54th American Association of Swine Veterinarians Annual Meeting 2023

Aurora, Colorado, USA
4 – 7 March 2023

ISBN: 978-1-7138-7102-6

Printed from e-media with permission by:

Curran Associates, Inc.
57 Morehouse Lane
Red Hook, NY 12571

Some format issues inherent in the e-media version may also appear in this print version.

American Association of Swine Veterinarians
830 26th Street
Perry, IA 50220-2328
USA

Phone: 515-465-5255
Fax: 515-465-3832

aasv@aasv.org

Additional copies of this publication are available from:

Curran Associates, Inc.
57 Morehouse Lane
Red Hook, NY 12571 USA
Phone: 845-758-0400
Fax: 845-758-2633
Email: curran@proceedings.com
Web: www.proceedings.com

Table of Contents

Howard Dunne Memorial Lecture

Alex Hogg Memorial Lecture

Sunday: Research Topics

Sunday Session #1: Student Seminar

Sunday Session #2: Industrial Partners

Sunday Session #3: Industrial Partners

Sunday Session #4: Industrial Partners

Poster Session: Veterinary Students

Poster Session: Research Topics

Poster Session: Industrial Partners

Monday: Be There!

Monday #1: Build Back Biosecurity and Make Health Great Again

Monday #2: Disease (PRRS and ...)

Monday #3: Foreign Animal Diseases: Be Prepared - Be Aware

Tuesday: Important Conversations: Be Engaged and Committed for a Common Goal

Howard Dunne Memorial Lecture

Be there. Be the leader for the pig, the client, the customer

Egan Brockhoff, DVM

In remembrance and honor of
Howard W. Dunne

Dr. Howard W. Dunne received his veterinary degree from Iowa State University in 1941. He was in engaged in private practice for one year and industry for four years before entering graduate school at Michigan State University. After receiving his PhD in 1951, he spent two years as deputy chief of veterinary microbiology at the Chemical Corps Biological Laboratories in Camp Detrick, MD. In 1953, he joined the faculty of Pennsylvania State University, where he headed the veterinary science research program and the large animal diagnostic laboratory. Dr. Dunne was internationally known for his work on swine diseases, particularly hog cholera and the swine reproductive viruses then referred to as the SMEDI syndrome. Dr. Dunne authored more than 90 scientific publications, and is perhaps best known for initiating and editing the textbook *Diseases of Swine*, which is still widely considered the "bible" of swine disease.

In 1971, the American Veterinary Medical Association presented Dr. Dunne with the XII International Veterinary Congress Prize to recognize his contributions to the international understanding of veterinary medicine. At the 1974 Congress of the International Pig Veterinary Society (IPVS), Dr. Dunne was elected president of the 1976 IPVS Congress, to be held in Ames, Iowa. His sudden death at the age of 61 in September of 1974 saddened the swine veterinary community, and the 1976 IPVS Congress was dedicated to his memory.

At the suggestion of Dr. Allen Leman, a financial gift made by the 1976 IPVS Congress organizers to the fledgling American Association of Swine Practitioners was combined with additional association funds to endow this prestigious annual lecture in remembrance and honor of Dr. Howard W. Dunne.

HOWARD DUNNE MEMORIAL LECTURERS

1978	Michael Muirhead	1990	Stanley Curtis	2002	Tim Blackwell	2014	Daryl Olsen
1979	W. J. (Bill) Smith	1991	Roy Schultz	2003	Lonnie King	2015	Gregory Stevenson
1980	N.C. Nielsen	1992	Joseph Connor	2004	Robert Desrosiers	2016	John Harding
1981	Allen Leman	1993	Camille Moore	2005	Scott Dee	2017	Jeff Zimmerman
1982	Michael Wilson	1994	Robert Morrison	2006	Gregg BeVier	2018	William R. DuBois
1983	Edward Bohl	1995	Rodney Johnson	2007	Tom Burkgren	2019	John T. Waddell
1984	Richard Ross	1996	Paul Yeske	2008	Timothy J. Loula	2020	Bret D. Marsh
1985	Kent Kislingbury	1997	Barbara Straw	2009	John U. Thomson	2021	Jerome O. Geiger
1986	Ralph Vinson	1998	Mike Daniel	2010	James F. Lowe	2022	Angela Baysinger
1987	Gary Dial	1999	Larry Rueff	2011	Locke A. Karriker		
1988	Thomas Alexander	2000	David Reeves	2012	R. Dean Boyd		
1989	Steven Henry	2001	Max Rodibaugh	2013	Craig Rowles		

https://doi.org/10.54846/am2023/1

Be there. Be the leader for the pig, the client, the customer

Egan Brockhoff, DVM

Introduction

It is truly an honour to present the Howard Dunne Memorial Lecture at this meeting of the AASV. A special thank you to Dr. Bill Hollis for inspiring me with his vision for this year's annual conference.

I don't remember studying policy development and analysis when I was a veterinary student, nor do I remember studying international commerce and trade. Yet here we are and here I am, serving a diverse set of clients with an ever-expanding set of demands on my time, my experiences, and my knowledge. Clients that push me to help them thrive in an increasingly complex global food system, pushing me to help them find new approaches to trade and market access, leveraging my expertise to ensure that their pigs are healthy, well and saleable in an incredibly competitive global animal protein sector. As swine veterinarians working at the farm level and beyond, we have a truly unique understanding and perspective of food models and food networks. We have a learned and genuine grasp on the pig itself, how our clients perceive the pig and shape their business around it, how the pig fits into a global food and commerce system and how our customers at all levels will identify with pork and pork products as a nutrient source. It is a remarkable space to find ourselves in, and one we must continue to work and prepare for to ensure we can be that leader. Leader for the pig, the client, and the customer.

It would seem almost cliché to me to address this esteemed body of professionals on the immense importance of being present, of being engaged, leaning in, accepting, and embracing that critical leadership. We all share this profound experience as swine veterinarians when we step back and examine the scope of the food system that we are working within. Swine health, swine welfare, how narrative influences policy, how policy impacts all aspects of our profession and sector, the importance of policy analysis, how policy influences and informs regulation at a local, regional, national, and global level. How all of this, in part and combined, will impact our opportunities for commerce and trade, the threats of market interruptions and border closures, success, and failure. The expanded necessity and ongoing demand for innovation in biosecurity, surveillance, traceability, nutrition, diagnostics, genetics, labour, and precision. We are a busy profession with a diverse set of demands placed upon us.

Dynamic clients

Our client base is changing at a pace and in a manner unlike anything we have seen before. Vertical integration, integrators, seems like such a simplistic way to describe these food systems. How they raise pigs, process and market pork is so diverse and only becoming increasingly so. Social media has opened doors, that has then opened other doors that then reveal something else that we could still call a door of some design, but only just. Expanding polarization of beliefs and attitudes, cancel culture, narrative, policy, nutrition, food security; This list can just go on and on. Our clients want to work with and produce a pig that will help them navigate through all these doors as best they can and in a way that still leaves them an opportunity for viability and sustainability. These are large models, constantly balancing risk and reward.

Client emergency preparedness: Hope for the best, prepare for the worst. There is extensive African swine fever (ASF) planning, preparedness, response, and recovery activity happening on a national and provincial level within government and accross the sector. This extensive work has not entirely trickled down to local, company level or farm level activities. It is a space that we need to engage our clients in with intention. A space that we simply cannot allow to remain sedentary, this is a space we must exercise with purpose. Unfortunately, there is more to be done than just foreign animal disease preparedness, this is emergency response planning in a broad sense. In the last year we have felt the impacts of climate change through unprecedented flooding and fires, the disruptions of COVID-19 to our workforce and critical infrastructure, and the overwhelming demand that Highly Pathogenic Avian Influenza has placed on our federal and provincial veterinary authorities and the poultry sector at large. All of these, along with an extensive list of others potential risks have made it evidently clear that our need to provide emergency response planning leadership, training and support will only continue to grow as a primary area of service and support.

Our clients are increasingly focused on the science and policy movement within the sustainability spaces. Climate change and carbon, animal welfare and enrichment, antimicrobials and one health. All these spaces are increasing demands on the executive, management, veterinary and production teams. By 2026 we will no longer be permitted to feed zinc oxide above nutritional levels in Canada. Concerns of antimicrobial resistance and environmental accumulation have driven these actions. The need for innovation within the post-zinc space is now. It reflects a need within the entire antimicrobial space. Access to non-traditional veterinary health products, new approaches to vaccine development and regional influenza vaccination are just a few of the areas we all need to be focusing our attention upon.

Well over half of the sow numbers that I have worked with during my career have now moved to loose-housing systems. Most of these transitions have been very successful, but not all started out that way. A lot of time was invested in understanding sow behaviour and why certain systems were so successful while others struggled, how genetics impacted success, new builds versus retrofits. Our clients have pushed us to provide value in this space in a very short amount of time. It reminds me a great deal of my early days in the cow-comfort space at the Summer Dairy Institute. There is certainly a real sense of pride when you finally get it dialed in, it very much changes your relationship, for the better, with the sows in your care.

From a veterinary perspective we are only now starting to spend time in the climate change space. I have had the real privilege of working with an insightful climate change policy specialist this past year. She has taught me a great deal about how valuable the veterinary team can be in identifying and influencing opportunities. The impact of antimicrobial use on soils has been an eye-opening experience. How health status translates to efficiencies that can drive down the carbon footprint per pig fed. I am very much looking forward to exploring this field of study with our clients. There is a lot of opportunity here to have a positive impact while at the same time affecting change within the narrative.

Risk management and the private-public partnership

Since 2018 we have been working extensively, sometimes it feels exclusively on ASF response planning. The work we have completed, and the work that still stands before us are both extremely noteworthy. We knew this work was going to be monumental in scope, what we have seen happen with ASF throughout the world has been nothing short of monumental. However, as we started bringing different minds together, exploring and tasking various reports and analytics on spaces from depopulation and disposal to novel approaches to business continuity a lot of valuable, intriguing, and quite frankly, useful ideas emerged. Ideas that could have real world impacts on our clients, our pigs, and our opportunities to mitigate risk from multiple paths.

Early in the ASF planning, preparedness, response, and recovery discussion we recognized that business continuity was going to be an extreme challenge in the face of a positive case within our borders. The ensuing market interruption was unquestionably going to be the more acute crisis versus control and eradication of the disease at the level of the infected premises. There was a very human story to all of this, the crisis was going to be overwhelming. As we looked at our existing business continuity and business risk management tools, we identified that there just wasn't enough on the upside opportunity for our sector in the recovery phase as we strive towards resuming "normalized" commerce and trade. Zoning arrangements are critically important, but when you look at the space closely from the perspective of an export driven sector you see that these tools are just not sufficient to avoid a large-scale welfare cull. What excited me most in this space was the discussions we began to have around private-public partnerships, the idea that industry could take on a larger role; leadership roles that were currently occupied exclusively by the competent veterinary authority. It was genuinely invaluable to inspire industry and government to work closely together to identify new and innovative approaches to managing this risk. One of the tools that emerged early in the discussion was that of African swine fever-free compartmentalization.

The idea of building an ASF-free Compartmentalization Program that could be utilized as the foundation of trade arrangements to be established in peacetime and then maintained in the face of disease was entirely instigated by the sectors farms and processors. The idea that a private-public partnership could be led and established by the pork sector took hold quickly and inspired a lot of great work, inspired new levels of cooperation and collaboration between government and the sector, and very much added some light to what looked like a very long dark planning and preparedness tunnel. Bringing together regulators, trade professionals, processors, producers, and veterinary colleagues to explore and develop this space was particularly rewarding. Seeing the response from within our client base was especially interesting as different minds found different ways to imagine how this could benefit their businesses. I remember in those early days of development, as we explained the concept of compartmentalization, how many people could only see this as an opportunity for the highly integrated. In some cases, getting downright grumpy that they didn't see an opportunity for themselves within the idea. It was truly energizing to see most of them later come back, finally uncovering an opportunity and recognizing the potential. Being a leader in this development process has been particularly empowering, regardless of what the immediate future holds for this concept. It has opened up new ways of thinking about trade and market access that will inform policy discussion for some time to come.

The pig

As many of you know, I also spend part of my time working with veterinary students at the University of Calgary. As you can well imagine, there is very little swine experience within the classes when they begin their program. An open book of opportunity. Regardless of this experience deficit, their curiosity for the pig is sincere. Canada has a Code of Practice for the care and handling of pigs, yet very few of the students have any knowledge of our on-farm Code system. They are unaware that pain management at castration and tail-docking is a requirement, that loose housing will soon become a requirement, and that there is a quality assurance program called Pig Safe and an animal care program called Pig Care. Regardless of how these conversations unroll, they are always interested in the pig and the state of its welfare, and this, I find so eye-opening and motivating. Seeing the pig from a veterinary student's vantage has helped me at times break down my own personal barriers, established notions and biases. An exercise I have enjoyed.

The pigs in our care work extremely hard for us and our clients alike. As their veterinarians we have a voice in their health, care, and handling. We are leaders in these fields, not just at the farm level but at the sector level as well. Our ability to advocate within this pork sector is unlike any others. When pain management at castration and tail-docking was introduced into the 2014 Pig Code we took on a leadership role in this space. There were no veterinary products labelled for pain, there were no established protocols or procedures. Discussions about pain management were largely unheard of beyond those of timely euthanasia. Our work to lead in this space is one that I am particularly proud of. We started working with various products, including NSAIDS, in various presentations We worked alongside the production teams with injectables, topicals, orals, most anything we could think of really. We trialed protocols to see how long it would take to process 1500 piglets with protocol a, then protocol b. And we kept trying until we found something that would work for each farm and system. The process was extremely rewarding, and the clients greatly appreciated the efforts we went to. At the end of it all, we had a manufacturer step up and register a product, we had protocols that had been barn-level tested, and we managed to publish on the welfare improvements. We have every occasion to be a leader and an advocate for the pig. Be there!

Closing

We can all grow as leaders within the veterinary profession and within our sector. My challenge is for all of us to move beyond our traditional areas of comfort. Now more than ever, veterinarians within the swine sector are needed to help shape and inform agriculture, food, and trade policy at all levels. To be leaders for the pig, our clients, and our customers. To be actively engaged in understanding the spaces where transboundary disease, animal health and one health can impact opportunities for sustainability in production and commerce. More is being demanded of us in a more diverse set of spaces than ever before. We have this opening before us right now, it is ours to take if we want to, if we have the courage and conviction to step out of our comfort zones and be that leader, that driver for innovation. A lot is coming at us faster than it ever has before. These are extremely exciting times, let's get ready.

Alex Hogg Memorial Lecture

Seizing opportunity within the challenges of modern swine veterinary medicine

Attila Farkas, DVM

In remembrance and honor of Alexander Hogg

Alexander Hogg
1920-2006

A native of Scotland, Alex (pronounced "Alec") Hogg immigrated to the United States with his family when he was six years old. After reaching adulthood, he served in the US Navy during World War II before entering veterinary school at Kansas State University. He received his DVM degree in 1950, and was a mixed animal practitioner in Iowa for twenty years. Dr. Hogg returned to college – this time, Iowa State University – to complete a Master's Degree in 1972. He spent the following years serving and educating swine veterinarians and producers in his roles as Nebraska Swine Extension veterinarian and professor at the University of Nebraska-Lincoln. Upon "retirement" from the university in 1990, Alex worked as a technical services consultant for MVP Laboratories. Ever an enthusiastic learner, he graduated from the University of Illinois Executive Veterinary Program in 1995.

An active and longstanding member of the AASV, Dr. Hogg joined what was then the American Association of Swine Practitioners shortly after it was formed. He served as secretary/treasurer of the association from 1974 to 1976, and was elected president in 1979.

The many awards and honors bestowed upon him over the years included several presented by the AASV: The Howard Dunne Memorial Award in 1985, the Meritorious Service Award in 1996, and in 2005, the first-ever AASV Foundation Heritage Award for a lifetime of achievement in veterinary medicine.

ALEX HOGG MEMORIAL LECTURERS

2011 Joseph F. Connor	2014 Mark J. Engle	2017 Matthew Turner	2020 William L. Hollis
2012 Lisa Tokach	2015 C. Scanlon Daniels	2018 Rodger Main	2021 Jeremy Pittman
2013 John T. Waddell	2016 Peggy Anne Hawkins	2019 Deborah Murray	2022 James A. Kober

https://doi.org/10.54846/am2023/2

Seizing opportunity within the challenges of modern swine veterinary medicine

Attila Farkas, DVM
Carthage Veterinary Service, Cartage Illinois

Introduction

Though I did not have the pleasure of knowing Dr. Alex Hogg, I am honored to give the seventeenth annual Alex Hogg Memorial Lecture. When I was first asked to give this lecture, I was very surprised. Honestly, the thought of talking about myself makes me uncomfortable. I do not want to ever do something that could be seen as bragging. Besides, talking to large audiences is really not in my blood. Not to mention, it is so hard to follow in the shoes of all the previous speakers. But I am going to give this my best.

Dr. Hogg was a man who wore many hats and seized many opportunities. He went to business college, joined the Navy, went to vet school, practiced mixed animal medicine, went back to school to earn a master's degree in veterinary pathology, served as a swine extension veterinarian, became a professor, and so many other things. He was a husband, a father, a son, and a friend. He lived a full life and helped others do the same. He had a heart for learning, which is crucial in an industry like ours where new challenges arise often.

As I am sure most of you know, Dr. Hogg was not from the US. And in case you could not tell from my accent, neither am I. Dr. Hogg immigrated with his family from Scotland as a child. My coming to America story is a little different, and it started later in my life.

After finishing vet school in Romania in 2006, I worked with Halav Love SRL, which is a dairy farm. They wanted to hire young, energetic veterinarians, and I fit the bill. I was one of four vets on their staff. My main responsibility was calf health management. I also worked monitoring milking data, looking for signs of mastitis.

One day in 2008, I got a call from a friend I made while visiting the US during college asking if I might be interested in working with swine in Texas. Smithfield's presence in Romania was starting to grow, and I anticipated that they were going to be the future of swine in my home country. The chance to go experience confined swine production firsthand was too big to pass up. I wanted to come and learn from the best in the world. I knew that the knowledge I gained from working with swine in the US would make me more marketable in Romania. So I interviewed for a US agriculture visa.

Ultimately, I was offered a swine technician position. I eagerly told them I had my vet license. They were pleased and told me the job would involve some veterinary tasks. I was excited to embark on this new adventure of learning, so I could come back home and be a better veterinarian in my country.

When I gave my two-week notice to my boss at the dairy farm, he tried to deny it. He thought I was joking or trying to get a pay raise. I explained I was moving to the US to work for a swine company. All these years later, his words still stay with me. He said, "In Europe, titles are important. They are respected. In the US, you will just be their buddy, and you will be scraping pig shit." I laughed and told him, "At least I will get paid."

I will never forget when I flew into Amarillo and then drove to the town where the farm was. Everything was so flat. There were no trees anywhere. It was just crops. I had never seen a landscape like that. When I arrived at the place I would be staying, it was quite the culture shock. It was a trailer park in a field. I had never even seen a trailer before, let alone lived in one.

When I arrived at orientation, it was the first time I had been to a sow farm. As I was shown around, I quickly realized no one working there spoke English ... or Romanian or Hungarian, which were the languages I knew. No one even knew my name. My start in the swine industry was not exactly what I had expected. But that was OK. I just needed to be patient and willing to learn.

In 2012, the company expressed an interest in having an on-site veterinarian and asked if I would be interested. I was eager to move back into a veterinary role, so I started the three-year program that foreign vets have to go through in order to be licensed in the US. And the rest is history, as they say.

Many of you are familiar with the story of the man who moved from Scotland as a child and ultimately became one of the swine industry's celebrated veterinarians. It is why we honor Dr. Hogg still today, seventeen years after his passing. And now you know the tale of a young, eager Romanian veterinarian, who moved to America to learn about confined swine production with the intentions of going back home.

Dr. Hogg had a special empathy with people of Eastern European countries. As a matter of fact, one of his favorite places to visit was the Pasteur Institute in Bucharest, Romania. He was not only concerned with his native homeland or the country he adopted as home. He cared about the world, which leads me to my next topic—global control measures.

Global control measures

As a person born and raised in another country, I have a unique perspective on the swine industry when it comes to examining it beyond the borders of the US. Over the years, the need for safe and efficient food production has continued to rise. The high demand for harvestable animal-sourced food both nationally and globally can create a lot of pressure on the professionals tasked with the health and wellbeing of that food source. The crisis brought about by the African swine fever (ASF) outbreak starting in 2018 is a perfect illustration of this.

While the twentieth century enjoyed the luxury of largely ignoring ASF, the twenty-first century has seen the rapid expansion of this devastating disease across most of the pork-producing countries in the world. It is hard not to consider the recent ASF outbreak in China as the largest domestic animal health disaster in the history of the world because half of the world's pig population was decimated. Diseases like ASF can have serious effects on worldwide swine production. It is a devastating, infectious disease that only infects swine and results in incredibly high mortality rates. African swine fever is one of the most economically important viral diseases in the swine industry. It is estimated that if the US experienced an ASF outbreak, it would cost us over $50 billion during the course of a ten-year period. Even a small scale ASF introduction into the US would result in devastating economic impacts to producers and the entire supply chain.

A lot of research has been devoted to trying to develop a commercial vaccine for ASF, but one does not exist yet. Recent advancements in vaccine technology have given hope to the ASF endemic regions of the world that disease control may be possible without depopulation. Even if a vaccine is only able to minimize the duration and volume of virus shedding, vaccination programs could go a long way toward decreasing the volume of ASF virus in the world, and thus slowing its creeping transmission across pig populations. Our industry should champion and support vaccine evaluation of safe vaccine candidates while demanding industries using "black market" ASF vaccines halt their use.

Currently, there are no effective treatment options for ASF either. Implementation of strict biosecurity measures is the best tool in our arsenal against the disease currently. African swine fever is largely a fomite-driven disease with contaminated meat products and swine carcasses being uniquely important components to its transmission. Producers generally do not need to change their biosecurity protocols to successfully prevent an ASF introduction, but they do need to execute them consistently. Nearly all pigs die once a herd is infected, and market access is lost to countries experiencing a break because no country is willing to risk bringing in contaminated pork. Depopulation of affected sites remains the gold standard for ASF control once it is introduced into a farm. Partial depopulation has been demonstrated to be effective on sites with regular testing, clinical sign monitoring, and constant internal biosecurity procedures to slow the spread of ASF inside the farm. Considerations for depopulation and carcass disposal methods should focus on timeliness of depopulation method, risk for carcass contamination in the disposal method, work safety, and mental wellbeing as well as animal wellbeing. Carcass disposal remains a challenge, particularly as it relates to regulatory compliance and practicality.

The rapid spread of ASF across China and other countries resulted in huge impacts to the global pig population, which created a deficit in the world's food supply chain. As of July 2022, over fifty countries had reported cases of ASF since the 2018 outbreak. Ten of those countries reported their first occurrence since 2020. So, while that first year of the outbreak was probably the most devastating, the problem is far from over or under control. The direct and indirect economic loss is estimated to be well into the billions. Finding out the exact number of pigs lost or the precise economic impact has proven to be challenging because the numbers are not always reported publicly, and some that are available are not considered accurate. It is likely that the disease is much more catastrophic than what we have heard.

There is a huge lack of crucial data available for proper research of this particular outbreak and its effects, which creates quite the challenge in analyzing the true impact and developing actionable steps to help prevent future outbreaks from having similar detrimental impacts.

Currently, the response to ASF varies widely depending on what part of the world we are talking about. For instance, in China, only pigs which test positive must be eliminated. While in Europe, once a positive case of ASF is confirmed, not only are infected animals euthanized but the entire farm must be depopulated. Then the site must go through two rounds of rigorous disinfecting, and then the site must sit empty for a significant period of time before repopulation is allowed.

The recent ASF outbreak highlighted some areas we need to work on as an industry. Initially, ASF moved across the whole country of China within three months. Only ninety days before the entire country, 3.7 million square miles, had been exposed to a disease that was first identified decades before. The situation caused many countries, the US included, to take a long, hard look at what needs to be done differently to learn from that situation. The answer so far seems to be to limit live movement, develop rapid ASF testing options, set up quarantines, and enforce the guidelines set.

For the US, one of the biggest hurdles is testing options. While there are diagnostic tests available for ASF, only a few sample types have been validated. They include whole blood, tonsils, spleen, and bone marrow for PCR testing and serum for ELISA testing. These all come with challenges, the primary one being the ease of quick collection if you need to survey a whole farm or an area. How many of you travel with the supplies in your vehicle to collect whole blood for an entire farm? Who is ready to necropsy a bunch of pigs in order to collect tonsils or spleens? Better yet, who wants to deal with the potential false-positive rate that is inherently part of ELISA testing when we are talking about a disease like ASF? Not me.

We need a better plan, a better method. But that takes a lot of work. It requires a lot of effort and collaborations between government agencies and producers to successfully tackle a disease like this. And with ASF, we have an even bigger challenge. Because so far we have been blessed to not have ASF within our borders, our access to samples for validation purposes is limited.

My connection with Eastern Europe has given me the opportunity to be part of a massive project to try to validate other sample types for monitoring for ASF. As part of the project, we were able to have samples collected of known ASF status for testing. We are trying to get data on the accuracy of population level samples like oral fluids versus individual samples like those that are currently validated. Many things must be considered when designing a project like this.

When we test a population sample, we accept a positive sample as positive. But how do we interpret a negative test result? Is it truly negative? Do we assume every animal chewed sufficiently on the rope, even those that might be sick? What if only one animal in the pen is sick, would we detect that? Would using oral fluids to survey a farm allow us to catch the index case, that first positive case, on a farm? These are things we do not know. And we will not know without a lot of testing.

In order to try to answer these questions, several organizations in the US are collaborating with organizations in Romania to sample ASF-positive farms to get us access to samples we have not had before. These samples will allow us to create scenarios with unvalidated sample types alongside validated ones to determine whether the unvalidated options are suitable for monitoring for ASF with accuracy.

Population type samples are always scrutinized initially before they become an acceptable option for surveillance of any disease. This is something we are used to. We went through it with PRRS for oral fluids and processing fluids. We went through this again with oral fluids for PED. Now we must do it for ASF.

With the help of our collaborators in Romania, we'll be able to create a known prevalence by collecting individual samples on the positive farms. Then we will hopefully be able to answer the big question at the early stages of an ASF break—when only one pig in a pen is positive, is that oral fluid for the pen also positive? Or does it take two pigs or five pigs? Where is the threshold of detectability?

I am not going to lie. This project has not been without its challenges. When the discussion first started about trying to validate other sample types, the first obstacle was actually gaining access to positive farms to collect the samples needed. Then the next challenge was getting the samples collected. In Eastern Europe, once a farm has a confirmed positive ASF case, the window for collecting samples is small because the farm must be depopulated within a few days. The next hurdle was getting the samples collected to a lab where all the confirmatory testing could be conducted at a level our US government would be comfortable with.

With teamwork, we were able to overcome these trials and get several thousand samples collected and delivered to a lab in the US for all the work to be done. Now comes the really hard part — waiting for the results.

It is one of the roles of swine veterinarians to help identify problems in swine herds, especially those that pose a risk to neighboring herds or the entire industry. Not only do vets help recognize any problems; they also try to implement solutions that ensure animal health under the heavy weight of public scrutiny. Swine vets may be called upon to utilize their skill set to help train humanitarian response to emergencies and disasters like disease outbreak, fires, storms, floods, and droughts. Diseases and disasters know no borders. Our communication and collaborations cannot either. We cannot be competitors in situations like these. We must be teammates.

Young vet perspective of industry opportunities

We have talked about some global issues facing our profession. Now it is time to shift our focus from a broader view to a more personal one.

Every new year brings reflection, analysis, and vision for the future for virtually every industry. The swine industry is no different. So how does the future look for us? How does it look for you? And how do we prepare for that future?

I would say the future looks pretty bright. Pork is the most consumed meat in the world. In general, pork production in the US has been on a steady increase for the past ten years, and it is not expected to decrease significantly. It is no secret that we like bacon ... and sausage and ham. The US is the world's third largest producer and consumer of pork and its products. In 2021, the US produced about 12.5 percent of the 100 million metric tons of pork produced.

The past few years, the US has been ranked first or second with regard to pork exports. In 2021, our pork exports surpassed 8.1 billion dollars in value. That is a whole lot of bacon.

The pork industry in our country continues to expand in spite of labor challenges and global swine health risks. Employment in the vet industry as a whole is expected to grow by 19 percent from 2021 to 2031, which outpaces many other sectors. For the upcoming decade, it is predicted that there will be 4800 openings for veterinarians each year. And while all of those will not be in the swine industry, when combining the job prospects for veterinary health and the growth of the pork industry, it stands to reason that the future looks very promising for swine vets in the US in the coming years. Regarding the future of swine veterinary medicine in the US, I would like to quote my colleague, Dr. Clayton Johnson. He said that the USA is still the best ride in the amusement park.

So how do young swine veterinarians and aspiring swine vets prepare to embrace the opportunities on the horizon to be able to enjoy the ride?

First, I would tell you do not shy away from getting your hands dirty. Swine medicine, like other livestock medicine, is not like working in a clinic where people bring their coughing pig in for an evaluation and an antibiotic. Swine medicine means going to the farm, examining the herd, surveying conditions and taking samples. You cannot keep your hands or boots clean when working with pigs. Embrace the uniqueness of swine medicine over that which is applied to small animals. Instead of treating an individual animal, you are practicing population medicine. You make health recommendations that can influence thousands of animals. As herds grow, the more vets must rely on population consideration and a holistic thought process for problem solving. This may mean seeking additional advice and relying on other experts' experiences. It is a hefty responsibility, but it is one you are capable of doing because you are here. You are willing to learn. You are willing to collaborate. You are willing to adjust and pivot when necessary.

I would also tell you to be pliable. As with everything, the only thing you can count on happening in the swine industry is that things will change. Change is never easy. Some changes will be for the good; some might not be. You have to be able to adapt as we learn more. Everything is subject to change—health risks, nutrition and feed ingredients, environmental conditions, genetics, diagnostics, and treatments. You need to be able to assess these items on any given farm and determine what do *these* pigs need and how can I help *this* producer to provide those things. I challenge you to view changes as opportunities not obstacles.

Society rewards professions that inspire innovation and embody an entrepreneurial mindset. The swine industry desperately needs these two skills in order to continue to thrive in a world where the demand for more food increases, while the societal understanding of the actual production and cultivation of that food decreases. Our industry has to deal with a perception that is not always accurate or positive. As swine veterinarians, this is part of the mantle you must carry.

As new technologies become more readily available, you will need to examine how those things can be implemented to improve pork production. Utilizing new tools will allow improved health monitoring, will increase the accuracy of diagnostics, and will change the way you are able to provide care for the farms you serve. You will need to effectively communicate these ideas with clients. You have to be flexible in order to properly address these new challenges and utilize the new solutions. We live in a world with rapidly changing technology. Human healthcare is driven by this fast-paced change. Veterinary medicine has to go along for the ride, and sometimes be prepared to drive if we want to stay in control of our own destiny in regard to animal healthcare.

Artificial intelligence and machine learning are poised to revolutionize the technology sector including animal health technologies. Virtual reality and telemedicine provide us with tools to connect with producers and farms in methods which were previously unfathomable. Because of biosecurity practices and other constraints that come with being a swine vet, you may have to utilize remote monitoring and care until you can visit a site in person. Use this capability as a way to be present when you cannot physically be present. Disease editing and resistance provide us with novel disease management tools that will forever change the way we think about disease control and elimination. Simply put, the practitioner who relies on outdated technologies and methods will quickly find themselves with outdated ideas and results. Embrace change, influence developments, and adapt to what the future brings.

Medicine is becoming more and more specialized all the time, and veterinary medicine is not immune from this transition. While we all graduate as veterinary "generalists" many of us choose to develop a formal or informal specialty. Swine veterinarian pioneers such as Dr. Ralph Vinson, Dr. Steve Henry, Dr. Joe Connor, Dr. Roy Schultz, Dr. Larry Rueff, Dr. Tim Loula, and many others paved the way for the current and future generations to swine veterinarians to focus 100 percent on commercial swine production. They broke from the mixed animal practice concept that dominated the early era of their careers to chart their own specialty course, and we should similarly pioneer additional specialty expansion. Boar stud specialists such as Dr. Chris Kuster, Dr. Darwin Reicks, and Dr. Doug Groth have followed this model, taking it one level deeper and specializing within a specific production class of swine production. Academics follow this model regularly, but swine practitioners have yet to fully exploit specialization, and this is an opportunity for all of us to carve out our own niches in the industry and share our knowledge with other practitioners when it is appropriate for a specific case.

Remember when I spoke briefly about my first swine job in the US? That job was a lesson in adaptability. I wound up in an entry level position even with a veterinarian license. That first week, I knocked down cobwebs and scraped hog manure from in front and behind the sows in the farrowing barn. My former boss had been right after all. But so was I. I got paid.

As time went on, I began to learn Spanish and became able to communicate better with my coworkers. Soon my responsibilities evolved from clean-up duty to vaccinating pigs at weaning and power washing the farrowing rooms. Eventually, I moved to day one processing of piglets. Over time, I worked through all the tasks to learn about confined swine production in the US. It was eye opening and quite educational to do tasks at each level on the farm.

Eventually, the company was trying to fill a GGP (great grandparent) sow farm position because the person in the position was being promoted. I decided to interview for the job. The interviewer ultimately became my wife. I got the position. And she became my boss before she became my wife. Then after some time, I wound up being her boss, that is until I wrote her up when I mistakenly thought she was wrong. Thankfully, someone in the company had the wisdom to change things up, and I was not ever in a position where I would need to make such a monumental mistake again.

An important skill that will serve you well as a swine vet is the ability to multitask. You will not just serve as a diagnostician. You are going to be a teacher. You will need to be able to teach farm managers things to look for so they can alert you to issues as soon as they become a problem. You cannot be on every farm every day. The people who are there will be your eyes and ears in your absence. You are going to need to be an effective communicator. You will often be the voice between farm staff and owner. You will need to explain your observations and diagnostic results to the clients and then be able to advise them of the best course of action. You will be a researcher. As new obstacles arise, you will need to tap into your inner scientist to come up with a way to evaluate and test solutions. The vet role in food animal production will continue to optimize animal health while keeping the goal of human nutrition in mind. Enhanced animal welfare practices will conform to consumer mindsets. You will need to balance your knowledge of sustainable farming practices with herd health management, not to mention, you will be balancing a personal life along with your professional one

Another important thing to remember is that you should never stop learning. Dr. Hogg got his master's degree when he was fifty-two years old and completed the Executive Veterinary Program when he was seventy-five. As I mentioned, the industry is constantly changing. Being a lifelong student creates the right mindset and attitude to be able to learn how to deal with these changes. When the opportunity arises for you to attend international veterinary programs, take it. Doing so will broaden your horizons, and it opens the lines of communication across borders and oceans. As discussed earlier and as we all understand even more now because of what we dealt with a couple of years ago, there are health crises that have far reaching impacts. Global understanding and teamwork are crucial to managing large scale problems.

As a swine vet, you are going to face challenges and obstacles. I will tell you a secret; you will not have all the answers, and it is OK. You will research and consult and find the answer, and the whole process will help you learn. This constant state of learning and seeking knowledge will make you better for your clients. As the future of swine veterinary medicine, it will become your job to ensure that swine veterinary medicine continues to evolve by developing new techniques, asking new questions, innovating new models of educating, recruiting new minds with a range of backgrounds and approaches to problem solving, and helping new vets gain knowledge.

The opportunities are here. The brightness of your future in swine veterinary medicine is up to you. Ted Engstrom said that the rewards for those who persevere far exceed the pain that must precede the victory.

What are you willing to do? What are you willing to learn? What are you willing to try?

So, what do I think the future of the swine vet field looks like? I think it looks like you.

Closing

While the last thirty years have seen the development of the swine veterinary systems of our current industry, there are significant risks to the model in place today. First and foremost, we face an immediate concern regarding incoming talent into the industry. While the current crops of students are extremely talented, recent salary surveys indicate the students interested in swine veterinary medicine are not being compensated as well as their peers in other species groups. Very few of us have the luxury to make career decisions independent of salary considerations, and for years, swine veterinarians enjoyed superior income rates than our peers in most other species groups. The last three years have seen this situation shift wildly out of our favor. We should be extremely concerned about how we can attract the top graduating veterinary talent given the new reality that our salaries are low compared to other opportunities. We must champion the value of our services and do everything we can to keep salaries as high as our competition in the companion animal and mixed animal worlds.

In addition to compensation concerns, we have to be concerned about the lack of pig farm exposure children are able to access. Biosecurity rules, no visitor policies, and the placement of farms in very remote locations have all combined to make pig farms inaccessible to everyone but our regular employees. Think about the last time you took your child or a young relative to a pig farm – I am guessing that experience happens rarely if ever, and that is a considerable risk to AASV members. If we cannot improve access to our farms and industry, how can we ever hope to inspire young generations to desire a career in swine medicine?

Veterinarians are considered trusted partners in protecting animal, human, and environmental health. We are stewards of animal welfare, research, food animal production, and species conservation. These roles will always matter, meaning veterinarians will always matter. Make sure the decisions you make build those partnerships. They will be vital to your success.

As a whole, swine vets have excelled at treating individual sick pigs. But to continue to be relevant in the industry, we must remember the scientific process. We cannot just observe and take note. We need to create controlled experiments to test theories and ideas. We must be proactive in seeking solutions, not just reactive. Years ago a saying was coined that you might have heard. The saying was that it takes a village to raise a child. These days, I say it takes a village to take care of the world's swine population. Being a vet encompasses knowledge in so many areas like immunology, pathology, virology, bacteriology, and epidemiology—areas we are not always experts in. Our individual success is limited by our ability to take clinical symptoms and relate them to the root biology of the issue. And then we have to conduct the proper diagnostic testing on the samples to support the diagnosis. Once an answer is found, treatment options become much more meaningful. Sometimes figuring all of this out takes the village. As we have seen with the current ASF break, we are going to need a village mindset to manage and control this issue.

When I came to the US, I was embarking on a mission to gain knowledge and perspective that I could take back home to build a career with. I never imagined I would gain so much more. I had not planned on finding my calling, creating a new home, finding my wife, and building an extended family. See what you can gain by taking risks, being pliable and choosing to embrace opportunities?

Our industry has certainly faced some challenges in recent years. But we are still here. In 1931, James Adams said America was the land of opportunity. And while many things have changed in the nine decades since then, America is still the land of opportunity. I leave you with this thought, Robert Collier said, "Success is the sum of small efforts repeated day in and day out." Make those small efforts. Do the work. Success will find you. I challenge you to think broadly. Think boldly. We make our own future. Are you ready to help feed the world?

https://doi.org/10.54846/am2023/3

Spatiotemporal relative risk distribution of porcine reproductive and respiratory syndrome virus in the southeastern United States

Felipe Sanchez[1,2]; Jason Galvis[1]; Nicolas Cardenas[1]; Chris Jones[2]; Cesar A. Corzo[3]; Gustavo Machado[1,2]
[1]Department of Population Health and Pathobiology, College of Veterinary Medicine, North Carolina State University, Raleigh, North Carolina; [2] Center for Geospatial Analytics, North Carolina State University, Raleigh, North Carolina; [3]Veterinary Population Medicine Department, College of Veterinary Medicine, University of Minnesota, St Paul, Minnesota

Introduction

Identifying areas and farms at greater risk for porcine reproductive and respiratory syndrome virus (PRRSV) transmission is pivotal for targeted surveillance and disease control strategies. The objectives of this study were to i) define the maximum spatial distance at which the risk of PRRSV transmission from infected to susceptible farms is increased, ii) describe how risk is distributed across different farm types, and iii) identify variables associated with the risk of PRRSV local transmission.

Methods

For this study, we used PRRSV status, geographical coordinates, and production type from 2,293 farms between the years 2018 through 2020 in the southeastern US. Farms were classified as cases or controls based on weekly PRRSV status records. An adaptive kernel density estimation approach was used to determine the maximum spatial distance of PRRSV local transmission, identify areas of significant risk, and estimate relative risk among the different farm types. Variables associated with PRRSV local transmission were identified using a Bayesian spatiotemporal hierarchical model approach. Variables considered in our model included, pig density, farm density, six environmental variables, three farm biosecurity features (site entry, line of separation access points (LOSAP), and perimeter buffer area access points (PBAAP), and six between-farm contact networks metrics for the year 2020.

Results

Our results indicate that the maximum spatial distance for which the risk of PRRSV transmission increased was 15.3 km in 2018, 17.6 km in 2019, and 18 km in 2020. Areas of significant risk were identified and consisted mostly of uninfected finisher farms (24.9%, 572/2,293), and infected sow farms (6%, 138/2,293). Relative risk estimates showed that sow farms consistently had higher relative risk values (27.8% - 32.7%), and finisher farms had the lowest (62.4% - 63.8%) among the different farm types. Variables found to be associated with PRRSV local transmission included out-degree (OR 1.11), line of separation access points (LOSAP) (OR 1.03), and the number of days of temperatures between 4°C - 10°C (OR 1.01).

Discussion

This study addresses an important gap in the understanding related to the spatial range of PRRSV local transmission by identifying the maximum distance at which PRRSV relative risk is increased. Our results suggest that downstream farms (ie, finisher farms) may play an important role in maintaining the circulation of PRRSV within the farm population, due to cases going undetected given the differences in biosecurity and surveillance in finisher and sow farms. We also demonstrated that the risk for local transmission increased for farms with more out-going movements (out-degree), a higher number of barn access points (LOSAP), and farms that experienced a higher number of days of temperatures between 4°C - 10°C. Information gathered from this study can be used to inform targeted surveillance and disease control strategies and calibrate future PRRSV transmission models.

Acknowledgment

Funding: College of Veterinary Medicine and the Center for Geospatial Analytics at North Carolina State University.

https://doi.org/10.54846/am2023/4

In-silico characterization of the relationship between PRRSV prevalence at the individual piglet level and prevalence at the litter level in a farrowing room

Onyekachukwu H. Osemeke[1], DVM, MS; Eduardo de Freitas Costa[2], DVM, PhD; Vinicius Weide[3], BS, PhD; Swaminathan Jayaraman[1], BE, MS; Gustavo S. Silva[1], DVM, MS, PhD; Daniel C. L. Linhares[1], DVM, MBA, PhD
[1]Veterinary Diagnostic and Production Animal Medicine, College of Veterinary Medicine, Iowa State University, Ames, Iowa;[2]Department of Epidemiology, Bioinformatics and Animal Models, Wageningen Bioveterinary Research, Lelystad, The Netherlands;[3]Instituto Federal de Educação, Ciência e Tecnologia do Rio Grande do Sul, Farroupilha, RS, Brazil

Introduction

Monitoring/surveillance remains an integral component of PRRSV control and elimination programs. Sampling for monitoring/surveillance is generally guided by assumptions, paramount of which is the assumed prevalence of the disease in the population to be sampled.[1,2] With the recent paradigm shift in US breeding herds' PRRSV surveillance, from individual-animal-based sampling to aggregate sampling,[3] the unit for which prevalence is estimated for these aggregate samples shifts from the individual animal to the animal group giving the sample, for example, the litter, in the case of family oral fluids (FOF).[4] To the best of the authors' knowledge, no published work describes how the mentioned prevalence types relate. Therefore, the objective of this study was to characterize the relationship between the proportion of viremic piglets in a farrowing room (PPV), the proportion of litters with at least one PRRSV-positive piglet (true litter prevalence, TLP), and the proportion of PRRSV-positive litters likely to be detected by FOF (apparent litter prevalence, ALP).

Materials and methods

Parameters from the referenced study

Based on the study of Almeida et al,[5] a predictive model was first built to characterize the relationship between within-litter prevalence (the proportion of viremic piglets within a litter) and the probability of a positive FOF sample from that litter. The degrees of clustering (heterogeneity) of PRRSV-positive pigs within sampled rooms from the reference study were scaled and measured, and the median value (0.61) was used as a baseline. The clustering factor used for this study ranged from 0 (meaning a completely random distribution of PRRSV-positive pigs between litters) to 1 (meaning PRRSV-positive pigs would be clustered within the fewest number of litters possible)). An empirical distribution of litter sizes from the sampled rooms across all the herds sampled in the reference study was also obtained.

Stochastic model

Farrowing rooms were simulated with a fixed number of litters (corresponding to the number of crates in the farrowing room) and using piglet-level PRRSV prevalence (PPV) ranging from 1 to 50%. The number of piglets within each litter was drawn from a discrete empirical distribution obtained from the referenced study. The number of viremic piglets within each litter was sampled from a recursive binomial model using a baseline clustering factor of 0.61.

The TLP per iterated room was obtained as the proportion of litters with at least one viremic pig, and the apparent litter prevalence was obtained as the predicted proportion of the litters in an iterated room that will be PRRSV-positive by FOF testing. A total of 5000 iterated rooms were obtained by Monte Carlo simulation, and the median values of TLP and ALP were obtained. All analyses were done on R statistical software.[6]

Results

Table 1 presents matched values of piglet-level prevalence and estimated litter-level prevalence for a 56-crate room.

Discussion and conclusion

The results of this study demonstrate how PPV, TLP, and ALP (by FOF) in a farrowing room compare. This study provides insight to swine practitioners on prevalence values to be used in estimating comparable sample sizes for either serum or FOF sampling for PRRSV surveillance in weaning-age pigs. For example, a PPV of 5% matches with an ALP of 6.48% (~4 crates), For a 56-crate room, using a conventional sample size calculator, this would mean sampling about 29 crates to have ≥ 95% confidence of detecting at least one positive litter by RT-rtPCR tests on FOF.

The results of this study also provide a framework for estimating the proportion of viremic piglets within a farrowing room, given the results of FOF testing (test positivity rate).

This study adds to the series of previous studies tailoring conventional sample size concepts to better fit peculiarities in typical US swine barns and the ecology of PRRSV.

References

1. Stevenson MA. Sample Size Estimation in Veterinary Epidemiologic Research. Front Vet Sci. 2021;7:539573. **doi:10.3389/fvets.2020.539573**

2. Fosgate GT. Practical sample size calculations for surveillance and diagnostic investigations. J Vet Diagnostic Investig. 2009;21(1):3-14. **doi:10.1177/104063870902100102**

3. Trevisan G, Linhares LCM, Crim B, et al. Macroepidemiological aspects of porcine reproductive and respiratory syndrome virus detection by major United States veterinary diagnostic laboratories over time, age group, and specimen. Shaman J, ed. PLoS One. 2019;14(10):e0223544. **doi:10.1371/journal.pone.0223544**

4. Osemeke OH, de Freitas Costa E, Almeida MN, et al. Effect of pooling family oral fluids on the probability of PRRSV RNA detection by RT-rtPCR. Prev Vet Med. 2022;206:105701. **doi:10.1016/j.prevetmed.2022.105701**

Table 1: Relationship between the piglet-level PRRSV prevalence in a 56-crate farrowing room and the true and apparent (by family oral fluids [FOF]) litter-level prevalence assuming a clustering level of 0.61

Piglet-level prevalence (%)	Litter-level prevalence (upper and lower 95% quantiles) (%)	Apparent litter-level prevalence by FOF (upper and lower 95% quantiles) (%)
1	5.36 (1.79, 7.14)	2.06 (1.07, 3.53)
5	8.93 (7.14, 12.50)	6.48 (5.30, 8.58)
10	14.29 (10.71, 17.86)	11.25 (9.31, 13.92)
15	19.64 (16.07, 23.21)	16.35 (14.47, 19.21)
20	23.21 (21.43, 26.79)	21.60 (18.73, 24.19)
25	28.57 (25.00, 32.14)	26.66 (23.50, 29.31)
30	33.93 (30.36, 37.50)	31.35 (28.77, 34.33)
35	39.29 (35.71, 42.86)	36.16 (33.49, 39.44)
40	44.64 (41.07, 48.21)	41.30 (38.05, 44.71)
45	48.21 (44.64, 53.57)	46.54 (43.10, 49.68)
50	53.57 (50.00, 57.14)	51.56 (48.34, 54.58)

5. Almeida MN, Zhang M, Zimmerman JJ, Holtkamp DJ, Linhares DCL. Finding PRRSV in sow herds: Family oral fluids vs. serum samples from due-to-wean pigs. Prev Vet Med. 2021;193:105397. **doi:10.1016/j.prevetmed.2021.105397**

6. R Core Team. R: A Language and Environment for Statistical Computing: R Foundation for Statistical Computing Vienna, Austria. 2019.

https://doi.org/10.54846/am2023/5

Refining PRRSV-2 genetic classification based on global ORF5 sequences and investigation of geographic distributions and temporal changes

Wannarat Yim-im[1], PhD; Tavis Anderson[2], PhD; Igor Paploski[3], DVM, PhD; Kimberly VanderWaal[3], PhD; Phillip Gauger[1], DVM, PhD; Karen Kreuger[1], PhD; Mang Shi[4], PhD; Rodger Main[1], DVM, PhD; Jianqiang Zhang[1], MD, PhD
[1]Veterinary Diagnostic and Production Animal Medicine, Iowa State University, Ames, Iowa; [2]Virus and Prion Research Unit, National Animal Disease Center, USDA-ARS, Ames, Iowa; [3]Department of Veterinary Population Medicine, University of Minnesota, St. Paul, Minnesota; [4]School of Medicine, Sun Yat-sen University, Shenzhen, China

Introduction

Porcine reproductive and respiratory syndrome virus (PRRSV) is an important swine pathogen affecting the global swine industry. The first ORF5-based genetic lineage classification system describing global PRRSV-2 genetic diversity was introduced a decade ago.[1] Recent studies proposed a refinement by dividing the lineage 1 into 9 sublineages based on US. PRRSV-2 ORF5 sequences.[2,3] However, PRRSV-2 genetic diversity at international levels has not been thoroughly explored after 2010 and the sublineages within other lineages have not been evaluated and updated. Thus, the classification system needs to be refined using more contemporary global sequences.

Methods

A dataset of 82,237 global PRRSV-2 ORF5 sequences obtained during 1989-2021 (57,260 sequences from the Iowa State University Veterinary Diagnostic Laboratory [ISU VDL] and 24,977 sequences from United States Swine Pathogen Database [US-SPD][4] that collects all sequences deposited in GenBank) were used in this study. For all ORF5 sequences, virus sequence ID, sample collection date, sample collection location, and RFLP (restriction fragment length polymorphism) typing information was compiled if such information was available. After redundant sequences were removed by mothur v.1.44.3, 40,601 PRRSV-2 ORF5 sequences from ISU VDL, 16,851 PRRSV-2 ORF5 sequences from US-SPD database, and 841 ORF5 reference sequences from the previous Shi et al study[1] were included to analyze and refine PRRSV-2 ORF5-based genetic classification system. Multiple sequence alignment was performed by the progressive method (FFT-NS-1) in MAFFT v7.407. Phylogenetic tree from the multiple sequence alignment was constructed using maximum likelihood with stochastic algorithm, general time-reversible nucleotide substitution with 10 categories of FreeRate heterogeneity model (GTR+F+R10) and 1,000 bootstrap replicates in IQ-TREE v1.6.12. Utilizing the classified sequences as anchors for lineage/sublineage classification, 1,063 reference sequences representing all lineages and sublineages were established and used to analyze all 82,237 sequences to determine their lineage/sublineage information.

As no standard cutoff of nucleotide (nt) identity has been defined to distinguish between vaccine-like and wild-type viruses, in the current study, we arbitrarily defined any sequence with ≥ 98%, 95 - < 98%, or < 95% ORF5 nt identity to a vaccine virus as a vaccine-like virus, vaccine-like suspect, or wild-type virus, respectively.

Results and discussion

In this study, based on analysis of 82,237 global PRRSV-2 ORF5 sequences reported during 1989-2021, we classified PRRSV-2 into ten genetic lineages (L1-L10) and 21 sublineages (L1A-L1F, L1H-L1J, L5A-L5B, L8A-L8E, and L9A-L9E). PRRSV-2 phylogenetic diversity at the inter-lineage levels was in the range of 8.54% (between L4 and L7) to 17.13% (between L3 and L6). Phylogenetic diversity at the intra-lineage level was in the range of 0.76% within lineage L7 to 11.90% within lineage L2.

At the lineage level, the lineages 1 to 9 are overall consistent with what Shi et al proposed.[1] However, the lineage 10 in our proposed classification system was previously undescribed and it is primarily detected in Thailand. The proposed refinement of L1 sublineages attempt to be consistent with what Paploski et al described.[2,3] However, some modifications are proposed. For example, the sublineages "L1B and L1G" proposed by Paploski et al[2,3] cluster together and form a monophyletic sublineage; therefore, "L1B and L1G" are combined into "L1B" in our new system and the use of "L1G" is discontinued. Similarly, "L1Dalpha and L1E" proposed by Paploski et al[2,3] cluster together and it is difficult to accurately differentiate them in the tree; therefore, "L1Dalpha and L1E" are combined into "L1E" in our new system and the use of "L1Dalpha" is discontinued. "L1Dbeta" proposed by Paploski et al[2,3] is therefore simplified to "L1D" in our new system and the use of "L1Dbeta" is discontinued. We also describe two new sublineages, L1I and L1J with the L1I sequences detected in the US, Thailand, and Canada and the L1J sequences detected primarily in South Korea. The proposed sublineages L5A and L5B correspond to L5.1 and L5.2 described by Shi et al,[1] respectively. Shi et al[1] described sublineages L8.1-L8.9 and L9.1-L9.17; however, we have refined them to sublineages L8A-L8E and L9A-L9E.

The proposed classification system is flexible for growth if additional lineages, sublineages, or more granular classifications are needed in the future. As an example, to meet epidemiological investigation needs, the sublineage L1C was further divided into five groups (L1C.1-L1C.5) with L1C.5 corresponding to recently emerged L1C variant strains.[5,6]

Comparison between RFLP typing and phylogenetic classification revealed that some RFLP patterns (eg, 2-5-2 and 1-7-4) were mostly detected in a particular lineage or sublineage while others (eg, 1-4-4, 1-8-4, 1-4-2, and 1-3-2) were distributed across multiple genetic lineages, reflecting the inaccuracy of using RFLP typing to determine PRRSV-2 genetic relatedness in most scenarios.

Geographic distribution analyses revealed that L1 (L1A-1C, L1D-1F, L1H, and L1I), L5 (L5A), L8 (L8A-L8C), and L9 (L9A and L9C) were the major lineages and sublineages in the US, which differed from the situation in Canada (mainly L1H, L1C, L5A and L8A), Mexico (mainly L1B, L1A, L5A and L8D), China (mainly L1C, L3, L5A, L8E [HP-PRRSV]), South Korea (mainly L1J, L2, L5A and L5B), Japan (mainly L4), Thailand (mainly L1I, L5A, L8E and L10), Europe (mainly L5A), and South America (mainly L1A and L1B).

Temporal dynamic changes of PRRSV-2 in the US were investigated by analyzing 73,092 PRRSV-2 ORF5 sequences collected during 1989-2021.

- At the lineage level (Fig 1A), lineage 9 was dominant before 2004 representing > 30% and decreased over years to less than 1% after 2013. During 1998-2001, lineage 1 was reported ~7.5% and increased to 29.6 % in 2002. Lineage 1 became a dominant lineage during 2004-2021, which dramatically increased from 33.3% to 74.4%. Lineage 5 represented 8.5%-16.6% during 2002-2009 and increased to ~20% during 2010-2021. Lineage 8 was reported approximately ~10% during 2002-2010 and 4.6%-10% during 2011-2021. Lineages 2, 6, and 7 were small lineages in which each of these lineages represented less than 1% or was rarely observed in recent years.
- At the sublineage level in lineage 1 (Fig 1B), expansion and contraction were observed in many sublineages. Sublineage L1F was dominant representing ~40% during 2002 - 2006 and declined from 33.8% in 2007 to 2.6% in 2014 then reduced to less than 1% in 2015 - 2021. Sublineage L1E was observed over 20% during 2002 - 2003 and reduced to less than 10% after 2005. Sublineage L1B represented ~10% during 2002 - 2005 and increased to 24.0% in 2006. Sublineage L1B became a dominant sublineage rising to over 40% during 2008 - 2009 and then decreased to less than 2% during 2019-2021. Sublineage L1C reported less than 7% during 2002 - 2007 was a dominant sublineage representing over 40% during 2010 - 2013 then decreased to less than 20% during 2015 - 2020 and increased to 28.4% in 2021. Sublineage L1A notably fluctuated from 2003-2013 and became a dominant sublineage during 2015 - 2021, increasing from 27.2% in 2014 to over 60% during 2015 -2018 and declining to ~40% in 2021. Sublinege L1H was increasingly observed from 1.69% in 2013 to over 20% during 2019 - 2021. However, expansion and contraction were rarely observed in sublineage L1D which was less than 8% during 2003 - 2021. Prevacent vaccine-like viruses in sublineage L1D were observed during 2019-2021 rising from 3.6% to 7.5%. Broadly, these temporal dynamics match trends in L1 diversity reported by Paploski et al,[2,3] which reinforces the robustness of these patterns across US PRRSV-2 databases despite varied geographic biases.
- At the sublineage level in lineage 5, sublineage L5A was a dominant sublineage from 2002 to 2021 in which sequences with < 98% nucleotide identity to Ingelvac MLV vaccine decreased over years while Ingelvac MLV-like virus with over 98% nucleotide identity to Ingelvac MLV vaccine increased over years.
- At the sublineage level in lineage 8, sublineage L8A was a dominant sublineage during 2003 - 2013, increasing from 49.3% to 60% during 2004 - 2012 and decreasing from 58.6% in 2013 to 12.3% in 2021. In sublineage L8A, wild-type virus and vaccine-like suspects with the respective < 95% and 95 - < 98% nucleotide identity to Ingelvac MLV ATP virus declined over year from 27.3% in 2002 to less than 1% during 2011 - 2021 for wild-type virus and less than 10% during 2002 - 2021 for vaccine-like suspects. Ingelvac PRRS ATP-like virus with ≥ 98% nucleotide identity to Ingelvac ATP vaccine represented 15.5% in 2002 and rose up to 50 - 70% during 2005 - 2013 then declined to less than 10% during 2002 - 2021. Sublineage L8B increased from 15.5% in 2002 to 42.8% in 2008 then reduced to 5.2% in 2015 and L8B sequences were not reported during 2016 - 2021. Sublineage L8C noticeably emerged in 2012 (15.7%) and became a dominant sublineage during 2014 - 2021. Fostera-like virus with ≥ 98% nucleotide identity to Fostera vaccine was first observed in 2012 (15.7%) in sublineage L8C and increased to 50 - 90% during 2014 - 2019 then dropped to 26.7% in 2021 while vaccine-like suspect with 95 - < 98% nucleotide identity to Fostera vaccine gradually increased to 60.1% in 2021.
- At the sublineage level in lineage 9, sublineage L9A was a dominant sublineage representing over 60% before 2006 and fluctuated from ~24% to 40% during 2008 - 2013 while sublineage L9C increased from 4.85% in 2002 to 44.53% in 2007 and fluctuated from ~33% to 55% during 2008 - 2013. Very few lineage 9 sequences were detected during 2013 - 2021.

This study refined PRRSV-2 ORF5-based phylogenetic classification after comprehensive analysis of global sequences and investigated the geographic distribution and dynamic changes of PRRSV-2 at the lineage/sublineage levels, including vaccine-like viruses. The refined classification system and molecular epidemiology data in this study will be invaluable for future characterization of PRRSV-2. In addition, reference sequences based on the new genetic classification are available for future epidemiological and diagnostic applications.

References

1. Shi M, Lam TT, Hon CC, Murtaugh MP, Davies PR, Hui RK, Li J, Wong LT, Yip CW, Jiang JW, Leung FC. 2010. Phylogeny-based evolutionary, demographical, and geographical dissection of North American type 2 porcine reproductive and respiratory syndrome viruses. J Virol 84:8700-8711.

2. Paploski IAD, Corzo C, Rovira A, Murtaugh MP, Sanhueza JM, Vilalta C, Schroeder DC, VanderWaal K. 2019. Temporal Dynamics of Co-circulating Lineages of Porcine Reproductive and Respiratory Syndrome Virus. Front Microbiol 10:2486.

3. Paploski I, Pamornchainavakul N, Makau DN, Rovira A, Corzo C, Schroeder DC, Cheeran M, Doeschl-Wilson A, Kao RR, Lycett S, VanderWaal K. 2021. Phylogenetic Structure and Sequential Dominance of Sub-Lineages of PRRSV Type-2 Lineage 1 in the United States. Vaccines 9:608.

4. Anderson TK, Inderski B, Diel DG, Hause BM, Porter EG, Clement T, Nelson EA, Bai J, Christopher-Hennings J, Gauger PC, Zhang J, Harmon KM, Main R, Lager KM, Faaberg KS. 2021. The United States Swine Pathogen Database: integrating veterinary diagnostic laboratory sequence data to monitor emerging pathogens of swine. Database 2021, baab078.

5. Kikuti M et al., Emergence of new lineage 1C variant of PRRSV2 in the United States. Frontiers in veterinary science 2021, 8, 752938-752938, **doi:10.3389/fvets.2021.752938**

6. Trevisan G et al., Complete coding genome sequence of a novel PRRSV2 restriction fragment length polymorphism 1-4-4 L1C variant identified in Iowa, USA. Microbiology Resource Announcements. 2021;10(21):e00448-21.

Figure 1: Temporal dynamics of PRRSV-2 ORF5 sequences at lineage level **A)** and sublineages within lineage L1 **B)** with samples collected in the US during 1989 - 2021. Percent distribution of each lineage and sublineage is indicated in the graph and the total number of sequences reported in a particular year is indicated in the table below the graph.

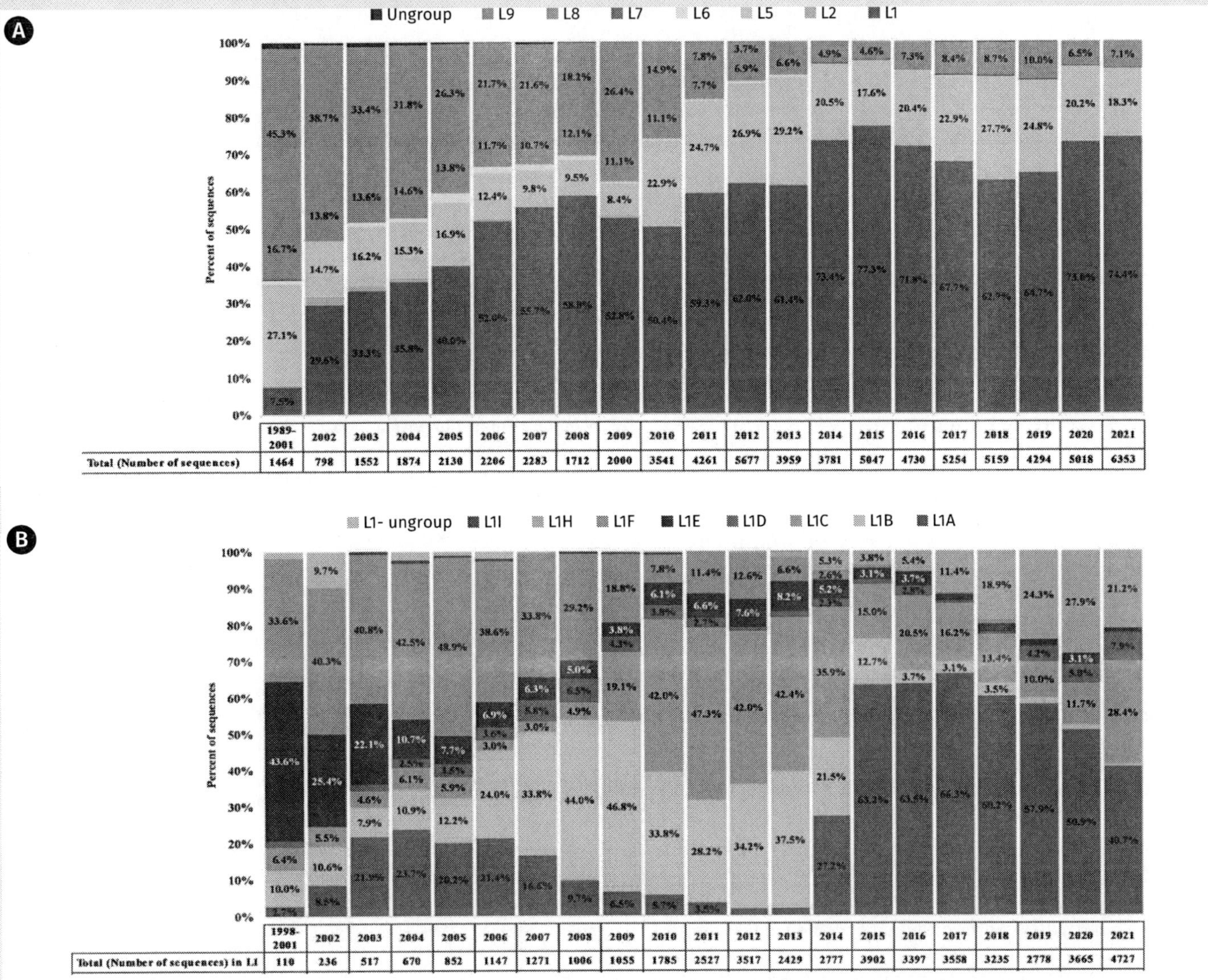

	1989-2001	2002	2003	2004	2005	2006	2007	2008	2009	2010	2011	2012	2013	2014	2015	2016	2017	2018	2019	2020	2021
Total (Number of sequences)	1464	798	1552	1874	2130	2206	2283	1712	2060	3541	4261	5677	3959	3781	5047	4730	5254	5159	4294	5018	6353

	1998-2001	2002	2003	2004	2005	2006	2007	2008	2009	2010	2011	2012	2013	2014	2015	2016	2017	2018	2019	2020	2021
Total (Number of sequences) in L1	110	236	517	670	852	1147	1271	1006	1055	1785	2527	3517	2429	2777	3902	3397	3558	3235	2778	3665	4727

https://doi.org/10.54846/am2023/6

Adsorption of PRRS virus strains to Minnesota soils, a possible transmission route

Joaquín Álvarez-Norambuena; Angie Quinonez; Sagar M. Goyal; Cesar A. Corzo
Veterinary Population Medicine, University of Minnesota, Saint Paul, Minnesota

Introduction

Porcine reproductive and respiratory syndrome (PRRS) has become the most significant disease in the US.[1] Data from the Morrison Swine Health Monitoring Project (MSHMP) indicate an increasing trend of PRRS cases between October and December, a time when manure is pumped and spread on soils near to swine farm premises.[2] It is known that PRRS virus can be found in feces and manure pits[3,4] but there is scarce data on whether PRRS virus present in the manure could reach groundwater after manure spread.[5-7] Therefore, the aim of this study was to compare different strains of PRRS virus for their percolating ability through different soil types from Minnesota.

Materials and methods

For the purpose of this study, we used three different PRRS viruses (RFLP 1-7-4, 1-4-4, and 1-26-2), which were grown and titrated in MARC-145 cells. A total of 13 soils were used; 6 from sites surrounding pig farms and 7 obtained from the UMN Agronomy Department. A vertical glass column model was developed and used in which 5g, 10g and 20g of each soil was placed inside the model with a filter paper in the bottom end. Enough water was added to soil columns to moisten the soil and remove air bubbles and mimic field conditions. Then a solution containing virus and water (1:1 v/v) were added to the glass column until the passage of solution by dripping. Then, percolated water was collected and subsequently titrated. Results were analyzed through ANOVA and multiple linear regression.

Results

We were able to isolate virus from the three different strains in the percolates regardless of the amount of soil added (5g, 10g, & 20g). With the 1-7-4 strain, the initial titter decreased from 6.50 to 3.83 $log_{10}TCID_{50}/100\ \mu L$ after being diluted with water (positive control). The virus was isolated from all 5g soil samples, 12 out of 13 10g soil samples and from 6 out of 13 samples. In the case of the 1-4-4 strain, the titter decreased from 5.50 to 3.17 $log_{10}TCID_{50}/100\ \mu L$ after being diluted with water. The virus was isolated from all the 5g soil samples, 6 out of 13 10g soil samples and 3 out of 13 20 g soil samples. As for the 1-26-2, the titter decreased from 5.17 to 3.50 $log_{10}TCID_{50}/100\ \mu L$. We were able to isolate the virus from all samples at 5g and 10g, and in 11 out of 13 20g soil samples (Figure 1).

Conclusions

1. All virus strains were able to percolate through all amounts and types of soil.
2. There is an inversely proportional relationship between viral titer and amount of soil.
3. Current findings agree with reports of other viruses (eg, poliovirus, echoviruses, coxsackievirus) having the capacity of percolation.

References

1. Holtkamp DJ, Kliebenstein JB, Neumann EJ, et al. Assessment of the economic impact of porcine reproductive and respiratory syndrome virus on United States pork producers. *J Swine Health Prod.*

2. Sanhueza, J. M., Stevenson, M. A., Vilalta, C., Kikuti, M., Corzo, C. A. (2020). Spatial relative risk and factors associated with porcine reproductive and respiratory syndrome outbreaks in United States breeding herds. *Preventive veterinary medicine, 183,* 105128.

3. Linhares DCL, Torremorell M, Joo HS, Morrison RB. Infectivity of PRRS virus in pig manure at different temperatures. *Vet Microbiol.* 2012; 160(1-2):23-28.

4. Montoya J, Vilalta C, Sanhueza J, Corzo CA.Prevalence of PRRSv, PEDv, PDCoV and TGE in pig farm manure pits. Allen D. Leman Conference. 2020.

5. Goyal SM, Gerba CP. Comparative adsorption of human enteroviruses, simian rotavirus and selected bacteriophages to soils. *Appl Environ Microbiol.* 1979; 38(2): 241-247.

6. Gerba CP, Goyal SM, Hurst CJ, LaBelle RL. Type and strain dependence of enterovirus adsorption to activated sludge, sopild and esturine sediments. *Water Res.* 1980. 14:1197-1198.

7. Goyal SM, Amundson D, Robinson RA, Gerba CP. Viruses and drug-resistant bacteria in groundwater of southeastern Minnesota. *J Minn Acad Sci.* 1989. 55:58-62.

Figure 1: Dot chart of the $TCID_{50}$ results by type of soil and virus.

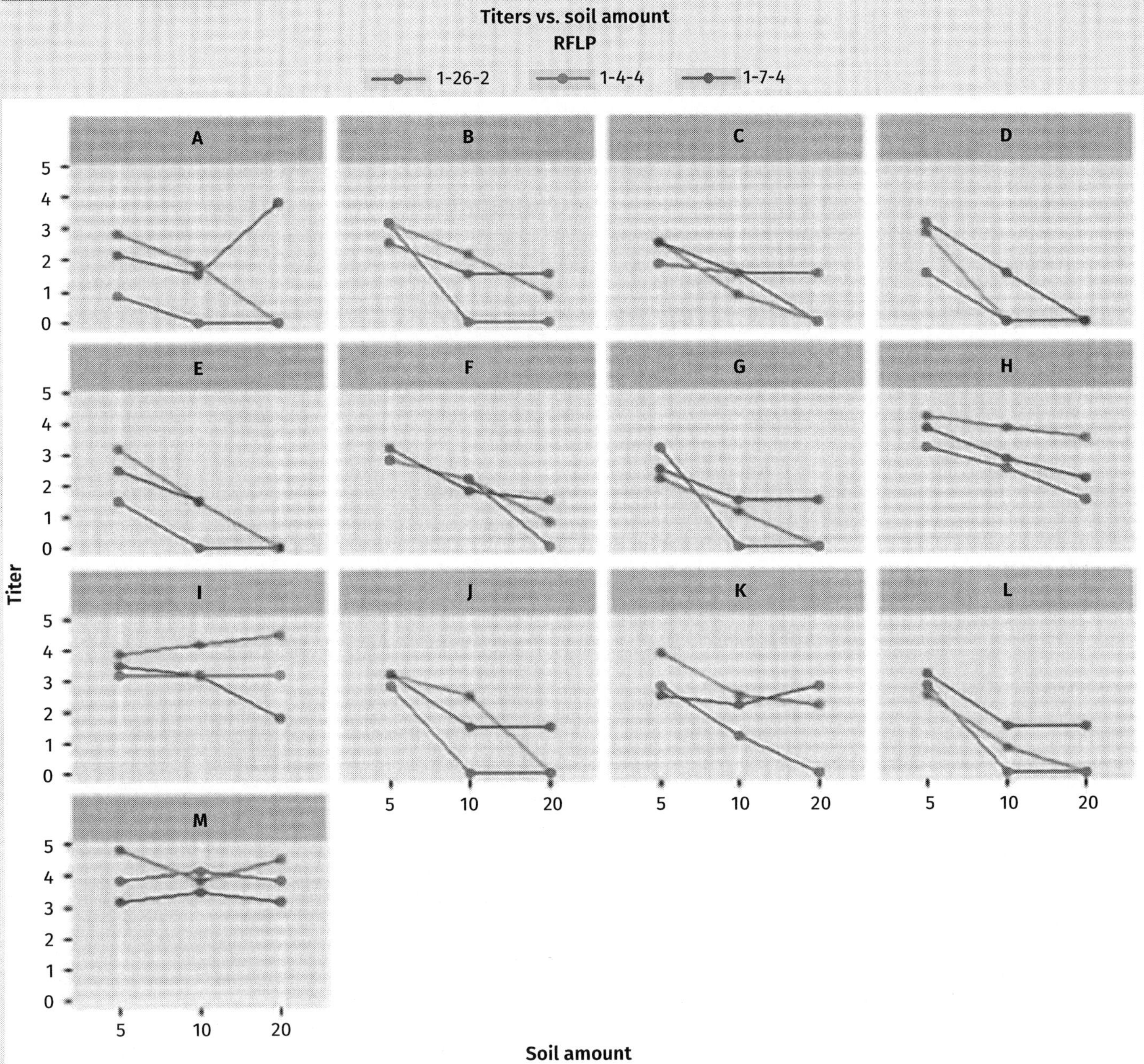

https://doi.org/10.54846/am2023/7

Hand sanitation protocols to decrease the risk of influenza transmission in pigs

Montse Torremorell; Laura Davis; Joaquin Alvarez-Norambuena; Marie Culhane; My Yang
Department of Veterinary Population Medicine, College of Veterinary Medicine,
University of Minnesota, St. Paul, Minnesota

Introduction

Influenza A virus (IAV) is one of the most significant respiratory pathogens in pigs and people.[1] Influenza is widespread globally and bi-directional transmission of IAV between pigs and people is well documented.[2,3] Proper hand hygiene protocols are important to minimize the risk of IAV transmission between pigs and people, in particular when considering prevention protocols at the pig and swine workers interface.[4-6] Previously, IAV has been cultured from people's hands,[7] therefore the CDC has recommended hand washing practices to decrease transmission risks.[8] Nonetheless, there is limited information on the effectiveness of hand sanitation protocols on the removal or inactivation of IAV on hands of swine farm workers contaminated with pig secretions. Furthermore, there is limited information on the duration of IAV viability on hands of farm workers. Respiratory pig secretions potentially increase the likelihood of viable IAV remaining on surfaces of the hands for unknown duration. Thus, we assessed four hand sanitation protocols and evaluated IAV viability on hands for up to 120 minutes after handling IAV-infected pigs.

Materials and methods

Pigs were experimentally infected with an H1N1 IAV. Participants were recruited to implement four hand hygiene procedures. Hands were sampled before and after handling infected pigs for 10 min and after the implementation of the specific hand hygiene procedure. Participants followed one of the following procedures: 1) hand washing with soap and water; 2) hand washing with water only; 3) hand sanitation with an alcohol-based hand sanitizer, and 4) wearing disposable gloves. Each procedure was evaluated in triplicate. A subset of participants also had their hands sampled for up to 120 minutes to evaluate IAV viability overtime. Hands were sampled using a moisten wipe and samples tested by IAV RT-PCR and virus isolation. All procedures were approved prior to the study by the University of Minnesota IACUC, IBC and IRB committees.

Results

Upon handling infected pigs, hands of all participants became contaminated with IAV. All hand hygiene treatments resulted in a reduction in the quantity of IAV genetic material detected by RT-PCR. However, the reduction observed in participants that washed hands with soap & water, or water only was limited. In contrast, samples from hands treated with an alcohol-based sanitizer and hands with gloves had larger reductions (Figure 1A, data not displayed). Notwithstanding, viable virus was isolated from multiple samples only after pig interaction and prior to hand sanitation, despite the positive RT-PCR results. We also observed a reduction in the amount of IAV genetic material present in the hands of workers for up to 120 minutes. The reduction observed ranged between 5.01 and 5.79 Ct values (Figure 1B, data not displayed).

Conclusions

We observed differences between handwashing protocols in their ability to remove or inactivate IAV in the hands of workers handling influenza-infected pigs. This study will be helpful to design protocols directed at mitigating IAV transmission in pigs and between pigs and people.

Acknowledgments

Funding, wholly or in part, was provided by The National Pork Board.

References

1. Dykhuis-Haden, C., Painter, T., Fangman, T., & Holtkamp, D. (2012). Assessing production parameters and economic impact of swine influenza, PRRS and *Mycoplasma hyopneumoniae* on finishing pigs in a large production system. *Denver: American Association of Swine Veterinarians,* 75–76.

2. Olsen, C. W., Brammer, L., Easterday, B. C., Arden, N., Belay, E., Baker, I., & Cox, N. J. (2002). Serologic evidence of H1 swine influenza virus infection in swine farm residents and employees. *Emerging infectious diseases, 8*(8), 814.

3. Newman, A. P., Reisdorf, E., Beinemann, J., Uyeki, T. M., Balish, A., Shu, B., Lindstrom, S., Achenbach, J., Smith, C., & Davis, J. P. (2008). Human case of swine influenza A (H1N1) triple reassortant virus infection, Wisconsin. *Emerging infectious diseases, 14*(9), 1470–1472.

4. Nelson, M. I., Schaefer, R., Gava, D., Cantão, M. E., & Ciacci-Zanella, J. R. (2015). Influenza A Viruses of Human Origin in Swine, Brazil. *Emerging infectious diseases, 21*(8), 1339–1347.

5. Diaz, A., Marthaler, D., Corzo, C., Muñoz-Zanzi, C., Sreevatsan, S., Culhane, M., & Torremorell, M. (2017). Multiple genome constellations of similar and distinct influenza A viruses co-circulate in pigs during epidemic events. *Scientific reports, 7*(1), 1-14.

6. Nirmala, J., Perez, A., Culhane, M. R., Allerson, M. W., Sreevatsan, S., & Torremorell, M. (2021). Genetic variability of influenza A virus in pigs at weaning in Midwestern United States swine farms. *Transboundary and Emerging Diseases, 68*(1), 62-75.

7. Lopez-Moreno, G., Garrido-Mantilla, J., Sanhueza, J. M., Rendahl, A., Davies, P., Culhane, M., ... & Torremorell, M. (2022). Evaluation of dam parity and internal biosecurity practices in influenza infections in piglets prior to weaning. *Preventive Veterinary Medicine, 208,* 105764.

8. Centers for Disease Control and Prevention (2021). Hand Sanitizer Use Out and About. **https://www.cdc.gov/handwashing/hand-sanitizer-use.html**

Figure 1: A) Total Ct-values after pig interaction and after each handwashing protocol. * Indicates statistical difference through Kruskal-Wallis between time of sampling per day. **B)** Total averages of Ct values for each sampling time according to day post infection.

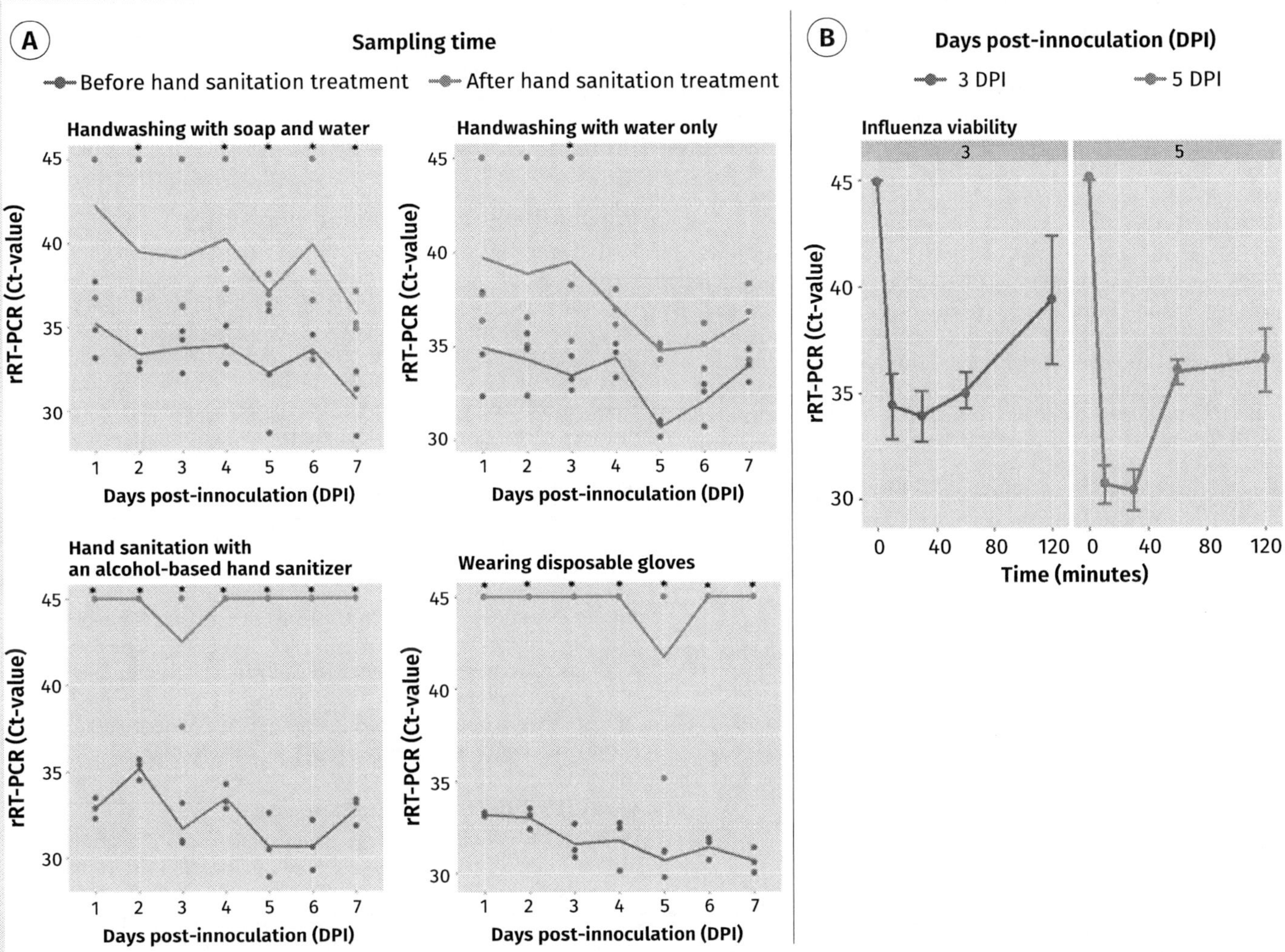

https://doi.org/10.54846/am2023/8

Divergent pathology between highly pathogenic avian influenza A (H5N1) strains in swine

B. Arruda[1], DVM, PhD; A. Buckley[1], DVM, PhD; M. L. Killian[2], PhD; N. L. Hines Bergeson[2], PhD; M. K. Torchetti[2], DVM, MS, PhD; A. L. Vincent Baker[1], DVM, PhD
[1]Virus and Prion Research Unit, National Animal Disease Center, Agriculture Research Service, US Department of Agriculture, Ames, Iowa; [2]National Veterinary Services Laboratories, Animal and Plant Health Inspection Service, US Department of Agriculture, Ames, Iowa

Introduction

Highly pathogenic avian influenza (HPAI) viruses pose substantial risk to poultry due to their negative agroeconomic impact and pandemic potential. Influenza A viruses occasionally cross species barriers, including H5 avian influenza virus (AIV) infections in humans, pigs, and marine mammals. Since February 2022, H5N1 HPAI belonging to the goose/Guangdong 2.3.4.4b hemagglutinin (HA) phylogenetic clade from a trans-Atlantic incursion caused outbreaks across the North American continent, resulting in more than 40 million poultry deaths, extensive mortality in wild bird species, and wild mammal infections.[1-3] Assessing the risk of 2.3.4.4b H5N1 strains from the current outbreak as a potential for infecting pigs is necessary due to its wide distribution in states with high populations of pigs and the potential for reassortment with endemic swine adapted IAV. If infectious and transmissible in swine, understanding the molecular mechanisms of how and why IAV cross species barriers and emerge in mammals to inform early warning systems is paramount to animal and public health.

Methods

Viruses

We evaluated two strains of the 2022 spring HPAI H5N1 2.3.4.4b clade outbreak, A/turkey/Minnesota/22-010654-002/2022 (A/turkey/MN/22) and A/bald eagle/Florida/22-006544-004/2022 (A/bald eagle/FL/22). These two strains resulted from Eurasian avian HPAI 2.3.4.4b H5N1 reassortment with North American low pathogenic AIV lineage internal genes and represented two different reassortment patterns frequently detected among H5N1 strains during the spring of 2022. Both strains retained the H5 and N1 from the 2.3.4.4b lineage. A/turkey/Minnesota/2022 contained North American LPAI PB2 and NP and A/bald eagle/Florida/2022 contained North American LPAI PB2, PB1, and NP. These two strains did not contain known mammalian adaptation mutations.

Virus propagation and titration

We conducted the study in compliance with the Animal Care and Use Committee of the USDA-ARS NADC under biosafety level 3 agriculture containment, including enhancements required by the Federal Select Agent Program.[4] Virus stocks were passaged in 10-day-old embryonating chicken eggs. The allantoic fluid from infected eggs was harvested, divided into aliquots, and stored at -70°C until use. Viral titers were determined in Madin-Darby canine kidney (MDCK) cells according to standard methods.

Pig challenge and transmission study

Forty-three pigs were obtained from a herd free of influenza A virus (IAV), porcine reproductive and respiratory syndrome virus, and *Mycoplasma hyopneumoniae*. Prior to challenge, pigs were confirmed seronegative to IAV by a blocking ELISA (IDEXX) and treated with tulathromycin (Zoetis) and ceftiofur crystalline free acid (Zoetis). Pigs were divided into one group of 3 (group 1; negative control) and two groups of 15 (one group per virus strain), housed in individual containment rooms. Fifteen pigs from groups 2 (A/turkey/MN/22) and 3 (A/bald eagle/FL/22) were inoculated intranasally with 1 ml of approximately 10^6 50% tissue culture infective dose ($TCID_{50}$) per mL. Five naive contact pigs were comingled with each of the virus-infected groups at 2 days post-inoculation (DPI). Nasal swabs (Minitip FLOQSwabs) samples were collected on 0, 1, 3, 5, and 7 DPI from principle pigs and 1, 3, 5 and 7 days post-contact (DPC) from contact pigs. Five principal pigs per group were necropsied at 3, 5 and 14 DPI. The five contact pigs were necropsied at 12 DPC. Bronchoalveolar lavage fluid (BALF) was collected from principle pigs at necropsy on 3 and 5 DPI. Sera were obtained from principal pigs at necropsy (3, 5, and 14 dpi) and contact pigs at necropsy on 12 DPC pigs.

Viral RNA detection and serology

Viral RNA was extracted from nasal swab and BALF samples using the MagMA™ CORE Nucleic Acid Purification Kit (Life Technologies) and subjected to RT-PCR assays that target a conserved regions of IAV (VetMAX™ Gold AIV Detection Kit, ThermoFisher Scientific) and the hemagglutinin gene of goose/Guangdong lineage influenza A virus (IAV) subtype H5, clade H5 2.3.4.4. Seroconversion was determined using the IAV NP blocking ELISA (IDEXX).

Macroscopic and microscopic scoring

At necropsy, the percentage of affected surface area per lung lobe was recorded from which a percent weighted lung score was calculated.[5,6] Tissue samples from the trachea and right middle or affected lung lobe were fixed in 10% buffered formalin for histologic examination. Microscopic lesions of the lung and trachea were evaluated by a veterinary pathologist blinded to treatment as previously described.[7,8] Five parameters were evaluated per lung section: necrotizing bronchitis and bronchiolitis, suppurative bronchitis and bronchiolitis, peribronchiolar and vascular lymphocytic infiltrates, alveolar septal inflammation and alveolar cellular exudate, edema, and hemorrhage. A composite score was computed with the sum of the five individual scores (0 – 20).[7] Two parameters were evaluated per tracheal section: epithelial degeneration and necrosis and inflammatory infiltrate for a possible composite score of 0-8.[8]

Immunohistochemistry

Heat-induced epitope retrieval was performed using citrate (pH 6) heated to 95°C for 2 cycles of 10 minutes. Slides were quenched with hydrogen peroxide, blocked with goat serum (1:10) for 60 minutes, and incubated with IAV NP antibody (1:4000; GeneTex GTX125989) for 60 minutes at room temperature, followed by a secondary antibody (EnVision+ anti-rabbit Poly-HRP-IgG) for 30 minutes. Diaminobenzidine chromogen detection was completed using Dako DAB Plus and Dako DAB Enhancer. Slides were counterstained using hematoxylin. Lung (conducting airways and alveolar lumen and septa) and tracheal immunohistochemical (IHC) score was recorded as previously described.[5]

Statistical analysis

All statistical analyses were performed using GraphPad Prism 8.1.2 software. A two sample Kolmogorov-Smirnov (nonparametric) test was used to compare cumulative distributions of pathologic assessments. RT-qPCR cycle threshold (Ct) values were compared using a Welch's t test. *P*-values < 0.05 were considered statistically significant.

Results

Neither HPAI H5N1 strain was detected in the upper respiratory tract nor transmitted to contact pigs

Viral RNA was not detected in any nasal swab sample in any primary or contact group and seroconversion was not detected in any of the contact pigs. Two of the five A/turkey/MN/22 and three of the five A/bald eagle/FL/22 principal pigs that survived to 14 dpi seroconverted.

Viral replication in the lower respiratory tract differed between strains

Both strains replicated in lungs of intranasally inoculated pigs at levels at or below those observed in swine-adapted influenza A virus (IAV). Viral RNA was detected in BALF from all principal pigs necropsied at 3 and 5 DPI in the A/bald eagle/FL/22 group with a mean Ct value of 30.14 and 23.22, respectively. Viral RNA was detected in BALF from 3 of the 5 principal pigs necropsied at both 3 (mean Ct value: 34.76) and 5 (mean Ct value: 34.70) DPI in the A/turkey/MN/22 group. The average Ct value of BALF samples were significantly lower (higher viral RNA quantity) from the A/bald eagle/FL/22 group compared to the A/turkey/MN/22 group at 5 but not 3 DPI.

Lesions and antigen labeling differed between strains

Primary inoculated pigs in each of the challenge strain groups developed gross and histologic lesions consistent with IAV, particularly the A/bald eagle/FL/22 challenged pigs (Figure 1). Yet, divergent pathology between HPAI strains was evidenced by statistically significant differences in pathologic assessments. The macroscopic lung lesion score, histologic lung lesion score, and lung IHC score of the A/bald eagle/FL/22 group compared to the A/turkey/MN/22 group at 5 but not 3 DPI were significantly higher. HPAI NP antigen was detected by IHC in the respiratory epithelium of conducting airways, macrophages, pneumocytes and endothelial cells of pigs inoculated with A/bald eagle/FL/22 (Figures 2 and 3). IAV NP antigen was not detected in the lung of any pig inoculated with A/turkey/MN/22.

Discussion

HPAI viruses can cause a range of asymptomatic or mild to severe infections in mammals.[9-12] Viral proteins including the HA, PA, PB2, and PB1 contribute to the pathogenicity, virulence, and transmissibility of these viruses and a combination of adaptive mutations and/or reassortment are likely necessary for efficient mammalian transmission.[13] The HPAI H5 clade 2.3.4.4b strains evaluated in this study did not contain previously identified molecular markers of mammalian adaption.

The receptor-binding specificity, glycosylation of the HA head, thermostability of the HA trimer, and pH of activation of the HA are important determinants in the host range of IAV.[13] The HA proteins of HPAI H5N1 preferentially bind to α2,3-linked sialic acids on the host cell which are at low abundance in the porcine upper respiratory system.[13] The low abundance of α2,3-linked sialic acids on epithelial cells of the upper respiratory tract of pigs may explain why the HPAI strains in this study did not transmit. Previous studies assessing the replication of H5N1 HPAI in swine demonstrated nasal shedding of some wild-type viruses,[14] although at lower titers than swine-adapted viruses. H5N1 containing 2009 pandemic H1N1 internal genes transmitted to direct contacts, demonstrating a potential risk if H5N1 HPAI reassorted with endemic swine strains.[15]

The quantity and distribution of α2,3-linked sialic acids is relatively higher in the lower respiratory tract of pigs compared to the upper respiratory tract and localized on type II pneumocytes lining the alveolar wall and on non-ciliated bronchiolar cells.[16-19] This distribution was consistent with the extent and distribution of IHC labeling in IAV antigen positive lungs in this study. IAV NP antigen was detected in pneumocytes and transudate/exudate in alveoli and suggests replication of A/bald eagle/FL/22 in the lower respiratory tract. This is further supported by a significantly lower Ct value (higher viral RNA quantity) in 5 out of 5 positive BALF samples from the A/bald eagle/FL/22 group compared to the 3 out of 5 positive BALF samples from the A/turkey/MN/22 group at 5 DPI.

It is not known if the different internal gene constellations between A/bald eagle/FL/22 and A/turkey/MN/22 played a role in the ability to replicate and induce lesions in the lungs of inoculated pigs. Future studies are needed to identify determinants of HPAI pathogenicity, virulence, and transmissibility in swine to inform early warning systems. The risk of reassortment with endemic swine IAV is a consideration; however, the risk is low in confinement operations with industry standard biosecurity. Bird proofing feed and facilities and not using untreated surface or well water that may be contaminated with wild waterfowl feces would be prudent to further reduce this risk.

References

1. Bird Flu Current Situation Summary. Centers for Disease Control and Prevention. September 15, 2022. Accessed September 19, 2022. **https://www.cdc.gov/flu/avianflu/avian-flu-summary.htm**

2. 2022 Confirmations of Highly Pathogenic Avian Influenza in Commercial and Backyard Flocks. Animal and Plant Health Inspection Service. September 15, 2022. Accessed September 19, 2022. **https://www.aphis.usda.gov/aphis/ourfocus/animalhealth/animal-disease-information/avian/avian-influenza/hpai-2022/2022-hpai-commercial-backyard-flocks**

3. National Wildlife Health Center. Distribution of Highly Pathogenic Avian Influenza in North America, 2021/2022. U.S. Geological Survey. September 16, 2022. Accessed September 19, 2022. **https://www.usgs.gov/centers/nwhc/science/distribution-highly-pathogenic-avian-influenza-north-america-20212022#overview**

Figure 1: A/bald eagle/Florida/22. Macroscopic lesions consistent with influenza A virus (IAV) in swine; multifocal to coalescing, cranioventral pulmonary consolidation (arrowhead).

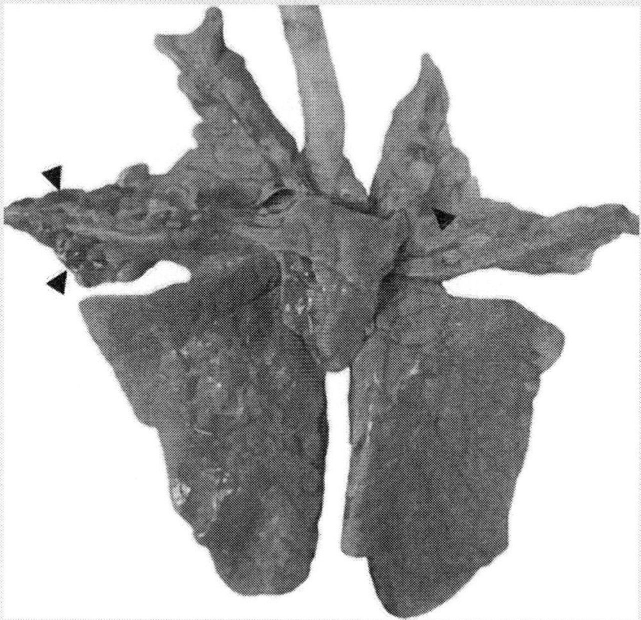

Figure 2: A/bald eagle/Florida/22. Detection of IAV by immunohistochemistry (IHC); labeling of respiratory epithelium lining a conducting airway (arrowhead) and macrophages (arrow).

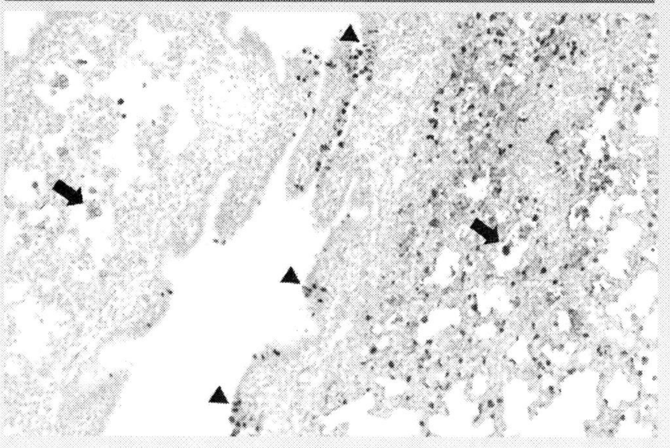

Figure 3: A/bald eagle/Florida/22. Detection of IAV by IHC; labeling of exudate/transudate in alveolar lumens (arrow) and pneumocytes (arrowhead).

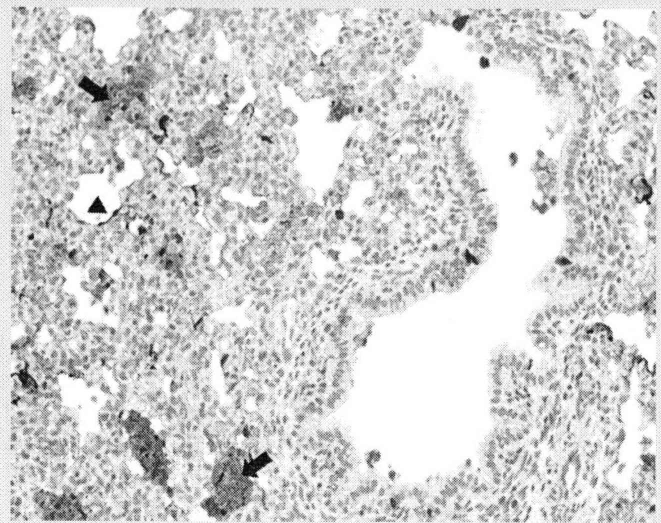

4. Richmond JY, Nesby-O'Dell SL. Laboratory security and emergency response guidance for laboratories working with select agents. Centers for Disease Control and Prevention. MMWR 2002;RR-19:1–6.

5. Gauger PC, Vincent AL, Loving CL, et al. Kinetics of lung lesion development and pro-inflammatory cytokine response in pigs with vaccine-associated enhanced respiratory disease induced by challenge with pandemic (2009) A/H1N1 influenza virus. *Vet Pathol.* 2012;49:900–912.

6. Halbur PG, Paul PS, Frey ML, et al. Comparison of the pathogenicity of two US porcine reproductive and respiratory syndrome virus isolates with that of the Lelystad virus. *Vet Pathol.* 1995;32:648–660.

7. Morgan SB, Hemmink JD, Porter E, et al. Aerosol delivery of a candidate universal Influenza Vaccine Reduces Viral Load in Pigs Challenged with Pandemic H1N1 Virus. *J Immunol.* 2016; 196:5014-5023.

8. Gauger PC, Loving CL, Khurana S, et al. Live attenuated influenza A virus vaccine protects against A(H1N1)pdm09 heterologous challenge without vaccine associated enhanced respiratory disease. *Virology.* 2014 Dec:471-473:93-104.

9. Shin DL, Siebert U, Lakemeyer J, et al. Highly Pathogenic Avian Influenza A(H5N8) Virus in Gray Seals, Baltic Sea. *Emerg Inf Dis.* 2019;25: 2295-2298.

10. Floyd T, Banyard AC, Lean FZX, et al. Encephalitis and Death in Wild Mammals at a Rehabilitation Center after Infection with Highly Pathogenic Avian Influenza A(H5N8) Virus, United Kingdom *Emerg Inf Dis.* 2021;27:2856-2863.

11. Postel A, King J, Kaiser FK, et al. Infections with highly pathogenic avian influenza A virus (HPAIV) H5N8 in harbor seals at the German North Sea coast, 2021. *Emerg Inf Dis.* 2022;219:725-729.

12. Rijks JM, Hesselink H, Lollinga P, et al. Highly Pathogenic Avian Influenza A(H5N1) Virus in Wild Red Foxes, the Netherlands, 2021. *Emerg Inf Dis.* 2021;27:2960-2962.

13. Neumann G. H5N1 influenza virulence, pathogenicity and transmissibility; what do we know? Future Virol. 2015;10:971-980.

14. Lipatov AS, Kwon YK, Sarmento LV, et al. (2008) Domestic Pigs Have Low Susceptibility to H5N1 Highly Pathogenic Avian Influenza Viruses. *PLoS Pathog* 4(7): e1000102.

15. Abente EJ, Kitikoon P, Lager KM et al. A highly pathogenic avian-derived influenza virus H5N1 with 2009 pandemic H1N1 internal genes demonstrates increased replication and transmission in pigs. *J Gen Virol.* 2017;98:18-30.

16. Shinya K, Ebina M, Yamada S, et al. Avian flu: influenza virus receptors in the human airway. *Nature.* 2006; 440(7083):435–436.

17. Baum LG, Paulson JC. Sialyloligosaccharides of the respiratory epithelium in the selection of human influenza virus receptor specificity. *Acta Histochem.* 1990; (Suppl. 40):35–38.

18. Nicholls JM, Chan MC, Chan WY, et al. Tropism of avian influenza A (H5N1) in the upper and lower respiratory tract. *Nat Med.* 2007; 13(2):147–149.

19. Yao L, Korteweg C, Hsueh W, et al. Avian influenza receptor expression in H5N1-infected and noninfected human tissues. *FASEB J.* 2008; 22(3):733–740.

https://doi.org/10.54846/am2023/9

Modeling the introduction and control of African swine fever within commercial swine populations in the United States

Abagael L. Sykes[1]; Jason A. Galvis[1]; Kathleen C. O'Hara[2]; Cesar Corzo[3]; Gustavo Machado[1]
[1]Department of Population Health and Pathobiology, College of Veterinary Medicine, North Carolina State University, Raleigh, North Carolina; [2]US Department of Agriculture, Animal and Plant Health Inspection Service, Veterinary Services, Strategy and Policy, Center for Epidemiology and Animal Health, Fort Collins, Colorado; [3]Veterinary Population Medicine Department, College of Veterinary Medicine, University of Minnesota, St. Paul, Minnesota

Given the proximity of African swine fever (ASF) to the US, there is an urgent need to simulate the dissemination of the virus within the US swine industry and to evaluate mitigation strategies. Here, we extended *PigSpread*, a spatially-explicit stochastic compartmental farm-level transmission model that incorporates six transmission routes including between-farm swine movements, vehicle movements, and local spread, to model the dissemination of ASF. We then examined the effectiveness of control and eradication actions listed in the ASF national response plan.

The average number of secondary infections during the first 60 epidemic days of the outbreak was 50 finisher farms, 17 nursery farms, 6 sow farms, and less than one infection in other farm types. The predominant route of ASF dissemination was the between-farm movements of swine with an average contribution of 71.4% to ASF dissemination, while local spread and movement of vehicles were less critical with an average of 14.5% and 14.2% (Figure 1). Additionally, we identified the potential for long distance spread of ASF up to a maximum of 494.24 km from the index farm, and dissemination of ASF between commercially unrelated companies.

We also demonstrated that the combination of quarantine and depopulation of detected farms, movement restrictions, contact tracing, depopulation of direct contacts and enhanced surveillance, was the most effective mitigation strategy, resulting in an average reduction of 74.0% of secondary cases within 140 days of the outbreak. In addition, such control strategies led to a median of 485,138 depopulated animals, 1,634,623 diagnostic tests, and 53,648 daily single-movement permits. Despite substantial reductions in cases, the above control strategy was only successful in bringing daily incidence to zero in 29% of simulations, suggesting that the successful elimination of an ASF outbreak is likely to require control actions to be implemented for more than four months or the intensification of the current control actions.

Nonetheless, our results provide a critical evaluation of the current national control and eradication actions with specific recommendations of the resources needed for depopulation, testing, and movement permits, while also highlighting the need to further study the implementation of control actions to limit the spread of a possible outbreak.

Acknowledgment

This project is funded by USDA's Animal and Plant Health Inspection Service, USDA-APHIS Award: AP22VSSP0000C004.

Figure 1: The contribution of the transmission routes to ASF infection over 60 days of epidemic, separated by the production type of the farm receiving infection. The y-axis represents the mean proportion of infection on a given day associated with the transmission route. Transmission of ASF through the movement of swine has been separated into the movement of exposed, infected and detected swine to reflect the all-in-all-out production system assumed in our model. Transmission starts at day two as day one is the time step in which the initial infection was seeded in the population.

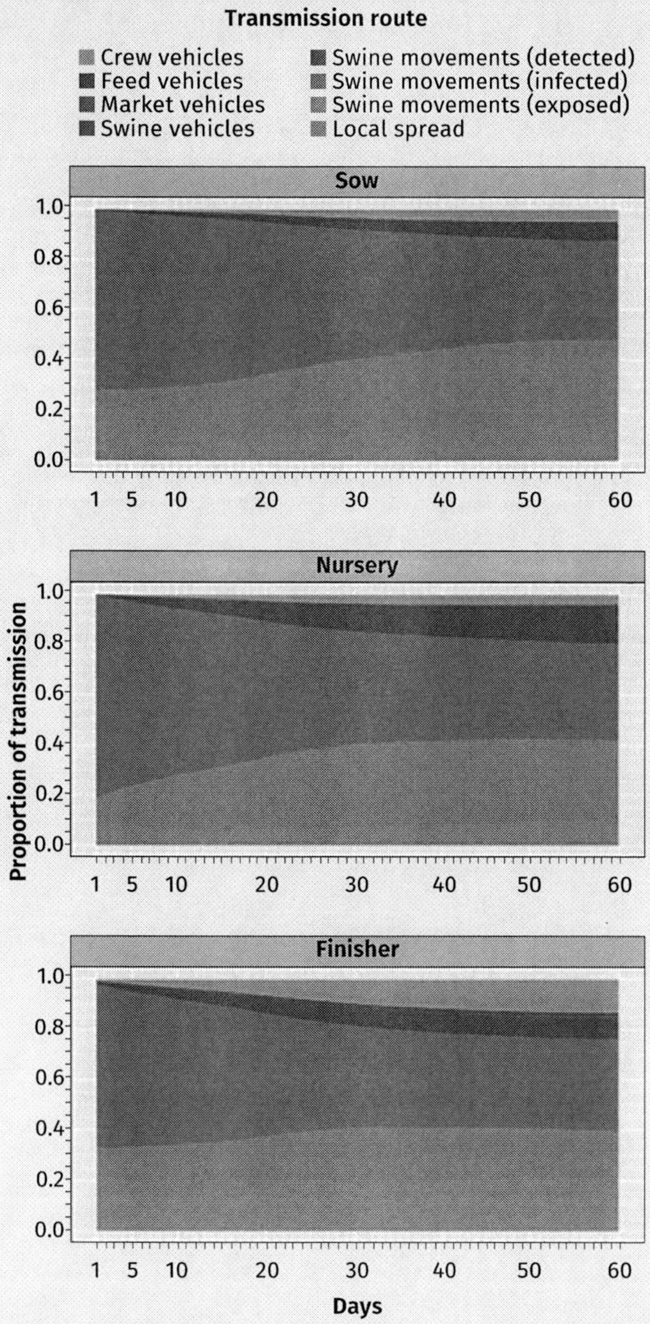

https://doi.org/10.54846/am2023/10

Actinobacillus pleuropneumoniae serotype 15 outbreak investigation: Comparative whole genome sequencing of pan-North American isolates

Antony Vincent[1], PhD; Alyona Michael[2], DVM, PhD, DACVP; Marcelo Gottschalk[3], DVM, PhD, DrHc; Jake Alsdurf[4], MS; Isadora Machado[2], DVM; Derald Holtkamp[2], DVM, MS; Marcelo Almeida[2], DVM, MS, PhD
[1]Département des sciences animals, Laval University, Québec, Canada; [2]Veterinary Diagnostic and Production Animal Medicine, Iowa State University, Ames, Iowa ; [3]Département de Pathologie et Microbiologie, University of Montreal, Québec, Canada ; [4]Cambridge Technologies, Worthington, Minnesota

Introduction

Actinobacillus pleuropneumoniae (App), the bacterial mediator of porcine pleuropneumonia, is subclassified into 19 known serotypes, which vary in pathogenicity depending on virulence factor expression. Highly virulent strains of App associated with high mortality outbreaks are uncommon in the United States, although sporadic losses can occur in the finisher phase within endemically infected systems. However, App's poor survival outside the living host limits its propensity for lateral transmission, which may contribute to the lack of routine vaccination against this agent in United States commercial operations.

In November 2021, swine finisher sites from 9 unrelated production systems in Iowa began reporting sudden severe App outbreaks, with peak weekly mortality exceeding 50% in some instances. Notably, affected sites were clustered within a narrow geographic radius in central Iowa, and isolates from tissues received by Iowa State University Veterinary Diagnostic Laboratory (ISUVDL) were identified as App serotype 15 (App15). Historically, the most commonly isolated App strains from diagnostic cases processed at ISUVDL belong to serotypes 7 and 8. App15 is associated with high virulence in Australia but is infrequently isolated and not commonly associated with high mortality outbreaks in the United States.

Methods

The sudden emergence of a readily laterally transmissible high-virulence strain of App15 triggered an epidemiologic investigation and genetic sequencing efforts aimed at identifying possible geographic and temporal origins for the prevalent strain and screening for genetic shifts in virulence factors that might account for its high pathogenicity. Fifty-two (52) App15 isolates were selected from the archives of ISUVDL and Groupe De Recherche Sur Les Maladies Infectieuses En Production Animale (GREMIP), the reference laboratory for App serotyping in Québec, Canada. Hybrid whole genome sequencing was performed on Illumina and Nanopore platforms, permitting genome reconstruction of isolates from the 2021 outbreak and comparison against historical pan-North American isolates derived from routine diagnostic cases received at ISUVDL and GREMIP.

Results and discussion

Preliminary analysis successfully characterized primary virulence factors, antibiotic resistance genes, and approximate genetic identity between App15 isolates. Sequence homology between 2021 outbreak isolates has confirmed strain identity and indicated clear phylogenetic divergence from historical Australian and pan-North American App15 isolates, including those from known endemic herds in Iowa. Ongoing sequence analysis has furthermore identified a regulator gene mutation that appears to correlate with the emergence of the virulent App15 outbreak strain. Additionally, insertional recombination events were documented in *Apx* toxin genes of the outbreak isolates. While the significance of these genetic permutations to the App15 clinical phenotype will require further exploration, these findings represent a promising avenue for investigating the genetic underpinnings of high-virulence App strain emergence.

https://doi.org/10.54846/am2023/11

Dynamics of infection of disease-associated *Streptococcus suis* (DASS) in the lactation phase

R. Mugabi[1]; A. P. Silva[1]; C. Haden[2]; T. Williams[2]; M. Rotolo[3]; J. Brown[1]; R. S. Jansen[2]; C. Scmidtt[2]; J. Nerem[2]; P. Harms[3]; L. Weinert[4]; G. Li[1]; O. Sahin[1]; C. Rademacher[1]; M. Gottschalk[5]; J. Wells[6]; L. Karriker[1]; A. Tucker[4]; M.J. Clavijo[1,3]
[1]Iowa State University, Ames, Iowa; [2]Pipestone Veterinary Services, Pipestone, Minnesota; [3]PIC North America, Tennessee; [4]University of Cambridge, Cambridge, United Kingdom; [5]Université de Montréal, Montréal, Canada; [6]Wageningen University, The Netherlands

Introduction

Streptococcus suis is an endemic Gram-positive facultative anaerobic bacterium that causes a disease characterized by meningitis, arthritis, endocarditis, septicemia, and pneumonia in pigs worldwide.[1] Although the bacterium can cause ailments in various pig age groups, infections are common in young pigs (5 - 10 weeks old), with mortality reaching 14%, causing economic losses to the farmers. *S suis* is part of the pigs' microbiota, and its natural habitat is the nasal and tonsils (upper respiratory tract). And has been isolated from the gastrointestinal and genital tracts.[2] Both commensal and pathogenic strains of *S suis* reside in the upper respiratory tract of pigs.

Disease-associated *Streptococcus suis* (DASS) is one of the most impacting bacterial diseases in swine worldwide. A key aspect to the control and prevention of DASS in swine farms hinges upon accurate detection of the disease-causing strains within the herd. This can be challenging due to its commensal nature, variability of on farm sampling methodologies, and lack of feasible diagnostic methods that predict virulence with high sensitivity. Developing improved sampling and testing methodologies allow veterinarians and producers to improve *S suis* surveillance programs, make timely adjustments to prevention programs. Therefore, novel diagnostic strategies are needed to understand the infection dynamics of virulent strains and the value of potential interventions. Understanding the infection dynamics of key agents that drive antimicrobial use in swine farms is paramount, and there is limited data on the infection dynamics of DASS strains in swine farms. The overall goal of this study was to describe the parity effect, optimal sampling site for DASS detection, and dynamics of infection of DASS in healthy dams, piglets, and the environment using a novel DASS qPCR assay.

Methods

In this study, we implemented a novel way of tracking DASS in a prospective cohort study in two pig flows, using an informative virulence-associated marker. A sensitive and specific real time PCR was validated, optimized and used to test antemortem samples from a longitudinal study to track DASS dynamics (eg, detects only disease associated strains), without the need to pursue bacteriological culture. In this study, 50 dams (25 P0 and 25 p4+) from the same farrowing week per farm were selected via stratified random for sampling by tonsil, nasal and vaginal swab collection at farrowing, and only tonsil swabs at weaning. In farm 2, additional samples were collected from sows, which included fecal swabs, farrowing crate swabs, and udder swabs. On the same day of farrowing, 14 piglets were randomly selected from each of the 50 litters, between 2 - 12 hours post birth. Piglets were ear-tagged, and one tonsil swab was collected from each piglet. A total of 631 and 629 piglets were sampled from farms 1 and 2, respectively. The piglets were sampled again at 7 and 21 days of age. All samples were also tested by a species-specific PCR (eg, detects all strains from the species by targeting the *recN* gene).

Results and discussion

All dams and piglets were *recN* positive in tonsil swabs, which highlights the high frequency of this species within pig population. Tonsil scrapings and nasal swabs revealed to be the samples of choice for detection of DASS in healthy dams compared to vaginal swabs. In one farm, nasal swabs provided the same or a better level of detection of DASS, compared to tonsil scrapings, suggesting the use of a more convenient sample type for surveillance of DASS. In terms of DASS carriage in dams from farm 1, the odds of 1130 (DASS marker) gene positivity was 82% lower in sows compared to gilts (odds ratio 0.17, 95% CI 0.35, 0.53, $P < 0.05$), a trend not seen in farm 2. The piglet DASS positivity followed a similar pattern between the two farms. In these herds between 20-70% of piglets were positive at birth, with a significant decrease in prevalence observed at day 7 (18-45%), followed by a significant increase in the prevalence of 1130 at day 21. The decrease at day 7 could reflect a transient detection at birth, or the effect of antibodies in colostrum. Furthermore, the DASS positivity was 83% and 63% lower in piglets from sows than piglets from gilts on farms 1 and 2, respectively ($P < 0.05$), highlighting a potential parity effect on DASS piglet carriage. At the litter level, the average litter prevalence was 32.7% and 64.97%, for farm 1 and 2, respectively. DASS was detected in farrowing crates, fecal, and udder samples, suggesting other relevant sources of infection for piglets, that should be further explored. Determining the timing and prevalence of colonization with virulent *S suis* strains allows for the refinement of control and/or elimination strategies, such as strategic medication and vaccination, early weaning or test and removal of individuals harboring DASS.

References

1. Marcelo, G. and Segura, M. (2019) 'Streptococcosis', *Diseases of Swine*. 11th ed: John Wiley & Sons, Inc, pp. 934-950.

2. Thongkamkoon P, Kiatyingangsulee T, Gottschalk M. Serotypes of Streptococcus suis isolated from healthy pigs in Phayao Province, Thailand. BMC research notes. 2017 Dec;10(1):1-7.

https://doi.org/10.54846/am2023/12

Antibody response to rotavirus A and C in gilts and their piglets after prenatal natural planned exposure

Deepak Kumar[1]; Amanda Anderson[2]; Jeremy Pittman[3]; Douglas Marthaler[4]; Waithaka Mwangi[1]

[1]Department of Diagnostic Medicine/Pathobiology, College of Veterinary Medicine, Kansas State University, Manhattan, Kansas; [2]Department of Veterinary Diagnostic and Production Animal Medicine, College of Veterinary Medicine, Iowa State University, Ames, Iowa; [3]Smithfield Foods, 434 E Main St., Waverly, Virginia; [4]Indical Bioscience, Kansas City, Kansas

Rotaviruses (RVs) are a significant cause of diarrheal disease in piglets globally. Due to difficulty in propagating rotavirus A (RVA) and lack of cell culture systems for rotavirus C (RVC), natural planned exposure (NPE) remains one of the most common method in swine production to boost lactogenic immunity through pre-farrow exposure to live virus. However, the efficacy of NPE protocols in generating lactogenic immunity against RVA and RVC, and association between RV fecal shedding and antibody response have not been investigated earlier. There is also a lack of serological tools to assess true effectiveness of NPE protocols against RVA and RVC in swine. A longitudinal study was conducted to investigate the dynamics of genotype-specific antibody response to Rotavirus A and C NPE in gilts and transfer of lactogenic immunity to their piglets. Pregnant gilts were randomly allocated into 4 groups. Group 1 received 3 doses of NPE at 5, 4, and 3 weeks pre-farrow (WPF), group 2 received 2 doses of NPE at 5 and 3 WPF, group 3 received one dose of NPE at 5 WPF, and group 4 received no NPE (control group). Blood (gilts and piglets) and colostrum/milk (gilts) samples were collected at multiple time points. VP7 (G4, G5 RVA and G6 RVC) and truncated VP4* (P[7], P[23] RVA and P[5] RVC) antigens were expressed in mammalian and bacterial cells, respectively. Purified antigens were used to optimize indirect ELISAs to determine antibody levels against RVA and RVC in gilts and piglets. RVA and RVC positive piglet fecal samples were (< Ct 26) were selected for next generation sequencing (NGS) from weeks 0 - 3 to assess RVA and RVC shedding in the presence of lactogenic immunity, and to compare genetic changes with the parent NPE RV strains. For RVA, we observed that treatment group 1 with 3 doses of pre-farrow NPE resulted in significantly higher IgG and IgA levels in colostrum compared to other groups except for G4 IgA. Control group gilts had colostrum lowest antibody levels in the absence of NPE. In piglet serum, group 1 piglets had higher IgG titers for all four antigens at day 0 than other groups. Piglet serum revealed higher IgA titers for group 3 piglets at day 0 for all antigens, however, the levels were not significantly different than group 1 piglets. Importantly, RVA NPE stimulated antibodies in all groups regardless of the treatment doses and prevented G4, G5, P[7] and P[23] RVA fecal shedding prior to weaning in piglets. The G11 and P[34] strains from pre-weaning piglets differed at multiple amino acid positions with parent NPE strains. For RVA G-types, some of these sites are also neutralizing escape mutation sites (NEMs) and high variability at these sites might have contributed in the detection of G11 genotype in the farrowing room.

For RVC, we found that group 1 had higher colostral IgG and IgA titers compared to other groups. Interestingly, group 1 RVC antibody levels in day 0 piglet serum were either significantly (P[5]) or numerically (G6) higher than other study groups. All 4 groups had significantly or numerically lower RVC colostrum antibody titers than RVA, irrespective of G and P-type. Piglet serum RVC IgA titers at day 0 were significantly lower than RVA titers. Our results prove that gilt colostrum and piglet serum contain significantly lower levels of antibodies to RVC than RVA, which could be the reason for higher prevalence of RVC in neonatal piglets. Higher number of RVC positive litters and high viral load at week 1 correlated with low colostrum G6 and P[5] IgA levels. NGS revealed a RVC genotype combination of G6P[5] in pre-weaning piglet fecal samples, which was similar to the parent NPE G6 and P[5] RVC strains and also shared very high nucleotide and amino acid percent identities.

In conclusion, results of this study expand our understating of the antibody response to porcine RVA and RVC and the role of NPE in providing lactogenic immunity to naïve piglets. To the best of our knowledge this is the first time a comparative study has been carried out to investigate the differences in antibody response to RVA and RVC in swine in a commercial setting.

https://doi.org/10.54846/am2023/13

Effect of rotavirus vaccination of gilts prior to farrowing following natural planned exposure at breeding on piglet growth, prevalence of diarrhea and pre-weaning mortality

L. Britton[1], BS; M. Johnson[2]; J. Pittman[3], DVM, MS, DABVP
[1]Department of Animal Sciences, North Carolina State University, Raleigh, North Carolina; [2]Ferguson College of Agriculture, Oklahoma State University, Stillwater, Oklahoma; [3]Smithfield Hog Production, North Region, Waverly, Virginia

Introduction

Rotavirus (RV) is a prevalent pathogen in suckling and post-weaning pigs worldwide, leading to diarrhea, decreased growth rate and increased mortality.[1] Rotavirus type A (RVA) and C (RVC) are most common in pigs. Piglets are highly susceptible at birth and with no effective treatment, manipulation of maternally derived antibodies (MDA) is a common strategy utilized by veterinarians to protect the piglets and mitigate the impact of the disease.[2] Natural planned exposure (NPE) is one method used to expose breeding age gilts to live RV, stimulating immunity and development of MDA.[3] The use of RV vaccination in dams after NPE and prior to farrowing can be utilized to booster and maintain higher levels of MDA passed to the piglets.[3] The objective of this study was to compare growth, prevalence of diarrhea, and mortality in piglets born to gilts exposed to NPE prior to breeding and subsequently vaccinated with one of two homologous RV vaccines, either once or twice prior to farrowing or not vaccinated.

Materials and methods

The study was conducted on a 3,400 head multiparous breed-to-wean sow farm. Gilts from four consecutive breeding groups were given NPE containing RVA and RVC orally twice prior to breeding. After pregnancy confirmation, gilts were housed in pens in groups of 6 to 8. Gilts were randomly assigned within pens to one of five treatments. Treatment 1 (TX1) were unvaccinated controls (CNT). TX2 received 2 mL of Medgene prescription platform (ppVx) vaccine IM at 4 and 2 weeks prior to farrowing (PF). TX3 received 2 mL of Medgene ppVx IM one week PF. TX4 received 1 mL of Vaccine B IM at 4- and 2-weeks PF. TX5 received 1 mL of Vaccine B IM one week PF. Refer to Table 1 for treatment breakdown. The ppVx and VXB were manufactured using the same RVA and RVC isolates contained in the NPE. At farrowing and during lactation, no piglet movement was allowed in study gilt litters. At 2 to 4 days of age (doa), whole litter weights were taken and each piglet received 1 ml of 1xPBS diluted (1:10) NPE orally to ensure RV exposure (RVA AVG CT: 17.2, RVC AVG CT 17.7). Litter diarrhea scores (0 = none, 1 = < 2 pigs, 3 = 2 or more pigs scouring) were taken at 1, 3, 5, 10, 15, and 20 doa. Whole litter weights were taken just prior to weaning, and visibly small pigs were individually weighed and categorized if 10 lbs or less. Intrinsic calculations within Microsoft Excel were used to calculate means, standard deviations, and standard errors for individual measurements for each treatment group. These data were used to calculate One-Way ANOVA significance for either 1-dose or 2-dose treatment groups, with significance of between-group variation tested using Tukey's modified t-test. Statistical significance was determined for all differences demonstrating $P < 0.05$.

Results

The number of lightweight weaned pigs was statistically lower for TX2 than TX4, and trended lower than TX1 ($P < 0.05$). Average diarrhea score for TX2 was significantly lower than TX1 ($P < 0.05$), but not compared to TX4. Refer to Table 2 for result comparison by treatment. Average daily gain (ADG) for TX2 and TX3 were numerically higher (0.242 kg) than TX1, TX 4 and TX5 (0.230 - 0.233 kg).

Discussion

The results of this study demonstrate that the prescription platform vaccine resulted in less lightweight pigs and diarrhea than VXB or CNT. The lack of difference between the CNT and all vaccine groups, as has been demonstrated in other studies,[3] could be due to an already effective RV control program. Pooled litter fecal PCR confirmed RV within study litters, suggesting adequate challenge of pigs. The less excluded pigs and potential higher ADG could lead to a potential return on investment by producers but must be weighed with the cost and implementation of vaccination. Further research on effective RV vaccination as a control method needs to be done.

Table 1: Treatment specifications

Treatment number	1	2	3	4	5
Treatment type	Control No vaccination	2 mL Medgene 2 + 4 WPF	2 mL Medgene 1 WPF	1 mL Vaccine B 2 + 4 WPF	1 mL Vaccine B 1 WPF
Gilts enrolled	50	49	49	49	50

WPF = weeks prior to farrowing

Table 2: Results by treatment. All treatments were compared to control results for this analysis. Results listed include the Mean ± Standard Deviation.

	Control	Medgene 2 + 4 WPF	Medgene 1 WPF	Vaccine B 2 + 4 WPF	Vaccine B 1 WPF
Mortality	0.9 ± 1.3	1.0 ± 1.3	1.0 ± 1.2	1.2 ± 2.0	1.0 ± 1.6
Scour mortality	0.1 ± 0.3	0.2 ± 0.7	0.1 ± 0.3	0.0 ± 0.0	0.1 ± 0.6
Avg wean weight	13.9 ± 1.7	14.3 ± 1.9	14.2 ± 2.3	13.9 ± 2.4	14.2 ± 2.5
Weight gain	119.1 ± 22.9	121.6 ± 23.8	115.6 ± 27.0	118.8 ± 32.6	118.2 ± 31.4
Num lightweight	0.5 ± 0.7^{a}	0.2 ± 0.4^{b}	0.4 ± 0.8	0.8 ± 1.0^{a}	0.5 ± 0.9
Avg scour score	$0.7 \pm 0.4^{a,x}$	0.4 ± 0.4^{b}	0.5 ± 0.4^{y}	0.6 ± 0.5^{ab}	0.5 ± 0.4^{xy}

One dose regimens were compared to each other and controls, and two-dose regimens were compared to each other and controls, but one- and two-dose comparisons were not compared.

Values with differing superscripts are statistically significant ($P < 0.05$). Values without superscripts were not significantly different from controls or each other.

WPF = weeks prior to farrowing

References

1. Vlasova AN, Amimo JO, and Saif LJ. 2017. Porcine rotaviruses: Epidemiology, immune responses and control strategies. Viruses 9(3): 48.

2. Groth D. 2014. Clinical management of rotavirus. American Association of Swine Veterinarians. 561-562.

3. Pittman J. 2022. Rotavirus: Current experiences and thoughts with prevention and control. American Association of Swine Veterinarians. 389-396.

https://doi.org/10.54846/am2023/14

Testing the efficacy of different combinations of time and temperature to inactivate PRRSV and PEDV

Mafalda Mil-Homens; Ethan Aljets; Rodrigo Paiva; Isadora Machado; Guilherme Cezar; Onyekachukwu Osemeke; Daniel Moraes; Swaminathan Jayaraman; Mckenna Brinning; Lauren Tidgren; Madison Durflinger; Mallory Wilhelm; Vivian Flores; Jolie Frenier; Daniel Linhares; Jianqiang Zhang; Derald Holtkamp; Gustavo Silva
Veterinary Diagnostic and Production Animal Medicine, College of Veterinary Medicine, Iowa State University, Ames, Iowa

Introduction

The use of foggers is an established method to decontaminate fomites in supply entry rooms on swine farms. Recent research has begun to question this method of inactivating pathogens, especially in complex situations where pathogens may be shielded by organic material or blind spots.[1] Different combinations of temperature and time (eg, 60°C [140°F] for 10 minutes or 65°C [150°F] for 5 minutes) have been shown to be effective for inactivating pathogens in contaminated trailers.[2] Therefore, the objective of this study was to evaluate different combinations of time and temperature at inactivating in porcine reproductive and respiratory syndrome virus and porcine epidemic diarrhea virus decrease in concentration on contaminated surfaces commonly found in supply entry rooms in swine farms.

Materials and methods

Two surfaces were tested, diamond plate aluminum and cardboard, at four temperatures (20°C [68°F], 30°C [86°F], 40°C [104°F] and 50°C [122°F]) in combination with six holding-times (15 minutes, 60 minutes, 6 hours, 12 hours, 24 hours, and 36 hours). The PRRSV MN184 and 1-4-4 L1C variant strain, and the NC46469/2013 PEDV isolates were used. Three replicates for each treatment were performed and each coupon was inoculated with 2 ml of virus for positive controls and test groups, or 2 mL of minimum essential media for negative controls. Once the surface temperature reached the desired temperature, the coupons were held for the designated holding time. Linear models were built for each combination of virus, surface and temperature and half-lives were estimated.

Results

PRRSV 1-4-4 L1C Variant was still present on both the aluminum and the cardboard surfaces after 36 hours at 20°C. However, no virus was detected after 12 hours at 30°C, after 6 hours at 40°C, or after 15 min at 50°C on the aluminum surfaces. In contrast, no virus was detected after 6 hours at 30°C, 40°C, or 50°C on the cardboard surfaces. For PRRSV MN184 both surfaces had virus after 36 hours at 20°C. At 30°C, no virus was detected after 36 hours on both surfaces. On the aluminum surface, no virus was detected after 24 hours at 40°C and after 6 hours at 50°C. On the cardboard surface, no virus was detected after 12 hours at either 40°C or 50°C. Regarding PEDV, the virus was present after 36 hours at 20°C on both the aluminum and the cardboard surfaces. On the aluminum, no viable virus was detected at 30°C and 40°C after 6 hours, and 50°C after 15 min. On the cardboard, no viable virus was detected after 12 hours at 30°C, 40°C, and 50°C. At 30°C, the infectivity half-life was 50 minutes in both surfaces for PRRSV MN184, 5 hours on cardboard and 2 hours on aluminum for PRRSV 1-4-4 L1C variant, and 36 minutes on cardboard and 47 minutes on aluminum for PEDV. At 40°C, the infectivity half-life was 16 minutes on aluminum and 20 minutes on cardboard for PRRSV MN184, 4 hours on aluminum and 47 minutes on cardboard for PRRSV 1-4-4 L1C variant, and 46 minutes on aluminum and 52 minutes on cardboard for PEDV. Finally, at 50°C the half-life was 21 minutes on aluminum and 16 minutes on cardboard for PRRSV MN184, less than 15 minutes on aluminum and 47 minutes on cardboard for PRRSV 1-4-4 L1C variant, and less than 15 minutes on aluminum and 36 minutes on cardboard for PEDV.

Conclusion

Under the conditions of this study, there seems to be a difference between PRRSV isolates, with PRRSV 1-4-4 L1C variant having increased half-lives at higher temperatures, it was also observed that aluminum surfaces took longer to reach the desired temperature. The minimum time that surfaces should be held at 30°C is 24 hours, at 40°C is 12 hours and at 50°C is 6 hours, considering that the surface should reach the desired temperature, and that difference in materials can impact the time that the surface should be held at.

References

1. Leuck, Kristen. 2020. "Automatic Foggers Fall Short in Disinfecting Hog-Farm Supplies." https://www.Thepigsite.Com/Articles/Automatic-Foggers-Fall-Short-in-Disinfecting-Hog-Farm-Supplies. May 14, 2020. **https://www.thepigsite.com/articles/automatic-foggers-fall-short-in-disinfecting-hog-farm-supplies**

2. Dee, Scott A., John Deen, Satoshi Otake, and Carlos Pijoan. 2004. "An Experimental Model to Evaluate the Role of Transport Vehicles as a Source of Transmission of Porcine Reproductive and Respiratory Syndrome Virus to Susceptible Pigs." Canadian Journal of Veterinary Research 68 (2): 128–33.

https://doi.org/10.54846/am2023/15

Attenuation phenotypes and protective efficacy of cell culture adapted PEDV non-S INDEL strain

L. Schumacher, DVM, PhD; Qi Chen, DVM, PhD; Holly Salzbrenner, DVM; Phillip Gauger, DVM, PhD; Jinhui Zhu, BS; Wannarat Yim-im, PhD; Mallory Wilhelm, BS; Ethan Aljets, BS; Gaurav Rawal, DVM, MS; Joseph Thomas, DVM, MS; Michael Welch, DVM, PhD; Drew Magstadt, DVM, MS; Jianqiang Zhang, MD, PhD
Department of Veterinary Diagnostic and Production Animal Medicine, College of Veterinary Medicine, Iowa State University, Ames, Iowa

Introduction

Porcine epidemic diarrhea virus (PEDV) continues to be an important driver of neonatal disease and economic loss to the US pork industry.[1] Currently, two genetically distinct strains are circulating in US herds and are designated as non-S INDEL (G2) and S-INDEL (G1b) PEDVs.[2,3] However, a safe and effective live attenuated vaccine is lacking. We previously demonstrated the non-S INDEL PEDV is overall the better choice of virus strain for PEDV MLV vaccine development in various age of pigs.[4] Our aims in the present research are to: 1) compare the virulence/attenuation of five lower passages (P7, P25, P50, P75 and P100) of a non-S INDEL PEDV isolate USA/IN19338/2013 in neonatal piglets (Study 1), 2) further compare the virulence/attenuation of higher passages (P100 and P200) of the same isolate in neonatal piglets followed by challenge to evaluate protective efficacy (Study 2), and 3) compare virus attenuation phenotypes with virus genetic changes after serial passages in cell culture.

Methods

A non-S INDEL PEDV isolate USA/IN19338/2013 obtained in our laboratory[5] and characterized as a virulent strain[6] was serially passaged in Vero cells (ATCC CCL-81).

In Study 1, 5-day-old PEDV naïve piglets (n = 60) were divided into six groups (10 pigs/group) and orogastrically inoculated with PEDV P7, P25, P50, P75, P100, or virus-negative culture medium, respectively (10 mL per piglet; all isolates had a titer of 10^4 $TCID_{50}$/mL) for the duration of 0-7 days post inoculation (DPI).

In Study 2, twelve PEDV negative pregnant sows were divided into 4 groups (3 sows/group) to farrow piglets. The piglets (3 to 7-day-old) were orogastrically inoculated with PEDV P7 (10^3 $TCID_{50}$/piglet), PEDV P100 (10^4 $TCID_{50}$/piglet), PEDV P200 (10^4 $TCID_{50}$/piglet), or virus-negative culture medium, respectively. Approximately 5 piglets from each group were randomly necropsied at 4 DPI for pathological examinations. At 21 DPI (4-week-old), remaining piglets were weaned, and sows euthanized. At 28 DPI, all piglets were orogastrically challenged with PEDV P8 (10^5 $TCID_{50}$/piglet) and necropsied one week later (35 DPI).

In both studies, clinical signs of diarrhea, lethargy, and body condition were assessed following the criteria described in the previous study.[6] Gross lesions, histopathological lesions, and immunohistochemistry staining of small intestines were conducted following previously described procedures.[6-8] PEDV RNA levels in rectal swabs were determined using a previously described quantitative PEDV nucleocapsid gene-based real-time RT-PCR.[9] Virus neutralizing antibody was tested as described previously.[9] PEDV antibody in serum was also determined using a PEDV S1-based multiplex fluorescent microsphere immunoassay (FMIA).[10] Whole genome sequences of selected serially passed viruses were determined using next-generation sequencing (NGS) technology as previously described.[5] All statistical analyses were performed using GraphPad Prism 8.0.

Results

In Study 1, clinical observations, virus shedding in rectal swabs, histopathologic lesions, and IHC staining data suggested that the P75 and P100 viruses were more attenuated than the P7, P25 and P50 viruses. However, the P75 and P100 viruses still caused severe diarrhea, suggesting insufficient attenuation.

In Study 2, P7-inoculated piglets had severe diarrhea, lethargy, and poor body condition as well as 75% mortality. P100- and P200-inoculated piglets had almost normal body condition and no lethargy; however, P100 inoculation caused severe diarrhea whereas P200-inoculated piglets had minimal diarrhea. P200-inoculated piglets shed significantly lower level of virus in fecal swabs at 1-3 DPI compared to P7- and at 1 DPI compared to P100-inoculated pigs. P200-, P100-, and P7-inoculation all significantly reduced or prevented fecal virus shedding after challenge at 28 DPI. P200- and P100-inoculated piglets developed similar levels of anti-PEDV IgG antibody (PEDV S1-based FMIA assay) and neutralizing antibody in serum although the antibody levels were lower than that induced by P7-inoculation at 28 DPI.

Whole genome sequencing via NGS indicated that the P7, P25, P50, P75, P100 and P200 PEDVs all had 28,038 nucleotides in length, excluding the 3' poly(A) tail. These viruses had > 99.84% nucleotide (nt) identity between each other at the whole genome level. Sequence alignment indicated that no nucleotide deletion or insertion occurred across these passages of PEDV but nucleotide and amino acid substitutions were detected at multiple positions. For example, amino acid changes were identified in nsp2, nsp3, nsp4, nsp5, nsp6, nsp7, S, E, M and N proteins. In addition, one nucleotide substitution was identified in ORF3 (24908A > T) that resulted in a truncated ORF3 protein in P25 - P200 viruses due to early stop codon of deduced amino acid translation.

Discussion

Based on previous data from our laboratory[4], non-S INDEL PEDV overall appears to be a better choice of virus strain for development of a PEDV MLV vaccine for use in various age of pigs. In the present research, we aimed to attenuate a highly virulent US non-S INDEL isolate towards development of a MLV vaccine. Specifically, we serially passed one non-S INDEL isolate, PEDV USA/IN/2013/19338, up to 200 passages in Vero cells and then characterized the *in vitro* growth kinetics and *in vivo* virulence/attenuation phenotypes and protective efficacy of viruses at different passages together with determination of whole genome sequences of these viruses. In Study 1, selected passages up to 100 were not attenuated in 5-day-old pigs based on severity of clinical signs. Therefore, the virus isolate was further propagated up to passage 200. In Study 2, when compared to piglets inoculated with P100 and P7, neonatal pigs inoculated with PEDV P200 showed significantly milder clinical signs, lower fecal viral shedding early on, and prevented shedding when challenged at weaning. Overall, the non-S INDEL isolate at passage 200 was attenuated, capable of providing protective immunity, and appears to be a promising MLV vaccine candidate. Whole-genome analysis of PEDV P7, P25, P50, P75, P100 and P200 revealed some amino acid mutations and truncation of ORF3 which may be associated with virus attenuation during serial passages in cell culture. However, functionality of these observed amino acid changes and their contributions to PEDV attenuation need to be assessed using reverse genetics approach in future studies and these may help elucidate the molecular mechanisms of PEDV attenuation. Overall, PEDV P200 is a safe and promising vaccine candidate. These data warrant further investigation of PEDV P200 for its potential reversion to virulence after serial passages in pigs, for its passive protection to suckling piglets after vaccinating pregnant sows, and for its benefits of being used in weaned and grower-finisher pigs.

References

1. Jung K, Saif LJ, Wang Q. 2020. Porcine epidemic diarrhea virus (PEDV): An update on etiology, transmission, pathogenesis, and prevention and control. *Virus Res* 286:198045.

2. Vlasova AN, Marthaler D, Wang Q, Culhane MR, Rossow KD, Rovira A. 2014. Distinct characteristics and complex evolution of PEDV strains, North America, May 2013–February 2014. *Emerg Infect Dis* 20:1620-8.

3. Wang Q, Vlasova AN, Kenney SP, Saif LJ. 2019. Emerging and re-emerging coronaviruses in pigs. *Curr Opin Virol* 34:39-49.

4. Schumacher, Loni L. 2022. Characterization of Virus Strain- and Pig Age-Dependence of Porcine Epidemic Diarrhea Virus: Implications for Vaccine Development. Order No. 29258562, *Iowa State University.* **https://www.proquest.com/dissertations-theses/characterization-virus-strain-pig-age-dependence/docview/2728053286/se-2**

5. Chen Q, Li G, Stasko J, Thomas JT, Stensland WR, Pillatzki AE, Gauger PC, Schwartz KJ, Madson D, Yoon KJ, Stevenson GW, Burrough ER, Harmon KM, Main RG, Zhang J. 2014. Isolation and characterization of porcine epidemic diarrhea viruses associated with the 2013 disease outbreak among swine in the United States. *J Clin Microbiol* 52(1):234-243.

6. Chen Q, Gauger PC, Stafne MR, Thomas JT, Madson DM, Huang H, Zheng Y, Li G, Zhang J. 2016. Pathogenesis comparison between the United States porcine epidemic diarrhea virus prototype and S-INDEL-variant strains in conventional neonatal piglets. *J Gen Virol* 97:1107-1121.

7. Madson DM, Magstadt DR, Arruda PHE, Hoang H, Sun D, Bower LP, Bhandari M, Burrough ER, Gauger PC, Pillatzki AE, Stevenson GW, Wilberts BL, Brodie J, Harmon KM, Wang C, Main RG, Zhang J, Yoon KJ. 2014. Pathogenesis of porcine epidemic diarrhea virus isolate (US/Iowa/18984/2013) in 3-week-old weaned pigs. *Vet Microbiol* 174(1-2):60-68.

8. Chen Q, Gauger P, Stafne M, Thomas J, Arruda P, Burrough E, Madson D, Brodie J, Magstadt D, Derscheid R, Welch M, Zhang J. 2015. Pathogenicity and pathogenesis of a United States porcine deltacoronavirus cell culture isolate in 5-day-old neonatal piglets. *Virology* 482:51-59.

9. Thomas JT, Chen Q, Gauger PC, Giménez-Lirola LG, Sinha A, Harmon KM, Madson DM, Burrough ER, Magstadt DR, Salzbrenner H, Welch MW, Yoon KJ, Zimmerman JJ, Zhang J. 2015. Effect of porcine epidemic diarrhea virus infectious doses on infection outcomes in naïve conventional neonatal and weaned pigs. *PLoS ONE* 10(10):e0139266.

10. Giménez-Lirola LG, Zhang J, Carrillo JA, Chen Q, Magtoto R, Poonsuk K, Baum DH, Piñeyro P, Zimmerman J. 2017. Reactivity of porcine epidemic diarrhea virus structural proteins to antibodies against porcine enteric coronaviruses: diagnostic implications. *J Clin Microbiol* 55(5):1426-1436.

https://doi.org/10.54846/am2023/16

Impact of sow farm PED outbreaks on the downstream nursery performance in the absence of PRRS acute herds

E. S. Magalhães[1]; T. Donovan[2]; D. Rosero[2]; G. Trevisan[1]; J. J. Zimmerman[1]; D. J. Holtkamp[1]; G. S. Silva[1]; D. C. L. Linhares[1]
[1] Veterinary Diagnostic and Production Animal Medicine Department, Iowa State University, Ames, Iowa;
[2] The Hanor Company, Enid, Oklahoma

Introduction

Porcine epidemic diarrhea virus (PEDV) in sow farms causes severe economic losses in naïve herds due to high mortality rates (up to 100%) in suckling piglets. Piglet survivability increases as sows are increasingly able to provide adequate levels of lactogenic immunity. Nevertheless, Alvarez et al,[1] reported an 11% increase in nursery mortality for the first batch of weaned piglets that survived the outbreak compared to the performance of piglets weaned before PEDV introduction. However, the overall impact on the nursery phase of PED positive sow farms after the acute phase and undergoing elimination has not been yet reported. Thus, the objective of this study was to describe changes on nursery mortality and average daily weight gain (ADWG) in groups originating from sow farms after the outbreak period until reaching stability.

Methods

A retrospective cohort study was conducted on historical data streams for 471 nursery groups from July 2021 to July 2022 in one US swine production system. The dataset included information on sow farm productivity, sow farm PRRS and PED weekly status, nursery performance reports, seasonality, and data from reports of disease(s) diagnosed by a pathologist for the groups analyzed. This report utilized a coding system developed by the pathologists at the ISU-VDL.[2] The dataset was then analyzed to compare nursery mortality and average daily weight gain (ADWG) of pig groups originating from PEDV positive unstable sow farms to that of groups originating from PED negative sow farms. For the purpose of this study, PEDV positive unstable was defined as the period of outbreak until achieving no clinical signs and 28 days of PCR-negative results on suckling pigs. The data were analyzed using a Linear Mixed Model (SAS PROC GLIMMIX) with the sow farm(s) originating the nursery cohort as the random effect, adjusting then the impact of PED by the groups` source. Tukey-test was utilized to quantify the effect of each variable on the outcome in the final model, and multicollinearity was assessed.

Results

The nursery mortality for this study population was 4.48% and 0.95 pounds (lbs.), respectively. Overall, the difference between PED positive unstable and negative flows was 4.4% higher for nursery mortality (6.8%a vs 2.4%b), and 0.08 lbs. lower for ADWG (0.90 lbs.a vs 0.98 lbs.b). The results of the overall PED status and the interaction with other variables are described in Table 1. Even though no sow farms for this study population were in PRRS category I-A (unstable-high prevalence), all farms were categorized as positive stable, either 2vx or 2fvi.

Discussion and conclusions

Contradictory to the common understanding that the impact of PEDV on the growing pigs` performance is relatively mild, especially for groups after the acute phase, this study demonstrated that nursery mortality increased by 4.4% for groups originating from positive unstable PEDV sow farms, particularly during the fall/winter season, for young average weaning age groups, for groups with a disease diagnosis assigned by a pathologist, and for both PRRS categories within this study population. Lastly, ADWG decreased by 0.08 lbs. for cohorts originating from positive unstable PED sow farms, especially during fall/winter seasons, and for nursery groups with a disease diagnosed record.

References

1. Alvarez, J., Sarradell, J., Morrison, R., Perez, A., 2015b. Impact of Porcine Epidemic Diarrhea on Performance of Growing Pigs. PLoS One 10, e0120532.

2. Derscheid RJ, et al. Disease diagnostic coding to facilitate evidence-based medicine: current and future perspectives. J Vet Diagn Invest 2021;33(3):419–427. Crossref. PubMed.

Table 1: Average nursery mortality (%) and ADWG for PED status and its interactions with other risk factors

Variables analyzed	Nursery mortality (%)		ADWG (lbs.)	
PED status & PRRS status	**2vx**	**2fvi**	**2vx**	**2fvi**
PED unstable	7.3%[a]	7.8%[a]	0.94[ab]	0.86[b]
PED negative	2.4%[b]	2.9%[b]	0.97[b]	1.01[a]
PED status & season	**Fall/Winter**	**Spring/Summer**	**Fall/Winter**	**Spring/Summer**
PED unstable	10.2%[a]	6.3%[b]	0.88[a]	0.91[a]
PED negative	2.4%[c]	2.1%[c]	0.98[b]	0.99[b]
PED status & weaning age	**19.5 days**	**22.5 days**	**19.5 days**	**22.5 days**
PED unstable	7.5%[a]	5.3%[a]	0.89[b]	0.94[ab]
PED negative	2.4%[b]	2.4%[b]	0.96[a]	1.01[a]
PED status & dx code*	**Positive**	**Negative**	**Positive**	**Negative**
PED unstable	9.3%[a]	6.9%[b]	0.87[a]	0.91[bc]
PED negative	3.9%[c]	2.1%[d]	0.95[b]	1.03[c]

* Dx code positive refers to groups with submission of tissue to the VDL for diagnostics, while negative means "no submission".

[abc] Different letters indicated a significant statistical difference between each of the four combinations of PED status and the individual variables analyzed (Tukey-Kramer test).

https://doi.org/10.54846/am2023/17

Evaluation of on-farm point-of-need testing using influenza A virus in swine (IAV-S) as a model

A. Tatnall[1], BS; C. LeFevre[2], DVM; D. Marthaler[3], PhD; A. Lower[2], DVM; N. Eddy[2] BS, MS
[1]University of Illinois College of Veterinary Medicine; [2]Carthage Veterinary Service, Ltd.; [3]INDICAL Biosciences

Introduction

Rapid and accurate diagnostics are an essential component of modern swine production. The diagnostic lab systems and overnight shipping have produced rapid, next-day results for most US swine production. However, development of on-farm, point-of-need (PON) testing would revolutionize the response speed to diagnostic results. We have recently experienced the progression of COVID-19 testing from PCR to point-of-need in home testing which dramatically influenced the cost of diagnostics and successful quarantine of infected individuals. Development of PON testing would allow quick assessment of farm disease status with the ability to adjust management practices, protocols, and pig flow. Additionally, the continual threat of foreign animal diseases, including the recent pressure of African swine fever virus entering the United States emphasize the importance of PON testing. However, there is a lack of knowledge surrounding the ease of use and accuracy of on-farm testing on swine farms.

Objective

The purpose of this study is to assess the ease-of-use and accuracy of the INDICAL Biosciences 'Easy Lab' model on a sow farm, using influenza A virus (IAV) as a model.

Materials and methods

The study took place on a 6000 head sow farm with endemic IAV. Sampling occurred 2 times per week, on consecutive days, for 4 weeks resulting in a total of 8 sampling periods. Six nasal swabs from wean age piglets were collected in each sampling period by sampling 5 piglets within a single nasal swab. To determine feasibility, a trained veterinary student oversaw the samplings by the farm staff and on-farm testing. This student trained the farm staff in running the test and subjectively assessed their ability to run the testing. The farm staff was also surveyed on their experience. For PON testing, nasal swab samples were placed in a microcentrifuge tube with PBS and manually agitated. The samples were extracted using the prefilled IndiMag Pathogen IM48 cartridge on the IndiMag 48s. The nucleic acid samples were pipetted into the appropriate PCR strip filled with INDICAL's virotype Influenza A Primers/Probes and IndiMix JOE and run in the IndiField. Results were automatically uploaded to Cloud storage for analysis. The samples and RNA were frozen and sent to a diagnostic lab for testing to compare the on-farm results. In the diagnostic lab, the nasal swabs were extracted in using the prefilled IndiMag Pathogen KF96 cartridge. The on-farm and diagnostic lab RNA were tested using INDICAL's virotype Influenza A Primers/Probes and IndiMix JOE.

Results

On samples with a Ct < 35, there was 100% agreement between the on-farm and lab testing with negative (n = 21) and positive samples (n = 19) Eight samples had disagreements with Ct values > 35 (Figure 1). The farm staff was able to complete on-farm sampling and testing during a normal day, and the procedure was easy. The farm had an error with one sample in which the RNA was not loaded into the on-farm PCR. The process required 20 minutes of hands-on time. Total run time from sample processing to RNA extraction was 30 minutes. Samples were then loaded into the PCR machine and results were available in 70 minutes. The total time from sample processing to results was 120 minutes. Which allowed testing to be completed within the confines of a normal farm day.

Discussion

The INDICAL's 'Easy Lab' model required only two-time points requiring focus and manual input, proving to be easy and feasible for farm staff. There was high equivalency between the on-farm and lab results, proving this is an accurate test for IAV in this study. The farm staff had some initial trouble with the proper pipetting technique. Proficiency occurred rather quickly with observation and practice. Early sampling during a wean event generated on-farm results by 10 AM. Point-of-need testing allows for altered management decisions or pig flows on an expedited schedule. Overall, this on-farm point-of-need pilot study on a sow farm demonstrates the testing is easy to use and accurate. This technology leads to potential use in all swine production phases. Furthermore, this technology has the potential to promote farm readiness in the face of a foreign animal disease outbreak.

Figure 1: Represents the samples with differing results based on testing location and modality. The black marker represents all IAV negative samples (n = 21) that were in agreement across all locations and modalities.

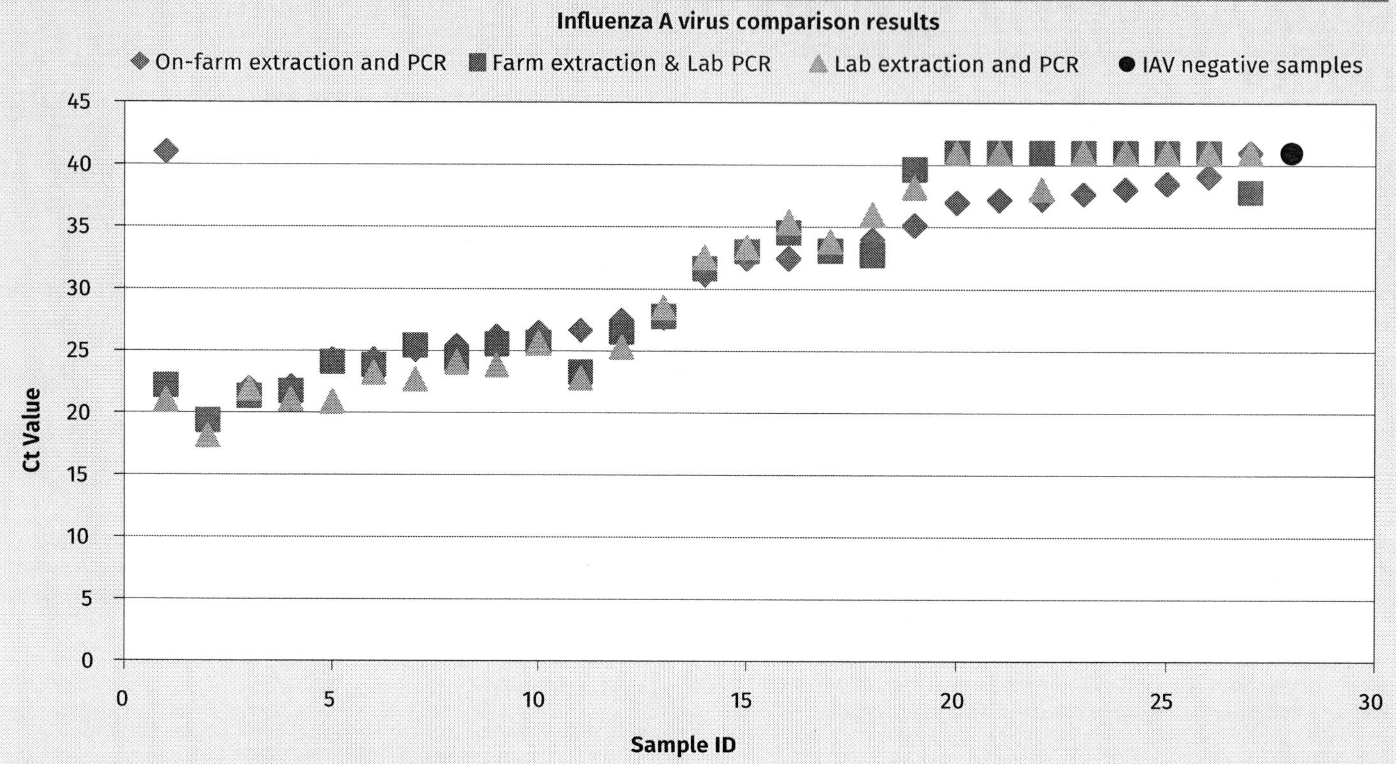

https://doi.org/10.54846/am2023/18

Comparison of IFA versus ELISA for assessing maternal and post-vaccination PCV2 antibody levels

E. Russell[1], MS; E. Larsen[1], MS; B. O'Brien[2], MS, VMD; B. Thacker[1], DVM, PhD, MBA
[1]College of Veterinary Medicine, Lincoln Memorial University, Harrogate, Tennessee
[2]Merck Animal Health, Lenexa, Kansas

Introduction

Measuring PCV2 serum antibody levels is performed in the US to assess potential vaccination interference by maternally derived antibodies (MDA) in order to properly time vaccinations and to evaluate post-vaccination antibody responses with vaccines that consistently induce a predictable, measurable titer.[1] Post vaccination antibody responses are often used to monitor staff compliance with an operation's vaccination protocol. The indirect immunofluorescent antibody assay (IFA) is generally considered to be the "gold standard" for measuring porcine circovirus type 2 (PCV2) serum antibodies. More recently, an ELISA test has become commercially available (Ingezim Circovirus IgG ELISA, Ingenasa) and compared to IFA, is less expensive and more easily performed in a diagnostic laboratory as existing equipment and staff can be utilized.

The objective of this study was to evaluate the PCV2 ELISA with regard to assessing MDA levels in young pigs and post-vaccination antibody responses in pigs vaccinated (or not) with Circumvent® PCV G2 (Merck Animal Health, Lenexa, Kansas) using a two-dose regimen.

Methods

For MDA assessment, 120 sera from non-vaccinated, 3-week-old pigs in a field survey were used. For post-vaccination responses, sera from 20 vaccinated and 20 non-vaccinated control pigs from an experimental study, and at the time of challenge (3 weeks post second vaccination), were used. IFA testing, using a PCV2b antigen, was performed at the Kansas State Veterinary Diagnostic Laboratory and ELISA testing was performed at the Merck Animal Health Diagnostic Laboratory. The IFA titer cut-offs for MDA that could interfere with vaccination and indicate a positive vaccination response were ≥ 640. For the ELISA, the test kit generates a "titer" by calculation and three cut-off titers were evaluated for MDA (160, 320, 640) and post vaccination responses (320, 640, 1280).

The five performance parameters (see Table 1) of the ELISA were based on the assumption that the IFA provided the "true" status of the sample.

Results

The results are presented in Table 1. For the MDA samples, the ELISA cut-off used impacted all five parameters and especially test specificity. For all parameters, the test results were in order based on the ELISA cut-off value used for the calculations except for negative predictive value and percent agreement where the ≥ 320 cut-off value was lowest. For test specificity, calculated values ranged from 89.2% (≥ 160 cut-off) to 58.7% (≥ 640 cut-off). For specificity, the values ranged from 17.5% to 78.9%; for positive predictive value, 71.9% to 84.0%; negative predictive value, 81.4% to 68.3% and percent agreement 66.1% to 68.3%.

For the post-vaccination samples., the calculated values for all 5 parameters were at least 95.2% and the calculated values for each parameter were the same for all three cut-off values. The calculated values were 100.0% for test sensitivity; 95.2% for specificity; 95.0% for positive predictive value; 100.0% for negative predictive value; and 97.5% for percent agreement.

Discussion

The performance of the ELISA with regard to assessing maternal antibody levels that might interfere with vaccination does not support the routine use of the test for that purpose. None of the five performance parameter values exceeded 90%, regardless of the titer cut-off.

Conversely, for assessing post-vaccination antibody responses, the ELISA results were aligned with the IFA in that all parameters exceeded 95%, regardless of the titer cut-off. Accordingly, the ELISA appears to be an acceptable alternative to the IFA for assessing post-vaccination antibody responses to two, 1 mL doses of Circumvent® PCV G2.

Table 1: The comparison between ELISA and IFA for measuring PCV2 maternally derived and post vaccination antibodies

Antibody type	IFA cut-off	ELISA cut-off	Sensitivity	Specificity	Positive predictive value	Negative predictive value	Percent agreement
Maternal	≥ 640	≥ 160	89.2%	17.5%	71.9%	81.4%	66.1%
		≥ 320	66.7%	49.1%	80.5%	58.8%	61.0%
		≥ 640	58.7%	78.9%	84.0%	68.7%	68.3%
Post-vaccination	≥ 640	≥ 320	100.0%	95.2%	95.0%	100.0%	97.5%
		≥ 640	100.0%	95.2%	95.0%	100.0%	97.5%
		≥ 1280	100.0%	95.2%	95.0%	100.0%	97.5%

Reference

1. Thacker, B., Creel, J., Blomme. PCV2 vaccination update: Field experiences and serological test comparisons. In: Proc. 46th Annual Meeting of the American Association of Swine Veterinarians, Orlando, Florida, 2015, pp. 205-208.

https://doi.org/10.54846/am2023/19

Prevalence of IAV-S in unvaccinated post-weaning piglets

Alyssa Ruston; Suzanna Storms, DVM
Applied Livestock Epidemiology Laboratory, College of Veterinary Medicine, University of Illinois, Urbana, Illinois

Introduction

Influenza A virus in swine (IAV-S) is a zoonotic pathogen that is endemic in the United States commercial swine herd. IAV-S is a major contributing pathogen to porcine respiratory disease complex (PRDC), has significant financial impacts on the pork industry due to declines in production, as well as the potential to cause public health concerns.[1] The weaned pig population is a particularly sensitive group to infection, reinfection, and transmission due to mixing of pigs in nursery and wean-to-finish housing. Weaned pigs are often undergoing the transition from acquired, passive immunity to the development of adaptive immunity. This, along with the stress associated with the weaning process, such as change in diet and adaptation to a new environment, makes the weaned pig population especially susceptible to infection.[2] Influenza is a negative sense, single stranded RNA virus, and is dependent on host cell machinery to replicate. Due to the lack of a proofreading mechanism and the presence of a segmented genome, influenza is particularly prone to point mutations resulting in antigenic drift.[3] Influenza is also capable of undergoing antigenic shift, resulting from reassortment events following coinfection of different influenza subtypes.[3] Because of the genetic variability between Influenza A viral strains, control and prevention of disease within swine populations can be difficult. Disease surveillance is an important aspect of detecting IAV-S and has traditionally been completed using nasal swabs or oral fluids. Our study aims to describe the prevalence of IAV-S infected piglets in unvaccinated post-weaning swine populations using nasal swabs. Understanding the prevalence of IAV-S in these sensitive populations will allow us to better understand and potentially mitigate the infectious challenges pigs face during maturation and development.

Methods

The sow farm from which the pigs were derived has an outbreak of H1N1 influenza during early gestation of these pigs, after being negative for several years. The sows were routinely immunized with the Zoetis FluSure XP vaccine. We followed 60 weaned pigs in a gilt development unit (GDU) that was connected by a hallway to the sow unit. The gilts were tagged at weaning and swabbed weekly for then weeks to monitor influenza shedding in nasal secretions. Nasal swabs were collected using commercial aluminum handled, polyester tipped swabs (Puritan). One swab was used for both nares, and the swabs were placed into 1 mL of Brain-Heart Infusion viral transport media and held on ice for transport to the lab. Samples were stored at -80°C until RNA extraction and RT-PCR could be performed. RNA extraction was completed via the KingFisher Flex Automated Purification instrument using the MagMAX Viral/Pathogen Nucleic Acid Isolation protocol. RT-PCR was completed using a QuantStudio 3 according to the VetMAX Gold SIV Detection kit. The cycle threshold values were analyzed according to the VetMAX Gold SIV Detection kit protocol, with manually normalized CT values below 38 being considered positive.

Results

Over the course of the ten weeks, 50/60 of the gilts were sampled for all periods. Of the 50 gilts that were sampled, 48 (96%) had at least one positive event and 12 of those had reinfection events, with three or more weeks between infections. 17 (34%) of the 50 gilts tested positive over two or more consecutive sampling periods. As shown in Figure 1, overall prevalence of IAV-S peaked during week one, with 87.5% prevalence, and with 32.7% at week two. After week two, the percentage of positive events declined progressively, which is expected given the time required to reach recovery from IAV-S. At week six however, the percentage of positive events increased to 11.8% compared to the 1.96% from the previous week. At week eight, the percentage declined again, eventually reaching 0% prevalence of IAV-S within this sample population.

Discussion

During an IAV-S infection, clinical signs can be present within one to two days post-infection and can resolve four to eight days post-infection due to rapid immune clearance of the virus.[4] In this sample population, we saw a very high prevalence of influenza in these post weaning pigs within the first one to two weeks after weaning. Reinfection events were also seen within the sample population, which leads us to believe that either our nasal swab technique was imperfect, or that the pigs recovered and were reinfected very soon after their first infection. To understand these results, further study, beginning with viral sequencing, is needed to understand the infection dynamics or to identify potential mutations in the virus. Understanding the shedding, infection, and reinfection dynamics will help veterinarians and producers update vaccination strategies in post weaning pigs to improve herd health.

Collection of nasal swabs and oral fluids are common sample types for the detection of IAV-S. However, oral fluids have been shown to have a higher probability of detecting IAV-S by RT-PCR.[5] Compared to nasal swab sampling, oral fluid sampling does not have the granularity to describe the number of infected animals, the length of time they are shedding the virus, or how the infection is moving around a group of animals.

Vaccination of sows is a common method of IAV-S control in weaning/weaned pigs. This method has been shown to prolong maternal antibody presence and partially protect against clinical disease, although piglets may still become infected and shed virus as maternal antibodies wane.[1] As previously mentioned, the sows of the pigs in this GDU were routinely vaccinated with the FluSure XP vaccine, which contains one strain of H1N1, one stain of H1N2, and two strains of H3N2. This would lead us to believe that maternal antibody transfer should provide protection against the H1N1 outbreak present on the farm. However, based on our results, we do not see evidence of protective maternal antibodies, as nearly all pigs were infected and shedding at the time of weaning, earlier than expected for maternal antibody decay. Alternatively, the sows may be producing adequately protective

Figure 1: Prevalence of influenza A virus (IAV-S) in nasal swabs collected weekly from post-weaning piglets.

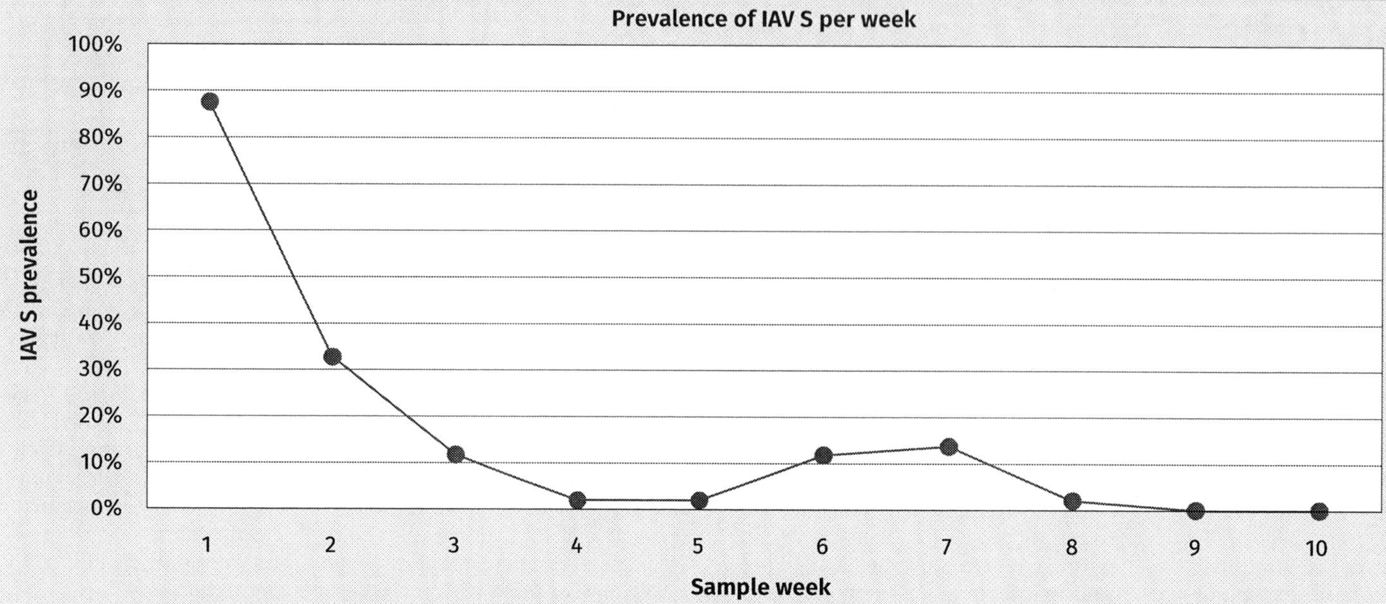

antibodies, and the amount of colostrum received is not enough to confer protection. Further study would be required to measure antibody titers within the vaccinated sows and transferred maternal antibodies within the gilts at time of weaning.

In summary, this study demonstrates a longer viral shedding period compared to the previously reported shedding period of five to seven days post-infection,[6] and that the post-weaning swine population is susceptible to repeated infections and for longer periods of time. Together, this leads to decreased profitability for producers. Possible interventions that would need more research could be to include IAV-S vaccines in post-weaning piglets or ensuring that sow vaccination provides adequate antibody responses. Both interventions may prove helpful to producers, lead to shorter shedding periods, and improve herd health.

References

1. Sandbulte MR, Spickler AR, Zaabel PK, Roth JA. Optimal use of vaccines for control of influenza a virus in swine. *Vaccines (Basel).* 2015;3(1). **doi:10.3390/vaccines3010022**

2. Campbell JM, Crenshaw JD, Polo J. The biological stress of early weaned piglets. *Journal of Animal Science and Biotechnology 2013 4:1.* 2013;4(1):1-4. **doi:10.1186/2049-1891-4-19**

3. Krammer F, Smith GJD, Fouchier RAM, et al. Influenza. *Nat Rev Dis Primers.* 2018;4(1):1-21. **doi:10.1038/s41572-018-0002-y**

4. Detmer SE. The clinical features, pathobiology, and epidemiology of influenza infections in pigs. *Animal Influenza.* Published online November 4, 2016:435-451. **doi:10.1002/9781118924341** CH16

5. Goodell CK, Prickett J, Kittawornrat A, et al. Probability of detecting influenza A virus subtypes H1N1 and H3N2 in individual pig nasal swabs and pen-based oral fluid specimens over time. *Vet Microbiol.* 2013;166(3-4):450-460. **doi:10.1016/J.VETMIC.2013.06.029**

6. Larsen DL, Karasin A, Zuckermann F, Olsen CW. Systemic and mucosal immune responses to H1N1 influenza virus infection in pigs.

https://doi.org/10.54846/am2023/20

Water-based foam depopulation in swine: Evaluating brain activity, animal behavior and logistical aspects

Jack Korenyi-Both[1]; Jorge Vidaurre[2]; Tim Held[2]; Justin Kieffer[3]; Magnus R. Campler[1]; Ting-Yu Cheng[1]; Steven Moeller[3]; Andrew S. Bowman[1]; Mark Flint[1]; Brad Youngblood[1]; Andreia G. Arruda[1]
[1]Department of Veterinary Preventive Medicine, The Ohio State University, Columbus, Ohio;
[2]Pediatric Neurology Department, Nationwide Children's Hospital, Columbus, Ohio;
[3]Department of Animal Sciences, The Ohio State University, Columbus, Ohio

Introduction

Foreign animal diseases (FADs) are a threat to the United States' swine industry. African swine fever, for example, which has spread in China, Europe, and in countries near the United States, can cause morbidity and mortality in swine populations and result in economic losses. The use of mass-depopulation to control and eradicate FADs will be needed in case of incursions to prevent additional spread and to mitigate effects on the national supply chain. In poultry, water-based foam (WBF) is listed as a preferred method by the American Veterinary Medical Association (AVMA) and is utilized in the event of disease outbreaks. While several depopulation methods are currently recommended for swine by the AVMA,[1] WBF is not currently one of them. However, recent work supports that it could be an attractive option especially when considering concerns associated with currently preferred methods or those permitted under constrained circumstances, including cost, logistics, animal welfare, user mental health impacts, and stockpiling capability. However, the lack of research on the use of WBF in swine challenges its consideration as a depopulation option.

Objectives

There were two objectives for this project, which had the overarching goal of investigating the effectiveness of WBF as a depopulation method for swine under field conditions. The goal of objective 1 was to use electroencephalogram (EEG) data to determine time to unconsciousness and brain death and subsequently investigate the effect of animal weight and sex on each of these outcomes. The goal of objective 2 was to investigate the effect of container fill rate (slow, medium, fast) on time to cessation of movement (COM), time to cardiac failure (CF), and to describe and quantify swine behavior during the foaming process.

Materials and methods

Objective 1

Twelve nursery pigs were individually fitted with six subdermal electrodes and a handheld electroencephalogram to track and collect brain wavelengths. Pigs were placed into a sling attached to a chain on a tractor and lowered into a 1.46 m3 plastic container and kept elevated to prevent pigs from having contact with the floor of the container. Baseline EEG data was collected 5 minutes prior to WBF application. WBF was applied and animals were kept submersed in the foam for 15 minutes. EEG data was analyzed by a clinical neurologist to determine times in each of four phases based on the definitions from a previous EEG study[2] (movement, normal EEG, HALF EEG, transitional EEG - as a measure of unconsciousness - and isoelectric EEG - as a measure of brain death). In brief, movement was defined as movement artifact produced by muscular activity, not representative of brain wavelengths. Normal EEG was determined using baseline EEG measurements prior to intervention. HALF EEG was defined as high amplitude, low frequency wavelengths. Transitional EEG was defined as EEG with amplitude < 1/2 of that of normal EEG; and finally, isoelectric EEG was defined as EEG with amplitude < 1/8 of that of normal EEG or little to no measured brain activity. A linear regression model was fit using Stata 17 to determine the effect of weight and sex (and their interaction) on time to unconsciousness and brain death; and significance was declared at $P < 0.05$.

Objective 2

The effect of fill rate on time to COM and time to CF, was tested in 2 replicates of 13 finisher pigs for each fill rate group (total 78 pigs). Slow, medium, and fast fill rate pertained to one, two, and three foam producing units, respectively. A subcutaneous logger (ST-Centi-HRT ACT Star-Oddi) capable of recording activity and electrocardiogram (ECG) was placed in the right or left caudal triceps of each animal using a 2 - 3-inch surgical incision closed by surgical staples. Animals were loaded into the front half of a custom-fabricated 64.5 m^3 rendering trailer, WBF was applied, and the container was left undisturbed for 7.5 minutes. Time to COM and CF began at the start of foam filling. ECG tracings were reviewed to determine time of death and type of cardiac failure (clinical asystole, third-degree atrioventricular block (AVB), atrial standstill or ventricular fibrillation (VF), and persistent or pulseless electrical activity (PEA) that occurred for each pig. Additionally, an animal welfare specialist observed all foaming events and performed a behavior assessment, which included quantification of foam surface breaks, escape attempts and vocalizations. A linear mixed effect model was used to compare time to COM and time to CF across treatment groups. Data were analyzed using Stata 17, and statistical significance was determined at $P < 0.05$.

Results

Objective 1

Average (mm:ss ± SD) time to unconsciousness and time to brain death was 01:53 ± 00:36 and 03:03 ± 00:56, respectively. Time in each EEG phase for animal subjects and an average can be seen in Figure 1. There was no effect of sex on time to unconsciousness ($P = 0.733$) or brain death ($P = 0.611$); and no effect of weight on time to unconsciousness ($P = 0.358$) or brain death ($P = 0.694$). Interactions between the two variables were also not significant for both time to unconsciousness ($P = 0.524$), and time to brain death ($P = 0.376$).

Figure 1: Results from objective 1, showing average and individual time (mm:ss) in each EEG phase. Blue = movement artifact. Yellow = Transitional EEG. Red = Normal EEG. Green = Isoelectric EEG. Not shown: 5-minute pre-intervention normal EEG, delay in onset of movement, and HALF EEG which is hidden in the intermittent component of transitional EEG.

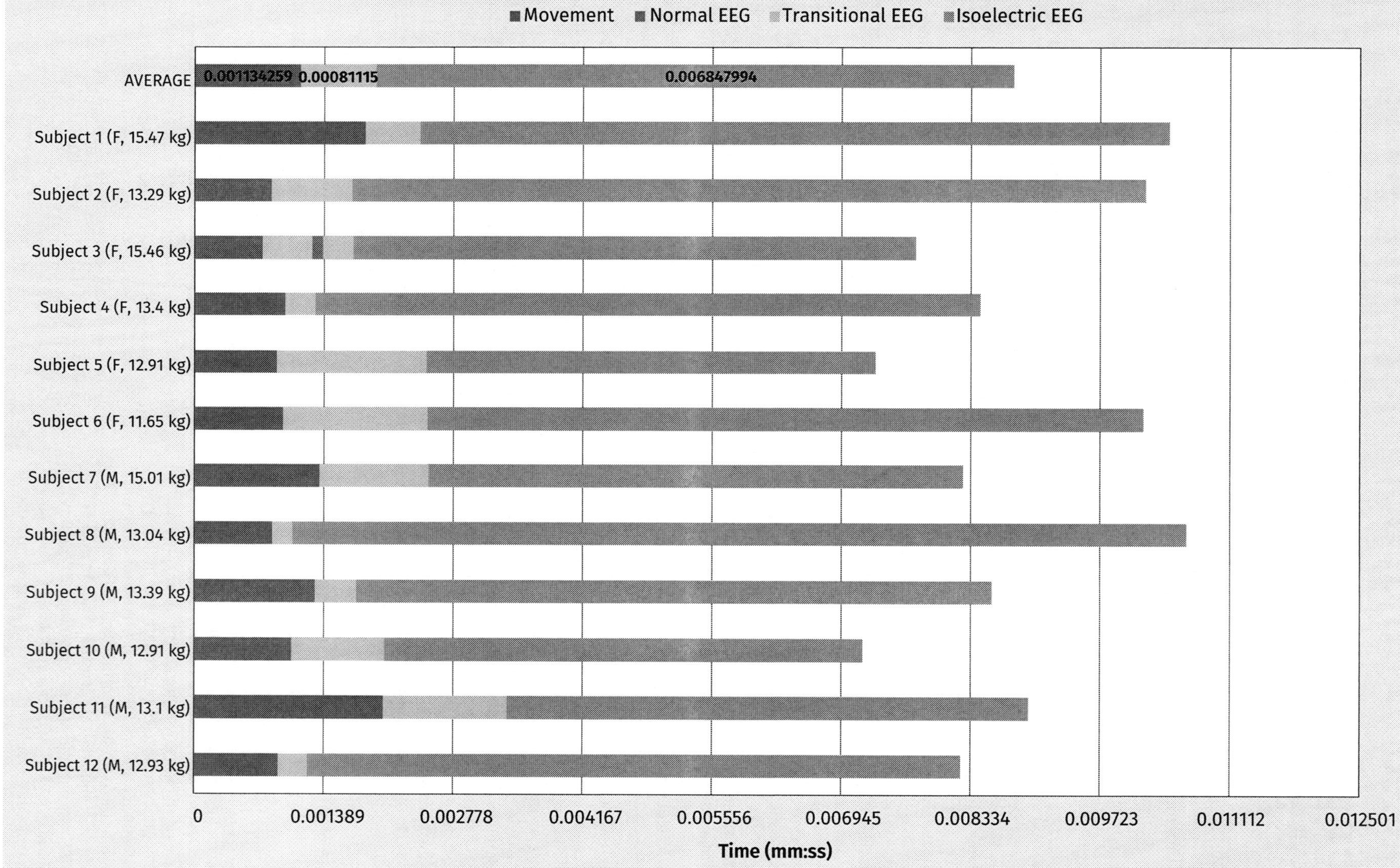

Objective 2

Average (mm:ss ± SE) time to COM for slow, medium, and fast foam rates were 05:22 ± 00:21, 03:32 ± 00:14, and 03:11 ± 00:13, respectively. The time to COM for slow fill rate groups was significantly longer when compared to both medium and fast groups ($P < 0.001$), as well as when compared to fast fill rate groups alone ($P < 0.001$). The average (mm:ss ± SD) time to CF was 09:18 ± 01:53, 10:50 ± 02:38, and 10:24 ± 01:49 for fast, medium, and slow foam rate groups, respectively. Time to CF was faster for pigs in the fast fill rate group compared to the slow and medium fill rate groups ($P < 0.001$). Behavior assessments showed 35 responses (35 foam surface breaks, 0 escape attempts), 5 responses (0 surface breaks, 5 escape attempts) and 11 responses (6 escape attempts, 5 surface breaks) for slow, medium, and fast fill rate groups, respectively. No vocalizations were heard for any of the fill rate groups. A summary of findings for objective 2 can be found in Table 1, which includes types of cardiac failure for each fill rate group.

Discussion

Objective 1 demonstrated that time to unconsciousness, an important consideration for the AVMA for depopulation method consideration, was under 2 minutes on average. This was the case regardless of animal weight and sex. This information is useful when preparing recommendations for the use of WBF; however, it is important to recognize that, given this was a terminal study using an experimental set-up with limited animal age (nursery pigs), generalization needs to be done with caution. Objective 2 demonstrated that a slow fill rate was associated with a longer time to COM, time to CF, and a higher number of responses compared to faster fill rates. These results highlight the importance of immersing pigs quickly in foam to minimize stress and to reduce the time to unconsciousness prior to death. This information also contributes to forming a recommendation for the use of WBF; however, limitations to be considered include the difficulty in assessing animal behavior due to the foam restricting full view of animals during the process, and this study being applied to similarly aged and sized pigs (finisher pigs). This work collectively adds scientific evidence for the use of WBF for rapid depopulation of swine and has the potential to contribute to the swine industry's need for multiple efficacious depopulation methods to implement in different scenarios.

References

1. American Veterinary Medical Association. AVMA Guidelines for the Depopulation of Animals: 2019 Edition. (2019).

2. Gibson, T. J., Oliveira, S. E. O., Costa, F. A. D. & Gregory, N. G. Electroencephalographic assessment of pneumatically powered penetrating and non-penetrating captive-bolt stunning of bulls. *Meat Sci.* 151, 54–59 (2019).

Table 1: Table depicting results for objective 2. For each fill rate group, average time (mm:ss ± SE) to cessation of movement (COM), average time (mm:ss ± SD) to cardiac failure (CF), type of arrythmia causing CF: asystole, ventricular fibrillation (VF), third-degree atrioventricular block (3rd AVB), and pulseless electrical activity (PEA), and responses (surface breaks and escape attempts) are depicted.

Group	Average time (mm:ss ± SE) to COM	
Slow	05:22 ± 00:21	
Medium	03:32 ± 00:14	
Fast	03:11 ± 00:13	
	Average time (mm:ss ± SD) to CF	**Type of fatal arrythmia**
Slow	10:24 ± 1:49	12 asystole, 5 VF, 9 PEA
Medium	10:50 ± 2:38	16 asystole, 8 VF, 2 PEA
Fast	9:18 ± 1:53	7 asystole, 1 3rd AVB, 15 VF, 2 PEA
	Surface breaks	**Escape attempts**
Slow	35	0
Medium	0	5
Fast	5	6

https://doi.org/10.54846/am2023/21

Impact of reduction of coccidiosis with Kenoco cleaner on piglet performance

E. Gibbs[1], BS; L. Britton[2], BS; J. Pittman[3], DVM, MS, DABVP
[1]University of Missouri College of Veterinary Medicine, Columbia, Missouri; [2]North Carolina State University, Raleigh, North Carolina; [3]Smithfield Hog Production – North Region, Waverly, Virginia

Introduction

Coccidiosis is an enteric disease of pigs caused by the protozoa, *Cystoisospora suis*. Pasty to watery diarrhea starts as early as five days of age (DOA) and continues through the post-weaning period.[1] The disease is characterized by high morbidity, low mortality, and reduced piglet weight gain. Coccidiosis continues to be a challenging disease due to the lack of availability of effective and legal antiprotozoals in swine, the resistance of the oocysts to many chemical disinfectants, and their survivability in the environment.[2,3] Breeding herd operations are left to combat coccidiosis with intense cleaning protocols and disinfectants that are labor intensive, unsafe, and often ineffective against inactivating *C suis* oocysts in the US. Kenoco (CID LINES; St. Paul, Minnesota) is a broad spectrum, amine-based cleaner shown to effectively inactivate *C suis* oocysts in other countries has recently become available in the US.[4]

Objectives

The purpose of this study was to evaluate the effectiveness of an amine based cleaner Kenoco, on environmental oocysts when applied to farrowing crates, and evaluate the effect on piglet coccidiosis.

Materials and methods

This study was conducted on a 1,500 head multiparous breed-to-wean sow farm in July and August in Eastern North Carolina. This site had a recent history and diagnosis of coccidiosis. The farrowing rooms enrolled in this study consisted of four rooms with twenty-four crates each with plastic/cast iron flooring and five rooms with twenty crates each with metal tri-bar flooring. Prior to loading sows, the rooms were soaked with a degreaser detergent and disinfected with bleach at 2oz/gal, and Virocid (CID LINES; St. Paul, Minnesota) at 1oz/246gal and let dry.[5] Each crate was randomly assigned to control (CNT) or treatment (TX) groups. Once crates dried, approximately 0.52 gal of Kenoco cleaner was applied to each TX crate at the labeled rate of 4% using a pump sprayer with a foam tip.[4] At loading, sow ID, parity, due date, and room/crate location were recorded. Within 24 hours of loading, fecal samples were collected from each sow and evaluated for *C suis* via fecal flotation by authors and confirmed by Iowa State University Veterinary Diagnostic Laboratory (ISU VDL) using semi-quantitative fecal flotation. ISU VDL speciated oocysts to determine the presence of *C suis* vs *Eimeria sp* oocysts. Litters were allowed to be fostered between the same treatment crates in the first 24 hours post farrowing. Litters born into enrolled crates from four sequential weaning groups were enrolled in the study between 2 to 5 DOA. At enrollment, litters were weighed, counted, and an average piglet weight was calculated using the litter's data. To prevent contamination, all team members wore gloves when handling piglets and piglets were placed in totes lined with clean trash bags when weighed. Trash bags and gloves were changed between each litter. Pooled litter fecal samples were collected twice weekly to obtain at least one sample at 10 to 14 DOA and 18 to 21 DOA. Samples were tested by a modified Wisconsin fecal floatation at the Merck Animal Health Laboratory in Lawrence, Kansas. Each litter was monitored bi-weekly for diarrhea severity using a standardized fecal scoring system. The fecal scoring system ranged from a severity of 1 to 4 with 1 being firm, 2 being pasty diarrhea, 3 being creamy diarrhea, and 4 being watery diarrhea. The fecal score also included a secondary classification following the number describing the prevalence of diarrhea in the crate. One plus sign was a few piglets exhibiting diarrhea where three plus signs indicated high prevalence of diarrhea throughout the entire litter. At weaning, each litter was weighed, counted, and an average weight and daily gain was calculated for each piglet using the litter's data. Piglets that were 4.54 kg and lighter were individually weighed and documented as light weight. Litters where a sow aborted, died, or was replaced were removed from the study. For univariate statistical analyses, continuous variables were analyzed by One-way or Kruskal-Wallis ANOVA depending on data distribution normality. Linear regression was also performed on selected continuous variables. For binary variables, the Chi Square test was used. Besides analyzing for main treatment effects, the impacts of floor type, parity and crate location within room were also determined. For all tests, a *P*-value $\leq$ 0.05 was considered significant. All statistical analyses were done using Statistix™ 10 (Analytic Software, Tallahassee, Florida).

Results

There were 196 sow/litter combinations enrolled in this study and ninety CNT and TX litters (n = 180) were included after removal of sows/litters that aborted or had contaminated litters from piglet mishandling between treatment crates. There were no differences in starting weight or fostering procedures between groups. There was no difference in weight gain between the groups at the end of the study. There were statistically lower oocysts shed between 10-14 DOA for TX groups than for CNT, but that was no longer observed at 18-21 DOA. There was a statistical difference between floor type and oocyst concentration such that litters with higher oocyst counts in fecal samples were observed more on plastic flooring than metal tri-bar as seen in Table 1. Piglets with higher diarrhea scores also had a higher prevalence of oocysts shed in their feces. Of the initial 196 dams loaded pre-farrow, 85 were confirmed positive for coccidia by authors and sent off for further testing at ISU VDL. Of the 85 dams, 72 were gilts, and 14 were sows. Out of 72 gilts, 21 (29%) were positive for *C suis* and 3 (4%) were positive for *Eimeria* sp. Of the 14 sows, 3 (21%) were positive for *C suis* and 1 (7%) was positive for *Eimeria* sp.

Discussion

This study aimed to describe the impact of the amine based cleaner, Kenoco on environmental oocyst presence within farrowing crates and its effect on piglet performance. Prior to loading sows, normal cleaning protocols were followed. It is crucial that all organic debris is cleared, surfaces are washed with

Table 1: Average diarrhea and oocyst scores per floor type

	Plastic flooring		Metal tri-bar flooring	
	Average diarrhea score (+/- SD)	Average oocyst score (+/- SD)	Average diarrhea score (+/-SD)	Average oocyst score (+/- SD)
Treatment	2.60 (+/- 0.61)	1.95 (+/- 1.15)	2.62 (+/- 0.57)	1.1 (+/- 1.05)
Control	2.48 (+/- 0.50)	2.25 (+/- 1.0)	2.78 (+/- 0.62)	1.45 (+/- 1.1)

normal sanitation protocols, allowed to fully dry, and then the cleaner applied and allowed to dry for best results. During this study, it was observed that embedded organic debris remained more with plastic flooring when compared to metal tri-bar flooring after power washing. Diarrhea caused by coccidia is a similar color to the plastic interlocking flooring used in this farm making it difficult for personnel to visualize trapped manure. Plastic flooring has more crevices than metal tri-bar which results in retained organic debris in the crates. Ultimately it was observed that plastic flooring is more difficult to effectively clean because it can harbor more oocysts after power washing than compared to metal tri-bar. This lessens the effectiveness of Kenoco as this protects oocysts from contact with the cleaner. It is possible that multiple applications of the cleaner may be required before adequate elimination of environmental oocysts results in reduction of disease. Effective sanitation with an amine-based cleaner is crucial for control of coccidiosis. Other current products are ineffective at inactivating *C suis* oocysts unless using ammonia or bleach at high, unsafe levels for animals and farm staff.[6,7]

The dam's influence, particularly younger parities, on environmental oocyst presence should be considered as well for potential piglet exposure to *C suis*. Dams shedding various species of coccidia were observed on this farm including *C suis* and *Eimeria* species. It was concluded that more gilts were shedding oocysts prior to farrowing compared to sows, and the most common species shed was *C suis*. When sows defecate into the crate, their feces may contain oocysts which will sporulate into an infective stage as early as 1-2 days later depending on temperature and humidity.[1] Newborn piglets may ingest these sporulated oocysts and show a clinical infection with coccidia as early as 5 days once the oocysts release sporozoites in the intestines leading to infiltration of enterocytes, causing pasty diarrhea as described previously.[1,2]

Kenoco has been proven to effectively inactivate *C suis* oocysts in the environment and works to decrease the level of disease in the herd.[4] Currently, there is not an oral antiprotozoal treatment product for gilts and sows that could be given prior to loading into farrowing crates that would reduce the number of oocysts shed in feces in the farrowing crate. Because of this interruption of the life cycle of coccidia is not yet possible making it difficult to eliminate the disease. If Kenoco was used in conjunction with an oral anticoccidial such as toltazuril (Baycox; Bayer, Whippany, New Jersey), it is believed that the overall presence of oocysts in the environment as well as the piglets would be decreased leading to reduced disease instance.[8] Baycox is not currently approved for use in the US, but efficacy of ponazuril (Marquis; Bayer, Whippany, New Jersey) have been described and are successful.[3,9] When combined with an amine based cleaner such as Kenoco, there could be significant reduction of the presence of coccidia.[8,9]

Coccidiosis continues to be a frustrating disease for veterinarians and producers within the swine industry due to a lack of oral antiprotozoal treatments for gilts and sows which allows dams to enter farrowing crates and contaminate their environment without ever breaking the lifecycle of coccidia. Along with this, limited products exist to inactivate *C suis* environmental oocysts safely for those applying it as well as the animals on farm. The only options to combat this disease are treating piglets with an oral antiprotozoal product such as Marquis and using anticoccidial cleaners such as Kenoco to reduce both the environmental and in vivo presence of *C suis* oocysts.[8,9] A study showing both of these product used in conjunction has not been done, but similar studies have shown the positive impacts of treatments and sanitation on coccidiosis.[8] Future investigation is required to understand the impact of use of an oral antiprotozoal in piglets in conjunction with an environmental product that inactivates *C suis* oocysts and their overall effects on reducing coccidiosis.

References

1, Neumann, E. J., Ramirez, A., & Schwartz, K. J. (2020). Coccidia and Other Protozoa. In *Swine disease manual* (11th ed., pp. 1015–1019). essay, American Association of Swine Veterinarians.

2. Christensen, J. P., & Henriksen, S. A. (1994). Shedding of oocysts in piglets experimentally infected with Isospora Suis. *Acta Veterinaria Scandinavica, 35*(2), 165–172.

3. American Association of Swine Veterinarians. (2022, August 18). *Use of ponazuril, diclazuril, and Toltrazuril in swine and cattle in ...* Retrieved November 5, 2022, from **https://www.aasv.org/documents/azuril.pdf**

4. Safiullin, R. T., & Skachkov, D. P. (2021). Measures against exo- and endogenous stages of pig coccidia in Industrial Farms. *IOP Conference Series: Earth and Environmental Science, 677*(4), 042002.

5.*Virocid®*. Virocid® | CID LINES. (n.d.). Retrieved November 8, 2022, from **https://www.cidlines.com/en-US/products/virocidr-0**

6. Stuart, B., Bedell, D., & Lindsay, D. (1982). Coccidiosis in swine: A search for extraintestinal stages of Isospora Suis. *Veterinary Record, 110*(4), 82–83. **https://doi.org/10.1136/vr.110.4.82**

7. Straberg, E., & Daugschies, A. (2007). Control of Piglet coccidiosis by chemical disinfection with a cresol-based product (Neopredisan 135-1®). *Parasitology Research, 101*(3), 599–604. **https://doi.org/10.1007/s00436-007-0521-z**

8. Hinney, B., Cvjetković, V., Espigares, D., Vanhara, J., Waehner, C., Ruttkowski, B., Selista, R., Sperling, D., & Joachim, A. (2020). Cystoisospora suis control in Europe is not always effective. *Frontiers in Veterinary Science, 7*. **https://doi.org/10.3389/fvets.2020.00113**

9. McCallister, S. et al. (2021). Impact of coccidiosis and treatment with ponazuril on piglet performance. *52nd Annual Meeting of the American Association of Swine Veterinarians*, 84.

https://doi.org/10.54846/am2023/22

Everything that grows is not gold: Evaluating the effect of sampling techniques on culture of disease-associated *Streptococcus suis* from diseased nursery-aged pigs

Amber Vegter[1], BS; Edgar Diaz-Estrada[4], MBA, DVM; Rebecca Robbins[4], PhD, DVM; Maria Clavijo[3,2], PhD, DVM; Alyona Michael[3,2], DVM, PhD, DACVP
[1]Iowa State University, College of Veterinary Medicine, Ames, Iowa; [2]Iowa State University, College of Veterinary Medicine- Department of Diagnostic & Production Animal Medicine, Ames, Iowa;
[3]Iowa State University, College of Veterinary Medicine - Veterinary Diagnostic Laboratory, Ames, Iowa;
[4]Ceva Animal Health, Lenexa, Kansas

Introduction

Streptococcus suis is a gram-positive commensal bacterium of swine. Disease-associated *S suis* strains (DASS) enter systemic circulation, resulting in a variety of clinical syndromes, including septic arthritis and meningitis. Additionally, compromised upper respiratory defenses can allow opportunistic infection of the lung with commensal or less virulent strains and contribute to localized pneumonia as a secondary agent. The strain variation from commensal or opportunistic, to highly pathogenic, complicates accurate diagnosis and hinders the success of any clinical intervention. The objective of this study was to evaluate the ability of different sampling protocols in successfully isolating DASS.

Methods

Eight 7-week-old untreated pigs from a farm with a history of *S suis* diagnosis and displaying neurological and/or lameness signs each had sequential samples taken for two different protocols. Protocol A was an "optimal" approach with minimal contamination risk. Protocol B was a "routine" approach that utilizes common sampling/euthanasia techniques that may increase the risk of bacterial contamination. Each pig was euthanized via electrocution, then CSF, a brain swab, and individually bagged lung, heart and spleen were collected as Protocol A. After samples for Protocol A were collected, sections of lung, heart and spleen were bagged together (pooled) for Protocol B to replicate routine shipping practices. A captive bolt was applied to each head to mimic a common euthanasia practice. All tissue, the brain swab, and the heads were packaged with ice and allowed to sit for 48 hours to simulate typical shipping time. After 48 hours, another brain swab was aseptically taken from the foramen magnum as part of Protocol B. Culture was attempted on all sample types for each pig. A summary of all samples taken per protocol is shown in Table 1. DASS isolates were defined as those PCR positive for the *mrp* gene (a gene associated with pathogenic *S suis* strains). Next- generation sequencing (NGS) was used on all *S suis* isolates to determine serotype and sequence type (ST). All tissue samples were evaluated histologically for evidence of bacterial lesions by a board-certified pathologist. Joint fluid from two pigs with acute arthritis was pooled and tested by PCR for *Mycoplasma hyorhinis* (MHR) and *Glassarella parasuis* (GPS) at the discretion of the study's diagnostician. Association of protocol and sample type was made by using a Fisher's exact test and strength of association reported as an odds ratio.

Results

Bacterial meningitis, serositis, or arthritis were present in only four, one, and seven of the total eight pigs, respectively. While bacterial arthritis was the most frequently observed disease process, lesions were chronic in five out of seven affected animals (Table 3). Overall, despite the history of *S suis* at this site, *S suis* was isolated from only three of four pigs with bacterial meningitis and two of seven pigs with arthritis. GPS was detected in pooled joint fluid of the single non-*S suis* meningitis case, thus accounting for the remaining lesions (Table 3). Regardless of protocol, eight of a total 72 culture attempts were *S suis* positive (11%). Protocol A yielded 6/8 isolates, with 4/8 being DASS positive and 2/8 being DASS negative. Two DASS isolates were recovered from Protocol B brain swabs. No DASS was isolated from other tissues collected using Protocol B (Table 2). The odds of isolating *S suis* from CNS samples was 12.37 times higher than isolating *S suis* from other tissues (P-value > 0.05). This is consistent with histologic lesions, as three of four pigs with suppurative meningoencephalitis were positive for DASS. While numerically higher, there was no statistical difference between Protocol A and B on DASS isolation (P-value > 0.05).

Protocols A and B were compared for isolation rate of contaminant bacteria, which were defined as non-pathogenic, environmental, or cutaneous and fecal host commensals that were inadvertently introduced into samples during necropsy or tissue commingling during simulated shipping conditions. These included the following genuses: *Aerococcus, Enterococcus, Pseudomonas, Staphylococcus, Enterobacter, Rothia, Escherichia*. When comparing contamination in DASS positive samples, Protocol B yielded more pigs with contaminant bacteria compared to Protocol A (P-value > 0.05). A total of 2/8, 0/8, 8/8 pigs had other bacterial agents grown from CSF, Protocol A brain swab and Protocol B brain swab, respectively. Five and 24 contaminant bacterial species were obtained from the CSF (Protocol A) and from Protocol B brain swabs, respectively (Table 4).

NGS revealed the traditional North American strain serotype ½ (or 2) ST28 in 7 of the 9 isolates. The other strains detected were serotype 3 ST117, considered a pathogenic type, and an untypeable strain by serotyping and MLST, considered commensal/ opportunistic.

Table 1: Description of sample types and number of tissues cultured by protocol type

	Protocol A	Protocol B
CNS samples	CSF Brain swab (Day 0)	Brain swab (Day 2)
Tissue samples	Individual heart Individual lung Individual spleen	Pooled heart Pooled lung Pooled spleen
Total cultured	5	4

CNS = central nervous system; CSF = cerebrospinal fluid

Table 2: Summary of the number of DASS positive isolates per sample type

	Protocol A DASS	Protocol B DASS
CSF	2	NA
Brain swab	1	2
Tissue	1	0
Total	4	2

DASS = disease-associated *S suis*; CSF = cerebrospinal fluid

Table 3: Summary of DASS isolation and diagnostic interpretation by histologic lesion. Diagnostic interpretation of disease was done by agreement of DASS isolation from the tissue type associated with the histologic lesion.

		DASS isolation - # of pigs				Diagnostic interpretation		
Histologic lesion:	**Lesions (# of pigs)**	**Protocol A**	**Protocol B**	**GPS (+)?**	**MHR (+)?**	**DASS-disease**	**Non-DASS bacterial disease**	**No bacterial disease**
Bacterial meningitis	Yes: 4	2	2	1	0	3	1	0
	No: 4	0	0	0	NA	0	0	0
Bacterial serositis	Yes: 1	1	0	1	0	0	1	0
	No: 7	0	0	0	NA	0	0	0
Bacterial arthritis	Yes- Acute: 2	0	0	2	NA	0	2	0
	Yes- Chronic: 5	2	2	0	NA	0	0	0
	No: 1	1	0	0	NA	0	0	1
No significant lesions	1	0	0	0	NA	0	0	1

DASS = disease-associated *S suis*; GPS = *Glassarella parasuis*; MHR = *Mycoplasma hyorhinis*

Table 4: Number of contaminant bacteria species in samples by protocol. Genuses included: *Aerococcus*, *Enterococcus*, *Pseudomonas*, *Staphylococcus*, *Enterobacter*, *Rothia*, *Escherichia*

	Protocol A	Protocol B
CSF	5	NA
Brain swab	0	24
Tissue	22	16

CSF = cerebrospinal fluid

Discussion

In pigs, bacterial meningitis, serositis and arthritis are most frequently associated with *S suis, G parasuis* or *M hyorhinis.* All three of these agents can present with similar clinical course, gross and histologic findings. Effective control of disease- associated *S suis* depends on successful isolation of relevant DASS, which can be compromised by prior antimicrobial treatment, chronic disease stage, or premature assumptions regarding causative bacterial etiology. Success can be maximized by establishing a case definition that includes consistent identification of acute systemic bacterial disease (meningitis, arthritis, or serositis), selection of untreated pigs, as well as comprehensive diagnostics to help rule out GPS and MHR involvement. Furthermore, commensal *S suis* strains can convolute bacterial culture results, so aseptic and targeted sampling is critical.

This study highlighted the challenges of successful isolation of *S suis* in the field. Only eight *S suis* strains were isolated from a total of 72 samples. While the low rate of *S suis* isolation was unexpected given the herd history, the non-homogenous lesion distribution, preponderance of subacute to chronic disease, and GPS involvement in selected pigs, all contributed to the low rate of DASS isolation. In this study, a CNS sample (CSF or brain swab) was 12.37 times more likely to have a *S suis* isolate, compared to other sample types, making it a vital sample type to include when *S suis* is a differential for pigs presenting with CNS signs (paddling, nystagmus). This was corroborated by histologic findings, as three of the four pigs with suppurative meningoencephalitis consistent with *S suis* infection, were positive for DASS. A fourth pig with *S suis* isolation from CSF failed to demonstrate histologic evidence of meningitis; thus, the isolate was interpreted as a likely contaminant. This latter animal did feature GPS-mediated arthritis. This re-enforces the need for a complete sample set, as the cause could have been inappropriately attributed to S suis if synovium had not been submitted for histology and GPS/MHR PCR were not completed to help rule out these fastidious organisms. Finally, while both protocols yielded DASS, the copious amounts and various types of contaminant bacteria obtained with samples in Protocol B, is a typical finding that hinders identification of the clinically relevant strains within a herd (1), and underlines the importance of a methodical and aseptic approach to sample collection. In addition, finding a bacteria within tissues but in the absence of pathology could also signal contamination (Table 3).

When potential DASS is isolated from appropriate sample types, Next- Generation Sequencing (NGS) can be a powerful tool for guiding clinical and management decisions for DASS control that include source segregation, selection of isolates for autogenous vaccines, and antibiotic therapy. The predominant strain detected in the study was ST 28, a strain considered pathogenic and frequently associated with disease in North America. However, other strains of likely lesser clinical role or potentially commensal were also detected, highlighting the importance of in-depth genetic characterization of strains, and submission of multiple pigs and tissues to identify the primary disease-causing strains in a herd. Based on previous data,[1] ST28 would be a viable candidate for an autogenous vaccine production for this system.

Effective DASS control, such as vaccine candidate selection, pig flow management, and antimicrobial selection, hinges on the successful culture of appropriate disease-associated strains present in a herd. However, polymicrobial interactions (ie, opportunistic bacterial agents), lack of proper sampling techniques, suboptimal animal selection (ie, treated or chronically affected pigs) and failure to rule out GPS/MHR involvement can hinder optimal identification of DASS isolates. This study demonstrated all of these challenges and explored the effect that they have on the end diagnosis. Overall, Protocol A, an optimal approach, was more conducive to DASS isolation and had significantly less bacterial contamination compared to Protocol B, a routine approach. CNS samples were found to yield a DASS isolate more often compared to other tissues. This emphasizes one of the biggest takeaways of this study, which is the need for a complete set of diagnostic samples from the correct pigs.

Reference

1. Clavijo et al., 2021. Friend or foe: what next generation sequencing can tell you about the endemic agents in your herd. AASV Proc. 2021.

https://doi.org/10.54846/am2023/23

Evaluation of the IgA response to attenuated-live oral *Lawsonia intracellularis* vaccine

Kendall Sattler[1]; Melissa Farber Billing[2], DVM, MVS; Luis Gimenez-Lirola[3], MS, PhD; Ronaldo L. Magtoto[3], PhD; Juan Carlos Mora-Diaz[3], DVM, MS, PhD; Fernando Leite[2], DVM, MS, PhD
[1]Purdue University, College of Veterinary Medicine, West Lafayette, Indiana; [2]Boehringer Ingelheim Animal Health USA Inc., Duluth, Georgia; [3]Iowa State University College of Veterinary Medicine Veterinary Medicine Diagnostic Laboratory, Ames, Iowa

Introduction

Porcine proliferative enteropathy (PPE), commonly referred to as ileitis, is an infectious enteric disease leading to decreased production performance of growing pigs. The etiological agent of this disease, *Lawsonia intracellularis*, is a Gram-negative bacterium found in swine worldwide.[1] An avirulent live oral vaccine, Enterisol® Ileitis (Boehringer Ingelheim Animal Health USA Inc.), is currently available to immunize pigs to prevent this disease and subsequent economic loss from mortality, decreased feed efficiency, and poor weight gain.[2] Recent research has found that anti-*L intracellularis* immunoglobin A (IgA) and immunoglobin G (IgG) can be observed in oral fluids and serum following experimental *L intracellularis* challenge.[3] Given the role that IgA plays in protecting the intestine and mucosal surfaces, we elected to evaluate IgA.[4,5] The objective of this study was to investigate if IgA against *L intracellularis* could be detected in serum and in oral fluids following vaccination with Enterisol® Ileitis.

Materials and methods

On study day zero, thirty 49-day old gilts from a population of 200 same age gilts were randomly selected, tagged, had serum collected, and received 1 mL of Enterisol® Ileitis via oral drench. The total population of 200 gilts was administered the vaccine via the drinking water per label instructions. On study day 21, serum was collected from the tagged gilts, they were administered an additional 1mL of Enterisol® Ileitis via oral drench, and the vaccine was administered to the entire population via the drinking water per label instructions to elicit a recall immune response. Serum was collected from tagged gilts again on study day 42. Three oral fluid samples were collected on study days 0, 7, 14, 21, 28, 35, and 42. Oral fluid and serum samples were tested using the commercial *L intracellularis* ELISA (Svanova) which was adapted for the assessment of anti-*Lawsonia intracellularis* IgA. Serum was also tested for *Lawsonia intracellularis* by ELISA (Svanova) at Iowa State University Veterinary Diagnostic Laboratory. To analyze the differences in the detection of IgA antibodies (S/P value) against *Lawsonia intracellularis* in oral fluids between each timepoint of collection, a linear mixed model (PROC Mixed and LSMeans Statement) was applied with "timepoint" as fixed effect and the "pen ID" as random effect. A linear mixed model (PROC Mixed and LSMeans Statement) was also used to analyze the differences in IgA antibodies detected against *L intracellularis* and the average percentage of inhibition in serum between the three collection points with "time" as fixed effect and "pig ID" as the random effect. For all the analyses, an adjusted P-value (Tukey-Kramer) of 0.05 was considered statistically significant. Statistical analyses were performed using SAS version 9.4 (SAS, Cary, NC, USA, SAS Institute, Inc).

Results

The IgA level measured in oral fluids increased considerably over time, having increased from an average S/P value of 0.28 to 1.19 from the first to the last collection timepoint, respectively ($P < 0.05$, Figure 1). While the increase in serum IgA observed from the first to last time point was also significant ($P < 0.05$), this increase was of much lower magnitude, as gilts had an average S/P value of 0.13 prior to vaccination and 0.22 at the last time point (Figure 1). The serum *L intracellularis* ELISA also increased over time as the tagged gilts had an average percent inhibition of 0.743 prior to vaccination, 9.786 prior to second vaccination, and 15.731 at the last sample collection. The increase in percent inhibition was significant between each time point ($P < 0.05$), however only three of the thirty gilts were considered seropositive at 42 days post initial inoculation considering the commercial test threshold of percent inhibition greater than or equal to 30. These results were expected as Enterisol® Ileitis does not usually lead to seroconversion post vaccination, and serum ELISA values do not correlate to protection to PPE.[2]

Discussion

In this pilot study, anti-*L intracellularis* IgA could be detected in both serum and oral fluids following vaccination with Enterisol® Ileitis. A drastic increase in IgA level was observed in oral fluid samples over time, while only a slight increase in serum was observed. This is likely because IgA can be secreted by salivary glands into oral fluids and better represent a measure of mucosal immunity, while serum measurement may mostly only represent systemic responses.[4,5] As serum results were less remarkable compared to oral fluids, it could be concluded that oral fluids are a more favorable diagnostic sample. Oral fluid samples offer a non-invasive sampling method that require minimal time and skill to collect. The major implication of this study is that with further validation, *L intracellularis* IgA levels in oral fluids could allow swine veterinarians to determine the immune status of a herd, including the evaluation of vaccination status, to ensure herd health and prevent economic losses due to poor growth performance caused by *L intracellularis*.

Figure 1: Average anti-*Lawsonia intracellularis* IgA S/P value in oral fluid and serum samples following vaccination.

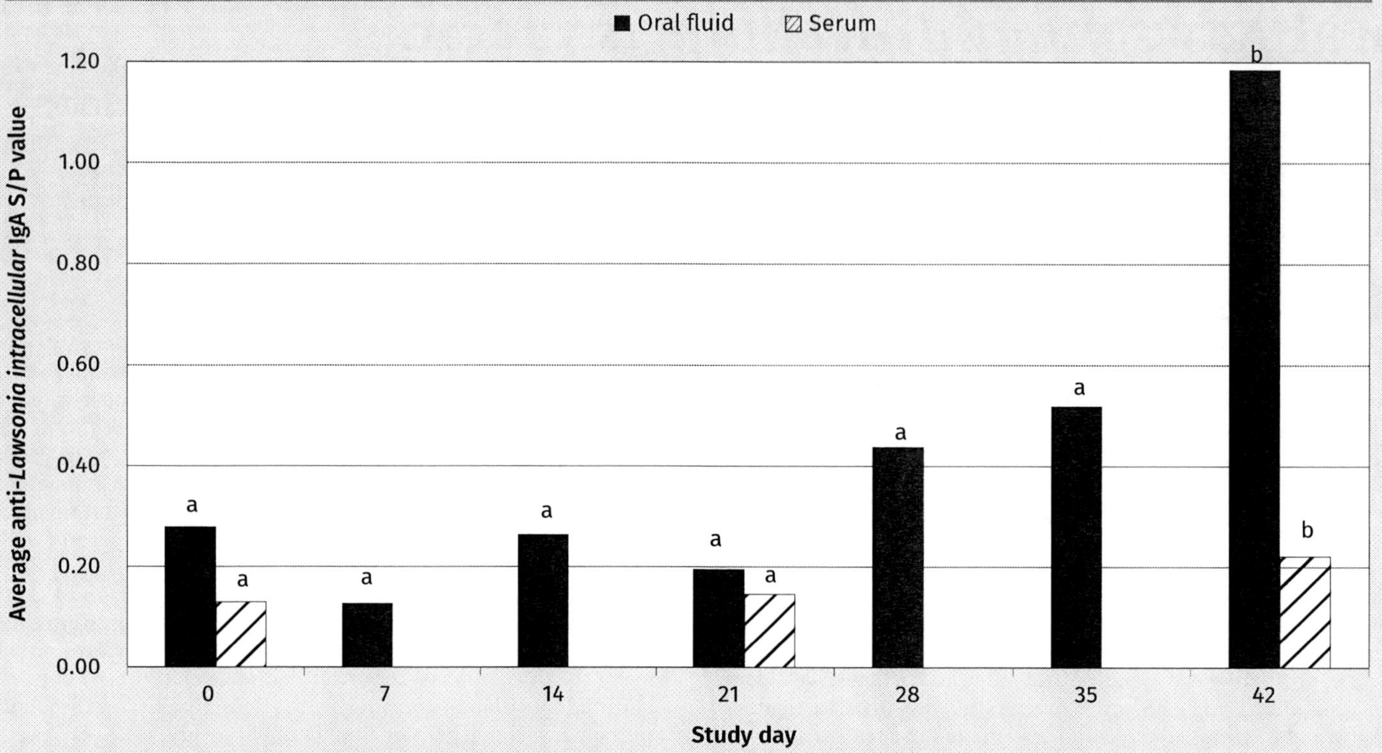

[a,b] Different superscripts within sample type are statistically different ($P < 0.05$)

References

1. Vannucci FA, Gebhart CJ, McOrist S (2019) Proliferative enteropathy. In: Zimmerman JJ, Karriker LA, Ramirez A, Schwartz KJ, Stevenson GW (eds) *Diseases of swine,* 11th ed. (pp 898–911). West Sussex, Wiley-Blackwell.

2. Leite F, Cline G, Waddell J, et. al. Enterisol® Ileitis, what we have learned in twenty years controlling *Lawsonia intracellularis.* Proc AASV. 2021:157-160.

3. Gabardo MDP, Sato JPH, Rensende TP, et. al. Use of oral fluids to detect anti *Lawsonia intracellularis* antibodies in experimentally infected pigs. *Pesq. Vet. Bras.* 2020;40(12):970-976. **doi:10.1590/1678-5150-pvb-6679**

4. Brandtzaeg P. Secretory immunity with special reference to the oral cavity. *J Oral Microbiol.* 2013;5. **doi:10.3402/jom.v5i0.20401**

5. de-Sousa Pereira P, Woof JM. IgA: Structure, Function, and Developability. *Antibodies (Basel).* 2019 Dec 5;8(4):57. **doi:10.3390/antib8040057**

https://doi.org/10.54846/am2023/24

Evaluation of various grueling protocols on nursery pig performance and survivability

C. Bates[1], BS; B. Burton[2], DVM; A. Reever[2], MS, DVM; J. Okones[3], DVM
[1]Iowa State University CVM, Ames, Iowa; [2]Suidae Health and Production, Algona, Iowa;
[3]Pharmgate Animal Health, Wilmington, North Carolina

Introduction

Proper post-weaning nutrition is critical, especially when considering pigs that have undergone transport stress and environmental changes. Previous work has shown that weaning causes morphological and functional changes of the small intestine of pigs, where majority of the nutrients are digested and absorbed, and these changes can result in severe diarrhea and possibly mortality.[1] Grueling is the technique of mixing water into feed and has long been used to aid weaned pigs in the transition from a milk-based diet to a solid feed diet. Grueling is commonly performed in pull pens, small pens, or health challenged groups that have been removed from the general population to allow a lower stocking density and more intensive care. According to a study from 2014, gruel feed is best fed multiple times throughout the day to mimic the activity of nursing.[2] Many producers have adopted gruel feeding as a daily chore post-placement; however, it is likely that it is under-fed by most caretakers, and grueling protocols are not well documented in the industry today. To best utilize this practice, it is important to understand proper gruel volume, ratios, frequency, feeder space, and therefore, production impact, of different techniques that can be used by producers to ensure proper gruel efficacy. The objective of this study was to evaluate performance as measured by average daily gain (ADG) and mortality rate in pull pen pigs that received gruel according to two protocols compared to a negative control group.

Materials and methods

The study was conducted in a 3,600 head nursery with three rooms in Eastern Iowa. There were 3 treatment groups; treatment 1, 2 and 3 (control). After each room filled, pigs were thoroughly evaluated at 1-, 3-, 7-, and 10-days post placement to determine which pigs needed pulled from the general population and enrolled into the study. Pigs were included in the study if they showed signs of severe lameness, had a low body condition score (spine/hip bones exposed), were significantly smaller than their pen mates, or appeared unthrifty. On each pull day, pigs were identified and then randomly allotted to a treatment using a random list generator application. Pigs were then tagged with individual RFID tags, weighed, and placed in their appropriate pens. At least six pens per room were designated for the trial (one per pull day for treatment 1, one pen for treatment 2, and one pen for treatment 3).

Treatment 1 pull pens contained only pigs that were pulled on a single day (one pen per pull day per room) with no continuous addition of other pigs after the original group was allotted and placed. The larger space allowance was derived from previous studies that have shown ADG improves with a smaller stocking density.[3] These pigs were gruel fed 0.1 pounds of feed per pig per day split between three feedings each day with proper water ratios that started at 75% water and declined to 25% water over seven days (Table 1). Both water and feed were measured in pounds, using a portable hanging scale. Feed was pulled directly from the dry feeders to match the appropriate diet phase at that time. The gruel was fed out of a four-foot trough that ensured at least three inches of space per pig.

Treatment 2 pull pens had continuous addition of pigs to the pen on all pull days (one pen per room). They received 0.1 pounds of feed per pig with equal amounts of water one time each day for seven days. The gruel was fed in a circular feeder with a 52-inch circumference with 10 head hole spaces. If the pen exceeded 40 pigs, another identical feeder was added. Treatment 2 varies from treatment 1 in the decreased frequency of gruel feed and the decreased feeder space.

Treatment 3 pull pens had continuous addition of pigs to the pen on all pull days (one pen per room). They received no gruel and acted as the negative control group.

All pigs in the study had continuous access to water and a dry mash nursery diet. If any single pen exceeded 55 pigs, an additional pen was added for that group. If any pig enrolled in the study died, its ear tag was cut out and date and death reasons were recorded. Pigs that were determined to be a cull pig were individually identified as well. All treatments, vaccinations, and daily production protocols were performed in accordance with the system's standard operating procedures. At the end of the nursery phase, pigs were weighed and ADG was calculated along with mortality percentages for each treatment. Pigs were followed through the finisher phase where pre-market weights, mortalities, and cull pigs at one hundred-days placed were recorded to better understand the lasting impact gruel feeding may have. Numerical observations, rather than statistical, were analyzed due to the design of the study.

Table 1: Amount of feed to water ratio in gruel fed to treatment group 1, amount of water decreased as the days in pull pen increased

Days after pull	Feed amount	Water amount
1	25%	75%
2	25%	75%
3	33%	66%
4	40%	60%
5	50%	50%
6	60%	40%
7	75%	25%

Results

No ADG differences were shown to correlate between treatment group. However, there were large numerical differences identified between piglet treatments and mortality percentage (Figure 1). The difference in mortality percentages between treatment 1 and the control group suggests that grueling several times per day with proper ratios can drastically increase survivability in nursery pigs. Treatment 1 resulted in a mortality percentage that was 13.54% less than that of the control and 5.23% less than treatment 2. This study successfully generated data to support two grueling options that will have immediate impact on nursery pigs in many systems and environments. No observable differences were seen for one-hundred-day culls based on treatment group. Just prior to marketing, a final weight measurement will be performed on all surviving pigs in the study. Additionally, overall mortality percentages will be calculated.

Discussion

Since the criteria for pulling a pig or leaving it in the general population is quite subjective, there may be differences between producers, managers and caretakers. Setting guidelines for each operation is the first step in implementing proper gruel management. It is important to train caretakers to accurately identify pull pigs within a system, while not overstocking pull pens and negating some of the advantages. Additionally, this study was completed on one farm and extrapolation to other farms/systems may be limited. Considering the prior lack of documentation regarding gruel feeding in a nursery setting, this study helped to better define the reductions to mortality that can be achieved. Although this trial did not show any significant data for ADG, it shows a strong indication that the addition of gruel feeding can reduce nursery mortality. These results are highly applicable to both nursery and wean-to-finish producers due to the ease of implementation and return on investment.

References

1. Zheng, Lan, et al. "Intestinal Health of Pigs upon Weaning: Challenges and Nutritional Intervention." Frontiers, Frontiers, 1 Jan. 1AD.

2. Bertram, Mark J. "Implications of Early Post-Weaning Management on Subsequent Performance of Piglets." Swine Information Library Access Restricted, Nov. 2014, **https://www.aasv.org/library/swineinfo/item.php?ISU%2F2014%2F125.pdf.**

3. Smith, Laura F, et al. "The Impact of Feeder Adjustment and Group Size-Floor Space Allowance on the Performance of Nursery Pigs." Swine Information Library Access Restricted, June 2004, **https://www.aasv.org/library/swineinfo/item.php?SHAP%2F2004%2Fv12n3p111.pdf.**

Figure 1: A comparison of mortality percentages and the assigned treatment groups.

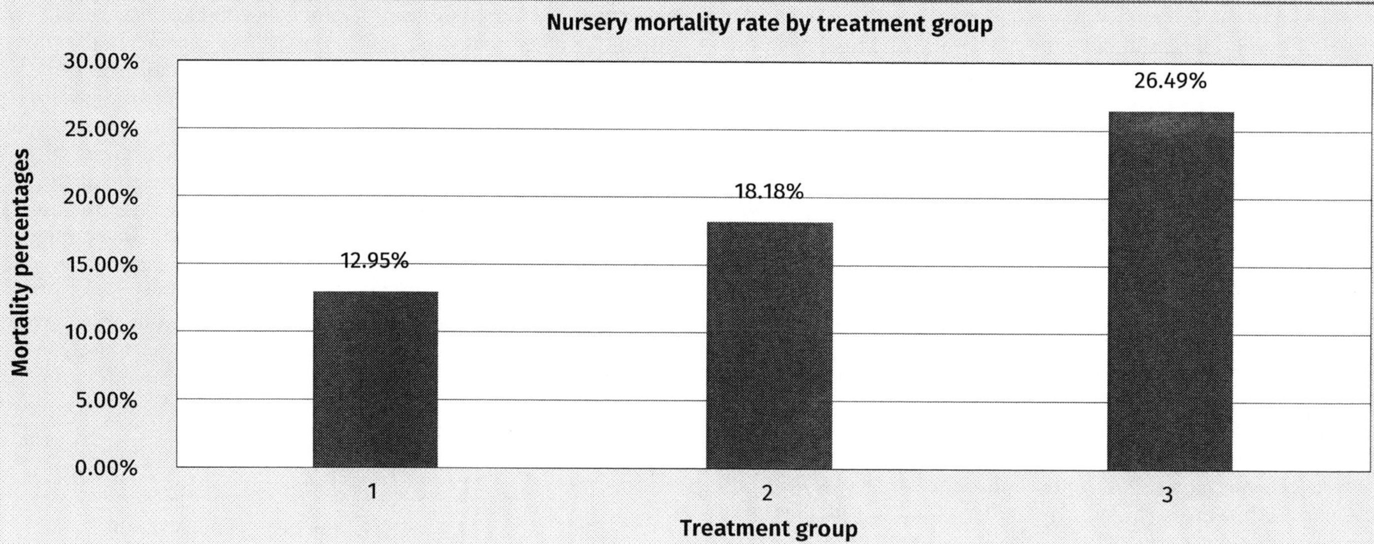

https://doi.org/10.54846/am2023/25

Mucin 4 is a marker of necrotizing bronchiolitis in influenza A virus infection

R. A. Kanefsky[1]; T. K. Anderson[2], PhD; A. L. Vincent Baker[2], DVM, PhD; B. L. Arruda[2], DVM, PhD
[1]Tufts Cummings School of Veterinary Medicine, Grafton Massachusetts; [2]Virus and Prion Research Unit, National Animal Disease Center, USDA Agricultural Research Service

Introduction

Influenza A virus (IAV) causes significant morbidity and similar pathology in the human and swine host. IAV infects epithelial cells in the respiratory tract of mammals causing a necrotizing bronchitis and bronchiolitis. Protecting these mucosal surfaces are large glycoproteins called mucins. Mucins can act as serum biomarkers for neoplastic progression and respiratory infections.[1] Mucins can be divided into two categories: gel-forming mucins secreted by goblet cells and transmembrane mucins that localize to apical surfaces of epithelial cells.[2] Transmembrane mucins can inhibit microbial invasion, serve as decoy receptors, and activate signal transduction pathways. Mucin 4 (MUC4) is a transmembrane mucin that consists of an alpha subunit responsible for surface protection and intracellular beta subunit involved in signal transduction to repress apoptosis and stimulate epithelial proliferation. [3] This study was designed to determine the expression and potential role of MUC4 during IAV infection.

Materials and methods

We used immunohistochemistry (IHC) in combination with a machine learning algorithm to quantify differential expression of MUC4 subunits in IAV-infected and uninfected porcine lung. Pigs were blocked by litter and randomized to one of six groups (8 per group), intranasally inoculated with 3 mL of H1N1 (gamma; 1A.3.3.3) or sham-inoculated with Opti-MEM™ and euthanized day post-inoculation (DPI) 1, 3 or 5. At necropsy three sections of formalin-fixed lung (right cranial, left cranial and caudal) were collected per pig and routinely processed. Heat-induced epitope retrieval was performed using citrate (pH 6) heated to 95°C for 2 cycles of 10 minutes. Slides were quenched with hydrogen peroxide, blocked with goat serum (1:10) for 60 (MUC4-beta subunit; 1G8) or 90 (MUC4-alpha subunit; 5B12) minutes, incubated with either 1G8 (1:3000) or 5B12 (1µg/mL) primary antibody for 60 minutes, followed by a secondary antibody (EnVision+ anti-mouse Poly-HRP-IgG) for 30 minutes. Diaminobenzidine chromogen detection was completed using Dako DAB Plus and Dako DAB Enhancer. Slides were counterstained using hematoxylin. HALO™ was optimized for each primary antibody, with parameters set to best capture the alpha subunit (clone 5B12) and beta subunit (clone 1G8) labeling independently. To determine basilar protein expression of MUC4 in bronchi, a layer that consisted of the nuclei of the respiratory epithelium lining a bronchus was generated using HALO™. To evaluate MUC4 protein expression in bronchioles, we selected 4 to 5 bronchioles per section from IAV inoculated animals from DPI 1, 3, and 5 with necrotizing bronchiolitis. For each affected bronchiole selected, a paired normal bronchiole within the same section was also selected. Statistical analyses were performed using GraphPad Prism. A P-value less than 0.05 was considered significant using a Kruskal-Wallis with Dunn's multiple comparisons test, Brown-Forsythe and Welch ANOVA with Dunnett's T3 multiple comparisons test, and Komogorov-Smirnov test.

Results

Both subunits of MUC4 were commonly detected by IHC at the apical surface of histologically normal bronchi, with some differential expression among healthy pigs, but were minimally detected in normal bronchioles. However, abundant differential MUC4 expression was observed in the conducting airways of pigs infected with IAV. Increased MUC4 expression was seen at the basilar aspect of respiratory epithelium lining bronchi (Figure 1A and 1B) and bronchioles with necrotizing bronchiolitis (Figure 2A and 2B; $P < 0.05$ DPI 1, 3, & 5) of IAV inoculated pigs.

Discussion

MUC4 expression was significantly increased in bronchioles with necrotizing bronchiolitis and likely represents an epithelial regenerative response to restore the integrity of the conducting airways. These data can be used to train machine learning algorithms to facilitate the diagnosis of bronchitis and bronchiolitis. Additionally, and with refinement, MUC4 serum concentrations and expression could act as a proxy for disease severity. Understanding the impact of differential MUC4 expression among healthy individuals and during IAV infection will facilitate control strategies by elucidating mechanisms associated with susceptibility to IAV that can be therapeutically or genetically regulated and may extrapolate to other respiratory diseases.

References

1. Chatterjee M, van Putten JPM, Strijbis K. Defensive Properties of Mucin Glycoproteins during Respiratory Infections-Relevance for SARS-CoV-2. *mBio*. 2020;11:e02374-20.

2. van Putten JPM, Strijbis K. Transmembrane Mucins: Signaling Receptors at the Intersection of Inflammation and Cancer. J *Innate Immun*. 2017;9:281-299.

3. Huguette A and Carraway KL, III. MUC1 and MUC4: switching the emphasis from large to small. *Cancer Biother Radiopharm*. 26;2011: 261-271.

Figure 1: MUC4 (beta subunit; 1G8) protein expression. **A)** Normal bronchi with minimal, weak (yellow) basilar labeling. **B)** Abundant, weak (yellow) and rare, strong (red) basilar labeling representative of influenza A virus inoculated pigs necropsied day post-inoculation 5. Labeling identified by HALO™.

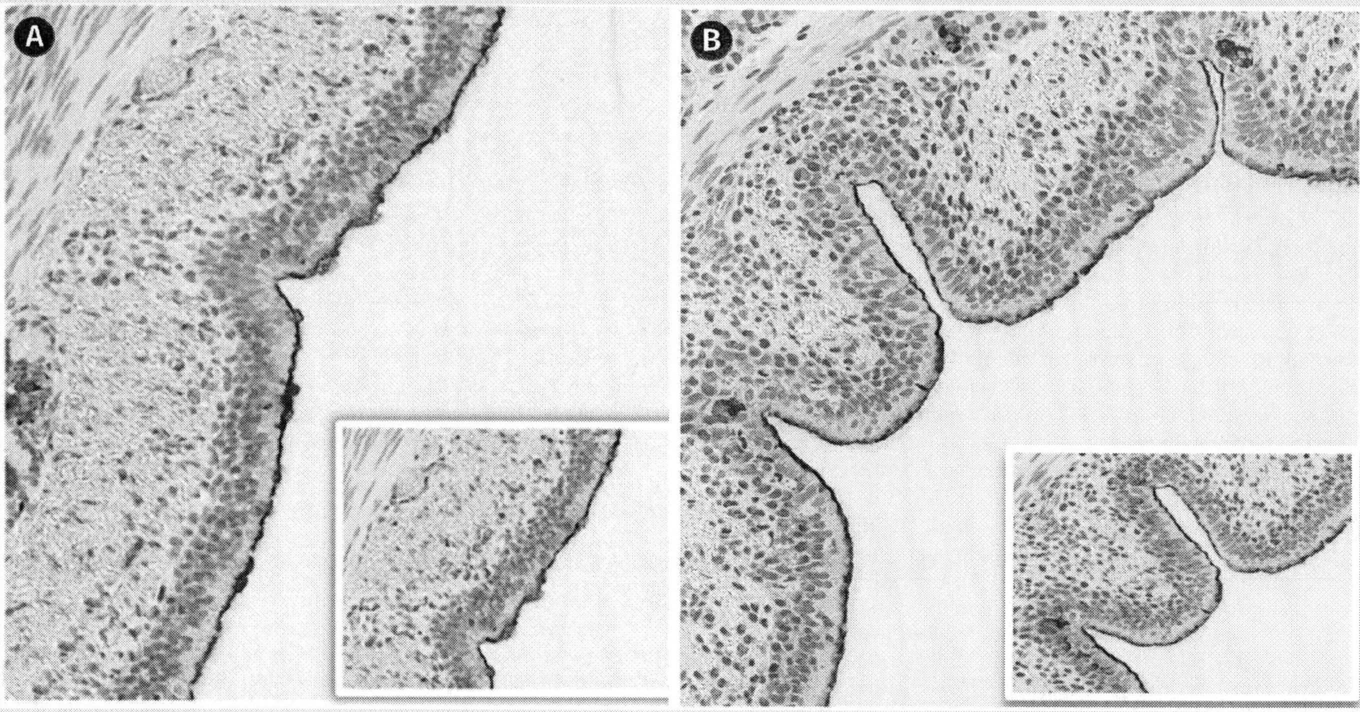

Figure 2: MUC4 (beta subunit; 1G8) protein expression. **A)** No labeling of histologically normal bronchiole representative of influenza A virus inoculated pigs. **B)** Abundant, weak (yellow) and strong (red) and rare, moderate (orange) labeling of a bronchiole with necrotizing bronchiolitis representative of influenza A virus inoculated pigs necropsied day post-inoculation 3. Labeling identified by HALO™.

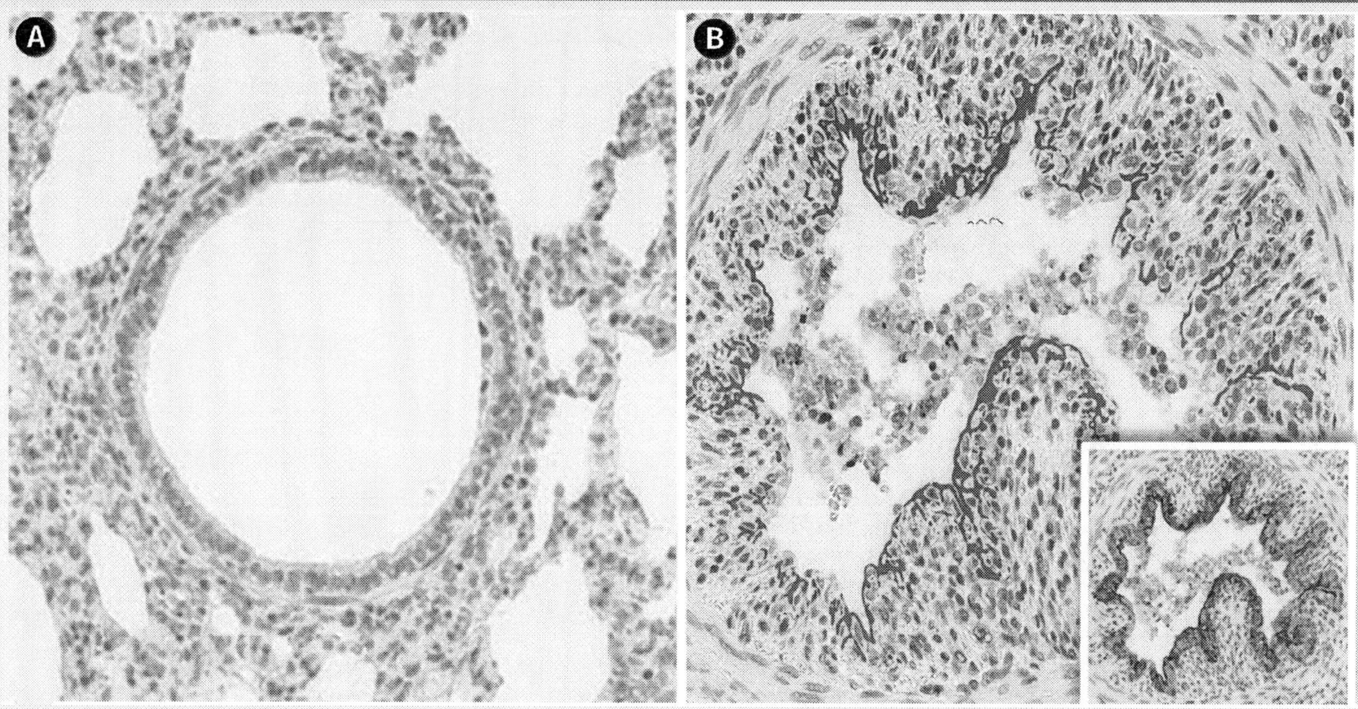

https://doi.org/10.54846/am2023/26

Comparison of tongue tips, serum, and processing fluids for PRRSV monitoring in neonates

Taylor Jansen[1], BS; John Waddell[2], DVM, MBA; Daniel Linhares[3], DVM, MBA, PhD; Isadora Machado[3], DVM; Claire LeFevre[4], DVM
[1]Purdue University, West Lafayette, Indiana; [2]Boehringer Ingelheim Animal Health USA Inc., Duluth, Georgia; [3]Iowa State University, Ames, Iowa; [4]Carthage Veterinary Service, Ltd., Carthage, Illinois

Introduction

Porcine reproductive and respiratory syndrome virus (PRRSV) is a pathogen that costs the United States swine industry $664 million per year.[1] PRRSV is commonly reported with reproductive losses and deaths in breeding herds, such as increased abortion rates, mummies, stillbirths and neonate mortality. Multiple sample types are used for PRRSV monitoring, such as blood samples, processing fluids,[2] oral fluids,[3] and family oral fluids.[4] Blood samples can be collected at all stages of production. However, the blood collection process requires adequate training, and may cause distress to the piglets. Processing fluids is another sample type that may be used for PRRSV monitoring, which is more time, cost, and labor efficient and involves collection from a larger population.

Recently, studies have evaluated the use of tongue tips from stillborns and dead pigs for PRRSV RNA detection, providing another surveillance option that would not require sophisticated training. Because tongue tips are obtained from dead pigs, they offer a risk-based approach for pathogen monitoring. The objective of this study was to compare the positivity rate and efficacy of different sample types used in PRRSV surveillance programs by comparing the RT-qPCR cycle threshold (Ct) values of stillborn tongue tips with stillborn serum samples, liveborn littermate serum samples, and processing fluids.

Materials and methods

This study was conducted at a commercial 5,600 breed-to-wean PRRSV positive unstable farm.[5] Litters containing a minimum of one stillborn were enrolled on trial, with a total of 30 litters enrolled per cohort. For each cohort, 30 stillborn tongue tips, 30 stillborn blood samples, 30 liveborn littermate blood samples (< 24 hours of age), and processing fluids from the same litters were collected. The smallest or most unthrifty pig of each litter was targeted for the liveborn littermate blood samples. Tongue tips and stillborn blood samples were collected from the same stillborn pigs. Tongue tips (approximately 2 cm) were removed from each stillborn using forceps and scissors, and placed into a clean Ziplock bag. The bags of tongue tips were frozen and subsequently thawed, and the exudate was collected. Processing fluids were collected at 3-5 days of age. The RT-qPCR Ct values were compared from each sampling method and the time commitment and ease of collection of each sampling type were noted for practitioner consideration.

Results and discussion

Tongue tip exudate had a lower average Ct value (29.4) when compared to processing fluids (33.8). Similarly, stillborn sera had a lower average Ct value (27.0) when compared to liveborn littermate sera (29.9) (Table 1). By utilizing risk-based sampling (tongue tips and stillborn sera), lower Ct values could be obtained due to targeting the diseased animals that are more likely to harbor the PRRSV. Tongue tips have proven to be an efficient and practical alternative surveillance method that can be used by practitioners and staff to evaluate PRRSV status.

References

1. Holtkamp DJ, Kliebenstein JB, Neumann EJ, et al. Assessment of the economic impact of porcine reproductive and respiratory syndrome virus on United States pork producers. *J Swine Health Prod.* 2013;21(2):72-84.

2. Lopez WA, Angulo J, Zimmermann JJ, et al. Porcine reproductive and respiratory syndrome monitoring in breeding herds using processing fluids. *J Swine Health Prod.* 2018;26(3):146-150.

3. Marisa L. Rotolo, Yaxuan Sun, Chong Wang, Luis Giménez-Lirola, David H. Baum, Philip C. Gauger, Karen M. Harmon, Marline Hoogland, Rodger Main, Jeffrey J. Zimmerman. Sampling guidelines for oral fluid-based surveys on group-housed animals. *Veterinary microbiology,* 209, 20-20. **https://doi.org/10.1016/j.vetmic.2017.02.004**

4. Almeida, M. N., Rotto, H., Schneider, P., Robb, C., Zimmerman, J.J, Holtkamp, D. J., Rademacher, C. J., & Linhares, D. (2020). Collecting oral fluid samples from due-to-wean litters. Preventive veterinary medicine, 174, 104810. **https://doi.org/10.1016/j.prevetmed.2019.104810**

5. Holtkamp D, Torremorell M, Corzo CA, Linhares DCL, Almeida MN, Yeske P, Polson DD, Becton L, Snelson H, Donovan T, Pittman J, Johnson C, Vilalta C, Silva GS, Sanhueza J. Proposed modifications to porcine reproductive and respiratory syndrome virus herd classification. J Swine Health Prod. 2021;29(5):261-270.

Table 1: Ct averages of percent positive samples for stillbirth serum, liveborn littermate serum, tongue tip exudate, and processing fluids

	Stillbirth serum		Liveborn littermate serum		Tongue tips exudate		Processing fluids	
Cohort	Positive samples (%)	Ct average (min-max)	Positive samples (%)	Ct average (min-max)	Positive samples (%)	Ct average (min-max)	Positive samples (%)	Ct average (min-max)
#1	1/30 (3.3%)	32.9	0/30 (0%)	–	1/1 (100%)	29.6	1/1 (100%)	30.4
#2	3/30 (10%)	27.9 (19.9-35.9)	1/30 (3.3%)	19.1	1/1 (100%)	27.4	1/1 (100%)	32.9
#3	2/30 (6.7%)	21.1 (20.0-22.1)	0/30 (0%)	–	1/1 (100%)	27.0	0/1 (0%)	–
#4	4/30 (13.3%)	28.5 (18.9-32.8)	4/30 (13.3%)	31.3 (30.0-32.8)	1/1 (100%)	27.5	1/1 (100%)	36.8
#5	1/30 (3.3%)	30.9	1/30 (3.3%)	33.4	1/1 (100%)	32.4	1/1 (100%)	35.2
#6	1/30 (3.3%)	24.5	1/30 (3.3%)	30.7	1/1 (100%)	29.9	0/1 (0%)	–
#7	1/30 (3.3%)	23.1	2/30 (6.7%)	35.1 (34.8-35.4)	1/1 (100%)	31.7	1/1 (100%)	33.7
Overall	**6.2%**	**27.0**	**4.3%**	**29.9**	**100%**	**29.4**	**71.4%**	**33.8**

https://doi.org/10.54846/am2023/27

Pharmacokinetics of oral clonazepam and its effect on selected stress markers in growing commercial pigs (*Sus scrofa domesticus*)

Gregory A. Shanks[1]; Chiara E. Hampton[2], DVM, MS; Stephanie Kleine[3], DVM, PhD; Joseph Smith[2], DVM, PhD; Zhu Xiaojuan[4], PhD; Pierre-Yves Mulon[2], DVM; Christopher Smith[3], DVM; Luca Giori[5], DVM, PhD; Sherry Cox[5], MS, PhD; Reza Seddighi[2], DVM, PhD

[1]College of Veterinary Medicine, University of Tennessee, Knoxville, Tennessee; [2]Department of Large Animal Clinical Sciences, University of Tennessee, Knoxville, Tennessee; [3]Department of Small Animal Clinical Sciences, University of Tennessee, Knoxville, Tennessee; [4]Office of Information and Technology, University of Tennessee, Knoxville, Tennessee; [5]Biomedical and Diagnostic Sciences Department, University of Tennessee, Knoxville, Tennessee

Introduction

The management of stress associated with various origins is an important aspect of swine welfare. Research for determining effective, readily administered, and clinically applicable tranquilizers is relevant to swine veterinarians seeking strategies to minimize stress and improve medical standards of care in swine populations. Clonazepam, a benzodiazepine, is commonly used in people and dogs in the treatment of stress, anxiety, and panic disorders. Benzodiazepines decrease hypothalamic-pituitary-adrenocortical axis activity[1] and may cause sedation and psychomotor impairment.[2] The objectives of this study were to determine the pharmacokinetic (PK) profile of oral clonazepam in pigs and to quantify its effect on serum cortisol as a biomarker of stress.[3]

Methods

Eight healthy, growing Landrace/Yorkshire cross pigs (3 barrows, 5 gilts; avg age 91 days; mean weight 43.8 ± 2.4 kg) were randomly assigned by day to receive oral treatment after fasting for 12 hours. For all pigs, on the first day of the experiment, capsules containing placebo were administered, and serial blood samples obtained from an indwelling jugular catheter implanted 2 weeks prior to the start of the experiment. Serial sampling at pre-established timepoints across 8 hours was performed to establish the baseline for cortisol concentrations. On the second day, the experiment was repeated after oral administration of clonazepam (0.5 mg/kg). Serial sampling was repeated, and additional blood samples were collected for 96 hours for the measurement of plasma concentrations of clonazepam as part of the PK study. Analysis of plasma clonazepam concentrations was performed using reverse-phase high-performance liquid chromatography. PK modeling was conducted via commercial software using a non-compartmental model to characterize time-concentration PK data. Serum cortisol was analyzed with an Immulite 2000 Xpi cortisol immunoassay. A mixed effect ANOVA was used to determine the effect of sex, treatment, time, and the treatment by time interaction on cortisol.

Results

Maximum plasma concentration (C_{max}) of clonazepam (mean, 97 ng/mL) occurred at 3 hours (T_{max}). Terminal elimination half-life and mean resident time were 5.6 and 10.5 hours, respectively. A large volume of distribution (V_z/F; 4.2 L/kg) was observed. Sedation was noted for a period of 2 to 4 hours after administration of clonazepam; no adverse effects were noted. When compared to placebo, clonazepam decreased overall serum concentrations of cortisol (mean, 2.27 and 1.95 µg/dL, respectively; $P = 0.0463$). No treatment by time interaction was detected.

Discussion

Stress reduction is desirable for swine populations subjected to stressors, such as on-farm procedures and transportation. Based on the results of this study, oral administration of clonazepam was effective in decreasing serum cortisol compared with a placebo. Oral clonazepam is promising as a useful tranquilizer for pigs and as a strategy for decreasing stress for clinically relevant periods of time. We anticipate that this drug could aid in managing stress associated with handling and minor medical/surgical procedures and provide an opportunity to improve welfare for pigs.

References

1. De Souza, E. B. Neuroendocrine effects of benzodiazepines. *Journal of Psychiatric Research* 24, 111–119 (1990).

2. dos Santos, F. M. et al. Pharmacokinetic/pharmacodynamic modeling of psychomotor impairment induced by oral clonazepam in healthy volunteers. *Therapeutic drug monitoring* 31, 566–574 (2009).

3. Martínez-Miró, S. et al. Causes, consequences and biomarkers of stress in swine: an update. *BMC Veterinary Research* 12, 1–9 (2016).

https://doi.org/10.54846/am2023/28

Comparison of environmental sampling methods to detect PRRSV on contaminated surfaces

H. D. Dohlman[1]; E. M. Kettelkamp[2], DVM; A. M. Betlach[2], DVM, PhD
[1]Iowa State University College of Veterinary Medicine, Ames, Iowa; [2]Swine Vet Center, St. Peter, Minnesota

Introduction

Environmental sampling is an important method for assessment of pathogen contamination and validation of surface cleanliness after sanitation of a given area. Detection of viral genetic material in the environment can be challenging, as diagnostic sensitivity may vary due to sample technique, handling, or potential RNA degradation prior to testing. While many techniques are available,[1,2] a standardized method for environmental sampling has not been established. Therefore, the objective of this study was to determine the optimum sampling method for detection of PRRSV genetic material on contaminated surfaces under in-vitro and field conditions.

Materials and methods

Under in-vitro conditions, three pilot studies were conducted to compare the detection of PRRS genetic material from inoculated surfaces utilizing different sampling methods and conditions: 1) swab type and surface type (3 × 3 factorial arrangement); 2) media and swab types (2 × 2 factorial arrangement); 3) swab type and viral dose (3 × 4 factorial arrangement). Moreover, the diagnostic sensitivity of different swab types for PRRS detection was assessed under field conditions.

Comparison of PRRS genetic material detection based on swab and surface type

Twenty-one sterile substrate coupons (4″ × 4″) made of aluminum, rubber, and plastic were used to represent commonly contaminated surface materials in the swine industry. Coupons were inoculated via administering 0.25 mL of Ingelvac PRRS® MLV (MLV) with a sterile syringe equipped with a rayon tipped applicator and drying at room temperature for 20 minutes. Coupons were then randomly assigned to undergo one of three aseptic sampling methods according to the following swab types: 1) Swiffer® Dry Sweeping Cloths soaked in 50 mL of phosphate-buffered saline (PBS; Swiffer); 2) Gauze 4 × 4 squares soaked in 20 mL of PBS (Gauze); and 3) BBL™ culture swabs (BBL). In total, 63 samples were assessed with 7 replicates per treatment.

Comparison of PRRS genetic material detection based on swab and media type

Evaluation of liquid media types as either PBS or minimal essential medium (MEM) were compared between Swiffer and Gauze sampling treatment methods. Swiffer® Dry Sweeping Cloths were soaked in either 50 mL of PBS or MEM and gauze 4 × 4 squares were soaked in either 20 mL of PBS or MEM. Sixty aluminum coupons were randomly assigned to one of four aseptic sampling treatments (n = 15 replicates per treatment) in a 2 × 2 factorial arrangement. Coupons were inoculated with 0.1 mL of MLV, dried at room temperature for 20 minutes, and then swabbed.

Comparison of PRRS genetic material detection based on swab type and viral dose

Thirty-six aluminum coupons were randomly assigned to one of twelve treatments that varied based on swab type (ie, Swiffer, Gauze, and BBL) and viral dose (n = 4) as a 3 × 4 factorial. Based on assigned treatment, coupons were either 1.0 mL, 0.5 mL, 0.25 mL, or 0.1mL of MLV, dried at room temperature for 20 minutes and then swabbed. A total of three replicates per treatment were achieved.

Diagnostic sensitivity of swab type under field conditions

Eight farrowing crates were randomly selected from a confirmed PRRSV positive sow farm located in southwest Minnesota. Family oral fluids (FOF) were collected to assess PRRS shedding status of pigs (approximately 18 d of age). For each crate, plastic crate dividers and rubber mats were sampled with the three aforementioned swab types. Two Swiffer, Gauze, and BBL swabs were collected across each measured crate surface. Moreover, sampling treatments were repeated in three randomly selected crates of weaned piglets present in an on-site nursery holding room. To limit potential impact of environmental sample processing on PCR results,[3] the liquid recovered from all Swiffer and Gauze samples were centrifuged for 10 minutes and the supernatant was submitted for testing.

All samples were submitted for PRRS testing via real-time PCR. A PCR cut-off value of 40 was used. Relative viral loads represented by Ct-value based on swab type, media type, viral load and their interactions were assessed within a generalized linear model with the glm function in R version 1.4.1. Significance was deemed at a $P < 0.05$.

Results and discussion

In field and in-vitro settings, PRRSV genetic material recovery was dependent upon swab type. No difference in sensitivity was detected between MEM and PBS liquid media types ($P > 0.10$; Table 1). There were no differences in relative viral load detection based on surface type ($P = 0.06$; Table 2). Across 0.25 mL - 1.0 mL of doses of MLV, BBL swabs yielded the average lowest visual Ct-values (20.81 - 21.89) compared to Gauze and Swiffer samples.

Table 1: Average Ct-values for PRRS detection based on swab and media type

	PBS	MEM
Swiffer swab	27.13	27.45
Gauze swab	23.39	23.98

Table 2: Average Ct-values for PRRS detection based on swab and coupon surface type

	Aluminum	Rubber	Plastic
Swiffer swab	26.8	26.12	25.99
Gauze swab	25.02	23.41	22.83
BBL swab	24.44	23.35	24.00

At 0.1 mL, the highest diagnostic sensitivity was achieved using Gauze swabs compared to Swiffer ($P < 0.001$). Gauze and BBL swabs yielded the lowest average in-vitro Ct-value across all coupons ($P < 0.001$).

In the field, 75% of the sampled farrowing crates resulted PRRSV positive using FOF. For farrowing and nursery room crate surfaces, Swiffer swabs yielded the average relative viral load (35.9 and 34.5 respectively) Ct-value compared to other swab types. In contrast to in-vitro coupon sampling, BBL culture swabs only yielded 4.5% positive PRRS PCR samples in the field. The field Swiffer swab treatments consistently detected PRRS material compared to other swab treatments. BBLInconsistency in the gauze swab detection may be due to the ratio of swab surface area and volume of PBS used. Additional research utilizing a larger sample size is warranted to validate findings observed under field conditions.

References

1. McCarty, M., Petznick, T., Kuster, C. 2015. Case report describing the clinical course of porcine epidemic diarrhea in a commercial boar stud and return of the stud to service after whole-herd inoculation with porcine epidemic diarrhea virus. J Swine Health Prod. 2015;23(5):264–271.

2. Garrido-Mantilla, J., Alvarez, J., Culhane, M. 2019. Comparison of individual, group and environmental sampling strategies to conduct influenza surveillance in pigs. BMC Vet Res 15, 61 (2019). **doi:10.1186/s12917-019-1805-0**

3. Elijah, G. C., A. K. Blomme, O. L. Harrison, J. Bai, J. C. Woodworth, C. K. Jones, E. G. Poulsen-Porter, C. B. Paulk, and J. T. Gebhardt. 2021. Evaluating the impact of presence of organic matter on environmental samples and sample processing technique on RNA detection of PEDV. Kansas Ag. Exp. Stat. Res. Report, Manhattan. Vol. 7: Iss. 11. **doi:10.4148/2378-5977.8221**

https://doi.org/10.54846/am2023/29

Efficacy of PRRSV vaccination regimens to overcome maternal immunity in PRRSV-positive weaned pigs

A. C. Steffensmeier[1], BS; J. P. Gooding[2], MS; C. L. Puls[2], PhD;
R. W. Evelsizer[2], DVM; A. C. Szczotka[3], DVM; N. C. Lauterbach[3], DVM
[1]Iowa State University, College of Veterinary Medicine, Ames, Iowa; [2]Elanco, Greenfield, Indiana;
[3]Smithfield Hog Production, Algona, Iowa

Introduction

Porcine reproductive and respiratory syndrome virus (PRRSV) is the most economically detrimental disease in the swine industry. PRRSV costs the United States swine industry more than $640 million annually.[1] PRRSV is an enveloped RNA virus that replicates in the alveolar macrophages and lymphoid tissues allowing the virus to persist in the host after the adaptive immune system has stimulated an adequate immune response against the virus.[2] The hematogenic spread of the virus by the phagocytic cells allows for dissemination to the lungs and other organ systems.[3] Ultimately, PRRSV infection increases the susceptibility to other pathogens leading to the further decreased performance of the herd. PRRSV vaccination has been shown to significantly reduce death loss and shedding of PRRSV by infected pigs.[4] The efficacy of PRRSV vaccinations in weaned pigs can vary based on several factors, including the level of maternal antibody interference originating from the dams.[4] The goal of this study was to determine if two full doses of a modified-live (MLV) PRRSV vaccine in wean pigs, originating from a PRRS-positive sow system, would overcome maternal antibody interference, resulting in reduced viremia and/or increased neutralizing antibody production when compared to a single full dose of MLV PRRSV vaccine at weaning.

Materials and methods

Pigs were sourced from a I-B PRRSV positive unstable, low prevalence sow unit[5] and placed in a 2,400-head, two-barn, naturally ventilated wean-to-finish facility in North Central Iowa. The source sow unit previously broke with a wild-type strain of PRRSV. The pigs for this trial were vaccinated for porcine circovirus 2, *Mycoplasma hyopneumoniae*, and *Lawsonia intracellularis* prior to arrival. Thirty robust pigs were allocated to one of two vaccine regimen groups and separated by barn. Each enrolled pig was ear tagged 1-60 and divided into 6 pens with no more than 10 enrolled pigs in each pen. Barn 1 received one full dose (1 mL) of a MLV PRRSV (Prevacent, Elanco Animal Health) vaccine on day 0. Barn 2 received one full dose (1 mL) of a MLV PRRSV (Prevacent, Elanco Animal Health) vaccine on day 0 and a second full dose on day 21. The animal caretakers were instructed to perform normal standards of care following set protocols during the course of the trial. Serum samples were collected on days 0 (24 hours after placement), 7, 14, 21, 28, 35, and 43. Three oral fluids were collected per vaccine regimen group on days 7, 14, 21, 28, 35, and 43. Serum samples were analyzed for fluorescent focus neutralizing antibody (FFN) specific to the Prevacent MLV PRRSV strain at South Dakota State University Animal Disease Research and Diagnostic Laboratory. Fluorescent focus neutralizing assays determines the amount of neutralizing antibodies specific to the PRRSV strain found in Prevacent in a serum dilution series. The samples were diluted in 2% horse serum and incubated to evaluate the fluorescent foci. The serum dilution series were 1:4-1:512 and were evaluated manually. The serum dilution series endpoints were determined at the 90% or greater reduction in fluorescent focus units.[6] Serum samples were also tested for PRRSV enzyme-linked immunosorbent assay (ELISA) at Iowa State University Veterinary Diagnostic Laboratory (ISU VDL), and PRRSV polymerase chain reaction (PCR) (ISU VDL). Oral fluid samples were analyzed for PRRSV PCR (ISU VDL).

Results and discussion

Diagnostic sampling confirmed the presence of PRRSV 1-2-4 lineage, 1C was present prior to vaccination. This strain originated from the sow source and was identified throughout the population during the trial. There was a 1% variation in all PRRSV sequences identified throughout the trial. Mortality was < 1% in both barns, while we lost seven-head from the study populations during the trial. The results for the mean serum PCR cycle time (Ct) for the presence of PRRSV genetic material are displayed in Figure 1 and Table 1. The mean serum PCR Ct reached the lowest point for two-dose on day 14 at 22.1 and on day 21 for the one-dose regimen at 22.5. ELISA results are found in Figure 2. The two-dose regimen saw a faster increase in ELISA-positive pigs compared to the one-dose. By day 42, 98% of all the pigs in the trial were positive for serum antibodies against PRRSV. Results of the Prevacent FFN are displayed in Figure 3. Day 0 FFN testing prior to vaccination showed a 40% positivity of neutralizing antibodies against the Prevacent strain for the two-dose compared to 73% positivity for the one-dose regimen. It is possible that there is a high level of neutralizing antibodies at day 0 due to the cross-reactivity of the wild-type strain to the Prevacent strain in the FFN test. Most of the titers are low but a few had higher titers that were likely due to the individual response to the previous PRRSV exposure. FFN results demonstrated an increase in mean titers for both groups by day 14. There was an anamnestic response following the second dose of the two-dose regimen by day 28. Both vaccine regimen groups saw mean titers fall on day 35. The two-dose regimen FFN results remained higher on day 42 with a mean of 0.165 vs. the one-dose regimen FFN at 0.125. Neutralizing antibody titers within the two-dose regimen were higher from day 21 to day 42 with a mean of 0.04. The *P*-value for the difference in FFN was > 0.05. While this trend was not significant for this population, this does demonstrate an ability to enhance the production of neutralizing antibodies to a MLV PRRSV strain in the face of maternal PRRSV antibodies. When a second full dose was given, the neutralizing antibody titers measured in this trial were found to be stronger and lasted longer. Further studies are warranted to evaluate if this same effect can be achieved in pigs originating from a

Figure 1: Mean PRRSV PCR by day.

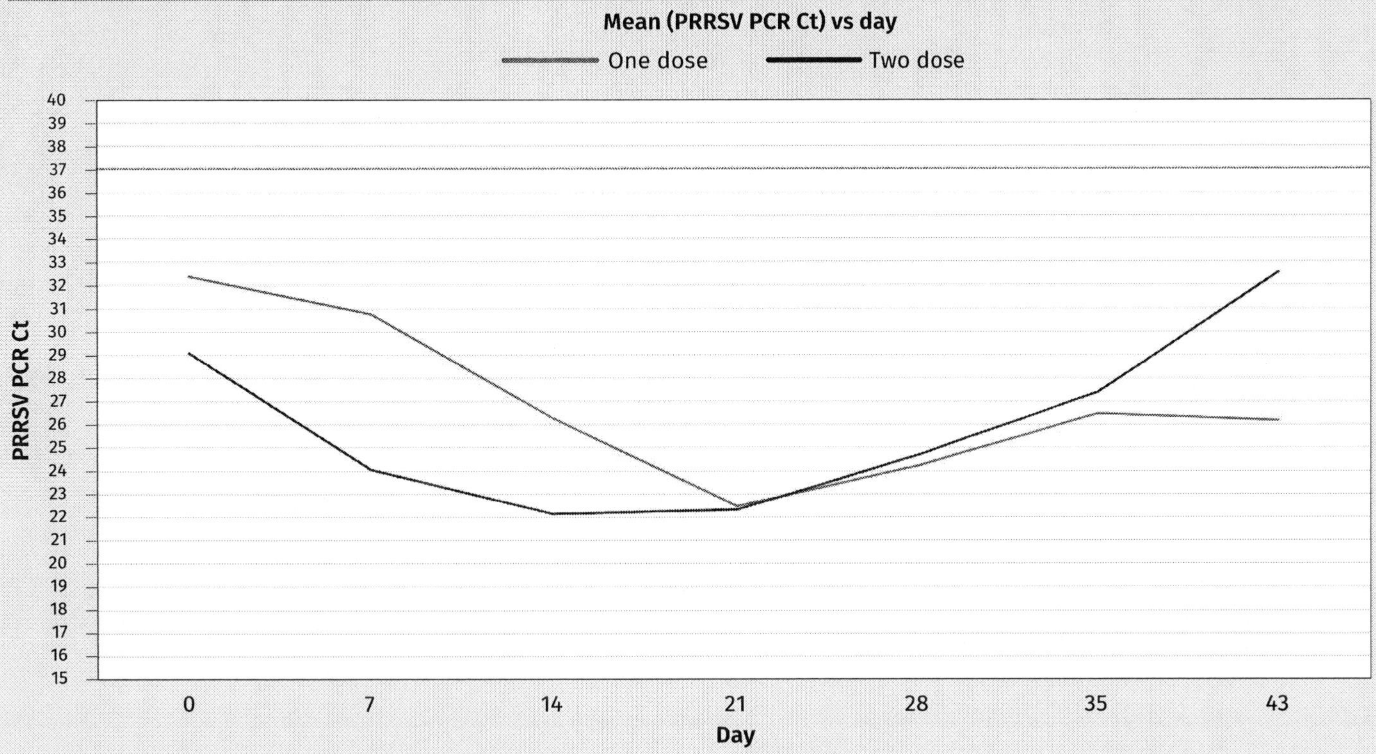

Table 1: PRRSV PCR Ct by day

Treatment	Animal ID	Day 0	Day 7	Day 14	Day 21	Day 28	Day 35	Day 43
1 Dose	GG [31 - 35]	23.7	22.3	20.3	20.1	22.8		
1 Dose	GH [36 - 40]	37.0	37.0	37.0	24.2	24.4	23.8	24.0
1 Dose	GI [41 - 45]	37.0	30.2	24.4	21.0	22.8	25.0	27.1
1 Dose	GJ [46 - 50]	22.6	22.0	21.2	22.4	27.4	27.4	27.0
1 Dose	GK [51 - 55]	37.0	37.0	21.0	22.0	25.8	29.1	24.6
1 Dose	GL [56 - 60]	37.0	36.0	33.7	25.1	22.1	27.0	28.1
2 Dose	GM [61 - 65]	37.0	27.7	21.4	23.0	24.9	28.5	26.4
2 Dose	GN [66 - 70]	33.4	22.9	20.2	22.0	27.9	25.6	29.8
2 Dose	GO [71 - 75]	23.9	24.8	22.0	20.9	20.9	28.6	37.0
2 Dose	GP [76 - 80]	23.0	22.7	24.6	22.4	24.3		36.1
2 Dose	GQ [81 - 85]	20.3	21.0	22.3	21.6	25.4	27.5	34.7
2 Dose	GR [86 - 90]	37.0	25.2	22.3	24.0		26.7	31.5

I-B PRRSV sow source by utilizing a split-dosing program instead of the two-dose conducted in this trial. It would also be beneficial to replicate this study in PRRSV naïve animals and evaluate what, if any, production impact the increased neutralizing antibody titer levels and duration of immunity would have if exposed to a wild-type PRRSV.

References

1. Holtkamp, D., et al. "Assessment of the economic impact of porcine reproductive syndrome virus on United States pork producers." *Journal of Swine Health and Production*, 2013;21(2):72-84.

2. Dubovi, Edward J., et al. "Arteriviridae and Roniviridae." *Fenner's Veterinary Virology*, Fourth ed., 2011, pp. 420–422.

3. McGavin, M. Donald, and Alfonso Lopez. "Respiratory System." *Pathologic Basis of Veterinary Disease*, Fourth ed., 2006, pp. 535–536.

4. Tizard, I. "The Use of Vaccines." *Veterinary Immunology.* 2018; 278-280.

5. Holtkamp, D., et al. "Proposed modifications to porcine reproductive and respiratory syndrome virus herd classification." *Journal of Swine Health and Production* 29.5 (2021): 261-270.

6. Robinson, Sally R., et al. "Broadly Neutralizing Antibodies against the Rapidly Evolving Porcine Reproductive and Respiratory Syndrome Virus." *Virus Research*, vol. 203, 2015, pp. 56–65.

Figure 2: ELISA results by day.

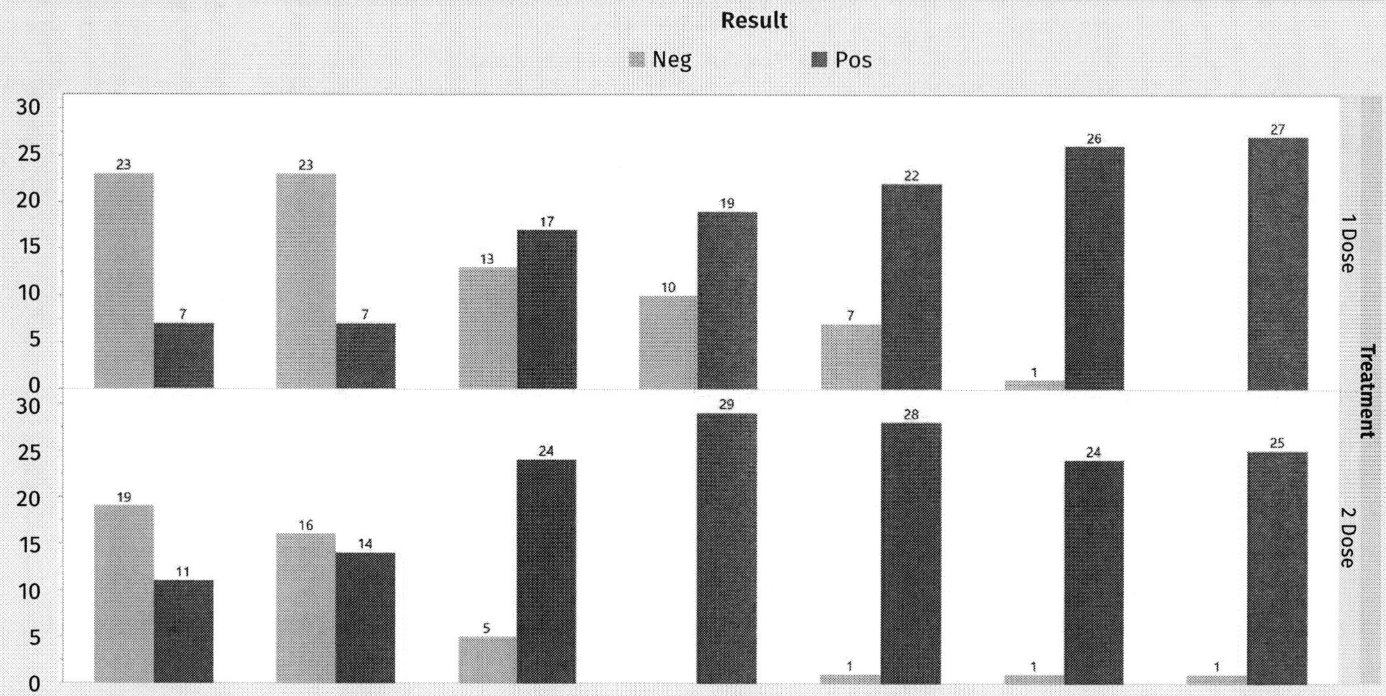

Figure 3: Prevacent FFN.

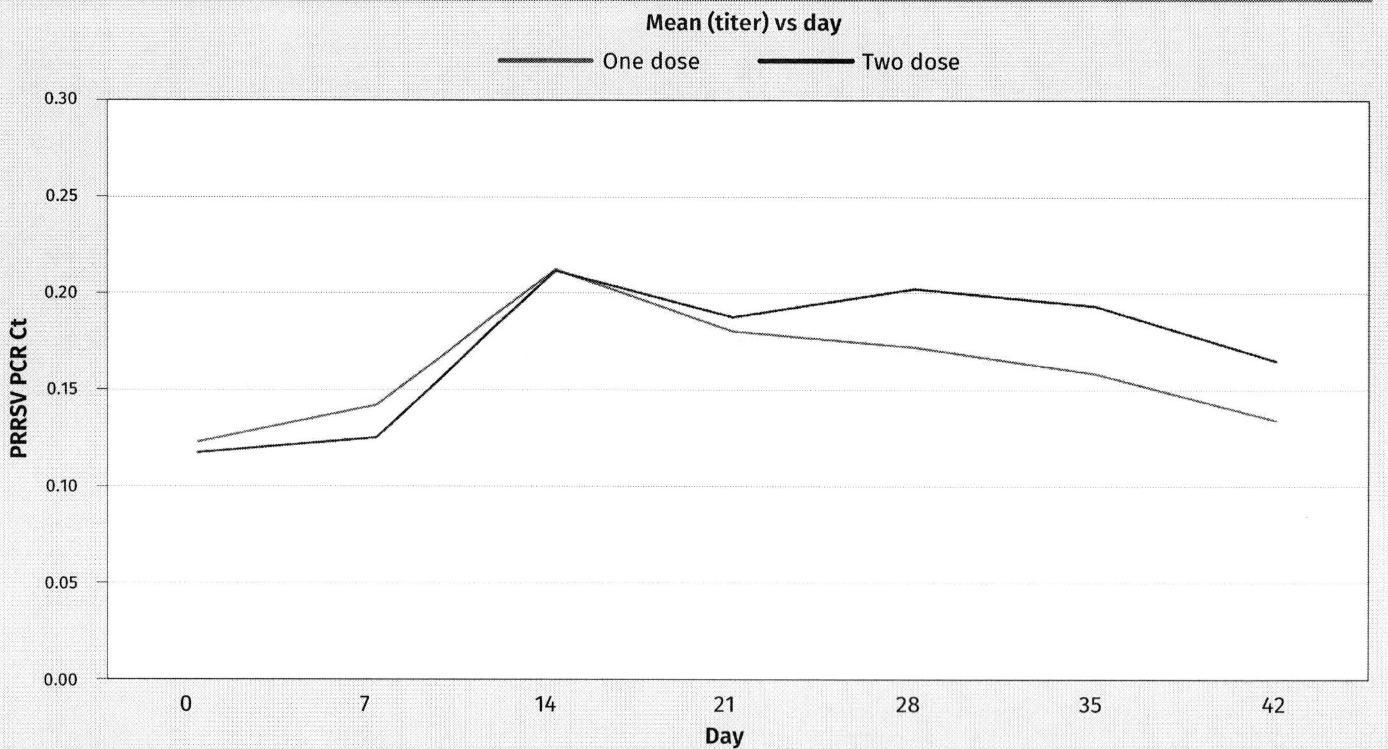

*no differences (*P*-value > 0.05)

https://doi.org/10.54846/am2023/30

Evaluation of porcine reproductive and respiratory syndrome virus maternal antibody interference in commercial wean pig populations using a PRRS modified live virus vaccine

Mallory Wilhelm[1], BS; Brandi Burton[2], DVM; Amanda Reever[2], MS, DVM;
Robert Evelsizer[3], DVM; Christopher Puls[3], PhD
[1]Iowa State University, College of Veterinary Medicine, Ames, Iowa; [2]Suidae Health and Production, Algona, Iowa;
[3]Elanco Animal Health, Greenfield, Indiana

Introduction

Porcine reproductive and respiratory syndrome virus (PRRSV), an RNA virus, has been estimated to cost the industry $664 million annually.[1] Efforts have been made to minimize the effect this virus has on the industry. However, in recent years, producers have incurred increasingly substantial losses due to PRRSV outbreaks. As a result, some producers are choosing to vaccinate their herd against PRRSV using a modified live or killed virus vaccine to minimize potential losses from this disease. Vaccinating sows with a modified live vaccine (MLV) has demonstrated benefit in improving outcomes when animals are subsequently exposed to wild type PRRSV.[2] The objective of this project was to observe differences in maternal antibody levels based on varying time since mass herd vaccination in sows. This trial may shed light on more effective timing and vaccine protocols in regard to sow immunity. Additionally, it may help to determine if piglet vaccination protocols should be adjusted over time relative to last mass-herd vaccination on the sow farm.

Materials and methods

This study was conducted in a 2,500 head commercial sow farm in northwest Iowa and three wean-to-finish sites (one 2,400 head site in southern Minnesota and two 2,400 head sites in southeast Iowa). All sows on the farm received a mass herd vaccination on a single day. Eight sows with at least 12 piglets were selected for each treatment group: Group A piglets were born 17 weeks post-previous mass vaccination (pre-herd vaccination), group B piglets were born 1-week post-herd vaccination, and group C piglets were born 3-weeks post herd vaccination. Piglets were all weaned between 18 and 21 days of age. At weaning, piglets within a litter were randomly allotted into four different treatment groups via a random number generator. Group 1 received a half dose of a modified live PRRSV vaccine at weaning. Group 2 received a half dose of a modified live PRRSV vaccine at weaning and a half dose booster 3-weeks post weaning. Group 3 received a full dose of a modified live PRRSV vaccine at weaning. Lastly, group 4 remained unvaccinated at weaning, receiving a half dose of a modified live PRRSV vaccine 3-weeks post weaning, as seen in Table 1. Piglets were tagged at birth, serum was collected, and piglets received the appropriate treatment, if necessary, after being allocated to a treatment group. Serum was collected from all pigs at 0 (pre-wean), 1, 3, 5, and 7-weeks post-weaning and stored in a -80°C ultra-low freezer until being submitted at one time to the Iowa State University Veterinary Diagnostic Lab to reduce variability. Samples were tested by PRRSV enzyme-linked immunosorbent assay (ELISA) to determine antibody response to vaccine. Blood samples collected from 90 pigs (30 pigs per mass vaccination timing group) at both pre-weaning and 7-weeks post weaning were sent for Prevacent fluorescent-focus neutralization assay (FFN) testing at South Dakota State University.

Results

Pigs from sows vaccinated 3 weeks post-vaccination had greater incidence of ELISA positives at pre-wean, 1-week, and 3-week post wean time points compared to pigs from sows vaccinated 17- or 1-week post-herd vaccination. Figure 1 displays the percent ELISA positives between each of the three treatment groups by time point. For piglets from sows 17-weeks and 1-week post-vaccination, almost 30% of piglets were ELISA positive prior to weaning/vaccination, suggesting sustained maternal antibody presence. At 1-week post weaning, pigs from sows 17-weeks and 1-week post-vaccination had decreased ELISA positive rates, while the 3-week post vaccination pigs had an increased percent ELISA positive rate. Through 5- and 7-weeks post weaning, piglets from sows 17-weeks and 1-week post-vaccination had increased ELISA positive rates compared to the 3-week post vaccination piglets. None of the piglet populations reached 100% ELISA positive rates. Regardless of the amount of time post-vaccination, ELISA positivity rates reached similar levels.

Figure 2 displays the average ELISA S/P ratio by piglet treatment group over the 7-week study. Each time point (pre-wean, 1-week post wean, etc) is presented by piglet treatment group (1-4) and sow treatment (A, B, and C). Evaluating specific piglet treatment groups, piglets born to sows 3-weeks post herd vaccination (Group C) that received a half dose of PRRSV vaccine treatment showed the greatest immune response (the greatest average ELISA S/P ratio) in comparison to the other treatment groups within this study.

The FFN analysis revealed that group A (pre-herd vaccination), treatment group 4 (unvaccinated at weaning, ½ dose 3 weeks post wean) had the highest FFN-specific antibody response at around 87% at 7-weeks post weaning. All treatment groups within group A also saw an increase in percent positives except for treatment group 1 (½ dose at weaning), which saw a decrease in percent positives over the 7-week period. Within group B (1-week post herd vaccination), treatment group 2 (½ dose at weaning and ½ dose booster at 3 weeks) had the highest FFN-specific antibody response, however, it was the same at about 70% for both pre-wean and 7-week post weaning time points. Group C (3-weeks post herd vaccination) had the overall had the highest percent FFN-positives from all treatment groups. As time advanced, the total FFN-percent positives from all treatment groups within group C

Table 1: Sow treatment groups (A-C) on the left and piglet treatment groups (1-4) per sow treatment group on the right

Sow treatment	Piglet treatment
Group A - 17-weeks pre-herd vaccination	Group 1 - ½ dose of modified live PRRSV vaccine at weaning
	Group 2 - ½ dose of modified live PRRSV vaccine at weaning and a ½ dose 3-weeks post weaning
	Group 3 - full dose of modified live PRRSV vaccine at weaning
	Group 4 - unvaccinated at weaning and ½ dose of modified live PRRSV vaccine 3-weeks post weaning
Group B - 1-week post herd vaccination	Group 1 - ½ dose of modified live PRRSV vaccine at weaning
	Group 2 - ½ dose of modified live PRRSV vaccine at weaning and a ½ dose 3-weeks post weaning
	Group 3 - full dose of modified live PRRSV vaccine at weaning
	Group 4 - unvaccinated at weaning and ½ dose of modified live PRRSV vaccine 3-weeks post weaning
Group C - 3-weeks post herd vaccination	Group 1 - ½ dose of modified live PRRSV vaccine at weaning
	Group 2 - ½ dose of modified live PRRSV vaccine at weaning and a ½ dose 3-weeks post weaning
	Group 3 - full dose of modified live PRRSV vaccine at weaning
	Group 4- unvaccinated at weaning and ½ dose of modified live PRRSV vaccine 3-weeks post weaning

decreased. Further study is warranted to evaluate maternal antibody incidence and decay and subsequent effects on piglet ELISA response.

Discussion

This study will enable producers, supervisors, and veterinarians to make better vaccination decisions through the addition of new knowledge on the role that maternal antibody interference plays in both sow and piglet immunity. By understanding the maternal antibody decay timeline within the piglet immune system, depending on when sows have been vaccinated, producers will be able to optimize vaccine scheduling practices. This information is important and relevant for all stages of production; including sow farms, nurseries, finishers, and wean-to-finish operations, since it will result in better protection against PRRSV downstream.

References

1. Holtkamp, D. J. & Kliebenstein, J. B. & Zimmerman, J. J. & Neumann, E. & Rotto, H. & Yoder, T. K. & Wang, C. & Yeske, P. & Mowrer, C. L. & Haley, C., (2012) "Economic Impact of Porcine Reproductive and Respiratory Syndrome Virus on U.S. Pork Producers", *Iowa State University Animal Industry Report* 9(1). **https://doi.org/10.31274/ans_air-180814-28**

2. Lowe JF, Zuckermann FA, Firkins LD, Schnitzlein WM, Goldberg TL. Immunologic responses and reproductive outcomes following exposure to wild-type or attenuated porcine reproductive and respiratory syndrome virus in swine under field conditions. J Am Vet Med Assoc. 2006 Apr 1;228(7):1082-8. **doi: 10.2460/javma.228.7.1082** PMID: 16579789.

Figure 1: Percent positive pigs in each of the three treatment groups displayed at varying time points post wean.

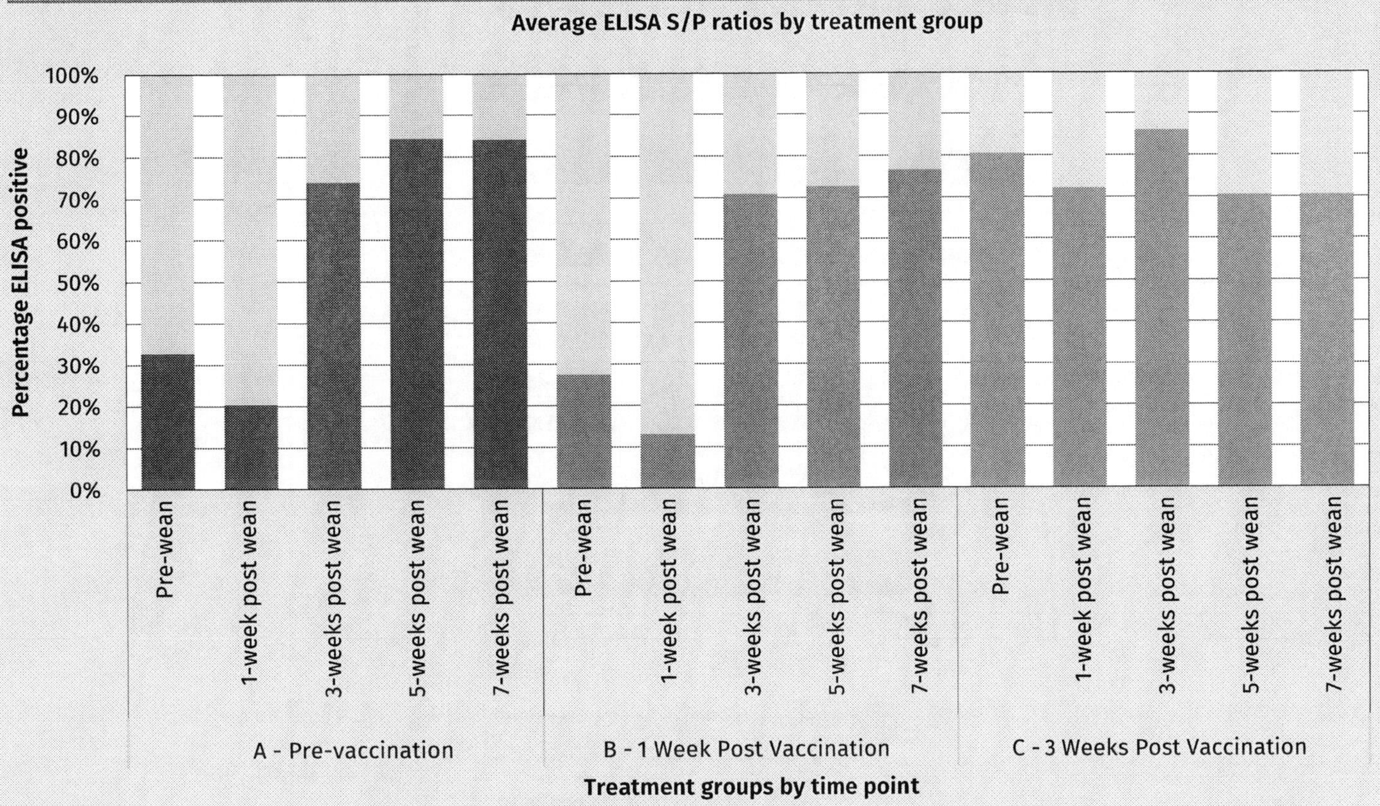

Figure 2: Sow Treatment Group

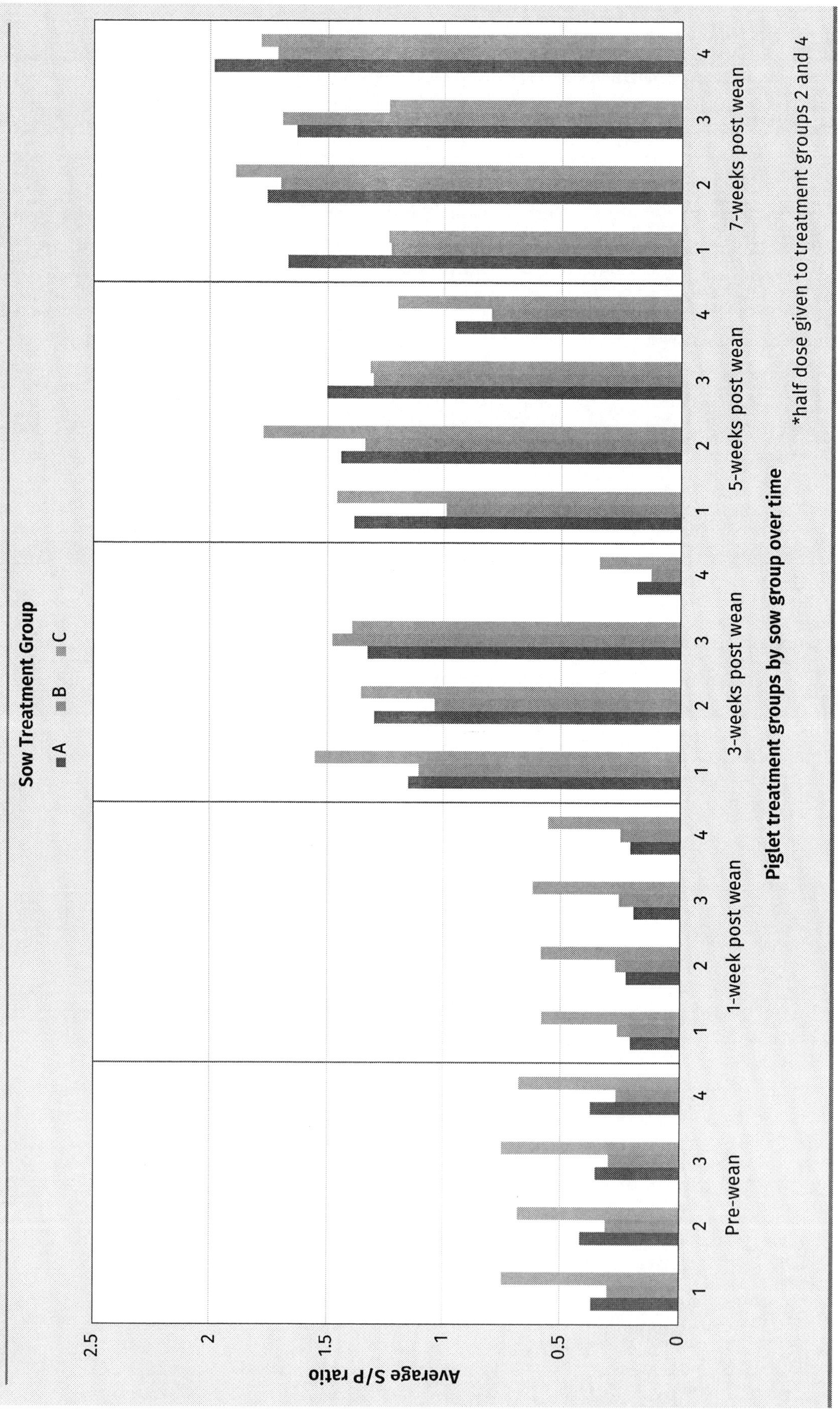

https://doi.org/10.54846/am2023/31

Assessing tongue samples from mortalities as additional PRRSV and IAV-S surveillance on protected nursery and finishing sites

Dylan Wulfekuhle[1], BS; Kathleen Wood[2], DVM; Jeff Okones[3], DVM; Gustavo Silva[1], DVM, MS, PhD; Marjorie Schleper[2], DVM
[1]Iowa State University, College of Veterinary Medicine, Ames, Iowa; [2]Christensen Farms, Sleepy Eye, Minnesota; [3]Pharmgate Animal Health, Wilmington, North Carolina

Introduction

Protected sites are nursery and finishing barns located within three miles of commercial sow barns, multipliers, or boar studs. These sites require regular testing using oral fluids (OF) or serum samples to detect the presence of pathogens such as porcine reproductive and respiratory syndrome virus (PRRSV) +/- influenza A virus of swine (IAV-S). The objective of this study was to determine whether sampling tongue homogenate or exudate would serve as a useful diagnostic tool for pathogen detection and whether this process would be a feasible testing measure to add to the surveillance protocols currently implemented for protected sites.

Materials and methods

Tongues were collected from all mortalities across four nurseries and four finishing sites. Tongues were cut at their base with a knife or scissors, placed in a clean Ziploc® bag and frozen in a commercial grade freezer. A minimum of five tongues were collected per airspace. On a bi-weekly basis, the tongue samples would be thawed and processed by two treatment groups: Tongue homogenate obtained from samples processed in a countertop blender and tongue exudate obtained by straining the homogenate through pantyhose to collect a fluid. Oral fluids were collected at the end of the two-week period of tongue collection and would serve as the control group to be compared to the tongue sample groups collected during the prior two weeks. The control and treatment group samples would be submitted for the detection of PRRSV and IAV-S by polymerase chain reaction (PCR). Contingency tables (2x2) were built to calculate the relative sensitivity and specificity to the control group. The agreement between groups was determined using the kappa statistic. Agreement was substantial if the kappa value (K) was 0.61 to 0.8, moderate if K was 0.41 to 0.6, and fair if K was 0.21 to 0.4. All analyzes were performed using the R program.

Results

Comparison of treatment group samples to current OF testing protocol samples reveal a fair agreement for PRRSV detection between OF and tongue homogenate (K = 0.3741) with a relative sensitivity of 0.73 and a relative specificity of 0.67. There was substantial agreement for PRRSV detection between OF and tongue exudate (K = 0.6489) with a relative sensitivity of 0.82 and a relative specificity of 0.83. For detection of IAV-S, there was substantial agreement between OF and tongue homogenate (K = 0.6972) with a relative sensitivity of 0.82 and relative specificity of 0.88. There was also substantial agreement for IAV-S detection between OF and tongue exudate (K = 0.6402) with a relative sensitivity of 0.88 and relative specificity of 0.75. On a generalized sample basis, both tongue samples showed a higher positivity rate and lower CT value for PRRSV when compared to oral fluids. Tongue exudate generated a higher positivity rate for IAV-S and higher average CT value. Tongue homogenate had a lower positivity rate and higher CT value when compared to oral fluids for IAV-S detection (Table 1).

Discussion

Tongue samples are suggestive of being a diagnostically sensitive test for PRRSV and IAV-S. Tongue exudate showed the highest positivity rate for both PRRSV and IAV-S detection with a typically lower cycle threshold (CT) value when compared to homogenate. Disadvantages of tongue sampling include ability to acquire an adequate sample size and the time required to collect and process samples. This sampling style works well on sites requiring extra surveillance when utilizing a pooled sample.

Reference

1. Baliellas J, Novell E, Enric-Tarancón V, Vilalta C, Fraile L. Porcine Reproductive and Respiratory Syndrome Surveillance in breeding Herds and Nurseries Using Tongue Tips from Dead Animals. Vet Sci. 2021 Nov 2;8(11):259.

Table 1: Results from different sample types for PRRSV AND IAV-S

Sample type	Samples tested	Samples positive	Positivity rate	CT average
PRRSV oral fluids	30	12	40.00%	32.95
PRRSV tongue homogenate	29	14	48.28%	31.61
PRRSV tongue exudate	29	17	58.62%	29.78
IAV-S oral fluids	30	17	56.67%	28.13
IAV-S tongue homogenate	29	16	55.17%	32.68
IAV-S tongue exudate	29	18	62.07%	29.28

https://doi.org/10.54846/am2023/32

The competency impact after initiating a new training program in the boar stud

M. Hood[1], DVM; S. Miller[2,3], DVM; S. Hornback[2,3], MBA; D. Reicks[1], DVM
[1]RVRC; [2]AgCreate Solutions, Inc; [3]Pork Avenue Training Portal

Introduction

High staff turnover in the swine industry has led to difficulties in training and retaining staff. This is especially challenging in boar studs, where staff are expected to master a specialized series of technical competencies alongside basic care. This can be further complicated by language barriers. In previous studies, training programs using the *'See It., Do It., Teach It.'* model were successful at raising baseline expressed competency levels of staff[1] and maintaining that competency over time.[2] A program using this method was developed for boar stud staff and implemented within a group of studs.

Methods

Two boar studs with entirely Spanish-speaking staff were selected to participate in the study. Each stud employed 5 staff members. Only two participants had worked on their site for greater than 6 months. Prior to initiating the study all new employees were trained by on-farm staff mentoring. Before implementing the new program, 30 competencies were selected from five learning experiences to evaluate the staff – 18 competencies on collection technique, 6 competencies on boar safety, and 6 competencies on boar training. Baseline competency measurements were taken using the 'teach it' method with the use of prompting questions by the evaluator if necessary. The staff were asked to teach the evaluator how to perform tasks including donning gloves for collection, collecting a boar, moving boars around the barn, choosing proper PPE, and training boars. Competencies were marked as yes/no present. Employees were given a percent expressed competency at the end of the session. Post baseline, employees were enrolled in the Pork Avenue Boar Stud Curriculum, including three learning experiences on collection technique with Spanish subtitles, plus one learning experience on boar behavior and safety and one on boar training, both with Spanish voiceover. All training videos ended with a testing simulation of desired end skills. After seeing the videos, mentors took employees to the barn to perform the task together. Finally, staff were re-evaluated by repeating the 'teach it' back phase with the same evaluator marking the competencies chosen as yes/no present.

Results

The employees had an average baseline expressed competency rate of 70.5% (Stud 1) and 73.6% (Stud 2). The areas where staff struggled the most were boar training and boar safety. Post external verification of training, the employees averaged 94.5% (Stud 1) and 95.5% (Stud 2) competency expression rate. The average employee improved by 31.1% and the overall improvement ranged from 10 to 66% between employees – see Figure 1. The area of greatest improvement was in boar training – the average employee improved by 88%.

Conclusions

The training program was successful at increasing staff competency expression rates for boar stud specific topics. In addition, the program was consistently able to bring all employees, no matter their background, to similar high levels of competency.

References

1. Probst Miller, Sarah and Krantz, Seth. Inspiring a verified culture of care post-identification of abuse on an undercover video using a baseline assessment of expressed competencies compared to post training measurement of expressed competencies.

2. Probst Miller, Sarah and Krantz, Seth. Culture of care longevity measurements post-training.

Figure 1: Pre- and post-assessment scores by individual participant.

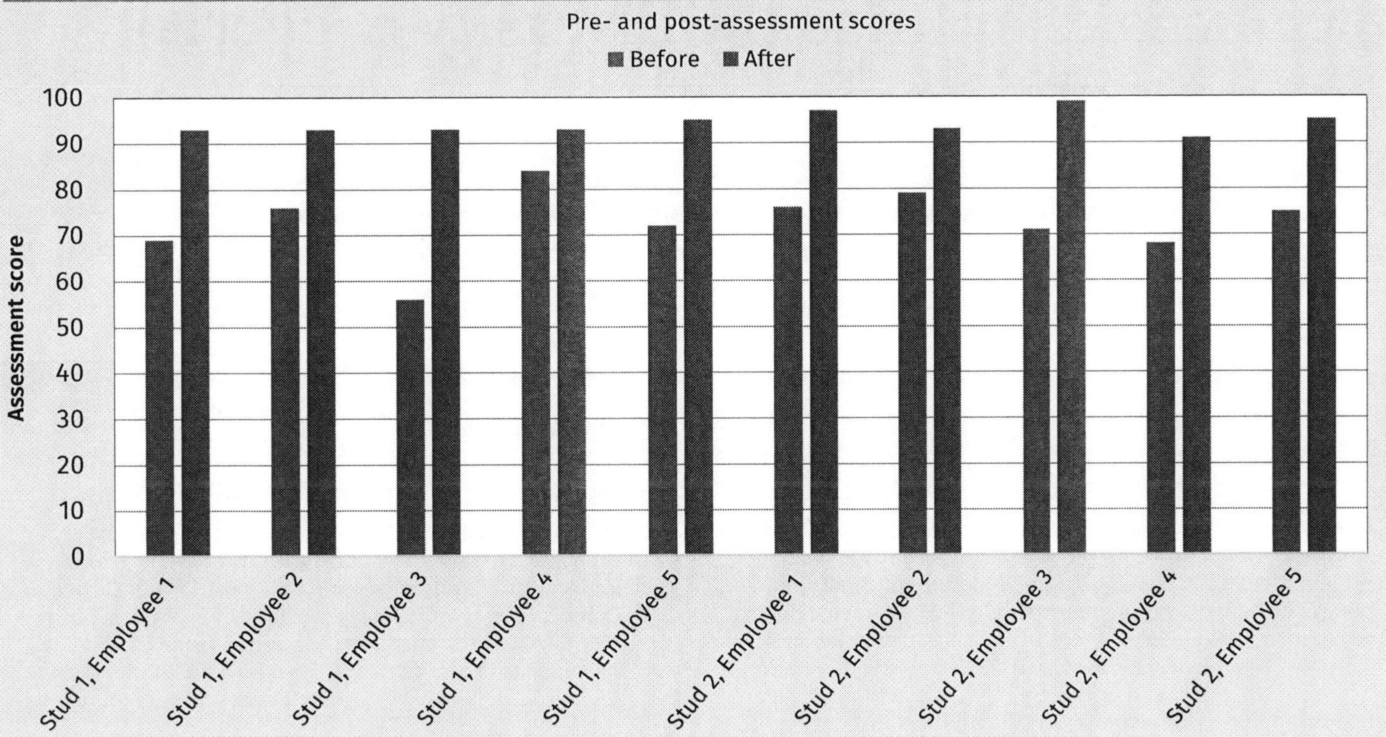

https://doi.org/10.54846/am2023/33

Epitope analysis confirms a bivalent PCV2 vaccine is relevant to the diversity of current, global strains

Meggan Bandrick[1]; Dennis Foss[1]; Andres H. Gutierrez[2]; Sudeep Perumbakkam[1]; Anne S. De Groot[2,3]; William D. Martin[2]; Terry E. Frances[2]; Alvaro Aldaz[4]; James R.D. Allison[4]; Jose Angulo[3]
[1]Zoetis, Global Biologics, Veterinary Medical Research and Development, Michigan; [2]EpiVax Inc., Providence, Rhode Island; [3]Center for Vaccines and Immunology, University of Georgia, Athens, Georgia; [4]Zoetis, Global Commercial Development, New Jersey

Introduction

Porcine circovirus type 2 (PCV2) is a worldwide swine pathogen with an almost ubiquitous presence. There are 9 proposed PCV2 genotypes labeled PCV2a-i of which PCV2a, b, and d are clinically relevant.[1] The economic impact of PCV2 and associated disease is substantial and vaccination against PCV2 rapidly became routine once vaccines were available. Until recently all vaccines were based on PCV2a, reflecting the predominant genotype when they were developed. The genomic shift to PCV2b started before vaccine introduction, but it seems likely that vaccination has encouraged genetic drift.[2,3] Along with a decrease in revalence, an increased diversity is seen within PCV2a viruses, focused on antigenic epitopes shared between field and vaccine strains.[2] In experimental studies, homologous vaccine / challenge combinations provided greater reductions in post-challenge PCV2 viremia than heterologous combinations[4,5] although cross-protection was still substantial, and in-field use PCV2a-based vaccines is generally judged effective against any wild genotype.

Whether current vaccines will remain effective, however, is obviously a concern and, given the often non-specific nature of PCV2's impact on health and productivity, it may be difficult to determine whether the protection obtained is optimal. In a recent field trial, a vaccine containing PCV2a and PCV2b reduced PCV2 viremia and improved weight gain compared to a PCV2a-only vaccine, although both were superior to non-vaccination. T cell Epitope Content Comparison (EpiCC) was used to predict antigenic similarity among the field strain isolated and the vaccines tested. EpiCC predicted the PCV2a-PCV2b vaccine increased coverage to the PCV2d field strain's capsid protein T cell epitopes by 33% compared to the PCV2a vaccine.[6]

T cell epitopes, such as from PCV2 vaccine and field strains, predicted to bind putative class I and class II Swine Leucocyte Antigen (SLA) can be expressed as an EpiCC score, reflecting the degree of antigenic relatedness of those sequences which might actually be recognized by the pig.[7] We've previously demonstrated EpiCC scores for specific monovalent PCV2 vaccines were highest when compared to field strains of the same genotype. The bivalent vaccine not only improved coverage across genotypes, but also gave higher scores than monovalents when the latter were compared to a homologous genotype, reflecting the value of an additional vaccine strain in compensating for antigenic diversity within as well as between genotypes. A bivalent PCV2a and PCV2b vaccine also gave biologically superior protection to monovalent PCV2a or PCV2b vaccines in a series of experimental challenge studies.[4] In this study, we use EpiCC to compare vaccine sequences to a larger and more recent population of field strains, all associated with herds affected by PCVAD and analyzed by geographic origin.

Materials and methods

746 PCV2, ORF2 nucleotide sequences were obtained directly from veterinary diagnostic laboratories (n = 684), or from GenBank and typically originated from investigations of suspected PCVAD or ongoing monitoring of herds with PCVAD. All were from 2017 to 2021. Sequences were obtained from 24 countries and were classified into Asia (n = 185), Europe (n = 269), North America (n = 235) and South America (n = 57) for presentation of results.

The field strain ORF2 DNA sequences were used to derive predicted amino acid sequences for their encoded PCV2 capsid proteins using EditSeq, Version 15.1.0 (DNASTAR, Madison, WI). The same was done for DNA sequences representing several PCV2 vaccines commercially available at the time of the study: the PCV2a and PCV2b ORF2 sequences from a bivalent PCV1-PCV2 chimeric vaccine (VacAB), the PCV2a sequence from this vaccine on its own (VacA), and two PCV2a ORF2 sequences from vaccines produced using baculovirus expression systems (VacAlt1, VacAlt2). The sequences were classified by genotype and mapped to known sequence clusters as defined.[8]

Using methodology fully described previously,[9] all field strain protein sequences were individually compared to the four vaccines in a series of two-way comparisons, using the EpiCC algorithm to quantify the degree of antigenic relatedness based on T cell epitope content. Briefly, potential T cell epitopes were identified for a panel of 11 SLA class I alleles and 5 SLA class II alleles using EpiVax's iVAX toolkit and SLA-binding matrices derived from the PigMatrix prediction algorithm. EpiCC was then used to compare the epitope content of the sequences, with a higher EpiCC score representing greater commonality or epitope coverage. For comparisons to the bivalent vaccine (VacAB), the EpiCC score reflects the matching of field strain T cell epitopes to epitopes in either or both vaccine strains with no additional score if the latter. The EpiCC algorithm does not account for any differences in dose.

EpiCC scores were converted to percent T cell epitope content coverage by dividing the EpiCC scores by the corresponding field strain baseline score. The non-parametric Wilcoxon test for paired samples was used to determine whether T cell epitope content coverage for each PCV2 genotype was significantly different between pairs of vaccines. *P*-values < 0.05 were considered significant.

Results

Overall, 36% of the tested sequences were from Europe, 32% from North America, 25% from Asia and 8% from South America, with 17% being PCV2a, 15% PCV2b and 68% PCV2d. Sequences were collected based on availability rather than as part of a prospectively designed survey, so the distribution may not be an accurate reflection of field epidemiology. PCV2d was the dominant genotype in all geographies and nearly all PCV2d sequences were cluster 4 or PCV2d-2. It is also noteworthy that PCV2a showed an apparent difference by continent, with all isolates from Europe and South America being in cluster 10, whereas 76 of the 77 North American isolates were in cluster 6.

The mean baseline T cell epitope score was 10.55, equivalent to 221.58 (177.24 class I and 44.34 class II) distinct T-cell epitopes within the capsid protein. Variation in baseline scores between geographies and genotypes was small, although, of the three genotypes PCV2a showed the greatest between-region variation; variation was smallest for PCV2d. The mean EpiCC score for VacAB against field sequences was 8.73, compared to 7.36, 6.74 and 7.07 for VacA, VacAlt1 and VacAlt2, equivalent to a mean of 82.69% T cell epitope coverage for VacAB, compared to 69.72%, 63.78% and 66.93% for the other vaccines (Figure 1). The differences in mean EpiCC score between VacAB and the other vaccines were statistically significant ($P<0.0001$). Mean EpiCC scores for PCV2b as well as for PCV2d field strains between VacAB and the monovalent vaccines were also highly significant ($P < 0.0001$).

For PCV2a, PCV2b and PCV2d, 75.4, 74.3 and 77.2% of the shared epitopes, respectively, were found in both PCV2a and PCV2b vaccine strains (Figure 2). Against PCV2a field viruses most of the additional score was provided by the PCV2a component, which was uniquely responsible for 22.6% of the overall coverage. Interestingly, 2.1% of the score against PCV2a field viruses was provided by the PCV2b component. For PCV2b field strains the PCV2b vaccine component provided 24.7% of the coverage, compared to only 1% from the PCV2a. Against PCV2d, the PCV2b vaccine contribution was the highest and uniquely responsible for 17.7% of the shared T cell epitope content, compared to 5.1% from PCV2a.

Figure 1: EpiCC scores of tested PCV2 sequences split by region and genotype.

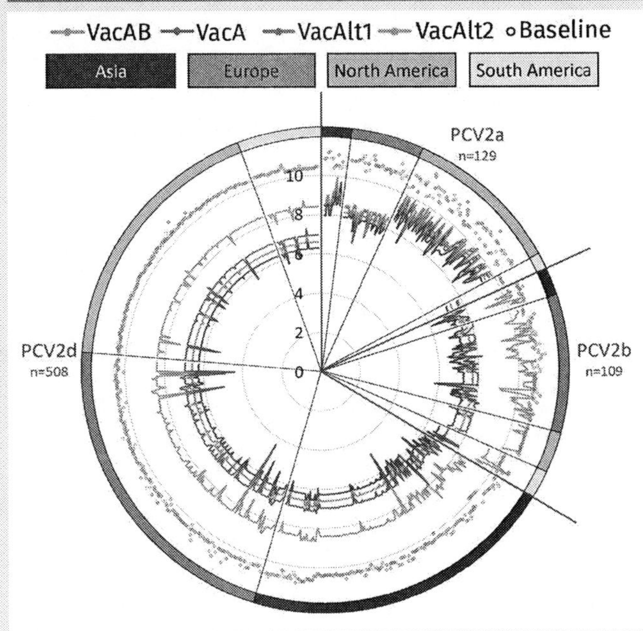

Figure 2: Relative contributions of the PCV2a and PCV2b components of VacAB to its overall epitope coverage against field viruses by PCV2 genotype.

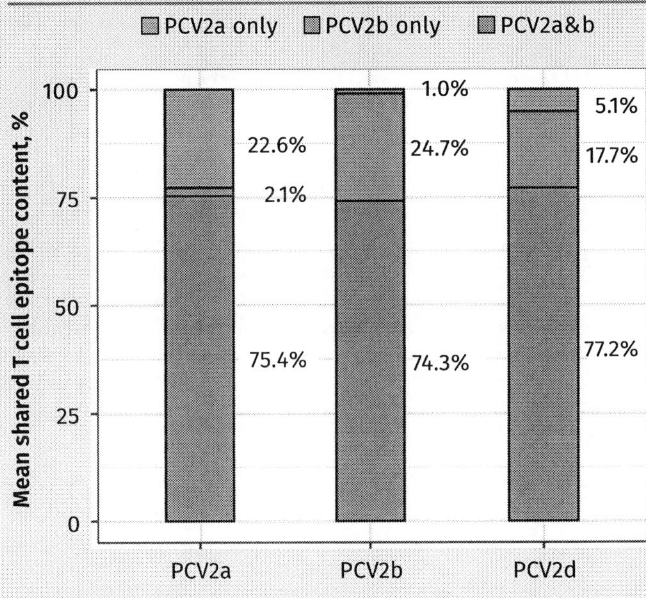

Conclusion

The study compared the T cell epitope content of PCV2 vaccines to a large and geographically diverse population of contemporary PCV2 field strains. Field strain sequences are biased towards laboratories and production systems routinely conducting PCV2 sequencing. At a high level, however, the results qualitatively confirm the continuing global relevance of all 3 of the major genotypes, with a predominance of PCV2d in all regions. Predominance of PCV2d with co-circulation of PCV2a and PCV2b agrees with recent surveys including from the US.[10] Even after allowing for regional differences, in agreement with previous work, PCV2a shows the greatest within-genotype genetic variation of the 3 genotypes.[2] PCV2d showed the least intragenotype variation, which may reflect its more recent spread.

The EpiCC scores confirm the conclusions of Bandrick et al[9] that the addition of a PCV2b strain to VacA, to create VacAB, increases mean T cell epitope coverage against all three of the tested genotypes. The major impact is on PCV2b, as would be expected, and there is a substantial increase for PCV2d, which is more closely related to PCV2b than PCV2a. For PCV2b and PCV2d, the addition of PCV2b improved mean epitope coverage by 23.4% and 13.7% respectively, representing increases of 33% and 21% over the coverage supplied by PCV2a alone. Even against PCV2a strains, however, the addition of PCV2b resulted in a modest but statistically significant increase in EpiCC score compared to PCV2a alone. Taken together, the results for this large and contemporary set of field isolates show that, compared to monovalent PCV2a vaccines, VacAB conferred broader T cell epitope coverage across different PCV2 genotypes and regions of origin.

The PCV2 EpiCC results reported here show immunologically relevant differences between PCV2 vaccines in their relationship to field strains. Work with influenza A viruses in swine has shown that such differences can be linked to differences in vaccine efficacy. Comparisons of EpiCC scores for an inactivated flu vaccine with the results of in vivo challenge experiments allowed estimation of EpiCC thresholds for full and partial protection, with higher scores predicting efficacy.[7] With influenza, however, vaccine failure due to a poor antigenic match is common, facilitating this type of analysis. In contrast, while studies with PCV2 have suggested that vaccination with a homologous genotype might be more successful than with a heterologous genotype[4,5] field efficacy from PCV2a-based vaccines is generally considered satisfactory and a recent analysis of isolates from possible cases of vaccine failure did not reveal any obvious difference to the general PCV2 population.[11] Nonetheless, vaccine failures have been observed, especially where vaccine and field strains are mismatched,[12] and the efficacy of vaccination in preventing production losses from subclinical PCV2 infection is difficult to assess and may be suboptimal. A closer match between vaccine and field strain has resulted in a meaningful improvement in growth performance under field conditions, together with a reduction in PCV2 viremia.[6]

Theoretically, higher EpiCC scores not only suggest a greater probability of vaccine success if failures due to evolutionary drift are occurring, but also a lower risk of vaccine failure in the future if the virus continues to change. PCV2 has changed over the past two decades with both optimization of receptor binding[13] and escape from natural or vaccine-acquired immunity[14] being suggested as the drivers. As described in a recent, comprehensive review[1] immunological relationships between and within PCV2 genotypes are complex, with some overall cross-reactivity but also qualitative differences in the breadth and efficacy of immunity. Given the capacity of PCV2 to mutate and recombine, and given that PCV2 vaccines can be described as leaky, in that they do not produce a sterilizing immunity and allow some viral replication and spread,[5] evolution will almost certainly continue. In this case, the more heterogeneous immune response resulting from the inclusion of a second genotype may not only preserve efficacy but also hinder the emergence of specific-immunity escaping variants, as has been speculated for other viral pathogens.[15]

References

1. Franzo, G., Segalés, J., 2020. Porcine circovirus 2 genotypes, immunity and vaccines: Multiple genotypes but one single serotype. Pathogens.

2. Franzo, G., Cortey, M., Segalés, J., Hughes, J., Drigo, M., 2016. Molecular Phylogenetics and Evolution Phylodynamic analysis of porcine circovirus type 2 reveals global waves of emerging genotypes and the circulation of recombinant forms. Molecular Phylogenetics and Evolution 100, 269–280.

3. Reiner G, Hofmeister R, Willems H (2015) Genetic variability of porcine circovirus 2 (PCV2) field isolates from vaccinated and non-vaccinated pig herds in Germany. Vet Microbiol 180:41-48.

4. Bandrick, M., Balasch, M., Heinz, A., Taylor, L., King, V., Toepfer, J., Foss, D., 2022. A bivalent porcine circovirus type 2 (PCV2), PCV2a-PCV2b, vaccine offers biologically superior protection compared to monovalent PCV2 vaccines. Veterinary Research 53, 12.

5. Karuppannan AK, Opriessnig T (2017) Porcine circovirus type 2 (PCV2) vaccines in the context of current molecular epidemiology. Viruses 9:99.

6. Um, H., Yang, S., Oh, T., Park, K., Cho, H., Suh, J., Min, K.D., Chae, C., 2021. Comparative evaluation of growth performance between bivalent and trivalent vaccines containing porcine circovirus type 2 (Pcv2) and mycoplasma hyopneumoniae in a herd with subclinical PCV2d infection and enzootic pneumonia. Vaccines (Basel).

7. Gutiérrez, A.H., Rapp-Gabrielson, V.J., Terry, F.E., Loving, C.L., Moise, L., Martin, W.D., de Groot, A.S., 2017. T-cell epitope content comparison (EpiCC) of swine H1 influenza A virus hemagglutinin. Influenza and other Respiratory Viruses 11, 531–542.

8. Xiao, C.T., Halbur, P.G., Opriessnig, T., 2015. Global molecular genetic analysis of porcine circovirus type 2 (PCV2) sequences confirms the presence of four main PCV2 genotypes and reveals a rapid increase of PCV2d. Journal of General Virology 96, 1830–1841.

9. Bandrick, M., Gutiérrez, A.H., Desai, P., Rincon, G., Martin, W.D., Terry, F.E., de Groot, A.S., Foss, D.L., 2020. T cell epitope content comparison (EpiCC) analysis demonstrates a bivalent PCV2 vaccine has greater T cell epitope overlap with field strains than monovalent PCV2 vaccines. Veterinary Immunology and Immunopathology 223, 110034.

10. Wang, Y., Noll, L., Lu, N., Porter, E., Stoy, C., Zheng, W., Liu, X., Peddireddi, L., Niederwerder, M., Bai, J., 2020. Genetic diversity and prevalence of porcine circovirus type 3 (PCV3) and type 2 (PCV2) in the Midwest of the USA during 2016–2018. Transboundary and Emerging Diseases 67, 1284–1294.

11. Sibila, M., Rocco, C., Franzo, G., Huerta, E., Domingo, M., Núñez, J.I., Segalés, J., 2021. Genotyping of porcine circovirus 2 (Pcv-2) in vaccinated pigs suffering from pcv-2-systemic disease between 2009 and 2020 in Spain. Pathogens 10.

12. Gava D, Serrao VHB, Fernandes LT, Cantao,ME, Ciacci-Zanella JR, Mores N, Schaefer R (2018) Structure analysis of capsid protein of Porcine circovirus type 2 from pigs with systemic disease. Braz J Microbiol 49:351-357.

13. Wei, R., Xie, J., Theuns, S., Nauwynck, H.J., 2019. Changes on the viral capsid surface during the evolution of porcine circovirus type 2 (PCV2) from 2009 till 2018 may lead to a better receptor binding. Virus Evolution 5, 1–12.

14. Franzo, G., Tucciarone, C.M., Cecchinato, M., Drigo, M., 2016. Porcine circovirus type 2 (PCV2) evolution before and after the vaccination introduction: A large scale epidemiological study. Scientific Reports 6, 1–10.

15. Franzo, G., Legnardi, M., Tucciarone, C.M., Drigo, M., Martini, M., Cecchinato, M., 2019. Evolution of infectious bronchitis virus in the field after homologous vaccination introduction. Veterinary Research 50.

https://doi.org/10.54846/am2023/34

Grow-finish model assessing ceftiofur crystalline free acid intervention for control of umbilical infections and hernias in a commercial production system

L. Glowzenski[1], VMD; N. Garbes[2], DVM; J.E. Guzman[2], MS
[1]TriOak Foods, Oakville, Iowa; [2]Zoetis, Parsippany, New Jersey

Introduction

Umbilical swellings, including infections and hernias, have been a frustration to the swine industry for years. Economic losses can occur from reduced performance, mortality, and alternative market sales. Since frequency of occurrence is variable, attention to this problem tends to gain focus when the prevalence tops an arbitrary new threshold. The most cited reference on umbilical hernias reports a prevalence range of 0.4% to 1.5%.[1]

Previous studies have tried to determine if umbilical hernias and abscesses are a result of genetics or management issues. One publication found no genetic risk factors for the development of umbilical hernias, but noted that castration was the most significant risk factor found and strongly recommended focus on day 1 pig care, aseptic surgical technique, and the use of antimicrobials to reduce incidence of herniation associated with castration.[2] More recent publications have addressed risk factors involved with umbilical swellings. One study looked at risk factors that impact umbilical defects and found that batch farrowing along with disinfection practices were significant for predicting umbilical defects. When placing batch and disinfectant in a model, the odds of umbilical defects were 16.90 times greater when disinfectant was not used. Umbilical defects observed in this study were necropsied between 3 - 12 weeks of age and equal numbers of umbilical defects were true umbilical hernias or true abscesses.[3] Further work on sow farms has led to the development of a model for evaluating judicious use of antibiotics for control of umbilical infections and hernias while documenting value of an antibiotic intervention within 24 hours of birth.[3,4] One study also demonstrated that ceftiofur crystalline free acid (CCFA; EXCEDE® for Swine, Zoetis, Parsippany, NJ) treated pigs had significantly lower navel scores and navel diameters compared to the negative control group at weaning, but failed to show a difference in the percent abdominal herniation at weaning.[5] This supported previous work that described omphalitis, or external inflammation at the umbilicus, at weaning not being associated with subsequent hernia development.[6]

The objective of this study was to use a grow-finish model to assess umbilical swelling prevalence before and after a CCFA intervention given to piglets within 24 hours of birth. An additional objective was to seek out infectious pathogens found within umbilici as part of a responsible use of antibiotics focus in this production system.

Materials and methods

Four 5000-head sow farms were selected to assess prevalence of umbilical swellings before and after CCFA intervention. Diagnostic tissue samples from pigs showing clinical signs of systemic disease were submitted to Iowa State University Veterinary Diagnostic Laboratory along with umbilicus stump samples from 10 dead pigs following the judicious use model described previously.[4]

A grow-finish model that evaluated all pigs in 20% of the pens within a wean-finish barn was used. Every fifth pen within a barn was evaluated unless the assigned pen was housing animals that had been pulled for other health reasons. If the fifth pen was a sick or pull pen, an adjacent pen was evaluated. This approach assessed three to four weeks of production from each sow farm flow of pigs 15 to 18 weeks of age. This age group was selected based on clinical history and published work which indicated the majority of hernias are present in pigs by nine to 14 weeks of age.[6] Animals older than 18 weeks of age can be assessed but tend to occupy more pen space increasing the difficulty of visualizing umbilical swellings.

Once the initial evaluation was completed, the CCFA intervention was initiated on the four sow farms, and pigs were again evaluated 7 months later. CCFA was given one time at 0.1 cc/pig intramuscularly within 24 hours of birth. Diagnostic results from umbilical swabs and tissues yielded *Streptococcus suis, Staphylococcus hyicus, Glaesserella parasuis,* and *Trueperella pyogenes.* All had antimicrobial susceptibility results supporting the use of CCFA.

After 7 months of intervention with CCFA within 24 hours of life, the same grow-finish model was used to evaluate the new prevalence. Prevalence for each site was calculated by dividing the number of umbilical swellings observed in 20% of the pens by the total number of pigs evaluated.

Results and discussion

The grow-finish model used in this study produced data that supports a previous evaluation of CCFA for umbilical swellings when given early in life. Results (Figure 1) demonstrated a reduction in umbilical swelling prevalence of 50% or greater in 3 of 4 farms, with all sow farm flows observing a reduction in prevalence. These results were similar to a previous published work demonstrating a significant reduction in umbilical hernias (4.0% non-treated vs 2.4% CCFA-treated) when CCFA was given within 16 hours of birth.[7]

Overall, a grow-finish model to estimate prevalence of umbilical swellings in a sow farm flow was needed to uncover intervention outcomes more accurately. This model was robust enough to warrant the use of CCFA as an antibiotic intervention reducing the incidence of umbilical swellings. It can be applied to other sow farms/pig flows as described here and by previous authors.[4,5] As producers and veterinarians looks to ways to improve the prevalence of umbilical defects, and the health and profitability of pigs, this is an approach worth considering and evaluating with proper diagnostic evidence.

Figure 1: Boxplots and mean percent umbilical swellings before and after CCFA intervention on four sow farm flows (mean = *).

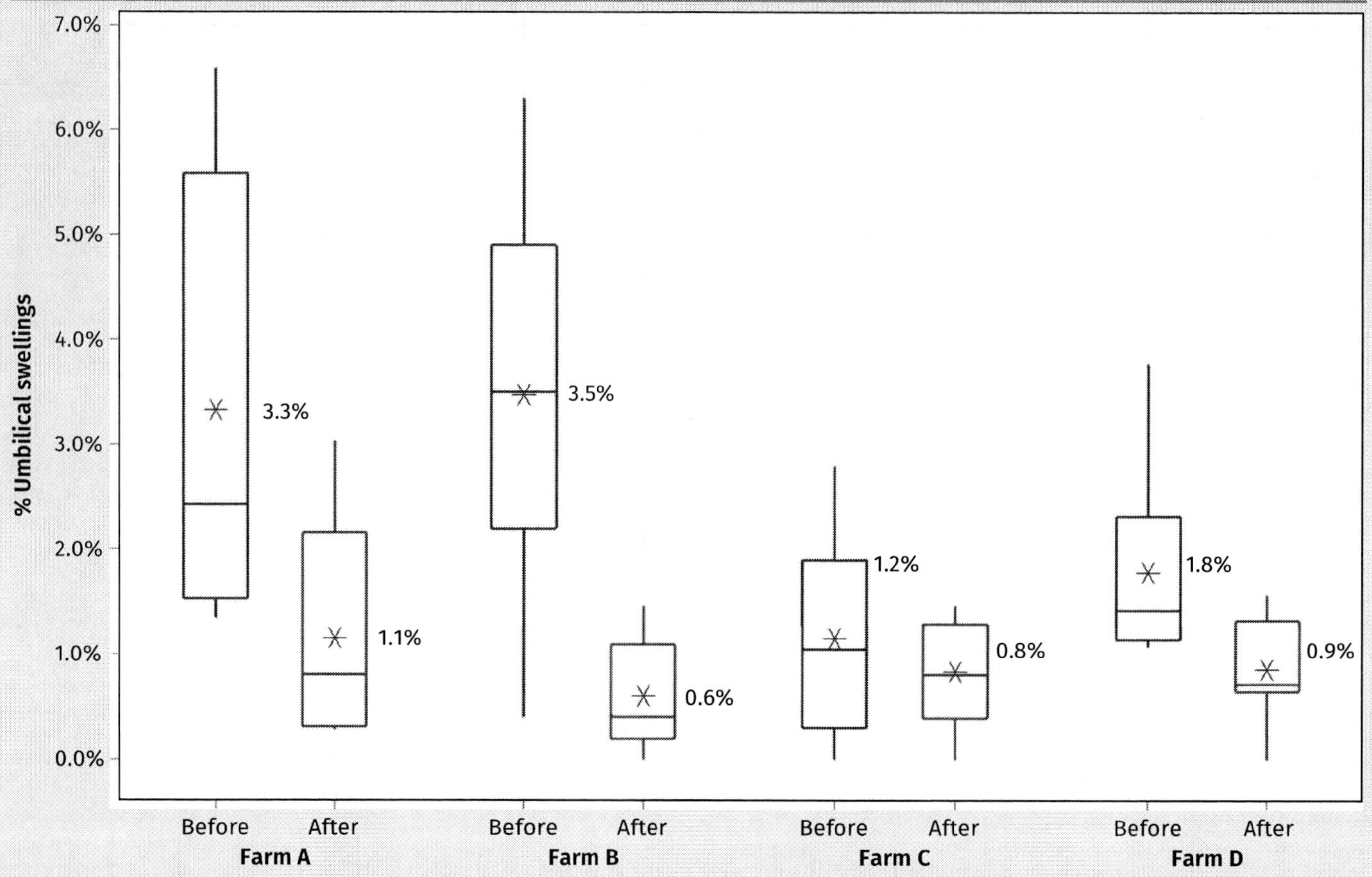

References

1. Straw B, Bates R, May G. Anatomical abnormalities in a group of finishing pigs: prevalence and pig performance. J Swine Health Prod. 2009; 17 (1): 28-31.

2. James H, Harding J, Waldner C. Umbilical hernias and abscesses: a result of management issues, not genetics. AASV. 2007.

3. Bracy M, Blake K, Wensley M, Madson D, Coble K, Davenport J, Baker K. Risk factors for umbilical defects in commercial pigs. AASV. 2021.

4. Fitzgerald C, Thomas P, Swalla R, Amodie D. Developing a useful model to evaluate the judicious use of a ceftiofur crystalline free acid to control umbilical infections and hernias in a commercial production system. AASV. 2018.

5. Szczotka A, Hayman K, Ratliff B, Rademacher C, Swalla R, Hoogland M, Brown J, Kittrell H, Skoland K, Breuer M, Forseth A, Wang C, Karriker L. Effects of antibiotic and non-antibiotic interventions applied to swine umbilici within the first 24 hours of life on the incidence of umbilical infection, growth, and mortality. AASV. 2019.

6. Searcy-Bernal R, Hind D. Effects of and factors associated with umbilical hernias in swine herd. JAVMA. 1994; 204:1660-1664.

7. Reiman J, Greiner L, Lowe J, Connor J. Bacterial infection of the navel cord in neonatal piglets: A follow up on prevalence and clinical outcomes. AASV. 2012.

https://doi.org/10.54846/am2023/35

Medium chain fatty acids fed to sows during gestation and lactation and the subsequent effects on the health and performance of the progeny

Stacie Crowder[1], PhD; Kobe Lannoo[2]
[1]PMI, Arden Hills, Minnesota; [2]Agrimprove, Ghent, Belgium

Introduction

Sow productivity has been a focus of the swine industry with genetic selection for highly prolific sows. The increased number of piglets per sow also increases the risk for an insufficient supply and quality of colostrum and milk for the piglets. Colostrum provides nutrients and immunological protection to the newborn piglets[1] and growth factors to support intestinal development and integrity.[2] This is of critical importance as the sow and farrowing environment are primary sources of pathogenic bacteria to the newborn piglets. Colostrum intake is negatively correlated with pre-weaning mortality[3] and nursery mortality.[4] Piglets with increased colostrum and milk intakes have better developed immune systems and are considered to be less susceptible to infections.[5] Therefore, the objective of this research was to determine the effects of feeding medium chain fatty acids (MCFA) to sows on the colostrum quality and piglet health and performance.

Materials and methods

Trial 1 MCFA was fed to sows at the rate of 2lbs/ton for 4 months. Piglets were weaned at 21 days of age. MCFA treatment sow performance measurements were collected for 3 months following 2 months of supplementation and compared to the 3 months pre-trial Control. Colostrum samples were collected from 20 sows at two time points pre-trial and after 4 months of MCFA supplementation. Piglet immunocrit[6] was measured from the same 20 sow litters with blood samples taken from 2, 2-day old piglets from each litter (40 samples/treatment).

Trial 2 MCFA was fed to sows at the rate of 2lbs/ton for 2 months. Piglet blood samples were taken from 30 2-day of piglets at three time points (1. pre-trial, 2. 1 month on MCFA, and 3. 2 months on MCFA) for immunocrit analysis.

Trial 3 was conducted on a 1600 sow farm that had a history of *Streptococcus suis* in the farrowing barn with increased pre-weaning mortality and increased arthritis at weaning. MCFA was fed to sows at 2lbs/ton for 5 weeks. Piglet immunocrit was at two timepoints pre-trial and after 5 weeks of MCFA supplementation using 30 2-day old piglets. Clinical arthritis scoring was done by a veterinarian at weaning each week of the trial. Sow performance control was pre-trial average and MCFA was the average of the 5 weeks during supplementation measurements were liveborn, mortality, and number weaned per litter.

Results

Results from these trials demonstrate that MCFA fed to sows increased the immunocrit level in colostrum samples and carried over in improved blood immunocrit levels of piglets. MCFA supplementation to sows also reduced pre-weaning mortality by 17% after 4 months of MCFA supplementation. Clinical observation of arthritis was reduced from 30% pre-trial to less than 10% after 2-5 weeks of MCFA supplementation. These results suggest that MCFA supplementation of sows is one in which to support sow and litter performance. Further research is needed to evaluate the duration of feeding needed to see the results in commercial production.

References

1. Rooke and Bland 2002 The acquisition of passive immunity in the newborn piglet. Livest. Prod. Sci.78:12-23.

2. Devillers et. al., 2007 Variability of colostrum yield and colostrum intake in pigs. Animal 1:1033-1041.

3. Declauwe et. al., 2014 Piglets' colostrum intake associates with daily weight gain and survival until weaning. Livest. Sci. 162:185-192.

4. Declerck et. al., 2016 Effect of energy supplementation to neonatal low birth weight piglets on mortality, weaning weight, and colostrum intake. Livest. Sci. 183:48-53.

5. Le Dividich et. al., 2005 Nutritional and immunological importance of colostrum for the newborn pig. J. Agric. Sci. 143:469-485.

6. Vallet et. al., 2012 A simple novel measure of passive transfer of maternal immunoglobulin is predictive of pre-weaning mortality in piglets. The Veterinary Journal.

https://doi.org/10.54846/am2023/36

Top financial planning tips for veterinarians

Andrew Kleis, AIF®, Insight Wealth Group

Introduction

Contrary to the advice of many investment professionals, the best investment owners of veterinary practices will make in their lifetime will be their commitment of time and capital into their practice. The business itself has been the primary tool for growing wealth and finding tax efficiencies available to the owner. Yet running a business is rarely the reason a veterinarian became a veterinarian. Our goal is to identify key areas within your practice and your experience as a business owner that can have the largest impact on your ability to turn your practice into an engine for wealth generation and generational estate planning.

Tax planning

The single largest expense in the lifetime of a business owner is taxes. Simple compliance with the complicated tax code at the state and federal level is itself a challenging task. Too often business owners do not have a tax planning strategy to accompany their business goals. Simple decisions such as which form your business takes (LLC, Partnership, C-Corporation, etc) can have significant and long-lasting impacts on the tax impact on your business and you are personally. Oftentimes these decisions are made at the formation of the business and are never adjusted for changes in tax code. Identifying a reliable partner to guide you through proper tax planning is crucial.

Retirement planning

Retirement planning goes hand-in-hand with tax planning. The benefits of business ownership are substantial in this arena. Whereas employees can contribute up to $6,500 in an IRA account or $22,500 in a 401(k) in 2023 (Table 1), business owners can contribute up $66,000 in profit sharing contributions. Utilizing the correct tools for retirement planning can be both a tax and retirement benefit that can have lifelong benefits.

Business succession planning

As it is likely your business will be your largest asset as you move to retirement, planning for your succession is a vital task that can never begin too early. Goals will be different for everyone. Some may wish to pass the practice along to family members, others may wish to sell to employees, or simply maximize the potential sale price for the practice. No matter the preferred outcome, structuring the business appropriately is a process that begins years ahead of any potential sale.

1031 exchange opportunities

Often the largest asset within a veterinary practice is the real estate owned by the practice or the practice owner. Properly structuring the sale of the real estate can result in lifelong tax benefits for the practice owner and their heirs. The utilization of a 1031 exchange (Figure 1) means the owner will pay no tax on the sale of the real estate and heirs will benefit in the step-up in basis upon the owner's passing. Liquidating a practice's largest asset without paying any tax is one of the most important opportunities for business owners to capitalize upon.

Estate planning

Proper estate planning is vital for any business owner. It ensures that, in case of an untimely death, the business can continue to operate, and the owner's heirs are properly protected. This process should proceed in conjunction with the Business Succession Planning process as it will inform how the business should be disposed of (or run) by your heirs. It will also determine the tax consequences of your estate and the final amount of assets available to your heirs.

Table 1: 2023 Contribution limit increases

Contribution type	2022 limit	2023 limit	Change
401(k)/403(b) Employee contribution	$20,500	$22,500	$2,000
401(k)/403(b) Catch-up contribution	$6,500	$7,500	$1,000
401(k)/403(b) Total contribution < 50	$61,000	$66,000	$5,000
401(k)/403(b) Total contribution 50+	$67,500	$73,500	$6,000
457(b) Contribution	$20,500	$22,500	$2,000
Traditional IRA contribution	$6,000	$6,500	$500
Traditional IRA catch-up contribution	$1,000	$1,000	$0
Roth IRA contribution	$6,000	$6,500	$500
Roth IRA catch-up contribution	$1,000	$1,000	$0
SEP-IRA contribution	$61,000	$66,000	$5,000
SIMPLE IRA/SIMPLE 401(k) contribution	$14,000	$15,500	$1,500
HSA Contribution (single)	$3,650	$3,850	$200
HAS Contribution (family)	$7,300	$7,750	$450
Healthcare FSA contribution	$2,850	$3,050	$200

Figure 1: Comparison of tax benefit of 1031 exchange.*

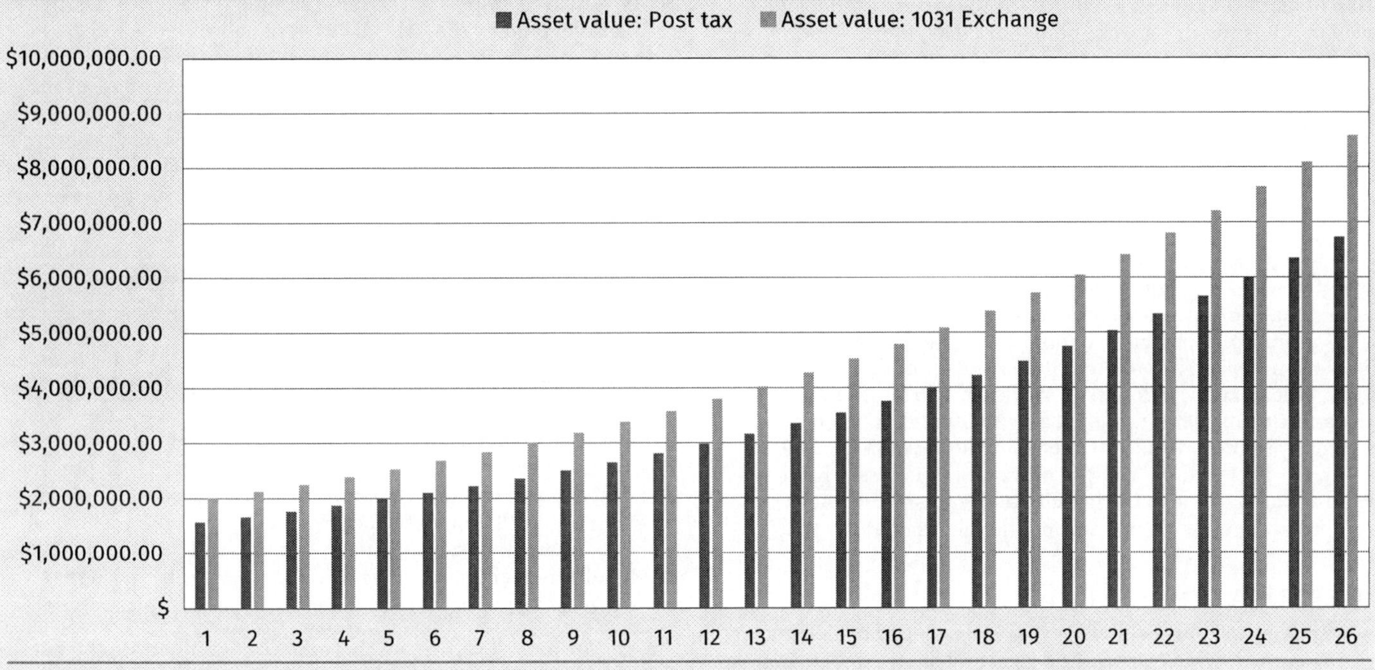

* Assumes 6% annual return on proceeds; total tax of 28.8%; cost basis of $500,000; sale proceeds of $2,000,000

https://doi.org/10.54846/am2023/37

Utilizing hybridization-based next generation sequencing for swine-related pathogen detection

B. Hause[1], PhD; G. Temeeyasen[1], DVM, PhD; E. Nelson[1], PhD; J. Christopher-Hennings[1], DVM, MS; A. Shin[2]
[1]Department of Veterinary and Biomedical Sciences, South Dakota State University, Brookings, SD;
[2]Celemics, Seoul, South Korea

Introduction

Diagnostic testing is the foundation of animal and herd health management. Traditional bacterial culture, virus isolation and PCR have long been mainstays for the detection of bacteria and viruses. These assays are relatively rapid and affordable, leading to widespread usage. Limitations of these assays include difficulties in culturing fastidious microorganisms and the highly specific nature of PCR, the latter of which requires a priori knowledge of a pathogen's genetic composition. Additionally, PCR testing for multiple pathogens is often cost-prohibitive, leading to a myopic view disease causation.

Next generation sequencing is revolutionizing diagnostic medicine. Increasingly affordable equipment and consumables has led to widespread adoption of new technologies for pathogen detection. In veterinary diagnostics, full genome sequencing is becoming more common for viruses such as porcine reproductive and respiratory syndrome virus (PRRSV) and influenza A virus (IAV). Metagenomic sequencing, which allows for comprehensive detection of all pathogens in a clinical sample, is also widely used. However, it has the drawbacks of higher costs, lower sensitivity and longer turnaround time as compared to PCR.

Hybridization-based sequencing is a technology that overcomes limitations inherent for PCR and metagenomic sequencing. Oligonucleotide probes complementary to pathogen sequences and coupled to magnetic beads are used to bind and purify pathogen DNA randomly amplified from a clinical sample. Following elution, pathogen nucleic acids are detected by next generation sequencing. Besides detection, pathogen genome sequences can be assembled and used for subsequent genetic analysis. Celemics (**https://celemics.com/en/index.asp**) produces hybridization sequencing assays for multiplex detection of targets such as human respiratory pathogens, inherited disease mutations and other swine pathogens such as African swine fever virus. Working with South Dakota State University, Celemics has designed a swine hybridization panel to detect many known swine pathogens, including more than 30 viruses and 15 bacteria. Therefore, this commercial assay has a significant advantage due to the breadth of organism detection and the ability to generate the pathogen sequence concurrently.

Materials and methods

A swine pathogen hybridization assay was developed by Celemics, targeting known swine bacteria and viruses. For viruses, oligonucleotide probes spanning the complete genome were developed. For bacteria, oligonucleotide probes were tiled across genes used for genetic analysis via methods such as multi locus sequence typing.

Performance of the hybridization assay was assessed on 50 archived clinical samples submitted to the SDSU Animal Disease Research and Diagnostic Laboratory for diagnostic testing and included both swine respiratory and enteric disease submissions of various sample types (serum, tissue homogenates, swabs). Diagnostic test results were compiled from the laboratory information management system and were conducted per the submitter's requests. Samples were processed for hybridization-based sequencing per the manufacturer's instruction. In brief, nucleic acids were isolated from tissue homogenates or swabs using a QiaAmp Viral RNA minikit or a MagMax nucleic acid isolation kit. Samples were then reverse transcribed, adapter-ligated, and randomly amplified by PCR, followed by hybridization of the amplified adapter-ligated DNA to the swine pathogen probes. Following amplification and elution, captured DNA was sequenced on an Illumina MiSeq. Sequencing reads were mapped to reference genomes corresponding to those represented by the hybridization panel. An organism was counted as detected when mapped sequence reads spanned greater than 20% of the genome.

Results and discussion

Influenza A virus was detected in 14 samples. Using sequence derived from the hybridization, positive samples were concurrently subtyped as H1N1, H1N2 or H3N2. For comparison, PCR detected 19 IAV-positive samples in the same sample set. All IAV-positive samples detected by the hybridization panel were positive by PCR. Of the 16 PRRSV PCR-positive samples, 10 were detected using the hybridization panel. For eight of these samples, over 50% of the genome was assembled, enabling further genetic and epidemiological analyses. Of the 50 samples, only nine were tested for porcine circovirus 2 (PCV2) by PCR; three were positive. The hybridization panel identified 15 PCV2-positives, suggesting undiagnosed or subclinical infection of PCV2 is common. The most prevalent viruses detected in the sample set were parvoviruses, with detection of porcine parvovirus (PPV) species 1-7 in 2 (4%), 28 (56%), 3 (6%), 22 (44%), 21 (42%), 5 (10%) and 14 (28%) of the 50 samples, respectively. While PPV1 is an established swine pathogen, the etiological significance of PPV2-7 is less well understood, though PPV2 has been associated with systemic disease. Numerous additional viruses were detected in the sample set that are rarely identified by traditional diagnostic approaches, including porcine respiratory coronavirus (n = 1), hemagglutinating encephalomyelitis virus (n = 1), porcine parainfluenza virus 1 (n = 1), porcine sapelovirus (n = 1), porcine enterovirus (n = 2), porcine sapovirus (n = 7), porcine astrovirus 4 (n = 12), atypical porcine pestivirus (n = 1), and PCV1 and PCV3 (n = 4 and 3, respectively). In addition, numerous bacteria that cause disease in swine were co-detected in the samples.

The preliminary results from this experiment illustrate the potential of hybridization-based sequencing to fundamentally change our understanding of swine disease etiology. This concurrent detection and sequencing assay would be useful for simultaneous pathogen discovery and routine monitoring of mutational changes within known pathogens. Many diseases are multifactorial and detection of the complement of microorganisms involved is critical to understand disease etiology and subsequent control measures.

https://doi.org/10.54846/am2023/38

Reduction of antibiotic usage in commercial boar semen production through accurate dosage of antibiotics (ADA)

R. Grossfeld[1], PhD; J. Quackenbush[2]; D. Zaritskiy[1]; E. Biesendorfer[1];
K. Simmet[1], DVM; C Simmet[1], DVM; M. Schulze[3], PD, DVM
[1]Minitüb GmbH, Tiefenbach, Germany; [2]Minitube USA Inc., Verona, Wisconsin
[3]IFN Schoenow, Bernau, Germany

Introduction

Freshly collected, native boar ejaculates commonly contain multiple bacterial strains.[1] These bacteria can damage the spermatozoa in the diluted semen and impair the fertility of the inseminated sows. Therefore, antimicrobial agents are added to the semen extender to prevent bacterial growth and disease transmission.

Commercial boar semen extenders are typically supplied as a powder with the antimicrobials included.[2] The powder is dissolved in purified water to prepare the ready-to-use extender, a process during which powder-dust containing antibiotics can build up. This exposes the laboratory environment to antibiotics, thus increasing the risk of building up antibiotic resistance in bacteria due to constant exposure to small doses of antibiotics.

Further, the initial antibiotic concentration in the prepared semen extender is reduced as the diluted semen is a mixture of extender with antibiotics and native ejaculate without antibiotics. The dilution ratio is variable as it depends on the ejaculate volume, sperm concentration and the desired sperm content in the semen dose. The dilution ratio can range from 1 + 4 up to 1 + 19 (v/v). A safety margin in antibiotic content in the semen extender is necessary, to assure that the finally diluted semen contains sufficient antibiotics to meet the highest inhibition concentration that is not known to have resistance for all relevant bacteria typically found in boar semen (epidemiological cut-off value: ECOFF). This is especially important, when the dilution ratio is low, and the diluted semen contains a relatively high amount of native semen without antibiotics.

Both a possible exposure of the semen laboratory environment to antibiotics and the imprecise antibiotic dosage increase the risk of the formation of bacteria strains that are resistant against antibiotics.

Accurate dosage of antibiotics (ADA) in the semen laboratory

Accurate dosage of antibiotics (ADA) is a new protocol for applying antibiotics in boar semen processing that prevents the exposure of antibiotics to the laboratory environment and that allows for a precise dosage of antibiotics in the diluted boar semen doses.[3]

For this, the amount of antibiotics that are added to the semen is based on the end-volume of the finally diluted boar semen. The antibiotics are not part of the main extender but will be added with an antibiotic concentrate in a two-step dilution process. The amount of the antibiotic-free main extender and the antibiotic concentrate can be calculated with the formulas according to Figure 1.

In addition to the accurate dosage of antibiotics according to this formula (Figure 1), a water-soluble packaging was developed, which allows the completely dust-free preparation of the antibiotic concentrate which contains 33fold of the desired antibiotic concentration in the diluted semen. Either main extender or antibiotic concentrate can first be added to the raw ejaculate (Figure 2).

Due to the accurate dosage of antibiotics based on the end-volume of the extended semen, a safety margin for the antibiotic content is no longer necessary. Therefore, the commonly applied antibiotic concentration in the standard procedure can be reduced by 15% without the risk that the ECOFF is not reached for bacteria found in boar semen.

Conclusion

With the application of the ADA protocol, all semen doses precisely contain the desired amount of antibiotics, independent of the dilution rate of the native ejaculate. Especially post-cervical artificial insemination semen doses with low volume but high sperm content will benefit from this new protocol. ADA will increase biosecurity in the swine artificial insemination (AI) industry, significantly minimize the selection pressure for resistant bacteria in laboratories and minimize the exposure of laboratory personnel to antibiotics. Less semen extender containing antibiotics is discarded and together with an obsolete safety margin for antibiotics, the overall usage of antibiotics in swine AI is significantly reduced.

References

1. Althouse, G. C., Kuster, C. E., Clark, S. G. & Weisiger, R. M. Field investigations of bacterial contaminants and their effects on extended porcine semen. *Theriogenology* 53, 1167–1176 (2000).

2. Schulze, M., Nitsche-Melkus, E., Hensel, B., Jung, M. & Jakop, U. Antibiotics and their alternatives in Artificial Breeding in livestock. *Anim Reprod Sci* 220, 106284 (2020).

3. Schulze, M. et al. Dose rates of antimicrobial substances in boar semen preservation—time to establish new protocols. *Reproduction in Domestic Animals* 52, 397–402 (2017).

Figure 1: Formula for the calculation of the amount of antibiotic concentrate and antibiotic-free main extender to be added to a boar ejaculate to reach the desired antibiotic concentration and final end-volume.

$$\textbf{antibiotic concentrate (mL)} = \frac{ND \times VO}{ACC}$$

$$\textbf{main extender (mL)} = ND \times VO - EJV - \text{antibiotic concentrate (mL)}$$

ND = number of semen doses that can be produced from a raw ejaculate
VO = volume per semendose (mL)
ACC = n-fold antibiotic concentration in AB concentrate (eg, 33fold)
JV = ejaculate volume (mL)

Figure 2: Workflow of accurate dosage of antibiotics (ADA).

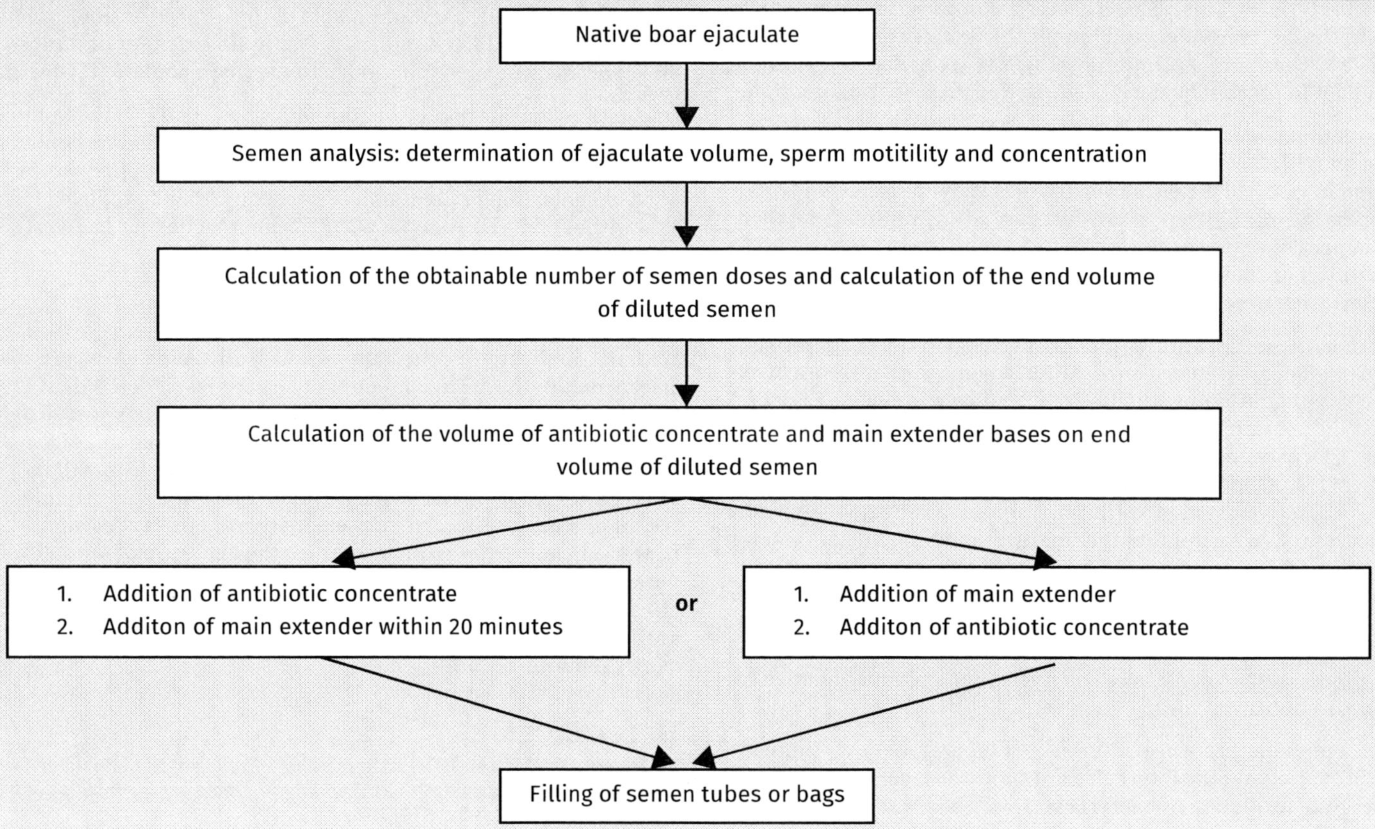

https://doi.org/10.54846/am2023/39

Improved performance feeding NutriQuest Baseline™ in sows and related impact on inflammation biomarker levels

Tom Petznick[1], DVM; Petra Chang[2], PhD; Ran Song[2], PhD
[1]ArkCare, Omaha, Nebraska; [2]NutriQuest, Mason City, Iowa

Introduction

Pigs face constant inflammatory challenges from environmental stress, production, and health challenges; activities like day-one pig care, vaccination and weaning contribute significantly to the inflammatory load on a pig. These environmental stresses and health challenges often increase inflammation in the body resulting in performance and profit loss. Reduced productivity due to inflammation that occurs during a pig's lifetime has significant impact to the swine industry, but quantifying its cost has been difficult, as there has not been a reliable biomarker for inflammation that is cost-effective and performance indicating to measure. After screening for several fecal and blood biomarkers, NutriQuest has successfully identified a serum biomarker of inflammation that is performance-indicating, consistent and low cost to evaluate. The data indicated a positive correlation between the identified inflammation biomarker and pig body weight measured from wean to finish in pigs with high and low health status (Figure 1). These data were used to build a model to predict pig lifetime productivity based on their inflammation biomarker status at weaning.

NutriQuest Baseline™ is a technology that has been shown to improve sow reproductive performance and post-weaning pig performance and quality. Sows fed Baseline™ had piglets with improved levels of the inflammation biomarker, which suggested a lower level of inflammation, and heavier weaning weights compared to control sows.[1]

The objective of the evaluations presented in this paper was to assess piglet performance and inflammation biomarker status before and after different durations of feeding Baseline™ to sows.

Methods

Four commercial sow farms evaluated the impact of feeding Baseline™ to sows on piglet inflammatory biomarker levels and weaning weights. All farms were located in the state of Nebraska, across three production companies, and differed in herd size and health status (Table 1).

Sows were fed Baseline™ in gestation and lactation diets at the inclusion level of 1.0 lb/ton. Piglet serum inflammation biomarker levels and weights were measured a few days prior to weaning and before feeding Baseline™, and subsequently at varying times after feeding Baseline™ to sows at each farm (Table 2). Six full litters were sampled and weighed at each time point; litters were representative of the health status of the farm at time of sampling and balanced for parity.

Data was analyzed by farm. Each piglet was considered the experimental unit within each farm for analyses of the inflammation biomarker level and weight. Pig age at collection was used as a covariate and the sow was used as a random effect when analyzing the impact of feeding Baseline™ on pig weight. A multivariate analysis was conducted to determine correlations between the inflammation biomarker and pig weights. For this analysis, all data points within each farm were used to determine Pearson correlation coefficients (r) and probabilities. Data was analyzed using JMP software (v. 16.1).

Feeding Baseline™ to sows improved inflammatory biomarker levels in piglet serum and weaning weights

Inflammation biomarker levels and weight results are presented as least square means in Figure 2. In Farm 1, inflammation biomarker levels improved ($P < 0.0001$) after 58 d and 198 d of feeding Baseline™, with no difference in levels observed at d 162 of feeding. Piglet weights increased significantly ($P < 0.01$) after 58 d of feeding and pigs were heavier at d 162 and d 198 compared to d 0 and d 23. In Farm 2, inflammation biomarker levels improved by 15% ($P < 0.01$) after 68 d of feeding Baseline™ to sows, and pigs tended to be heavier ($P < 0.1$). Farm 3 showed improved inflammation biomarker levels ($P < 0.0001$) and weights ($P < 0.01$) after 93 d of feeding compared to d 0 and d 51. Farm 4 had decreased levels of the inflammation biomarker ($P < 0.0001$) after 51 or 93 d of feeding, and piglet weights ($P < 0.1$) did not show a significant improvement after Baseline™ was fed to sows.

Correlation coefficients and probabilities between the inflammation biomarker concentration and piglet weights are presented in Figure 3. All farms but Farm 4 showed positive and significant correlations between the inflammation biomarker and weights.

Conclusions

Feeding NutriQuest Baseline™ to sows improved piglet inflammation biomarker levels and weights in all farms but one. Farm 4 had higher initial (d 0) inflammation biomarker levels compared to all other farms, which indicates low inflammation status of the farm. Therefore, such status may reduce the opportunity to enhance pig productivity by use of a technology like Baseline™. The improvements found in the inflammation biomarker and piglet weights in Farms 1, 2 and 3 were detected after over 50 d of feeding Baseline™ to sows, suggesting that sows need to consume the product for at least that duration to be able to detect an improvement in piglet inflammation biomarker levels and weights. The positive and significant correlations found between the inflammation biomarker and weights corroborate with NutriQuest's previous findings, highlighting the potential of the inflammation biomarker as a useful tool in predicting pig lifetime performance and profitability.

Figure 1: Piglet inflammation biomarker levels are positively correlated with pig body weight measured from wean to finish in pigs with **A)** high and **B)** low health status.

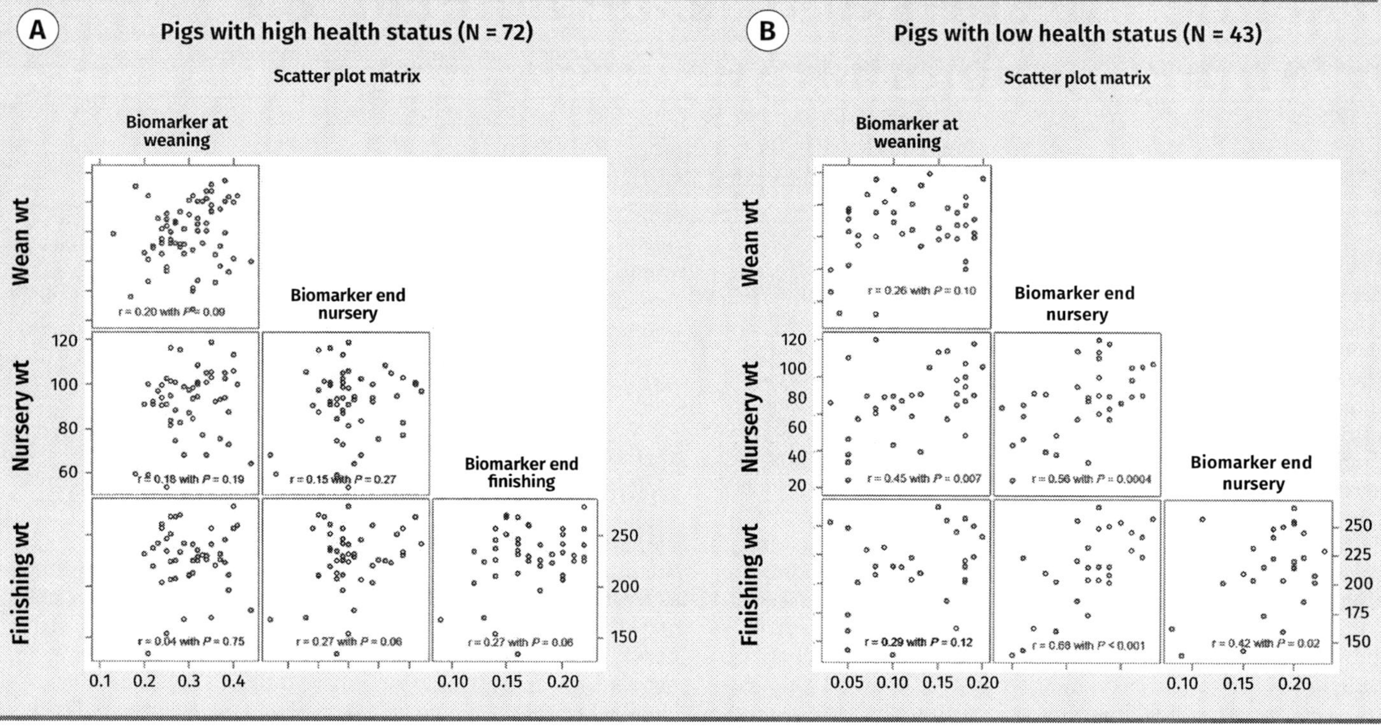

Table 1: Description of farms and timeline of evaluations

Farm	Herd size	Health status	Date initial (d 0) samples taken
1	5,000 sows	PRRSV, M hyo naive, coronavirus naive, flu positive	September 2021
2	650 sows	PRRSV break in May 2022 before initial samples collection, M hyo naïve, coronavirus naïve, flu positive	May 2022
3	7,350 sows	PRRSV positive, PEDV break in April 2022	June 2022
4	3,000 sows	PRRSV positive, PEDV break in April 2022	June 2022

Reference

1. Song R, Friesen K, Musser R. Using sow technologies to drive sow and subsequent pig performance associated with stress induced inflammation. *53rd AASV Proc*. 2022;326-327. **doi: 10.54846/am2022/149**

Table 2: Sampling schedule and average piglet age at collection

Farm	Time on Baseline™	Total pigs sampled[1], N	Pig age[2], days
1	Initial, d 0	70	19.0
	d 23	71	18.5
	d 58	66	19.2
	d 162	77	18.2
	d 198	70	17.7
2	Initial, d 0	74	27.3
	d 68	74	25.5
3	Initial, d 0	81	18.3
	d 51	79	17.2
	d 93	76	19.3
4	Initial, d 0	73	15.8
	d 50	78	19.4
	d 92	76	19.4

1 Six full litters were selected and balanced by parity.

2 Average age of all 6 litters sampled at each time point.

Figure 2: Piglet inflammation biomarker concentrations and weaning weights before and after feeding Baseline™ at each farm (values are presented as least square means).

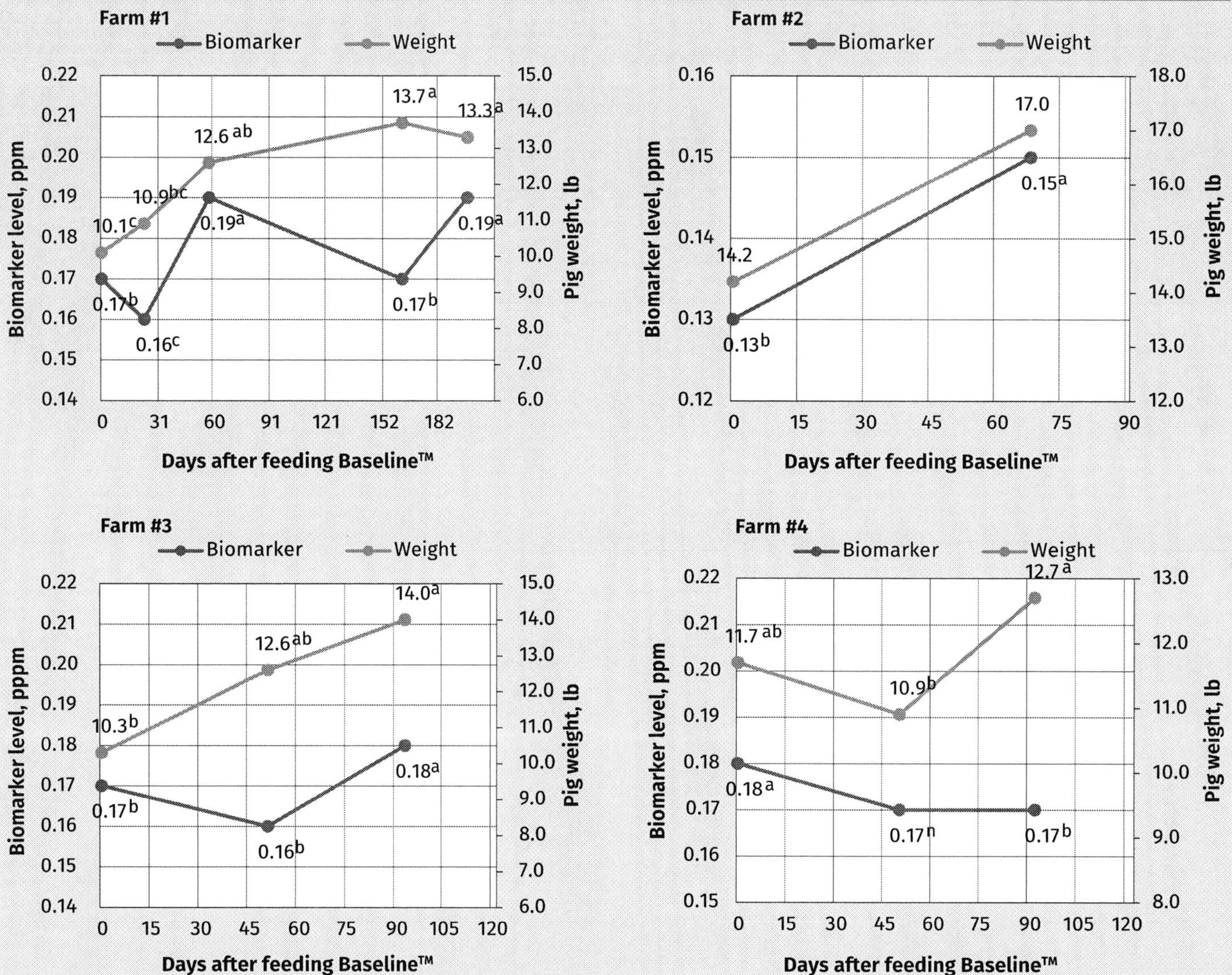

abc Means with different superscripts differ; P < 0.05.

Figure 3: Correlations between piglet inflammation biomarker levels and weights for each farm

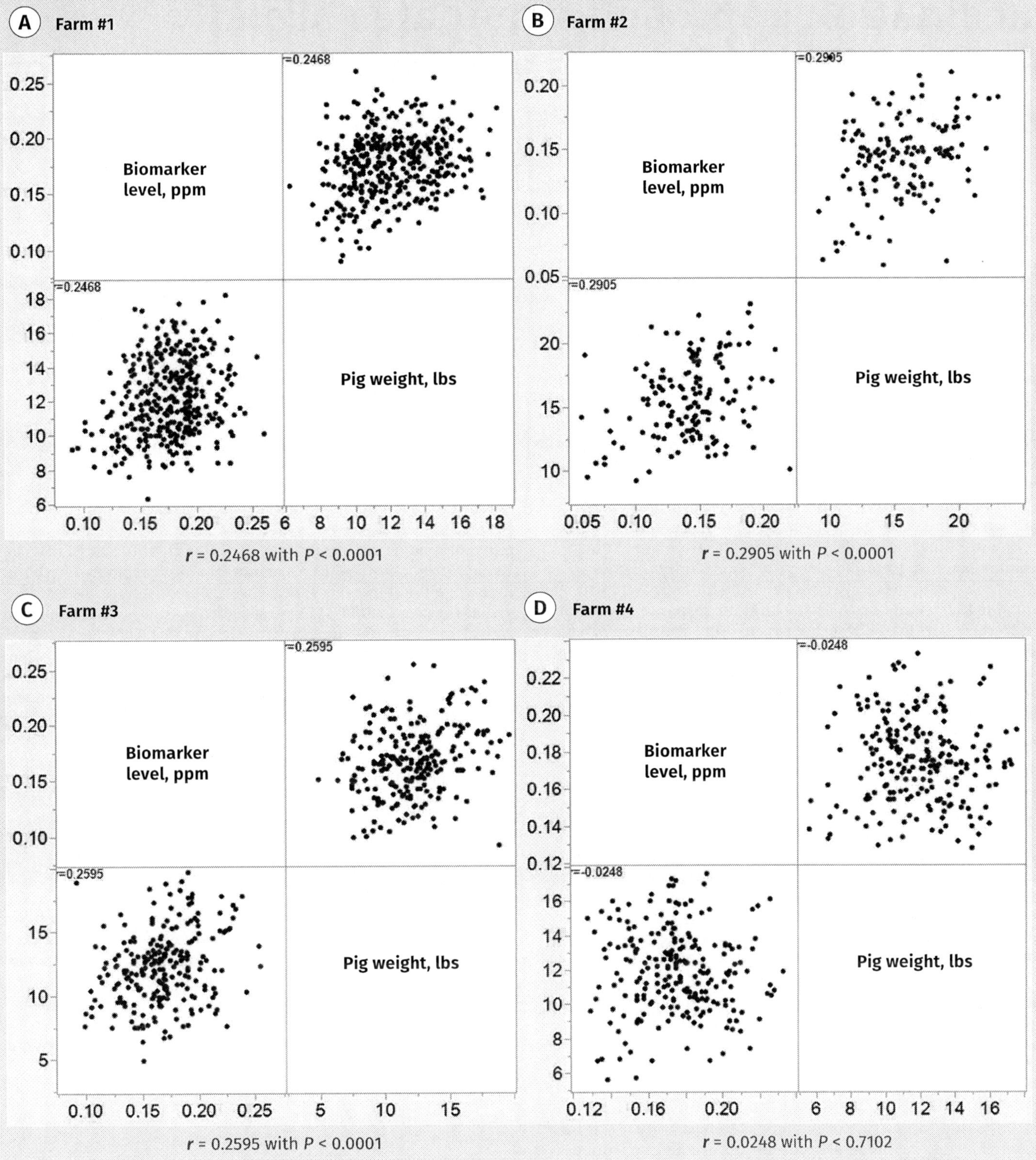

https://doi.org/10.54846/am2023/40

A modified live virus prime/autogenous boost vaccination approach stimulates broadly neutralizing antibodies and improved protection to PRRSV

J. Park, DVM, PhD; E. Bumgardner; E. Collin, PhD; D. Stine, DVM, PhD; B. Hause, PhD
Cambridge Technologies, Worthington, Minnesota

Introduction

Despite decades of research, porcine reproductive and respiratory syndrome (PRRS) remains the most significant disease affecting pigs worldwide. A variety of fully licensed products, including modified live and inactivated vaccines, as well as autogenous vaccines are available. Yet largely, these are perceived as inefficacious. Protection conferred by modified live virus (MLV) vaccines is generally considered superior to inactivated vaccines for genetically similar viruses. However, rapid virus evolution via mutation and recombination limits vaccine effectiveness. Live virus vaccines additionally pose significant safety concerns as recombination between vaccine and wild-type viruses are well-documented and can lead to novel, virulent viruses. Inactivated vaccines are inherently safe yet often fail to stimulate a robust immune response. Heterologous prime-boost immunization strategies have increasingly been explored for many viruses as a method to bolster and broaden the immune response to vaccination. Here, we evaluated whether concurrent administration of a commercial MLV vaccine with either a monovalent or a multi-strain autogenous vaccine could bolster immunity and broaden protection as compared to administration of MLV alone.

Materials and methods

Thirty-two approximately 21-day old specific pathogen free pigs were confirmed negative for influenza A virus, PRRSV and *Mycoplasma hyopneumoniae* by ELISA and PCR. Pigs were randomly divided into four groups of eight pigs and allowed one week to acclimate. On study day 0, group 1 was vaccinated with a commercial MLV per label instructions. Group 2 was co-administered MLV and a monovalent autogenous PRRSV vaccine with a restriction fragment length polymorphism (RFLP) of 1-4-4 that was homologous to the challenge virus. Group 3 was co-administered MLV and a trivalent autogenous PRRSV vaccine (RFLP types 1-8-4, 2-6-2, 1-4-2) heterologous to the challenge virus. Group 4 pigs were mock vaccinated. On study day 28, pigs in groups 2 and 3 received a booster dose of autogenous vaccine. On study day 42, pigs were challenged intranasally and intramuscularly with a wild type PRRSV (RFLP type 1-4-4). Neutralizing antibodies were measured on serum collected prior to challenge using homologous virus and a panel of heterologous viruses using serum neutralization (SN) assays. Induction of cell mediated immunity (CMI) was evaluated by measuring interferon gamma production from peripheral blood mononuclear cells (PBMCs) isolated on the day of challenge following stimulation with one of the vaccine strains. Viremia was assessed by quantitative RT-PCR (qRT-PCR) on serum collected beginning every third day post challenge until day 56. Pigs were euthanized on day 56 and gross lung lesions were assessed by a blinded observer. Fixed lung tissues were scored for microscopic lesions by a blinded pathologist using hematoxylin and eosin staining as well as immunohistochemistry (IHC). Differences between group means were assessed by ANOVA and post-hoc Tukey's HSD.

Results and discussion

Mock and MLV-vaccinated pigs were negative for neutralizing antibodies to the challenge virus on the day of challenge. In contrast, pigs vaccinated with the homologous virus had a SN geometric mean titer (GMT) of 26.9, while pigs vaccinated with the heterologous trivalent autogenous vaccine has a SN GMT of 7.3. Besides the challenge virus, SN titers were assessed against a panel of nine heterologous field strains with differing RFLP types. Using a protective titer of ≥ 8, sera from pigs vaccinated with the homologous autogenous vaccine were able to neutralize 9 of 10 viruses assayed with GMT's ranging from 8-128, all of which were significantly greater than control pig sera ($P < 0.05$). Sera from pigs vaccinated with the trivalent autogenous vaccine were able to neutralize 5 of the 10 viruses assayed with a GMT ranging from 10-64 which were significantly greater than mock vaccinated control sera ($P < 0.05$). In contrast, sera from mock and MLV-vaccinated pigs failed to neutralize any of the viruses. These results demonstrate that co-administration of MLV and autogenous vaccine leads to a rapid, robust and broadly heterologous neutralizing antibody response which is absent with administration of MLV vaccine alone. Additionally, PBMCs collected from monovalent autogenous vaccinated pigs showed significantly greater IFN-γ production following stimulation with vaccine virus as compared to MLV-only or mock vaccinated pigs.

Viremia was assessed on days 3, 6, 10 and 13 post challenge by qRT-PCR. There were no significant differences between mock and MLV-vaccinated pigs at any time point. In contrast, pigs vaccinated with either homologous or trivalent heterologous autogenous vaccines had significantly lower viremia as compared to mock vaccinated pigs on day 10 post challenge ($P < 0.05$). These results demonstrate that co-administration of MLV and autogenous vaccine can significantly reduce PRRSV viremia.

Lung samples were collected on day 56. Similar to PRRSV viremia, significantly lower PRRSV titers were measured in pigs co-administered MLV and autogenous vaccines as compared to mock vaccinated pigs ($P < 0.05$). Vaccination with MLV alone had no impact on PRRSV titers in the lung. These results demonstrate that co-administration of MLV and autogenous vaccine can significantly reduce PRRSV lung titers.

Lung lesions were assessed macroscopically and microscopically. There were no statistically significant differences between experimental groups as assessed by gross lung lesion scores. In contrast, microscopic lung lesions were significantly lower for pigs co-administered MLV and the trivalent autogenous vaccine than all other experimental groups ($P < 0.05$). While all vaccine groups had numerically lower lung IHC scores as compared to mock vaccinated pigs, only pigs co-administered MLV and the trivalent autogenous vaccine were significantly lower ($P < 0.05$). These results demonstrate that co-administration of MLV and autogenous vaccine can significantly reduce lung lesions associated with PRRSV infection.

Together, these results convincingly demonstrate that two doses of autogenous vaccine administered to MLV-primed animals can stimulate robust cell mediated and neutralizing antibody responses that are absent from MLV-only vaccinated pigs. These antibodies are able to neutralize a broad range of genetically divergent viruses. Following challenge, pigs receiving autogenous vaccines showed lower PRRSV viremia and lung titers, as well as reduced lung lesions, when compared to mock vaccinated pigs. Vaccination with MLV alone failed to show significant protection to the RFLP 1-4-4 challenge virus.

https://doi.org/10.54846/am2023/41

Viral isolation and sequencing demonstrate flies and gnats serve as vectors of swine rotaviruses and sapovirus in nurseries

Allison Knox[1], DVM; Gene Spellman[2], BS; Miranda Capper[3]
[1]Walcott Veterinary Clinic, Walcott, Iowa; [2]Central Life Sciences, Schaumburg, Illinois;
[3]University of Iowa, Iowa City, Iowa

Introduction

The integrity of swine facility biosecurity measures are continuously challenged by non-blood sucking insects, such as the house fly, *Musca domestica*, and small gnats of the suborder Nematocera. Their small stature and ability to fly allows them to easily enter and exit barns and vehicles while carrying viral, bacterial, and parasitic pathogens. For example, recent bioassay studies have shown the capability of house flies carrying PEDV to transmit live virus resulting in clinical disease in pre-weaning pigs.[1] Similarly, porcine rotaviruses and sapovirus pose a clinical disease risk to pre- and post-weaning pigs.[2,3] Rotaviruses and sapovirus infections are leading causes of enteritis in young pigs and result in extensive morbidity and mortality in severe cases. The potential for non-blood sucking insects to carry and transmit swine rotaviruses and sapovirus has not yet be evaluated. In this study, PCR and genetic sequencing technologies were used to assess flies and gnats for their ability to serve as vectors for swine rotaviruses and sapovirus in nursery pigs. Additionally, viral isolation was used to evaluate non-bloodsucking insects' ability to harbor live rotavirus serotype A.

Materials and methods

A sow farm and downstream nursery in southeast Iowa were chosen for the study due to a history of rotaviral and sapoviral challenges in pre- and post-weaning pigs. A pilot study was conducted with these farms in 2021 evaluating proof of concept focused on PCR testing and viral isolation of rotavirus serotype A. The farms were re-enrolled for the expanded study in 2022, with additional samples collected focusing on rotaviruses A, B, C and sapovirus. Two replicates were performed in the expanded study, with one week between the start of replicate 1 and replicate 2. Sample collection began at the sow unit three days pre-weaning, where rectal swabs were collected from five nursing pigs out of a group of approximately 500. Sampled pigs were chosen at the discretion of the farm manager, with scouring or unthrifty pigs being the primary candidates. These rectal swabs were pooled and PCR tested for rotavirus A, B, C, and sapovirus. Two days pre-weaning, environmental samples and flies were collected from the clean and disinfected downstream nurseries. Environmental swabs consisted of 4 × 4" gauze squares soaked in sterile saline. These were used to wipe pig-contact areas in the nursery, such as the pen flooring, gates, feeders, and water cups. The saline and swabbed materials were wrung out and poured into falcon tubes. Environmental samples were PCR tested for rotaviruses and sapovirus on an individual basis. Flies and gnats were collected using two Captivator® liquid traps (Starbar®, Central Life Sciences) per room. Traps were used with the goal of reducing potential cross-contamination from the environment. Flies were allowed to enter the traps on their own volition, at which point they would become suspended in the water and attractant mixture. A control fly trap was hung in addition to the two active traps. The control trap consisted of a fly trap that was open to the air but used cheese cloth over the openings prevent flies and gnats from entering the trap. Fly traps were allowed to hang for 24-48 hours before the collected flies or control liquid were collected into conical tubes. Flies and control liquid were PCR tested on an individual basis for rotaviruses and sapovirus. Once weaned for approximately five days, pig rectal swabs and fly traps were recollected in the nursery. A third sampling at twelve days post weaning was also performed. All samples collected were submitted to the Iowa State Veterinary Diagnostic Lab (ISU VDL) for testing. Once preliminary PCR testing was completed, positive samples were selected for follow up sequencing and viral isolation. Environmental, control trap, or fly samples that were PCR positive for rotavirus A with a Ct around 25 or lower were chosen for attempted viral isolation. As viral isolation of rotavirus B, C, and sapovirus was not yet feasible at the time of the study, RNA sequencing was relied on to identify viral strains. Dendrograms of generated sequences were then assembled to assess percent homology between strains found in pig rectal swabs, the environment, and in the insect samples. Sequences were generated from the VP7 proteins of the rotaviruses and from the VP1 protein of the genogroup III sapoviruses.

Results

In the pilot study, pigs fecal swabbed post wean were PCR positive for rotavirus serotypes A, B, and C (Table 1). Viral isolation attempts on the rotavirus A were successful, with the lab providing a positive growth Ct of 9.6, 10.2, and 10.3 in the three groups of flies harvested. The control trap had no PCR positives for rotavirus A, B, or C.

In replicate 1 of the follow up study, pigs in the sow farm tested positive for only rotavirus A pre-weaning at relatively low Cts. (Table 2). The empty nursery had evidence of A, B, & C rotavirus serotypes at high Cts, and flies that were carrying rotaviral A RNA. The control trap was also positive for rotavirus A, but at a Ct higher than that found in the flies. No fly or environmental samples were positive for sapovirus pre-weaning. Samples at five days post-weaning found that pigs began to shed rotavirus A, B, and C serotypes, and that the flies were carrying the three serotypes as well. The control trap was also positive for serotypes A, B, & C, again at higher Cts compared to the fecal or fly samples but was negative for sapovirus. The rotavirus A found in fly samples and in the control trap was confirmed live via positive viral isolation, with isolated Cts of 9.5 and 13.0 in the flies and 18.5 in the control trap. The rotavirus A found in the pre-weaning fecal samples, the five day-post weaning fecal samples, and in the flies had 100% sequence homology. Rotavirus B in pig feces at five days post-wean and in the flies likewise had 100% sequence homology. Sequencing results for the rotavirus C was more unique as environmental swabs from the empty nursery had a 99.4% homology

Table 1: Diagnostic summary, pilot study

Flow/Site	Date	Sample	Rotavirus PCR (Ct)	Rotavirus sequence	Rotavirus A viral isolation (Ct)	Sapovirus PCR	Sapovirus sequence
Nursery (Post-wean)	Jul. 1, 2021	Fecal swab	A: 21.9 B: 25.3 C: 15.7	– – –	–	–	–
		Flies	A: 22.5 B: 27.7 C: 20.5	– – –	Positive 9.6	–	–
Nursery (Post-wean)	Jul. 15, 2021	Fecal swab	A: 14.1 B: 31.5 C: 24.4	– – –	–	–	–
		Control trap	A: Negative B: Negative C: Negative	– – –	–	–	–
		Flies 1	A: 27.0 B: Negative C: Negative	– – –	Positive 10.2	–	–
		Flies 2	A: 26.1 B: Negative C: 30.8	– – –	Positive 10.3	–	–

to the strain found in the flies at five days post-weaning. However, the rectal swab samples had only 79.8% homology to the environmental swabs and 80.3% homology to the post-weaning flies. Pigs had also become positive for sapovirus by five days post-weaning, and sapovirus was identified in the flies with a 99.6% match in sequence homology. Samples at twelve days post-weaning similarly had pigs shedding and flies carrying rotaviruses A, B, C, and sapovirus. The control trap also tested positive for rotavirus A, B, C, and sapovirus at higher Cts than those found in the fly samples at twelve days post-weaning. Sapoviral sequencing from flies at twelve days post-weaning showed a 99.7% homology to the fecally shed sapovirus from five days-post weaning and a 99.8% homology to the strain carried by the flies at five days post-weaning.

In replicate 2, the pre-weaning pigs were only positive for rotavirus serotype A, while negative for rotavirus B, C, and sapovirus (Table 3). The empty pre-weaning nursery had PCR detectable levels of rotavirus A, B, and C, but was negative for sapovirus. The control trap was also positive for rotavirus A, but at a Ct higher than the flies. Once weaned for five days, pigs were positive for rotaviruses A, B, C, and sapovirus. In the same time frame, flies and the control trap became positive for the three rotavirus serotypes, while only the flies were sapovirus positive. Virus isolation of the rotavirus A from the two fly traps were positive for live virus, and the sequence homology was 100% similar to the sequences from pre-weaning pig feces, the five-day post-weaning pig feces, and both fly samples. Rotavirus B sequenced from the five-day post-weaning pig feces and the five-day post-weaning fly samples also had 100% sequence homology. While a rotavirus C was sequenced from the pre-weaning environmental samples, no other samples tested produced a positive rotavirus C sequence for comparison. Sapovirus sequences proved to be 99.9% homologous between pig feces at five-days post-wean and in flies at the same point in time. At twelve-days post-weaning, pigs were still PCR positive for the three rotavirus serotypes and sapovirus. The control trap was still positive for rotavirus A and C at high Cts, but had become negative for rotavirus B and sapovirus. As the Cts on the rotavirus A from the flies had gotten higher, viral isolation failed. However, the Cts for sapovirus had gone lower, and sequencing showed a 99.6% homology of the pig feces and the flies at twelve-days post-weaning.

Discussion

The pilot study showed proof of concept that in post-weaning nurseries where pigs were fecally shedding rotavirus serotype A, rotaviral RNA was detectable on or in flies and gnats via PCR testing. Furthermore, it was possible to identify the rotavirus A as living virus via positive viral isolation. In the continued study, both replicates found that it was possible to isolate rotavirus A from flies at five days post-weaning. Sequencing of rotavirus A, B, and sapovirus showed homologies close to or achieving a 100% match between the strains found in pigs compared to those found in the flies. This would suggest that the flies were carrying strains shed from the pigs, or that the pigs had been infected by strains carried by the flies and gnats. The exception to this was the

Table 2: Diagnostic summary, replicate 1

Flow/Site	Date	Sample	Rotavirus PCR (Ct)	Rotavirus sequence	Rotavirus A viral isolation (Ct)	Sapovirus PCR	Sapovirus sequence
Sow unit (Pre-wean)	Aug. 15, 2022	Fecal swab	A: 19.1 B: Negative C: Negative	Positive – –	–	Negative	–
Nursery (Pre-wean)	Aug. 16, 2022	Environmental swab 1	A: 27.1 B: 35.0 C: 30.9	Negative – Positive	–	Negative	–
		Environmental swab 2	A: 30.8 B: Negative C: 33.3	– – –	–	Negative	–
		Control trap	A: 33.6 B: Negative C: Negative	– – –	–	Negative	–
		Flies 1	A: 30.7 B: Negative C: Negative	Negative – –	–	Negative	–
		Flies 2	A: 34.2 B: Negative C: Negative	– – –	–	Negative	–
Nursery (Post-wean)	Aug. 23, 2022	Fecal swab	A:15.0 B: 22.3 C: 22.9	Positive Positive Positive	–	16.1	Positive
		Control trap	A: 27.5 B: 34.5 C: 32.3	– – –	Positive 18.5	Negative	–
		Flies 1	A: 23.3 B: 30.4 C: 27.0	– – Positive	Positive 13.0	23.0	–
		Flies 2	A: 23.3 B: 30.4 C: 27.0	Positive Positive –	Positive 9.5	22.0	Positive
Nursery (Post-wean)	Aug. 30, 2022	Fecal swab	A: 24.8 B: 34.1 C: 27.0	– – –	–	24.2	–
		Control trap	A: 28.8 B: 32.4 C: 28.5	– – –	–	30.7	–
		Flies 1	A: 27.0 B: 31.9 C: 27.7	– – –	–	22.3	–
		Flies 2	A: 25.0 B: 30.0 C: 24.3	– – Negative	–	22.3	Positive

Table 3: Diagnostic summary, replicate 2

Flow/Site	Date	Sample	Rotavirus PCR (Ct)	Rotavirus sequence	Rotavirus A viral isolation (Ct)	Sapovirus PCR	Sapovirus sequence
Sow unit (Pre-wean)	Aug. 22, 2022	Fecal swab	A: 14.4 B: Negative C: Negative	Positive – –	–	Negative	–
Nursery (Pre-wean)	Aug. 23, 2022	Environmental swab 1	A: 31.9 B: Negative C: Negative	– – –	–	Negative	–
		Environmental swab 2	A: 26.3 B: 35.3 C: 30.8	Negative – Positive	–	Negative	–
		Control trap	A: 34.2 B: Negative C: Negative	– – –	–	Negative	–
		Flies 1	A: 33.0 B: Negative C: Negative	– – –	–	Negative	–
		Flies 2	A: 32.0 B: Negative C: Negative	– – –	–	Negative	–
Nursery (Post-wean)	Aug. 30, 2022	Fecal swab	A: 16.3 B: 25.3 C: 19.1	Positive Positive Negative	–	23.6	Positive
		Control trap	A: 27.8 B: 34.8 C: 33.0	– – –	–	Negative	–
		Flies 1	A: 17.9 B: 23.7 C: 24.7	Positive Positive Negative	Positive 8.6	19.8	Positive
		Flies 2	A: 18.0 B: 27.1 C: 24.3	Positive – –	Positive 8.8	21.7	–
Nursery (Post-wean)	Sept. 8, 2022	Fecal swab	A: 22.6 B: 31.1 C: 21.9	– – –	–	19.4	Positive
		Control trap	A: 32.8 B: Negative C: 32.3	– – –	–	Negative	–
		Flies 1	A: 23.5 B: 29.4 C: 23.3	– – –	Negative	18.9	Positive
		Flies 2	A: 25.8 B: 33.9 C: 22.5	– – –	Negative	22.5	–

rotavirus C of replicate 1, which demonstrated that the flies were carrying strains isolated from the nursery environment, but not the strain present in the pigs at that time. This may have been a result of having multiple rotavirus C strains involved, where only the dominant strain in each sample were sequenced, or that not enough time had passed for pigs to acquire the second strain from the environment or from the flies. Another interesting finding in the course of the study was that several of the control traps were positive for rotavirus A, B, C, and sapovirus at one or more time points in each replicate. As these traps were covered to prevent insect entry, but still open to the air, it insinuates that rotavirus and sapovirus can be found in barn dust and particulates in the air. Generally, these PCR values from the rotavirus and sapovirus found in the control traps were higher (lower amount of virus) as compared to what was identified in the pig samples or in the fly samples. This suggests that flies and gnats in the barn are able to carry more virus or allow the virus to survive longer compared to virus suspended in barn dust or particulates. In summary, flies and gnats were demonstrated to be vectors of swine rotaviruses and sapovirus in post-weaning swine nurseries. As these viral pathogens tend to be ubiquitous and hardy in the environment, and current control methods have variable efficacy, new methods for prevention and control should be sought out. Practitioners and producers should consider including non-biting insect control protocols into farm biosecurity measures, as they may contribute to the reduction of rotavirus and sapovirus infections and transmission in swine nurseries.

References

1. Allison, G., Gauger, P., Zhang, J., Spellman, G. PEDV positive bioassay reveals houseflies (Musca domestica) can transmit infectious PEDV to pigs. *Proceedings 50th AASV Annual Meeting 2019.* 206-208.

2. Dufrense, L. Rotavirus: Field challenges of prevention and control of a reemerging disease. *Proceedings 42nd AASV Annual Meeting 2012.* 433-434.

3. Thacker, B., Petznick, T., Creel, J., Mogler, M., Gauger, P., Crawford, K., Sebo, C., Maurer, J. Porcine sapovirus: Field experiences with a potentially emerging pathogen. *Proceedings 52nd AASV Annual Meeting 2022.* 199-202.

https://doi.org/10.54846/am2023/42

Effect of various levels of stabilized fish oil source on sperm quality and production in boars

M. Parsley[1], BS; M. Wilson[1], MS, PhD; T. Gall[2], MS; M. R. M. Ballard[3], MS
[1]Feedworks USA, Ltd., Cincinnati, Ohio ; [2]Gall Consulting, Burlington, Iowa; [3]The Ballard Group, Inc., Cincinnati, Ohio

Introduction

Boars are selected for breeding based on estimated breeding values (EBV) and structural soundness. Real success of boars in a boar stud depends on quality of sperm as well as quantity of total sperm cells per ejaculate. Dietary needs for a boar play a significant role in maintaining adequate requirements for energy, amino acids, minerals, and vitamins that have been established by the National Research Council (NRC). Too often, comparisons are made about dietary nutrient responses when there are differences in boar genetic lines, grain sources differ in regions around the globe, amount of fiber in the diet and quantity and quality of feed stuffs depending on storage and handling. Experiments conducted utilizing essential fatty acids, including those sourced from fish oils, have not succinctly determined their requirement levels and impact on sperm production and quality. The benefits of essential fatty acids, particularly docosahexaenoic acid (DHA), on spermatogenesis, sperm quality, sperm maturation, and male reproductive system maintenance are well known.

Metabolic processes of Sertoli cells can actively convert 18 and 20 carbon poly unsaturated fatty acids (PUFAs) into 22 and 24 carbon metabolites. Sertoli cells react in response to signals such as AMP-activated energy status and the hypoxia inducible factors are sensitive to oxygen concentrations. The peroxisome proliferator-activated receptors (PPARs) are influenced by fatty acid availability in Sertoli cells. DHA is concentrated in the head or the tail of the sperm and levels vary among different species. If DHA is deficient, acrosome biogenesis is halted after the release of proacrosomal granules. In simple terms, Sertoli cells have a great deal of involvement in spermatogenesis, from the spermatogonium up through the process of making spermatids and maturation of the sperm cells. In addition, Sertoli cells signal Leydig cells to increase testosterone production and androgen receptors sights.

The purpose of this study was to evaluate the inclusion of a stabilized marine lipid (Salmate® DHA+) in diets provided to mature boars in a commercial stud at 3 levels (dose titration) and compare Salmate® DHA+ to a traditional source of marine polyunsaturated fatty acids. In addition, two different genetic lines were evaluated, PIC 337 (Pietrain cross) and PIC 800 (Duroc), to determine if the responses differed by breed.

Methods

The experiment was conducted at the Geode Gene Transfer Center boar stud (TriOak Foods, Inc., IA, USA). Boars were housed in individual stalls on slatted floors over a deep pit in an air-conditioned barn with a positive pressure air filtration system. Fresh water, sourced from a rural public water supply, was provided by individual nipple waterers and feed was supplied individually each morning based on the body condition score (BCS) of each boar. Ideal BCS is 3+ on a scale of 1 - 5 with 1 being thin and 5 being over conditioned. Boars were fed approximately 3.2 kg of feed per day.

PIC 337 (Pietrain cross) and PIC 800 (Duroc) lines were utilized in this study (PIC, Hendersonville, TN, USA). Three groups of 87 mature boars per group were assigned to treatment 1, 2 or 3, with each group made up of 54 PIC 337 and 33 PIC 800 boars. Boars were equalized within treatment for age, average sperm production in the previous 12 weeks and BCS.

All boars received a corn-soybean meal diet that was balanced for net energy and formulated to meet or exceed NRC requirements for amino acids, minerals and vitamins (Table 1).

Trial 1

Part 1 of the trial (dose titration) compared three levels of DHA intake per boar per day: 1.83 g/b/d (treatment 1), 2.38 g/b/d (treatment 2) and 2.94 g/b/d (treatment 3) provided from Salmate® DHA+ (The Ballard Group, Inc., Cincinnati, OH, USA) on semen production and quality parameters. Treatments 2 and 3 were achieved by top-dressing Salmate® DHA+ over the treatment 1 diet each day. Boars began receiving the treatment diets on 10-August-2020 and were fed for a 9-week pre-feeding period prior to data collection for comparison, which began on 12-October-2020 and was completed on 02-November-2020.

Trial 2

Production data from 77 of the 87 boars on the control diet (previous DHA source) were compared to the data from the same boars on treatment diet 1 with both diets providing 1.83 g DHA per boar per day. Data collected from 14-June-2020 through 05-July-2020 were used to represent the control feeding period and compared to data from semen collected between 12-October-2020 and 02-November-2020.

Semen collection

Semen was collected by the gloved hand technique from boars previously trained to mount dummy sows. Ejaculate volume was determined by weight. Within 2 minutes of completion of ejaculation, a sample of the raw semen was diluted at 1:14 ratio with AndroStar Plus® semen extender (Minitube USA, Inc., Verona WI USA). The sample was processed through an eFlow® chamber to determine sperm concentration and quality (Androvision® Computer Assisted Sperm Analysis System (CASA), Minitube USA, Inc., Verona, WI, USA). The CASA analysis organized data on sperm concentration, motility, both progressive and non-progressive, and gross morphology, which included percent proximal and distal cytoplasmic tail droplets as well as translocated tail abnormalities. Findings were automatically transferred into Prism® software and recorded.

Ejaculates were required to meet the following criteria to be accepted for use:

1. Combined percent of morphological defects (proximal and distal cytoplasmic droplets and translocated tails) must be < 25%.
2. Total motility must be > 80% and progressive motility must be > 70%.
3. Minimum doses per ejaculate must be > 5.

Table 1: Composition of diets (kg)

Ingredient name	Control	Test diet 1
Corn grain	665.15	670.25
Soybean meal	162.75	162.75
Soybean hulls	100	100
Monocalcium phosphate	19.2	19.2
Choice white grease	9.7	9.7
Calcium carbonate	9.1	9.1
Omega 3 supplement[a]	13.3	0
Salmate®-Omega 3 supplement[a]	0	8.2
Salt	6.05	6.05
Vitamin- mineral premix[b]	14.75	14.75
Kcal net engergy	2,358	2,344

[a] Omega 3 sources supplied 1.83 g DHA per boar per day for each diet

[b] Vitamin-mineral premix, synthetic amino acids, choline chloride and phytase were isogenous for both diets

Chemical analysis

Feed samples for each of the diets were analyzed by Eurofins, Des Moines, IA, USA to determine the level of fatty acids supplied by the diets. The average value of DHA level in the feed samples was within 7% variance of calculated levels of DHA.

Statistical analysis

For sperm quality parameters, statistical analysis was conducted using a general linear model determined through stepwise regression and included the fatty acid treatment, the genetic lines of boars, the age of the boars and the number of days rest between the collection of semen. The analysis was carried out using Minitab 16 Statistical Software®. Chi-square analysis was conducted to assess differences between ejaculate rejection rates. A two tailed test with Yates corrected was used.

Results

Results trial 1

Results showing the effects of the 3 dietary levels of DHA (dose titration) are provided in (Table 2). There were noteworthy differences between the 2 genetic lines for this phase of the research. Sperm concentration was greater for the PIC 800 line ($P < 0.001$), while ejaculate volume was greater with the PIC 337 line ($P < 0.001$). Other differences, including total sperm per ejaculate, were not different ($P > 0.2$) between the lines. There were no differences in the individual sperm quality parameters measured that could be attributed to the dietary treatments.

Findings for semen rejection parameters are provided in semen collection criteria. The overall rejection rate was low for all treatments. While there were no differences in the individual semen quality parameters, the intermediate treatment level resulted in fewer ejaculates being rejected due to low motility ($P < 0.026$) and fewer total ejaculates being discarded ($P < 0.045$) (Figure 1). This overall increase in percent of ejaculates accepted for treatment 2 equates to 7.5% and 6.4% more ejaculates processed compared to the lowest and highest levels of DHA inclusion respectively.

Results trial 2

Ten of the originally assigned boars were not available at the end of the trial and were not included in either data set. The remaining boars for this comparison were 49 PIC 337 and 28 PIC 800 boars. Results showed that there were no differences in the extent of sperm abnormalities between the 2 dietary fatty acid sources. However, the concentration of sperm was lower ($P < 0.023$) for the boars fed Salmate® DHA+ compared to the control diet. This was offset by a significantly greater volume of semen per ejaculate during the test period ($P < 0.001$), resulting in a greater number of total sperm cells ($P < 0.001$) and significantly more insemination doses being produced per ejaculate ($P < 0.001$) (Figure 2).

Conclusions

Salmate® DHA+ significantly reduced the percent of rejected ejaculates when fed at a rate of 2.38 g DHA per boar per day, compared to 1.83 g and 2.94 g, by 7.5% and 6.4%, respectively, in mature boars in a commercial boar stud in two terminal genetic lines. Salmate® DHA+ also significantly increased total sperm output without altering sperm cell quality or the percent rejected ejaculates when compared to another commercially available product when both provided 1.83 g DHA per boar per day. Although sperm concentration was reduced, total volume, and thus the total number of sperm cells increased per ejaculate by 17%.

References

Available upon request.

Table 2: Results obtained from ANOVA General Linear Model (Trial 1)

CASA measurements	Treatment			Line		P-value		
	Test 1	Test 2	Test 3	L337	L800	Treatment	Line	T*L
Motility, %	83.1	83.6	82.8	83.5	82.5	0.869	0.347	0.788
Progressive motility, %	76.1	76.2	76.2	77.2	75.6	0.951	0.318	0.898
Proximal droplets, %	7.0	8.6	8.1	8.1	7.6	0.345	0.259	0.062
Distal droplets	9.6	9.5	9.5	9.4	9.9	0.971	0.583	0.446
Translocated tails	0.3	0.7	0.3	0.5	0.3	0.156	0.274	0.450
Volume, mL	231.6	219.6	227.4	255.3	210.4	0.553	0.001	0.187
Concentration	0.391	0.412	0.394	0.347	0.450	0.570	0.001	0.424
Total sperm, $*10^9$	90.6	90.5	89.6	88.6	94.7	0.765	0.486	0.194
Doses possible[a]	40.2	40.2	39.8	39.4	42.1	0.872	0.824	0.168

[a] Mean of 2.25×10^9 cells/dose

Figure 1: % Rejects due to motility and all reasons.

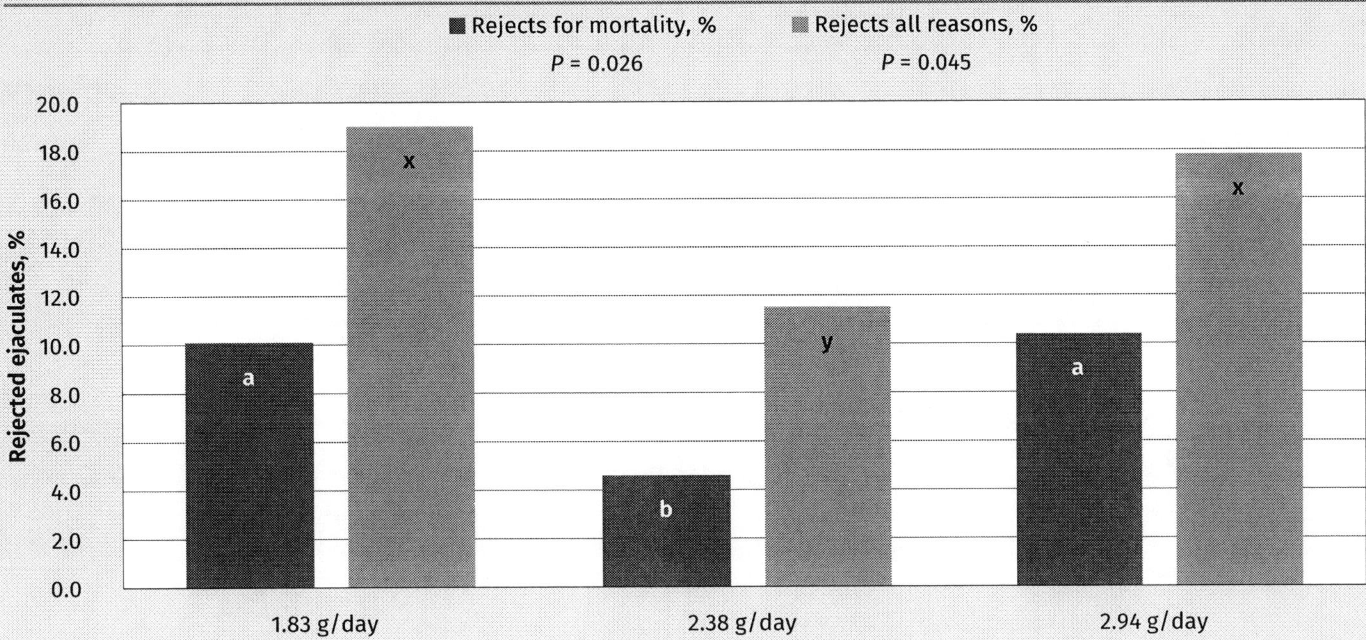

Figure 2: Sperm production and possible doses.

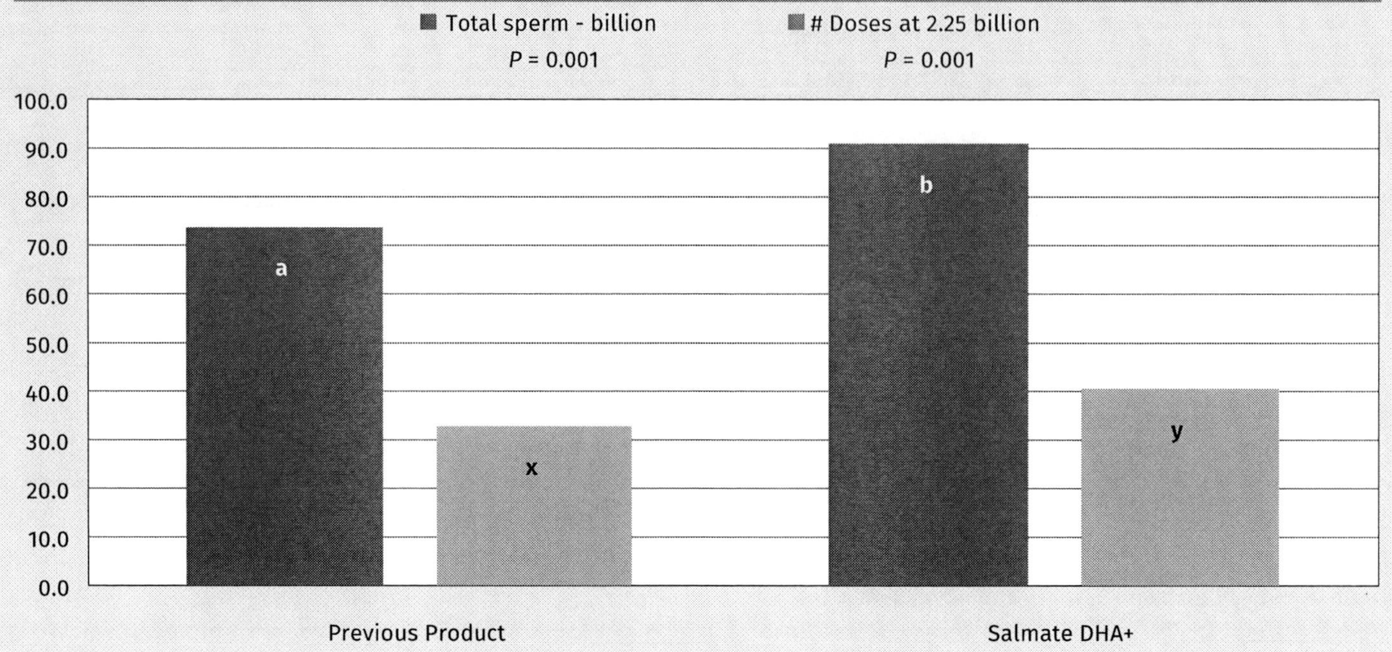

https://doi.org/10.54846/am2023/43

Conditions of storage and preincubation of boar seminal doses and its effect on motility and kinetic parameters measured with Magavision SCI

Raquel Ausejo[1], Jacob Stratton[2], Victor Marqués[1], Pedro Soler-Lloréns[1]
[1]Magapor S.L. PCTA Valdeferrin-Aula Dei. Ejea de los Caballeros, Zaragoza, Spain
[2]MB Swine reproduction LLC 666 Walnut St., Suite 2302 Des Moines, Iowa

Introduction

Artificial insemination (AI) is the most common method for sow fecundation nowadays.[1–3] More than 90% of swine AI worldwide are carried out with extended liquid semen stored up to 5 days.[4,5] The quality of the doses can be affected by several factors, eg, the boar, processing of the sample, transportation conditions, and preservation temperature, among others. AI centers take measures to minimize negative effects by evaluating the quality of the ejaculate: color, smell, motility, concentration and abnormal shapes. Besides, the temperature is controlled during the production of the doses[4] and most of boar studs store samples per batch to evaluate the motility of the doses at one or several checkpoints, being the last one at the expiration date

Computer assisted semen analyzer (CASA) systems provide user information for motility, concentration and abnormal shapes and do the calculations needed for the production of doses.[6] Accurate analysis of the ejaculates relies on the correct handling of the volume of the liquids, the slide used, and the temperature of the elements or the correct focusing of the sample, among others. When evaluating stored seminal doses, the user has to decide how to store them (small vs large volumes) and how long the sample is warmed up prior to evaluation, which can alter the results.

The aim of the present study was to evaluate two factors affecting the motility analysis: the time of warming up incubation of the samples and the recipient used to store the samples. Motility (total and progressive) and the kinetic parameters, curvilinear velocity (VCL), straight line velocity (VSL), average path velocity (VAP), amplitude of lateral head displacement (ALH), and beat cross frequency (BCF), were compared to highlight benefits and limitations when analyzing samples.

Materials and methods

Semen samples collection and storage

All samples were obtained from one AI center located in the province of Zaragoza, Spain. The collection and dilution was done under standard operation procedures of the AI Center. Ejaculates were diluted with Vitasem® extender (Magapor, Spain).

Analysis of sperm motility and morphology

Motility, morphology and concentration of the seminal doses were evaluated using Magavision SCI® CASA system (Magapor, Spain). Aliquots of 1.5 mL of each sample were pre-warmed at 37°C, and 3.5 µL were transferred into disposable 20 µm depth counting chambers (Magapor, Spain) pre-warmed to 37°C and analyzed by capturing at least 5 different fields and 500 sperm. The report of the analysis included: Total motility (%), progressive motility (%), VCL (µm/s), VSL (µm/s), VAP (µm/s), LIN (%), STR (%), WOB (%), ALH (µm), BCF (Hz), total concentration (M/mL) and total abnormal shapes (%).

Effect of incubation of the sample, and storage container

To assess the effect of the warming up of the samples prior the analysis, seminal doses of 50 different boars were measured after 1, 5, 10, 15, 20 and 30 minutes of incubation at 37°C. In a second experiment, the effect of the recipient used to store the seminal doses on motility was evaluated. In this case, 10 samples of different boar semen ejaculates were used. Ejaculates were analyzed under production conditions and after final dilution, samples were separated in different recipients: insemination bottles (80 mL), Post-cervical AI blisters (45 mL), 15 mL tubes (12 mL), 6 mL tubes (5 mL) and 1.5 mL microtubes (1.5 mL). Three samples were stored and at day 1, 3 and 7 post extraction and one replicate was measured after 10 minutes of incubation at 37°C.

In a third experiment, the effect of adverse environments during the storage on motility was evaluated. For this, 10 samples of different boar semen ejaculates were used. Ejaculates were analyzed under production conditions and after final dilution, control samples were prepared on insemination bottles (80 mL), 6 mL tubes (5 mL) and 1.5 mL microtubes (1.5 mL), while air chamber samples were prepared by adding half of the volume (40 mL, 2.5 mL and 750 µL, respectively). Three samples were stored and at day 1, 3 and 7 post extraction and one replicate was measured after 10 minutes of incubation at 37°C.

Results

The results for the effect of the incubation time of the sample on the motility are summarized in Figure 1. Total motility increases during the first five minutes (76.6% after one minute and 85.5% after five minutes) and then remains stable for the rest of the experiment. Progressive motility shows a continuous increase with the incubation time, from 22.4% after one minute and 65.03% after thirty minutes. The kinetic parameters are summarized in Figure 2. VCL, VAP show maximum values after five minutes of incubation (167.0 µm/s, 86.2 µm/s respectively and ALH shows maximum value after 10 minutes of incubation ALH (1.3 µm). VSL and BCF show a progressive increase with the incubation time (from 37.5 µm/s after one minute to 53.4 µm/s after 30 minutes for VSL; from 21.6 Hz after one minute 35.5 Hz after 30 minutes for the BCF).

Table 1 shows the motility results for doses stored in different containers under standard procedures. No significant differences were observed among the different containers within the same

Figure 1: Evolution of motility with the incubation time.

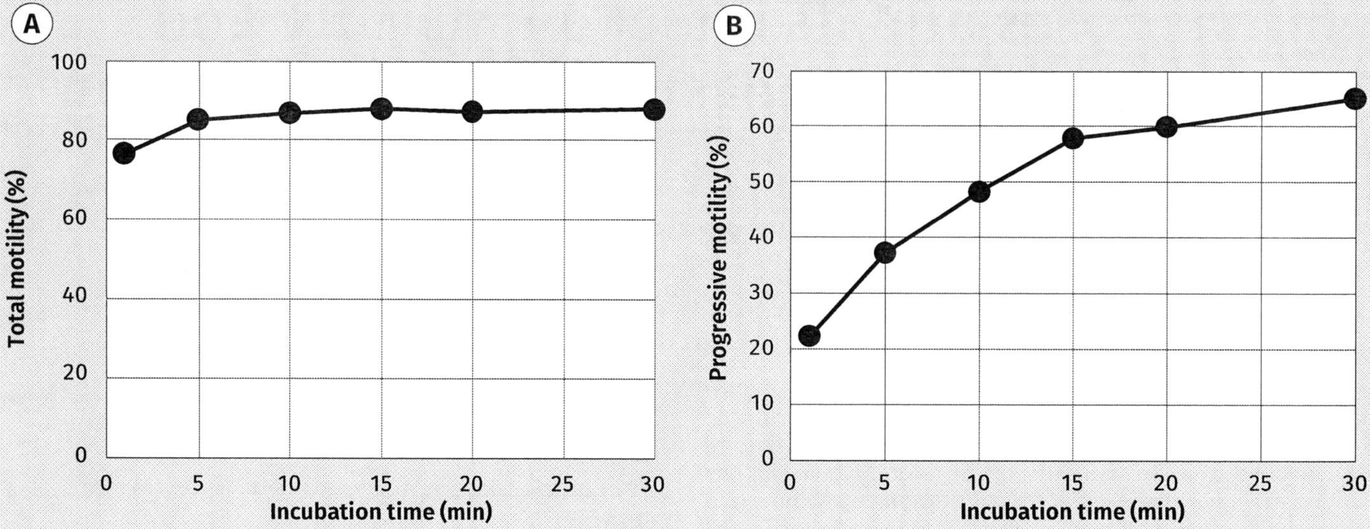

Figure 2: Evolution of the kinetic parameters with the incubation time. Each graph represents the effect of time incubation to **A)** VCL, **B)** VSL, **C)** VAP, **D)** ALH and **E)** BCF.

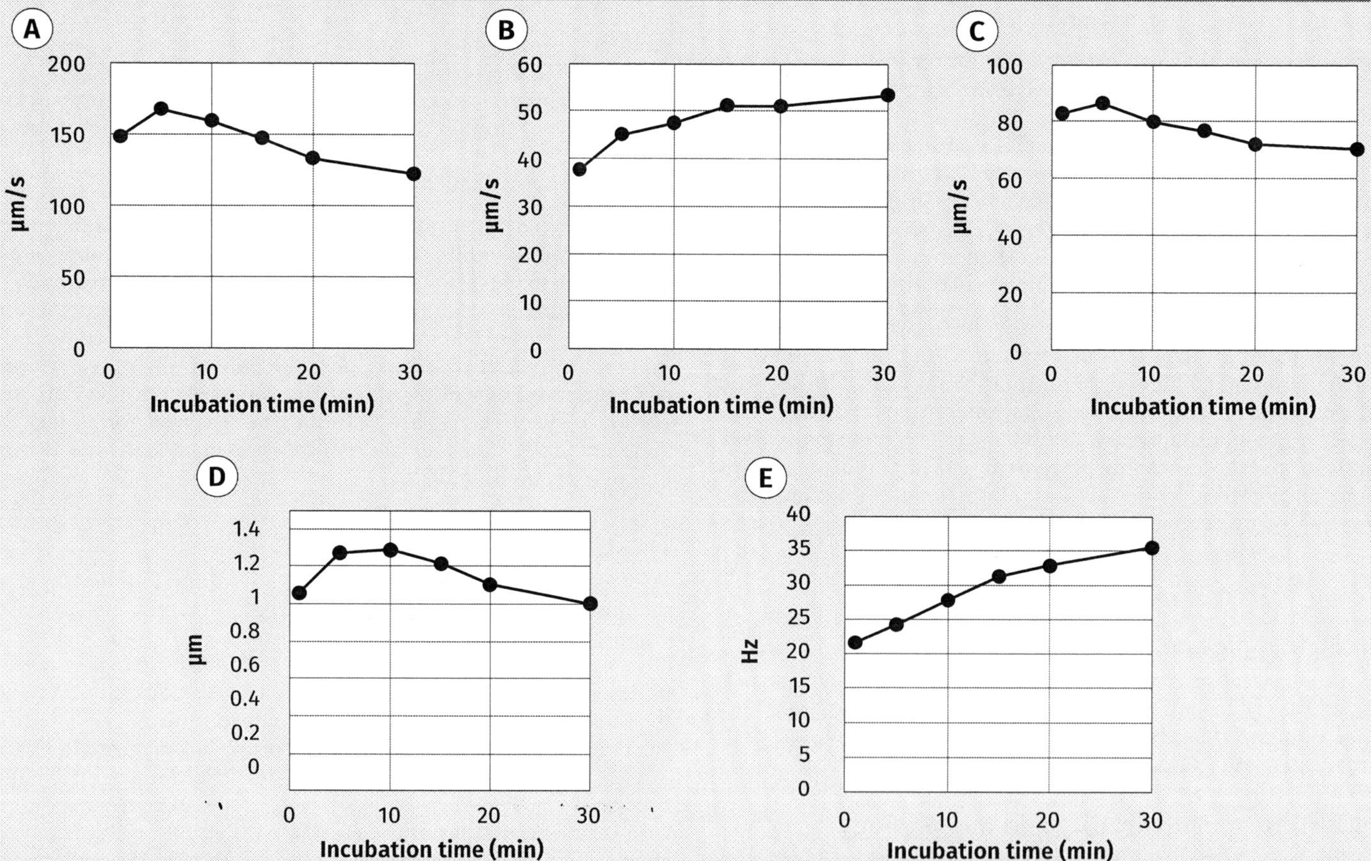

VCL = curvilinear velocity; VSL = straight line velocity; VAP = average path velocity; ALH = amplitude of lateral head displacement; BCF = beat cross frequency

day. At day seven values of TM, VCL, VSL, VAP and ALH significantly differ ($P < .05$) from day one. Motility results for the doses stored in different containers under adverse conditions are shown in Table 2. The Air chamber doses showed differences in TM and VSL at day seven compared to control ($P < .05$). 1.5 microtubes showed lower values compared to bottles at day seven for TM, VCL, VSL and VAP ($P < .05$).

Discussion

The implementation of CASA systems to evaluate the quality of the ejaculate and the doses offer several advantages: measurement is fast, limits human bias in the selection of fields, miscount or misinterpretation; allow standardization among AI centers and guarantee the repeatability of these results.[6] Complete standardization requires taking care of the conditions outside the software such as the volume handling, storage and warming up conditions.

The time of incubation is an important feature when measuring the quality of a sample, and AI centers want to implement quality controls that are cost effective. If the center is only taking into account total motility, results show that it reaches maximum value within the first 5 to 10 minutes. Progressive motility, on the other hand, is an incubation time dependent value, and thus 10 or more minutes would be required to have an acceptable value. As for the kinetic parameters, we observe that values such as VCL, VAP and ALH reach a maximum at 5 - 10 minutes, while VSL and BCF are also time dependent.

The container where the sample is stored is the other factor studied in this work. Under ideal conditions, no differences were observed among any of the tubes-blisters used. Motility parameters were not affected in this case (Table 1); however, differences can be observed when samples are stored in the presence of a high amount of air (Table 2). TM was above 70% in all the conditions tested, but differences could be observed between bottles and 1.5 mL microtubes. As seen in previous studies, air chamber can change the pH of the media reducing motility, and these factors are less affected when a long-term extender was used.[7]

This study did not take into account parameters such as the straightness (STR), linearity (LIN) or wobble (WOB) indexes, since these are derivations from VCL, VSL and VAP. Although they are useful parameters, they can mask the differences found in the velocities. Future studies should take into account also other factors such as viability, acrosome degradation or early apoptosis signs, since internal damage can be masked under a correct motility.

In conclusion, AI centers should take into account the time of incubation, the extender and the recipient used when measuring the samples in order to achieve accurate results.

Table 1: Effect of the storage container on natural conditions

	TM (%)	PM (%)	VCL (μm/s)	VSL (μm/s)	VAP (μm/s)	ALH (μm/s)	BCF (μm/s)
Day 1							
Boteclip	78.8 ± 6.9	55.6 ± 9.2	148.7 ± 29.4	49.8 ± 12.1	81.1 ± 17.4	1.2 ± 0.2	27.2 ± 3.9
Blister	80.8 ± 6.7	57.4 ± 9.0	148.6 ± 31.1	50.8 ± 11.7	81.5 ± 1.5	1.2 ± 0.2	28.3 ± 3.6
15 mL tube	79.4 ± 6.2	59.3 ± 10.7	145.6 ± 33.1	50.2 ± 11.8	79.5 ± 18.9	1.2 ± 0.3	28.5 ± 3.7
6 mL tube	79.1 ± 5.0	59.4 ± 11.2	145.7 ± 31.6	50.9 ± 10.7	80.5 ± 17.3	1.2 ± 0.3	29.7 ± 3.1
1.5 mL microtube	80.0 ± 6.2	63.5 ± 11.4	143.1 ± 30.1	52.4 ± 10.8	79.7 ± 16.8	1.2 ± 0.2	29.7 ± 3.1
Day 3							
Boteclip	80.6 ± 8.3	57.5 ± 11	148.3 ± 31.1	50.2 ± 11.6	80.8 ± 17.4	1.2 ± 0.3	27.6 ± 4.1
Blister	78.9 ± 11.8	56.7 ± 11.4	150.8 ± 33.5	50.7 ± 13.4	80.8 ± 19.9	1.2 ± 0.3	28.1 ± 3.9
15 mL tube	79.6 ± 9.4	62.1 ± 9.8	139.8 ± 27.1	49.1 ± 7.9	75.2 ± 14.0	1.1 ± 0.2	29.6 ± 3.0
6 mL tube	77.9 ± 8.8	59.8 ± 9.9	141.9 ± 33.5	48.9 ± 8.6	76.4 ± 16.7	1.2 ± 0.3	29.7 ± 2.8
1.5 mL microtube	79.3 ± 7.9	60.9 ± 9.3	137.9 ± 28.8	48.5 ± 9.9	75.2 ± 15.6	1.1 ± 0.2	28.9 ± 2.9
Day 7							
Boteclip	71.5 ± 15	58.5 ± 11.6	128.6 ± 30.9	45.5 ± 11.9	70.2 ± 17.1	1.0 ± 0.2	26.9 ± 2.7
Blister	74.1 ± 8.9	60.7 ± 9.3	123.8 ± 33.1	44.0 ± 12.5	67.2 ± 18.7	1.0 ± 0.3	26.9 ± 2.5
15 mL tube	73.0 ± 8.4	64.7 ± 10.4	116.3 ± 26.8	43.2 ± 9.8	63.8 ± 14.5	0.9 ± 0.2	27.8 ± 2.6
6 mL tube	73.2 ± 8.6	65.6 ± 10.5	117.7 ± 28.1	44.2 ± 8.9	64.6 ± 14.6	1.0 ± 0.2	29.3 ± 2.5
1.5 mL microtube	75.0 ± 9.7	66.2 ± 9.3	121.3 ± 26.9	45.2 ± 8.5	66.6 ± 1.1	1.0 ± 0.2	28.6 ± 2.2

Table 2: Effect of the storage recipient on adverse conditions

	TM (%)	PM (%)	VCL (μm/s)	VSL (μm/s)	VAP (μm/s)	ALH (μm/s)	BCF (μm/s)
Day 1							
Control							
Bottle	88.7 ± 3.8	66.3 ± 15.0	153.2 ± 26.8	53.3 ± 6.0	84.8 ± 12.2	1.2 ± 0.2	31.4 ± 2.5
5 mL tube	89.4 ± 3.1	59.6 ± 16.2	162.5 ± 28.8	50.9 ± 7.0	87.3 ± 15.6	1.3 ± 0.2	31.8 ± 2.5
1.5 mL microtube	87.3 ± 3.8	65.9 ± 12.5	150.1 ± 25.9	51.6± 7,1	81.2 ± 12.4	1.2 ± 0.2	32.2 ± 2.7
Air chamber							
Bottle	85.3 ± 4.6	62.3 ± 16.7	146.2 ± 18.7	47.6 ± 8.3	77.9 ± 10.8	1.2 ± 0.2	32.2 ± 3.4
5 mL tube	84.0 ± 4.2	58.8 ± 15.9	153.5 ± 20.4	46.1 ± 6.6	80.4 ± 12.2	1.3 ± 0.2	31.6 ± 3.4
1.5 mL microtube	83.6 ± 5.4	62.3 ± 13.9	139.9 ± 23.9	43.7 ± 6.1	72.5 ± 13.3	1.2 ± 0.2	31.9 ± 3.5
Day 3							
Control							
Bottle	88.1 ± 4.3	67.2 ± 17.3	147.3 ± 28.2	50.1 ± 9.4	80.8 ± 16.1	1.2 ± 0.2	29.,5 ± 4.0
5 mL tube	87.3 ± 3.5	65.5 ± 18.1	153.7 ± 26.3	49.6 ± 9.6	81.7 ± 14.2	1.3 ± 0.2	30.5 ± 3.6
1.5 mL microtube	86.0 ± 3.1	67.9 ± 12.6	139.9 ± 19.6	48.2 ± 7.5	75.2 ± 10.7	1.2 ± 0.2	30.5 ± 4.0
Air chamber							
Bottle	85.2 ± 4.4	65.6 ± 20.9	147.8 ± 22.9	47.5 ± 10.6	77.4 ± 14.1	1.2 ± 0.2	31.0 ± 4.2
5 mL tube	83.3 ± 4.9	62.3 ± 19.2	150.5 ± 28.9	45.5 ± 9.9	77.7 ± 16.8	1.3 ± 0.3	30.6 ± 4.5
1.5 mL microtube	79.0 ± 7.3	58.3 ± 15.7	140.0 ± 32.5	42.2 ± 9.3	71.1 ± 18.4	1.2 ± 0.3	30.5 ± 3.6
Day 7							
Control							
Bottle	83.4 ± 5.7	61.7 ± 16.1	131.1 ± 30.9	50.3 ± 10.9	73.5 ± 15.4	1.1 ± 0.3	30.7 ± 3.7
5 mL tube	83.9 ± 3.9	60.0 ± 19.9	131.8 ± 27.8	46.2 ± 4.4	71.7 ± 14.2	1.1 ± 0.3	31.7 ± 3.6
1.5 mL microtube	80.6 ± 4.8	56.7 ± 14.8	120.4 ± 30.9	437.8 ± 77.8	66.6 ± 16.5	1.0 ± 0.3	31.1 ± 4.0
Air chamber							
Bottle	78.9 ± 6.4	63.0 ± 14.4	112.9 ± 26.3	43.8 ± 9.6	63.0 ± 14.9	0.9 ± 0.2	31.4 ± 2.8
5 mL tube	70.8 ± 15.8	56.1 ± 22.6	116.7 ± 28.7	41.9 ± 9.8	62.7 ± 15.5	1.0 ± 0.3	31.5 ± 4.0
1.5 mL microtube	70.2 ± 7.9	54.7 ± 12.7	94.9 ± 17.8	34.2 ± 6.8	50.3 ± 9.2	0.8 ± 0.2	30.9 ± 3.7

References

1. Riesenbeck A. Review on International Trade with boar semen. Reprod Domest Anim 2011;46:1–3. **Doi:10.1111/j.1439-0531.2011.01869.x**

2. Bortolozzo FPF, Menegat MB, Mellagi APG, Bernardi MLM, Wentz I. New Artificial Insemination Technologies for Swine. Reprod Domest Anim 2015;50:80–4. **Doi:10.1111/rda.12544**

3. Knox R V. Artificial insemination in pigs today. Theriogenology 2016;85:83–93. **Doi:10.1016/j.theriogenology.2015.07.009**

4. Johnson LA, Weitze KF, Fiser P, Maxwell WMC. Storage of boar semen. Anim Reprod Sci 2000;62:143–72. **Doi:10.1016/S0378-4320(00)00157-3**

5. Khalifa T, Rekkas C, Samartzi F, Lymberopoulos A, Kousenidis K, Dovenski T. Highlights on Artificial Insemination (AI) Technology in the Pigs. Maced Vet Rev 2014;37:5–34. **Doi:10.14432/j.macvetrev.2013.09.001**

6. Amann RP, Waberski D. Computer-assisted sperm analysis (CASA): capabilities and potential developments. Theriogenology. 2014 Jan 1;81(1):5-17.e1-3. **Doi: 10.1016/j.theriogenology.2013.09.004**

7. Balogun, KB, Stewart, KR. Effects of air exposure and agitation on quality of stored boar semen samples. Reproduction in Domestic Animals. 2021; 56(9), 1200– 1208. **Doi:10.1111/rda.13975**

https://doi.org/10.54846/am2023/44

Influenza strategies: From serology to vaccination

Pravina Kitikoon[1], DVM, PhD; Susan Knetter[1], PhD; Channing Sebo[1], DVM;
Tara Donovan[2], DVM; Mark Mogler[1], PhD; Chandra Morgan[1], BS; Allison Hoehn[1], MS;
Heather Dempsey[1], BS; Ruizhong Shen[1], PhD
[1]Merck Animal Health, De Soto, Kansas; [2]Hanor, Spring Green, Wisconsin

Introduction

Serological evaluation of the immune response to influenza A virus (IAV) in swine herds is a valuable tool for production systems to monitor historical and existing virus presence, and when possible, also to evaluate and make informed vaccine decisions. However, the two most common practices used to combat swine IAV (IAV-S) are monitoring for clinical signs and the presence of live virus in the herd. The process is based on genetic confirmation of the virus via IAV-specific quantitative real time polymerase chain reaction (qRT-PCR) test on nasal swabs collected from sick animals, oral fluids, udder and snout wipes. Positive results are further subtyped and, in most cases, sequenced to identify the hemagglutinin (HA) and neuraminidase (NA) genetic subtype and clades. This information is important for IAV-S vaccine decisions involving strain-matched vaccine formulations, especially when autogenous or veterinary prescribed vaccines are used in the herd. The measurement of successful IAV control is based on production parameters such as reduced morbidity, mortality, treatment rate, and viral load in wean pigs or replacement gilts by a negative qRT-PCR test. While viral surveillance and characterization is necessary for vaccine decisions, using this approach alone may not truly reflect active virus circulation and cannot be used to evaluate the immune response to infection or vaccination. Virus shedding in infected animals typically lasts less than a week and may diminish by the time of sampling. In addition, endemic IAV-S may cause reduced clinical signs in some animals and thus shedding animals may remain undetected. Lastly, qRT-PCR positive samples may not necessarily reflect live virus circulation in the herd but rather detect inactive virus contamination. As such, virus surveillance, while important for vaccine decisions and monitoring, may not accurately reflect the true prevalence of IAV-S impacting a herd.

Serological evidence of infection has a wider window of detection and can be used to study current and historical exposure to IAV in the swine herd and in some cases, the immune response to IAV-S vaccination. The viral nucleoprotein (NP) and two surface envelope glycoproteins, hemagglutinin (HA) and neuraminidase (NA), are known targets for IAV-S serological diagnostic assays. Key characteristics of these assays are compared in Table 1.

Nucleoprotein (NP) ELISA

The NP wraps the RNA-dependent RNA polymerases around the viral RNA to form the ribonucleoprotein (RNP) complex, to facilitate transport from the cytoplasm into the nucleus for virus replication. This protein is highly conserved within each influenza type, A, B and C and therefore can be used to detect pigs exposed to all IAVs independent of strains and subtypes. Antibody titer to NP is frequently detected after IAV infection, and after immunization with killed virus vaccines (KV).[1] Antibody to NP alone is not protective, however sows vaccinated with IAV-KV can transfer NP-specific maternally derived antibody (MDA) to their progeny. Currently an IDEXX swine influenza virus antibody test kit (IDEXX Laboratories, ME) is commercially available to detect NP antibodies to the IAV of any subtype. It is a blocking ELISA where the presence of the conserved NP-specific antibodies in the serum will block the anti-NP conjugated antibody from binding to the NP antigen coated on the plate. The color development is inversely proportional to the amount of anti-NP antibodies in the test samples and calculated as a sample-to-negative (S/N) value. This is a qualitative assay with a positive result set at an S/N of < 0.60. It has been shown to have a sensitivity of 95.5% and specificity of 99.6% and can detect IAV antibodies regardless of the viral subtype.[2] It is suited for indicating exposure of pigs to any IAV strains through infection (as early as 7-14 days of initial infection) or vaccination with KV or modified live virus (MLV) vaccines. It can also be used to determine the presence of IAV-specific MDA, especially when kinetics are followed in the same animal born from an IAV-vaccinated or IAV-infected sow. This assay is easy and fast to perform and can therefore be used to screen a large number of samples at a given time. However, other assays must be used in conjunction with this test if antibody subtyping is needed. This test cannot differentiate infected from vaccinated animals (DIVA), limiting result interpretation when an autogenous or commercial KV or MLV vaccine is used in the herd. However, vaccines that contain HA or NA only, such as Sequivity® Veterinary Prescription HA (Sequivity® HA) or Sequivity® IAV-S NA (Sequivity® NA), makes these vaccines DIVA compatible with NP ELISA.

Hemagglutinin inhibition (HI) assay

The HA and NA glycoproteins are the major antigenic determinants of IAV subtypes (H1N1, H1N2 and H3N2), clades within the subtypes, and strains within each clade. The functional role of HA occurs at the entry point of virus infection and replication. The HA initiates infection by binding to the host cell receptor and triggering endocytosis of the virion. The HA protein then goes through a conformational change to fuse with the endosomal membrane and uncoats the viral RNP into the cell cytoplasm to start the replication process.[3] As such, specific antibodies to the HA protein can neutralize the virus to prevent infection.

The HI assay is based on the ability of certain viruses (ie, IAV, new castle disease virus, parvovirus) to agglutinate red blood cells (RBC) of certain animal species. This phenomenon is called "hemagglutination" and occurs via binding of the viral hemagglutinin protein to the sialic acid receptor molecules expressed on RBC membranes.[4] Prior to performing the HI assay, an IAV must first be isolated and enriched (through culture in eggs or cell culture) and quantified to a standard amount (4-8 hemagglutination units, (HAU) per µL) to enable visual detection of lattice formation (agglutination) with RBC. The presence of HI antibody in serum can bind to the HA protein of the virus and inhibit hemagglutination activity (thereby diminishing the RBC lattice formation). Test serum is typically prediluted 1:10 followed by serial two-fold dilution to determine the functional levels of HI antibodies specific to the test virus. The reciprocal of the highest serum dilution in which HA activity is completely inhibited

Table 1: Serological assays for influenza A virus in swine

	Nucleoprotein (NP) ELISA	Hemagglutinin inhibition (HI) assay	Neuraminidase inhibition enzyme-linked lectin assay (NI-ELLA)
Type of detection	Binding antibody	Functional antibody	Functional antibody
Subtype or clade specific	No	Yes	Yes
Requires pre-serum treatment	No	Yes	No, for pigs ≥ 7 days of age
Confounding factor	None	Can cross-react with different HA strains	Can cross-react with different NA strains HI antibody interference if wild type virus with matching HA is used as test antigen
Correlate to protection	No	Yes	Yes
DIVA (differentiate infection from vaccination assay)	Yes for Sequivity® HA, Sequivity® NA or subunit vaccines	No	No
Available in the diagnostic laboratory	Yes	Yes	No

(demonstrated by a button of pooled RBCs instead of agglutinated, lattice formation) is considered the HI titer. The HI assay has been established since the early 1940s and is viewed as the "gold standard" for measuring strain-specific protective antibodies. In humans, an HI antibody titer of ≥ 40 following IAV vaccination is generally accepted to represent the protection threshold of clinical efficacy.[5] The assay is simple, inexpensive and does not require expensive equipment. Disadvantages to the assay include (1) results are subjective to the operator's readout; (2) non-specific inhibitors of hemagglutination require treatment/removal; (3) control sera and virus antigen must be standardized each time a test is performed and; (4) careful interpretation is required due to cross-reactivity between the strains of the subclades (but not between different subtypes, ie, H1 vs H3). Overall, the HI antibody titer correlates proportionally to the serum neutralization antibodies, thus the HI assay continues to be used for vaccine formulations and evaluation of immune response to vaccination.

Neuraminidase inhibition (NI) enzyme-linked lectin assay (NI-ELLA)

The functional role of NA protein occurs at the exit point of virus replication. The enzymatic function of the NA protein contributes to the release of newly produced virions from the infected host cells and thereby assists virus spreading in the host.[3] Specific antibodies to the NA protein can limit virus spread in the host.

The NI-ELLA is a modified colorimetric assay to quantify free sialic acid, performed in a 96-well plate format and used to measure functional NA-specific antibodies in serum.[6,7] The assay utilizes fetuin protein that is heavily glycosylated with sialic acids. The NA enzyme cleaves sialic acid, and in this assay, that cleavage is detected with a peanut agglutinin (PNA)-horse radish peroxidase (HRP) conjugate followed by an addition of TMB substrate to obtain a colorimetric outcome. To perform the assay, test serum samples are incubated with the target NA antigen or virus and added to the fetuin coated plates. If NI antibodies are present in the serum, they will inhibit the NA activity to cleave sialic acid on fetuin and negate the positive signal from the PNA-HRP and TMB substrate. The assay quantifies the level of NI antibody activity by pre-diluting the serum 10-fold followed by a series of 2-fold dilution prior to incubation with NA antigen. Proper controls including positive and negative serum and no serum-antigen only wells are required on every plate and used to determine the percent inhibition level for each dilution of each serum sample. The NI titer is the reciprocal of the highest dilution that results in a percent inhibition that is ≥ 50%. Similar to HI antibodies, NI antibodies are shown to correlate to protection and reduction of IAV-associated disease.[8] One drawback for this assay is the interference of HI antibody that matches to the HA protein on the virus, if whole virus is used as the test antigen. This HI antibody can block NI antibody binding to the NA protein through steric hindrance.[9] Any animal infected with IAV or vaccinated with KV or MLV is expected to have some level of HI antibodies, and therefore likely to confound the outcome. To overcome this, genetically reverse virus with the target NA in combination with a HA subtype that is not naturally found in the swine population (ie, H6 or H9) can be used as the test antigen. Additionally, a recombinant NA protein that maintains its native tetramer and enzymatic activity can also be used (as demonstrated in this study). Inversely, an endemic IAV can still be used as test antigen in the NI-ELLA assay if serum does not contain matched HA antibodies, as when evaluating the immune response to IAV-S NA only vaccines such as Sequivity® IAV-S NA, recombinant NA protein vaccines, or NA-subunit vaccines.

Serological data should be interpreted with caution due to cross-reactivity of antibodies within virus subtypes and clades. Interpreting a change in antibody levels between two samples from the same animal (longitudinal sampling) is considered more reliable than one based on a single serum sample, as the individual pig serves as their own control. However, cross-sectional sampling at different ages provides value when performed routinely, to provide a consistent profile of the disease and response to vaccination regimen. The HI and NI assays measure strain-specific antibody function that correlates to disease protection against the homologous strain.[8,10] The presence of HI and NI antibodies indicate exposure to IAV through infection or vaccination and can indicate the presence of HI- or NI-MDA titers in young pigs. Since IAV can infect all ages of pigs it is important to understand the principle and limitations of these different serological tests so they can be used as tools to strategically plan an effective IAV-S control program to fit each production system's goals. This study provides a production system example of how these three serological assays can be used in concert and what interpretations can be made, in addition to an assay overview and limitations. Data generated by using these assays to monitor IAV disease outcome in swine herds vaccinated with Sequivity® based IAV vaccines is pending and will be disclosed at the conference.

References

1. Potter CW, Oxford JS. Determinants of immunity to influenza infection in man. *Br Med Bull* 1979;35:69-75.

2. Goodell CK, Prickett J, Kittawornrat A, et al. Evaluation of Screening Assays for the Detection of Influenza A Virus Serum Antibodies in Swine. *Transbound Emerg Dis* 2016;63:24-35.

3. Dou D, Revol R, Ostbye H, et al. Influenza A Virus Cell Entry, Replication, Virion Assembly and Movement. *Front Immunol* 2018;9:1581.

4. Kitikoon P, Gauger PC, Vincent AL. Hemagglutinin inhibition assay with swine sera. *Methods Mol Biol* 2014;1161:295-301.

5. Hannoun C, Megas F, Piercy J. Immunogenicity and protective efficacy of influenza vaccination. *Virus Res* 2004;103:133-138.

6. Eichelberger MC, Wan H. Influenza neuraminidase as a vaccine antigen. *Curr Top Microbiol Immunol* 2015;386:275-299.

7. Kaplan BS, Vincent AL. Detection and Titration of Influenza A Virus Neuraminidase Inhibiting (NAI) Antibodies Using an Enzyme-Linked Lectin Assay (ELLA). *Methods Mol Biol* 2020;2123:335-344.

8. Memoli MJ, Shaw PA, Han A, et al. Evaluation of Antihemagglutinin and Antineuraminidase Antibodies as Correlates of Protection in an Influenza A/H1N1 Virus Healthy Human Challenge Model. *MBio* 2016;7:e00417-00416.

9. Chen YQ, Lan LY, Huang M, et al. Hemagglutinin Stalk-Reactive Antibodies Interfere with Influenza Virus Neuraminidase Activity by Steric Hindrance. *J Virol* 2019;93.

10. Cox RJ. Correlates of protection to influenza virus, where do we go from here? *Hum Vaccin Immunother* 2013;9:405-408.

https://doi.org/10.54846/am2023/45

An evaluation of the use of Sequivity® PPV2 prescription vaccine to reduce the impact of lameness in growing pigs from a high health commercial farm

Deborah Murray[1], DVM; Levi Johnson[1], DVM; Brett O'Brien[2], VMD, MS; Stephanie Rossow[3], DVM, PhD
[1]New Fashion Pork, LLC, Jackson, Minnesota; [2]Merck Animal Health, Lenexa, Kansas;
[3]Minnesota Veterinary Diagnostic Laboratory, University of Minnesota, St. Paul, Minnesota

Introduction

Porcine parvovirus 2 (PPV2) has historically not been recognized as a clinically relevant cause of lameness in the US swine industry. However, repeated diagnostic surveillance from a high health commercial farm experiencing a high incidence of lameness has yielded high viral loads of PPV2 with the absence of any other pathogens typically associated with lameness. Given that PPV2 and PPV1 have fairly low VP2 homologies (less than 50%), there was a low expectation that commercially available PPV1 vaccines would help mitigate the high viral loads of PPV2. Therefore, the Sequivity prescription platform was used to evaluate a PPV2 vaccine administered to either dams pre-farrowing or growing pigs to determine if it could help mitigate the incidence of lameness in this commercial herd.

Methods

Two studies were conducted at a high health, GAP certified, commercial herd located in Wisconsin.

Study A (Sow vaccination)- A total of 180 dams were selected from three breed groups staggered three weeks apart at six weeks pre-farrowing, blocked by parity group (gilt, P1-2, P2+) and randomly assigned to treatment (T01: Controls- Sterile saline and T02: Sequivity PPV2). Each dam received 1 mL of their assigned treatment administered intramuscular in the right side of the neck and were bled at 6 and 3-weeks pre-farrow. A colostrum sample and processing fluids (litter testicles) were collected from all enrolled sows. Cross-fostering piglets off of a study litter was permitted within the first 24 hours but no piglets were allowed to be fostered onto a study litter. Any study pig that was moved to a fallback litter was removed from study. Piglets born to enrolled sows received one colored tag in the left ear at birth (different colors were used per treatment group and all tags were uniquely numbered) and had their gender recorded. At approximately 3 weeks of age (WOA), 127 litters (T01 = 67, T02 = 60) still remained on test in that the litters had remained intact and had at least 6 piglets left within the litter. A subset of one piglet from thirty T01 and from thirty T02 litters were randomly selected for bleeding throughout the study and received a second, different colored right ear tag. The remaining piglets also received a second ear tag in the right ear matching the original color. At weaning (4 WOA), littermates were placed in pens together in the wean-to-finish (wtof) barn to maintain litter integrity since litter was the experimental unit. An equal number of litters from T01 and T02 were placed into each pen.

Study B (Growing Pig vaccination)- A total of 1804 suckling piglets were enrolled at 3 WOA from a cross section of parities from three breed groups staggered three weeks apart. Within litter, piglets were enrolled in equivalent body size (S/M/L) pairs, sexed, tagged in each ear (different color per treatment group), and each pig within the pair was randomly assigned to either T01: Controls - 1 mL sterile saline IM at 3 and 6 weeks of age or T02: 1 mL Sequivity PPV2 given IM at three and six weeks of age. Thirty pairs of piglets were randomly assigned to be bled throughout the study and these pigs received a unique colored tag in their right ear. Pigs were weaned into pens at 4 WOA such that piglets from the same litter were placed into the same pen.

Study A and B- Pigs were weaned into one room of a wtof barn set up for 120 pigs per pen. Pigs were maintained on study until 18 WOA. Pens were observed daily and documented for general health. Any pigs that were moved to a hospital pen remained on study but this move was recorded. All study personnel, caretakers, and the statisticians were blinded to treatment assignments. Serum from the 60 subset pigs and oral fluids (OF) from all pens containing test animals were collected at 4 WOA and every three weeks until 18 WOA. All serum and OF samples were tested individually by PCR for PPV2 at the University of Minnesota Veterinary Diagnostic laboratory (UMN VDL). All pigs were vaccinated for *E coli* and *Salmonella* in the farrowing house and post-weaning, for PCV2 immediately prior to weaning and for Erysipelas and Ileitis in the nursery period. Pigs in the nursery period received Oregano treatment twice and Emcel (Vit D/E) or Vitamin D monthly. Lame pigs were treated with injectable Dexamethasone per veterinary prescription. All individual treatments were recorded. Any animal failing to show recovery from treatment was humanely euthanized and their cause and date of death was recorded. Mortalities attributable to lameness between 12 and 16 WOA had a complete diagnostic work-up performed on tissues and the intact affected limb at UMN VDL by Dr. Stephanie Rossow. Lameness scoring was performed by two trained scorers and was performed every three weeks post-weaning until 12 weeks of age. From 12-16 WOA, pigs were score for lameness weekly and then a final lameness score was obtained at 18 WOA. Lameness score assignment were either a "0" for pigs with no lameness or a "1" for those exhibiting mild to moderate lameness (i.e. shifting weight from affected limb, stiffness while walking, reluctant to stand/walk, limping). Percent mortalities, hospital pen removals, and sera PCR data were statistically analyzed using Statistix 10. Data was checked for normality using the Wilkes-Shapiro test. PPV2 CT's were analyzed using one-way ANOVA and mortality and hospital pen removal were analyzed by Chi^2 in Study B and the rates by Kruskal-Wallis one-way ANOVA in Study A. A P-value less than 0.05 was considered statistically significant while those between 0.2 and 0.5 were considered a trend. Lameness scores and necropsy results were summarized using descriptive statistics.

Results

Study A (Sow Vaccination)- All groups were PPV2 positive on OF at 6 WOA with peak levels observed at 12 WOA (See Table 1). On sera, peak PPV2 virus levels were seen at 9 WOA at which time the mean CT of the controls (24.1) was significantly lower than mean CT of vaccinates (28.8) ($P < 0.05$). (Table 2) There was a significant reduction in the hospital pen removal rate in litters s from vaccinated dams (8.7%) compared to the controls (11.2%) ($P < 0.05$). The mean percent lame (combined from mortalities and hospital pen removals) was significantly lower for the litters from vaccinate sows (7.7%) versus the control litters (11.0%) ($P < 0.01$). Overall mortality rates were low for both groups and were not significantly different (T01 = 5.2% and T02 = 5.3%). There were a total of nine pigs with severe lameness that were necropsied (T01 = 4 and T02 = 5) and there was a PPV2 diagnosis in 8 of these. Average tissue PPV2 CT in the necropsied piglets from control litters was 25.1 and 26.0 for piglets from litters from vaccinated sows. Lameness scores were on average 0.7% lower for the vaccinated group.

Study B (Growing Pig vaccination)- All pens were PPV2 positive on OF and sera by 9 WOA with peak levels observed at 12 WOA (Table 1). On sera, peak PPV2 virus levels were not different between the treatment groups at 12 WOA. However, at 18 WOA PPV2 viral loads were significantly lower in vaccinated pigs (mean CT = 35.7) compared to the controls (mean CT of 31.2). There was a trend towards reduction in the mortality rates of vaccinated pigs (5.1 %) compared to controls (7.2%) ($P = 0.05$). Of total mortalities, 1.1% were due to lameness in vaccinates compared with 2.8% for controls ($P = 0.19$). There were no differences in the hospital pen removal rate nor the overall recorded lameness rate. There were a total of 13 pigs with severe lameness that were necropsied and 62.5% of the controls had a PPV2 diagnosis while only 40% of the vaccinates did. Lameness scores were on average 0.5% lower for the vaccinated pigs.

It is important to note that lameness is multifactorial and in both studies there was evidence in the necropsied animals of other factors contributing to lameness such as growth plate, bone formation and OCD lesions in 27% of those submitted, evidence of prior injury in 10%, arthritis in 64%and muscle necrosis in 45%.

Discussion

In this study, Sequivity PPV2 prescription vaccine helped reduce the incidence of lameness in piglets that died or were pulled to the hospital pens. This was more pronounced in the study in which dams were vaccinated pre-farrowing. Offspring from vaccinated dams demonstrated a reduced rate of lameness and pulls to the hospital pen. While there were notable differences in the pre-farrowing vaccination study on mortality rates this might be attributable to the fact that the study ended at 18 WOA and that mortalities attributable to lameness would most likely come from the lame animals still held in the hospital pen. Vaccinating growing pigs did not reduce the overall incidence of lameness within hospital pens but did show reduced lameness scores and a trend of reduced mortality. Given what we know about parvovirus transmission dynamics, it makes sense that a control strategy in dams where the infection source probably lies might have the larger impact versus vaccinating growing pigs. The necropsy results are a reminder that lameness is multifactorial (ie, management, pen design, genetics, nutrition) play a role but the results from these studies indicate that PPV2 appears to contribute to higher incidence of lameness in some farms.

Table 1: PPV2 percent positive PCR and average Ct oral fluid results

Group	6 WOA	9 WOA	12 WOA	15 WOA	18 WOA
Sow- Rep 1	60% (Ct = 35.0)	100% (Ct = 26.7)	100% (Ct = 29.6)	40% (Ct = 36.1)	60% (Ct = 35.7)
Sow- Rep 2	20% (Ct = 38.5)	40% (Ct = 36.0)	100% (Ct = 26.0)	80% (Ct = 34.7)	20% (Ct = 35.9)
Sow- Rep 3	20% (Ct = 38.7)	0% (Ct = 40)	100% (Ct = 24.0)	100% (Ct = 32.5)	100% (Ct = 33.5)
Growing pig- Rep 1	N/A	80% (Ct = 29.0)	100% (Ct = 26.8)	60% (Ct = 34.1)	60% (Ct = 35.7)
Growing pig- Rep 2	40% (Ct =36.3)	60% (Ct = 30.6)	100% (Ct = 29.0)	100% (Ct = 34.3)	60% (Ct = 35.4)
Growing pig- Rep 3	0% (Ct = 40)	80% (Ct=31.8)	100% (Ct = 31.1)	80% (Ct = 35.4)	0% (Ct = 40)

Table 2: PPV2 percent positive PCR and average Ct sera results by treatment group

Group	3 WOA	6 WOA	9 WOA	12 WOA	15 WOA	18 WOA
Sow- controls	15% (Ct = 37.7)	N/A	100% (Ct = 24.1)	100% (Ct = 28.7)	72% (Ct = 34.5)	48% (Ct = 35.2)
Sow- PPV2 vx	0% (Ct = 40)	N/A	70% (Ct = 28.8)	100% (Ct = 29.0)	48% (Ct = 35.2)	39% (Ct = 35.2)
Growing pig- controls	0% (Ct = 40)	0 (Ct = 40)	48% (Ct = 36.0)	100% (Ct = 26.6)	N/A	100% (Ct = 31.7)
Growing pig- PPV2 vx	0% (Ct=40)	0% (Ct = 40)	26% (Ct = 36.8)	100% (Ct=26.9)	N/A	66% (Ct = 35.7)

https://doi.org/10.54846/am2023/46

Circumvent® CML: A newly licensed commercial vaccine for the protection against PCV2a, PCV2d, *Mycoplasma hyopneumoniae*, and *Lawsonia intracellularis*

B. O'Brien, VMD, MS; M. Allen, MA; F. Roerink, MS; A. Crowley, MS; S. Knetter, PhD; C. Morgan, BS; H. Wei, PhD; R. Segers, PhD
Merck Animal Health, DeSoto, Kansas

Introduction

Porcine circovirus type 2 (PCV2), *Mycoplasma hyopneumoniae* (M hyo), and *Lawsonia intracellularis* (LI) are three of the most important pathogens in the swine industry worldwide today. PCV2 is the causative agent of porcine circovirus-associated diseases, with postweaning multisystemic wasting syndrome and involvement in the porcine respiratory disease complex (PRDC) being the most economically important.[1] Currently, the most prevalent genotype of PCV2 in the United States is PCV2d.[2] *Mycoplasma hyopneumoniae* is also one of the main pathogens in PRDC and contributes to financial losses in the pig industry due to chronic, non-productive cough, reduced weight gain and worsened feed conversion.[3] *Lawsonia intracellularis* multiplies in gut enterocytes and results in significant economic losses worldwide by causing ileitis which reduces feed conversion and average daily weight gain.[4]

For over ten years, Merck Animal Health has marketed Circumvent® PCV, PCV G2, PCV M and PCV M G2, a family of vaccines targeted at PCV2 and M hyo. Merck Animal Health has also marketed Porcilis® Ileitis, the first adjuvanted *L intracellularis* bacterin with a label for intramuscular (IM) application. Due to the widespread use of these vaccines, pork producers have been requesting an updated version of Circumvent, which includes LI. This new trivalent vaccine, Circumvent® CML, reduces the number of vaccinations given to a piglet and, in turn, reduces labor cost and animal stress. The objective of these studies was to evaluate the safety and efficacy of Circumvent CML against diseases caused by PCV2a, PCV2d, *M hyopneumoniae,* and *L intracellularis* under laboratory and field conditions.

Materials and Methods

Vaccines

Test vaccines were produced containing a baculovirus-expressed ORF2 antigen of PCV2, inactivated M hyo cells, inactivated LI and the Microsol Diluvac Forte® adjuvant. Microsol Diluvac Forte® is an oil-in-water emulsion containing mineral oil and Vitamin E. The vaccines were given IM in one, 2 mL dose to pigs three weeks of age or older according to the product label. The placebo vaccine contained all of the components of the CML vaccine except for the antigen fraction being tested.

Efficacy studies

Both onset of immunity (OOI) and duration of immunity (DOI) for each of the three antigen fractions were determined in experimental challenge studies. Depending on the challenge organism, three-week-old pigs from herds seronegative and free of M hyo, LI or PCV2 were used for the OOI and DOI studies, with the exception that cesarean derived colostrum deprived (CDCD) pigs were used for the PCV2d DOI study. For all studies, pigs were randomly divided into two groups (vaccine and placebo) prior to vaccination.

For the PCV2a OOI study, 25 vaccinated and 25 control pigs were enrolled at 3 weeks of age (WOA) and intranasally challenged at 7 weeks post vaccination (wpv) with a mixture of PCV2a and porcine reproductive and respiratory syndrome virus (PRRSV) given 3 mL per nostril at a viral dose of 2.5-2.6 $\log_{10}$TCID50 and 7.1-7.2 $\log_{10}TCID_{50}$, respectively. Both the PCV2a and PRRSV challenge viruses were obtained from the University of Nebraska at Lincoln. Blood samples were taken weekly after challenge and assayed for PCV2 viremia by qPCR at the Iowa State University Veterinary Diagnostic Laboratory (ISU-VDL) and for antibody titers via immunofluorescence assay (IFA) performed at the Kansas State University Veterinary Diagnostic laboratory (KSU-VDL). Five weeks after challenge, all pigs were humanely euthanized and a necropsy was performed to collect the tonsil, mesenteric and bronchial lymph nodes. The tissues were analyzed for lymphoid depletion by immunohistochemistry (IHC) at the ISU-VDL. The blinded pathologist scored the tissues as 0 (normal), 1 (mild lymphoid depletion with loss of overall cellularity), 2 (moderate lymphoid depletion) or 3 (severe lymphoid depletion with loss of follicle structure).

For the PCV2d studies, 25 vaccinated and 25 control pigs were enrolled at 3 WOA and then intranasally challenged at 4 wpv (OOI) or 16 wpv (DOI) with 3 mL per nostril of a U.S. PCV2d isolate procured from Dr. Richard Hesse at the KSU-VDL at a final dose of 2-6 $\log_{10}TCID_{50}$. Blood samples were taken weekly after challenge and assayed for PCV2 viremia by qPCR at the ISU-VDL and for IFA at the KSU-VDL. Five (OOI) and four (DOI) weeks after PCV2d challenge, all pigs were humanely euthanized and a necropsy was performed to collect the tonsil, mesenteric lymph node, bronchial lymph node and Peyer's Patch. The tissues were analyzed and scored for lymphoid depletion by IHC at the ISU-VDL by blinded pathologists, as described above. Statistical analysis of the PCV2 viremia and lymphoid depletion of tonsil, mesenteric lymph node, bronchial lymph node, and Peyer's Patch (PCV2d studies only) data were conducted in SAS® using the R code PF package RRtosst module to perform the litter-coalesced prevented fraction analysis. Additionally, the litter-stratified prevented fraction analysis was conducted by calculating the 95% confidence interval (CI) for the complement of the stratified risk ratio 1- (Pv/Pc) from the Gart-Nam and/or Mantel-Haenszel estimators in the PF package in R-3.5.0. PCV2 serology data were analyzed by one-way ANOVA in Minitab®.

For the M hyo OOI study, 35 vaccinated and 35 control pigs were enrolled at 3 WOA and challenged intratracheally with 10 mL of a US isolate strain 232 lung homogenate (procured from the Iowa State University College of Veterinary Medicine), diluted to contain 4.8 to 5.6 log_{10} color changing units (CCU)/mL at 5 wpv. For the M hyo DOI study, 70 vaccinated and 70 control pigs were challenged with 20 mL of the same lung homogenate at 10 wpv. Blood samples were taken prior to vaccination, one day prior to challenge and 28 (OOI) or 30 (DOI) days after challenge. Sera were tested with a commercial ELISA (IDEXX or DAKO M hyo Ab test) at the ISU-VDL. Twenty-eight or 30 days after challenge, necropsies were performed on all pigs and the lungs were collected and evaluated for lung lesions by a blinded pathologist. Lung lesions were scored by shading the approximate size and shape of the lesion onto a gridded outline of a lung and its associated lobes. The number of squares shaded were counted for the apical, cardiac, diaphragmatic, and intermediate lobes on both the dorsal and ventral sides. The total number of shaded squares was divided by the total number of squares in the lung grid and then multiplied by 100 to get a percent lung lesion score (LLS). The final LLS was the average LLS of two scorers. Lung lesion scores of the vaccinate group were compared to the placebo group using the R code MFBoot module to estimate the litter-coalesced mitigated fraction (MF) and associated 95% CI (highest density). The R code MFClusBoot module was used to estimate the litter-stratified (with resampling within the litter) MF and associated 95% CI (highest density).

For the LI OOI study 40 vaccinated and 40 control pigs and for the LI DOI study 50 vaccinated and 50 control pigs were enrolled at 3 WOA and challenged by orally inoculating pigs with 25 mL of a U.S isolate containing 4.3-4.4 $log_{10}TCID_{50}$ at 5 wpv (OOI) or 20 wpv (DOI). Blood samples were taken prior to vaccination, one day prior to challenge and 21 days after challenge. In the DOI study, blood samples were also taken at regular intervals between vaccination and challenge. An in-house IFA was performed for the DOI samples whereas for the OOI, an immuno-peroxidase monolayer assay (IPMA) was performed at the University of Minnesota Veterinary Diagnostic Laboratory (UMN-VDL). For the in-house IFA assay, serial dilutions of the sera were tested on microtiter plates containing McCoy cells infected with LI. After incubation with sera, the plates were stained with a FITC-conjugated anti-swine IgG and the titers were determined using a fluorescent microscope. The serum titer was calculated as the reciprocal of the highest serum dilution that gave a positive signal. For the IPMA assay, results were expressed as negative or positive according to the test kit parameters. Three weeks after challenge, pigs were humanely euthanized and a necropsy was performed to evaluate the intestine. The length of the affected ileum was measured and the severity of the gross lesions was scored as 0 (normal mucosa), 1 (slight mucosal edema or slight hyperemia), 2 (moderate PPE, mucosal edema, hyperemia, some mucosal thickening), 3 (severe PPE, mucosal edema, severe hyperemia and mucosal corrugation) or 4 (severe PPE, score 3 plus hemorrhaging and/or necrosis, blood clots or yellowish pseudomembrane) by two scorers blinded to treatment group. In addition, the mucosa proximal to the ileo-cecal junction was collected for histopathological testing. These tissues were scored as 0 (no diagnostic lesions), 1 (mild individual crypt proliferative change) or 2 (marked proliferative enterocolitis). Lastly, samples from the ileo-cecal junction were collected and assayed for bacterial load which was performed by qPCR at the ISU-VDL and was reported as cycle threshold (Ct) or log_{10}DNA copies/mL. Incidence of ileitis and colonization due to *L. intracellularis* were analyzed (where appropriate) using the litter-stratified prevented, or mitigated, fraction and the associated 95% CI. The litter-stratified prevented fraction analysis was conducted by calculating the CI for the complement of the stratified risk ratio 1 - (Pv/Pc) from the Gart-Nam and/or Mantel-Haenszel estimators in the PF package in R. The R code MF package MFClusBoot routine was used to perform the litter-stratified (boot.cluster=TRUE and boot.unit=TRUE) mitigated fraction (MF)analysis with associated bootstrap highest density confidence interval.

Field safety study

A field safety study was performed in one herd each from the states of Minnesota, Iowa and Texas for a total of three herds. Six hundred and eighty-one healthy, 17 to 24 day old, weaned pigs were vaccinated, in the order of handling, alternating between two separate Circumvent CML serials. Two hundred and twenty five to two hundred and thirty pigs per farm were enrolled in the study with approximately half of the pigs vaccinated with each serial. Medication was recorded and pigs that died during the study were examined post-mortem to establish the cause of death. The investigator routinely observed the animals at vaccination, 1 hr. after vaccination, and at least once daily for 14 days for general health abnormalities and any systemic or local reactions suspected to be related to vaccine administration. In addition, the injection site for each pig was palpated on study days 3, 4 and 5. All visible injection site reactions were followed daily until resolution.

Results

Efficacy studies

None of the PCV2 challenge infections resulted in clinical signs except for one pig in a control group that had an occasional cough. Nevertheless, qPCR data clearly showed infection as demonstrated by viremia, as well as lymphoid depletion (Figures 1 and 2). Median virus titers were generally 2-7 log_{10} lower in the vaccinated pigs in all studies and the differences between the groups met the statistical criteria for prevention. Lymphoid depletion scores in the vaccinate groups were lower than the control groups (Figure 2). Vaccination also resulted in significantly higher antibody levels against PCV2 at 7 wpv (PCV2a OOI), 4 wpv (PCV2d OOI) as well as for the 16 wpv timepoint (PCV2d DOI (Figure 3]), compared to the control groups which were serologically negative or had low antibody titers (median $\leq 5.2\ log_2$) until the time of challenge. In all studies, following challenge, vaccinates developed an anamnestic response and the pigs in the placebo group started to seroconvert.

In both of the M hyo efficacy studies, the median M hyo induced lung lesions in the vaccinated groups were 50% (OOI) and 54% (DOI) lower than in the placebo groups at necropsy. The differences in the lung score data between groups met the statistical criteria supporting the reduction of lung lesions caused by M hyo. Seroconversion after vaccination was also seen in both studies with 65% of vaccinated animals seropositive at both 5 wpv and 10 wpv. Almost all the placebo animals responded serologically to the challenge infection, indicating a successful challenge.

In both of the LI efficacy studies, the median LI induced ileum gross lesion scores in the vaccinated groups were 74% (OOI) and 44% to 47% (DOI [Figure 4]) lower than in the placebo groups, depending on the scorer. In addition, the incidence of LI-induced histopathological lesion scores in the vaccinated groups were 48% (OOI) and 57% (DOI [Figure 4]) lower than in the placebo groups. Finally, median bacterial loads at the ileo-cecal junction were 3.1 log_{10} (OOI) lower or 6.4 Cts (DOI) higher in the vaccinated pigs.

Figure 1: PCV2d viral load in serum and percentage pigs positive from duration of immunity (DOI) study.

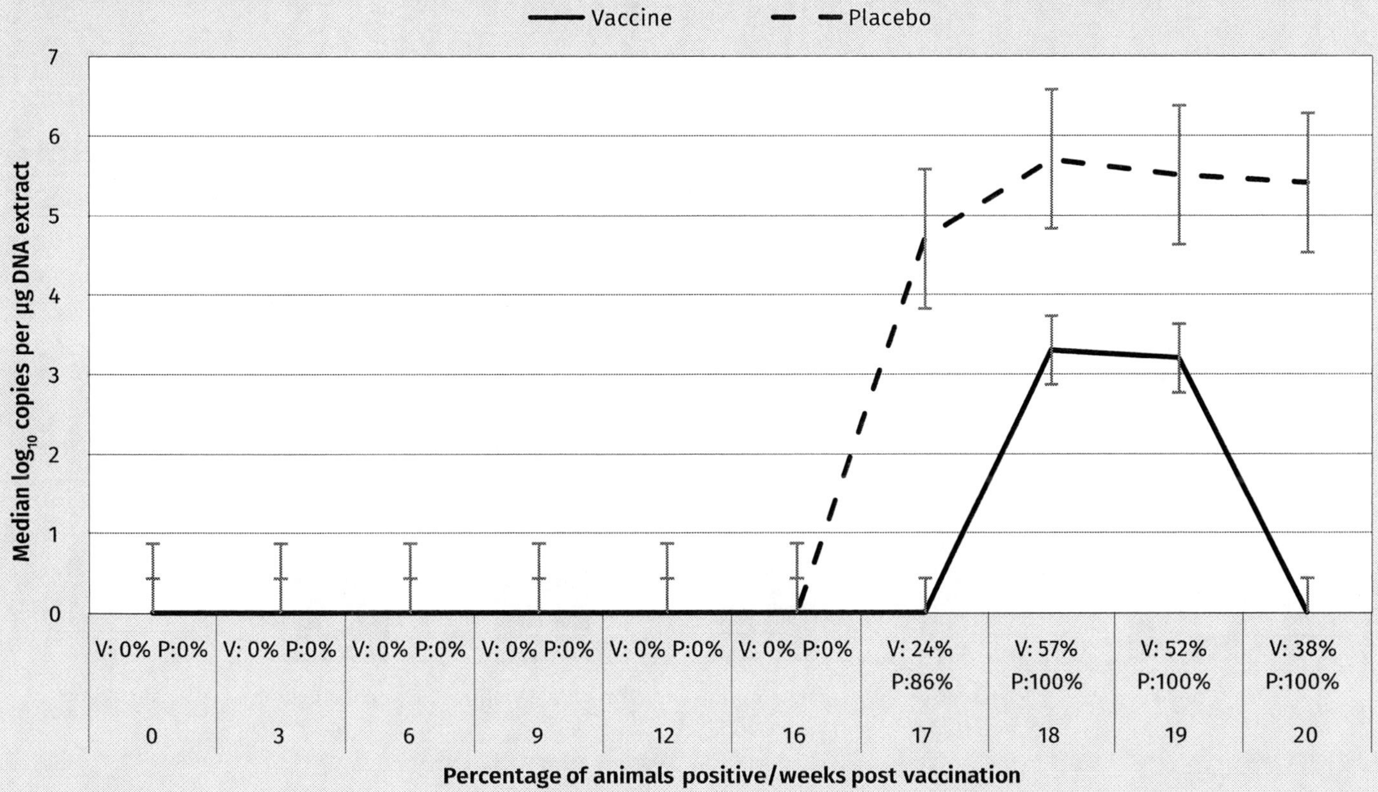

Seroconversion to LI was observed in both studies with 85% of the vaccinated animals seropositive at 5 wpv and 100% seropositive at 20 wpv. Nearly all the placebo animals serologically responded to the challenge infection in the OOI study and all of the placebo animals serologically responded in the DOI study, indicating a successful challenge (Figure 5).

Safety

In the efficacy studies, no animals developed systemic reactions and injection site swellings were observed in 44% of the vaccinate and 40% of the placebo pigs. Mean swelling size in the vaccinates was 4.6 cm and 4.1 cm for the placebo groups. The maximum duration was 14 days for the vaccinate and 13 days for the placebo groups. In the field safety study, 0.59% of the pigs fainted after vaccination and 61% developed injection site swellings which measured between 5 and 10 cm. The fainting pigs were alert within a few minutes and the swellings lasted for a maximum of 10 days. None of the injection site swellings progressed to abscesses and the pigs did not show any sign of discomfort during the post vaccination period.

Discussion

These studies demonstrated the safety and efficacy of Circumvent CML, the first trivalent ready-to-use vaccine for PCV2, M hyo and LI given once to pigs 3 weeks of age and older. The efficacy studies indicate that OOI occurs as early as 4 (PCV2d) and 5 weeks (M hyo and LI) after vaccination and lasts for at least 16 weeks (PCV2d), 10 weeks (M hyo), and 20 weeks (LI), respectively.

Circumvent CML resulted in a significant reduction in PCV2 viremia and lymphoid depletion against both PCV2a and PCV2d infections in the OOI studies and against PCV2d in the DOI study. Currently, the most prevalent PCV2 genotype in the U.S. is PCV2d with PCV2a still circulating in the population. These studies demonstrate that Circumvent CML cross protects against genotype 2d, despite the fact that it is based on a baculovirus-expressed ORF2 antigen of PCV2a.

M hyo induced lung lesions were lower in both OOI and DOI studies by at least 50%. Sixty-five percent of piglets receiving one IM dose were seropositive to M hyo at the time of challenge.

LI induced ileitis was greatly reduced after vaccination with a significant reduction in the percentage of pigs with ileum lesions (74%) and colonization (58%) in the DOI study. OOI and DOI PCR results of ileum scrapings further supported the reduction in LI colonization as vaccinated animals had a 10 to 1000-fold reduction in the median amount of LI at the ileo-cecal junction. This reduction implies that Circumvent CML can reduce the shedding of LI and, in turn, reduce the spread of the organism in a herd.

Circumvent CML is satisfactorily safe with moderate local reactions that completely resolved on average within 2 weeks. No systemic reactions were observed in the laboratory studies, and the low incidence of fainting pigs in the field safety study may be related to the genetics or vaccination practices of swine operations rather than the vaccine itself.

In conclusion, the studies demonstrated the safety and efficacy of the first ready-to-use trivalent PCV2, *M hyopneumoniae*, and *L intracellularis* vaccine that can be given as a single 2 mL dose to

Figure 2: Lymphoid depletion: Percent positive animals and median score from duration of immunity (DOI) study.

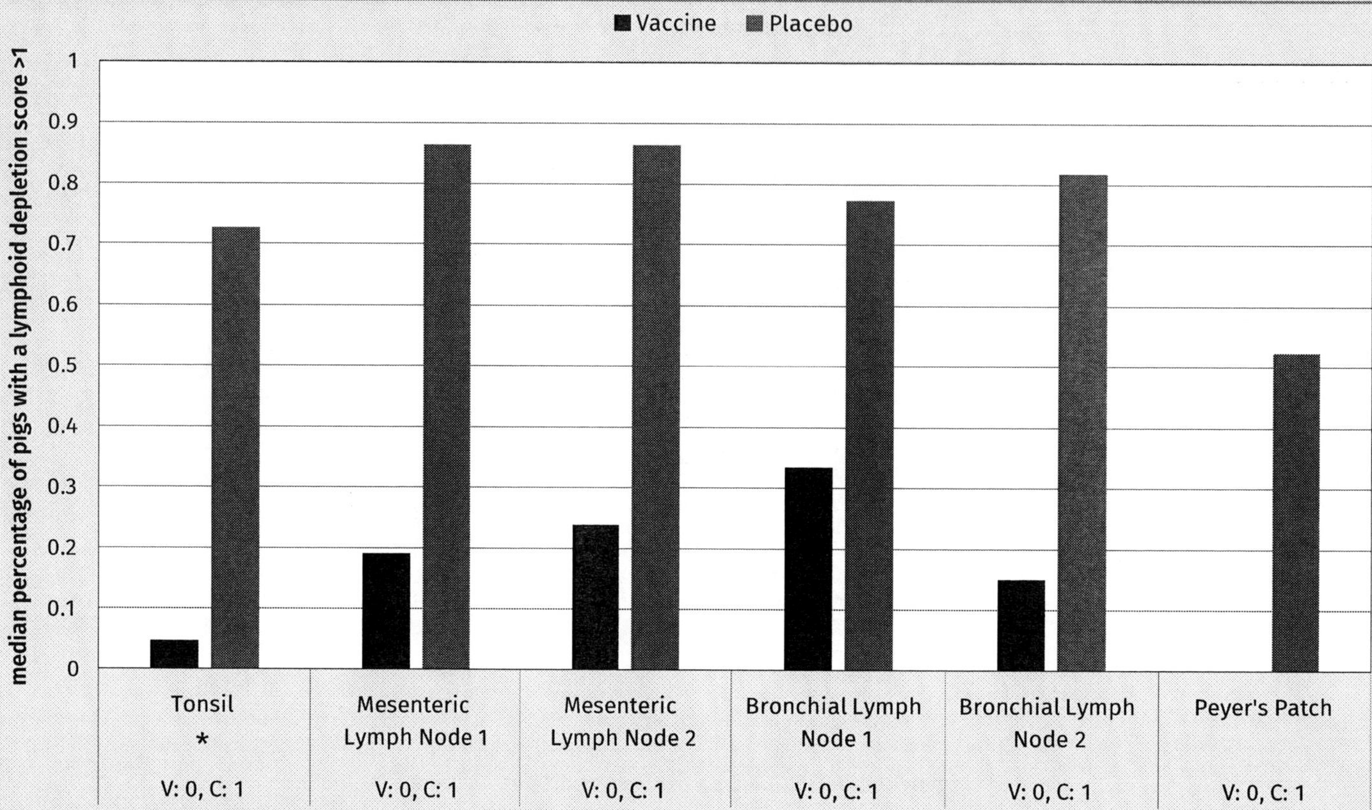

* denotes the median severity score

piglets three weeks or older. Vaccination with Circumvent CML can meet the pork producer's need to conserve valuable nutritional resources, to reduce labor inputs, and to improve animal health and welfare.

Acknowledgment

The authors wish to thank Lijuan Deng for the statistical analyses.

References

1. Segales. J. Porcine Circovirus Type 2 Infection: Postweaning Multisystemic Wasting Syndrome and Other Conditions. *Proceedings of IPVS mtg.* 2002. 035.

2. Madson. D, Baker, K, Blake K, and Davenport J. PCV2 Vaccine Considerations. *Proceedings of 2021 ISU James D. McKean Swine Disease Conference,* 25-28.

3. Sibilia, M. Pieters, M., Molitor, T., et al., Current perspectives on the diagnosis and epidemiology of *Mycoplasma hyopneumoniae* infection. *The Veterinary Journal.* 2009. 181. 221-231.

4. McOrist, S. Defining the full costs of endemic porcine proliferative enteropathy. *The Veterinary Journal.* 2005. 170(1); 8-9.

Figure 3: PCV2d serological response from duration of immunity (DOI) study.

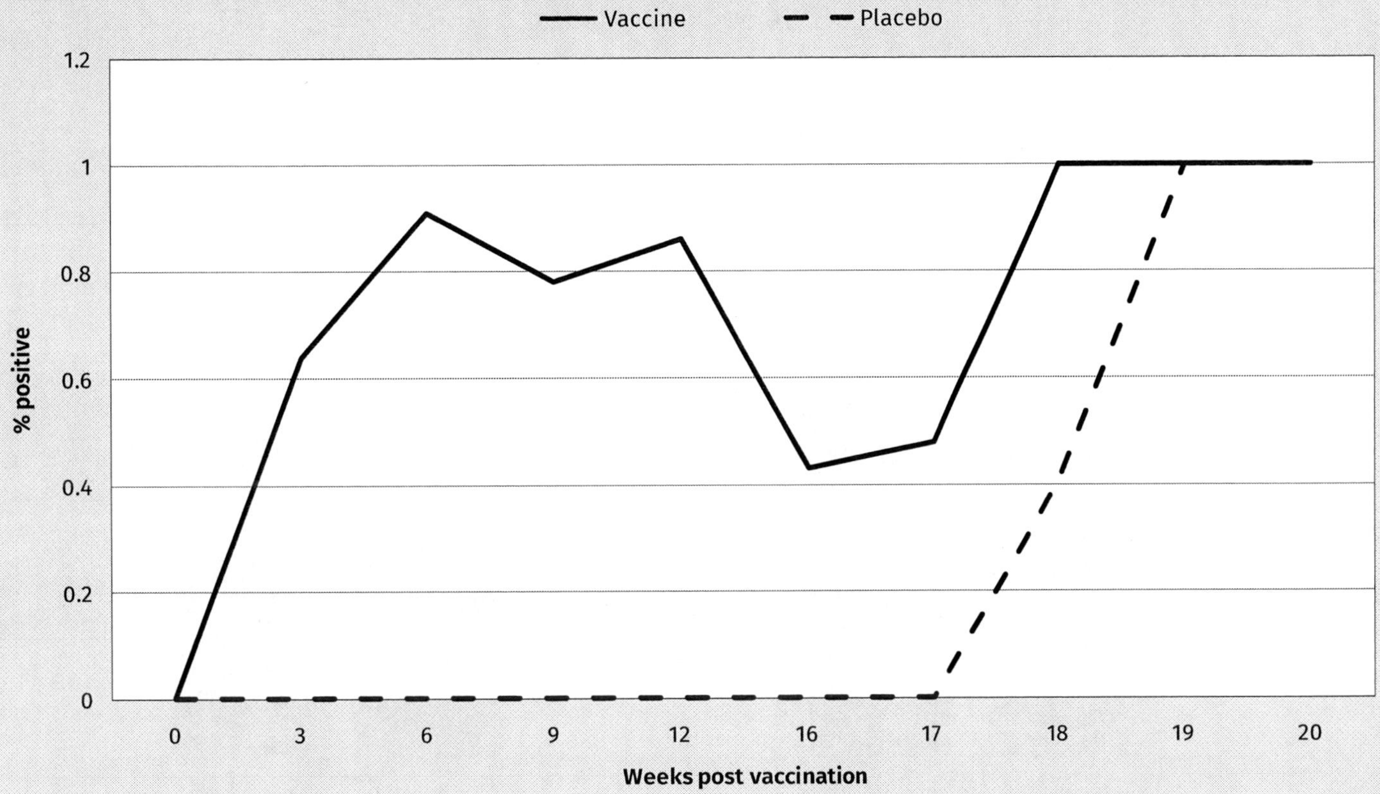

Figure 4: LI induced ileum gross lesions and histopathology scores for duration of immunity (DOI) study.

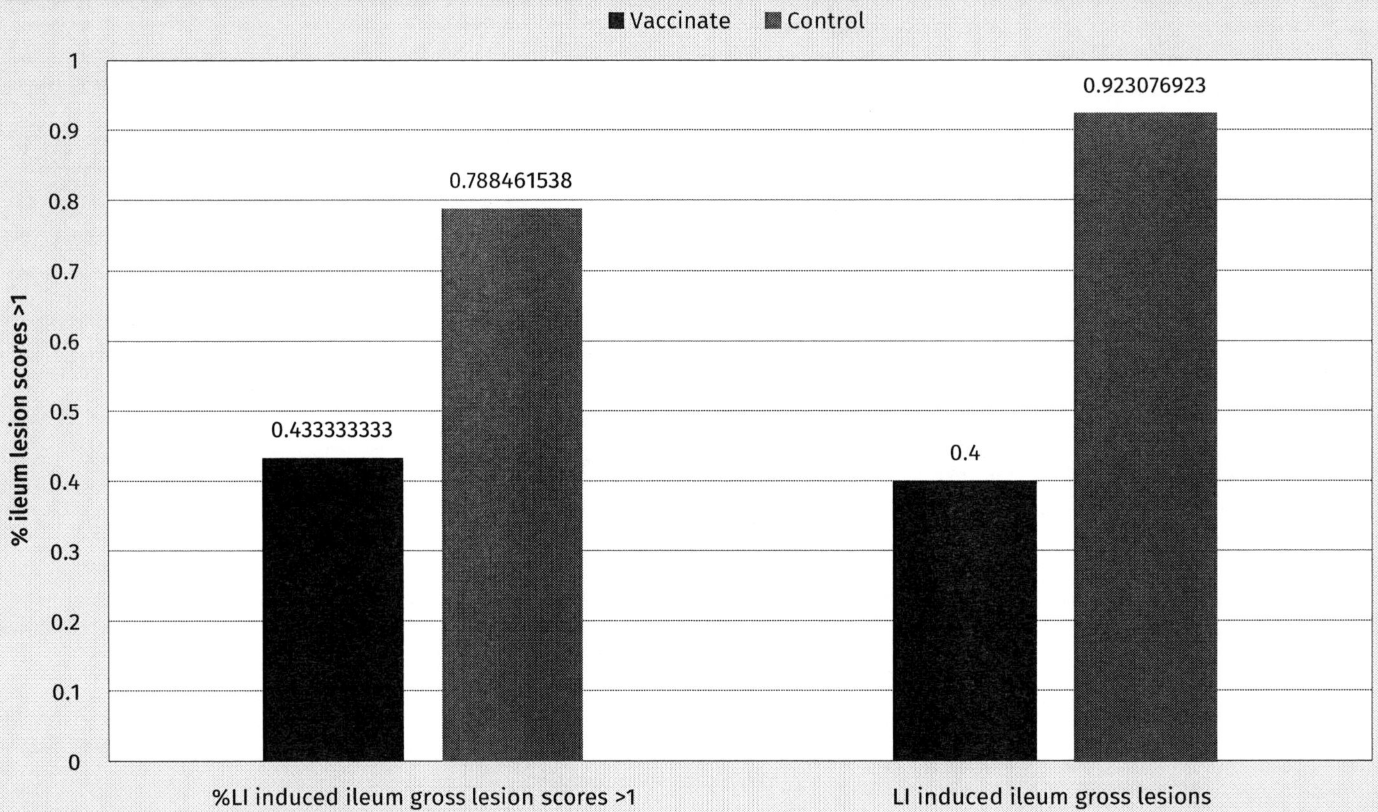

Figure 5: Seroconversion in LI onset of immunity (OOI) study.

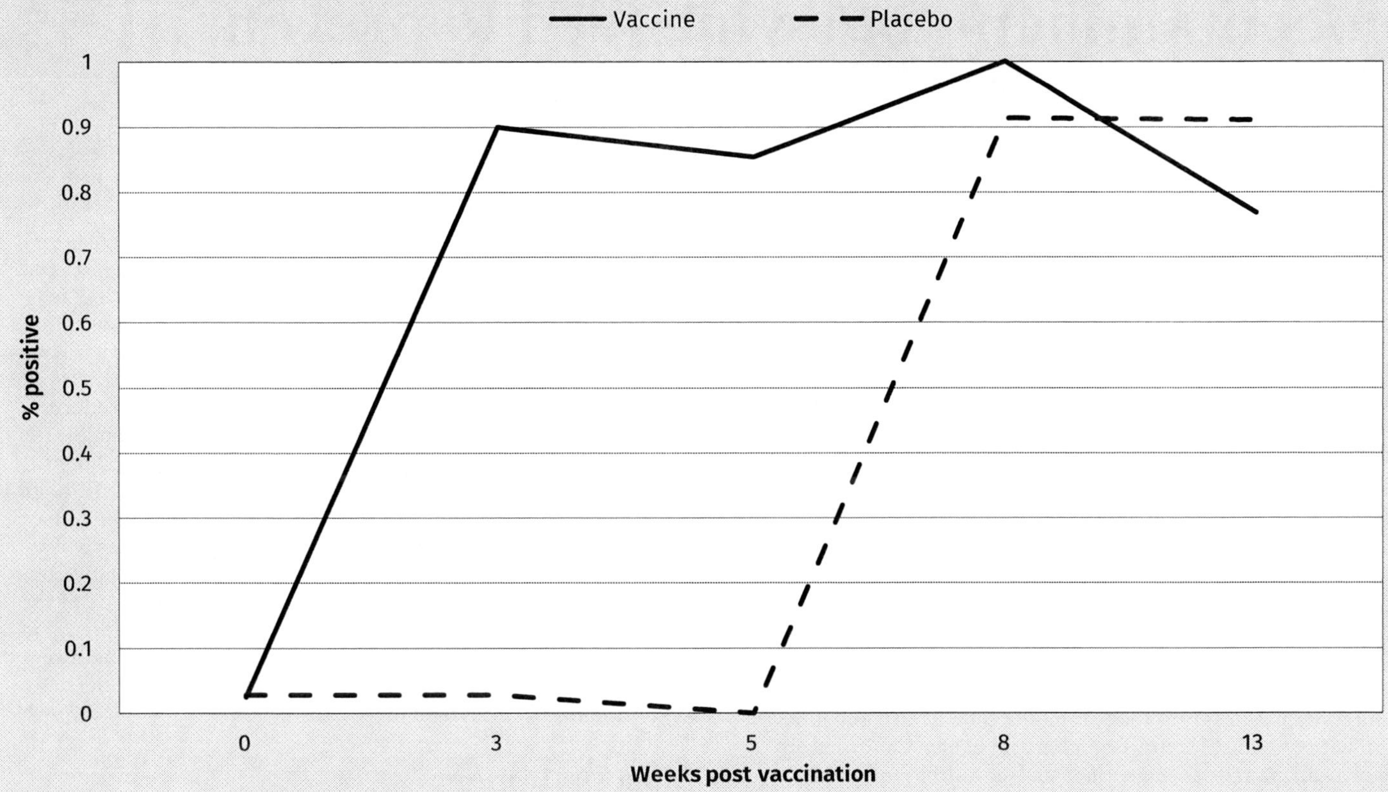

https://doi.org/10.54846/am2023/47

Biosecurity methodology including biocontainment for swine feed to prevent animal disease

P. Yeske
Swine Vet Center, St Peter, Minnesota

Introduction

The main concept of biosecurity is to break the cycle of disease infection by separation or prevention of external infectious contaminates or fomites entering a farm with susceptible or naïve animals and prevent transmission between sites. Biosecurity is made of Bioexclusion (keeping animals from becoming infected) and biocontainment (process of preventing pathogens to move between sites). Handling procedures for feed products to properly avoid contamination as well as the use of mitigants are a good method of Bioexclusion from feed.

Practices that can help to reduce bacterial or viral load of infected animals should also be part of a biocontainment plan as well, while providing the best inputs for swine health and viability. The innate immune status of the pig to resist infection and reduce viraemia are an important aspect of a biocontainment plan, and a sound nutrition program that promotes good health and growth, along with good health management and a farm environment that minimizes stress, can support better animal resistance to many infectious agents.

Nutritional, environmental, pharmacological, and immunological practices that improve the innate immune status of the pig to resist infection can help reduce viral or bacterial load within the farm and potentially reduce risk of farm-to-farm transmission and improve biocontainment, especially considering aerosol transmitted diseases like PRRS.

Bioexclusion

As an example of biosafety standards used by a feed ingredient manufacturer, the bioexclusion portion of control is done through multiple control strategies in the manufacturing process for spray dried porcine plasma. The first precaution is to separate the manufacturing plant into unprocessed and processed areas with dedicated one-way flow through the plant and sanitary barriers located throughout the plant. Employees are dedicated to the specific areas as well as assigned color coded uniforms to demonstrate their status. Equipment is dedicated to each area of the plant. There are strict polices and protocols that must be followed in the plant. No one that works in the plant can be around livestock outside of work and visitors must spend 24 hours away from livestock to enter. In the US, specific plants are dedicated entirely to either pure porcine or pure bovine raw materials and are dedicated by specie of origin throughout the collection and processing manufacturing system. In other plants with both specie of origin raw material, incoming blood within a plant is separated by species by having different sized connectors and pipes to unload and pump into separate storage if more than one species is processed in the plant. In specific plants within the USA, the plasma is passed through a computer-controlled UV-C photo purification system and is then concentrated. Then the concentrated plasma is spray dried at 80° C (has been recognized to inactivate OIE list disease including FMDV, CSFV and ASFV). The dry product is packaged in new containers only. Final storage in a temperature-controlled environment at 20°C for a minimum of 14 days is certified with strategically placed probes to ensure the proper storage temperature. A final quality control set of testing for product nutrient guarantee, certified processing, and storage temperatures is completed before product is released for sale. Product either in totes or 25 kg bags on pallets is wrapped in plastic prior to shipment. Such procedures implemented by spray dried blood product manufacturers minimize risk of post-processing contamination and enables an effective management of supply chain feed safety from manufacturer to the feed mill using state of the art practices. Once transported to a feed mill, the ingredient biosafety standards should be maintained at the feed mill using good manufacturing practices and, in some cases, including the use of feed mitigants in final feed product.

Biocontainment

As a form of biosecurity, biocontainment is a means to reduce numbers and levels of viremia in pigs and reduce the opportunity for lateral spread within a farm or to neighboring farms. Some of the components including immunoglobulins, albumen, transferrin, peptides, cytokines, and growth factors present in spray dried plasma affect the innate immune response to reduce viral levels. Furthermore, spray dried plasma is a high-quality, highly digestible protein that modulates the immune system to work more efficiently to combat pathogens including viruses to minimize the impact of pathogen aggression. Animals fed diets with spray dried plasma have been shown to maintain and support intestinal barrier function and integrity[1] as well as systemically maintain and support a positive TH-1 to TH-2 inflammatory cytokine balance in the pulmonary[2,3] and reproductive mucosal systems.[4,5] Collectively research has demonstrated that animals fed diets with spray dried plasma have a more efficient immune system response against pathogens, thereby conserving more nutrient resources for productive purposes.

For pigs fed diets with spray dried plasma, several studies have shown a reduction in viral load or pathogen shedding under experimental challenge conditions. Research has demonstrated that pigs fed spray dried porcine plasma under a PRRSV challenge had a more rapid clearance of serum viral load[6] and this was proposed to be due to enhanced proinflammatory cytokine (IFN-γ, IL-6) activity in lung fluid.[7] Furthermore, commercial pigs seropositive for PRRSV at weaning fed nursery diets with spray dried bovine plasma compared to complex alternative diets with multiple specialty proteins and additives had reduced mortality and heavier body weight at the end of the nursery.[8] Also, an unstable, chronic PRRS sow herd had improved farrowing rate, reduced repeat matings, and more pigs born alive and weaned per sow served after including 0.5% spray dried plasma in gestating and lactating sow feed.[9]

Weaned pigs under challenge with PEDV and fed a diet with bovine spray dried plasma had a more rapid increase in PEDV antibody development and a more rapid clearance rate of PEDV fecal excretion compared to a control diet without spray dried plasma.[10]

Recently, pigs fed a diet with UV-C treated spray dried porcine plasma had a 63% improvement in survival, reduced fecal score, and a 1.5 kg heavier average final body weight compared to pigs fed a diet with soy protein isolate when subjected to a rather harsh F18 ETEC challenge under a study protocol that prohibited any medical interventions.[11]

Furthermore, recent evidence showed that naïve contact pigs fed a diet with spray dried porcine plasma that were intranasally vaccinated with a candidate ASFV vaccine were 100% protected against a wild-type ASFV (Georgia 2007 strain) challenge, whereas vaccinated pigs fed a diet without spray dried plasma became infected.[12]

Conclusions

Such evidence clearly demonstrates that spray dried plasma manufactured under state of the art, good manufacturing processes, is not only safe, but highly efficacious in support of swine herd health management. Feed programs utilizing spray dried plasma produced under good manufacturing processes can promote swine health and reduce pathogen (bacterial or viral) load, thereby facilitating both preventative and biocontainment biosafety practices.

Starting with good bioexclusion practices for feed ingredients and additional use of feed mitigants can help to ensure that we don't bring the virus to the farm in the first place. However, if the herd becomes infected with virus from a different source; nutritional and immunological practices that improve the innate immune status of the pig to resist infection can help reduce viral loads and help to act as an additional form of biocontainment for the site. Realizing that this is just one piece of the complex puzzle for control.

References

1. Peace et al., 2011. Spray-dried porcine plasma influences intestinal barrier function, inflammation, and diarrhea in weaned pigs. J. Nutr. 141:1312-1317.

2. Maijo et al., 2011. Dietary plasma proteins attenuate the innate immunity response in in a mouse model of acute lung injury. Brit. J Nutr. 107:1-9. **doi:10.1017/S007114511003655**

3. Maijo et al., 2012. Dietary plasma proteins modulate the adaptive immune response in mice with acute lung inflammation. J Nutr. 142:264-270.

4. Song et al., 2015. Spray-dried plasma attenuates inflammation and improves pregnancy rate of mated female mice. J. Anim. Sci. 93:298-305.

5. Liu et al., 2018. Spray-dried plasma attenuates inflammation and lethargic behaviors of pregnant mice caused by lipopolysaccharide. PLoS ONE 13(9):e0203427. **https://doi.org/10.1371/ journal.pone.0203427**

6. Pujols et al., 2011. Commercial spray-dried porcine plasma does not transmit porcine circovirus type 2 in weaned pigs challenged with porcine reproductive and respiratory syndrome virus. Vet. J. 190:e16-e20.

7. Diaz et al, 2010. Potential positive effect of commercial spray-dried porcine plasma on pigs challenged with PRRS virus. Proc. 21st Intl. Pig Veterinary Society, Vancouver, Canada. p 560.

8. Crenshaw et al., 2017. Effects of a nursery feed regimen with spray-dried bovine plasma on performance and mortality of weaned pigs positive for porcine reproductive and respiratory syndrome virus. J. Swine Health Prod. 25(1):10–18.

9. Campbell et al., 2006. Use of statistical process control analysis to evaluate the effects of spray-dried plasma in gestation and lactation feed on sow productivity in a PRRS-unstable farm. Amer. Assoc. Swine Vet. p 139-142.

10. Duffy et al., 2018. Impact of dietary spray-dried bovine plasma addition on pigs infected with porcine epidemic diarrhea virus. Transl. Anim. Sci. 2018.2:349–357. **doi:10.1093/tas/txy088**

11. Campbell et al., 2022. Survival of pigs challenged with F-18 Escherichia coli was improved when fed diets containing ultraviolet light treated spray dried porcine plasma. Allen D. Leman Swine Conference, St. Paul, MN.

12. Blázquez et al., 2022. Improving the efficiency of a candidate ASFV vaccine through feeding spray-dried plasma to pigs. Proceedings IPVS, Rio de Janeiro, Brazil.

https://doi.org/10.54846/am2023/48

Sow herd performance following vaccination of PRRSV naïve sows with a modified live PRRSV vaccine

C. Fruge[1]; J. Gooding[2]; C. L. Puls[2]; R. Evelsizer[2]
[1]The Maschoffs LLC, Arnold, Nebraska; [2]Elanco, Greenfield, Indiana

Introduction

Porcine reproductive and respiratory syndrome virus (PRRSV) is an RNA virus that causes reproductive failure in gestating sows and systemic disease in the growing herd. PRRSV is transmitted vertically, resulting in weaned piglets with respiratory disease, usually weak and underweight.[1] Lateral spread also occurs from area spread, transportation of infected stock, and biosecurity lapses.[2] The sow performance impact of PRRSV is significant with the literature reporting 1.7 pigs lost/sow/year in infected herds.[3] The estimated total cost of productivity losses due to PRRSV in the US in the national breeding and growing-pig herd was estimated to be $664 million annually.[4] As managing PRRSV is a difficult and a serious matter for veterinarians and producers, difficult and complicated decisions must be made. These decisions encompass financial evaluations, best care for the animal's welfare, and are typically emotional to caretakers. Although mitigation of a clinical outbreak is difficult to predict, some producers will vaccinate their herds to mitigate the high production loss associated with PRRSV. In this case study, four sow farms were vaccinated to mitigate PRRSV losses. In this case study, area PRRSV pressure or spread influenced the decision to use a modified live virus PRRSV vaccine (Prevacent) in 4 naïve sow herds to mitigate possible losses in the (increasingly likely) event of sow herd exposure to the circulating area wild-type PRRSV.

Materials and methods

Study herds

Numerous wild-type PRRSV confirmed breaks in the area surrounding 4 PRRSV-naïve (AASV category IV[5]) sow farms influenced a decision to vaccinate these farms to mitigate production losses in the event of sow herd exposure and subsequent clinical disease.

PRRSV vaccination history

Farms 1 and 2 sows received a full dose of Prevacent at single point of time regardless of production cycle phase. Farms 3 and 4 sows received a full dose of Prevacent excluding the sows in their last trimester of pregnancy (weeks 1-12). Sows in their last trimester of pregnancy (weeks 13 - 16) received a full dose of Prevacent following farrowing.

Productivity impact

Herd productivity data was evaluated to compare the 8-week time frame prior to vaccination to the 8-week time frame following vaccination.

Results and discussion

Comparisons in sow herd productivity measures in the 8 weeks prior to receiving Prevacent to the 8 weeks following Prevacent are presented in Table 1.

No significant differences in the listed productivity measures were observed in the 8 weeks prior to vaccination with Prevacent compared to the 8 weeks following Prevacent. However, there was a trend towards increased combined stillborn and mummified in Farms 1 and 2 following vaccination. The impact to combined stillborn and mummified appeared to be neutralized in Farms 3 and 4 by avoiding vaccination of 3rd trimester animals, resulting in more liveborn.

Producers must balance the high production loss associated with PRRSV infection with immunization with PRRSV vaccination and the potential to mitigate a clinical outbreak. Vaccination of naive sows is not without risk. This experience indicates Prevacent had little to no impact on farrowing performance parameters as previously reported.[6] A recent report indicates modified live virus (MLV) vaccination of PRRSV naïve herds can potentially affect farrowing parameters leading to a decrease in pigs weaned per litter for a six-week period and increase repeat services creating lower farrowing rate.[7] As with all MLV vaccines and wild-type PRRSV infections, a persistent lymphoid period post vaccination needs to be considered in decision making about PRRSV status and transmission risk.

References

1. Karniychuk, U. U. and H. J. Nauwynck. 2013. Pathogenesis and prevention of placental and transplacental porcine reproductive and respiratory syndrome virus infection. Vet Res. 44(1):95. doi: 10.1186/1297-9716-44-95. PMID: 24099529; PMCID: PMC4021427.

2. Pileri, E. and E. Mateu. 2016. Review on the transmission porcine reproductive and respiratory syndrome virus between pigs and farms and impact on vaccination. Veterinary Research. 47(1):108.

3. Nieuwenhuyse, N., et al. 2012. Economic analysis of outbreaks of porcine reproductive and respiratory syndrome virus in nine sow herds. Vet Rec 170(9): 225.

4. Holtkamp DJ, Kliebenstein JB, Neumann EJ, et al. 2013. Assessment of the economic impact of porcine reproductive and respiratory syndrome virus on United States pork producers. J Swine Health Prod. 21(2):72-84.

5. Holtkamp DJ, Polson DD, Torremorell M, et al. 2011. Terminology for classifying swine herds by porcine reproductive and respiratory syndrome virus status. J Swine Health Prod. 19(1):44-56

6. Trevisan, G., et al. 2021. Description of changes of key performance indicators and PRRSV shedding over time in a naïve breeding herd following a PRRS MLV exposure. Transbound Emerg Dis: 1-6.

7. Petznick, T., A. Witt, C. L. Yoder, R. Evelsizer, J. Risser, D. Linhares, and J. M. Hammer. 2022. Performance and diagnostic evaluation of a lineage 1 (1-8-4) PRRSV vaccine in naïve sows with subsequent field exposure. Pg. 101-104 in Proceedings 53rd Annual Meeting of the AASV.

Table 1: Effects of vaccination on sow herd performance

Farm name	Parameter	8 weeks prior	8 weeks post	Delta (post vs previous)
Farms 1 & 2	Mean (total born)	14.75	15.12	0.37
	Mean (stillborn)	0.72	1.06	0.34
	Mean (mummified)	0.38	0.52	0.14
	Mean (stillborn, mummified)	1.10	1.58	0.48
	Mean (liveborn)	13.65	13.54	-0.11
	Mean (piglets weaned)	11.84	11.35	-0.49
	Stillborn, %	4.88%	7.00%	2.12%
	Mummified, %	2.58%	3.45%	0.87%
	Stillborn, mummified, %	7.46%	10.45%	2.99%
	Service failures	51	28	-23
Farms 3 & 4	Mean (total born)	14.44	14.82	0.38
	Mean (stillborn)	1.10	0.98	-0.12
	Mean (mummified)	0.24	0.30	0.06
	Mean (stillborn, mummified)	1.35	1.28	-0.06
	Mean (liveborn)	13.10	13.54	0.45
	Mean (piglets weaned)	10.69	11.01	0.32
	Stillborn, %	7.64%	6.55%	-1.09%
	Mummified, %	1.66%	2.00%	0.34%
	Stillborn, mummified, %	9.30%	8.55%	-0.75%
	Service failures	16	13	-3

https://doi.org/10.54846/am2023/49

Pulmotil used for *Mycoplasma* eradication mid-herd closure

P. Yeske[1]; R. Strobel[1]; H. Johnson[1]; A. Betlach[1]; R. Evelsizer[2]; J. Gooding[2]
[1]Swine Vet Center, St Peter, Minnesota; [2]Elanco Animal Health, Greenfield, Indiana

Introduction

Many farms are moving to *Mycoplasma* elimination to simplify management of the breeding herd, improve production and growth in finishing.[1-3] There are many different methods for elimination. The majority of herds have used the traditional approach of exposure and herd closure with good success.[1,2] Some producers have tried whole herd medication with no closure and have success, but not as good as herd closures.[1,2] The objective of this project evaluated the use of Pulmotil AC for elimination near the middle of herd closure rather than the end to observe if gains in downstream finishing pig performance with elimination could be accelerated.

Methods

A farm had experienced a recent outbreak of *Mycoplasma hyopneumoniae.* The entire sow herd was exposed with lung homogenate from *M. hyopneumoniae* positive sows within the herd to allow for consistent exposure and to set time zero of the herd. Exposure was confirmed using tracheal swabs in the sow herd and in the gilt developer. The herd was closed for a traditional elimination (240 days). Whole herd medication in water using Pulmotil AC including gilt developer, gestation, and farrowing was initiated following confirmation of exposure. Pulmotil AC was administered at 200 ppm for 7 days, followed by 7 days non-medicated water, and again at 200 ppm for an additional 7 days. Piglet treatments began on the same day as initiation of water medication. All pigs were injected on site with a single dose of Baytril (7.5 mg/kg) on approximately day 7 and 14 of age. Data was summarized for the 16-wk period prior to Pulmotil AC initiation vs. 16-wk, 32-wk, and 48-wk periods after Pulmotil AC initiation.

Results

Treatment and elimination were successful by testing of gilts in the gilt developer unit, sow herd and downstream pigs with tracheal swabs. Downstream performance metrics were evaluated, as well as clinical signs. Herd closure was completed, but no additional treatment was required since all indications were that the *M. hyopneumoniae* had been eliminated. Table 1 shows the impact of Pulmotil on farrowing house performance as measured by preweaning mortality. Table 2 demonstrates the differences in performance in nursery phase following treatment. Table 3 demonstrates the differences in performance in finishing phase following treatment. The use of Pulmotil AC reduced percent preweaning mortality (PWM), increased nursery and finishing average daily gain (ADG) and reduced overall nursery and finishing mortality.

Table 1: Pre-weaning mortality

Parameter	Prior 16 weeks	Pulmotil 16 weeks	Delta
Pre-wean mortality %	12.96%	12.31%	-0.65%

Table 2: Nursery performance

Parameter	Prior 16 weeks	Pulmotil 16 weeks	Pulmotil 32 weeks	Pulmotil 48 weeks
Number of closeouts	5	9	10	8
Placement weight, lbs	13.0	12.7	13.4	12.6
Out weight, lbs	49.1	43.2	45.5	43.2
Mortality, %	1.1%	0.9%	0.8%	1.0%
ADG, lbs	0.81	0.76	0.85	0.84
FC, lbs feed/lbs live gain	1.64	1.52	1.55	1.50
DOF	43.8	39.7	37.5	35.9

Table 3: Finishing performance

Parameter	Prior 16 weeks	Pulmotil 16 weeks	Pulmotil 32 weeks	Pulmotil 48 weeks
Number of closeouts	6	8	9	7
Placement weight, lbs	46.3	41.6	47.0	43.1
Out weight, lbs	272.8	281.2	282.7	276.6
Mortality, %	5.2%	6.2%	3.7%	3.4%
ADG, lbs	1.87	1.90	2.05	1.94
FC, lbs feed/lbs live gain	2.86	2.85	2.86	2.90
DOF	113.0	116.7	109.5	114.2

Discussion

Pulmotil AC can be used for elimination of *M. hyopneumoniae* following exposure of the breeding herd at confirmation of exposure rather than waiting until the end of closure and obtain additional performance benefits in the finishing animals prior to actual re-opening of the herd. Pulmotil AC can be used in herds with an acute *M. hyopneumoniae* outbreak to stabilize the herd until it can be either stabilized for long term control or move on to elimination.

References

1. Yeske, P. 2020. Eradication of *Mycoplasma hyopneumoniae* from pig herd. Mycoplasma in Swine. Chapter 12 pg. 229-247.

2. Holst S. 2015. Elimination of *Mycoplasma hyopneumoniae* from breed to wean farms: a review of current protocols with emphasis on herd closure and medication. *J of Swine Health Prod.* 24:321-330.

3. Linhares D. 2017. Economics of *Mycoplasma hyopneumoniae* elimination. Swine health and research section Allen D Leman Conference. St Paul, MN.

https://doi.org/10.54846/am2023/50

The effects of different dietary levels of supplemental vitamin D, 25-OH-D_3, and vitamin E on nursery pig performance and vitamin status

J. Bergstrom[1], PhD; S. Hough[1], DVM; K. Herkelman[2], PhD
[1]DSM Nutritional Products, Parsippany, New Jersey; [2]Cooperative Research Farms, Richmond, Virginia

Introduction

In population surveys, research studies, and diagnostic cases, the vitamin D status of pigs is often found to be relatively low at weaning and during the nursery period, and this has sometimes become a source of controversy. Most domestic pigs are now born and raised indoors without exposure to the UVB rays of sunlight (to allow for synthesis of vitamin D_3 via skin), and sows naturally provide very low levels of vitamin D in colostrum and milk. Therefore, it is not surprising that the vitamin D status of many young pigs can be low. Although low plasma levels of 25-OH-D_3 are often found in pigs suspected of having "metabolic bone disease" or vitamin D deficiency symptoms, plasma levels below the currently suggested minimum (of 18 ng/mL of 25-OH-D_3) for 3- to 4-week-old pigs can also be found in pigs without any obvious compromises in their growth or health. Nevertheless, an increased vitamin D status reduces the likelihood of vitamin D deficiency and can improve immunological functions and optimize bone development and health. However, even with common industry levels of vitamin D_3 fortification at 7- to 14-times the NRC (2012) requirement estimate of 220 IU/kg for nursery diets, the vitamin D status of nursery pigs is frequently considered to be low.

Vitamin E is another essential nutrient for swine, that is required to maintain reproductive, muscular, circulatory, nervous, and immune system functions.[1] Although sows' colostrum and milk can provide a substantial amount of vitamin E to their suckling litters, the vitamin E status of pigs typically declines sharply within the first week post-weaning. As a result, young pigs are more prone to vitamin E deficiency and associated hepatosis dietetica, mulberry heart, and muscular degeneration.[2,3] Consequently, early nursery diets are typically fortified with vitamin E at 3- to 6- times the estimated requirement suggested by the NRC (2012, 16 IU/kg diet). However, the 2022 DSM Optimum Vitamin Nutrition (OVN™) Guidelines for dietary vitamin E supplementation of nursery pigs indicates that 105 – 160 IU/kg diet is recommended to support the greater needs for growth and optimum immune function in modern genetic lines of pigs housed indoors.

Therefore, the objective of this experiment was to evaluate the effects of different levels of supplemental vitamin D, 25-OH-D_3 (from HyD®; DSM Nutritional Products, Parsippany, NJ), and vitamin E on the growth performance and vitamin status of nursery pigs.

Experimental procedures

A total of 720 PIC (800 sire × 1050 sow) weanling pigs, averaging 6.49 kg and approximately 21.5 days of age, were allotted by weight to 8 dietary treatments for the 42-day nursery experiment. A description of the treatments is presented in Table 1. Briefly, Treatment A was the "standard" nursery program with 2,000 IU/kg of vitamin D_3 and 70 IU/kg of vitamin E. Treatment B was as Treatment A with vitamin E increased to 140 IU/kg. Treatments C and D were as Treatment A and B, respectively, with HyD (providing 25-OH-D_3) fully replacing vitamin D_3 at 2,000 IU/kg. Treatment D and E were as Treatment A and B, respectively, with the vitamin D_3 levels increased to 3,500 IU/kg. Treatments F and G were as Treatments C and D, respectively, with vitamin D_3 added at 1,500 IU/kg to result in a total of 3,500 IU/kg of vitamin D from the combination of vitamin D_3 and HyD.

There were 3 diet phases during the experiment. Feed-grade antibiotics were not used. Aside from the intended differences in vitamin D and vitamin E levels between the treatments, all the diets within each phase were formulated identically to meet or exceed the NRC or PIC requirement estimates for all other nutrients. Phase 1 diets were typical, complex corn-soybean meal-based diets containing 2500 ppm added Zn (from ZnO) and were fed for the initial 7 days post-weaning. The Phase 2 diets were less-complex diets containing 2500 ppm added Zn and were fed from day 7 to 21 post-weaning, The final, Phase 3 diets were typical corn-soybean meal-based diets with 1000 ppm added Zn and were fed from day 21 to 42 post-weaning. Diets for Phase 1 and Phase 2 were manufactured as 3.175 mm pellets. Diets for Phase 3 were manufactured as a mash.

Pen was the experimental unit with 10 pigs per pen and 9 replications per treatment. Each pen within a replication contained an equal proportion of barrows and gilts. Pens had totally slatted metal floors, were equipped with a 4-hole self-feeder, and one combination nipple-cup waterer. Pen density was approximately 0.279 m^2 per pig. The pigs were weighed individually on d 0 and by pen on days 7, 14, 21, 28, 35, and 42 of the experiment. Feed consumption for each pen was determined by weighing and recording all the feed offered and subtracting the remaining feed in the feeders at the end of each weekly period. Pen weights and feed consumption were used to calculate the average daily gain (ADG), average daily feed intake (ADFI), and gain:feed (G:F) on a per pig basis for each pen and diet phase. To assess the serum status of 25-OH-D_3 and vitamin E in pigs during the experiment, 1 average weight gilt and 1 average weight barrow in each pen were initially identified for blood sampling on days 1, 21, and 42 of the experiment. After blood sampling at each time-point, the serum from each sample was kept frozen until analysis.

At the end of the experiment, the growth performance and serum data were statistically analyzed using pair-wise comparisons between treatments and as a 2 × 2 × 2 factorial using the LME procedure of R (R Core Team, 2021), with block used in the random statement. Average pen body weight on d 0 was used as a covariate for all performance parameters. The model was: u = mean + Day 0 Body Wt + Vitamin D + HyD + Vitamin E + Vitamin D*HyD + Vitamin D*Vitamin E + HyD*Vitamin E + Vitamin D*HyD*Vitamin E + Residual Error. Significance levels were reported at ($P < 0.05$) and a tendency ($P < 0.10$).

Table 1: Description of treatments

Treatment	Total vitamin D, IU/kg	Vitamin D source		Vitamin E, IU/kg
		Vitamin D_3, IU/kg	25-OH-D_3 (HyD), IU/kg	
A	2000	2000	None	70
B	2000	2000	None	140
C	2000	None	2000	70
D	2000	None	2000	140
E	3500	3500	None	70
F	3500	3500	None	140
G	3500	1500	2000	70
H	3500	1500	2000	140

Results

Growth performance

Phase 1 & 2 (day 0 to 21 post-weaning). There were no treatment interactions for growth responses during the first 21 days post-weaning (Tables 2 and 3). Pigs fed 3500 IU/kg of total Vitamin D had greater ($P \leq 0.05$) average daily feed intake (ADFI), average daily gain (ADG), and average body weight (BW) on day 21 compared to pigs fed 2000 IU/kg total Vitamin D. HyD and Vitamin E levels did not affect performance during this period.

Phase 3 (day 21 to 42 post-weaning). From day 21 to 42, pigs fed 3500 IU/kg of total Vitamin D were heavier ($P < 0.05$) at the end of the nursery period. Pigs fed 3500 IU/kg of total Vitamin D consumed more feed and grew faster ($P < 0.05$) than pigs fed 2000 IU/kg total Vitamin D, but these responses were dependent on HyD and Vitamin E levels.

Pigs fed 2000 IU/kg total Vitamin D had reduced ADG and ADFI when HyD was the source of Vitamin D fed in combination with 140 IU/kg added Vitamin E, especially when compared to pigs fed 3500 IU/kg of Vitamin D_3 combined with 140 IU/kg added Vitamin E or 3500 IU/kg total vitamin D that included HyD with either 70 or 140 IU/kg vitamin E. Pigs fed 3500 IU/kg Vitamin D_3 with 70 IU/kg Vitamin E had ADG that was similar to that of pigs fed any of the treatments with 2000 IU/kg Vitamin D (Vitamin D × HyD, $P \leq 0.07$; Vitamin D × Vitamin E, $P < 0.05$; HyD × Vitamin E, $P < 0.05$).

A three-way, Vitamin D × HyD × Vitamin E interaction ($P = 0.04$) occurred for the efficiency of feed utilization (G:F) during Phase 3. Pigs fed 3500 IU/kg of Vitamin D_3 with 140 IU/kg of Vitamin E had greater G:F than pigs fed 3500 IU/kg of Vitamin D_3 with only 70 IU/kg of Vitamin E, and had greater G:F than pigs fed 2000 IU/kg of total Vitamin D using HyD with 140 IU/kg of Vitamin E.

Overall (day 0 to 42 post-weaning). Feeding 3500 IU/kg of total Vitamin D increased ($P < 0.05$) ADG, ADFI, and BW on d 42 when compared to feeding 2000 IU/kg of total Vitamin D. Feeding 3500 IU/kg of total Vitamin D also tended ($P = 0.09$) to improve G:F; however, pigs fed 3500 IU/kg Vitamin D_3 with 70 IU/kg had performance that was intermediate to those fed any of the treatments with 2000 IU/kg total Vitamin D and those fed 3500 IU/kg Vitamin D_3 with 140 IU/kg Vitamin E or 3500 IU/kg total Vitamin D that included 2000 IU/kg Vitamin D from HyD.

Serum 25-OH-D_3 and Vitamin E status

There were no treatment interactions for serum responses during the experiment (Table 4). Feeding 3500 IU/kg total Vitamin D increased serum 25-OH-D_3 in pigs on d 21 ($P < 0.05$) and d 42 ($P < 0.10$) when compared to feeding 2000 IU/kg total Vitamin D. Supplementation using 2000 IU/kg Vitamin D from HyD resulted in further increases ($P < 0.0001$) in the serum levels of 25-OH-D_3 on d 21 and d 42.

Increasing dietary Vitamin E supplementation from 70 to 140 IU/kg improved ($P \leq 0.05$) serum Vitamin E status on d 21 and d 42. Interestingly, on d 42, pigs supplemented with 2000 IU/kg Vitamin D from HyD tended ($P = 0.08$) to have an improved serum Vitamin E status when compared to pigs fed without HyD supplementation.

Discussion and conclusions

Few studies, if any, have been performed to simultaneously evaluate different levels and forms of dietary Vitamin D with different levels of Vitamin E supplementation in nursery pig diets. In this experiment, feeding 3500 IU/kg of total Vitamin D for 42 days post-weaning improved the serum 25-OH-D_3 status of pigs when compared to feeding 2000 IU/kg total Vitamin D. However, supplementing 2000 IU/kg 25-OH-D_3 from HyD as the source of dietary Vitamin D resulted in a larger magnitude of improvement in serum 25-OH-D_3 status. Not surprisingly, feeding pigs 140 IU/kg Vitamin E resulted in improved serum Vitamin E status when compared to feeding 70 IU/kg Vitamin E. Regardless, this is likely one of the first experiments to demonstrate that supplementing Vitamin D using HyD may improve the Vitamin E status of nursery pigs, whereas, simply increasing the level of Vitamin D supplementation using Vitamin D_3 had no effect on the Vitamin E status.

Another interesting finding in this experiment was that growth performance was improved by feeding 3500 IU/kg total Vitamin D from day 0 to 42 post-weaning, but this primarily occurred when Vitamin E was increased from 70 IU/kg to 140 IU/kg if Vitamin D_3 was the sole source of Vitamin D. However, feeding 3500 IU/kg total Vitamin D using the combination of 1500 IU/kg Vitamin D_3 and 2000 IU/kg 25-OH-D_3 from HyD resulted in improved growth performance with either 70 IU/kg or 140 IU/kg Vitamin E. In a similar experiment that compared feeding 2000 IU/kg Vitamin D_3 to feeding 4000 IU/kg total Vitamin D from adding either 2000 IU/kg Vitamin D_3 or 2000 IU/kg 25-OH-D_3, with all diets formulated to contain

Table 2: Main effects means and factorial analysis of pig performance[1]

Item	Vitamin D, IU/kg (D)			HyD, IU/kg (HD)			Vitamin E, IU/kg (E)			*P*-values						
	2000	3500	SEM	0	2000	SEM	70	140	SEM	D	HD	E	DxHD	DxE	HDxE	DxHDxE
Phase 1&2 (d 0 to 21)																
Day 0 BW, kg	6.49	6.49	0.43	6.49	6.50	0.43	6.50	6.50	0.43	0.73	0.72	0.93	0.99	0.98	0.95	0.79
ADG, g	289.5[a]	305.8[b]	6.2	296.4	298.9	6.2	296.4	298.8	6.2	0.02	0.72	0.73	0.64	0.20	0.76	0.98
ADFI, g	343.3[a]	358.4[b]	5.6	349.6	352.1	5.6	350.9	350.8	5.6	0.01	0.68	0.99	0.58	0.29	0.16	0.59
G:F	0.844	0.854	0.009	0.849	0.848	0.009	0.844	0.853	0.009	0.43	0.91	0.51	0.19	0.41	0.18	0.47
Day 21 BW, kg	12.66[a]	12.95[b]	0.12	12.80	12.80	0.12	12.80	12.80	0.12	0.05	0.97	0.96	0.43	0.23	0.90	0.82
Phase 3 (d 21 to 42)																
ADG, g	552.9[a]	601.5[b]	19.8	581.2	573.2	19.7	576.0	578.4	19.7	< 0.001	0.41	0.80	0.07	0.005	0.002	0.43
ADFI, g	865.3[a]	927.2[b]	12.7	899.0	893.4	12.7	892.9	899.5	12.7	< 0.001	0.62	0.56	0.004	0.02	0.02	0.31
G:F	0.637	0.648	0.013	0.645	0.641	0.013	0.643	0.642	0.013	0.14	0.57	0.86	0.59	0.05	0.02	0.04
Day 42 BW, kg	24.28[a]	25.60[b]	0.50	25.03	24.85	0.50	24.90	24.98	0.50	< 0.001	0.53	0.78	0.11	0.12	0.01	0.41
Overall (d 0 to 42)																
ADG, g	418.8[a]	452.7[b]	12.4	436.5	435.0	12.4	433.9	437.6	12.4	< 0.001	0.81	0.57	0.12	0.18	0.009	0.66
ADFI, g	600.1[a]	641.1[b]	9.2	620.3	621.0	9.2	617.9	623.4	9.2	< 0.001	0.93	0.46	0.05	0.23	0.01	0.21
G:F	0.697[x]	0.706[y]	0.010	0.703	0.700	0.010	0.701	0.702	0.010	0.09	0.55	0.88	0.99	0.31	0.20	0.007

[1] Initial body weight used as a covariate for all performance parameters.
[a,b] Treatment difference within a main effect ($P < 0.05$).
[x,y] Treatment difference within a main effect ($P < 0.10$).

Table 3: Pig performance by individual treatment

Vitamin D_3, IU/kg:	**2000**	**2000**	**0**	**0**	**3500**	**3500**	**1500**	**1500**		
HyD, IU/kg equiv.	**0**	**0**	**2000**	**2000**	**0**	**0**	**2000**	**2000**		
Vitamin E, IU/kg:	**70**	**140**	**70**	**140**	**70**	**140**	**70**	**140**		
Treatment:	**A**	**B**	**C**	**D**	**E**	**F**	**G**	**H**	**SEM**	***P*-value**
Phase 1&2 (d 0 to 21)										
Day 0 BW, kg	6.49	6.49	6.49	6.49	6.49	6.49	6.49	6.49	0.43	0.28
ADG, g	283.1	296.7	284.5	293.7	305.2	300.5	312.9	304.3	10.5	0.38
ADFI, g	331.6	349.2	348.8	343.5	359.5	358.2	363.8	352.3	9.1	0.18
G:F	0.857	0.849	0.811	0.857	0.851	0.841	0.859	0.865	0.018	0.55
Day 21 BW, kg	12.63	12.78	12.51	12.69	12.95	12.82	13.12	12.89	0.21	0.51
Phase 3 (d 21 to 42)										
ADG, g	566.6^{bc}	564.9^{bc}	564.9^{bc}	515.2^{c}	561.7^{bc}	631.5^{a}	610.7^{ab}	602.0^{ab}	23.0	< 0.0001
ADFI, g	875.6^{bc}	894.1^{ab}	874.6^{bc}	816.7^{c}	889.1^{ab}	937.3^{ab}	932.4^{ab}	949.9^{a}	18.7	< 0.0001
G:F	0.644^{ab}	0.630^{ab}	0.645^{ab}	0.629^{b}	0.629^{b}	0.674^{a}	0.653^{ab}	0.635^{ab}	0.016	0.03
Day 42 BW, kg	24.53^{abc}	24.65^{abc}	24.38^{bc}	23.56^{c}	24.75^{abc}	26.17^{a}	25.94^{ab}	25.53^{ab}	0.60	0.0002
Overall (d 0 to 42)										
ADG, g	419.9^{cd}	429.5^{abcd}	422.8^{bcd}	403.1^{d}	432.0^{abcd}	464.7^{a}	460.9^{ab}	453.1^{abc}	14.7	0.0001
ADFI, g	594.8^{b}	619.3^{ab}	608.8^{ab}	577.7^{b}	621.7^{ab}	645.4^{a}	646.4^{a}	651.0^{a}	12.7	< 0.0001
G:F	0.704	0.692	0.694	0.697	0.694	0.720	0.711	0.697	0.012	0.07
Morbidity + mortality, %	6.67	2.22	2.22	2.22	2.22	3.33	1.11	0.00	1.72	0.25

[abcd] Means within a row without a common superscript are different ($P < 0.05$).

Table 4: Main effects means and factorial analysis of serum 25-OH-D_3 and vitamin E levels

Item	Vitamin D, IU/kg (D)			HyD, IU/kg (HD)			Vitamin E, IU/kg (E)			P-Values						
	2000	3500	SEM	0	2000	SEM	70	140	SEM	D	HD	E	DxHD	DxE	HDxE	DxHDxE
25-OH-D3, ng/mL[1]																
Day 1	4.30	4.09	0.24	4.05	4.34	0.24	4.07	4.32	0.24	0.52	0.37	0.47	0.54	0.17	0.82	0.70
Day 21	22.57^{a}	25.95^{b}	0.86	14.53^{a}	34.00^{b}	0.85	24.86	23.67	0.84	0.002	< 0.0001	0.30	0.70	0.63	0.62	0.26
Day 42	30.22^{x}	32.99^{y}	1.14	20.33^{a}	42.89^{b}	1.12	32.22	31.00	1.14	0.10	< 0.0001	0.43	0.47	0.88	0.44	0.94
Vitamin E, mcg/mL[2]																
Day 1	1.24	0.95	0.11	1.10	1.10	0.11	1.13	1.07	0.11	0.03	0.99	0.65	0.55	0.60	0.42	0.15
Day 21	1.12	1.19	0.06	1.12	1.19	0.06	1.06^{a}	1.25^{b}	0.06	0.40	0.35	0.03	0.82	0.67	0.60	0.16
Day 42	0.92	0.89	0.08	0.81^{x}	1.00^{y}	0.08	0.79a	1.02^{b}	0.08	0.73	0.08	0.05	0.46	0.70	0.69	0.61

[1] Initial Vitamin D level used as a covariate.
[2] Initial Vitamin E level used as a covariate.
[a,b] Treatment difference within a main effect ($P < 0.05$).
[x,y] Treatment difference within a main effect ($P < 0.10$).

only 30 IU/kg Vitamin E, performance of pigs was only improved when the higher level of total Vitamin D supplementation was achieved using the combination of Vitamin D_3 and 25-OH-D_3 from day 0 to 42 post-weaning.[4]

In conclusion, this research indicates that increasing the total Vitamin D from 2000 to 3500 IU/kg in nursery diets may improve performance of nursery pigs, but an increased supplementation of Vitamin E may simultaneously be required to improve performance unless 2000 IU/kg 25-OH-D_3 is included to achieve a total Vitamin D level of 3500 IU/kg. Using 25-OH-D_3 from HyD as the source of dietary Vitamin D will improve serum 25-OH-D_3 status when compared to feeding only Vitamin D_3 and may also improve the Vitamin E status of pigs. More research is needed to understand the relationship of 25-OH-D_3 and Vitamin E in pigs.

References

1. Combs, G. F., Jr. The Vitamins, 4th Edition, Academic Press, 2012.

2. Ullrey, D. E. Vitamin E for Swine. Journal of Animal Science, 1981, 53(4):1039-1056.

3. Nafstad, I; and S. Tollersrud. The Vitamin E-Deficiency Syndrome in Pigs: I. Pathological Changes. Acta Veterinaria Scandinavica, 1970, 11, 452-480.

4. Zhou, X.; et al. Dietary Supplementation of 25-Hydroxyvitamin D3 Improves Growth Performance, Antioxidant Capacity and Immune Function in Weaned Piglets. Antioxidants 2022, 11, 1750. **https://doi.org/10.3390/antiox11091750**

https://doi.org/10.54846/am2023/51

Clinical manifestations of mycotoxin contaminated feed in swine

S. Hough[1], DVM; L. Zheng[1], PhD; P. Gott[1], PhD; J. Bergstrom[1], PhD; D. Johnson[2], PAS
[1]DSM Animal Nutrition and Health; [2]Hueber Foods LLC

Introduction

Mycotoxins are toxic secondary metabolic products of molds, which damage crops. With feed accounting for 70% of the cost of production, mycotoxin contamination can cause huge economic losses, both in losses in feed quality and animal performance.[1] This makes mycotoxins one of the many challenges pig producers have to overcome. In addition to impairing animal performance, mycotoxins can increase susceptibility to disease and reduce vaccine efficiency.[2] Diseases related to consuming contaminated feed are labeled as mycotoxicosis, and the systems affected vary by toxin, with potential impacts on liver or kidney function, gastrointestinal and/or respiratory health, and reproductive performance (Figure 1).[3] (Mycotoxins can also have subclinical implications and can play as predisposing factors of disease. Certain toxins, alone or in combination, have been shown to increase disease severity of pigs infected with porcine epidemic diarrhea (PEDV), increase intestinal colonization of bacterial pathogens, perpetuate the effects of porcine reproductive and respiratory sydrome (PRRSV) and reduce reproductive efficiencies in sows.[4-7] With their potential influence on pig health and welfare, veterinarians need to be aware of clinical expression when present.

Biomin, now part of DSM Animal Nutrition and Health, has conducted an annual global mycotoxin survey since 2004. As part of the global survey, corn grain and other corn-based ingredients are reported for the United States. Samples are screened at Romer Labs (Union, MO) via liquid chromatography with tandem mass spectrometry (LC-MS/MS) method for six major mycotoxin groups: aflatoxins (Afla), type A trichothecenes (A-Trich), type B trichothecenes (B-Trich), fumonisins (FUM), zearalenone (ZEN), and ochratoxin A (OTA). The 2022 crop year data is just becoming available (n = 44) as harvest is still in progress in the writing of this abstract. To date, 97% of the samples evaluated have contained at least one mycotoxin, compared to 90% that was observed in 2021. The percentage of Afla contaminated samples increased in 2022 corn compared to the 2021 harvest years (42 vs. 7%). The percentage of FUM contaminated samples in 2022 corn is higher compared to 2021 (88 vs. 64%). The preliminary results of the 2022 US corn crop survey indicate mycotoxin occurrence and contamination levels are similar or higher (Afla and FUM) to the prior crop year and will continue to challenge animal health and performance as these ingredients are fed out.

Over 90% of the mycotoxins in the feed are produced mainly in the field (ex, FUM, Trichothecenes, ZEN) and the rest are results of storage conditions (ex, Afla).[1] There are optimum growing temperatures and humidity levels for each type of mold and some feed types/cereals are more prone to be contaminated versus others. Several factors affect the clinical signs and severity of mycotoxin toxicity in pigs, like type of mycotoxin, level of mycotoxin in the feed, duration of exposure, age of the pig and production phase.[8] Young pigs and breeding swine are the most susceptible and combinations of certain mycotoxins may be more harmful than individual mycotoxins.[9] This should be considered when evaluating levels and associated clinical issues. Pigs are also a species that is very sensitive to mycotoxins, especially due to their high consumption of potentially contaminated cereal grains.[2]

Case examples

Case 1

In late November 2021, a nutritionist was visiting a 5600-sow farm during his quarterly visit and noticed a substantial number of weak pigs post farrowing. After the farm walk through, he opted to take corn samples (whole and ground) from the producer-owned feed mill. The mill operator said he had a sample of corn he received 3 days prior to the visit. The corn sample was poor quality with damage and had the appearance of black mold. Samples were sent out for mycotoxin testing. The initial results in whole corn were 135.96 ppb Afla, 75.76 ppm FUM and 2.43 ppm DON. The farm had just completed a *Mycoplasma hyopneumoniae* elimination project and was also considered PRRS and swine influenza A negative. No clinical issues were apparent in the sows, but piglet mortality started to increase within 48 hours of birth and the herd veterinarian visited within a week. A workup for infectious causes of an increase in pre-wean mortality yielded no etiologies and complete feed samples from lactation, which included corn oil, were found to contain 7.2 ppb Afla and 20.4 ppm FUM. An in-feed mitigation strategy was started within 2 weeks of discovery at a low dose and corn oil was replaced with choice white grease. With continued piglet losses and clinical issues in farrowing, the mitigation additive inclusion was doubled at about 4 weeks post-exposure. After 32 days of exposure, the contaminated corn was gone, and new corn was added to the bin. The testing results from the new corn were 0 ppb Afla and 0.72 ppm FUM. It took just over 50 days for the piglet livability to return to normal for that particular site from initial clinical issues in pigs until resolution. Looking back at the timeline, an increase in pre-wean mortality started with sows that were exposed the levels of mycotoxin contamination listed in the original corn sample at approximately 105 days of gestation. Losses were reduced dramatically when sows, that were exposed at day 56 of gestation and less, farrowed. The only pathological finding was idiopathic embryonic liver damage in piglets.

Case 2

A nursery in Austria that makes its own feed onsite had an issue with mycotoxins in 2021. The case report was put together by DSM colleagues in Europe. There were 1000 pigs on site that were about 60lbs, and the farm ran out of feed. Pigs (500) remained on the onsite feed, while the other 500 were fed purchased feed. Within one day of the diet change, there was restlessness noted, widespread tail biting and some occasional ear biting in the pigs with the purchased feed. Thirty percent clinical incidence tail biting and 4% mortality was observed within days. The complete feed was tested for mycotoxins and yielded 1030 ppb DON. They

Figure 1: Mycotoxicosis in swine.

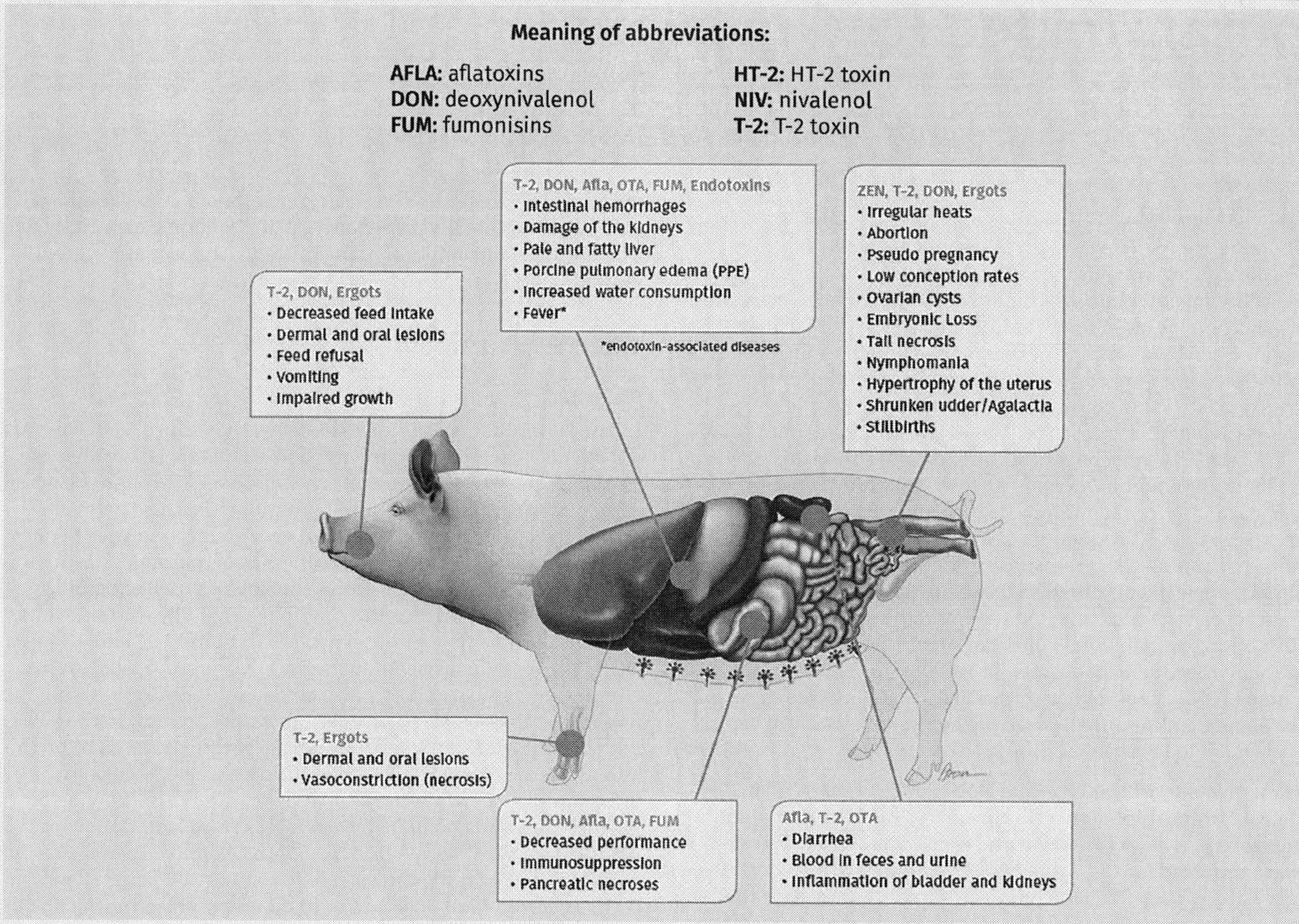

were able to salvage the pigs by treating some with antibiotics and changing the feed. Clinical incidence of the vices mentioned above went away almost immediately with a change in feed. The purchased feed was made with cereal grains harvested a year prior and had been in storage. No tail biting was observed in the pigs that did not receive any of the purchased feed.[10] The pathogenesis of tail biting is not well understood and causes multifactorial.[11] It does appear that off feed events can lead to various vices so there is a relationship of mycotoxin contamination and clinical expressions are seen in this case.[1,12]

Discussion

Consumption of mycotoxins causes clinical signs such as feed refusal and diarrhea, impacts the immune system, and makes pigs more susceptible to diseases as outlined above. The co-contamination with multiple mycotoxins in feed can cause more adverse effects than a single mycotoxin due to synergism. Furthermore, limited practice on mitigating low-dose mycotoxins may cause chronic exposure and impact production efficiency. Due to the continued challenge of mycotoxins on animal health and performance, understanding the prevalence of mycotoxins and applying mitigation strategies is critical.

DSM has a new resource that you can pick up at the exhibit booth titled Differential Diagnosis for Mycotoxicosis in Swine. This booklet outlines mycotoxins' potential role in swine health by covering nine conditions (reduced feed intake, immune modulation, vomiting, rectal prolapse, ear/tail/skin necrosis, conjunctivitis, icterus, reproductive failure, and sudden death). Practitioners should know when to consider mycotoxins, and how to workup some of these perplexing cases.

Considering the frequent occurrence of mycotoxins in feeds and the potential negative impact these mycotoxins have on animal production following exposure, mitigation options should be considered. Veterinarians need to work with their nutrition team to determine the best program for their herds. Strategies for mitigation will be covered in the in-person presentation, but one must consider the chemical structure of each mycotoxin and the potential to be bound with conventional products at the relatively harsh pH of the gastrointestinal tract. Biotransformation or degradation is an alternative detoxification method which can address mycotoxins that have limited binding potential. Fumonisin esterase, FUMzyme®, is a novel, purified enzyme that was developed to specifically and irreversibly degrade fumonisins into non-toxic metabolites, thus limiting the potential impact of fumonisins on animals. FUMzyme is the first mycotoxin degrading ingredient to go through FDA's Food Additive Petition process and is the first product ever to be approved for degrading fumonisins in poultry and swine feed in the US market. Fumonisin esterase is now available in the US as part of the products Biofix® Plus with FUMzyme® and Biofix® Select with FUMzyme®.

Acknowledgments

Authors would like to acknowledge Romer Labs for feeding mycotoxin testing and thank those willing to share case examples.

References

1. Biomin Mycotoxin Compendium (2018). Biomin Holding GmbH, Erber Campus 1, Getzersdorf, Austria

2. A. Pierron, A., I. Alassane-Kpembi, I.P. Oswald. Impact of mycotoxin on immune response and consequences for pig health *Animal Nutrition*. 2 (2016) 63-68.

3. Iowa State University College of Veterinary Medicine Veterinary Diagnostic and Production Animal Medicine Disease Topics: Mycotoxins. Accessed November 8, 2022. **https://vetmed.iastate.edu/vdpam/research/disease-topics/toxicology/mycotoxins**

4. Liu, D., L. Ge, Q. Wang J. Su, X. Chen, C. Wang, K. Huang. Low-level contamination of deoxynivalenol: A threat from environmental toxins to porcine epidemic diarrhea virus infection. *Environment International*. 142 (2020) 105949.

5. Oswald, I.P, C. Desautels, J. Laffitte, S. Fournout, S.Y. Peres, M. Odin, P. Le Bars, Le Bars P, J.M. Fairbrother. Mycotoxin fumonisin B1 increases intestinal colonization by pathogenic Escherichia coli in pigs. *Applied Environmental Microbiology*. 69:10 (2003) 5870-5874.

6. Rückner, A., L. Plagge, K. Heenemann, M. Harzer, B. Thaa, J. Winkler, S. Dänicke, J. Kauffold, T. W. Vahlenkamp. The mycotoxin deoxynivalenol (DON) can deteriorate vaccination efficacy against porcine reproductive and respiratory syndrome virus (PRRSV) at subtoxic levels. *Porcine Health Management*. (2022) 8:13.

7. Young, L.G, H. Ping, G.J. King. Effects of feeding zearalenone to sows on rebreeding and pregnancy. *Journal of Animal Science*. 68:1 (1990) 15-20.

8. Whitney, M. University of Minnesota Swine Extension Website. Swine Nutrition: Mycotoxins harm swine health and reproductive performance. Accessed November 8, 2022. **https://extension.umn.edu/swine-nutrition/mycotoxins-harm-swine-health-and-reproductive-performance**

9. Andretta, I. M. Kipper, L. Hauschild, C.R. Lehnen, A. Remus and R. Melchior. Meta-analysis of individual and combined effects of mycotoxins on growing pigs. *Scientia Agricola*. 73:4. (2016) 328-331.

10. Behler-Woechtl and Nies (2021): Schwanzbeißen in einem spezialisierten Ferkelaufzuchtbetrieb im Zusammenhang mit Deoxynivalenol ein Fallbericht, KTP (29) 131-134– Translated - Tail biting in a specialist piglet rearing farm in connection with deoxynivalenol - a case report.

11. Widowski, T. (2002): Causes and prevention of tail biting in growing pigs: a review of recent research. In London Swine Conf., 47.

12. Baker, S., S. Hough, E. Jablonski, J. Angulo, E. Nemechek. Development of a scoring system to determine the prevalence, severity, and associated risk factors of tail-biting and lameness. *AASV Proceedings* (2019).

https://doi.org/10.54846/am2023/52

Improving sow livability through selection for better structure and reproductive health

J. R. Dunkelberger[1,2], PhD; T. Stevens[1], MSc; E. F. Knol[1], PhD
[1]Topigs Norsvin Research Center B.V., Beuningen, The Netherlands;
[2]Topigs Norsvin USA, Burnsville, Minnesota

Introduction

Sow mortality has drastically increased for swine breeding herds in recent years. The top three reasons for sow mortality are: 1) unknown/sudden death; 2) feet/leg structure; and 3) uterine prolapse.[1] Previous research conducted by our group provides evidence of a genetic component for susceptibility to uterine prolapse. This conclusion is based on analysis of over 16,000 removal records and three generations of pedigree data, for which heritability of uterine prolapse was estimated at 22%.[2,3] Results from a recent follow-study show that, when using genomic (rather than pedigree) data to account for relatedness, heritability was even higher than previously realized (35%).[4] This finding indicates that genetic selection can be used as part of a multidisciplinary approach to reduce the incidence of this complex issue.

While genetic variation in susceptibility to uterine prolapse presents an encouraging opportunity to address a major reason for sow death loss through the breeding program, previous investigation by our group provides evidence that feet/leg structure and sow longevity in general, can also be improved through genetic selection. For example, heritability of osteochondrosis has been estimated at 31%,[5] longevity to parity 2 at 8%,[6] and longevity to parity 5 at 13%[6]. However, pairwise correlations between these traits have not yet been reported, nor have results to demonstrate the efficacy of breeding values for these traits to predict removal in the field. Therefore, the objectives of this study were to: 1) estimate genetic correlations between osteochondrosis and longevity; and 2) to validate that breeding values for these traits are predictive of removal at the commercial level.

Materials and methods

Data

Removal records collected on 6,487 purebred Topigs Norsvin females were used for analyses (Dataset A). Records were collected between 2012 and 2020 from a single commercial multiplication facility located in the US. Available phenotypes included removal status and reason for removal. Removal was analyzed as a binary trait, where the phenotype for removal due to a leg problem was coded as "1", vs. removal due to any other reason as "0". In this dataset, a "leg problem" was defined as any one of the following removal reasons: arthritis, joint infection, damaged/bruised legs, broken leg, abscesses on the legs, leg position (front or hind legs), or leg deformity. Only sows removed between parity 1 and 4 were used for analyses. Using all 6,487 available records, data was parsed into the following subsets, consisting of sows removed: between parity 1 and 2 (P2, n = 2,853), parity 1 and 3 (P3, n = 4,776); and parity 1 and 4 (P4, n = 6,487).

Estimated breeding values (EBVs) for each sire of the 6,487 sows in Dataset A were available for the following traits: osteochondrosis (OCD), longevity to parity 2 (LGY2), and longevity to parity 5 (LGY5). Sire EBVs were made available from routine genomic evaluation. During this process, EBVs can be estimated for an individual even when phenotypic information is lacking, so long as the individual's genotype information is available. Therefore, although phenotypes for LGY2 and LGY5 are only collected on females, EBVs for these traits can be estimated for sires based on their genomic information. A breeding value for a given trait represents the genetic merit of the individual for that trait, relative to the population mean. Thus, an EBV can be expressed as either a positive or negative value and is represented using the same units in which it is recorded. Phenotypes for OCD are assigned based on computerized tomography-derived images, according to the approach described by Aasmundstad et al.[5] For this trait, a score of "0" indicates no evidence of osteochondrosis, vs. a score of "23", which is indicative of severe osteochondrosis. Therefore, a higher EBV for OCD indicates increased susceptibility to OCD. For the longevity traits, a phenotype of "1" is assigned if the sow survives to parity 2 or 5 (for LGY2 and LGY5, respectively) or "0" if the sow was removed from the herd prior to that parity. Accordingly, a higher EBV for LGY2 or LGY5 is representative of increased chance of survival.

Genetic correlations

Pairwise correlations between EBVs for OCD, LGY2, and LGY5 were estimated to assess the magnitude and direction of the genetic relationship between traits. All available EBVs from routine genomic evaluation were used for this analysis.

Regression analysis

A regression analysis was conducted to assess the ability of sire EBV for osteochondrosis and longevity to predict offspring performance for sows in Dataset A, which were reared and housed in a commercial environment. For a given sire, offspring performance was calculated as the % of his daughters removed due to a leg problem, out of the total number of his daughters removed from the herd (due to any reason). Three values for offspring performance were calculated per sire, based on including sows removed between parity 1 and 2, 1 and 3, or 1 and 4 in the analysis, as described above. Offspring performance was regressed on sire EBV for OCD, LGY2, and LGY5 to estimate the correlation between offspring performance with sire EBV for each of these traits. Only sires with 30 or more daughters (aka "Reliable Sires", n = 112) were used for analyses.

Results and discussion

The correlation between EBVs for osteochondrosis and longevity were estimated at -0.24 for OCD/LGY2 and -0.28 for OCD/LGY5. These results indicate that OCD is a genetically distinct trait from LGY2 and LGY5, sharing only a moderately similar genetic basis. These results also indicate that selection for lower osteochondrosis score is expected to increase longevity to parity 2 and longevity to parity 5. The correlation between EBVs for LGY2 and LGY5 was high and positive (0.93), as expected, indicating that LGY2 and LGY5 are genetically similar traits. Therefore, selection for increased longevity to parity 2 is also expected to increase longevity to parity 5.

Overall percent removal due to a leg problem was 17.1%, 12.2%, and 13.7% for analysis of datasets P2, P3, and P4, respectively. Similar values (18.2%, 12.7%, and 13.8%) were obtained when analyzing datasets P2, P3, and P4 including records collected on daughters of Reliable Sires only (Table 1). The average sire EBV for OCD was also calculated for sows removed due to a leg problem vs. another reason (Table 1). Regardless of the number of records used for analyses (all records vs. only records for daughters of Reliable Sires) or timeframe of removal, the average EBV for OCD was consistently higher (less favorable) for sows removed due to a leg problem vs. sows removed due to another reason, as expected.

Results of the regression analysis show that the correlation between offspring performance with sire EBV for OCD was 0.18 (P2), 0.20 (P3), and 0.27 (P4). These results indicate that a higher (less desirable) sire EBV for OCD is associated with a higher rate of removal due to a leg problem among his daughters, as expected. The correlation between offspring performance and EBV for LGY2 was -0.22 (P2), -0.21 (P3), and -0.18 (P4). Similar estimates were obtained for the correlation between offspring performance and EBV for LGY5 at -0.15 (P2), -0.13 (P3), and -0.08 (P4). These results indicate that a higher sire EBV for longevity (either LGY2 or LGY5) is associated with a lower rate of removal due to a leg problem among his daughters, also as expected.

In summary, results from this study show that genetic selection for reduced osteochondrosis and better longevity is associated with lower removal rate due to a leg problem among sows reared and housed in a commercial environment. Correlations between EBVs show that, while OCD and LGY2/LGY5 are genetically distinct traits, they are favorably correlated. Therefore, selection for reduced osteochondrosis is expected to increase sow longevity to parity 2 and longevity to parity 5. Results of the regression analysis provide evidence of the efficacy of these EBVs in the field. For instance, sires of sows removed due to a leg problem had a higher (less favorable) average EBV for OCD than sires of sows removed due to another reason. In summary, results from this exercise show that the less desirable a sire's EBV for OCD, LGY2, or LGY5, the higher the probability that his daughter will be removed due to a leg problem.

Taken together, results from this study, in combination with results from previous research conducted by our group, address the genetic basis for sow longevity, including both of the top two specific reasons (uterine prolapse and feet/leg structure) for sow mortality in the US today. We have demonstrated that uterine prolapse, osteochondrosis, and sow longevity are heritable traits, indicating that these traits can be improved via genetic selection. Results from this study, and previously published work from our group, demonstrate the efficacy of EBVs to predict susceptibility to uterine prolapse, or removal due to a leg problem for sows reared and housed in a commercial US environment. Therefore, genetic selection is an important tool that should be used to address this critical, multifactorial issue in modern sow production.

References

1. Ross J. Identification of putative factors contributing to pelvic organ prolapse in sows. Improving Pig Survivability. 2019.
2. Stevens T, Dunkelberger JR, and Knol EF. Heritability and Validation of Sow Uterine Prolapse in the United States. *J Anim Sci.* 2021;99(E-Suppl. 3).
3. Dunkelberger JR, Stevens T, and Knol EF. Heritability of sow uterine prolapse in a commercial maternal line. *Proc of the 53rd Annual Meeting of the AASV.* 2021.
4. Bhatia V, Stevens T, Dunkelberger JR, et al. Confirmation of the Genetic Basis of Sow Uterine Prolapse Using Genomic Data. *Proc of the Leman Swine Conference.* 2022.
5. Aasmundstad T, Kongsro J, Wetten M, et al. Osteochondrosis in pigs diagnosed with computed tomography: heritabilities and genetic correlations to weight gain in specific age intervals. *Anim.* 2013;7(10):576–1582.
6. Iversen MW, Nordbø Ø, Gjerlaug-Enger E, et al. Predicting survival and longevity of sows using pure- and crossbred data. *Trans Anim Sci.* 2020;4(2):993-1005.

Table 1: Average sire estimated breeding value (EBV) for osteochondrosis (OCD) by timeframe of removal

Data used for analyses	**P[a]2**		**P3**		**P4**	
	N	**Mean[b]**	**N**	**Mean**	**N**	**Mean**
All sows[c]						
Leg problem[d]	488	1.02	234	1.09	234	1.04
Other reason	2,365	0.81	1,689	0.86	1,477	0.92
Daughters of reliable rires only[e]						
Leg problem	471	1.01	224	1.08	223	1.03
Other reason	2,119	0.81	1,534	0.84	1,395	0.92

[a] P: Sows removed between parity 1 and 2 (P2), 1 and 3 (P3), or 1 and 4 (P4).

[b] Mean: Average sire EBV for OCD.

[c] All sows: All 6,487 available removal records used for analyses.

[d] Leg problem: Any one of the following removal reasons: arthritis, joint infection, damaged/bruised legs, broken leg, abscesses on the legs, leg position (front or hind legs), or leg deformity.

[e] Daughters of Reliable Sires only: Only removal records for daughters of Reliable Sires (sires with 30 or more progeny in the dataset) used for analyses.

https://doi.org/10.54846/am2023/53

The PigFlow® individualized pig care and herd diagnostics platform

Matthew Rooda; Ron Ketchem
SwineTech, Inc., Solon, Iowa

Introduction

Significant losses have been attributed to sow and piglet mortalities. The global swine industry is continuing to see a rise in overall unexpected sow mortality at an average of ~15%, according to Chris Rademacher, DVM, extension swine veterinarian, Iowa State University. Additionally, 1.08 piglets per litter are found to be dead due to complications during the birthing process (stillborns). Once born, the first three days of the piglet's life represent the greatest danger of death. Approximately 70% of these piglet deaths occur during this time frame and according to the MetaFarms Production Index roughly 13.5% of all pigs born alive will not survive.

Furthermore, labor turnover and access to skilled labor are plaguing the US Swine Industry. The National Pork Board 2016 - 2017 "Employee Compensation & HR Practices in Pork Production" report shows the average reported turnover rate for animal caretakers on large and small to midsize producers was 35.03% and 20.34%. In many cases employee turnover will exceed 100%. This is a very costly problem for producers representing 4.05 billion US dollars in annual losses. This high turnover rate is also resulting in an annual loss of on-site knowledge, experience, and skill sets. It is leading to greater variability in processes and pig care.

Our greatest challenge is that as an industry we are unaware and unable to measure what is being completed daily, who is completing it, and whether it was completed correctly as indicated by company SOPs. A solution is needed to bring more organization, communication, and visibility to daily operations so that farms can efficiently and effectively train new team members, focus on what matters most, and improve workflow efficiencies, process compliance, and daily pig care.

Research and methods

Industry interviews were conducted with 33 of the top 40 producers in the US, and 93.9% of these producers agreed that there is a significant lack of visibility into daily workflows, process compliance, and pig cares. In response to these interviews, 167 interviews were conducted with on farm personnel to understand the key stressors and issues that mattered to them. There were 103 English speaking participants and 64 Spanish speaking participants. Each participant was interviewed in their native tongue. The answers were consistent from farm to farm and fell into the following categories:

- Desire a more organized and timely onboarding experience
- Communication throughout the workplace is poor due to language barriers
- It is not clear throughout the day where people are and what needs to happen next
- Many tasks were inefficient in both preparation and execution
- The use of notepads causes information to remain invisible to the greater team
- Traditional handhelds offer no value to employees on the farm
- A strong desire to see key performance indicators tied to their individual performance

These responses indicated an opportunity to improve the quality of life of employees while also driving a more efficient workplace with improved pig cares. It was deemed that in order to properly address process compliance in sow farms a solution was needed for these employees that actually makes their quality of life better.

Results and discussion

In response to these findings a group of 11 team members at SwineTech, Inc. and a couple dozen industry members which included owners, veterinarians, production supervisors, farm managers, and on farm employees collaborated in the development of the PigFlow Platform. Inspired by workforce management solutions in long-term healthcare facilities, PigFlow is a specialized communication tool that enables an optimized flow of information that delivers actionable intelligence to the person most able to act in a timely enough manner to achieve the best outcome at every level of business. It also provides a record of both the communication and the resulting action in order to evaluate the effectiveness of decisions and processes. This has led to improvements in labor time to productivity, pig survivability, and the implementation and management of precision technologies. PigFlow is currently managing over 150,000 sows in the US and Canada on sow farms from 400 sows to 12,000 sows. Producers have reported a reduction in labor needs, stillborns, pre-wean mortality, and sow mortality. In addition to these benefits, teams are more connected than ever and have reported to be more organized and less stressed throughout the day. It is our expectation that over time we will continue to see sustained improvements in employee onboarding, development, and retention.

In addition to these impacts, this presentation will be primarily focusing on the pig care feature developed in collaboration with many reputable swine veterinarians, Johnsonville Foods, and Genus PIC to help producers accomplish the following:

- Creation of custom treatment regimens for both medical and non-medical interventions.
- Complex yet easy to use workflows for farm staff.
- Measurement of treatment and vaccination compliance.
- Ability to highlight if a selected anti-microbial isn't working for the identified disease process.
- Prompt identification of sudden mortality spikes to minimize ongoing losses.
- Early identification of mortality trends that may be indicating early signs of disease.
- Monitor the introduction of autogenous vaccines to replace medical treatments.

References

1. Ketchem, Ron. "70% Of Stillborns Come from 20% of the Sows; Who Is at Risk?" *National Hog Farmer,* 5 June 2019, **https://www.nationalhogfarmer.com/reproduction/70-stillborns-come-20-sows-who-risk**

2. "MetaFarms Production Index Analysis." MetaFarms, 2020.

3. Wilson, Keith. "Sow Mortality Frustrates Experts." *National Hog Farmer,* 4 Feb. 2016, **https://www.nationalhogfarmer.com/mag/farming_sow_mortality_frustrates**

https://doi.org/10.54846/am2023/54

Precision pig weights using LeeO™

A. Jenkins[1], MS; Z. Rao[1], MS; J. Gebhardt[2], DVM, PhD; J. Woodworth[1], PhD; M. Tokach[1], PhD; J. DeRouchey[1], PhD; R. Goodband[1], PhD; T. Holck[3], DVM, MS, MBA; Joel Stave[4], BA
[1]Department of Animal Sciences and Industry, Kansas State University, Manhattan, Kansas; [2]Department of Diagnostic Medicine/Pathobiology, College of Veterinary Medicine, Kansas State University, Manhattan, Kansas; [3]Feed His People LLC, Gilbert, Iowa; [4]Prairie Systems, Spencer, Iowa

Introduction

LeeO is an animal technology system focused on traceability and individualized data collection. This technology utilizes radio frequency identification (RFID) tags to identify animals and subsequently record individual weights, injections, and a plethora of other actions using a LeeO O reader and Bluetooth technology. LeeO scales can be adjusted based on system goals for two different stabilization settings, one setting is measured in grams and the other is measured in milliseconds. The pig weight that is accepted by the technology is only locked in after staying within the set weight range (g) for the set amount of time (mSec). In theory, a higher stabilization weight or a shorter stabilization time can shorten the time required for the weight to be locked in. Conversely, increasing the stabilization time or lowering the stabilization weight may allow for a more accurate and/or precise weight collection.

Accuracy refers to how close a measurement is to its true value. While precision refers to how close a set of measurements is to one another. Accuracy can be difficult to assess in real-world settings because it is often not known what the true animal weight is. Precision is easier to obtain, as it is simply the measure of variation as measurements are collected repeatedly on the same animal. Increasing precision is paramount in animal research as reducing between-animal variation is one of the underlying goals of animal research. The primary objective of this trial was to determine the most beneficial combination of scale head settings for an intensive research system to balance accuracy, precision, and efficiency of data collection in nursery and finishing pigs.

Materials and methods

Nursery LeeO scales are preprogrammed with a 100 g stabilization weight setting and a 500 mSec stabilization time setting. In the nursery the following settings were used: 50, 100, or 200 g for stabilization weight and 250, 500, or 1,000 mSec for stabilization time. Finishing LeeO scales are preprogrammed with a 500 g stabilization weight setting and a 1,000 mSec stabilization time setting. In the finisher the following settings were used: 250 or 500 g for stabilization weight and 1,000 or 2,000 mSec for stabilization time. There is a third LeeO scale setting which can be adjusted, this is the weight confirmation delay. The weight confirmation delay is the amount of time that the recorded weight stays on the screen during which time the operator can choose to confirm the weight or reweigh the pig. LeeO scales come preprogrammed with a 3 second weight confirmation delay, this setting was used for all weights taken in both the nursery and finishing portions of this trial.

Before research with animals began, the accuracy and precision of each combination of settings was determined using a 25-pound test weight in the nursery and a 50-pound test weight in the finisher. To accomplish this, the test weight was weighed 5 times on each of the predetermined combinations of scale settings. The coefficient of variation for that setting was calculated as the measure of precision. For accuracy, the absolute difference of each of the weights obtained by the scale for that combination of settings from the true weight of that test weight was determined. Since there were only 1 replication for calculating CV, statistics were not ran for precision. A total of 30 nursery pigs (5.17 + 0.2 kg) and 18 finishing pigs (86.97 + 2.0 kg) were used in this study. Each pig in the study was weighed 5 times on each of the combinations of stabilization settings for a total of 45 weights for each nursery pig and 20 weights for each finishing pig. The order of settings was randomized for each animal. All weights within each setting combination were collected within 20 minutes of beginning that setting in order to minimize variation in weights due to factors unrelated to the LeeO scale settings. In order to obtain repeated weights for each pig, the weight survey function in the LeeO app was utilized. Between each new scale setting, the scale was tared, and between each weight taken the scale was allowed to zero prior to the pig being weighed again. Coefficient of variation (CV) for each setting on each pig was calculated by dividing the standard deviation of those 5 weights by the average of the weights collected for that setting. Average time required to collect each weight in nursery pigs was calculated by dividing the elapsed time between when the first weight was recorded for that setting until the fifth weight was recorded by the number of weighing events. Therefore, this average time includes the time required to remove the pig, begin another weight survey, allow the scale to zero, scan the pig again, place the pig back on the scale, weigh the pig, wait for the 3 second weight confirmation delay, and save the survey. In finishing pigs, the time from when the gate was closed on the scale until the weight locked in was timed to determine the average time required to weigh each pig on each setting. The accuracy of each setting was estimated by calculating the absolute difference between the average weight of that setting from the overall average of all weights collected for that pig using all settings. Experimental data were analyzed using the RStudio environment with each pig serving as the experimental unit. The interaction of stabilization time and weight was tested, along with the main effects of stabilization time and stabilization weight. Differences were considered significant at $P \le 0.05$ and marginally significant at $0.05 < P \le 0.10$.

Results

When looking at the precision of weighing the 25-pound and 50-pound test weights, the overall precision of all settings was very good with CV estimates of 0.1% or less when the test weight was measured multiple times within each combination of settings. For accuracy, there was no difference ($P > 0.10$) in the absolute difference of the weights collected from the true value of the test weight based on stabilization weight for both the nursery pig and finishing pig scale settings. There was a significant effect of stabilization time on accuracy for the 25-pound test weight with longest stabilization time resulting in lower differences in the weights collected from the true value of the test weight ($P < 0.001$; Table 1).

Table 1: Main effects of LeeO scale stabilization settings on precision and accuracy in weighing 25-pound and 50-pound test weight

	No motion, g[1]					No motion, mSec[2]				
25-pound test weight	**50**	**100**	**200**	**SEM**	***P* =**	**250**	**500**	**1,000**	**SEM**	***P* =**
Precision										
CV, %[3]	0.082	0.085	0.094	---	---	0.107	0.106	0.048	---	---
Accuracy										
Difference from true, g[4]	4.20	7.53	3.53	2.735	0.547	10.87[a]	9.53[a]	-5.13[b]	2.735	< 0.001

	No motion, g[1]				No motion, mSec[2]			
50-pound test weight	**250**	**500**	**SEM**	***P* =**	**1000**	**2000**	**SEM**	***P* =**
Precision								
CV, %[3]	0.020	0.036	---	----	0.032		---	---
Accuracy								
Difference from true, g[4]	-11.60	-10.60	6.364	0.930	-8.60	-13.60	6.364	0.677

a,b Means within row with different superscripts differ ($P < 0.05$).

1 No motion, g is the terminology used in the scale settings portion of the LeeOTM app to denote stabilization weight which is the range that the weight has to stay within during the specified stabilization time in order for the weight to be locked in.

2 No motion, mSec is the terminology used in the scale settings portion of the LeeOTM app to denote stabilization time which is the length of time that the weight has to stay within the specified weight range in order for the weight to be locked in.

3 Calculated by dividing the standard deviation of the 5 weights taken for the test weight on each unique setting by the average of those 5 weights.

4 The difference of each weight recorded utilizing that combination of stabilization settings from the true weight of the test weight.

For precision in nursery weights, there was a tendency for an interactive effect of stabilization time and stabilization weight settings ($P = 0.082$; Table 2). Within each stabilization weight setting, pairwise comparisons revealed that there was no evidence of a difference in CV based on stabilization time ($P > 0.05$). Within each stabilization time setting, pairwise comparisons revealed that there were no differences in CV based on stabilization weight ($P > 0.05$). There was a significant main effect of stabilization weight setting ($P = 0.023$; Table 3) on CV with the stabilization weight setting of 50 g having a lower ($P < 0.05$) CV than the stabilization weight setting of 200 g with the 100 g setting intermediate. There was also a significant effect of stabilization time ($P = 0.003$) on CV with the stabilization time setting of 250 mSec having a higher ($P < 0.05$) CV than the stabilization time setting of 1,000 mSec with the 500 mSec setting intermediate. There was a tendency for an interactive effect of stabilization time and stabilization weight settings ($P = 0.091$) on absolute difference of the average of that setting from the overall mean, but no evidence of pairwise differences were observed. There was no significant main effect ($P > 0.10$) of stabilization time or stabilization weight setting on absolute difference of the average of that setting from the overall mean. It took less time ($P < 0.001$; Table 3) to weigh nursery pigs utilizing the stabilization weight settings of 100 or 200 g compared to the 50 g setting. Average time was lower ($P < 0.001$) when weighing nursery pigs with a stabilization time setting of 250 or 500 mSec compared to 1,000 mSec.

For finishing pigs, data indicate there were no significant interactive or main effects ($P > 0.10$; Table 4) of stabilization time and stabilization weight settings in finishing weight precision or accuracy. There was an effect of stabilization time setting on average time required to collect each weight with a setting of 1,000 mSec requiring half the time to lock in a weight as compared to the setting of 2,000 mSec ($P = 0.001$; Table 5). Average time to collect a weight was more than 10 seconds shorter ($P < 0.001$) when weighing finishing pigs with a stabilization weight setting of 500 g compared to 250 g.

Discussion

This work is the first to describe the effects of different scale settings on the precision, accuracy, and efficiency of weight collection in an intensive swine research setting using the LeeO system. Using any of the scale settings, the precision was very high when using the test weights. For accuracy, there was no effect of stabilization weight or stabilization time on the difference from true weight for the finishing scale settings as well as no effect of stabilization weight on the accuracy of the nursery scale settings. However, for the accuracy of the nursery scale settings with the test weight the longest stabilization time (1,000 mSec) resulted in the closest accuracy to the true weight of the test weight. It is important to keep in mind, however, that these estimated accuracy results were very good across all scale settings with the accuracy generally within 10 g of the true weight (representing approximately 99.9% accuracy) when measuring the accuracy and precision of an object of known mass.

The optimal settings for LeeO scales in the nursery depend on the operation's priorities. In a system where minimizing variability of weights is the first priority, a lower stabilization weight and higher stabilization time setting would be the best fit. Conversely, for an operation that places more of a priority on efficiency of time to collect weights, a higher stabilization weight value and shorter stabilization time would be recommended. For a research setting with the goal to balance precision and efficiency of data

Table 2: Interactive effects of LeeO scale stabilization settings on precision, efficiency, and accuracy in weighing nursery pigs

No motion, g[1]:	50			100			200				
No motion, mSec[2]:	250	500	1,000	250	500	1,000	250	500	1,000	SEM	Weight × time, *P* =
Precision CV, %[3]	0.247^{ab}	0.169^{b}	0.204^{b}	0.261^{ab}	0.306^{ab}	0.152^{b}	0.373^{a}	0.296^{ab}	0.205^{ab}	0.0410	0.082
Accuracy Difference from mean, g[4]	10.91	8.96	14.53	10.81	14.09	9.43	13.68	14.63	11.80	2.308	0.091
Efficiency Avg. time, sec[5]	11.30^{bc}	11.39^{bc}	13.36^{a}	10.23^{c}	10.84^{bc}	12.27^{ab}	10.23^{c}	10.53^{c}	11.74^{a}	0.431	0.780

a-c Means within row with different superscripts differ ($P < 0.05$).

1 No motion, g is the terminology used in the scale settings portion of the LeeO™ app to denote stabilization weight which is the range that the weight has to stay within during the specified stabilization time in order for the weight to be locked in.

2 No motion, mSec is the terminology used in the scale settings portion of the LeeO™ app to denote stabilization time which is the length of time that the weight has to stay within the specified weight range in order for the weight to be locked in.

3 Calculated by dividing the standard deviation of the 5 weights taken for each pig on each unique setting by the average of those 5 weights.

4 Calculated by averaging the absolute difference of the average weight recorded utilizing that combination of stabilization settings from the overall average weight for that pig.

5 Calculated by dividing the elapsed time from when the first weight was recorded to when the final weight for that setting combination was recorded by the number of weighing events to obtain an average time. This time measurement includes the time required to remove the pig from the scale, begin another weight survey, allow the scale to zero, set the pig back on the scale, weigh the pig, wait for the 3 second weight confirmation delay, and save the survey.

Table 3: Main effects of LeeO scale stabilization settings on precision, efficiency, and accuracy in weighing nursery pigs

	No motion, g[1]					No motion, mSec[2]				
	50	100	200	SEM	*P* =	250	500	1,000	SEM	*P* =
Precision CV, %[3]	0.206^{b}	0.240^{ab}	0.291^{a}	0.0275	0.023	0.294^{a}	0.257^{ab}	0.187^{b}	0.0275	0.003
Accuracy Difference from mean, g[4]	11.47	11.45	13.37	1.679	0.380	11.80	12.56	11.92	1.679	0.876
Efficiency Avg. time, sec[5]	12.02^{a}	11.11^{b}	10.83^{b}	0.327	< 0.001	10.59^{b}	10.92^{b}	12.46^{a}	0.327	< 0.001

a,b Means within row with different superscripts differ ($P < 0.05$).

1 No motion, g is the terminology used in the scale settings portion of the LeeOTM app to denote stabilization weight which is the range that the weight has to stay within during the specified stabilization time in order for the weight to be locked in.

2 No motion, mSec is the terminology used in the scale settings portion of the LeeOTM app to denote stabilization time which is the length of time that the weight has to stay within the specified weight range in order for the weight to be locked in.

3 Calculated by dividing the standard deviation of the 5 weights taken for each pig on each unique setting by the average of those 5 weights.

4 Calculated by averaging the absolute difference of the average weight recorded utilizing that combination of stabilization settings from the overall average weight for that pig.

5 Calculated by dividing the elapsed time from when the first weight was recorded to when the final weight for that setting combination was recorded by the number of weighing events to obtain an average time. This time measurement includes the time required to remove the pig from the scale, begin another weight survey, allow the scale to zero, set the pig back on the scale, weigh the pig, wait for the 3 second weight confirmation delay, and save the survey.

Table 4: Interactive effects of LeeO scale stabilization settings on precision, efficiency, and accuracy in weighing finishing pigs

No motion, g[1]:	250		500			
No motion, mSec[2]:	1,000	2,000	1,000	2,000	SEM	Weight × time, P =
Precision						
CV, %[3]	0.253	0.222	0.259	0.207	0.0490	0.818
Accuracy						
Difference from mean, g[4]	163.96	225.03	162.00	217.69	40.250	0.942
Efficiency						
Avg. time, sec[5]	15.20	28.57	6.95	11.94	2.881	0.103

1 No motion, g is the terminology used in the scale settings portion of the LeeOTM app to denote stabilization weight which is the range that the weight has to stay within during the specified stabilization time in order for the weight to be locked in.

2 No motion, mSec is the terminology used in the scale settings portion of the LeeOTM app to denote stabilization time which is the length of time that the weight has to stay within the specified weight range in order for the weight to be locked in.

3 Calculated by dividing the standard deviation of the 5 weights taken for each pig on each unique setting by the average of those 5 weights.

4 Calculated by averaging the absolute difference of the average weight recorded utilizing that combination of stabilization settings from the overall average weight for that pig.

5 Calculated by averaging the elapsed time from when the scale door was shut until a weight was locked in.

Table 5: Main effects of LeeO scale stabilization settings on precision, efficiency, and accuracy in weighing finishing pigs

	No motion, g[1]				No motion, mSec[2]			
	250	500	SEM	P =	1,000	2,000	SEM	P =
Precision								
CV, %[3]	0.238	0.233	0.0361	0.917	0.256	0.214	0.0361	0.374
Accuracy								
Difference from mean, g[4]	194.50	189.85	30.559	0.901	162.98	221.36	30.559	0.121
Efficiency								
Avg. time, sec[5]	21.89[a]	9.45[b]	2.264	< 0.001	11.08[b]	20.26[a]	2.264	0.001

a,b Means within row with different superscripts differ ($P < 0.05$).

1 No motion, g is the terminology used in the scale settings portion of the LeeO™ app to denote stabilization weight which is the range that the weight has to stay within during the specified stabilization time in order for the weight to be locked in.

2 No motion, mSec is the terminology used in the scale settings portion of the LeeO™ app to denote stabilization time which is the length of time that the weight has to stay within the specified weight range in order for the weight to be locked in.

3 Calculated by dividing the standard deviation of the 5 weights taken for each pig on each unique setting by the average of those 5 weights.

4 Calculated by averaging the absolute difference of the average weight recorded utilizing that combination of stabilization settings from the overall average weight for that pig.

5 Calculated by averaging the elapsed time from when the scale door was shut until a weight was locked in.

collection, the optimal settings appear to be the combination of 50 g for stabilization weight and 500 mSec for stabilization time. For finishing pigs, there appears to be minimal effect of stabilization time or stabilization weight on precision or accuracy. However, the time required to collect data does differ substantially between finishing scale head settings. When utilizing the 250 g stabilization weight setting and 2,000 mSec stabilization time combination, time to collect one weight had the largest range, with the longest time being over 4 minutes for the scale to lock in a weight. For this reason, it is recommended to use a shorter stabilization time and larger stabilization weight to optimize efficiency of data collection. When utilizing the most efficient nursery scale setting (200 g × 250 mSec), the total time that it took to weigh nursery pigs under standard weighing conditions was 8.4 seconds, this time included picking up the animal, scanning their ear tag, weighing the pig, and putting it back down. This translates to 428 pigs per hour that can be weighed utilizing the LeeO scale using the current setting recommendations. When utilizing the most efficient finishing scale setting (500 g × 1000 mSec), the total time that it took to weigh finishing pigs under standard weighing conditions was 31.7 seconds equating to a rate of 113 finishing pigs per hour.

In summary, the data herein helps guide decisions pertaining to initial stabilization settings for LeeO nursery and finishing scales. Within all settings utilized in this study accuracy was very high. However, the recommended combination of settings when factoring in precision and efficiency for the LeeO nursery scale are 50 g for stabilization weight and 500 mSec for stabilization time and for the LeeO finishing scale, 500 g for stabilization weight and 1000 mSec for stabilization time. Future research will focus on the inclusion of more setting combinations and more replications with the existing setting combinations to gain a better understanding of this new technology and better cater it to an intensive research setting.

https://doi.org/10.54846/am2023/55

PCR detection of ASFV nucleic acid in oral swabs from clinical and nonclinical sows of three swine farms in Vietnam

C. Goodell[1], DVM, MS, PhD, DACVPM; J. Torrison[2], DVM, PhD, DACVPM; V. Tran[1], DVM; H. Nguyen[1], DVM; Q. Tran[1], DVM; C. Maala[1], DVM; C. Helm, MBA[2]; D. Lai[3], DVM; T. Nguyen[3], DVM; H. Le[3], DVM, MSc, PhD; N. Nguyen[3], DVM, MSc; D. Do[3], DVM, MSc, PhD; T. Nguyen[3], DVM, MSc, PhD; N. Nguyen[4], PhD; J. Zimmerman[5], DVM, PhD, DACVPM; O. Gomez-Duran[1], DVM, MS, PhD, MRCVS, Certificate in Pig Medicine, Diplomate ECPHM

[1]Boehringer Ingelheim Vetmedica GmbH, Ingelheim, Germany; [2]Longhorn Vaccines and Diagnostics LLC, Bethesda, Maryland; [3]Animal Sciences and Veterinary Medicine, Nong Lam University, HCM City, Vietnam; [4]Research Center for Genetics and Reproductive Health, School of Medicine, Vietnam National University, HCM City, Vietnam; [5]Iowa State University, Ames Iowa

Introduction

Vietnam first reported the first outbreak of African swine fever virus (ASFV) in February 2019.[1] Since this time, after losing over 20% of their swine herd, Vietnam has worked to return to pre-ASFV level of production, albeit with continued, sporadic ASFV breaks.[2] In 2022, Vietnam began use of the first ASFV vaccine in the field and results appear promising.[3] Until now, and until vaccine use is more widespread, ASFV control has focused on test and remove, or partial depopulation programs. These methods, although imperfect, were implemented because producers desired to retain as much of their herd as possible when ASF was detected.[4] Individual blood samples have been the accepted sample type for detection of ASFV by real-time PCR to identify ASFV in clinical animals and their non-clinical direct and indirect neighbors. ASFV is readily detected in blood, however this method risks shedding and spreading ASFV given ASFV viability in blood and stability outside the host. Additionally, blood sample collection requires higher technical skills which might be limiting for rapid, large-scale implementation compared to collecting oral swabs. Therefore, this study was performed to assess the feasibility of oral swabs as a sample type for ASFV detection when submitted to a diagnostic laboratory in a novel molecular transport media (PrimeStore® MTM, Figure 1). Prime store® MTM stabilizes and inactivates infectious agents particularly for molecular diagnostics.[5]

Materials and methods

Three of the 52 suspected ASFV events in a previously reported study were also sampled by oral swabbing.[6] "Suspected ASF event" was defined as a sow exhibiting OIE-established clinical signs as determined by a farm caregiver (OIE Technical Disease Card: African swine fever - OIE - World Organization for Animal Health). For each suspected ASFV event, whole blood samples were collected from the index sow plus approximately 14 contact animals (direct and indirect) located in gestation stalls on each side of the index sow (Figure 2). The oral swab procedure entailed a single use sterile swab inserted into the sow's mouth along the buccal surface of the lower jaw, which collected crevicular fluid and saliva. The end of the swab was placed in the PrimeStore® MTM media tube and sealed before shipment to the diagnostic laboratory. Samples were tested for ASFV DNA by an OIE-approved real-time ASFV PCR within 24-hours of arrival to the laboratory. ASFV PCR results from oral swab results were compared to results of blood from the same sows.

Results

a. 45 oral swab samples were collected alongside 45 blood samples from the 3 suspected ASF events.
b. The oral swabs detected 10 of the 16 sows positive for ASFV by PCR in blood (Table 1).
c. Oral swabs detected 3 of 3 clinical sows positive for ASFV by PCR, 100% agreement with blood.
d. Success of ASF nucleic acid detection by PCR in oral swabs was related to sow blood Ct values (Figure 3 & Table 2).

Discussion

As used in this project, oral swabs did not provide the level of sensitivity required for individual animal testing for presence of ASFV as part of a partial depopulation strategy. This is consistent with observations from controlled infection experiments previously reported.[7,8] Other sample types such as whole blood in this study or blood swabs in other studies have been shown to provide greater sensitivity for characterizing the antemortem ASFV status of an individual pig at a point in time.[9]

Figure 1: PrimeStore MTM media tube.

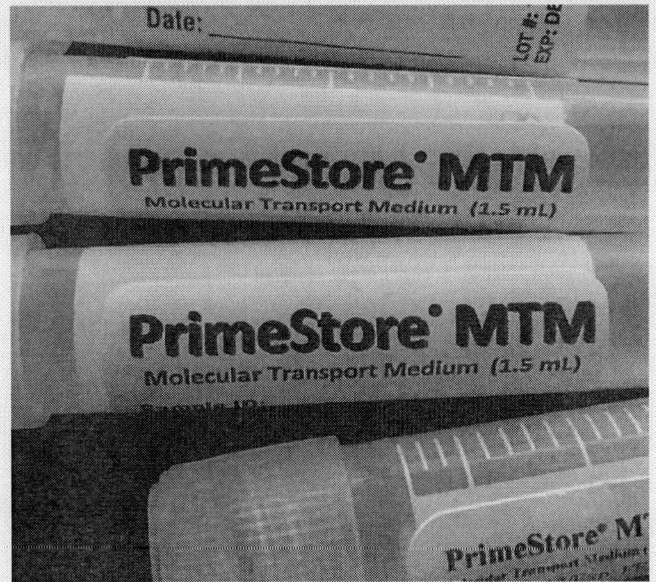

Figure 2: Example of blood sampling protocol around a suspected ASF clinical sow in gestation.[†‡]

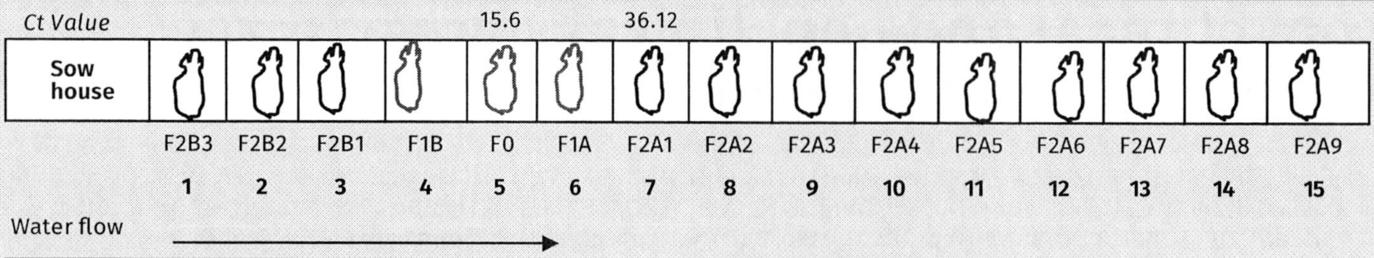

† F0 = index sow, F1 = direct/closest contact neighbors, F2 = indirect contact neighbors; A = "down" row, B = "up" row

‡ The rationale for the sampling distribution was due to an assumption drinking water flowed down a common water trough

Table 1: Comparison of ASFV PCR by sample type

Sample type	OIE PCR test positive	OIE PCR test negative	Total tested
Blood	16	29	45
OS	10	35	45

The low number of farms (3) and samples (45) evaluated for the comparison is a limitation of this study. However the higher Ct values for the oral swab vs. whole blood samples in addition to the lower number of positive test results is compelling evidence that oral swabs are less sensitive for detecting ASFV (Tables 1 & 2).

Oral swab samples from individual pigs should not be confused with oral fluids collected from a population of pigs, which was not evaluated in this study as the status of individuals was the objective here.

PrimeStore® MTM should continue to be investigated for ASFV field sampling.[9] There is high value in a transport media which will inactivate infectious agents but not to the detriment of diagnostic detection, particularly in foreign animal and transboundary disease investigations. Additionally, the benefit of not having to maintain a cold chain for clinical samples from the time of collection through to testing has advantages of cost, convenience and, in some cases, feasibility.

Conclusions

Oral swabs were 62.5% sensitive and 100% specific in detecting ASFV by PCR compared to blood; 69% sensitive when comparing blood samples with Ct < 25. Oral swabs should not be a preferred sample type for individual antemortem animal testing in ASFV partial depopulation programs. PrimeStore® MTM warrants consideration as an inactivating transport media for safe and effective use for high consequence animal disease testing.

Acknowledgment

Funding for this project was provided by the Swine Health Information Center, Ames, IA.

References

1. Report # VM2019-0067, Foreign Agricultural Service, USDA, December 06, 2019. Available at: **DownloadReportByFileName (usda.gov)**

2. Report # VM20221-0042, Foreign Agricultural Service, USDA, May11, 2021. Available at: **https://www.fas.usda.gov/data/vietnam-african-swine-fever-update**

3. Swine Health Information Center. SHIC- ASF Vaccine Status in Vietnam Update. September, 2022. SHIC – ASF Vaccine Status in Vietnam Update – Swine Health Information Center, accessed Nov 1, 2022.

4. Bui, Thi To Nga, et al. "Implications of partial culling on African swine fever control effectiveness in Vietnam." *Frontiers in Veterinary Science* (2022): 1289.

5. **Longhorn-Molecular-Transport-Medium.pdf (cdc.gov)**

6. Goodell et al., Efficacy of the "Tooth Extraction" (test-and-remove) protocol in commercial swine farms in Vietnam". Proceedings of the 53rd AASV Annual Meeting. P 43. **https://doi.org/10.1089/vim.2017.0050**

7. Guinat, Claire, et al. "Dynamics of African swine fever virus shedding and excretion in domestic pigs infected by intramuscular inoculation and contact transmission." *Veterinary research* 45.1 (2014): 1-9.

8. Beemer, Oriana, et al. "Assessing the value of PCR assays in oral fluid samples for detecting African swine fever, classical swine fever, and foot-and-mouth disease in US swine." *PloS One* 14.7 (2019): e0219532.

9. Pikalo, J., et al. "African Swine Fever Laboratory Diagnosis—Lessons Learned from Recent Animal Trials. Pathogens 2021, 10, 177." (2021).

Figure 3: Distribution of ASFV PCR results by sow location.

Farm A

Sow location	F0	F1	F1	F2	F2	F2	F2	F2	F2
Blood PCR Ct	17.66	19.39	18.37	18.2	17.88	17.18	-	19.15	-
Oral swab PCR Ct	25.81	-	35.90	36.64	35.16	33.18	-	-	-

Farm B

Sow location	F0	F1	F1	F2	F2	F2	F2	F2	F2	F2	F2	F2	F2
Blood PCR Ct	16.37	19.25	19.58	-	-	-	-	-	-	-	-	-	-
Oral swab PCR Ct	27.97	-	36.06	-	-	-	-	-	-	-	-	-	-

Farm C

Sow location	F0	F1	F1	F2	F2	F2	F2	F2	F2	F2	F2	F2	F2	F2	F2	F2	F2	F2	F2	F2	F2	F2	F2
Blood PCR Ct	17.82	36.1	-	-	30.89	24.82	-	-	-	-	17.29	-	-	32.75	-	-	-	-	-	-	-	-	-
Oral swab PCR Ct	31.34	-	-	-	-	-	-	-	-	-	30.1	-	-	31.94	-	-	-	-	-	-	-	-	-

F0: clinical sow; F1: direct contact sows; F2: indirect contact sows

Table 2: ASFV PCR positive sows sorted by blood Ct values

Location	ASFV PCR Ct blood	ASFV PCR Ct oral swab
F0	16.37	27.97
F2	17.18	33.18
F2	17.29	30.1
F0	17.66	25.81
F0	17.82	31.34
F2	17.88	35.16
F2	18.2	36.64
F1	18.37	35.9
F2	19.15	-
F1	19.25	-
F1	19.39	-
F1	19.58	36.06
F2	24.82	-
F2	30.89	-
F2	32.75	31.94
F1	36.1	-

F0: clinical sow; F1: direct contact sows; F2: indirect contact sows

https://doi.org/10.54846/am2023/56

Evaluation of airborne shedding and production setback post-weaning from Pharmgate PRRSGard® vaccine in commercial conditions

E. Kettelkamp[1]; A. Betlach[1]; P. Yeske[1] L. McCuiston[2]; J. Okones[2]
[1]Swine Vet Center P.A., St. Peter, Minnesota; [2]Pharmgate Animal Health, Wilmington, North Carolina

Introduction

Vaccinating post-weaning pigs for PRRSV from susceptible populations has become common across the swine industry. Modified live PRRS vaccine and unidirectional pig flow structures have been shown to help protect PRRSV-negative growing pigs placed in high pig-dense regions. Additionally, vaccination helps minimize the area spread of PRRSV by reducing the amount of shedding that occurs in pigs if field strain infection occurs. Previous work has described the value of vaccinated pig performance when faced with field strain PRRSV introductions.[1-4] Without PRRSV field exposure, the use of various PRRS vaccines may have a negative impact or no difference in growth performance outcomes.[5-7] The objectives of this study are to define the amount and duration of viremia in the population and tested aerosol samples, both inside and outside of a typical commercial barn, and to evaluate the effect of vaccination with Pharmgate PRRSGard® on pig performance.

Materials and methods

A source of PRRSV-negative pigs was identified from a single source to use in the trial. This farm is a PRRSV naïve herd and had a negative diagnostic history supporting the status of the pigs. A site designated to this flow was identified to receive these pigs and then vaccinate them in the first 1-2 weeks post-placement. This was a typical two-room, tunnel-ventilated commercial wean-to-finish barn with a large pen (150 head per pen) finishing design. Each room was designed to 1,800 head capacity with pit fans for minimum ventilation.

Room 1 was assigned as the vaccinated room and room 2 was assigned to the negative controls. Four pens in the center of each room were allotted to the study to control for performance variation based on location within the barn. Within each pen, 100 pigs were randomly selected and individually tagged.

On Day 0, vaccinated pigs were inoculated with Pharmgate PRRSGard® vaccine intramuscularly at the labeled dose (1mL/pig). Control pigs were administered 1.0 mL of saline.

Individual body weights were collected on days 0, 49, and 119 of the study.

Thirty pigs per room in each treatment were randomly identified. The selected pigs were ear-tagged with a unique identification number and blood tested at days 0, 3, 7, 14, 21, 28, 35, and 49-days post-vaccination (dpv). Blood samples were tested for quantitative PRRS PCR and PRRS ELISA. Oral fluids were collected from each pen while the blood samples are collected, followed by collection every two weeks throughout the finishing period and tested for quantitative PRRS PCR.

A Cyclonic Air Collector (Midwest Micro-Tec, Brookings, South Dakota) and Cub Air Collector (Innovaprep, Drexel, Missouri) were used to assess the amount of aerosol shedding post-vaccination on days 0, 7, 14, 21, 28, and 35 of the study. Paired air samples were collected from the following locations: inside the vaccinated room of the barn, directly outside one of the pit fans of room 1, and 1 mile away from the site in a down-wind direction for the day. Air samples were collected by running the two air collectors side-by-side for 30 minutes at each location. All samples were tested for PRRSV rt-PCR.

To assess the likelihood of PRRS shedding to naïve pigs when comingled with Pharmgate PRRSGard® vaccinated pigs (at room level), two pens located on one side of the room remained unvaccinated to assess the potential for PRRS transmission. In the two unvaccinated pens within room 1, a subset of pigs (n = 10 total) was randomly selected and ear-tagged with a unique identification number and sampled at days 0, 3, 7, 14, 21, 28, 35, 49-dpv and tested for quantitative PRRS PCR.

Results

There was no PRRSV aerosol shedding detected during the first 35-dpv at any of the locations: in the room above the vaccinated pens, outside the nearest pit fan and 1 mile downwind of the site.

All pigs serologically converted to vaccine showing an immune response within 14 to 28-dpv.

The pigs turned viremic promptly, becoming 100% positive 3-dpv, and the viremia was sustained for 21-dpv before beginning to drop off at 28 - 35-dpv. The 10 negative control pigs that remained in adjacent pens to vaccinated pigs in the room never tested positive within 5 weeks post-vaccination.

The oral fluids testing went positive very early in the vaccinated pens and the ct levels minimally varied over time within the first 35 days of the study. Unvaccinated pigs tested PRRSV negative until day 91 of the study. At approximately day 100 in the trial, the site broke with clinical influenza confirmed on oral fluid PCR. When the influenza virus went through the population, this correlated with increased shedding of PRRSGard® across the site and detection of the vaccine in the unvaccinated control room.

Growth performance and mortality were very similar through day 119 of the trial between the vaccinates and controls.

Discussion

In this study, there was no detection of the vaccine virus in the aerosol testing for the first 35 days following vaccination. Including the testing in the barn, outside the nearest located pit fan, and downwind 1 mile. Other vaccines have been reported to have virus detected at the pit fan and downwind for up to 14 days following vaccination.[6] This is promising for using PRRSGard® vaccine next to negative sites with less concern of regional spread while protecting the site from the effects of field virus spread into the site.

The Pharmgate vaccine developed a sustained viremia 3 – 21-dpv before shedding began to reduce through the end of the nursery period. Comparable to other PRRS vaccine studies,[7] nonvaccinated controls became infected in the neighboring room at the end of the study following the clinical outbreak of influenza virus.

There was no significant loss of growth performance either at the end of the nursery phase, finishing phase, or wean-to-finish all the way through when the vaccine was applied to the negative pigs as compared to a sham vaccination with saline. This is different than shown work with other vaccines.[4]

This study demonstrates that PRRSGard® may be a useful tool to control PRRS infection with reduced risk of vaccine area spread without negative effects on pig performance.

References

1. Oppressing T. et. al.; Demonstration of Protection Against Heterologous Challenge; J Swine Health and Production 13:246-253.

2. Roof M.; 2007 Cross Protection in 16 Independent Trials; Data presented at a meeting in Poland.

3. Johnson, C., Kliebenstein, J., Mabry, J., Neumann, E., 2004. The Impact of PRRS on the Cost of Pig Production. Allen D. Leman Swine Conference Recent Research Reports. Volume 31. pg: 11.

4. Polson, D., et.al, 2000. Evaluating the Beneficial Performance and Financial Effects of Pig Vaccination with Ingelvac® PRRS MLV in a PRRS-Positive Multi-Site Production System. Proceedings of the 16th IPVS. Pg: 152.

5. Pretzer S., et al. 1996. The Effects of Porcine Reproductive and Respiratory Syndrome (PRRS) vaccination on postweaning growth performance. KSU Swine Day: 83-86.

6. Yeske P. 2011. A study to evaluate airborne shedding of PRRS virus in pigs following vaccination with Ingelvac® PRRS ATP in commercial production systems. Proceedings AASV 43rd Annual Meeting.

7. Yeske P. 2021. Evaluation of shedding and effect on pig performance of Prevacent® PRRS Vaccine in commercial conditions. Proceedings AASV 52nd Annual Meeting.

https://doi.org/10.54846/am2023/57

Comparative efficacy of Aivlosin® and tilmicosin for treatment of swine respiratory disease in nursery-age pigs

C. Sexton[1], DVM; C. Smith[2], DVM; N. Lauterbach[2], DVM; J. Okones[2], DVM
[1]Southern Swine Research, North Carolina; [2]Pharmgate Animal Health, Wilmington, North Carolina

Introduction

Swine respiratory disease (SRD) refers to the component of acute respiratory disease in swine associated with bacterial pathogens such as *Actinobacillus pleuropneumoniae, Bordetella bronchiseptica, Haemophilus parasuis, Pasturella multocida*, or *Streptococcus suis*.[1] Pigs impacted with SRD can display numerous clinical signs such as coughing, lethargy, weight loss, and mortality. Aivlosin® (62.5% w/w Tylvalosin as Tylvalosin Tartrate) is a water soluble macrolide labeled for the control of SRD associated with *Bordetella bronchiseptica, Haemophilus parasuis, Pasturella multocida, Streptococcus suis*, and *Mycoplasma hyopneumoniae* in groups of swine intended for slaughter in buildings experiencing an outbreak of SRD. Aivlosin® is also labeled for the control of porcine proliferative enteropathy (PPE) associated with *Lawsonia intracellularis* infection in groups of swine intended for slaughter in buildings experiencing an outbreak of PPE. The objective of this study was to compare the efficacy of Aivlosin® and tilmicosin, another water soluble, macrolide antibiotic, for the treatment of swine respiratory disease in nursery-age pigs.

Materials and methods

Study design

This study was conducted on a multi-room nursery site in North Carolina. Three rooms within this site were utilized for this trial. In room one of the nursery, the pens were 20 square feet in size, while pens within rooms two and three were 70 square feet in size. In all three rooms, the pigs were placed at a similar stocking density within the pens. 770 total pigs, averaging 22.5 days old, were placed to fill a room per weaning event, consistent with normal pig flow within this production system. These pigs originated from a porcine reproductive and respiratory syndrome virus (PRRSV) stage 2vx sow source.[2]

Pigs were weighed and allocated for the trial by pen (n = 25), with the pens on one side of the nursery room set to receive Aivlosin® during the trial and the pens on the other side of the nursery room set to receive tilmicosin. Following allocation, pigs were monitored for a minimum of 14 days prior to the administration of any water soluble medication. After the initial 14 days, water medication was started to the room at the onset of clinical signs consistent with SRD, as noted by the attending veterinarian. The pens received either Aivlosin®, administered at 50 ppm, or tilmicosin, administered at 200 ppm, for five consecutive days. Individual stock solutions containing either Aivlosin® or tilmicosin were mixed daily. Water volume administered to each set of pens was measured and recorded daily. 14 days after initial administration of Aivlosin® or tilmicosin, pen weights were collected and the trial was complete.

During the trial, diagnostics were collected and submitted to Iowa State University Veterinary Diagnostic Laboratory. Tissue samples and oral fluid samples were collected and tested for PRRSV, influenza A virus (IAV), porcine circovirus 2 (PCV 2), and *Mycoplasma hyopneumoniae* (MHP). Histopathology and culturing of tissue samples were also conducted on each individual tissue submission. Results from the oral fluid samples can be seen in Table 1. Tissue submissions revealed changes consistent with secondary bacterial bronchopneumonia throughout the trial, with *Glaesserella parasuis* and *Streptococcus suis* identified at multiple samplings.

Statistical analysis

Due to incorrect pig inventory at the end of the trial, four of the 25 pens within the Aivlosin® treatment group were disqualified and excluded from statistical analysis. This correction reduced the total number of pens in the trial treated with Aivlosin® (n = 21), but did not have an impact on the total number of pens treated with tilmicosin (n = 25). Within the pens remaining on trial, an ending inventory of 305 head were treated within the 21 pens that received Aivlosin® and 362 head were treated within the 25 pens that received tilmicosin. Analysis of variance (ANOVA) was performed to assess differences between treatments assuming the pen as the experimental unit. Once all assumptions were met, a Tukey-Kramer was performed to assess differences between the two groups. Two models were built to assess differences in starting weights and ending weights. Covariates to adjust for different barn layout was tested and removed from the model due to non-significance. A chi-squared statistic was computed to assess differences on the frequency of animals removed from the trial between treatments. The significance level of 5% was set to all analysis.

Table 1: Oral fluid diagnostic sample results

Timing	PRRSV PCR	IAV-S PCR	MHP PCR	PCV 2 PCR
Allocation (Average Ct)	8/15 positive (33.9 Ct)	0/15 positive	0/15 positive	9/15 positive (34.5 Ct)
Start of medication (Average Ct)	15/18 positive (32.2 Ct)	0/18 positive	0/18 positive	7/18 positive (35.4 Ct)
Conclusion of trial (Average Ct)	18/18 positive (29.5 Ct)	0/18 positive	0/18 positive	1/18 positive (36.3 Ct)

Results and discussion

Average pen weights

At allocation, there were no significant differences between the average starting weights in the pens, as shown in Table 2. The groups allocated to receive Aivlosin® averaged 16.39 pounds, while the groups allocated to receive tilmicosin averaged 16.21 pounds. At the conclusion of the trial, the average ending weight for the pigs within the pens treated with Aivlosin was 45.46 pounds, while the average ending weight for the pigs within the pens treated with tilmicosin was 45.89 pounds. This difference of 0.43 pounds was not found to be statistically significant, as shown in Table 2.

Overall mortality

As previously stated, mortality was tracked daily by treatment group for the duration of the trial. At the end of the trial, the pens treated with Aivlosin® had 13 total head of mortality. The pens treated with tilmicosin had 20 total head of mortality, resulting in a difference of seven head. While there was a numeric trend for mortality to be lower in the pens treated with Aivlosin® compared to the pens treated with tilmicosin, this difference was also not found to be statistically significant, as shown in Table 2.

Summary

There are various antimicrobials to consider when attempting to treat the bacterial pathogens associated with SRD. This study demonstrated that, when pens were treated with either Aivlosin® or tilmicosin at the onset of clinical signs associated with SRD, there was not a significant difference between the average end weight of the pens or the mortality within those pens. Further work should be done to continuously evaluate the antimicrobials available to the industry when facing health challenges.

References

1. Center for Veterinary Medicine. *CVM GFI #178 Recommended design and evaluation of effectiveness studies for swine respiratory disease claims.* FDA-2006-D-0479. October 2007.

2. Holtkamp, D., Clavijo, MJ. *New AASV Herd Classification Guidelines for PRRSV and Mycoplasma hyopneumoniae.* ISU James D. McKean Swine Disease Conference. November 2021.

Table 2 : Statistical analysis of starting weight, ending weight, and mortality

	Aivlosin	Tilmicosin	*P*-value	SEM
Number of pens on trial	21	25		
Average start weight (lbs)	16.15	16.02	0.860	0.543
Average end weight (lbs)	45.46	45.89	0.783	1.084
Total head mortality	13	20	0.592	–

https://doi.org/10.54846/am2023/58

Impact of nurse sow strategies on preweaning and nursery performance

John P. Sonderman[1], PhD; Brady M. McNeil[1]; Christine Mainquist-Whigham[2], DVM; Ethan Stephenson[2]; Emily D. Mauch-Swinford[1], PhD; Tom Rathje[1], PhD; Amanda J. Cross[1], PhD
[1]DNA Swine Genetics, Columbus, Nebraska; [2]Pillen Family Farms, Columbus, Nebraska

Introduction

Genetic and production improvements have increased born alive numbers in commercial swine farms. In 2004 the United States swine industry averaged 10.3 born alive[1] compared to top producers in 2022 averaging 15.0.[2] These improvements have challenged management's ability to effectively wean the highest number and quality of pigs for a wean group. Historically, farms may have over farrowed for the number of farrowing crates available to maximize the number of pigs weaned. Today, sow farms may farrow under the number of farrowing crates they have and move in nurse sows. Nurse sows are females added to a wean group, increasing the number of teats available for suckling. Nurse sows can be moved into a room at a variety of times including day 0/1 post farrow to account for high born alive days, day 3-5 to pull initial fall-behind pigs, or day 7-10 to pull late term fall-behind pigs. Farms using nurse sows reduce the total number of pigs born alive in a group with the goal of lowering their pre-wean mortality and thus weaning a higher percentage of the pigs born. However, the effects of utilizing nurse sows within a farrow to finish operation are not well understood. Therefore, the aim of this study is to determine the impact of utilizing a 7% nurse sow per farrow group compared to a 0% nurse sow on pre-weaning and nursery performance.

Materials and methods

Data were recorded on DNA L241 crossbred (Yorkshire × Landrace) sows (n = 618) representing parities 1 through 6. Sows were assigned to one of two treatments based on farrowing date. Treatments included a 0% nurse sow (NoNurse) treatment with all available farrowing crates in a room (n = 14) filled with sows due to farrow or a 7% nurse sow (Nurse) treatment that left 1 open farrowing crate per farrowing room with 14 crates. At birth, all pigs (n = 9,109) were individually tagged and weighed. In both treatments, pigs were cross-fostered in the first 24 hours after birth within treatment. Pig movement was minimized as much as possible for sows, with born alive between 13 and 15 pigs, to maximize the number of sows lactating her own pigs. If cross-fostering was necessary, pigs were placed on sows to maximize functional teat count utilization. In both treatments, fall-behind pigs, those pigs who were growing slower and at a disadvantage, due to size, compared to their littermates were identified and weighed between day 3 and 5 of life. Littermates to the fall-behind pigs were also identified and weighed. In the NoNurse treatment, fall-behind pigs were left in their litters. In the Nurse treatment, fall-behind pigs were moved to a nurse sow. Pigs were weaned at approximately 21.5 days of age and weaning weights were recorded at this time. At weaning, pigs (n = 7,928) were placed in commercial nurseries. Throughout the trial, mortality reason and date were recorded. At approximately 10 weeks of age, a subset of pigs (n = 2,169) were weighed.

Results and discussion

Birthweights of pigs between treatments were not significantly different ($P > 0.1$), however fall-behind pigs weighed on average 0.34 kg less than non-fall-behind pigs ($P < 0.05$) regardless of treatment (Figure 1). Fall-behind pigs in the NoNurse treatment weighed on average 0.75 kg ($P < 0.05$) less at weaning than the fall-behind pigs in the Nurse treatment (Figure 2). Average wean weight for NoNurse treatment pigs was 0.28 kg less than the pigs from the Nurse treatment ($P < 0.05$). NoNurse treatment sows weaned 0.73 more pigs ($P < 0.05$) than sows in the Nurse treatment. There was no difference in pre-wean mortality between the Nurse and NoNurse treatments ($P > 0.1$). Weights at 10 weeks of age showed fall-behind pigs were 1.51 kg ($P < 0.05$) for NoNurse and 1.66 kg ($P < 0.05$) for Nurse less than their contemporaries within treatment. Fall-behind pigs within the NoNurse treatment weighed 1.55 kg ($P < 0.05$) less than fall-behind pigs in the Nurse treatment (Figure 3). Nursery mortality was not significantly different at 2.34% compared to 2.64% for NoNurse and Nurse treatment ($P > 0.1$), respectively. The results of this study indicate that farms utilizing nurse sows' do not accomplish their goal of significantly reducing their pre-wean mortality, and farms not utilizing nurse sows are able to wean more total pigs per wean group. By utilizing nurse sows, farms can wean heavier pigs, by reducing competition within each crate. This extra weight at weaning was carried into the nursery phase resulting in heavier pigs at 10 weeks of age. While results were not significantly different for nursery mortality, the direction of the advantage does agree with internal Pillen Family Farms nursery data comparing nursery mortality for farms in McREBEL farrowing house protocols compared to farms that allow pig movement after day 1. Overall, producers need to evaluate their economic model to determine which is the best course of action for them. Based on these results, a sow farm producer could minimize the number of nurse sows to produce more total pigs.

References

1. PigCHAMP. (2004). PigCHAMP Benchmarking USA year summary. Accessed November 13, 2022. **https://www.pigchamp.com/benchmarking/benchmarking-summaries**

2. PigCHAMP. (2022). PigCHAMP Benchmarking USA quarterly summary. Accessed November 13, 2022. **https://www.pigchamp.com/benchmarking/benchmarking-summaries**

Figure 1: Average birth weights of pigs for NoNurse and Nurse treatments with fall-behind status of a pig moved to a nurse sow (1), a pig marked as fall-behind but remaining on their original dam (2), pigs from litters with fall-behind pigs marked (3), and litters with no fall-behind pigs (4).

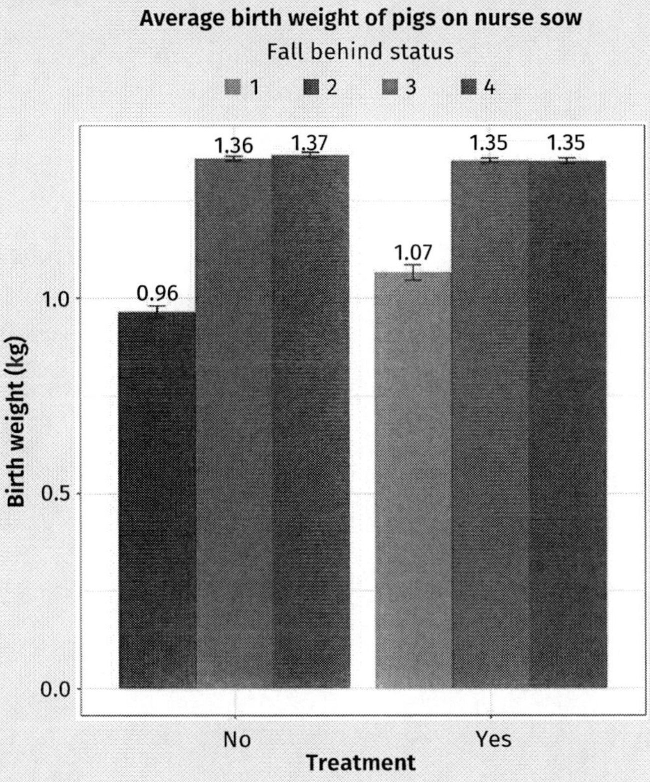

Figure 2: Average wean weights of pigs for NoNurse and Nurse treatments with fall-behind status of a pig moved to a nurse sow (1), a pig marked as fall-behind but remaining on their original dam (2), pigs from litters with fall-behind pigs marked (3), and litters with no fall-behind pigs (4).

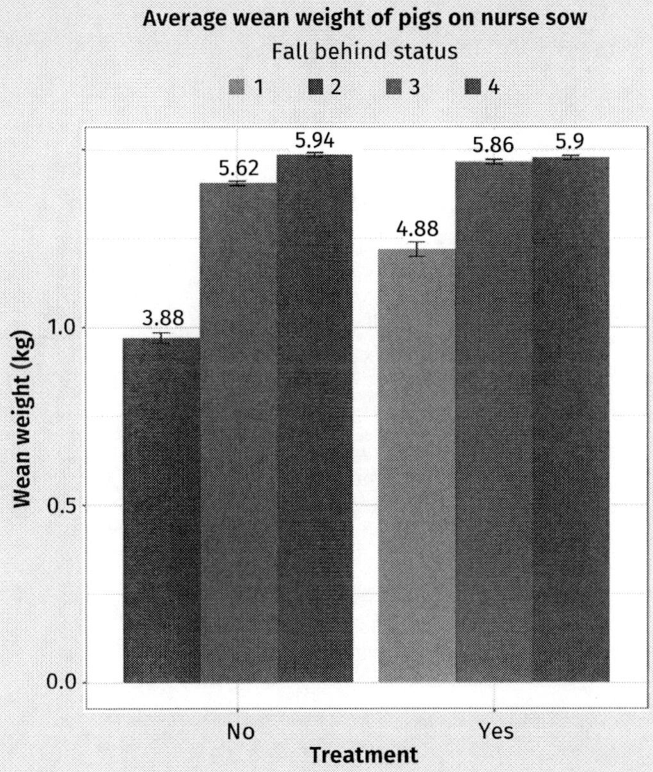

Figure 3: Average 10-week weights of pigs for NoNurse and Nurse treatments with fall-behind status of a pig moved to a nurse sow (1), a pig marked as fall-behind but remaining on their original dam (2), pigs from litters with fall-behind pigs marked (3), and litters with no fall-behind pigs (4).

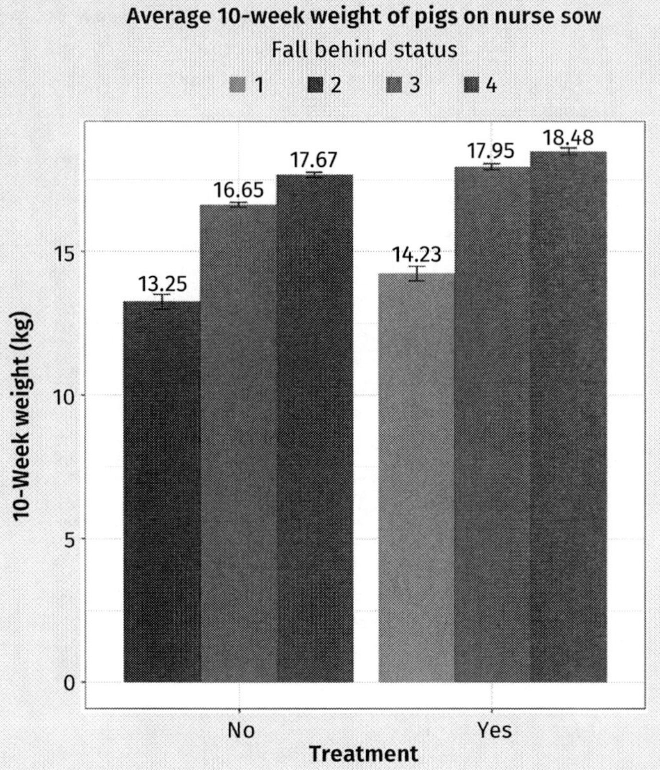

https://doi.org/10.54846/am2023/59

Enterotoxigenic *E coli* populations across sow herds and neonatal piglets were reduced after using a *Bacillus*-based product

S. Hernandez; S. Burmeister; R. Teal; A. Smith, PhD; E. Davis, PhD
Arm and Hammer Animal and Food Production, Waukesha, Wisconsin

Abstract

Enterotoxigenic *E coli* (ETEC) is a top concern across commercial swine industries. This group of *E coli* produce toxins which stimulate the lining of the intestines causing them to secrete excessive fluid resulting in diarrhea. ETEC challenges in the farrowing house and in the nursery impact the health of neonatal and nursery piglets causing economic losses for the swine industry. The purpose of this study was to determine the level of enterotoxigenic *E coli* across sow herds over a sampling period of one year and determine the efficacy of a *Bacillus*-based product introduced into the sow diet. Rectal swabs were collected from gestating sows, lactating sows and neonatal piglets from six different sow farms at five different time periods; September 2020 (Baseline 1), November 2020 (Baseline 2), January 2021 (Treated 1), June 2021 (Treated 2) and September 2021 (Treated 3). Swabs were processed to determine colony forming units (CFU) per swab on *E coli* selective media. Selected isolates (five per swab) were subjected to a multiplex PCR to determine the presence of nine targeted genes; STb, STaP, LTb subunit, K99, F18, 987P, K88, F41 and Stx2e.[1] Random amplification of polymorphic DNA (RAPD) PCR was performed on all isolates determined to contain at least one virulence gene and genetic diversity was determined based on the banding pattern. ETEC were higher before the *Bacillus*-based product was introduced into the system, on average the ETEC decreased from 5.8 Log CFU to 4.1 Log CFU ($P < 0.01$). The percent of swab samples that contained at least one pathogenic *E coli* isolate decreased from 69% to 47% after the *Bacillus*-based product was introduced ($P < 0.01$). The percentage of all *E coli* isolates determined to be ETEC decreased after implementation from 37.3% to 14.6% ($P < 0.01$). These results indicated a decrease in the number of potentially pathogenic *E coli* in the sow herds. The genetic populations of ETEC shifted at each time point of the study and genetic diversity increased after supplementation with the *Bacillus*-based product. According to the RAPD-typing, pathogenic *E coli* isolates that were found in the rectal swabs of neonatal piglets were genetically similar to those found in the sow showing the potential transmission of pathogenic *E coli* from sow to piglet. In conclusion, genetic populations of ETEC change over time, but the *Bacillus*-based product reduced the levels and the presence of pathogenic *E coli* among the sow herds.

Introduction

Enterotoxigenic *E coli* (ETEC) and shiga-toxin *E coli* (STEC) are a top-concerns across commercial swine industries and are important pathogens in post-weaning piglets.[2] Pathogenic *E coli* are the cause of several different diseases in swine that are collectively referred to as colibacillosis. Edema disease (ED) commonly occurs after weaning and is caused by pathogenic *E coli*.[1] There are some effective vaccines that are available that will protect neonatal piglets against diarrhea causing ETEC however, there are no current vaccines that are as effective against ETEC in the nursery.[3] ETEC challenges in the farrowing house and in the nursery impact the health of neonatal and nursery piglets causing economic losses for the swine industry. These pathogens cause decreases in weight gain and increases in mortality especially in the nursery phase of production. Many sows harbor a significant ETEC population within the gastrointestinal tract and could provide a critical control point to mitigate disease transfer to her offspring. Enterotoxigenic *E coli* often harbor specific genes that code for adhesion and colonization of the gastrointestinal tract (K88, K99, F18, F41 and 987P). When attached, these ETEC isolates can produce and excrete enterotoxins (Ltb, STaP and Stb) which cause diarrhea and dehydration. When the ETEC isolate also contains the shiga-toxin (Stx2e) the piglets can show signs of edema disease and suffer neurological signs with mortality rates of up to 90%.[2] The objective of the current study was to determine the level of enterotoxigenic *E coli* across sow herds over a sampling period of one year and determine the efficacy of a *Bacillus*-based product introduced into the sow diet.

Methods

Six sow farms were selected for a study that lasted one year. A total of 1204 rectal swabs were collected from gestating sows, lactating sows and neonatal piglets from the six farms. Samples were collected in September 2020 (Baseline 1), November 2020 (Baseline 2), January 2021 (Treated 1), June 2021 (Treated 2) and September 2021 (Treated 3). Each swab was processed to determine colony forming units (CFU) per swab on *E coli* selective media. A maximum of five isolates were selected from each sample and grown overnight at 37° in Tryptic Soy Broth (TSB). *E coli* isolates were subjected to a multiplex PCR to determine the presence of nine target genes: STb, STaP, LTb subunit, K99, F18, 987P, K88, F41 and Stx2e.[1] *E coli* isolates that contained the virulence genes were determined to be ETEC; these isolates were then subjected to random amplified polymorphic DNA (RAPD) PCR to determine the genetic diversity among the isolates. The diversity of the ETEC community was examined using the RAPD PCR. Principle coordinate analysis of the 1-completment of those similarities was used to visualize the communities in the two-dimensional plots. A group of genetically diverse isolates were selected to be screened against seven different Arm & Hammer™ proprietary *Bacillus* strains and a CERTILLUS™ targeted microbial solution was designed using the two strains that showed the best efficacy against the ETEC populations. ETEC counts were log transformed for statistical analysis. Samples with nondetectable counts were assigned a log CFU value of "1". Data generated from each treatment sampling period was analyzed using the Fit Model option in JMP (version 15.0.0) to generate least squares means and these means were separated using Tukey HSD when the F statistic was significant ($P \leq 0.05$).

Results

Over the course of this study, the ETEC populations shifted across both the sow samples and the piglet samples. Utilizing the five different timepoints, a total of 607 isolates (65 – Sept. 2020; 210- Nov. 2020; 79-Jan. 2021; 159-June 2021; 94-Sept. 2021) were screened against a panel of seven Arm & Hammer proprietary *Bacillus* strains. The first *Bacillus* combination was *Bacillus* strains 747 and 839 which was formulated based on the September 2020 sampling. After the January sampling, there was a shift in the ETEC microbial populations and *Bacillus* strains 839 and 4976 were determined to be the optimal formulation. In September 2021, we did our last sampling and determined the ETEC microbial populations shifted away from previous isolates and the formulation was adjusted to *Bacillus* strains 747 and 839.

Enterotoxigenic *E coli* (ETEC) counts were greater ($P < 0.05$) at the Baseline 2 sampling compared to all other timepoints (Figure 1), however, ETEC counts were greater ($P < 0.01$) before the CERTILLUS targeted microbial solution was implemented (Baseline 1 and 2; 5.8 ± 0.4 log CFU) compared to the timepoints after implementation (Treated 1, 2, 3; 4.2 ± 0.4 log CFU). The level of ETEC increased across the baseline samplings while showing the greatest increase in ETEC levels in the piglets at Baseline 2. After CERTILLUS was implemented into the diet, the levels decreased across sows and piglets (Figure 2).

The percentage of all *E coli* isolates determined to be ETEC was greater at the Baseline 2 sampling compared to all other timepoints ($P < 0.05$). The average percentage of *E coli* isolates that were determined to be ETEC during the baseline samplings was 37.3% compared to 14.6% after CERTILLUS was implemented ($P < 0.05$; Figure 3).

The percentage of samples containing at least one ETEC isolate (prevalence) was greater ($P < 0.05$) at Baseline 2 when compared to all other timepoints. ETEC prevalence was greater ($P < 0.01$) before CERTILLUS was implemented; on average, 69% of the baseline samplings contained at least one ETEC isolate whereas, 47% of swab samples contained at least one ETEC isolate after CERTILLUS was implemented.

The average number of ETEC genes detected within the *E coli* isolates decreased ($P < 0.01$) after CERTILLUS was implemented in sow diets compared to the average number at the baseline samplings (102 vs. 36 ± 19). The greatest number of ETEC genes, Stx2e, STaP and F41, was detected ($P < 0.05$) in the Baseline 2 sampling (Table 1). The number of Stx2e, Stb, STaP, K99, Ltb, and F41 genes detected in *E coli* isolates was less ($P < 0.06$) in the post-CERTILLUS sampling periods compared to the two baseline periods (Table 2).

According to the RAPD fingerprint typing, pathogenic *E coli* isolates that were found in the rectal swabs of neonatal piglets were genetically similar to those found in the sow showing the potential transmission of ETEC from sow to piglet. Visualization of the diversity of confirmed ETEC isolates showed how the populations shifted at each timepoint of the study and genetic diversity increased after supplementation with the CERTILLUS targeted microbial solution (Figure 4).

Discussion

In this study, the different farms showed varying levels of ETEC and the genes associated with pathogenicity across the time periods of these samplings. ETEC isolates were successfully inhibited by more than one Arm & Hammer proprietary *Bacillus* strains and a formulation was recommended. That formulation was updated when the microbial populations shifted. The levels of ETEC before the CERTILLUS targeted microbial solution was introduced was greater than after the product was formulated and added into the diet. This trend extended to the neonatal piglets across the three sampling timepoints after the CERTILLUS targeted microbial solution was introduced into the sow diet. The prevalence across both swab samples and across *E coli* isolates decreased after the CERTILLUS targeted microbial solution was added. The multiplex PCR was able to determine that all of the nine genes associated with pathogenic *E coli* were present in varying amounts across the five sampling timepoints. The greatest number of ETEC genes present was seen during the Baseline 2 sampling that was conducted in November of 2020. This sampling also had the highest CFU levels in swab samples and the highest prevalence of ETEC recovered from the swab samples. This increase could be associated with seasonal effects or an outbreak among the farms, however this sampling design prevented the ability to draw conclusions related to these effects. RAPD fingerprint typing of the samplings showed that isolates that were recovered from sows were genetically related to isolates that were found within the neonatal piglets in the farrowing house this suggests the transmission of those potentially pathogenic *E coli* from sow to piglet. In conclusion, genetic populations of ETEC change over time, but a customized CERTILLUS targeted microbial solution reduced the levels and the presence of pathogenic *E coli* among the sow herds. These findings could have implications related to the transmission of pathogenic *E coli* from sow to piglet.

References

1. Casey, T.A., Bosworth B.T. (2009). Design and evaluation of a multiplex polymerase chain reaction assay for the simultaneous identification of genes for nine different virulence factors associated with Escherichia coli that cause diarrhea and edema disease in swine. J Vet Diagn Invest.21:25-30.

2. Hrala M, Bosák J, Micenková L, Křenová J, Lexa M, Pirková V, Tomášťíková Z, Koláčková I, Šmajs D. 2021. *Escherichia coli* strains producing selected bacteriocins inhibit porcine enterotoxigenic *Escherichia coli* (ETEC) under both in vitro and in vivo conditions. Appl Environ Microbiol 87:e03121-20. **https://doi.org/10.1128/AEM.03121-20**

3. Nagy B., Fekete P. (2005). Enterotoxigenic *Escherichia coli* in veterinary medicine. Int J Med Microbiol 295: 443-454.

Figure 1: Levels of ETEC (log CFU) across the five different samplings ([a,b] $P < 0.05$).

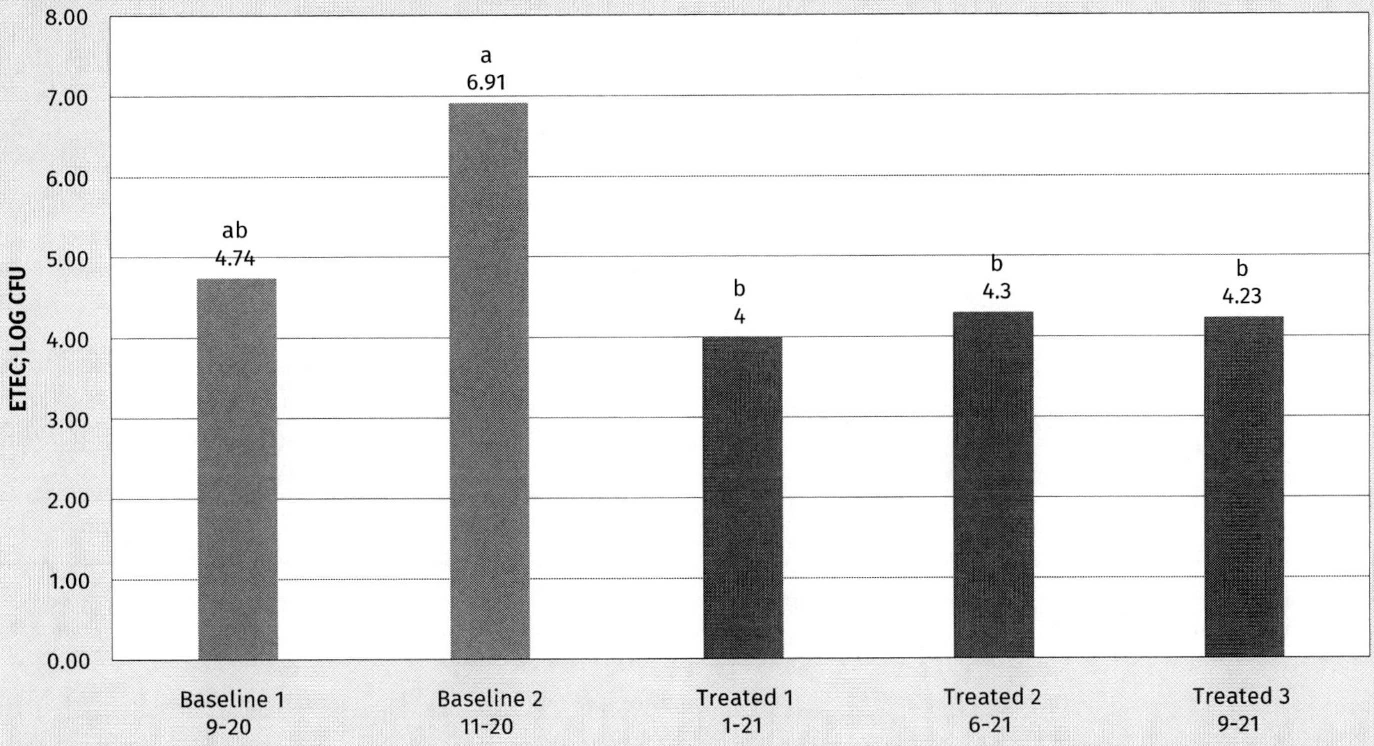

Figure 2: Levels of ETEC (log CFU) across timepoints and sample source.

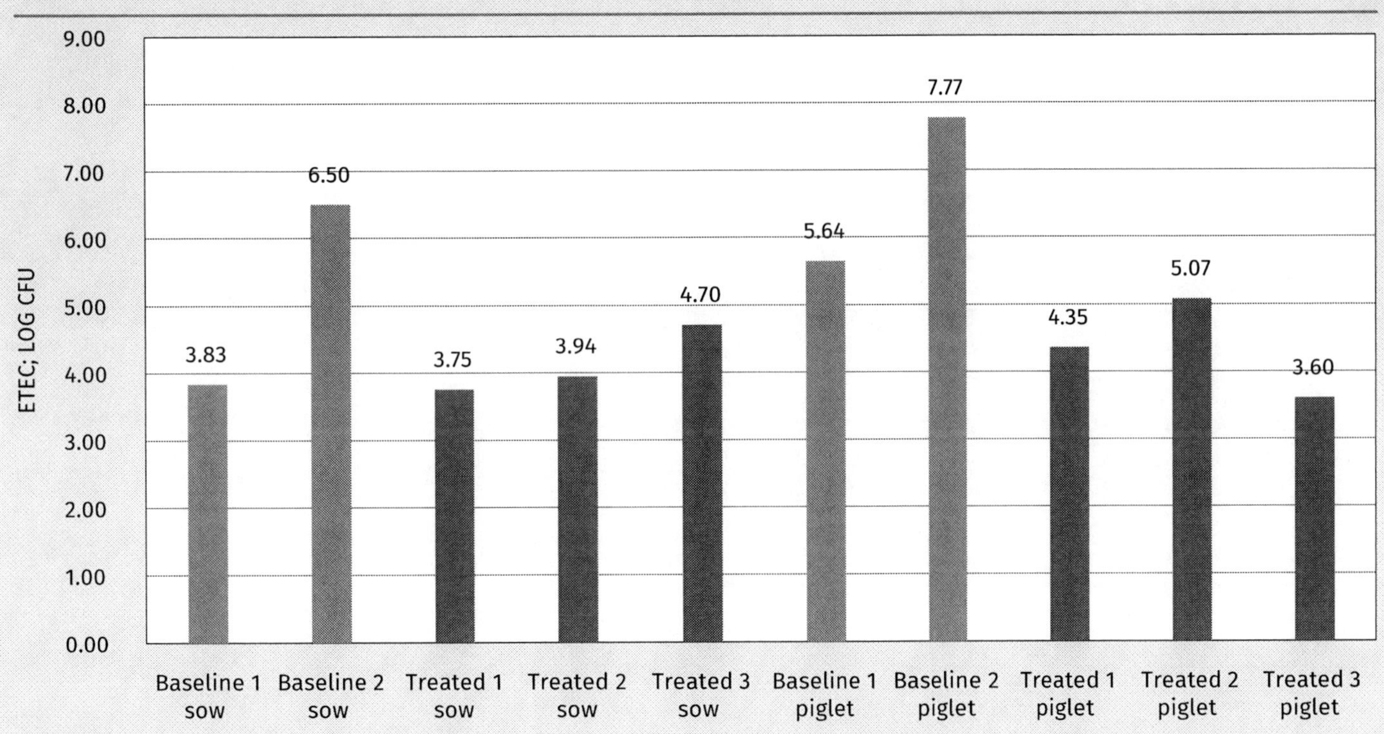

Figure 3: ETEC (%) prevalence before and after the implementation of CERTILLUS ([a,b] $P < 0.01$).

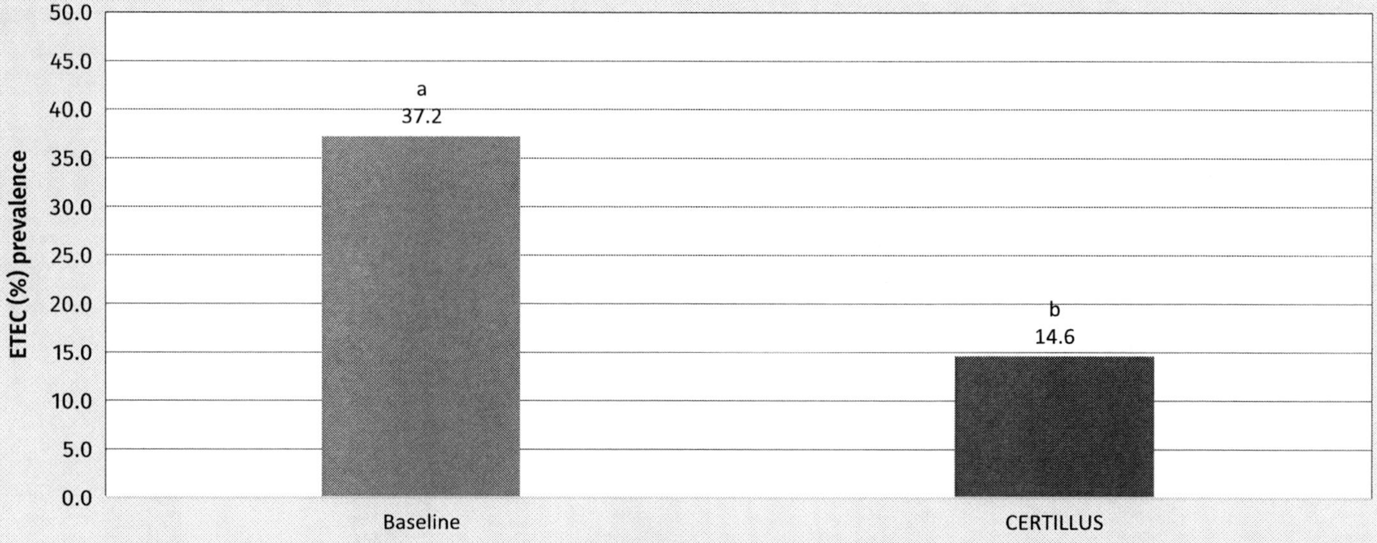

Table 1: Total virulence genes associated with each sampling timepoint ([a,b] $P \leq 0.05$)

	Stx2e	Ltb	Stb	STaP	K99	F18	987P	K88	F41
Baseline 1, 9-20	2.8[b]	8.4	16.4	11.0[a,b]	7.4[a]	2.4	0.2[b]	2.4	12.4[a,b]
Baseline 2, 11-20	16.2[a]	11.0	45.7	33.3[a]	2.3[a,b]	0.5	3.5[a,b]	2.5	27.0[a]
Treated 1, 1-21	2.3[b]	4.5	0.8	14.3[a,b]	2.3[a,b]	4.2	0.3[b]	2.0	2.0[b]
Treated 2, 6-21	4.0[b]	7.9	1.5	10.2[a,b]	1.5[b]	2.3	6.8[a]	4.5	7.5[a,b]
Treated 3, 9-21	4.0[b]	0.8	3.7	1.2[b]	1.5[b]	2.0	3.3[a,b]	4.0	8.0[a,b]

Table 2: Contrast statements showing the average virulence genes present in the baseline samplings compared to the treated sampling

Virulence gene	Baseline	Treated	SE	$P =$
Stx2e	9.49	3.44	2.4	0.02
Stb	31.04	2.00	11.6	0.02
STaP	22.15	8.73	6.9	0.06
K99	4.87	1.78	1.1	0.01
Ltb	9.70	4.22	2.7	0.05
F41	19.70	3.50	4.7	0.01

Figure 4: Visualization of the diversity of confirmed ETEC isolates. **Pane A:** all of the isolates obtained from each sampling time point; **Pane B:** Baseline 1 and Treated 3 sampling timepoints only; **Pane C:** the average location for each of the timepoints. Blue-Baseline 1; Green-Baseline 2; Red-Treated 1; Orange-Treated 2; Black-Treated 3.

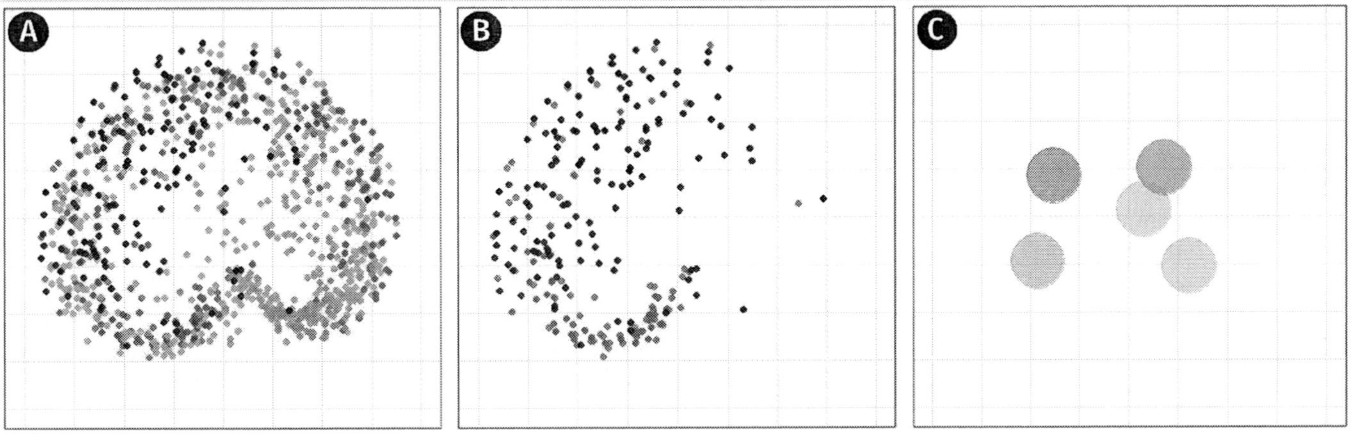

https://doi.org/10.54846/am2023/60

Efficacy of Tilmovet®AC for PRRS control by water medication in weaned pigs

Ulrich Klein[1]; Wouter Depondt[1]; Thomas Fangman[2]; David Nolan[2]
[1]Huvepharma® NV Belgium; [2]Huvepharma® Inc. Peachtree City, Georgia

Introduction

Porcine reproductive and respiratory syndrome (PRRS) is recognized as the most endemic and economically important disease of swine worldwide. PRRSV causing PRRS is responsible for respiratory disease in weaned and growing pigs, as well as productive failures in sows.

It is well known that pulmonary alveolar macrophages (PAMs) are the primary target of PRRSV infection and location of virus replication. It has been demonstrated that tilmicosin exhibits PRRSV-antiviral effects when concentrated in the pulmonary alveolar macrophages.[5] Rapid uptake of tilmicosin throughout the respiratory tract and highest intracellular accumulation in comparison to other macrolide antibiotics in lung macrophages, is the basis for its blocking effect of PRRSV replication and the high antibacterial effect of Tilmovet®.[3,9,10,12-14]

A reduction in clinical signs and performance improvements were shown in PRRS infected pigs when medicated with tilmicosin.[4] In most of these studies animals were experimentally infected with PRRSV and medicated with tilmicosin via feed.[2,7,8] Only a limited number of studies are published in which tilmicosin was administered via drinking water to control the PRRS infection after PRRSV inoculation[1] or PRRS naturally infected nursery pigs.[6]

The purpose of this paper is to describe two clinical cases of PRRSV-infected weaned pigs where Tilmovet AC (tilmicosin phosphate) was administered via water to control PRRSV infection and to present data on viremia, antibody titers, clinical and production parameters and severity of PRRS infection.

Materials and methods

Study No.1: The study was conducted in Taiwan with animals from a farrow-to-finish herd infected with PRRSV. 160 four-week-old-weaned piglets from a PRRSV-infected herd were equally divided into four treatment groups (40 piglets /group), (see Table 1).

Parameters assessed: blood samples (20 pigs/group) for viremia assessment (qPCR) and PRRSV antibodies (ELISA Herd Check, IDEXX Laboratories) determination before/after Tilmovet® water medication in different age groups (4,6,8,10,12 weeks of age). ELISA cut-off value: S/P ratio ≥ 0.4 = positive for antibody against PRRSV. DWG, survival rate (%), duration of PRRSV viremia, serum PRRSV load were determined. On the Taiwan farm diverse bacterial pathogens (*G parasuis, M hyopneumoniae, A pleuropneumoniae*) were present during the experiment.

Study No.2: The study was conducted in a transition farm in Spain. The farm was selected based on PRRSV outbreaks two months before study onset and the presence of other respiratory pathogens (*G parasuis, M hyopneumoniae*). 800 animals divided into two groups at arrival. Group 1: Tilmovet AC treatment at 3 time periods of 5 consecutive days (dose 15mg/kg bw). Between the three medication time periods, non-treatment periods of 14 days were established (see Table 2).

Parameters assessed: blood samples for assessment of viremia (qPCR) and PRRSV antibodies (ELISA); PRRSV clinical signs (nasal discharge, cough, behavior, body condition), DWG, FCR, health status, mortality rate. ELISA cut-off value: S/P ratio ≥ 0.4 = positive for antibody against PRRSV. On the Spanish farm diverse bacterial respiratory pathogens (*G parasuis, M hyopneumoniae*) were present during the study.

Table 1: Description of treatment groups in study No.1 (Taiwan)

Treatment groups	Dosage (mg/kg bw)	Treatment duration	Age medication period
Group 1 (Tilmovet AC)	20mg/kg bw	5 days (1-time)	4 weeks of age (start day 0)
Group 2 (Tilmovet AC)	15mg/kg bw	5 days (2-times)	4 & 6 weeks of age (start day 0 & day 14)
Group 3 (Tilmovet AC)	15mg/kg bw	5 days (3-times)	4, 6 & 8 weeks of age (start day 0, 14 & day 28)
Group 4 (control)	–	0 days	No medication

Table 2: Description of treatment groups in study No.2 (Spain)

Treatment groups	Dosage (mg/kg bw)	Treatment duration	Age medication period
Group 1 (Tilmovet AC)	15mg/kg bw	5 days (3-times)	4, 6 & 8 weeks of age (start day 0, 14 & day 28)
Group 2 (control)	–	0 days	No medication

Results and discussion

Study No.1

PCR results verify early PRRS infection in a low percentage of pigs. At the start of the study (4 weeks of age) and at 6 weeks of age lower percentage of PRRSV-positive pigs were found in Tilmovet®-medicated group 1 (10%/15%), group 2 (0%/5%) and group 3 (0%, 25%) in comparison to the non-medicated control group (30%/65%). A higher PRRS virus burden from trial onset until study end was determined in non-medicated pigs. This results in higher PRRSV antibody levels (S/P ratios) in the control group throughout the study (see Figure 1). Tilmovet® water medication led to depression of PRRS virus replication and to lower antibody titers in medicated pigs at 4, 6, 8, 10 and 12 weeks of age in comparison to the non-medicated pigs at these time periods.

A retardation effect on PRRS infection in medicated pigs were determined based on Tilmovet® water medication.

A higher survival rate from age week 4 to 12 was determined in the 2-time (95%) and 3-time (87.5%) Tilmovet®- medicated group in comparison to the control and the 1-time medicated group (82.5% each). Higher survival rate in tilmicosin medicated pigs is based on its PRRSV antiviral effect together with the very effective control of respiratory bacteria pathogens present on the farm.

A higher daily weight gain was determined in the pigs with 2-time and 3-time Tilmovet® medication in comparison to the control and the 1-time medication group (see Table 3).

Study No.2

A higher level of PRRSV-positive pigs at study start was identified in comparison to study 1. The percentage of pigs PCR-positive at the start of the study (3 weeks of age) was 51.25% in the control group and 45.57% in the Tilmovet® treatment group.

For the evaluation of the effect of Tilmovet® water medication three subgroups were created according to the piglet's qPCR results (ct value) at the enrolment of the study. The three subgroups were composed of those pigs that were negative to qPCR (ct value > 40), those that had low viral load at the entrance (ct value between 30 - 40) and those animals with high viremia on the day of the enrolment (ct value < 30).

In PCR-negative pigs at the beginning of the study (3 weeks of age) a positive effect of Tilmovet® water medication was determined. After the infection peak at 6 weeks of age the PRRS infection was reduced by 8% (week 8) and by 14% (week 10). In non-medicated pigs the percentage of positive pigs increased by 8% on week 8 and by 5 % on week 10 (see Figure 2).

Figure 1: PRRSV ELISA S/P ratio of the four treatment groups during the study.

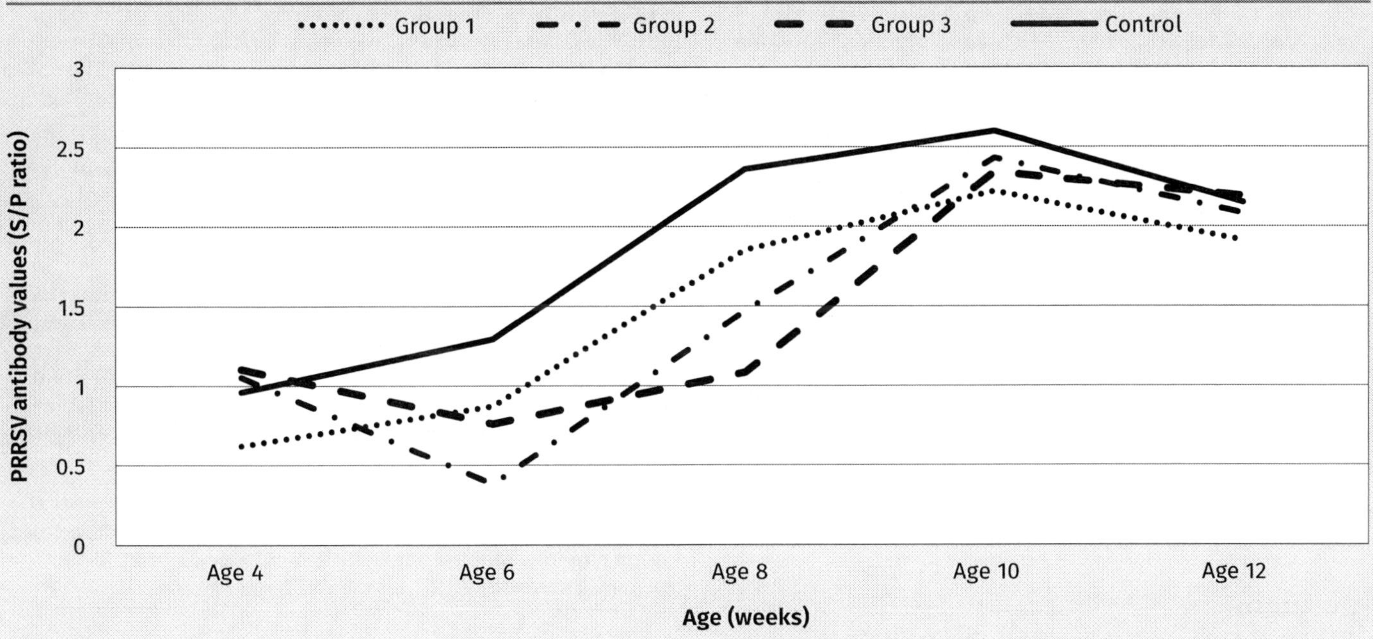

Table 3: Body weight and average daily gain (ADG) of pigs in the four treatment groups

Age (weeks)	Group 1	Group 2	Group 3	Control
4	5.60	6.89	6.72	6.61
12	23.0	28.64	27.33	26.57
ADG (kg)	0.31[b]	0.38[a]	0.36	0.35

[a,b] different letters indicate stat. sign. difference ($P < 0.05$).

Figure 2: Evaluation viral load of negative piglets at study start, from the beginning to the end of the experimental trial.

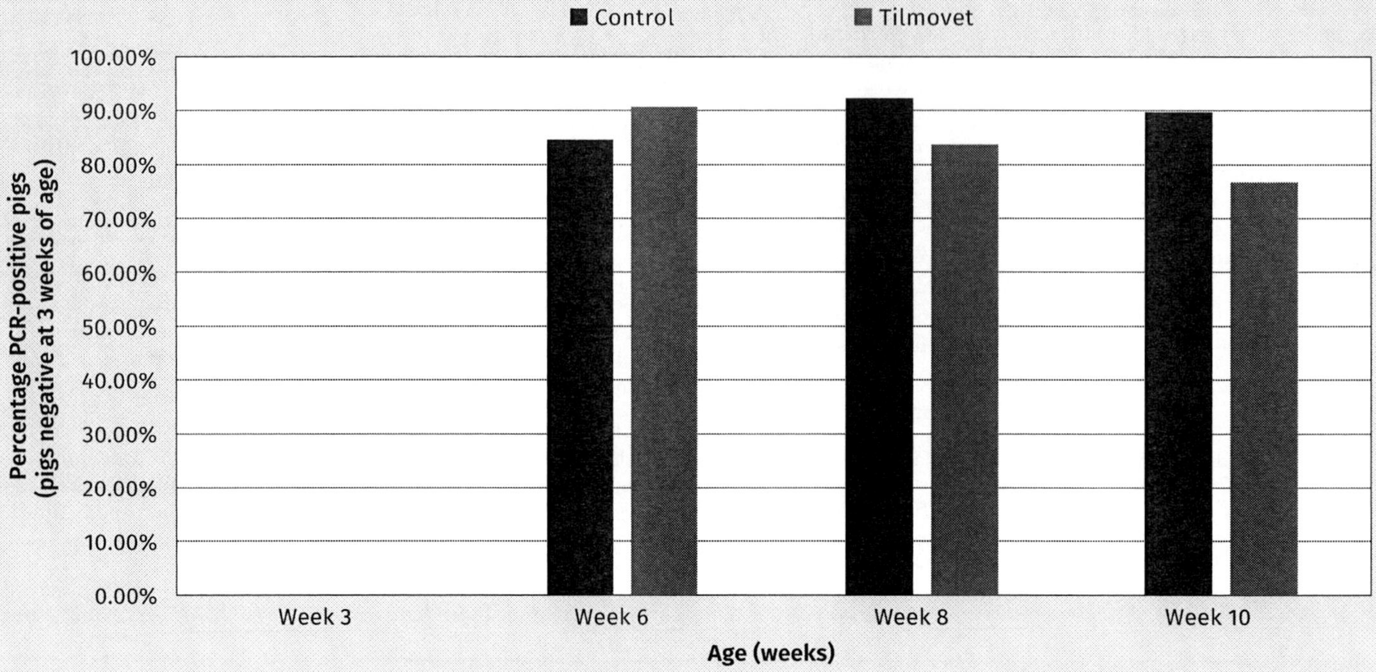

Low level of PCR-positive pigs at the beginning of the study in Tilmovet® medicated group showed a higher infection reduction by 25% (6 weeks of age) and by 47% at 8 weeks of age vs. the control group, in which the viremia decreased by 15% on age week 6 and 8% on age week 8 (see Figure 3).

In high level PCR-positive pigs at the beginning of the study, a similar infection reduction by 25% at 8 weeks of age and 40% at 10 weeks of age in the Tilmovet® group vs. the control group with 29% (week 8) and 40% (week 10) was found (see Figure 4).

A higher survival rate at the study end was determined in the Tilmovet® group (83%) in comparison to the control group (78%). Relative low survival rates in both groups were based on the highly pathogenic PRRSV strain and the first disease outbreak caused by this strain on the Spanish farm.

A higher daily weight gain was determined in pigs with Tilmovet® water medication in comparison to the control group (see Table 4).

A higher daily weight gain was determined in pigs with water medication in comparison to the control group.

Conclusion

Study results were generated in farms being naturally affected by PRRSV infections. Different infection levels at study start were determined in farm 1 (low infection level) and farm 2 (high infection level). In both farms secondary respiratory pathogens were present during the study.

3-time Tilmovet® AC administration had the best effect on the clinical expression of PRRS infection as verified by differences in the S/P ratio obtained with the ELISA test, the daily weight gain and higher survival rate.

The positive effect of 3-time Tilmovet® AC medication was shown after the viremia peak. In PCR negative pigs and low percentage of PCR-positives at the start of the study, a higher reduction of PRRSV infection was found in comparison to the control group after the viremia peak. This indicates a reduction of PRRSV replication and virus spreading in the pigs medicated post-weaning.

Tilmovet® AC can help reduce PRRSV load in farms after the PRRSV viremia peak has been reached. The effect on reduction in groups with low percentage of PCR-positive and viremic pigs seems to be higher in comparison to groups with high percentage of positive and viremic animals.

Tilmovet® AC medication programs in PRRS naturally infected herds with diverse viremia situations and the presence of respiratory bacteria pathogens (*G parasuis, M hyopneumoniae, A pleuropneumoniae*) can effectively stabilize the PRRS infection situation and prevent severe PRDC infections in weaned pigs. A better effect is expected in viremic situations with a lower percentage of positive pigs.

Consequently, the implementation of Tilmovet® AC 3-time water medication programs after farrowing can be considered as a highly effective tool to control PRRSV infections in PRRSV-viremic herds and with PRDC problems.

Figure 3: Evaluation of viral load of low positive piglets at study start, from the beginning to the end of the experimental trial.

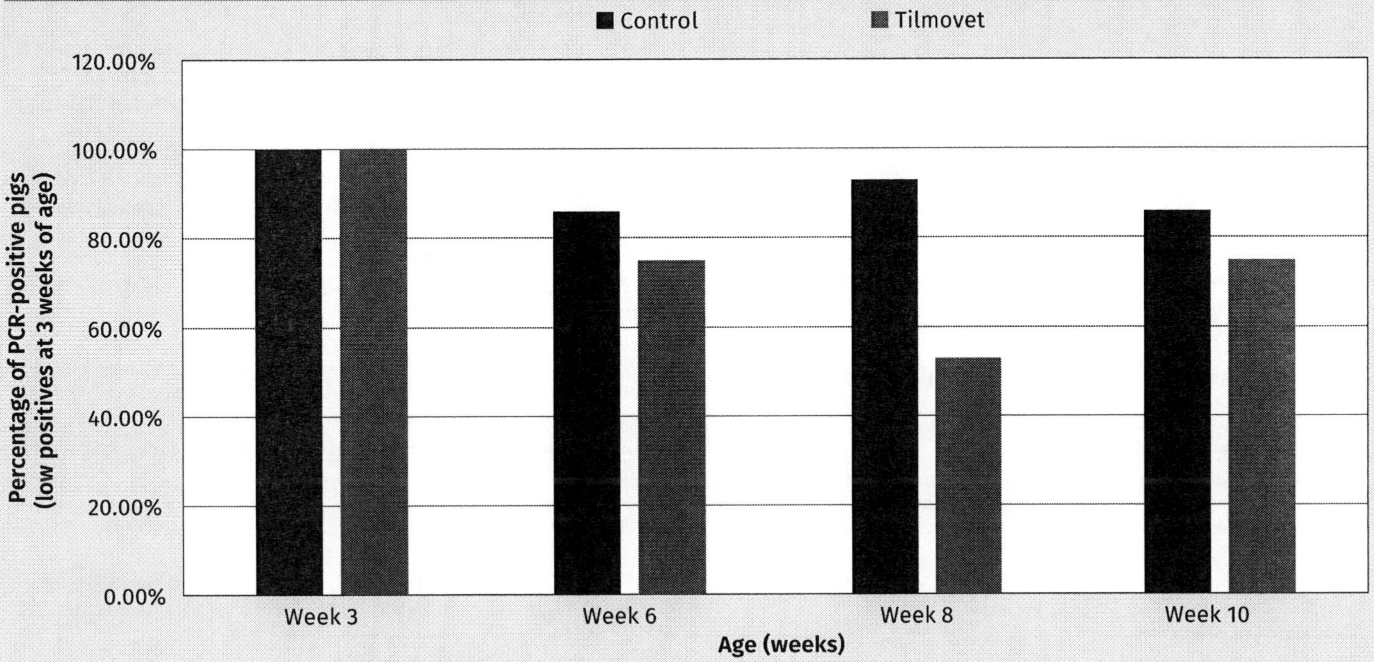

Figure 4: Evaluation of viral load of high positive piglets at study start, from the beginning to the end of the experimental trial.

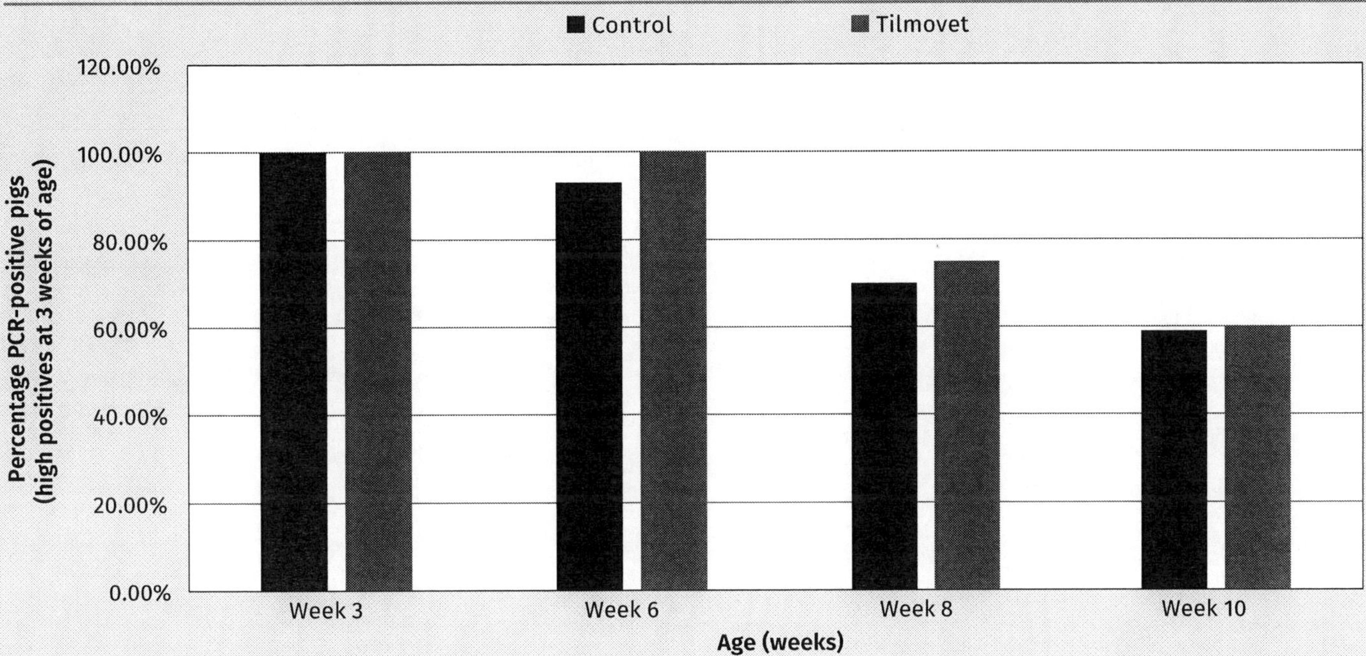

Table 4: Body weight and average daily gain (ADG)* of pigs in the two treatment groups

Age (weeks)	Group 1 (in kg)	Control (in kg)
3	5.56	5.30
12	19.47	17.97
ADG (kg)	0.30	0.27

* results on stat. evaluation not available at abstract submission date

References

1. Batista, L. *et al* (2009). Evaluation of the effect of tilmicosin administered in drinking water on nursery pigs inoculated with Porcine Reproductive and Respiratory Syndrome virus. Proc. American Association of Swine Practitioners, pp.173-178

2. Benfield, D.A. *et al.* (2002). An evaluation of the effects of tilmicosin in feed of nursery pigs inoculated with porcine reproductive and respiratory syndrome virus. Proc. American Association of Swine Practitioners, pp.87-91.

3. Blais, J. and S. Chamberland (1994). Proc. 13th IPVS Congress Bangkok, p.331.

4. Depondt, W. *et al.* (2014). Stabilizing PRRSV with Tilmovet®. Proc. 6th ESPHM Congress Sorrento, p.163.

5. Du, Y. *et al.* (2011). Antiviral activity of tilmicosin for Type 1 and Type 2 Porcine Reproductive and Respiratory Syndrome Virus in cultured Porcine Alveolar Macrophages. J.Antivir. Antiretrovir. Volume 3, pp.28-33. **doi :10.4172/jaa.1000031.**

6. Fraile, L. *et al.* (2004) Reduction of PRRSV and Respiratory Clinical Disease by Tilmicosin. Proceedings 18th IPVS Hamburg, Vol. 1, page 38.

7. Lehe, K. *et al.* (2003). Case study: effect of Pulmotil® on mortality and growth in nursery pigs with clinical PRDC initiated by PRRSV. Proc. American Association of Swine Practitioners, pp.129-130.

8. Molitor, T.W. *et al.* (2001). Tilmicosin affects Porcine Reproductive and Respiratory Syndrome Virus replication. Proceedings Leman Conference, 28, 31-32.

9. Hawser, S. (2022). European Collection of Veterinary Pathogens (VetPath V) – minimal inhibitory concentration (MIC) testing, IHMA Study Report 3638 September 2022.

10. Pridmore, A. (2018). European Collection of *Mycoplasma* Pathogens (MycoPath II): collection, storage, identification and MIC determination for *Mycoplasma* strains isolated from cattle, pigs and poultry. Don Whitley Scientific Study Report 015/16.

11. Lin, C.-N. et al. (2016). Tilmicosin reduces PRRSV loads in pigs in vivo. Journal of Agricultural Science, Vol.8, No.1 154-162.

12. Scorneaux, B., Shryock, T. (1998). Intracellular accumulation, subcellular distribution and efflux of tilmicosin in swine phagocytes. J. Vet. Pharmacol. Ther.21(4): pp.257-268.

13. Stoker, J. et al. (1996). Conc. Of tilmicosin in pig serum and respiratory tissue following administration via feed at level of 400g/tonne. Proc. 14th IPVS Congress Bologna, p.656.

14. Stuart, A.D. et al (2008). Tylvalosin a macrolide antibiotic inhibits the in-vitro replication of European and American PRRSV. The Pig Journal, 61, 42-48.

https://doi.org/10.54846/am2023/61

Impacts of dietary salinomycin and crude protein on coccidiosis burdened nursery pigs

M. Nisley[1]; C. Sparks[2]; D. Nolan[2]; E. R. Burrough[3]; N. K. Gabler[1]
[1]Department of Animal Science, Iowa State University, Ames, Iowa; [2]Huvepharma Inc., Peachtree City, Georgia; [3]Department of Veterinary Diagnostic and Production Animal Medicine, Iowa State University, Ames, Iowa

Introduction

Endemic enteric challenges antagonize nursery pig livability and performance. Recent limitations in the therapeutic control of coccidia has increased the prevalence of coccidiosis in suckling and nursery pigs. Enteric pathogens such as coccidia, in aggregation with rotaviruses, pathogenic *E coli*, and excess hind gut nutrient fermentation can induce intestinal inflammation and consequently compromise intestinal function and health. Feeding nursery pigs high crude protein (CP) diets can have negative implications on overall pathogen burden and increase scouring incidence.[2] Therefore, our objective was to evaluate the effect of salinomycin, an antimicrobial coccidiostat ionophore, and dietary crude protein levels on performance, livability, and coccidia shedding in endemically enterically challenged nursery pigs.

Methods

One hundred and twenty mixed sex pigs (5.63 ± 1.4 kg BW) were sourced from a coccidia, rotavirus, and hemolytic *E coli* positive sow farm and were randomly allotted to single sex pens in a 2 × 2 factorial design (10 pens/treatment) (Figure 1). Pigs were fed either a high CP (HCP; 24% CP) or Low CP (LCP; 18% CP) diets with or without salinomycin. All pigs were fed in 3 phases lasting 9, 12, and 21 days, respectively. Salinomycin was fed in phases 1 and 2 at 90 ppm, and at 60 ppm in phase 3. Pen bodyweights and feed disappearance were collected at the start and end of each phase to calculate average daily gain (ADG), average daily feed intake (ADFI), and gain to feed ratio (G:F). Fecal consistency was scored for diarrhea presence daily on a 0-3 scale, with 1 being normal and 3 being loose diarrhea. Within pen, pooled fecal samples were collected on days 9 and 20 for assessment of coccidia oocyst counts by fecal float. On day 9, one pig per pen, based on median ADG within pen, was euthanized for ileum and colon tissue collection and histopathology assessment of villous atrophy, coccidia presence, and colitis. Performance data, mortality/removals, and fecal coccidia presence were analyzed for the fixed effects of CP, salinomycin, and their interaction. Pen as the experimental unit and significance set at $P \leq 0.05$ and tendency set at $P \leq 0.10$.

Results

Fecal coccidia oocysts counts were reduced by 85% ($P < 0.05$) and 80% ($P = 0.066$) at day 9 and 20, respectively in salinomycin versus control fed pigs. The primary species of coccidia observed in the feces was *Cystoisospora suis*. LCP and salinomycin fed pigs both individually decreased diarrhea incidence during phase 2 ($P < 0.05$; Figure 2). During phase 2 and overall (d0 - 42), salinomycin treated HCP pigs observed reduced diarrhea compared to unmedicated pigs ($P < 0.05$ and $P = 0.09$, respectively; Figure 2). Pigs fed salinomycin had reduced removal rates of 10 and 0%, compared to 15 and 30% for non-medicated LCP and HCP pigs ($P < 0.05$). Irrespective of CP, salinomycin fed pigs had longer ileum villus length compared to untreated pigs (326 vs. 269 µm, respectively, $P < 0.05$). Histopathology analysis of ileum tissue revealed a 50% and 100% reduction in oocyte presence in LCP and HCP pigs fed with salinomycin relative to their non-salinomycin counterpart (Figure 3 and 4). These data corresponded to the fecal floatation analysis. No CP by salinomycin interactions for performance metrics were observed. In phase 1, no main effects were observed with pig performance. However, compared to HCP, LCP tended to increase ADFI in phase 2 (0.44 vs. 0.52 kg/d, respectively, $P = 0.064$), but no ADG and G:F differences were reported. During phase 3, compared to non-medicated pigs, salinomycin increased G:F ($P < 0.05$) and tended to increase ADG ($P = 0.080$). Overall 42-day nursery performance was not different between treatments (Table 1).

Figure 1: Pen of pigs on day 3 of study.

Discussion

Without labor and facility design constrictions, deep cleansing of farrowing facilities with the proper disinfectant can provide a valuable preventative measure in coccidial oocyte transfer from dam to piglet.[1] Unfortunately, with the state of labor for many farrowing facilities it is not always realistic to mitigate coccidiosis via preventative cleaning. Further, producers have limited regulatory approved pharmacological tools to mitigate coccidia in suckling and nursery pigs. Our results suggest the utilization of in-feed salinomycin in freshly weaned pigs is a viable way to improve livability and potentially decrease diet-dependent scouring. Salinomycin alone or dietary CP protein level did not influence nursery pig performance in this endemically enterically challenged nursery flow. In conclusion, although not approved for use in pigs in the US, if allowed, salinomycin can mitigate coccidia oocyte numbers and coccidiosis. Further, this in-feed pharmaceutical can improve pig livability and viability of enterically challenged nursery pigs.

References

1. Hinney, B., V. Cvjetković, D. Espigares, J. Vanhara, C. Waehner, B. Ruttkowski, R. Selista, D. Sperling, and A. Joachim. 2020. *Cystoisospora suis* control in Europe is not always effective. Frontiers in Veterinary Science 7.

2. Zhang, H., N. van Wielen, B. van Hee, J. Wang, W. Hendriks, and M. Gilbert. 2020. Impact of fermentable protein, by feeding high protein diets, on microbial composition, microbial catabolic activity, gut health and beyond in Pigs. Microorganisms 8:1735.

Figure 2: Treatment on fecal consistency.

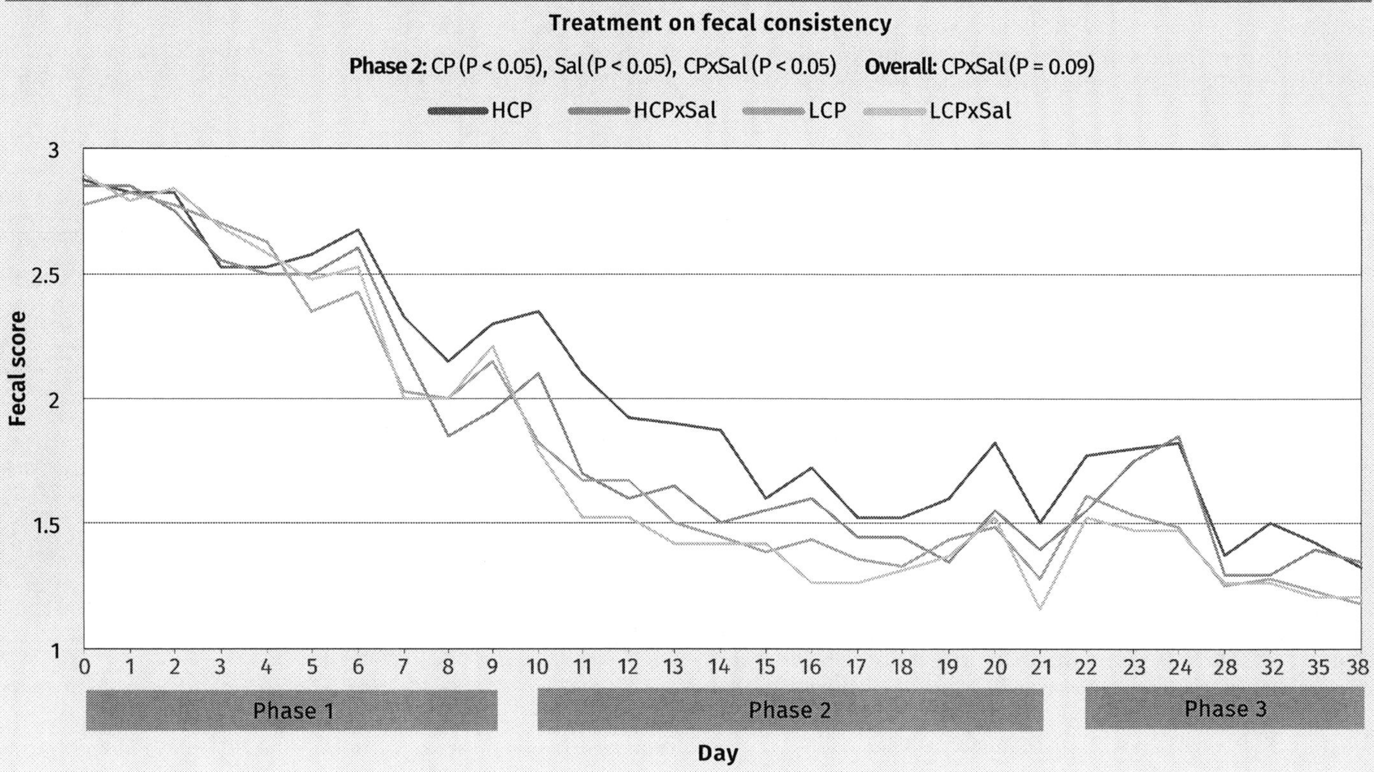

Figure 3: Ileum histology of salinomycin treated HCP pig.

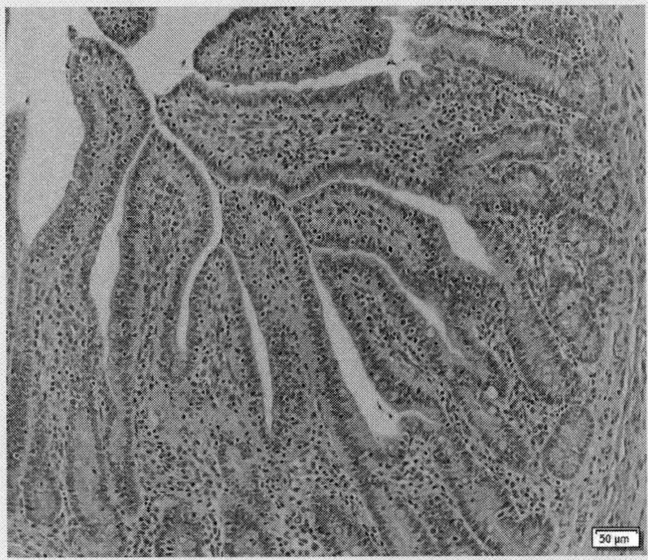

Figure 4: Ileum histology of untreated HCP pig infected with *Cystoisospora suis*.

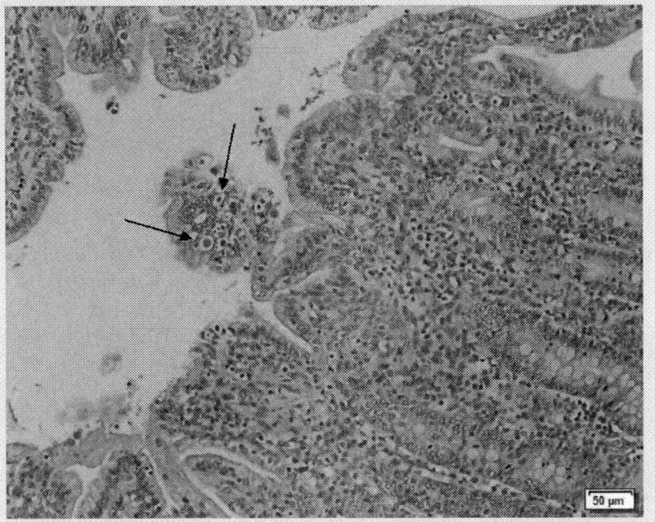

Table 1: Overall 42-day nursery performance (n = 10 pens per treatment)

Parameter		Low crude protein		High crude protein		SEM	*P*-value		
	Salinomycin	No	Yes	No	Yes		CP	Sal	CP x Sal
BW d0, kg		5.9	5.7	5.3	5.6	0.367	0.354	0.819	0.459
BW d9, kg		7.45	7.2	6.5	7.3	0.606	0.503	0.682	0.414
BW d14, kg		9.4	9.4	7.9	8.8	0.839	0.217	0.608	0.624
BW d21, kg		12.9	12.9	12.2	11.6	0.928	0.288	0.740	0.740
BW d42, kg		26.0	27.2	24.4	24.9	0.260	0.211	0.570	0.810
Phase 1, d0-9									
ADG, kg		0.17	0.16	0.14	0.19	0.035	0.921	0.620	0.488
ADFI, kg		0.22	0.19	0.17	0.20	0.035	0.573	0.910	0.399
G:F		0.69	0.91	0.79	0.91	0.114	0.667	0.146	0.660
Phase 2, d10-21									
ADG, kg		0.50	0.52	0.52	0.39	0.049	0.265	0.310	0.147
ADFI, kg		0.52	0.52	0.44	0.44	0.041	0.064	0.894	0.971
G:F		1.03	1.01	1.20	0.89	0.108	0.741	0.118	0.162
Phase 3, d22-42									
ADG, kg		0.62	0.68	0.58	0.64	0.032	0.169	0.080	0.915
ADFI, kg		0.94	1.01	0.95	0.91	0.063	0.473	0.802	0.392
G:F		0.66	0.68	0.63	0.70	0.019	0.834	0.026	0.100
Overall, d0-42									
ADG, kg		0.48	0.51	0.45	0.46	0.029	0.178	0.502	0.617
ADFI, kg		0.60	0.62	0.53	0.58	0.047	0.262	0.404	0.785
G:F		0.83	0.84	0.89	0.80	0.044	0.865	0.372	0.247

https://doi.org/10.54846/am2023/62

Factors affecting different claw lesions in sows: Lessons from a partial least square approach

T. Kramer[1], DVM, MSc; A. Klein[1], DVM; A. Cornelison[1], MS; W. Schweer[1], PhD; Z. Rambo[1], PhD; M. Socha[1], PhD; G. C. Alberton[2], DVM, PhD; L. Rodrigues[1], DVM, PhD
[1]Zinpro Corporation, Eden Prairie, Minnesota; [2]Federal University of Parana, Palotina Sector, Palotina, PR, Brazil

Introduction

In relevance to welfare and productivity, sow lameness, which is mainly impacted by claw lesions. is one of the greatest concerns faced by modern, intensive swine operations. The objective was to explore, using Partial Least Squares Analysis (PLS), the factors orchestrating occurrence and severity of specific claw lesions in sows.

Methods

A total of 33 Brazilian pig herds were selected. Sows [n = 7,650; 10% of inventory size within each herd, up to a maximum of 100 sows per herd) were randomly selected and scored from 0 (no lesions) to 3 (severe lesions) for claw lesions in the hind legs after parturition].[1] Lesions included heel overgrowth and erosion (HOE), heel-sole crack (HSC), separation along the white line (WL), horizontal (CHW) and vertical (CVW) wall cracks, and overgrown toes (T), or dewclaws (DC). A lesion index (L-Index) was calculated as the sum of the higher scores from each lesion, per sow. Information about region (Southeast, South, Midwest), mineral- (zinc, manganese, and copper) amino acid complex supplementation (ZN; supplemented or not) and duration of supplementation (ZN-T; months), housing type during gestation (individual vs. group), inventory size [(small, 250 - 650), (mid-small, 650 -1050), (mid-large, 1050 - 3250), (large, > 3250)], and parity (1 to 7+) were collected. The information collected was used to predict each lesion using PLS, by selecting variables with a Variable Importance in the Projection (VIP) greater than one.

Results

Heel overgrowth and erosion, CHW, T, and L-Index were mostly impacted by ZN (2.543 - 2.655) and ZN-T (2.096 - 2.294). Heel-sole crack was mostly impacted by ZN (2.341 - 2.599), ZN-T (1.465 - 1.761), and inventory size (0.820 - 1.552). Separation along the white line was mostly impacted by ZN (1.908 - 2.133), ZN-T (1.766 - 1.982), and housing type during gestation (0.916 - 1.186). Vertical wall cracks were mostly impacted by ZN-T (1.870 - 2.164), ZN (1.829 - 2.187), region (1.393 - 1.804), and inventory size (0.871 - 1.459). Dewclaws were mostly impacted by parity (1.755 - 1.990), ZN (1.696 - 1.917), ZN-T (1.413 - 1.603), and region (1.083 - 1.312).

Conclusions

Mineral- (zinc, manganese, and copper) amino acid complex supplementation and duration of supplementation had a marked influence on all the claw lesions. Furthermore, inventory size, housing type during gestation, region, and parity appear to influence specific lesions. Such information may help to establish appropriate monitoring and nutritional strategies to prevent and control lameness in sows.

Reference

1. Deen, J., Schuttert, M., Van Amstel, S., Ossent, P., & Van Barneveld, R. (2009). Feet first from Zinpro: lesion scoring guide. *Zinpro Corporation, Eden Prairie, MN, USA.*

https://doi.org/10.54846/am2023/63

Addressing the "why" and "how": Tips for integrating telemedicine into swine veterinary practice

Gilbert Patterson, VMD, MPH, DACVPM
VetNOW, Canonsburg, Pennsylvania

The use of telemedicine has increasingly become part of modern swine veterinary practice. As communication and data-capture technologies advance, both the quality and quantity of information that can be remotely extracted and shared among swine practitioners has grown significantly. While the future opportunity of telemedicine in swine practice appears promising, many questions are still being sorted out surrounding the effective, legal, and appropriate use of telemedicine. Furthermore, even though many swine veterinarians have been practicing telemedicine unofficially for some time, many would be reluctant to say that they provide telemedicine as an bona fide service offering to their clients. Despite relying heavily on digital communication, there remains a general lack of acumen among swine veterinarians when it comes to developing a fully operational, effective, and profitable telemedicine program for modern swine practices.

There are a number of key issues that are stymieing swine veterinarians from fully embracing telemedicine as a tool to increase efficiency and productivity. The first of these hurdles is a clear understanding of the legal framework that dictates how veterinarians are permitted to practice telemedicine. The State Boards of Veterinary Medicine across the US have the authority to determine the capabilities and limitations of telemedicine, and have adopted policies and recommendations into their practice acts regarding the use telemedicine in each state. There now exist widespread differences across the states when it comes to these telemedicine regulations, creating a potential source of confusion for the swine practitioner. This presentation will summarize all the relevant by-laws, practice acts, and recent press releases from the State Boards of Veterinary Medicine and/or Animal Health pertaining to telehealth, telemedicine, or virtual veterinary care. Special focus will be given to how some of these laws affect the intricacies of swine practice, with the purpose of equipping swine veterinarians with the knowledge and confidence to legally practice telemedicine.

Secondly, this presentation will cover some of the common strategic and operational stumbling blocks that hinder veterinarians from developing a fully functional telemedicine service within their practice. This will include important considerations for practices to evaluate before diving into telemedicine, as well as setting some reasonable goals for virtual care. Information will be provided on how to best select a telemedicine service to suit the practice's needs, as well as case examples for easy "starting" appointments in swine telemedicine. Finally, insight on how to best promote, market, and monetize a telemedicine service within a veterinary clinic will be covered.

By addressing both the legal technicalities, as well as some of the common logistical challenges of implementing telemedicine, this presentation will equip veterinarians with insights on how to develop the telemedicine offerings of their practice effectively and efficiently. The future of telemedicine offers many benefits to the swine practitioner, such as improved accessibility, task prioritization, biosecurity, client satisfaction, and profitability. As demand for high quality swine veterinary care increases, it is important that veterinarians continue to capture these benefits while also staying within the bounds of a valid Veterinary-Client-Patient Relationship (VCPR), and also maximizing their ability to improve animal health and profitable returns.

https://doi.org/10.54846/am2023/64

Effect of blends of organic acids and monoglycerides on growth performance, diarrhea, and systemic inflammation of *Escherichia coli* F18 challenged weaned pigs

Ying Chen[1]; Ross Wolfenden[1]; Sangwoo Park[2]; Yanhong Liu[2]
[1]Animal Nutrition, Eastman Chemical Company, Kingsport, Tennessee;
[2]University of California, Davis, California

Introduction

Enterotoxigenic E coli (ETEC) is reemerging and causing significant issues in post-weaning pigs in the US. Organic acids and monoglycerides of fatty acids have been reported to effectively prevent the proliferation of pathogenic bacteria like ETEC. In addition, organic acids acidify both the feed and stomach, subsequently improving nutrient digestibility, and attenuating the negative impact of post-weaning diarrhea. Monoglycerides of fatty acids can modulate immune response and alleviate inflammation, which contributes to improved overall health status of nursery piglets. In this study, we investigated dietary supplementation of an organic acid blend, a monoglyceride blend, and a combination of both on growth performance, diarrhea, and systemic inflammation of weaned pigs experimentally infected with pathogenic ETEC F18.

Methods

Forty post-weaning pigs (7.81 ± 0.84 kg BW) were individually housed in disease containment rooms and randomly allotted to one of four treatments with 10 replicate pigs per treatment. The four dietary treatments were: 1) Control: nursery basal diet, 2) Acids: basal diet supplemented with a blend of organic acids (formic acid, lactic acid, and sodium formate) at 0.3%, 3) Monoglycerides: basal diet supplemented with a blend of monoglycerides of fatty acids at 0.3%, 4) Combination of acids and monoglycerides: basal diet added with the acid blend at 0.2% and the monoglyceride blend at 0.2%. The experiment lasted 28 d [7 d before and 21 d after the first inoculation (d 0)]. All pigs were orally inoculated with *E coli* F18 10^{10} CFU/3 mL for 3 consecutive days from d 0. Growth performance was measured weekly. Diarrhea score (1, normal, to 5, watery diarrhea) was daily recorded for each pig. Fecal samples were collected on d 2, 5, 7, 10, 14, and 21 post-inoculation (PI) to confirm the presence of β-hemolytic coliforms in feces. Whole blood samples were collected on d 0, 5, and 14 PI to perform complete blood counting analysis. All data were analyzed by ANOVA using the PROC MIXED of SAS with pig as the experimental unit. Frequency of diarrhea was calculated and analyzed by Chi-square test.

Results

Supplementation of the acid blend, monoglyceride blend, or the combination of both significantly reduced ($P < 0.05$) frequency of diarrhea (25.5 to 29.8%) compared with control (39.3%). ETEC F18 infection increased ($P < 0.05$) the counts of total white blood cells, neutrophils, and lymphocytes on d 5 and 14 PI, compared with d 0. Pigs fed with the 0.3% monoglyceride blend had higher ($P < 0.05$) neutrophil count than pigs in the control group on d 14 PI. The combination treatment of acids and monoglycerides reduced ($P < 0.05$) the counts of total white blood cells, lymphocytes, and the ratio of neutrophils to lymphocytes, and tended ($P < 0.10$) to reduce neutrophil count on d 5 PI, compared with control. There was no significant difference on growth performance among the treatment groups.

Conclusions

A holistic approach is needed when addressing ETEC issues. Nutritional interventions using feed additives can effectively enhance intestinal and overall health status of nursery pigs, subsequently reducing the incidence of postweaning diarrhea caused by ETEC infection. Our results suggested dietary supplementation of the organic acid blend, monoglyceride blend, or the combination of both can reduce diarrhea frequency in ETEC F18 challenged weaned pigs. Furthermore, the combination of organic acids and monoglycerides alleviated systemic inflammation caused by ETEC F18 infection.

https://doi.org/10.54846/am2023/65

VetAssure™ ATP: Applications for the US pork industry

David Pyburn, DVM; Anne Calitz
BioChek BV, Reeuwijk, The Netherlands

Introduction

With the increased focus on biosecurity due to disease transmission risks and severe disease outbreaks veterinarians, farmers, and production companies need a tool to help verify and track the successful implementation of hygiene and sanitation protocols as part of their biosecurity plan. The VetAssure™ ATP biosecurity verification solution helps veterinarians and farmers to verify cleanliness and biosecurity procedures for barns, people, deliveries, equipment, and livestock trailers. The VetAssure™ ATP test uses bioluminescence technology to identify and measure adenosine triphosphate, commonly known as ATP. This measurement helps users to objectively determine from a microbiological level if a surface has been cleaned properly or if it requires a corrective action such as a recleaning.

This all-in-one swab bioluminescence test contains a pre-wetted swab bud for optimal sample recovery. A unique, liquid-stable reagent provides superior reproducibility and robustness of the ATP measurements.

VetAssure™ ATP

ATP is the basic energy component molecule of all plant and animal cells and is present in all microorganisms and organic residues. Due to this, its presence on a surface can be used as an indicator of organic contamination. When ATP reacts with luciferin in the presence of luciferase it produces a release of energy in the form of light, this is called bioluminescence. This is also the mechanism for firefly light in their abdomen. The presence of ATP can be detected as a measure of the light output and intensity being emitted. Therefore, the measurement of light intensity provides a visual indicator of ATP present and of the total organic contamination from the surface tested. ATP bioluminescence is used extensively in the food industry and in hospitals as a tool for monitoring and maintaining cleanliness.

ATP bioluminescence is a rapid and simple method to assess surface cleanliness and microbiological contamination.[1] ATP bioluminescence detects, all eukaryotic and procaryotic cellular ATP material including microbial cells, food residues, bacteria, fungi, and yeast, which might persist after inadequate cleaning.[2] It is measured in Relative Light Unit (RLU) values, which is directly proportional to the amount of ATP present. Several studies have demonstrated strong correlations between RLU values and bacterial microbe counts or surfaces which were judged to be dirty.[1,3-5] The reaction between ATP and the luciferin/luciferase is immediate, allowing results to be processed and read out in ten seconds by the luminometer. Results are expressed numerically on the screen in RLU values.

Studies have shown that the ATP bioluminescence method is a useful and efficient method to evaluate the cleanliness of a variety of surfaces in farm environments. Potential uses include:

- Determine the cleanliness of surfaces like floors, feeders, enrichment material, nipple drinkers, and walls from farms and animal transports.
- Highly accurate and cost-efficient alternative to microbiological culture plate methods for rapid on-farm/ on-trailer assessments.
- Useful to identify critical areas for subsequent further washing and disinfection.
- ATP bioluminescence assays are also a simple and reliable method to test showering/handwashing compliance.

Test results are displayed on the luminometer in ten seconds and then tracked over time in cloud-based software by location, plan, and user. This information allows users to clearly associate results with the specific location, or group of locations tested and who performed the tests. This allows users to track biosecurity performance over time and look for continual improvement.

Research on ATP bioluminescence in production agriculture

Several studies have shown that the ATP bioluminescence method is useful and efficient to evaluate the cleanliness of a variety of surfaces in farm environments. For example, to monitor the hygiene standards on dairy farms by taking samples of milking equipment.[3,4,6,7] Furthermore, ATP bioluminescence testing has been applied to determine the cleanliness of surfaces like floors, feeders, enrichment material, nipple drinkers and walls from farms and livestock trailers.[1,5,8] These studies also show that ATP testing is a highly accurate and cost-efficient alternative to culture plate microbiological methods for rapid on-farm or on-truck assessments. It can be used, in addition to visual inspection, to determine the cleanliness status of farms and livestock trailers and is useful to identify critical areas for subsequent washing and disinfection.

ATP bioluminescence assays are also a simple and reliable method to test handwashing/showering compliance.[9] ATP technology is used in several areas to monitor handwashing, as for example in hospitals or food-processing companies. It needs to be considered that skin is a living organ and therefore has natural levels of ATP, even when it is clean. It is thus impossible to achieve an ATP level of zero. Furthermore, due to variations of skin characteristics from person to person, it is not possible to define universal baseline ATP levels for clean hands/skin.

For all surfaces to be tested with ATP bioluminescence technology thresholds for the differentiation between "clean" and "dirty" vary depending on the surface and the environment where the sample is being taken. A collection of studies indicating threshold values and methods to define the thresholds are presented in Table 1. When starting out to use VetAssure™ ATP for biosecurity verification users must first take the time to collect initial measurements and establish testing/swabbing SOPs that enable the determination of thresholds and individual standard values for clean/dirty.

Many pathogens that pose a risk to pork production sites are viruses. ATP assays do not directly measure viruses. However, some studies suggest that reduced ATP measurements correlate

Table 1: Methods to define ATP RLU clean/dirty thresholds

Location	Tested surfaces	Threshold	Method to define thresholds	References
Dairy farms	Milking equipment	Clean: < 150 RLU Dirty: > 300 RLU Acceptable but not good: 150 – 299 RLU	Based on recommendations by the manufacturer.	Lindell et al., 2018[3] Irie et al., 2021[4] Costa et al., 2006[11]
Dairy farms	Milking equipment	Teat cup rubbers = 152 RLU Teat dip containers = 242 RLU Milk receivers = 282 RLU Pipeline joints – 1821 RLU	Reference values for each surface were established based on mean RLU values observed in class A farms. Class A farms are farms with a very low bacterial count (< 10 × 10^3) in bulk tank milk.	Vilar et al., 2008[7]
Swine livestock trailer	Floors, walls, ramps, partitions, and trailer exterior surfaces	Pass (clean): 430 RLU/100 cm^2 Critical: 431 – 850 RLU/ 100 cm^2 Fail: (dirty): > 850 RLU/ 100 cm^2	Defined by a microbial assessment through MacConkey agar contact plates. The thresholds for the definition of cleanliness through the microbiological assessment was taken from the guidelines of the Canadian swine health board. Based on those thresholds the researchers calculated the equivalent threshold values for RLU/ 100 cm^2	Cabahug et al., 2016[1]
Food processor company	Hands	Clean: < 600 RLU Dirty: > 600 RLU Or Drop of ATP level of 80% after hand washing	Not described in the paper	Bennion et al., 2013[12]
Dentistry students	Hands	Very clean skin: < 100 RLU Clean skin: 101 – 250 RLU Somewhat clean skin: 251 – 500 RLU Dirty skin: 501 – 750 RLU Very dirty skin: 751 – 1000 RLU	Not described in the paper	Bustamante et al., 2018[2]
Hospital	Hands	Clean: < 60 RLU Dirty > 60 RLU	Not described in the paper	Hygiena[13], n.d.
Hospital	Hands	Clean: < 1000 RLU/25 cm^2 Dirty: > 1000 RLU/25 cm^2	Based on clinical trials	Marena et al., 2002[9]

significantly with reduced viral recovery. ATP bioluminescence measurements can, therefore, be used to monitor the effectiveness of interventions taken to reduce viral contamination.[10]

Conclusions

Several studies have demonstrated that ATP bioluminescence is a fast and easy to use method to monitor microbial contamination and the effect of hygiene interventions for a variety of surfaces (eg, equipment, livestock haulers, pen surfaces, and skin). No general thresholds for the differentiation between clean and dirty exist, the differentiation needs to be defined based on the surface tested, the environment and the desired degree of cleanliness. There is no standard method for the definition of thresholds. Methods include reference values based on objects/surfaces with a low bacterial count, calculation of RLU-thresholds based on existing thresholds of microbiological assessments, or the relative drop of ATP levels after a hygiene intervention. Limitations of the method mainly concern the specificity of the assay, since high ATP-levels can also be caused by non-pathogenic or dead microbes, as well as interactions with certain disinfectant materials.

People and vehicles are major risk factors regarding the introduction of pathogens into a farm. This risk can be reduced by certain biosecurity measures, such as proper hand washing and showering or cleaning and disinfecting vehicles and livestock haulers before entering a farm. ATP bioluminescence is a suitable tool to monitor these measurements in real time and enable disease prevention actions to be taken.

The VetAssure™ ATP bioluminescence method of biosecurity verification can be utilized to demonstrate the organic material reduction effect of cleaning. Pre- and post-cleaning plus disinfection data tracking and analysis lead to improved biosecurity and labor behavioral changes towards improved biosecurity for the long-term.

References

1. Cabahug et al.: 2016, 2016 American Society of Agricultural and Biological Engineers Annual International Meeting p. 1-7.
2. Bustamante et al.: 2018, Transylvanian Review – 26:8909-8914.
3. Lindell et al.: 2018, Journal of Dairy Science – 101:2438-2447.
4. Irie et al.: 2021, The Journal of Veterinary Medical Science – 83.
5. Yi et al.: 2020, Journal of Animal Science and Technology – 62:682-691.
6. Reinemann et al.: 2000, ASAE Annual International Meeting – ASAE Paper No. 003009.
7. Vilar et al.: 2008, International Journal of Food Microbiology – 125:357-361.
8. Heinemann et al.: 2019, Individual Training for Farmers based on Results for Protein and ATP Rapid Tests and Microbiological Conventional Cultural Methods Improves Hygiene in Pig Fattening. Oxford University Press – 2:1-27.
9. Marena et al.: 2002, American Journal of Infection Control – 30:334-340.
10. Sifuentes et al.: 2017, Food and Environmental Virology – 9:14-19.
11. Costa et al.: 2006, Brazilian Journal of Microbiology – 37:345-349.
12. Bennion et al.: 2013, International Food Hygiene – 17:23-24.
13. Hygiena Internal Communication, no date.

https://doi.org/10.54846/am2023/66

Identifying and evaluating anti-inflammatory compounds that inhibit African swine fever virus

Charles C. Elrod[1,2], PhD; Joshua A. Jackman[3], PhD; Erik Arabyan[4]; Theresia A. Lavergne[1], PhD; Hovakim Zakaryan[4], PhD
[1]Natural Biologics Inc., Newfield, NY; [2]Department of Animal Science, Cornell University, Ithaca, NY;
[3]School of Chemical Engineering and Translational Nanobioscience Research Center, Sungkyunkwan University, Suwon, Republic of Korea; [4]Laboratory of Antiviral Drug Discovery, Institute of Molecular Biology of NAS, Yerevan, Armenia

Introduction

African swine fever virus (ASFV) is a major threat to the global swine industry and there is an outstanding need to identify economically feasible antiviral agents that can inhibit ASFV and reduce disease burden.[1] Among different options, naturally occurring compounds with both anti-inflammatory and anti-ASFV activities demonstrate particularly high potential for industrial use. To date, several anti-inflammatory compounds have been reported to inhibit ASFV and other viruses and exhibit various antiviral mechanisms. For example, we recently reported that kaempferol, a naturally occurring flavonoid, inhibits the entry and post-entry stages of the ASFV replication cycle, in turn affecting viral protein and DNA synthesis, and can also induce autophagy in infected cells.[2] Given these findings, there is strong motivation to further explore and comparatively evaluate anti-inflammatory compounds as potential anti-ASFV agents. In this study, we employed a cell-based, high-throughput screening approach to investigate the anti-ASFV activity of 125 anti-inflammatory compounds by measuring protection against virus-induced cytopathic effect (CPE) *in vitro* using the lab-adapted ASFV Ba71V strain.

Methods

High-throughput screening

Based on chemical and structural similarity to known antiviral compounds and documented anti-inflammatory activity, 125 compounds were selected for screening. These compounds included various classes of molecules, including alkaloids, anthraquinones, chalcones, coumarins, diterpenoids, flavonoids, iridois, lignans, phenols, quinones, and xanthones. All of the tested compounds can be found in natural plant sources. For this study, the compounds were purchased as a library from Chemfaces (Wuhan, China). The screening method measured the ability of compounds to protect cells from virus-induced CPE and involved testing with the ASFV Ba71V strain. Confluent Vero cells of African green monkey kidney origin were grown in 96-well cell culture plates at a seeding density of 2×10^4 cells per well in Eagle's Minimum Essential Medium and were infected with the ASFV Ba71V strain (10^5 $TCID_{50}$ per well, where $TCID_{50}$ denotes the 50% tissue culture infectious dose) and immediately treated with the respective compound at 50 μM compound concentration. After incubation for 72 h, the cell culture supernatant was collected and titrated by a CPE-based assay in order to quantify the infectious amount of virus. Compounds that exhibited > 50% CPE inhibition were selected for additional testing.

Cytotoxicity testing

The cytotoxicity of selected compounds was evaluated using Vero cells by the crystal violet staining method.[3] Confluent cells at a seeding density of 2×10^4 cells per well were exposed to 50 μM of the respective compound and incubated for 24 h at 37°C in 5% CO_2. After incubation, the cell culture medium was removed, and 4% crystal violet solution (in ethanol) was added to the wells and incubated for 40 min at room temperature. Then, the wells were washed with distilled water and 200 μL methanol was added to each well for 20 min. The optical density of each well was measured at 570 nm with a plate reader.

Viral yield reduction assay

Vero cells were grown in 24-well cell culture plates to a seeding density of 2×10^5 cells per well and then infected with the ASFV Ba71V strain (0.2 $TCID_{50}$ per cell) and treated with selected compounds at 6.25 to 50 μM compound concentrations. The infection was allowed to proceed for 24 h before the cell culture supernatant was collected and titrated by CPE-based assay.

Viral attachment and internalization testing

To evaluate the effects on virus attachment, Vero cells at a seeding density of 2×10^5 cells per well were grown in 24-well cell culture plates and then incubated with the ASFV Ba71V strain (0.2 $TCID_{50}$ per cell) and 50 μM of the test compound at 4 °C for 1 h. This temperature was used because it permits virus binding but prevents cellular internalization. Unbound virus and compound were then removed by thoroughly washing the cells with phosphate-buffered saline (PBS). Next, fresh medium containing 3% fetal bovine serum (FBS) was added. The cell culture supernatant was collected and titrated at 24 h post-infection.

In the internalization experiments, Vero cells at a seeding density of 2×10^5 cells per well were incubated with ASFV Ba71V at 4°C for 1 h. Then, the unbound virus was removed by thoroughly washing the cells with PBS. The test compound at 50 μM concentration was next added to the cells and incubated at 37°C to allow virus entry to proceed for 1 h. The compound was subsequently removed by washing with PBS to prevent the compounds from potentially interfering with later stages of infection as well. The infection was allowed to proceed for 24 h before the cell culture supernatant was collected and titrated.

Results

Vero cells seeded at a density of 2×10^4 cells per well were infected with 10^5 $TCID_{50}$ units of ASFV and then immediately treated with the respective test compounds at 50 μM compound concentration. After 72 h incubation, cell viability was measured in a CPE assay. Specifically, the assay was designed whereby viral infection caused loss of cell viability, i.e., CPE, in infected cells alone. Antiviral compounds could be identified by inhibition of CPE where cell viability was maintained above a defined threshold. Among the 125 screened compounds, five compounds (tetrandrine, berbamine, pinoresinol, curcumin, and paeonol) that prevented CPE by at least 50% were identified (Table 1). These five antiviral compounds were further assessed for cytotoxicity and found to be biocompatible (> 80% cell viability) at the tested concentration of 50 µM.

Next, an ASFV suspension of 2×10^5 $TCID_{50}$ was incubated with each of the lead compounds at 50 µM concentration for 1 h before being added to Vero cells and incubated for 72 h, followed by supernatant collection and titration by a CPE-based assay. Tetrandrine and berbamine demonstrated the greatest reductions in viral yield of 3.6 ± 0.38 log $TCID_{50}$/mL and 2.0 ± 0.25 log $TCID_{50}$/mL, respectively ($P < 0.05$ compared to virus-only control). Pinoresinol, curcumin, and paeonol caused smaller viral yield reductions of around 1.1 log $TCID_{50}$/mL.

In follow-up studies, tetrandrine demonstrated dose-dependent viral yield reductions of 0.8 to 2.0 log $TCID_{50}$/mL at compound concentrations ranging from 6.25 µM to 50 µM after 24 h incubation. Berbamine also showed dose-dependent viral yield reductions ranging from 0.50 to 1.38 log $TCID_{50}$/mL across the same concentration range.

To better understand the specific stages at which these compounds work, viral attachment and internalization studies were conducted and demonstrated that tetrandrine treatment resulted in 1.45 ± 0.53 log $TCID_{50}$/mL and 1.78 ± 0.29 log $TCID_{50}$/mL reductions in virus attachment and internalization, respectively. On the other hand, berbamine treatment resulted in 1.11 ± 0.51 log $TCID_{50}$/mL and 1.0 ± 0.19 log $TCID_{50}$/mL reductions in virus attachment and internalization, respectively.

Discussion

Several anti-inflammatory compounds were identified that exhibit antiviral activity against ASFV. Of the compounds studied, tetrandrine and berbamine demonstrated the greatest antiviral activity levels, which appear to be related to disrupting early stages of the virus life cycle such as virus binding to and internalization by susceptible cells. Quantitatively, tetrandrine had the greatest antiviral effect. In addition to ASFV, tetrandrine has been reported to inhibit the entry of other viruses such as Ebola virus.[4] Likewise, berbamine has been reported to inhibit the entry of severe acute respiratory syndrome coronavirus 2 (SARS-CoV-2) by blocking virus-cell fusion.[5] Interestingly, tetrandrine and berbamine are both calcium channel blockers and ongoing time-of-addition studies will further help to unravel the molecular mechanisms by which these compounds inhibit ASFV.

References

1. Arabyan E, Kotsynyan A, Hakobyan A, et al. Antiviral agents against African swine fever virus. *Virus Res.* 2019;270:197669. **DOI: 10.1016/j.virusres.2019.197669**

2. Arabyan E, Hakobyan A, Hakobyan T, et al. Flavonoid library screening reveals kaempferol as a potential antiviral agent against African swine fever virus. *Front Microbiol.* 2021;12:736780. **DOI: 10.3389/fmicb.2021.736780**

3. Feoktistova M, Geserick P, Leverkus M. Crystal violet assay for determining viability of cultured cells. *Cold Spring Harb Protoc.* 2016. **DOI: 10.1101/pdb.prot087379**

4. Sakurai Y, Kolokoltsov AA, Chen C-C, et al. Two-pore channels control Ebola virus host cell entry and are drug targets for disease treatment. *Science* 2015;347(6225):995-998. **DOI: 10.1126/science.1258758**

5. Zhang Z-R, Zhang Y-N, Zhang H-Q, et al. Berbamine hydrochloride potently inhibits SARS-CoV-2 infection by blocking S protein-mediated membrane fusion. *PLoS Neglected Trop Dis.* 2022;16(4):e0010363. **DOI: 10.1371/journal.pntd.0010363**

Table 1: Antiviral activity of selected anti-inflammatory compounds against ASFV in vitro, expressed as protection percentage against virus-induced CPE in top row and as mean log $TCID_{50}$/mL in other rows

Testing method	Virus only	Tetrandrine	Berbamine	Pinoresinol	Curcumin	Paeonol
CPE protection		101%	86%	54%	57%	50%
Antiviral	6.42	2.81*	4.42*	5.37*	5.26*	5.37*
Attachment	3.86	2.42*	2.76*			
Internalization	3.98	2.20*	2.98*			

CPE = cytopathic effect

* $P < 0.05$

https://doi.org/10.54846/am2023/67

Time to PRRSV negative response in live virus inoculated gilts injected with Ivomec

Grant Allison[1], DVM; Phillip Gauger[2], DVM, MS, PhD; Jianqiang Zhang[2], MD, PhD; Paul Lawrence[3], MSc, PhD
[1]Walcott Veterinary Clinic, Walcott, Iowa; [2]Iowa State University Veterinary Diagnostic Laboratory, Department of Veterinary Diagnostic and Production Animal Medicine, Ames, Iowa; [3]Bimeda Biologicals, Oakdale, Minnesota

Introduction

The goal of every swine producer is to wean PRRSV-negative pigs to reduce losses associated with disease. However, one problem is persistence of the virus and how long PRRSV will remain infectious within a population and by default, continue circulating on the farm.[1] One report states the virus will eventually be cleared in most animals by 150 days post inoculation (DPI).[2] Another report detected PRRSV by reverse transcription real-time PCR (RT-rtPCR) in blood in inoculated pigs from 7 to 77 DPI[3] and an additional report found viral RNA in tonsillar tissue up to 251 DPI.[4] It is challenging to reduce the length of viremia and detectable infection toward weaning PRRSV negative pigs from a PRRSV unstable farm. The two most common methods currently used are depopulation & repopulation with PRRSV naïve pigs or through a process called "load-close-expose" (LCE) for farms lacking resources and economic justification to depopulate. In most situations, exposure to PRRSV is through modified live virus (MLV) vaccine or the wild-type field strain endemic to the farm through live virus inoculation (LVI). MLV vaccines may not significantly reduce transmission of endemic strains but may reduce the duration and level of viremia.[5] These reports are consistent with additional references that have detected the PRRSV genome by RT-rtPCR in blood (viremia) up to 77 DPI but up to 251 DPI in tonsil tissue, a potential source of infection and ultimately a route of transmission. However, it has been reported that using MLV vaccine in LCE protocols shortens the time-to-PRRSV stability (TTS) when compared with LVI. Considering both protocols, the mean TTS defined as the number of weeks required to produce negative pigs at weaning was 26.6 weeks (~186 days) with a range of 21.6 -33.0 weeks.[6] The advantage of using a PRRSV strain endemic to the farm during gilt acclimation programs in lieu of commercial live vaccine virus is the difference in the efficacy of the immune response to a genetically similar virus. In the former, homologous immune responses induce more effective and longer lasting protection compared to the latter that represents heterologous exposure and suboptimal protection as they enter the breeding farm. In either situation, viremia and shedding occur with the risk that the inoculum used for live exposure is more likely to contain multiple strains of PRRSV, some potentially unknown, or contain other consequential but unknown pathogens at the time of exposure. Therefore, to reduce the risks associated with live virus exposure in breeding systems, primarily due to the induction of active viremia, viral shedding, exposure to more than one strain of PRRSV or exposure to other pathogens (PCV2, etc.), this project will evaluate the antiviral properties of Ivermectin that have been previously described for other pathogens in an attempt to shorten the duration of viremia and shedding of PRRSV in gilts or breeding females after LVI and prior to entry into the breeding farm in unstable production systems.[7-9]

Materials and methods

Twenty gilts from the same genetic source were placed in an offsite gilt developer unit (GDU) from a PRRSV negative source. The gilts were confirmed PRRSV negative upon arrival using RT-rtPCR and ELISA antibody tests, conducted at the Iowa State University Veterinary Diagnostic Laboratory (ISU VDL). The 20 PRRSV naïve gilts were divided into the Ivomec treated group (N = 10) and Control non-treated group (N = 10). The Control group of gilts were housed adjacent to the treated group. At 9 days post arrival, all 20 gilts were exposed via LVI to at least 5,000 copies of a wild type RFLP 1-4-4 and Lineage 1A (L1A) PRRSV based on RT-rtPCR quantitative analysis. The PRRSV L1A LVI originated from a clinical case of PRRSV at the farm, cycle threshold value 21.1, and RT-rtPCR quantity detected at 62,243,706 genomic copies per mL of serum. At 2 days post LVI exposure, blood was collected from 9 gilts and submitted in pools of 3 to confirm success of LVI exposure to PRRSV through RT-rtPCR. The 1st injection of Ivermectin, dose of 1 mL/11 kg of body weight (11 mL), occurred 6 days after LVI to 10 gilts in the Ivomec group. At approximately 2 weeks, 4 weeks, and 6 weeks after the 1st Ivomec treatment, the 2nd, 3rd, and 4th injection of 11 mL of Ivomec was administered to the 10 gilts in the Ivomec group, respectively. The 10 Control gilts housed adjacent to the Ivomec group did not receive any Ivomec throughout the study. At each time point, approximately 2, 4, and 6- weeks post Ivomec injection, and up to 61 days post LVI, serum and oral fluid samples were collected from the 20 Ivomec and Control gilts. The PRRSV status of Ivomec and Control gilts was monitored using RT-rtPCR (both serum and oral fluid samples) and PRRS X3 ELISA antibody test (serum samples) collected at each time point. All diagnostic testing, RT-rtPCR, PRRSV ELISA, and Sanger sequencing, was conducted at the ISU VDL. Bimeda Biologicals performed PRRSV ORF5 gene sequencing. Statistical analysis on log10 transformed genomic copies was performed using JMP® version 16 (JMP Statistical Discovery, Carey, NC).

Results

Gilts in both Ivomec and Control groups were bled on arrival and individually tested RT-rtPCR and ELISA negative. All subsequent quantitative RT-rtPCR data was log10 transformed to normalize the data for statistical analysis. At day 6 post LVI, 5/10 Ivomec and 6/10 Control gilts were PRRSV RT-rtPCR positive on individual testing at the time when the first dose of Ivomec was administered to the treated group. At day 16 post-LVI, all gilts in both Ivomec and Control groups were PRRSV RT-rtPCR positive. Collectively, there was a trend indicating the Ivomec treated gilts demonstrated lower mean log10 genomic copies of PRRSV per mL of serum compared to the Control gilts at day 16, 30 and 43 post LVI although values were not statistically significant. One gilt in the Ivomec group and one gilt in the Control group were viremic at 57 days post-LVI but at low concentrations of the virus based on quantitative RT-rtPCR. However, all 20 gilts were RT-rtPCR PRRSV negative at 61 days post-LVI. Interestingly, all 5 gilts PRRSV RT-rtPCR positive at day 6 post-LVI in the Ivomec treated group had a reduction in genomic copies/mL in serum at day 16 post-LVI compared to the Control group with 6 gilts PRRSV RT-rtPCR positive at day 16 post-LVI that demonstrated an increase in genomic copies/mL in 3 of 5 gilts suggesting Ivomec treatment may have reduced or prevented replication of PRRSV in the treated group. The mean Log 10 transformed genomic copies per mL of serum and geometric mean genomic copies per mL of serum for each time point post-LVI are reported in Table 1.

The time-to-negative status was defined as the first time point when a PRRSV positive gilt was detected RT-rtPCR negative and remained PCR negative for the remainder of the trial. One gilt in the Ivomec group and one gilt in the Control group were respectively determined PRRSV RT-rtPC R negative at day 43 post-LVI but became RT-rtPCR positive on the next collection date at day 57 post-LVI. However, both gilts were RT-rtPCR negative on the last date of collection at day 61 post-LVI. Considering the limitation that the exact date of negative status could not be determined, the median days between the date(s) of collection were used to define the time to negative (TTN) status for the 2 groups (Table 2).

Oral fluids were collected from the treated and non-treated pens prior to collection of blood to not disturb social order and limit interest in the rope(s). Sufficient time was allowed for all gilts to have access to the rope in both the Ivomec treated and non-treated Control pens. Post collection of oral fluids, the samples were placed on ice in a cooler and remained on ice until submission to the ISU VDL for RT-rtPCR. Oral fluid collected on day 48 post-LVI was taken the day gilts were moved to the sow unit. Therefore, samples collected on day 48 post LVI were from gilts in isolation while samples taken on day 57 post LVI were taken from gilts after placement in the sow unit. Since it would be unlikely that all gilts in a pen would chew on the rope, all 20 gilts were individually orally swabbed on day 57 and tested in pools of 5 to confirm they were truly PRRSV negative. Oral fluid PRRSV RT-rtPCR cycle threshold and quantitative results are reported in Table 3.

PRRSV ORF5 sequencing was conducted on serum samples with the lowest Ct value from the Ivomec and Control groups to confirm the virus was genetically similar to the PRRSV administered through LVI. A phylogenetic analysis using Geneious Prime (version 2022.2.2) MAFFT alignment and phylogenetic trees created in Geneious using RAXmL demonstrated the LVI strain was similar to the virus detected at days 6 and 30 post-LVI (Figure 1). Genetic homology ranged from 99.5% to 99.83% nucleotide similarity across PRRSV ORF5 sequenced from the LVI material and both experimental groups at both time points. Sequencing was not possible at day 43 post-LVI and thereafter due to the low concentration of the virus that was present in the sample.

Discussion

Antiviral drugs are used to treat viral infections and unlike most antibiotics, antiviral drugs do not destroy the pathogen, they inhibit its replication. The first antiviral drug (idoxuridine) was approved for human use in 1963. As of 2016, there were 90 antiviral drugs in 13 functional groups approved for use in 9 human diseases.[10] The first reference to using Ivomec as an antiviral drug to control PRRSV in pigs was published in 2016.[7] Since 2016, little has been done to investigate potential antiviral drugs of any type to

Table 1: Summary of serum samples tested by PRRSV quantitative RT-rtPCR in Ivomec and control gilts at different days post live virus inoculation

		Days post live virus inoculation					
Group	**Units**	**6**	**16**	**30**	**43**	**57**	**61**
Ivomec	Number of PCR-positive gilts	5/10	10/10	6/10	0/10	1/10	0/10
Control	Number of PCR-positive gilts	6/10	10/10	6/10	1/10	1/10	0/10
Ivomec	Mean Log10 genomic copies/mL	2.94[a*]	4.96[a]	2.16[a]	0[a]	0.41[a]	0
Control	Mean Log10 genomic copies/mL	3.55[a]	5.47[a]	2.42[a]	0.36[a]	0.36[a]	0
Ivomec	Geometric mean genomic copies/mL	877	92,097	144	0	2.6	0
Control	Geometric mean genomic copies/mL	3,573	298,291	264	2.3	2.3	0

* Superscripts within a column indicate statistical significance if different.

Table 2: Median PRRSV time-to-negative (TTN) status in serum based on RT-rtPCR

Range (days)	17 – 30	31 – 43	44 – 57	58 – 61	TTN (group)
Median (days)	23.5	37	50.5	59.5	
Treated (no. gilts)	4	5	0	1	
Treated (total days)	94	185	0	59.5	
Non-treated (no. gilts)	4	4	1	1	
Non-treated (total days)	94	148	50.5	59.5	
Treated median TTN (total days)					338.5
Non-treated median TTN (total days)					352.0

Table 3: PRRSV RT-rtPCR results and time-to-negative status based on oral fluid testing cycle threshold values or genomic copies/mL

Days Post-LVI	Ivomec RT-rtPCR (Ct)	Control RT-rtPCR (Ct)	Ivomec RT-rtPCR (log 10)	Control RT-rtPCR (log 10)	Ivomec geometric mean (genomic copies/mL)	Control geometric mean (genomic copies/mL)
6	33.2	29.3	4.07	5.26	11,733	179,916
16	26.5	28.5	5.82	5.20	657,459	160,262
30	31.3	32.2	4.33	4.07	21,609	11,821
43	36.2	33.5	2.98	3.80	946	6,304
48	37	35.2	0	3.27	0	1,866
57	37	37	0	0	0	0

Figure 1: PRRSV phylogenetic tree demonstrating the genetic relationship between the live virus inoculation PRRSV 1-4-4 lineage 1A compared to the ORF5 detected at day 6 and 30 post-live virus inoculation. Reference sequences are provided in blue and trial sequences are red.

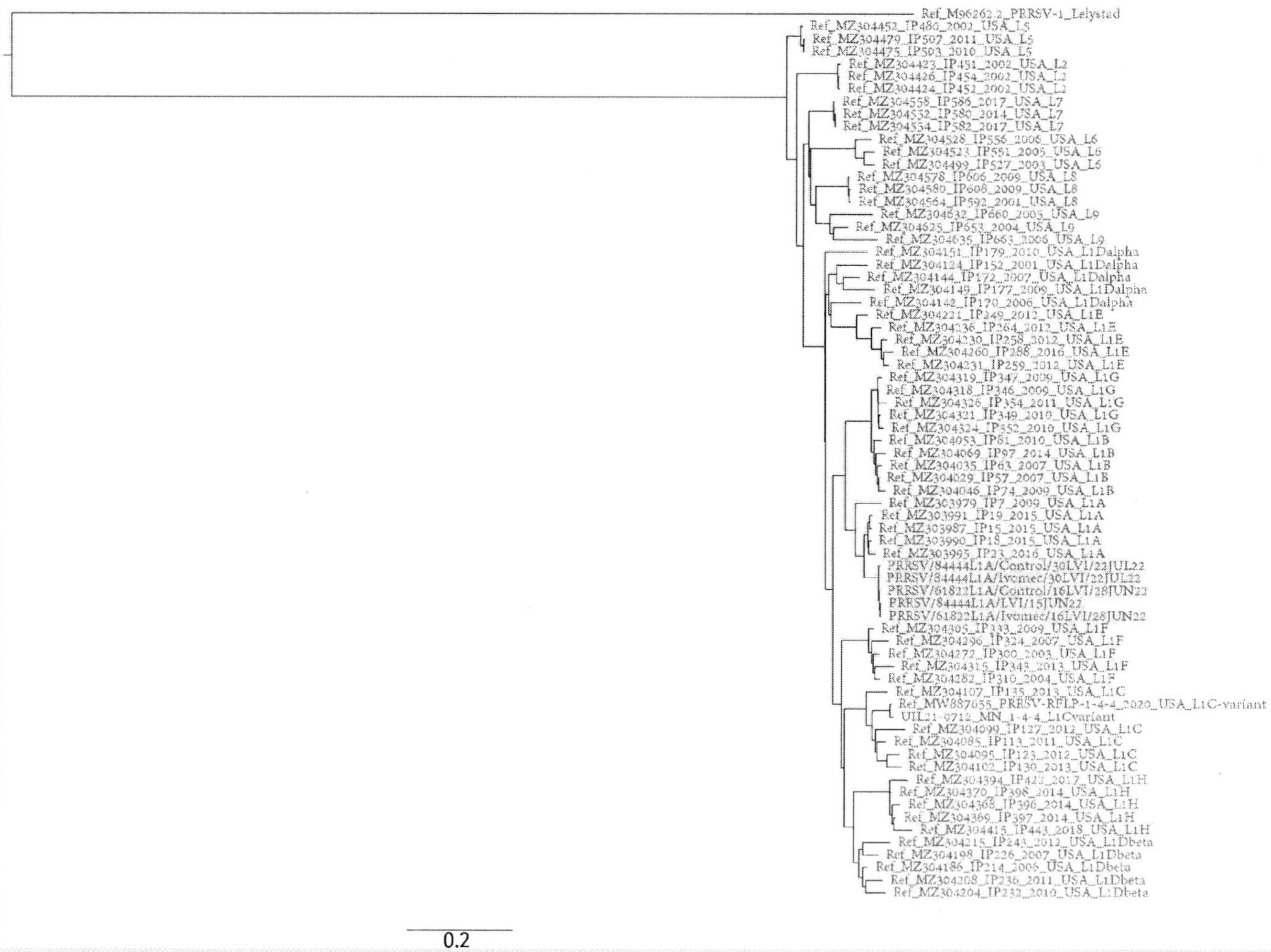

control PRRSV in pigs. The scheduled use of Ivomec in this pilot study shortened the time-to-negative status for PRRSV in serum; median TTN 33.8 days (Ivomec treated) and 35.2 days (Control non-treated), respectively. The single (serum) positive gilts on day 57 may have been a false positive result given all oral fluid samples were negative on day 57. The time-to-negative status for PRRSV in oral fluid was also shortened using Ivomec; 43 – 48 days (Ivomec treated) and 48 – 57 days (Control non-treated) respectively. Using median values for oral fluids the TTN was 45.5 days (Ivomec treated) and 52.5 days (Control non-treated), respectively. During the trial, no gilt required treatment with antibiotics and no mortality was recorded. In addition, gilts in the Ivomec group demonstrated a reduction in genomic copies post-treatment that was not observed in all Control gilts where 3 of 6 viremic animals demonstrated a subsequent increase in genomic copies at the next blood collection suggesting Ivomec may reduce replication and magnitude of viremia although this requires additional studies to confirm. Collectively, the goal to reduce viremia and subsequent exposure to the sow farm from gilts actively shedding PRRSV post introduction from isolation to the sow unit was achieved in 48 days using Ivomec.

Acknowledgment

Funding, wholly or in part, was provided by the Iowa Pork Producers Association

References

1. Trevisan G, Jablonski E, Angulo J, Lopez WA, Linhares DCL: Use of processing fluid samples for longitudinal monitoring of PRRS virus in herds undergoing virus elimination. *Porcine Health Manag* 2019, 5:18.

2. Allende R, Laegreid WW, Kutish GF, Galeota JA, Wills RW, Osorio FA: Porcine reproductive and respiratory syndrome virus: description of persistence in individual pigs upon experimental infection. *J Virol* 2000, 74(22):10834-10837.

3. Charpin C, Mahe S, Keranflec'h A, Belloc C, Cariolet R, Le Potier MF, Rose N: Infectiousness of pigs infected by the Porcine Reproductive and Respiratory Syndrome virus (PRRSV) is time-dependent. *Vet Res* 2012, 43:69.

4. Wills RW, Doster AR, Galeota JA, Sur JH, Osorio FA: Duration of infection and proportion of pigs persistently infected with porcine reproductive and respiratory syndrome virus. *J Clin Microbiol* 2003, 41(1):58-62.

5. Nodelijk G, de Jong MC, van Leengoed LA, Wensvoort G, Pol JM, Steverink PJ, Verheijden JH: A quantitative assessment of the effectiveness of PRRSV vaccination in pigs under experimental conditions. *Vaccine* 2001, 19(27):3636-3644.

6. Linhares DC, Cano JP, Torremorell M, Morrison RB: Comparison of time to PRRSv-stability and production losses between two exposure programs to control PRRSv in sow herds. *Prev Vet Med* 2014, 116(1-2):111-119.

7. Lee YJ, Lee C: Ivermectin inhibits porcine reproductive and respiratory syndrome virus in cultured porcine alveolar macrophages. *Arch Virol* 2016, 161(2):257-268.

8. Wang X, Lv C, Ji X, Wang B, Qiu L, Yang Z: Ivermectin treatment inhibits the replication of Porcine circovirus 2 (PCV2) in vitro and mitigates the impact of viral infection in piglets. *Virus Res* 2019, 263:80-86.

9. Lv CJ, Liu WK, Wang B, Dang RY, Qiu L, Ren J, Yan CQ, Yang ZQ, Wang XL: Ivermectin inhibits DNA polymerase UL42 of pseudorabies virus entrance into the nucleus and proliferation of the virus in vitro and vivo. *Antivir Res* 2018, 159:55-62.

10. DeClercq, E, Li, G.:Approved antiviral drugs over the past 50 years. *Clin Microbiol. Rev.*2016;Jul29(3):695-747.

https://doi.org/10.54846/am2023/68

Case report of a PCV2d-based vaccine in a clinical PCV2d challenged flow

Melissa Farber Billing[1], DVM, MVS; Terri Specht Benson[2], DVM; Eduardo Fano[1], DVM, PhD; Eric Bettinger[1]
[1]Boehringer Ingelheim Animal Health USA Inc., Duluth, Georgia; [2]Heimerl Farms LTD, Johnstown, Ohio

Introduction

Porcine circovirus type 2 (PCV2) is the primary causative agent behind porcine circovirus–associated disease (PCVAD), one of the most significant diseases affecting swine producers around the world. Since the introduction of PCV2 vaccines into the US market in 2006, the benefit of vaccination has been clearly demonstrated. Currently PCV2 genotype d is the predominant genotype of PCV2 circulating in the United States.[1] Existing PCV2 vaccines contain genotype a or b and provide cross-protection against PCV2d.[2,3] However in some situations where PCVAD is consistently confirmed in PCV2 vaccinated populations and vaccine compliance issues have been ruled out, then reviewing the vaccine choice should be considered.[3]

The pigs from the investigated flow were previously identified as having a natural PCV2d challenge plus had PCVAD confirmed cases in vaccinated pigs in early to mid-finishing.[4] The objective of this report is to describe a case where a PCV2d-based killed baculovirus vector vaccine was utilized in a selected number of groups in a flow that was previously identified as naturally PCV2d challenged.

Case description

This case report was documented in growing pigs from a single 5000-sow farm with PRRS and *Mycoplasma hyopneumoniae* (Mhp) negative status and conventional facilities. The group housed sow herd weans approximately 3000 piglets per week. Interestingly, despite the PRRS and Mhp negative status, the pigs from this flow had consistently PCVAD confirmed cases in vaccinated pigs at about 150-180 pounds.[4] To confirm natural challenge, tissues were taken from pigs with clinical signs of wasting and found to have a high viral load of PCV2 which was consistently sequenced as PCV2d, had lymphoid depletion on histopathology, and identification of PCV2 in lymphoid organs by immunohistochemistry. In addition, mortality in this flow was higher than expected. These conditions meet the standard criteria of PCVAD.[5] Historically pigs received a single dose of 2 mL of Fostera® Gold PCV (Zoetis), a PCV2 vaccine that contains genotypes a and b, at 3 weeks of age per label instructions. Due to the natural PCV2d challenge in this flow, d-FENSE® (Boehringer Ingelheim Animal Health, Inc), a PCV2d-based killed baculovirus vector vaccine was utilized and assessed for ten consecutive weeks.

To understand the PCV2 exposure during the assessment, six oral fluid and 30 random serum samples were collected longitudinally at approximately 6, 10, 14, and 20 weeks of age from three groups of pigs vaccinated with 1 mL of d-FENSE® at weaning per label instructions compared to two groups of pigs that received the historical PCV2 vaccine. Oral fluids were tested for PCV2 PCR, PRRS PCR, and Mhp PCR while serum was tested for PCV2 PCR, PCV2 ELISA (Ingezim), and PCV2 IFA 4 dilution at Iowa State University Veterinary Diagnostic Laboratory (ISU-VDL). During the assessment, surveillance for clinical signs of PCVAD was under observation of a veterinarian as well as caretakers. In PCVAD suspected cases, tissues were collected and sent to ISU-VDL for further confirmation. Finishing closeout market data was collected from eleven groups of historically vaccinated pigs as well as four groups of d-FENSE® vaccinated pigs to evaluate key performance indicators such as mortality, average daily gain, and percentage of pigs sold as standard market hogs.

Observations

At 10-20 weeks of age, the average oral fluid PCV2 PCR CT values were higher in the d-FENSE® vaccinated group and the d-FENSE® vaccinated pigs had a lower percentage of serum samples that were positive for PCV2 by PCR in the same age pigs. Also, at 10-20 weeks of age, the ELISA s/p ratios for PCV2 were lower in the d-FENSE® vaccinated pigs, while the percentage of pigs that had an PCV2 IFA 4 dilution of 320 or greater was increased in the historical vaccine group (Table 1). In addition, PCV2d was sequenced from oral fluids in the historical vaccine group at 6, 10, and 14 weeks of age while PCV2d could only be identified from oral fluids in the d-FENSE® group at 6 weeks of age. PCV2 genotype e was sequenced sporadically on oral fluids from both the historical and d-FENSE® groups (clinical relevance unknown). All oral fluids remained negative for PRRS and Mhp throughout the study. No PCVAD cases were confirmed during the assessment on the PCV2d vaccinated pigs. Production data shows an improvement in average daily gain, mortality, and percentage of pigs sold as standard market hogs in groups that received Boehringer Ingelheim's d-FENSE® compared to groups that received the historical PCV2 vaccine protocol (Table 2). Furthermore, d-FENSE® vaccine provided an observable benefit in this flow due to improved uniformity.

Conclusion

Under the conditions of this case report, diagnostic results suggest that pigs that received Boehringer Ingelheim's d-FENSE® vaccine had lower circulation of PCV2 compared to pigs receiving the historical vaccination protocol, based on oral fluid and individual serum samples. PCV2d was not identified in the d-FENSE® vaccinated pigs during the timeframe that was previously associated with PCVAD in this flow. Production parameters also showed an advantage to the groups that received d-FENSE® vaccine. In terms of diagnostics, pathogen exposure, and key performance indicators, the PCV2d-based vaccine provided a benefit in this PCV2d challenged flow compared to the PCV2a + PCV2b based vaccine.

References

1. Madson, D. Porcine circovirus: Are changes occurring? Proceedings AASV Annual Meeting. 2019;5-8.

2. Opriessnig T, Gerber PF, Xiao CT, Halbur PG, Matzinger SR, Meng XJ. Commercial PCV2a-based vaccines are effective in protecting naturally PCV2b-infected finisher pigs against experimental challenge with a 2012 mutant PCV2. *Vaccine*. 2014;32(34):4342-4348. **doi:10.1016/j.vaccine.2014.06.004**

Table 1: Results from serum and oral fluid diagnostics

	Age (weeks)	PCV2 PCR serum (% positive)	PCV2 PCR serum average CT	PCV2 ELISA serum (% positive)	PCV2 ELISA serum average s/p ratio	PCV2 IFA 320 or more (% positive)	PCV2 PCR oral fluids (% positive)	PCV2 oral fluids average CT
d-FENSE®	6	0.0	n/a	50.0	0.373	40.0	66.7	34.8
	10	2.2	33.6	35.6	0.433	6.8	77.8	34.3
	14	6.7	35.8	36.7	0.299	3.3	91.7	33.3
	20	13.3	35.6	30	0.313	16.7	83.3	34.7
Historical	6	0.0	n/a	60.0	0.391	40.0	66.7	30.7
	10	46.7	33.2	53.3	0.546	40.0	100.0	29.2
	14	43.3	34.28	70.0	0.703	53.3	100.0	27.3
	20	33.3	35.8	86.7	0.825	93.3	100.0	31.2

Table 2: Average of key performance indicators from group closeouts (6 weeks of age to market)

	Mortality (%)	Average daily gain (lbs.)	Standard sales (%)
d-FENSE®	1.19	1.99	98.18
Historical	2.23	1.94	97.44
Difference	1.04	0.05	0.74

3. Fano, E. Pineyro, P. Schwartz, K. Tousignant, S. Exploring the efficacy of a PCV2d-based vaccine under current severe PCVAD conditions. Proceedings AASV Annual Meeting. 2022;144-145. **https://doi.org/10.54846/am2022/53**

4. Taylor, A, Billing, MF, Specht, T, Fano, E. Longitudinal assessment of PCV2/PCV3 viremia in PCV2 vaccinated and non-vaccinated animals under PCV2d field exposure. Proceedings AASV Annual Meeting. 2020;274-275.

5. Sorden, SD. Update on porcine circovirus and postweaning multisystemic wasting syndrome (PMWS). *Swine Health Prod.* 2000;8(3):133-136.

https://doi.org/10.54846/am2023/69

DFM PAK™ improves manure characteristics to influence environmental and health benefits in the pit, the pen and in the air

J. Lee[1], M.S.; J. Spencer[1], PhD; E. Galbraith[2], MS; S. Son[2], MBA; M. King[2], PhD
[1]United Animal Health Inc, Sheridan, Indiana; [2]Microbial Discovery Group, Franklin, Wisconsin

Introduction

In pork production systems, manure composition heavily influences the production environment in barns and impacts factors associated with manure management, sanitation, and environmental sustainability. Naturally, diet is a primary contributor to manure composition and changes in major dietary ingredients can lead to new challenges in the production environment. Modern swine diets commonly contain high inclusions of dried distillers' grains with solubles (DDGS) and supplemental fat sources. Relative to manure from pigs fed traditional corn and soybean meal-based diets, modern manure contains higher levels of undigested fiber, protein, nitrogen, and long chain fatty acids. These shifts in dietary ingredients have led to increased solids accumulation, poor manure viscosity, increased foaming in pits, higher gas concentrations in facility air space, and a fecal consistency more prone to adhering to and harboring pathogenic bacteria on the slats.

Labor shortages and an apparent increase in pathogenic *E coli* abundance detected across the swine industry[1] have compounded these environmental challenges. This creates a need for easily applied, reliable products capable of improving the manure characteristics responsible for pen slat buildup and increased bioburden, thereby improving the ease and quality of sanitation.

Bacillus-based direct fed microbial products have been documented in the literature to improve manure characteristics when fed to swine.[2] However, those historical studies were conducted primarily with corn/soybean meal-based diets, and therefore, do not reflect the composition of modern manure and the associated challenges it causes in the production environment. It is well accepted that *Bacillus* strains show a wide range of genetic diversity between individual species and strains, creating vast differences in functional capabilities.[3] Therefore, selecting *Bacillus* with the appropriate attributes needed to improve modern manure challenges is needed.

To address a growing number of concerns related to new environmental challenges caused by modern manure, United® Animal Health leveraged its expertise in *Bacillus* strain screening, municipal wastewater management and unique environmental animal research capabilities to take a focused approach in the development of DFM PAK™ (United® Animal Health, Inc), a *Bacillus*-based direct fed microbial capable of improving the new challenges modern manure causes in the swine production environment.

Bacillus strain selection for environmental benefits

Rigorous *in vitro* strain selection criteria and genomic comparisons were employed by United® Animal Health to screen *Bacillus* from a proprietary library made up of strains sourced from the intestinal tract of high performing pigs. Genome comparisons (Figure 1) and functional screening methods were used to identify and understand individual strain attributes important for degrading manure components associated with the new environmental challenges caused by modern diets.[4]

Since higher dietary fiber and fat sources contribute to modern manure composition, identifying strains most capable of digesting DDGS, corn/soybean fiber and the long chain fatty acids in pit foam was an initial focus. Breaking down corn fiber and long chain fatty acids lend to a strain's ability to degrade manure solids in the pit, which in turn, increases storage volume, decreases the anaerobic sludge accumulation where H_2S, CH_4 gases are produced, improves manure viscosity/pumpability, and reduces pit foaming.

Additional screening was employed to candidate strains to identify which were most capable of utilizing ammonium for *Bacillus* growth and sporulation. *Bacillus* utilization of ammonium in the pit environment promotes nitrogen retention in the manure for land application and reduces the opportunity for the conversion of liquid ammonium to ammonia gas in stored pit manure.

Controlled environmental research

Following *in vitro* characterization and selection, the DFM PAK™ strain combination was validated *in vivo*, through pigs, in a commercial scale research facility, specifically designed with sixteen individual manure pits and two separate airspaces to determine manure impact on different areas of the swine production environment. In these controlled environmental trials, novel methods were developed to determine treatment differences in manure characteristics during pit storage, harmful gas production, manure buildup on slats and *E coli* bioburden in pens.[5]

Results from the environmental research trials supported the strain selection strategy and provided evidence that DFM PAK™ promotes positive environmental impacts in the pit, pen and airspace. In these controlled trials, treatment with DFM PAK™ was shown to reduce the amount of total solids in pit manure ($P \leq 0.05$), showed significant ($P \leq 0.05$) and numerical reductions in pit sludge fiber, numerically improved pit sludge viscosity, and reduced concentrations of ammonia ($P \leq 0.05$), hydrogen sulfide, carbon dioxide ($P \leq 0.05$) and methane in the airspace of the barn at pump out (Table 1). Further improvements included significant ($P \leq 0.05$) reductions in *E coli* bioburden in pens (Figure 2A) and reduced ($P \leq 0.05$) manure buildup on slats (Figure 2B) when compared to CONTROL environments. Collectively, improvements observed in controlled animal evaluations support DFM PAK™ as a novel tool for improving environmental challenges caused by the composition of modern manure.

Client evaluation

A field evaluation was conducted in a client's wean-to-market facilities to evaluate the ability of DFM PAK™ to improve manure consistency and reduce manure buildup in the pen environment.

Figure 1: Functional distribution of genes unique to two DFM PAK™ *Bacillus* strains relative to reference strain *B subtillus* 168, based on functional classification. Genome annotations, functional classifications and comparisons were performed using RAST[6] and SEED.[7]

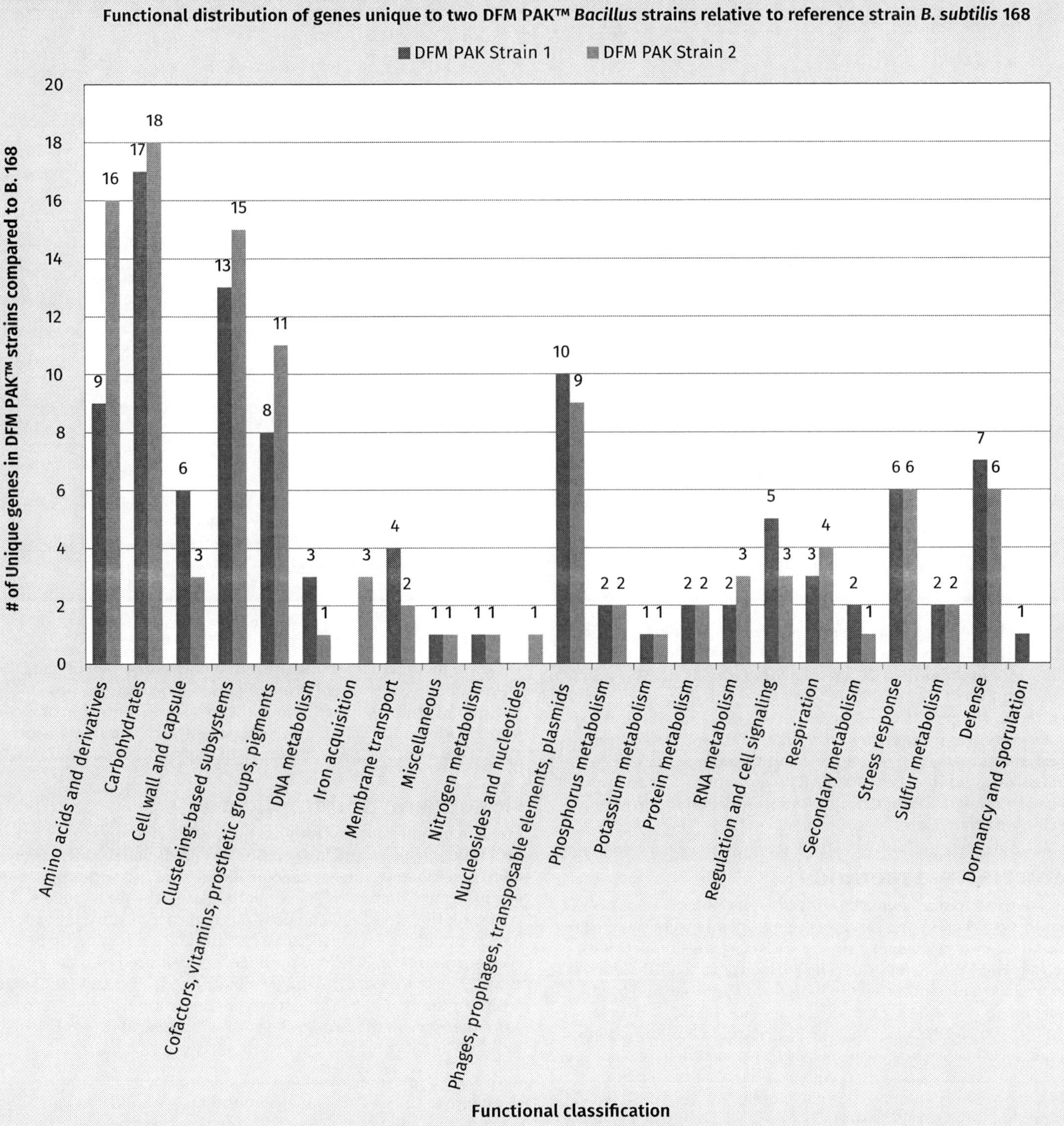

Table 1: Pit manure solids, pit sludge fiber, pit sludge viscosity, and gas concentration results from DFM PAK™ controlled environmental trial[5]

Measurement	Control	DFM PAK™	SEM, pooled	*P*-Value
Pit manure solids[a]				
Total solids %	8.86	5.54	1.13	**0.035**
Pit sludge fiber[b]				
ADF dry weight %	22.21	14.53	1.90	**0.038**
NDF dry weight %	28.61	24.17	1.78	0.225
Pit sludge viscosity[c]				
Viscosity, (cPs)	15,616.13	10,427.14	2,744.23	0.365
Gas concentrations,[d] (ppm)				
Ammonia	25.28	61.53	2.37	**0.048**
Hydrogen sulfide	21.77	18.42	1.03	0.114
Carbon dioxide	746.00	675.62	9.46	**< 0.001**
Methane	7.95	2.83	2.89	0.38

[a] Pit manure core samples were collected from between slats, with a sampling pipe. Six locations per pit were sampled and homogenized to achieve a 1-liter, representative sample from each pit (8 control, 8 DFM PAK™).

[b] Sludge samples collected from each pit (8 control, 8 DFM PAK™ treatment) after pit liquids were drained and analyzed using Ankom Tech. Methodology

[c] Sludge samples collected from each pit (8 control, 8 DFM PAK™ treatment) after pit liquids were drained and analyzed using Brookfield Rotational Viscometry.

[d] Average concentration of gas emissions in treatment airspace during individual manure pit drainage. Gas emissions measured using at the end of Trial 2 with a Drager X-am 5600 meter. Each individual manure pit (8 control, 8 DFM PAK™ Treatment) was drained separately, gas measurements collected during the initial 35 minutes of drainage time.

Diets included co-product ingredients, which consistently resulted in high amounts of manure accumulation in pens and strong adherence of manure to the slats, leading to considerably more time spent washing pens. This client was utilizing a third-party sanitation company for pen pressure washing and was notified that a large cost increase for washing their barns would be applied unless a solution to reduce manure caking was identified. United® Animal Health's DFM PAK™ was identified as an appropriate solution to evaluate for the environmental improvements desired in this system.

Material and methods

Twelve wean to market barns with a history of high amounts of manure build up in pens and washing difficulty were selected for this client evaluation and paired into TREATMENT (Diets with 1lb./ton DFM PAK™) and CONTROL (Diets devoid of DFM PAK™) barns, based on similarities in barn design, pig age and population density. DFM PAK™ was offered in TREATMENT diets for no less than 80 days prior to collecting manure buildup samples. Manure build up was measured by using a handheld scraper tool to remove the total manure contents adhered to the top of the slats, within a 4in × 12in rectangle template. Within each barn, manure buildup scraping samples were collected from 6 pens, at 5 standardized locations within each pen. Scraped manure was collected into a zip closure bag, allowed to air dry, and weighed. Total scraping weights per treatment were utilized to determine the difference in manure buildup in pens.

Results

A 33% reduction in total manure buildup was observed in pens where DFM PAK™ was fed through pigs to inoculate the pen environment (Figure 3). In addition to the measured reduction of manure buildup in DFM PAK™ influenced pens, the client feedback supported a noticeable reduction in manure adherence to the slats.

Conclusion and implications

Modern manure contains higher quantities of undigested fiber, proteins, nitrogen, and long chain fatty acids which collectively contribute to new, interconnected challenges in the pit, pen, and airspace environments. These challenges in the pit include solids accumulation, foaming and poor pumpability of stored manure. As a result of the increased solids accumulation in pits, higher concentrations of gases are produced and emitted into the airspace. The consistency of modern manure also creates challenges with sanitation and health in the pen environment, due to stronger adherence of manure to slats and a tendency for manure to build up and harbor biofilm forming, pathogenic bacteria.

A direct fed microbial intervention that can be fed through pigs to inoculate the pen and pit environments with *Bacillus* spores is desirable, due to ease of application. However, not all *Bacillus* strains possess the functional capabilities needed to efficiently degrade modern manure substrates. The *Bacillus* strains selected for DFM PAK™, an environmental direct fed microbial product, were rigorously screened for functional capabilities *in vitro*, and further validated *in vivo* at a controlled environmental research facility. Furthermore, a field evaluation in a client's facilities

Figure 2: Pen cleanliness results from controlled environmental trial.[5] As determined by quantification[a] of *E coli* genes **A)** and manure buildup[b] **B)** on slats in pens of pigs fed diets with or without DFM PAK.™

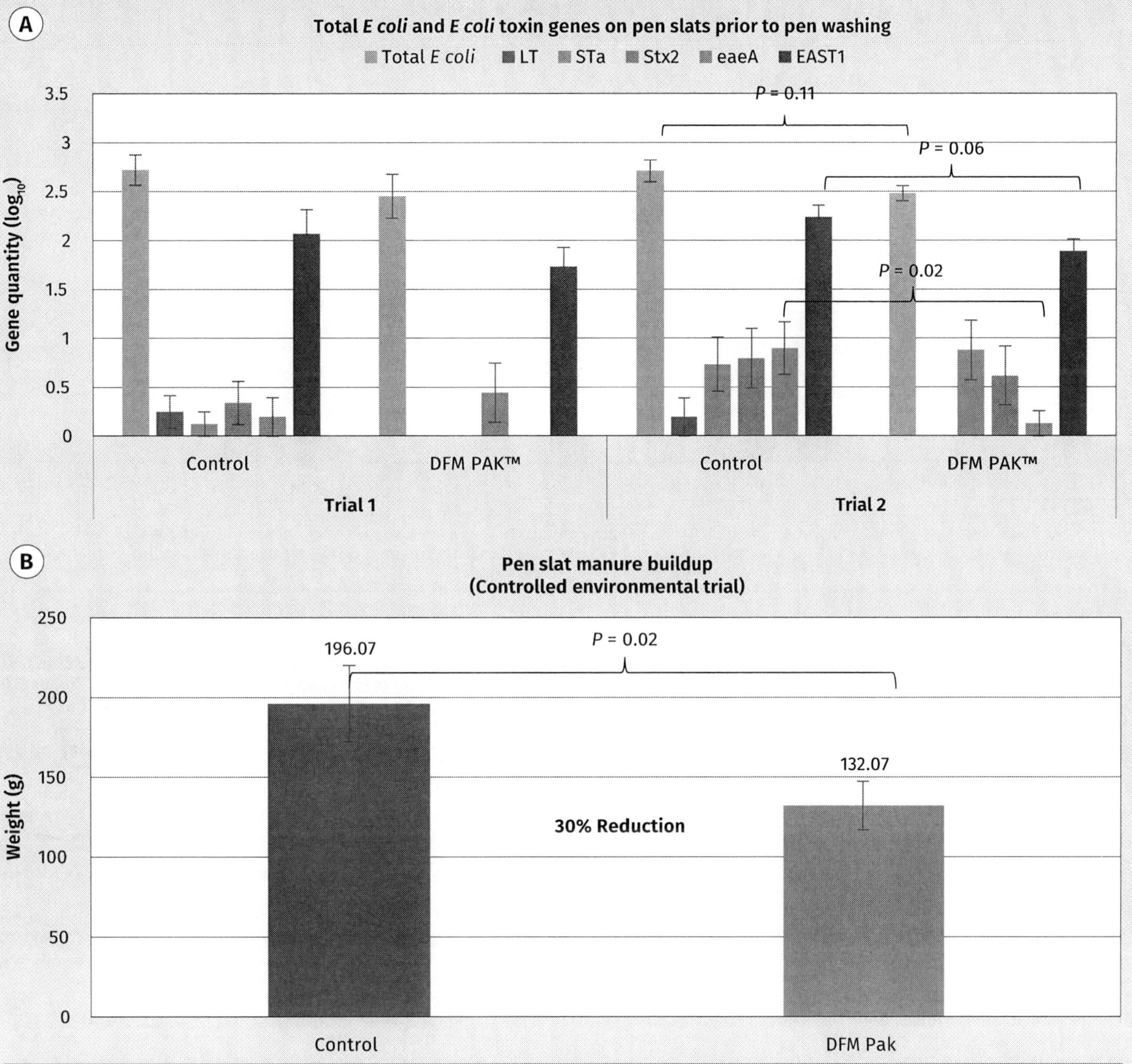

[a] Samples were collected at the end of each trial before washing pens. Slats were swabbed from a consistent location within each pen using a sterile foam swab, then transferred to and FTA card followed by DNA extraction and quantitative PCR analysis.

[b] Measured by hand scraping and weighing total manure contents accumulated in a consistent 6in. × 13in. rectangular area in the center of each pen (n = 24 pens per treatment).

Figure 3: Pen cleanliness as determined in a client field evaluation by scraping and weighing manure buildup from templates placed at standardized locations in 6 barns fed diets with DFM PAK™ and 6 barns fed Control diets (no DFM PAK™).[a]

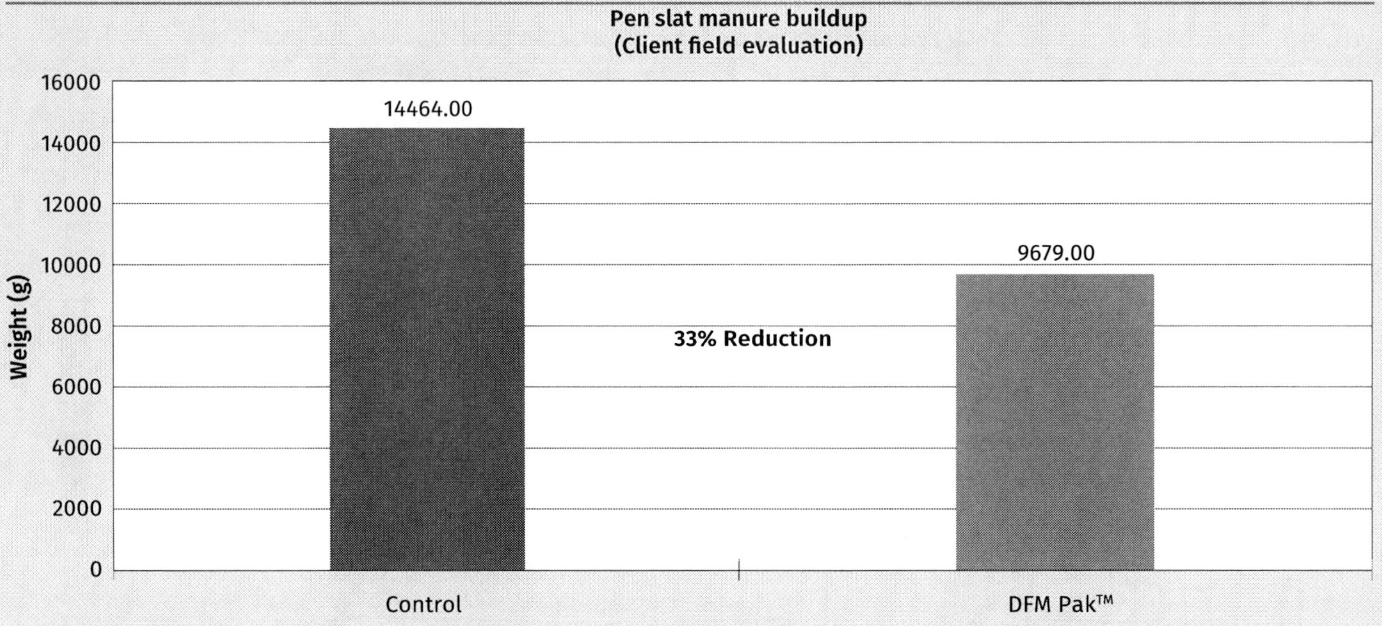

[a] Manure buildup was measured by hand scraping the total manure contents within a 4in × 12in rectangle template withing each barn (6 Control barns, 6 DFM PAK™ barns). Within each barn scrapings were collected from 6 pens at 5 standardized locations per pen, dried, and weighed.

showed a 33% reduction in slat manure buildup in DFM PAK™ influenced pens, mirroring a 30% reduction observed in the controlled environmental research trial. Consistencies in reducing manure buildup in a controlled environmental research facility and in a clients commercial finishing barns further support United® Animal Health's *Bacillus* strain screening model for identifying *Bacillus* strains most capable of improving fiber breakdown, manure adherence and other associated challenges modern manure causes in the pit, pen and in the airspace environments.

References

1. Magstadt, D., Murray, D., Kuecker, K. (2021, September 28). F18-Associated Gut Edema Management (webinar). Swine Health Information Center (SHIC) and American Association Swine Veterinarians (AASV).

2.Davis, M.E., T. Parrot, D.C. Brown, B.Z. de Rodas, J.B. Johnson, C.V. Maxwell, and T. Rehberger. (2008). Effect of a *Bacillus*-based direct fed microbial feed supplement on growth performance and pen cleaning characteristics of growing-finishing pigs. J. Anim. Sci. 86:1459.

3. Earl, A.M., R. Losick, and R. Kolter. 2007. *Bacillus subtilis* genome diversity. J Bacteriol. 2007;189(3):1163–1170.

4. Lee, J.; E. Galbraith; J. Spencer; S. Son; M. King. (2020). DFM Pak®, a novel combination of *Bacillus* strains selected for corn fiber utilization and biofilm activity, shows beneficial impacts on environmental and storage characteristics of manure. AASV Annual Meeting Poster Number 76.

5. Lee, J.; E. Galbraith; J. Spencer; S. Son; M. King. (2021) DFM Pak®, a novel combination of *Bacillus* strains reduces environmental *E. coli* bioburden, manure buildup, and odor compounds in swine housing environment. AASV Annual Meeting Poster Number 73.

6. The RAST Server: Rapid Annotations using Subsystems Technology. Aziz RK, Bartels D, Best AA, DeJongh M, Disz T, Edwards RA, Formsma K, Gerdes S, Glass EM, Kubal M, Meyer F, Olsen GJ, Olson R, Osterman AL, Overbeek RA, McNeil LK, Paarmann D, Paczian T, Parrello B, Pusch GD, Reich C, Stevens R, Vassieva O, Vonstein V, Wilke A, Zagnitko O. BMC Genomics, 2008.

7. The SEED and the Rapid Annotation of microbial genomes using Subsystems Technology (RAST). Overbeek R, Olson R, Pusch GD, Olsen GJ, Davis JJ, Disz T, Edwards RA, Gerdes S, Parrello B, Shukla M, Vonstein V, Wattam AR, Xia F, Stevens R. *Nucleic Acids Res.* 2014.

https://doi.org/10.54846/am2023/70

Post-weaning feed and water outage event with concurrent mycotoxin exposure additively impacts gut morphology and performance of nursery pigs

J. A. Mahoney, PhD; Y. O. Burden, PhD; N. L. Horn, PhD; A. D. Woodward, PhD
United Animal Health, Sheridan, Indiana

Introduction

Poor feed and water intake is a common issue during the first week post-weaning and is often linked to post-weaning growth lag. Reduced feed and water intake may cause microbial disruption, affect gastrointestinal regulation, and impact immune function, leading to short- and long-term effects on nursery pig performance and health. Additionally, feed and water disruption may impact intestinal morphology and nutrient uptake. While the implications of a post-weaning fast have been quantified by research,[1] many piglets still experience disrupted access to feed and water during the weaning process.

The presence of mycotoxins in many grain products used in livestock feed poses risks to animal health and performance by negatively impacting production and leading to reduced profitability.[2,3] Even low levels of mycotoxin contamination can depress feed intake, leading to reduced gain in nursery pigs.[4] Negative effects extend beyond obvious production metrics, as dietary mycotoxin presence can also impact immune response and damage the gastrointestinal tract.[5,6]

Either feed and water outage or mycotoxins alone can severely disrupt performance of nursery pigs[7]; however, the gut health implications of multiple stressors due to feed and water outage in conjunction with mycotoxin exposure have not been fully elucidated. We hypothesized that a combination of feed and water outage and sustained mycotoxin exposure would cause cumulative damage to the gut, leading to even more depressed performance than either feed and water outage or mycotoxin exposure alone. Therefore, we aimed to understand the intestinal integrity and immune changes responsible for reduced post-weaning growth performance in nursery pigs experiencing a combination of feed and water outage and mycotoxin exposure.

Materials and methods

Trial design and procedures were approved by the United Animal Health Animal Care and Use Committee. Mixed sex nursery pigs (n = 256) were weaned at 19 days of age and allotted to pens based on initial body weight (13.11 ± 2.49 lbs.) in a completely randomized block design arranged in a 2 × 2 factorial of 14 replicates of 4 to 5 pigs/pen for a 42-d trial. Treatments consisted of (1) Control, no Mycotoxin, no Feed and Water Stress, (2) Mycotoxin alone, (3) Feed and Water Stress ("Stress") alone, and (4) Mycotoxin + Stress. Mycotoxin exposure consisted of corn and soybean meal diets with the addition of 2.5 ppm deoxynivalenol (DON) via a naturally contaminated corn source (2020 crop) secured by United Animal Health; no other mycotoxins were present in high concentrations. Mycotoxin diets were fed to the respective treatments throughout the 42-d trial. On day 2 post-wean, feed and water stress was performed by turning pen feeders outward and shutting off water for 24 hours. After 24 hours, feed and water access was resumed for pigs undergoing stress.

Pen body weights and feed intakes were recorded at weaning, and 7, 21, and 42 d post-weaning to calculate body weight (BW), average daily gain (ADG), average daily feed intake (ADFI), and feed conversion ratio (F/G). Blood was collected from 1 average body weight pig per pen on d 21 and 42 and serum separated for analysis of markers of oxidative stress using ELISA kits. In the 8 replicates with 5 pigs per pen, 1 average body weight pig per pen was humanely euthanized on d 21 and 42 and jejunal samples were collected to measure intestinal morphology and gene expression.

All data were checked for normality, and any data point exceeding ± 3 standard deviations from the mean was removed from the data set. Data were then analyzed as a 2 × 2 factorial with the random effect of replicate using the Mixed Procedure of SAS to test the main effects of mycotoxin and stress and their interaction within each timepoint. Statistical significance was set at $P \leq 0.05$, with trends towards significance showing at $P \leq 0.10$. Results are presented as mean ± standard error (Tables 1 to 4).

Results

Growth performance

From d 0 to 7, both mycotoxin exposure and stress decreased ($P < 0.01$) body weight, ADG and ADFI. A significant interaction was not observed; however, the exposure to both mycotoxin and stress decreased ($P > 0.10$) BW by 0.5 lbs., ADG by 23.8%, and ADFI by 17.9% compared to either factor alone (Table 1).

By d 21, BW was still decreased ($P < 0.01$) by the ongoing mycotoxin exposure, but only tended to decrease ($P = 0.053$) due to the prior feed and water stress. Still, both ADG and ADFI were decreased by both mycotoxin exposure ($P < 0.01$) and by feed and water stress ($P < 0.05$) from d 0 to 21. The impact of mycotoxin and feed and water stress was additive (Mycotoxin × Stress, $P > 0.10$), with pigs experiencing both having numerically reduced ADG by 5.3 or 12.9% compared to pigs experiencing just mycotoxin or just feed and water stress, respectively (Table 1).

Cumulatively from d 0 to 42, mycotoxin exposure regardless of stress decreased ($P < 0.01$) BW by approximately 5 lbs. compared to no mycotoxin exposure. Additionally, mycotoxin exposure decreased ($P < 0.01$) ADG by 14.8% and ADFI by 11.5%, while feed efficiency was reduced by 4.3%. The impact of feed and water stress on end BW was not evident ($P > 0.10$, Table 1).

Table 1: Impact of mycotoxin (additional 2.5 ppm DON), feed and water stress (24-hr feed and water deprivation, d 2 post-wean), or the combination of mycotoxin and feed and water stress on growth performance of nursery pigs[1]

Stress	–	–	+	+		Level of significance		
Mycotoxin	–	+	–	+	**SEM**	**Mycotoxin**	**Stress**	**Mycotoxin × stress**
Body weight, lbs.								
D 0	13.12	13.14	13.11	13.09	0.68	0.980	0.211	0.382
D 7	15.11	14.60	14.59	14.17	0.68	0.001	0.001	0.722
D 21	27.13	25.10	26.14	24.75	0.97	< 0.001	0.053	0.341
D 42	52.04	45.57	50.04	45.75	1.19	< 0.001	0.307	0.226
D 0 – 7 Performance								
ADG, lbs.	0.29	0.21	0.21	0.16	0.02	0.001	0.001	0.559
ADFI, lbs.	0.35	0.29	0.27	0.23	0.01	< 0.001	< 0.001	0.546
F/G, lb./lb.	1.21	1.38	1.29	1.44				
D 0 – 21 Performance								
ADG, lbs.	0.67	0.57	0.62	0.54	0.02	< 0.001	0.036	0.417
ADFI, lbs.	0.75	0.62	0.71	0.59	0.02	< 0.001	0.013	0.846
F/G, lb./lb.	1.13	1.11	1.15	1.08	0.01	0.003	0.872	0.214
D 0 – 42 Performance								
ADG, lbs.	0.91	0.75	0.85	0.75	0.03	< 0.001	0.169	0.142
ADFI, lbs.	1.25	1.09	1.19	1.08	0.04	< 0.001	0.128	0.229
F/G, lb./lb.	1.38	1.45	1.39	1.44	0.02	0.003	0.956	0.592

[1] ADG = Average daily gain; ADFI = Average daily feed intake; F/G = Feed/Gain ratio.

Table 2: Impact of mycotoxin (additional 2.5 ppm DON), feed and water stress (24-hr feed and water deprivation, d 2 post-wean), or the combination of mycotoxin and feed and water stress on markers of oxidative stress[1] in nursery pigs

Stress	–	–	+	+		Level of significance		
Mycotoxin	–	+	–	+	**SEM**	**Mycotoxin**	**Stress**	**Mycotoxin × stress**
Day 21								
GPX, µg/mL	560.78	671.75	553.95	603.53	52.32	0.111	0.451	0.537
CAT, µg/mL	3.22^{ab}	3.57^{ab}	3.74^{a}	3.06^{b}	0.24	0.500	0.990	0.038
SOD, µg/mL	36.32	37.99	40.53	37.84	2.32	0.809	0.334	0.301
Day 42								
GPX, µg/mL	897.79	1002.85	1091.21	1042.87	105.28	0.702	0.121	0.304
CAT, µg/mL	8.24	8.42	9.56	9.26	0.83	0.927	0.078	0.687
SOD, µg/mL	44.12	43.60	50.42	49.31	2.73	0.765	0.033	0.914

[ab] Means with uncommon superscripts differ, $P \leq 0.05$.

[1] GPX = Glutathione peroxidase; CAT = Catalase; SOD = Superoxide dismutase

Markers of oxidative stress

On d 21, mycotoxin exposure numerically increased ($P = 0.111$) serum glutathione peroxidase activity. Additionally, serum catalase concentration appeared to be increased by mycotoxin exposure and by feed and water stress, but not when both factors were experienced (Mycotoxin × Stress, $P < 0.05$, Table 2).

On d 42, stress increased ($P < 0.05$) serum superoxide dismutase activity, tended to increase ($P = 0.078$) serum catalase activity, and numerically increased ($P = 0.121$) serum glutathione peroxidase activity.

Jejunal intestinal morphology

Villi height (VH) and crypt depth (CD) on d 21 were numerically decreased ($P = 0.179$ and 0.188, respectively; Table 3) by feed and water stress regardless of whether pigs had mycotoxin exposure. On d 42, crypt depth was numerically increased ($P = 0.120$) in pigs that had been feed and water stressed.

Six-weeks of mycotoxin exposure decreased ($P < 0.05$) villi height and numerically reduced ($P = 0.173$) VH:CD on d 42.

Jejunal gene expression

On d 21, the interactive treatment of mycotoxin exposure with feed and water stress increased jejunal MUC2 gene expression compared to either factor alone, with the control treatment having intermediate expression (Mycotoxin × Stress, $P < 0.05$; Table 4). Jejunal GLUT2 gene expression tended to decrease due to feed and water stress relative to control, but did not differ from control when both mycotoxin exposure and feed and water stress were experienced (Mycotoxin × Stress, $P = 0.084$). Mycotoxin exposure regardless of feed and water stress tended to decrease ($P = 0.096$) jejunal SGLT1 gene expression.

There were no differences in jejunal gene expression on d 42; however, the expression pattern of MUC2 ($P = 0.166$) and GLUT1 ($P = 0.183$) were similar on d 42 compared to d 21.

Conclusion

Both mycotoxin exposure and feed and water stress cause long-term impacts on growth during the nursery period, with mycotoxicosis being more severe. Exposure to both mycotoxin and feed and water deprivation caused cumulative losses to performance during the early nursery phase. Both mycotoxin and feed and water stress negatively influenced markers of oxidative stress, jejunal morphology, and jejunal gene expression, and the long-term effects of each were still evident by the end of the nursery period. Implementing strategies to minimize mycotoxins in diets and ensuring feed and water are continually available immediately post-wean will allow for optimal intestinal health and performance of nursery pigs.

References

1. Horn N et al. Impact of acute water and feed deprivation events on growth performance, intestinal characteristics, and serum stress markers in weaned pigs. *J Anim Sci.* 2014;92(10):4407.
2. Bryden WL. Mycotoxin contamination of the feed supply chain: Implications for animal productivity and feed security. *Animal Feed Sci and Tech.*2012;173(1-2):134.
3. Shull LR and Cheeke PR. Effects of synthetic and natural toxicants on livestock. *J Anim Sci.* 1983;57:330.
4. Diekman MA et al. Mycotoxins and Swine Performance. *Pork Industry Handbook.*1992;E2420.
5. Pierron A et al. Impact of mycotoxin on immune response and consequences for pig health. *Anim Nutr.* 2016;2:63.
6. Gao Y et al. The compromised intestinal barrier induced by mycotoxins. *Toxins.* 2020;12(10): 619.
7. Burden YO et al. P202. Effect of post-weaning acute feed and water deprivation and high DON diet on growth performance and intestinal health in piglets. *Animal-science proceedings.* 2022 Jul 1;13(2):220.

Table 3: Impact of mycotoxin (additional 2.5 ppm DON), feed and water stress (24-hr feed and water deprivation, d 2 post-wean), or the combination of mycotoxin and feed and water stress on jejunal intestinal morphology of nursery pigs

Stress	-	-	+	+		Level of significance		
Mycotoxin	-	+	-	+	**SEM**	**Mycotoxin**	**Stress**	**Mycotoxin × stress**
Day 21								
Villi height, µm	1036	996	969	932	46	0.415	0.179	0.976
Crypt depth, µm	950	858	929	905	49	0.256	0.790	0.500
VH:CD	1.204	1.250	1.113	1.111	0.083	0.767	0.188	0.783
Day 42								
Villi height, µm	1088	952	1123	951	56	0.015	0.769	0.755
Crypt depth, µm	719	732	842	784	55	0.684	0.120	0.525
VH:CD	1.609	1.402	1.464	1.353	0.109	0.173	0.398	0.671

Table 4: Impact of mycotoxin (additional 2.5 ppm DON), feed and water stress (24-hr feed and water deprivation, d 2 post-wean), or the combination of mycotoxin and feed and water stress on jejunal immune and nutrient transport gene expression cycle thresholds (CT) in nursery pigs[1,2,3]

Stress	-	-	+	+		Level of significance		
Mycotoxin	**-**	**+**	**-**	**+**	**SEM**	**Mycotoxin**	**Stress**	**Mycotoxin × stress**
Day 21								
IL-1β	9.05	10.11	9.12	7.94	0.86	0.946	0.251	0.220
IL-6	7.18	7.41	8.49	7.10	0.81	0.424	0.492	0.271
MUC2	8.47[ab]	9.75[a]	9.44[a]	7.65[b]	0.65	0.663	0.342	0.015
GLUT2	1.44[y]	2.33[xy]	2.62[x]	1.80[xy]	0.46	0.951	0.497	0.084
SGLT1	0.08	3.42	0.78	1.09	1.09	0.096	0.446	0.162
Day 42								
IL-1β	10.72	9.93	10.83	10.07	1.07	0.465	0.906	0.994
IL-6	9.10	9.58	9.57	9.12	1.02	0.989	0.999	0.648
MUC2	8.60	8.97	9.88	6.63	1.26	0.266	0.675	0.166
GLUT2	2.68	3.36	4.17	1.43	1.24	0.417	0.860	0.183
SGLT1	8.92	8.35	9.79	7.61	1.51	0.392	0.965	0.607

ab Means with uncommon superscripts differ, $P \leq 0.05$.

xy Means with uncommon superscripts differ, $P \leq 0.10$.

1 Data is expressed as means of CT_{target} - $CT_{housekeeping}$. Numerically higher means relate to lower gene expression.

2 GAPDH served as a housekeeping gene and was not different between treatments ($P > 0.10$).

3 IL-1β = Interleukin 1β; IL-6 = Interleukin 6; MUC2 = Mucin 2; GLUT2 = Glucose transporter 2; SGLT1 = Sodium/glucose cotransporter 1.

https://doi.org/10.54846/am2023/71

Effect of water enriched with Nextein APF® on weaned pigs after long duration transport under summer conditions

T. Bauman[2]; H. Lathom[2]; D. Shaffer[1]; D. McKilligan[1]; M. Almeida[3]; K. Coulson[1]; V. Utrera[1]
[1]TechMix Global Inc., Stewart, Minnesota; [2]The Maschhoffs; [3]Veterinary Diagnostic Laboratory, Iowa State University

Introduction

Transport stress of weaned pigs during elevated temperatures may exacerbate the effect of weaning on these animals with the negative consequence on health and performance. Dehydration and heat stress might be important predisposing factors leading to villous atrophy and mal absorption syndrome commonly observed during the first days post weaning into nursery. Water enrichment with nutrient dense liquids have been shown to improve growth performance and reduce morbidity and mortality in the early post-weaning period.[1,2] Nextein APF® is a nutrient dense liquid that supplies electrolytes, functional proteins and energy helping to mitigate the effects of nutrient disruption and dehydration after weaning.

The purpose of this field trial was to evaluate the effect of Nextein APF® provided in two different doses vs control non treated pigs, 2 days after weaning on several parameters after being weaned and transported for 6 or more hours under US mid-west summer conditions.

Materials and methods

Pigs

A total of 4500 pigs were divided in three groups: 1) control non treated, 2) pigs received water enriched with Nextein APF® at 1 oz/gal, 3) pigs received water enriched with Nextein APF® at 4 oz/gal. The trial was replicated in three wean to market sites. The study was carried out from weaning until 2 weeks post placement.

Heat index at time of arrival. For calculation of heat index, the air temperature and dew point temperature at times of arrival were obtained from online weather service and input into the heat index calculator provided online by the National weather service (**wpc.ncep.noaa.gov**). The average time of transport was 10 hours for each site.

Placement

Control: Animal husbandry performed in accordance with the grower's standard operating procedures was performed. Bottom 10% were not pulled off the truck. Water was off in the wet/dry feeder for first 48 hours. Water was available via cup waters. Mat feeding was provided.

Treatments: The target was flooding wet/dry feeders with a target volume of 0.1 gal/hd per flooding. Wet/dry feeder were turned off and Nextein APF® solution was delivered in feeder pans. Pigs had access to water cups and mat feed. Bottom 10% of pigs were not pulled off truck.

Protocol

Pigs entering the barn were gate cut. No sizing or pulling pigs upon arrival. Control pens had access to feed in feeder, water cups, and mat feed. The water bar in the wet/dry feeders was shut off. Twenty gallons Nextein APF® and water solution was added to the pan of the feeders in the first 12 hours in the treated pens. Pigs had access to cup waters and mat feed during this time. After 12 hours, feeders were opened and an additional 0.1 gallons/head of Nextein APF® solution was placed in the feeder pans. Fluid total per pen (2 feeders at 10 gallons each, twice in 18 hours). Target 3 sites within that flow to assess Nextein APF® delivered in the trough of the wet dry feeder for 24 hours post arrival. Mortality and treatments were recorded by pen. Total pulls were also recorded.

Hydration status was tested via blood urea nitrogen (BUN)/creatinine ratio and hematocrit. Five pigs from each treatment and five from the control group (15 total per barn) were tagged, weighed and blood sampled to determine hydration status. Test was done at arrival and 48 hrs later.

Enteric tissues were collected for bacterial culture and rotavirus PCR. Oral fluids were collected for PRRSv and PED 7 days after arrival.

Fecal scores were assessed at 10 days and 17 days post placement. Mortality and treatment records were recorded.

Rectal temperatures were recorded from tagged pigs.

Villous atrophy scoring

Histopathology was assessed at the Iowa State University Veterinary Diagnostic Laboratory. Scoring system was assessed according to Madson et al.[3] Briefly, three pairs of villous and crypts were measured in each section and average values were obtained. All measures were in micrometers. Three pigs per each group were euthanized and necropsies were performed to assess histopathological evaluation of small intestine.

Statistical analysis

The effect of water enriched with Nextein APF® on weaned pigs after long duration transport under summer conditions was assessed using analysis of variance (ANOVA). One model for each outcome (weight gain difference, temperature difference and fecal score) was built and the treatment remained as the independent variable. Post-hoc analysis to identify differences between treatments was conducted using t-test and Tukey-Kramer adjustment. The difference between 2 collection points for BUN, creatinine and hematocrit, were evaluated using a non-parametric test (Kruskal–Wallis test).

Results and discussion

Heat index at arrival: Site 1: June 13th, heat index: 105°F; pigs arriving under this high temperature weather showed signs of stress: labored breathing and reluctance to move. Site 2: June 27th, heat index: 81°F; Site 3: July 8th, heat index: 82°F. In general, pigs in both treated groups learned to operate the wet/dry feeder than pigs in control groups.

BUN/creatinine ratio: the highest levels of blood urea nitrogen were observed in all pigs arriving on June 13th when the heat index was 105°F. The BUN values increased in control pigs in 2 of the 3 sites 24 hrs after arrival. BUN can increase when muscle catabolism is occurring and can be interpreted as a consequence of weaning stress and nutrient disruption.[4] A decrease in BUN values was observed in all pigs treated with Nextein APF® 4 oz, regardless the weather conditions at arrival, suggesting a mitigating effect on dehydration. Creatinine values remained under normal ranges.[5]

Hematocrit: Hematocrit values were elevated at arrival and remained elevated 24 hrs later, above normal range, especially in those pigs arriving on June 13th when heat index was 105°F. Hematocrit values decreased in all groups of pigs that were drinking enriched water (Nextein APF® 1 oz and 4 oz). A decrease on hematocrit values can be interpreted as indicative of rehydration associated with water enrichment. The hematocrit values in control pigs were always the highest as compared with the Nextein APF® groups, suggesting that water consumption in these pigs was not effective for their rehydration in the first 24 hrs after arrival. These results suggest a positive effect of Nextein APF® on rehydration of pigs under the conditions of this trial.

Rectal Temperature: A reduction in average rectal temperature was observed in all pigs 24 hrs after arrival, however the average temperature was still above the normal range (103.2°F) in 83% of the pigs evaluated.[5] The highest average temperature was observed in pigs at arrival when the environmental heat index was 105°F. No difference on rectal temperatures reduction 24 hrs after arrival were observed between Nextein APF® treated and control groups.

Fecal scores, mortality, and morbidity: No differences in fecal scores, mortality, and morbidity were observed between treated and control groups after 2 weeks in the nursery. All the fecal samples evaluated were positive to Rotavirus A, B and C, 7 days after arrival.

Villous atrophy was detected in all the pigs evaluated at arrival in the sites 1 and 2. However only two control pigs showed moderate atrophy in the 3rd site. No treated pigs showed villous atrophy in this site 3. A numerical reduction in the severity of atrophy was observed in proximal jejunum samples evaluated at 48 hrs of water enrichment (3.61:1 control vs 3.73:1 Nextein APF® 1 oz, $P < 0.11$) and (3.61:1 control vs 3.93:1 Nextein APF® 4 oz, $P < 0.2$). Villous height to crypt depth ratio in 3-week-old pigs can vary, normal scores are 4.5:1, 3.6:1, and 3.5:1 for proximal jejunum, distal jejunum, and ileum, respectively,[6] lower scores are indicative of villous atrophy. Villous atrophy was observed in 89% (8/9) of control pigs as compared to 67% (6/9) pigs in Nextein APF® enriched groups, indicating a higher proportion of affected pigs.

In conclusion, these results suggest a reduction on the severity of villous atrophy of pigs when Nextein APF® was provided for 2 days to pigs immediately after arrival from a long-distance transport under summer conditions.

References

1. Morris, J. R., M. Ellis, J. Estrada, A. M. Gaines, C. M. Shull, and O. Mendoza. 2017. Effects of a liquid supplement and a nursery starter diet given immediately post-weaning on growth performance and morbidity and mortality of nursery pigs. J. Anim. Sci. 95(Suppl. 2):102.

2. Schmitt, R. L., M. Ellis, O. F. Mendoza, C. M. Shull, D. McKilligan, and N. Upah. 2018. The effect of administration of a nutrient dense liquid at weaning on growth performance and morbidity and mortality of pigs during the nursery period under commercial conditions. J. Anim. Sci. 85(Suppl. 2):115.

3. Madson DM, D.R. Magstadt, P.H.E. Arruda, H. Hoang, D. Sun, L.P. Bower, M. Bhandari, E.R. Burrough, P.C. Gauger, A.E. Pillatzki, G.W. Stevenson, B.L. Wilberts, J. Brodie, K.M. Harmon, C. Wang, R.G. Main, J. Zhang, K.J. Yoonet, 2014. Pathogenesis of porcine epidemic diarrhea virus isolate (US/Iowa/18984/2019) in 3-week-old weaned pigs. Veterinary Microbiology, 174(1-2):60-68.

4. Friendship, RM, Lumsden JH, McMillan I et al. 1984. Can J Comp Med 48:390-393.

5. Dewey C.E, Straw B.E. 2006. Herd examination. In Straw BE, Zimmerman JJ, D'Allaire S, et al., eds Diseases of Swine, 9th ed. Blackwell Publishing, pp 3-14.

https://doi.org/10.54846/am2023/72

Effects of zinc oxide and mannan-rich fraction (Actigen®) effects on nursery swine performance and removals

J. Soto[1], BS, PhD; H. Cemin[2], BS, MS, PhD; M. Hart[2], BS, MS; J. Pietig[2], BS, MS; S. Hansen[2], BS, MS; E. Hansen[2], BS, MS
[1]Alltech, Nicholasville, KY; [2]Hubbard Feeds, Mankato, MN

Introduction

Zinc (Zn) is a trace mineral with an essential role in multiple physiological processes. Whereas 50 to 100 ppm are enough to meet swine requirements, supplementing Zn at higher concentrations of 2,000 to 3,000 ppm as zinc oxide (ZnO) has been traditionally utilized in weanling pigs to positively impact growth, feed efficiency, and reduce post-weaning diarrhea.[1] However, there is growing public concern that high levels of ZnO may cause pollution when manure is spread on farm ground.[2] In addition, pharmacological use of Zn in nursery pigs could play a role in the selection and persistence of methicillin-resistant *Staphylococcus aureus*.[3] As a result, many countries have banned or restricted the use of pharmacological levels Zn in diets for nursery pigs. Thus, finding nutritional strategies offering similar benefits as ZnO are of high interest to the swine industry. Mannan-oligosaccharide (MOS), a yeast cell wall derivate, has been identified as a feasible alternative or a complement to pharmacological concentrations of ZnO in nursery swine diets. A mannan rich fraction (Actigen® [ACT]), has shown to positively impact intestinal morphology and permeability,[4] inhibit gut colonization of enteric pathogens,[5] and improve the pattern of immune responses with[6] and without[4] viral challenges. Most recently, it was reported that pigs fed ACT or ACT in combination with ZnO improved intestinal morphology and lowered diarrhea incidence.[7] Thus, we hypothesized that ACT could replace or complement ZnO to alleviate the weaning stress, positively impact growth performance, and decrease removal prevalence of weaned pigs reared under commercial conditions.

Methods

Two experiments were conducted to determine the interactive effects of ZnO and ACT on nursery pig performance and removals. In Exp. 1, 2,592 pigs (initial BW 6.9 ± 0.4 kg) were used in a 37-d trial. Pens of 36 mixed sex pigs were randomly assigned to 1 of 6 dietary treatments with 12 replications per treatment. Treatments were arranged in a 2 × 3 factorial with main effects of zinc (100 or 3,000 ppm) and ACT (0, 400, or 800 ppm). Diets were fed in two phases: phase 1 from d 0 to 21 and phase 2 from d 22 - 37, which represented the experimental and common diets, respectively. Phase 1 diets were corn-soybean meal based and contained 9% whey permeate, 5% rolled oats, and formulated to contain 1.38% SID Lysine. Phase 2 diets were corn-soybean meal based and formulated to contain 1.30% SID Lysine. In Exp. 2, 3,998 pigs (initial BW 5.8 ± 0.4 kg) were used in a 43-d trial. Pens of 46 to 53 mixed sex pigs were randomly allotted to 1 of 4 dietary treatments with 20 replications per treatment. Treatments were arranged in a 2 × 2 factorial with main effects of zinc (100 or 3,000 ppm) and ACT (0 or 800 ppm). Diets were fed in two phases: phase 1 from d 0 to 21 and phase 2 from d 22 - 43 and were comprised with similar ingredients and nutrients specification as indicated for Exp. 1. In both experiments, data was analyzed using the GLIMMIX procedure of SAS (SAS Institute, Inc., Cary, NC) in a randomized complete block design with pen serving as the experimental unit and initial BW serving as the blocking factor. In Exp. 1, preplanned linear and quadratic orthogonal contrast were built using coefficients for equally spaced treatment and used to determine the main effects of increasing ACT supplementation. Main effects of Zn as well as their interaction were tested. In Exp. 2, main effects of Zn and ACT as well as their interaction were tested. In both experiments, dietary treatments were the fixed effect and block served as the random effect in the analysis. Results from the experiments were considered significant at $P < 0.05$ and a marginally significant $P > 0.05$ and $P \leq 0.10$.

Results

In Exp. 1, no significant ZnO × ACT interactions were observed in growth performance or removals. From d 0 to 37, pigs fed diets with 3,000 ppm Zn had higher ($P < 0.0001$) ADFI, greater ($P < 0.0001$) ADG, and heavier ($P < 0.0001$) BW compared to pigs fed diets containing 100 ppm Zn (Table 1). Additionally, pigs fed diets containing 3,000 ppm of Zn had marginally increased ($P = 0.077$) removals prevalence compared to pigs fed diets with lower Zn concentration (Table 1). However, pigs fed diets with 3,000 ppm Zn and incremental levels of ACT marginally lower (linear, $P = 0.083$) dead & removals prevalence (Table 2). In Exp. 2, there was a significant ZnO × ACT interaction for removals ($P = 0.040$) where prevalence of removals was lower in pigs fed diets with ACT and 3,000 ppm Zn, however it was not impacted in pigs fed diets with 100 ppm Zn (Table 2). Similarly, there was a marginally significant ZnO × ACT interaction for ADFI ($P = 0.068$) where pigs fed diets with ACT and 3,000 ppm Zn had lower ADFI, however it was not impacted in pigs fed diets with 100 ppm Zn (data not showed). From d 0 to 43, pigs fed diets with 3,000 ppm of Zn had greater ($P < 0.0001$) ADG and heavier BW ($P < 0.0001$) (Table 1). Furthermore, pigs fed diets supplemented with ACT or 3,000 ppm Zn had marginally poorer ($P = 0.058$) or had poorer ($P = 0.014$) G:F, respectively (Table 1). Additionally, whereas pigs fed diets with 3,000 ppm Zn had higher ($P = 0.028$) incidence of dead & removals (Table 1), pigs fed diets supplemented with ACT had marginally lowered ($P = 0.093$) dead & removals incidence (Table 2).

Discussion

Limited voluntary feed intake and associated reduced growth are typically observed as a result of the weaning process. To lessen the impact of the weaning transition, nursery diet formulation frequently includes pharmacological levels of Zn (2,000 - 3,000 ppm), which has showed to positively influence the weanling pig. High levels of Zn reduce the incidence of diarrhea and increase weight gain of pigs.[1] The results of our experiments are in line with the previously published body of work. In both experiments, pigs fecd dicts supplemented with 3,000 ppm of Zn had greater ADG and heavier final BW compared to pigs fed diets containing

Table 1: Main effects of Zn supplementation on growth performance and removals of nursery swine

	Exp. 1*		SEM	Exp. 2†		SEM
Zn, ppm	**100**	**3,000**		**100**	**3,000**	
BW, kg						
Initial	7.0	7.0	0.27	5.8	5.9	0.12
Final	24.0[a]	24.6[b]	0.70	22.6[a]	23.4[b]	1.07
Overall						
ADG, g	443[a]	459[b]	4.6	436[a]	457[b]	12.8
ADFI, g	548[a]	572[b]	9.0	562[a]	592[b]	22.3
G:F, g/kg	810	804	8.8	780[a]	775[b]	8.2
Pigs removed						
Dead, %	0.61	0.54	0.329	0.41	0.35	0.202
Removals, %	2.06[x]	3.12[y]	0.853	0.55[a]	1.40[b]	0.304
Dead & removals, %	2.65	3.64	0.945	0.96[a]	1.77[b]	0.358

* A total of 2,592 pigs were used in a 37-d trial with 36 pigs per pen and 12 replicates per treatment

† A total of 3,998 pigs were used in a 43-d trial with 46 - 53 pigs per pen and 20 replicates per treatment

ab Means with a superscript indicates significant Zn main effect ($P \leq 0.05$)

xy Means with a superscript indicates marginally significant Zn main effect ($P > 0.05$ and $P \leq 0.10$)

Table 2: Interactive effects of Zn and ACT supplementation on nursery swine removals

	Exp. 1[a,b]					Exp. 2[c]				
Zn, ppm	**100**		**3,000**		**ACT Lin[d]**	**100**		**3,000**		**Zn × ACT**
ACT, ppm	**0**	**800**	**0**	**800**		**0**	**800**	**0**	**800**	
Dead, %	0.69	0.46	0.69	0.00	0.139	0.51	0.31	0.30	0.40	0.441
Removals, %	2.21	2.68	3.61	2.19	0.170	0.51	0.59	1.99	0.81	0.040
Dead & removals[e], %	2.88	3.11	4.27	2.17	0.083	1.03	0.89	2.31	1.22	0.188

a In Exp.1, results for treatment with 400 ppm ACT are not shown

b A total of 2,592 pigs were used in a 37-d trial with 36 pigs per pen and 12 replicates per treatment

c A total of 3,998 pigs were used in a 43-d trial with 46-53 pigs per pen and 20 replicates per treatment

d ACT linear effect within pigs supplemented with 3,000 ppm ZN

e Marginally significant main effect of ACT within Exp. 2

100 ppm of Zn. Whereas the mode of action of Zn is not clear, potential modes of action are the improvement of gut integrity and permeability, modulation of gut bacteria, development and functionality of the immune system, and downregulation of the expression of proinflammatory cytokines.[8] Additionally, high levels of Zn have been shown to stimulate voluntary intake. The results of both experiments validate the latter mode of action, where pigs feed 3,000 ppm Zn had increased voluntary feed intake compared to pigs fed diets containing 100 ppm Zn. Mechanistically, high levels of Zn have shown to stimulate stomach and plasma ghrelin, which plays an important role in appetite regulation.[9] Whereas ACT have shown to improve gut morphology or limit bacterial infection, improvements on growth performance has been inconsistent. For example, in two independent studies while pigs fed diet supplemented with ACT had lowered diarrhea incidence, ACT had modest or no effect on ADG, ADFI, or G:F.[4,7] The results of our experiments confirm these results, where pigs fed ACT had no impact on growth performance.

Although Zn at pharmacological levels has shown to consistently improve growth rate and reduce the incidence of diarrhea, there is limited evidence of the impact of pharmacological levels of Zn on morbidity and mortality incidence in weanling pigs. A case study reported higher nursery pig mortality (8.0% vs 0.96%) because of lower ZnO (612 vs 3,000 ppm) concentration due to a diet formulation error, and suggested that ZnO might be useful pharmacologically for preventing mortality associated with *E coli*.[10] A recent study reported that preweaning mortality was lowered (15.0 vs ≤ 13.2%) in sows supplemented with higher Zn concentration during late gestation (≤ 595 vs 125 ppm Zn).[11] Conversely, the results of our experiments suggested that pigs fed diets with 3,000 ppm Zn had increased removals and dead & removals prevalence compared to pigs fed diets containing 100 ppm Zn. While increases of morbidity and mortality in pigs fed high Zn in the present studies is not fully understood, recent work had suggested that certain diet ingredients bind more acid in the stomach,[12] which leads to increases in gastric pH and thus decreasing nutrient digestion and allowing pathogens to proliferate. Zinc oxide has one the highest acid-binding capacity (ABC) among ingredients, thus we speculate that increased removals and dead & removals might have been influenced by ZnO ABC. Feeding ACT has showed to improve the health outcomes in nursery swine. In PRRS infected pigs, ACT supplementation altered the pattern of immune responses, boosting innate and cell-mediate immunity.[6] In addition, ACT supplementation have improved intestinal morphology and lower permeability, resulting in lower diarrhea and mortality incidence.[4] The results of our experiments are in line with these findings. In Exp. 1, pigs fed diets with 3,000 ppm of Zn containing ACT, dead & removals prevalence was marginally lowered. Similarly, in Exp. 2, prevalence of removals was lowered in pigs fed diets with ACT and 3,000 ppm Zn. Furthermore, whereas the magnitude of dead & removals reduction appears to be higher in pigs fed diets with higher levels of Zn, ACT marginally lowered dead & removals prevalence in Exp.2, regardless of Zn concentration.

In summary, pigs fed pharmacological levels of Zn had increased overall growth performance, regardless of ACT supplementation. Furthermore, while pigs fed the highest levels of Zn had increased the prevalence of dead and removed pigs, ACT offset these effects, which was particularly evident in pigs fed high levels of Zn. Thus, there may be substantial value in utilizing ACT in nursery swine diets, particularly in the presence of pharmacological levels of Zn.

References

1. Jacela J, DeRouchey J, Tokach M, Goodband R, Nielssen J, Renter D, Dritz S. Feed additives for swine: fact sheet – high dietary levels of copper and zinc for young pigs, and phytase. J. Swine Health Prod. 2010;18:87-91.

2. Poulsen, H. Zinc and copper as feed additives, growth factors or unwanted environmental factors. J. Amin. and Feed Sci. 1998;7: 135–142.

3. Slifierz M, Frienship R, Weese J. Methicillin-resistant *Staphylococcus aureus* in commercial swine herds is associated with disinfectant and zinc usage. Appl. Environ. Microbiol. 2015;81: 2690-2695.

4. Min, S., Yijie, F., Han, S., Jiayi, Y., Fangfang, L., Xiaotong, Z., Lina, W., Ping, G., Gang, S., Zhonggang, W., Fang, Z., Gang, L., Yan, X., Qingyan, J., Songbo, W. Effects of Actigen, a second-generation mannan rich fraction, in antibiotics free diets on growth performance, intestinal barrier functions and inflammation in weaned piglets. Livest. Sci. 2019; 229: 4-12.

5. Browne, N., Traynor, A., & Horgan, K. Mannan rich fraction from yeast modulates inflammatory responses in intestinal cells (HT-29) exposed to Escherichia coli. J. of Appl. Anim. Nut. 2019;7;1-17.

6. Che, T., Song, M., Liu, Y., Johnson, R., Kelley, K., Van Alstine, W., Dawson, K. and Pettigrew, J. Mannan oligosaccharide increases serum concentrations of antibodies and inflammatory mediators in weanling pigs experimentally infected with porcine reproductive and respiratory syndrome virus. J. Anim. Sci. 2019.90:2784-2793.

7. Zhang, G., Zhao, J., Dong, W., Song, X., Lin, G., Li, D., Zhang, S. Yeast-derived mannan-rich fraction as an alternative for zinc oxide to alleviate diarrhea incidence and improve growth performance in weaned pigs. Anim. Feed Sci. and Tech. 2021; 281: 1-12.

8. De Mille, C., Burrough, E., Kerr, B., Schweer, W., Gabler, N. Dietary Pharmacological Zinc and Copper Enhances Voluntary Feed Intake of Nursery Pigs. Front. Anim. Sci. 2022;3: 1-12.

9. Yin, J., Li, X., Li, D., Yue, T., Fang, Q., Ni, J. Dietary supplementation with zinc oxide stimulates ghrelin secretion from the stomach of young pigs. J. Nutr. Biochem. 2009; 20(10): 783–790.

10. Tokach, L., Dritz, S., Tokach, M. Diagnosis and calculation of economic impact of incorrect pharmacological dosage of zinc oxide supplementation aided by record analysis of nursery performance. Swine Health Prod. 2000;8(5): 229-233.

11. Holen, J., Urriola, P., Schwartz, M., Jang, J., Shurson, G., Johnston, L. Effects of supplementing late-gestation sow diets with zinc on preweaning mortality of pigs under commercial rearing conditions. Transl. Anim. Sci. 2020, 4, 519-530.

12. Stas, E., Tokach, M., DeRouchey, J., Goodband, R., Woodworth, J., Gebhardt, J. Evaluation of the acid-binding capacity of ingredients and complete diets commonly used for weanling pigs. Transl. Anim. Sci. 2022, 6, 1-9.

https://doi.org/10.54846/am2023/73

High health herd response to activated phenolic molecules, APM™

A. Bents[1], DVM; C. Schmidt[2], DVM; R. Tims[2]
[1]Hubbard Feeds, Mankato, Minnesota; [2]Precision Health Technologies, Worthington, Minnesota

Background

Intestinal health in pigs continues to be a focus of producers and veterinarians. A growing environment of increased consumer concern and escalating difficulty in controlling enteric challenges demands that our industry must explore options beyond conventional use of antibiotics in addressing these challenges. Phytobiologics continue to gain interest amongst the veterinary community and may offer a valuable tool to meet these increasing challenges.

Polyphenols are a part of the innate immune system of many plants.[1] Disruption of plant cell integrity releases these polyphenols, which interact with intercellular fluids and enzymes to produce molecules that form protein cross-linkages to seal the injury site, interfere with microbial growth, and neutralize toxins that may delay the healing of the injured site.[2] These interactions also take place when certain plant polyphenols are placed in the presence of swine epithelial cells in the intestinal tract and skin. As a result, plant polyphenols offer a great potential alternative to address the injury, inflammation, microbial growth, and resulting presence of toxic compounds that develop as a result of enteric challenge in pigs.

The goal of this study was to evaluate the effect of a proprietary blend of plant polyphenols on the performance of piglets in a relatively high health farrowing environment. Precision Health Technologies in Brookings, SD continues to refine the extraction and quantification process, and evaluate implementation and delivery techniques to maximize the potential benefit of plant polyphenols in livestock production. GI Revive™ is a product that has emerged as a result of this continued evaluation of the needs of producers.

Material and methods

The trial was conducted at a 340-head farrow-to-finish family operation located in Southwest Minnesota. The operation retains ownership on a portion of the pigs through finishing and sells the remainder as premium-priced isoweans. The trial included all litters from two consecutive batches, weaned in September and October 2022. At processing, piglets from each litter were individually ear notched, weighed, sorted between males and females, and randomly assigned to treatment or control groups, both of which were present within every litter. Treated pigs were administered 1 mL of GI Revive orally, control pigs received no additional treatment. Following weighing and allocation into study groups, all piglets were then given an oral *E coli* vaccine, a 200 mg injection of iron fortified with ampicillin, all tails were docked, and boars were castrated. A small percent of piglets also displayed signs of mild scouring at this time, so rectal swabs were collected for further diagnosis of any pathogens present at time of processing. Farm staff was blind to the treated and control piglets in each litter.

Treatment protocol for scours on the farm entailed application of a proprietary product on the mat at the first sign of scours, continuing for up to three days or until the scours have subsided. After the fortified iron, no additional antibiotics were given to the piglets until they were creep fed for the last 5 days before weaning. Pigs were cross fostered at 48 - 72 hours of age using standard farm protocol to average the number of pigs per litter and were moved only once if at all. Creep feed was Hubbard Feeds' Powerstart Solo pellets containing 50g/ton Mecadox. Final weights were collected two days prior to weaning.

All sows followed the same feed curve: feed was dropped twice per day to all sows. All sows start at 6 lbs. total feed delivery at entry into the farrowing house, after farrowing feed is increased by 1 lb. per day as intake allows until they reach the maximum daily feed allowance of 16 lbs. There was no difference in sow feed intake at the end of either batch. All sows remained in adequate condition through weaning, and sow breed-back history is at an 85% average rate with 95+% of breed eligible sows inseminated by 5 - 7 days post-weaning. Herd average wean age is 18.5 days, averaging 14.4 lbs. at weaning., and average pigs weaned per litter has been 10.6 over the past one-year period.

Results and discussion

Diagnostic results on the rectal swabs collected identified multiple non-hemolytic, non-shiga producing *E coli*, along with Clostridium perfringens type A. PCR testing for Rotavirus types A and C were negative, and processing fluids tested for PRRS were also negative. Results can be seen in Tables 1 through 4. Table 1 shows the effect of GI Revive on pigs by their dam's parity; it appears that there is a slight benefit in the litters of both the younger and older sows, compared to mid-parity sows. The advantage appears greater when light birth weight pigs are evaluated separately, shown in Tables 2 and 3. Table 4 shows the effect of giving GI Revive to pigs scouring at processing, which showed some advantage as well. Based on the results in relatively healthy pigs, it can be hypothesized that results would be greater in farms with higher born alive litters, lighter birth weights, and greater scour challenges.

References

1. Simonson, R. "Alternative scour prevention and control: History and science of options" AASV Lecture, Hilton Orlando Buena Vista Palace, Orlando, FL, March 2015.

2. Huang, Alex. Harnessing the biology of plant immunity. June 20, 2014.

Table 1: GI Revive overall results by parity group

	Control		Treated		Δ ADG	Avg total gain, 14.5d	Total pigs
Parity	ADG, lbs.	n	ADG, lbs.	n	lbs.	lbs.	n
P1-P2	0.629	199	0.643	200	0.014	0.20	399
P3-P5	0.648	92	0.651	95	0.003	0.046	187
P6-P8	0.589	228	0.595	226	0.006	0.09	454
Total	**0.615**	**519**	**0.624**	**521**	**0.009**	**0.13**	**1040**

Table 2: GI Revive results in pigs started at 2.5 pounds and less, by parity group

	Control		Treated		Δ ADG	Avg total gain, 14.5d	Total pigs
Parity	ADG, lbs.	n	ADG, lbs.	n	lbs.	lbs.	n
P1-P2	0.498	4	0.500	10	0.002	0.029	14
P3-P5	0.514	2	0.719	1	0.205	2.970	3
P6-P8	0.396	14	0.454	11	0.058	0.835	25
Total	**0.428**	**20**	**0.487**	**22**	**0.059**	**0.849**	**42**

Table 3: GI Revive results in pigs started at 2.51 pounds and greater, by parity group

	Control		Treated		Δ ADG	Avg total gain, 14.5d	Total pigs
Parity	ADG, lbs.	n	ADG, lbs.	n	lbs.	lbs.	n
P1-P2	0.634	195	0.649	190	0.015	0.215	385
P3-P5	0.651	90	0.650	94	(0.001)	(0.009)	184
P6-P8	0.602	214	0.602	215	0.001	0.012	429
Total	**0.623**	**499**	**0.629**	**499**	**0.006**	**0.09**	**998**

Table 4: GI Revive results in pigs started with observed scouring

	Control		Treated		Δ ADG	Avg total gain, 14.5d	Total pigs
Parity	ADG, lbs.	n	ADG, lbs.	n	lbs.	lbs.	n
P1-P8	0.614	17	0.627	24	0.013	0.188	41

https://doi.org/10.54846/am2023/74

Efficacy of Enterisol Salmonella T/C® in reducing the shedding and colonization of multidrug resistant monophasic *Salmonella enterica* serovar I 4,[5],12:i:-

Fernando Leite[1], DVM, MS, PhD; Paulo Arruda[2], DVM, MS, PhD; Dianna Jordan[1], DVM, PhD; Shawn Bearson[3], PhD
[1]Boehringer Ingelheim Animal Health USA Inc., Duluth, Georgia; [2]VRI-AMVC, Ames, Iowa; [3]ARS-USDA, Ames, Iowa

Introduction

The monophasic *Salmonella* I 4,[5],12:i:- serovar is now the most common *Salmonella* serovar identified in US swine clinical cases.[1,2] This serovar has also become one of the most common to cause human foodborne salmonellosis and is also the most frequent multidrug resistant (MDR; resistance to ≥ 3 antimicrobial classes) serovar in the U.S.[3,4] Recently, it has been found that 83% of globally distributed monophasic *Salmonella* strains of swine origin are MDR;[3] with the most common resistances to the antimicrobial classes of tetracyclines, aminoglycosides, beta-lactams and sulfonamides.[3,4] A common resistance profile is R-type ASSuT or resistance to ampicillin, streptomycin, sulfisoxazole and tetracycline. The majority of monophasic strains (93%) also contain Salmonella Genomic Island 4 (SGI-4) which confers metal tolerance to copper, arsenic and silver.[3,4] In February of 2022, the Food Safety and Inspection Service proposed the implementation of performance standards for *Salmonella* in pork,[5] which adds to the need for strategies to reduce the colonization along with the shedding and transmission of this pathogen. It is recognized that the origin of the *Salmonella* that can contaminate the swine carcass is mainly the live pig itself, and lymph nodes which can be colonized by *Salmonella* can also lead to the direct contamination of pork products.[6,7] This study had the objective of evaluating a commercial vaccine, Enterisol Salmonella T/C®, in conferring protection against monophasic *S.* I 4,[5],12:i:- as measured by fecal shedding and colonization of ileal cecal lymph node (ICLN), ileal Peyer's patch (IPP), cecal tissue (CT) and cecal contents (CC).

Materials and methods

A randomized, blinded, controlled study was conducted evaluating pigs in three different treatment groups. Each treatment group consisted of 20 pigs, blocked for the effects of litter, sex and weight. The treatment groups were: 1) non-vaccinated non-challenged (NVNC); 2) non-vaccinated challenged (NVC) and 3) vaccinated with Enterisol Salmonella T/C® and challenged (EVC). Animals were challenged four weeks post vaccination at approximately 7 weeks of age with a dose of 2×10^9 *S.* I 4,[5],12:i:- strain SX 240. This strain has the R-type ASSuT resistance profile and was associated with a human foodborne outbreak from pork in the US.[8] Fecal samples were collected at 0, 2, 3, 7, 10 and 14-days post inoculation (dpi). All pigs were euthanized at 14 dpi, when different tissues (ICLN, IPP, CT, CC) which represent preferential sites of *Salmonella* colonization, were collected for *Salmonella* culture and quantification.

Results

Fecal shedding peaked in both challenged groups at 2 dpi with shedding levels of 6.3 and 5.6 $\log_{10}$ cfu/gram in the NVC and EVC groups, respectively (Figure 1). A significant ($P < 0.05$) reduction of shedding in the EVC group was observed beginning on 3 dpi and on each subsequent timepoint following challenge (Figure 1). At 14 dpi, the NVC group shed on average 3.5 $\log_{10}$ cfu/gram while the EVC group shed on average 1.4 $\log_{10}$ cfu/gram ($P < 0.05$; Figure 1). Vaccination led to a significant ($P < 0.05$) reduction in the colonization level of all measured tissues (Table 1). This vaccine also led to a significant ($P < 0.05$) reduction in the number of pigs colonized by MDR monophasic *Salmonella* I 4,[5],12:i:- in ICLN, IPP, and CT (Table 1).

Discussion and conclusions

We have previously shown that Enterisol Salmonella T/C® is able to significantly reduce the clinical impact of *S.* I 4,[5],12:i:- infection in swine, leading to a significant reduction of diarrhea and a significant improvement of weight gain.[9] Given the importance of this serovar to both swine and public health, we further evaluated the effect of this vaccine on parameters that can affect food safety as well as the environmental load and transmission of this pathogen. Enterisol Salmonella T/C® is the only commercially available vaccine in the U.S. that contains a live Typhimurium serovar vaccine strain, a serovar of the same serogroup as *S.* I 4,[5],12:i:-.[4,10] Given that higher *Salmonella* colonization levels have been associated with a greater likelihood of carcass contamination and that lymph nodes can directly contaminate final pork product, it is significant that Enterisol Salmonella T/C® can reduce both the level and number of colonized pigs.[7,11] These results suggest that the Enterisol Salmonella T/C® vaccine can be a preharvest tool to improve food safety and help prevent MDR monophasic *Salmonella* I 4,[5],12:i:-.

Figure 1: Fecal shedding of MDR monophasic *Salmonella* I 4,[5],12:i:- in pigs following challenge. NVNC = non-vaccinated non-challenged; NVC = non-vaccinated challenged; EVC = vaccinated with Enterisol Salmonella T/C® and challenged. Error bars represent standard error of the mean. Different superscript letters indicate statistical significance within each day post inoculation ($P < 0.05$).

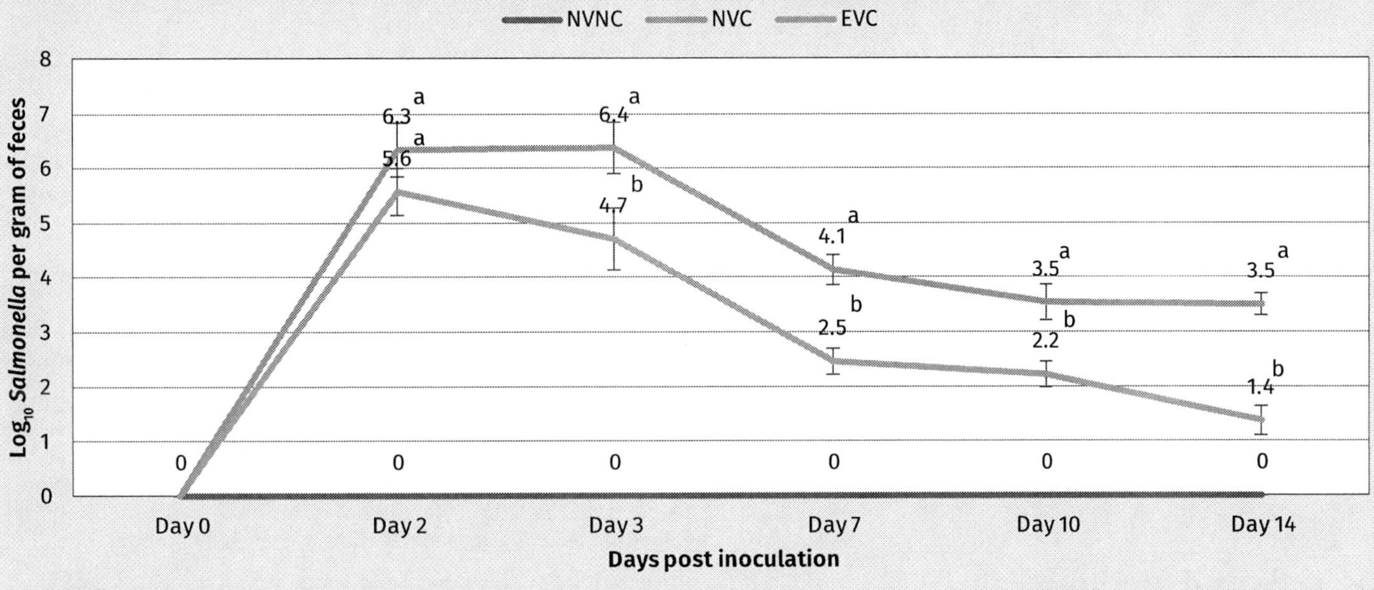

Table 1: Colonization of MDR monophasic *Salmonella* I 4,[5],12:i:- in different tissues at necropsy (14 dpi). Values represent average $\log_{10}$ *Salmonella* colony forming units per gram, followed by the percentage of culture positive pigs per treatment group.

Tissue	NVNC	NCV	EVC
Ileal cecal lymph node	0^a, $0\%^a$	2.50^b, $100\%^b$	1.32^c, $65\%^c$
Ileal Peyer's patch	0^a, $0\%^a$	3.12^b, $100\%^b$	0.79^c, $35\%^c$
Cecal tissue	0^a, $0\%^a$	2.94^b, $100\%^b$	1.19^c, $60\%^c$
Cecal content	0^a, $0\%^a$	4.33^b, $100\%^b$	2.33^c, $85\%^b$

NVNC = non-vaccinated non-challenged; NVC = non-vaccinated challenged; EVC = vaccinated with Enterisol Salmonella T/C® and challenged. Different superscript letters within tissue colonization level and percentage of positive pigs indicate statistical significance ($P < 0.05$).

References

1. Naberhaus SA et. al., Emergence of Salmonella enterica serovar 4,[5],12:i:- as the primary serovar identified from swine clinical samples and development of a multiplex real-time PCR for improved Salmonella serovar-level identification. J Vet Diagn Invest. 2019. 31(6):818-827.

2. SHIC. 2021. *Salmonella* 1,4,[5],12:i:-. p1-19.

3. Trachsel JM, Bearson BL, Brunelle BW, Bearson SMD. Relationship and distribution of Salmonella enterica serovar I 4,[5],12:i:- strain sequences in the NCBI Pathogen Detection database. BMC Genomics. 2022 Apr 6;23(1):268. **doi: 10.1186/s12864-022-08458-z**

4. Bearson SMD. Salmonella in Swine: Prevalence, Multidrug Resistance, and Vaccination Strategies. Annu Rev Anim Biosci. 2022 Feb 15;10:373-393. **doi: 10.1146/annurev-animal-013120-043304**

5. FSIS. Changes to the Salmonella Verification Testing Program: Proposed Performance Standards for Salmonella in Raw Comminuted Pork and Intact or Non-Intact Pork Cuts and Related Agency Verification Procedures. **https://www.fsis.usda.gov/policy/federal-register-rulemaking/federal-register-notices/changes-salmonella-verification-0**

6. Marin C, Chinillac MC, Cerdà-Cuéllar M, et al. Contamination of pig carcass with *Salmonella enterica* serovar Typhimurium monophasic variant 1,4[5],12:i:- originates mainly in live animals. Sci Total Environ. 2020 Feb 10;703:134609. **doi: 10.1016/j.scitotenv.2019.134609**

7. Miller M, Maher JM, Wiseman B, Gragg SE. *Salmonella* is Present in Multiple Lymph Nodes of Market Hog Carcasses at Slaughter. Food Protection Trends. 2022. Vol 42, No. 2, p. 100–106.

8. Shippy DC, et al., Porcine Response to a Multidrug-Resistant *Salmonella enterica* serovar I 4,[5],12:i:- Outbreak Isolate. Foodborne Pathog Dis. 2018. 15(5):253-261.

9. Leite F, Arruda P, Jordan D, Bearson S. Efficacy of Enterisol® Salmonella T/C in protecting against monophasic *Salmonella* I 4,[5],12:i:-. Published in: AASV Annual Meeting 2022. p. 142-143. **doi.org/10.54846/am2022/52**

10. Leite F. *Salmonella enterica* serovar I 4,[5],12:i:- update on risk factors and control. Published in: AASV Annual Meeting 2021. p.161-163.

11. Pesciaroli M, Cucco L, De Luca S, et al. Association between pigs with high caecal Salmonella loads and carcass contamination. Int J Food Microbiol. 2017 2;242:82-86. **doi: 10.1016/j.ijfoodmicro.2016.11.021**

https://doi.org/10.54846/am2023/75

A methodology for assessing the economic value of early disease detection in growing pigs

D. Polson[1], DVM, MS, PhD; C. Alonso[2], DVM, MS, PhD
[1]Boehringer Ingelheim Animal Health USA Inc., Duluth Georgia;
[2]Boehringer Ingelheim Vetmedica GmbH, Ingelheim, Germany

Introduction

Respiratory disease outbreaks continue to be a major pig production problem, impacting antibiotic use, welfare, productivity and profitability.[1] SoundTalks is an audio-based technology that continuously identifies and quantifies respiratory problems in pigs, generating alerts (yellow warnings, red alarms) when respiratory outbreak onset is detected. These alerts have enabled triggering earlier caregiver awareness than caregiver observations alone.[2] However, further research is needed to quantify the economic impact of this technology. The objective of this study was to evaluate the performance and economic differences resulting from earlier detection and intervention following the onset of a clinically detectable respiratory disease episode measured by SoundTalks.

Materials and methods

Eleven-week-old pigs (n = 1655) were placed in 72 pens throughout two rooms (airspaces). Each room contained three SoundTalks monitors, one monitor per 12 pens. Study groups were randomly allocated within each zone. In every pen, three randomly selected seeder pigs were challenged with *Mycoplasma hyopneumoniae* for two consecutive days followed by a PRRS virus challenge seven days later.

Continuous sensor data was recorded at the zone level (eg, audio, temperature, humidity) and pen level (eg, water use, temperature). Performance was measured at both the individual pig and pen levels. Individual pigs were weighed at the beginning of the study and every four weeks through the end of the study. Feed disappearance was recorded weekly for every pen. Daily individual pig treatments and mortality were also recorded. Pig carcass weights and the processing plant carcass discount-premium matrix were obtained.

There were three study groups, defined as: SoundTalks (ST) alert day zero (G0), day 5 (G5) and day 10 (G10). Alert day 0 was defined as the day that the first ST yellow/red alerts were reported post-challenge. All pigs received the same treatment protocol – differentiated only by the date the treatment protocol was started. Group ST-G0 received the intervention beginning the day of the first actionable SoundTalks alert. Subsequently, groups ST-G5 and ST-G10 received the same intervention at ST alert day 0 + 5 days and ST alert day 0 + 10 days, respectively.

Economic differences among treatment groups were calculated using a "Standardized Economic Index" (SEI) based on a partial budget spreadsheet model. The SEI is a function of finished pig performance measures, historical feed ingredient costs, historical market pig prices, historical inflation rates, processing plant carcass discount-premium grid and the calculated cost-to-operate (CTO) of the technology being evaluated. Pig performance measures utilized in the SEI were average daily gain (ADG), feed conversion rate (FCR), average daily feed (ADF), mortality, and individual pig treatment cost. Historical monthly market prices and feed ingredient costs were obtained for a roughly 31-year period from January 1992 through October 2022 (the most recent available USDA data). The most recent 12.8-year (154 month) time period (January 2010 through October 2022) was used as the focus of the economic analysis using the most current data available. In addition to hardware installation costs (labor and materials), a weighted SoundTalks hardware lifespan of 48 months was used to calculate the hardware cost on a per pig marketed basis for inclusion in the CTO.

To represent the impact of a more natural (contact) exposure and infection dynamic under routine operational conditions, performance data for contact challenged pigs that did not experience exceptional handling (eg, did not experience individual snaring, bleeding, tracheal catheterization) were used to model the SEI.

A sensitivity analysis was conducted by constructing a 3 × 3 table (nine combinations) where the X-axis was pig live market price (USD per kg live body weight) and the Y-axis was diet cost per metric ton (USD per 1000 kg complete diet). The pig price middle value was the calculated average of the 154 historical monthly market price values from January 2010 through October 2022 (12.8 years). The pig price lower and higher sensitivity matrix values used were calculated as the 99% lower control limit (LCL) and 99% upper control limit (UCL) from the 154 months of historical data for the same period.

The diet cost per metric ton middle value was the calculated average of the 154 historical monthly cost values from January 2010 through October 2022 for:

- Number two yellow corn as the primary dietary energy source
- 44% soybean meal as the primary dietary protein source

Non-energy and non-protein ingredient cost estimated were obtained[3] and were adjusted across the 12.8-year period using monthly US inflation rate (ie, CPI) data. The fee for finished diet preparation and delivery (GMD) was obtained[4] and was also adjusted across the 12.8-year period using monthly US inflation rate data. Typical grower-finisher pig diet formulations were obtained,[3] and utilized the aforementioned ingredient and GMD costs to calculate an average diet cost per metric ton representing the 154-month period. The diet cost lower and higher sensitivity matrix values used were calculated as the 99% lower control limit (LCL) and 99% upper control limit (UCL) from the 154 months of historical data for the same period.

Results

After the seeder pig dual challenge (*Mycoplasma hyopneumoniae* followed by PRRS virus), two respiratory outbreaks caused by swine A influenza virus (IAV-S) were documented during the study period. All pigs and pens were treated accordingly to study design.

Contact challenged pigs from G0 had 12.7 and 20.4 grams higher ADG compared to those from G5 and G10 respectively. Similarly, contact challenged pigs from G0 had a 23.4% and 10.1% decrease in individual treatments when compared to G5 and G10 respectively. Contact challenged pigs from G0 had a 0.26% higher and 1.22% lower percent mortality compared to those from G5 and G10 respectively. All production variables were introduced into the SEI model.

Figures 1 and 2 show the 12.8-year (154 month) time series and overall frequency distribution of the estimated benefit:cost ratio based on the aggregate performance differences observed in the study, monthly market price (including carcass discount-premium), diet cost and inflation data. Throughout the 154-month period from January 2010 through October 2022, based on a CTO of USD $0.254/pig marketed (inclusive of installation, hardware and software subscription for a 4800 head grow-finish site) the mean monthly B:C ratio was 6.51, ranging from 3.89 to 11.28. The B:C ratio (G0 vs G5) exceeded 2:1 for 154 of 154 (100%) months and exceeded 4:1 for 153 of 154 (99.4%) months. For the same 154-month period, the 48-month rolling B:C ratio (G0 vs G5) ranged from 4.91 to 7.45, exceeding both 2:1 and 4:1 for 154 of 154 (100%) intervals.

When a higher CTO of USD $0.303/pig marketed (inclusive of installation, hardware and software subscription for a smaller single barn 1200 head grow-finish site) is used for the same 154-month period, the mean monthly B:C ratio (G0 vs G5) was 5.47, ranging from 3.27 to 9.47. The B:C ratio (G0 vs G5) exceeded 2:1 for 154 of 154 (100%) months, and exceeding 4:1 for 138 of 154 (89.6%) months. For the same 154-month period, the 48-month rolling B:C ratio (G0 vs G5) ranged from 4.13 to 6.26, exceeding both 2:1 and 4:1 for 154 of 154 (100%) intervals.

For the sensitivity analysis of the nine combinations of pig market price and diet cost (based on the 99% confidence limits), the middle market price ($1.322/kg) and diet cost ($55.14/1000kg) generated a benefit:cost estimate of 4.17 where a denominator cost of $0.254 for SoundTalks was used, reflecting the technology and installation cost per pig marketed for a 4800 head grow-finish site. As would be expected, the lowest benefit:cost (3.79) occurred where the market price was lower ($1.269/kg) and the diet cost was higher ($270.77/1000kg), and the highest benefit:cost (4.55) occurs where the market price was higher ($1.376/kg) and the diet cost was lower ($239.52/1000kg).

When a denominator cost of $0.303 was used (reflecting the cost per pig marketed for a single-barn 1200 head grow-finish site), the middle market price ($1.322/kg) and diet cost ($255.14/1000kg) generated a benefit:cost estimate of 3.50, with the lowest benefit:cost (3.18) occurring where the market price was lower ($1.269/kg) and the diet cost was higher ($270.77/1000kg), and the highest benefit:cost (3.82) occurring where the market price was higher ($1.376/kg) and the diet cost was lower ($239.52/1000kg).

Conclusions and discussion

The results of this study and the standardized economic analysis suggests that there can be a consistent and long-term favorable economic impact based on aggregate performance differences using a technology that enables earlier detection and treatment intervention.

Earlier alerts of the onset of clinical disease episodes, in and of themselves, do not have a direct positive impact on the course of disease in affected pigs. Alerts can only provide the producer the opportunity to take the most appropriate action to mitigate and resolve the developing clinical disease episode, ie, knowledge requires informed action to generate value. As soon as meaningful alerts are observed, only immediate and optimal interventions taken by the pig producer will enable capturing the value potential enabled by any detection technology that provides early alerts of disease episode onset.

Figure 1: Monthly benefit:cost ratio (thin dashed and solid lines) as well as its 48-month rolling average (thick dashed and solid lines) for SoundTalks investment based on study group live performance and carcass premium differences over a 12.8-year period (January 2010 through October 2022).

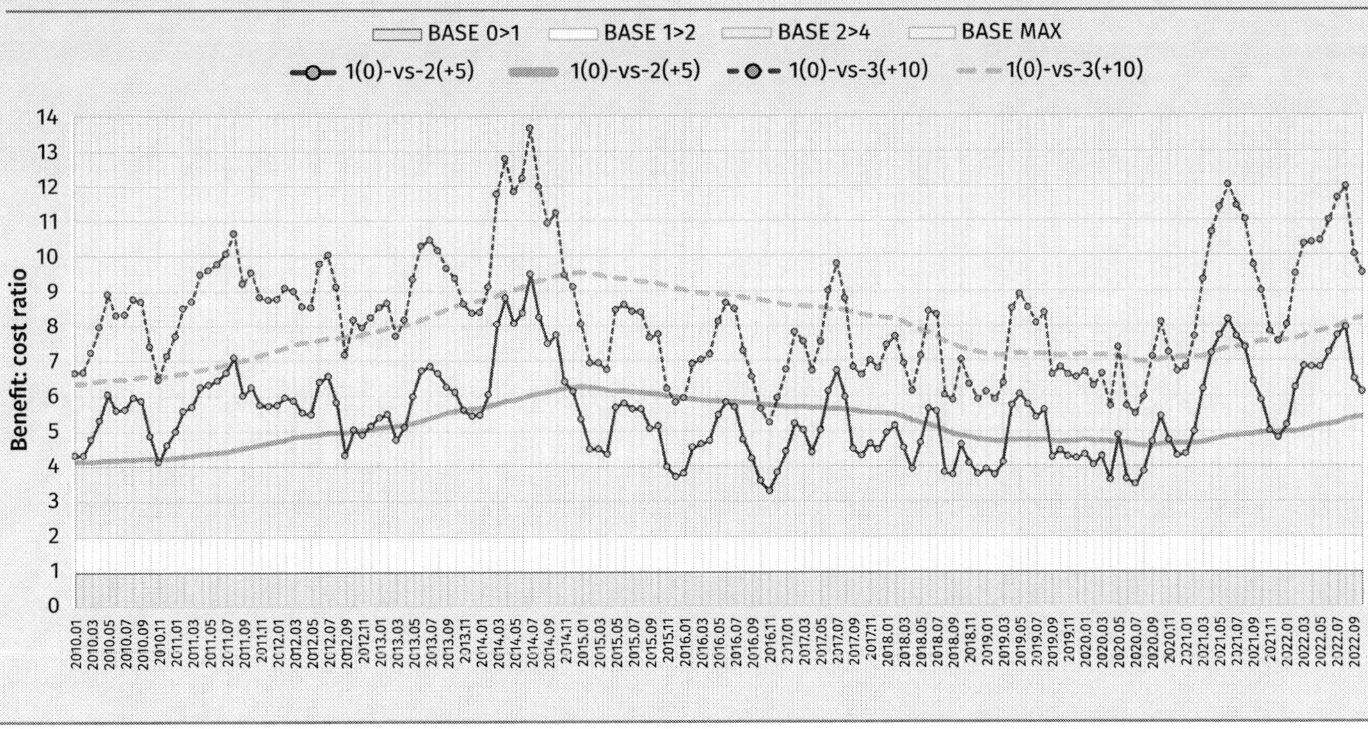

Figure 2: Monthly benefit:cost ratio frequency distribution for SoundTalks investment based on study group performance and carcass premium differences between Alert Day 0 and Alert Day +5 intervention groups over a 12.8-year period (January 2010 through October 2022).

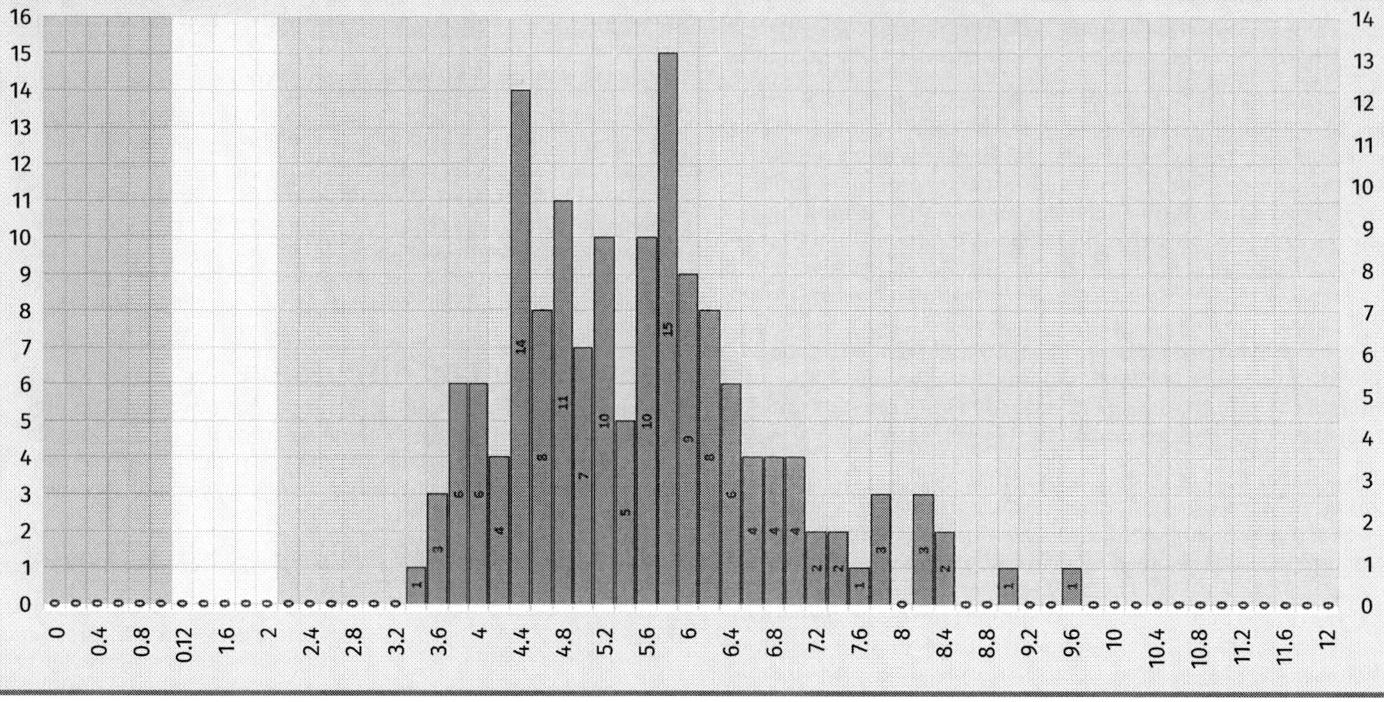

An obvious advantage of evaluating the benefit:cost using actual historical pig/pork market prices and major feed input costs instead of a traditional analysis using more subjective, artificially constructed market price and input cost combinations is that financial impact can be evaluated based on real (and realistic) combinations of cost inputs that reflect prior experience as well as current conditions. To account for expected and emerging market conditions, further development of the SEI method will include incorporation of forecasting based on available futures contracts for pork and the major feed ingredients as well as the ability to generate stochastic output distributions and confidence limits around future projected time series.

References

1. Lopes Antunes AC, Jensen VF, Jensen D (2019) Unweaving tangled mortality and antibiotic consumption data to detect disease outbreaks – Peaks, growths, and foresight in swine production. *PLoS ONE* 14(10): e0223250.

2. Polson D, Playter S, Berckmans D, et al., (2018) Precision livestock farming (PLF) for pig health and production: Sound as a diagnostic sample. *Proceedings of the 2018 American Association of Swine Veterinarians*, San Diego California, USA, 21-24.

3. Dr. Mike Tokach, Kansas State University, Manhattan Kansas, personal communication.

4. Jon Hoek, Summit Smart Farms, Remington Indiana, personal communication.

https://doi.org/10.54846/am2023/77

Evaluation of meloxicam in sows post-farrowing on piglet performance

B. Payne[1], DVM; D. Rosener[1], DVM; K. Stewart[2], PhD; A. Fisk[2], MS; K. Harlow[2], PhD
[1]Veterinary Pharmaceutical Solutions, St. Peter, Minnesota; [2]Purdue University, West Lafayette, Indiana

Introduction

Meloxicam is a nonsteroidal anti-inflammatory drug (NSAID) which has been widely used in human and companion animal medicine to treat arthritis. Internationally, it has been approved for use in swine for lameness, to relieve post-operative pain associated with minor soft tissue surgery such as castration, and for supportive therapy together with antibiotics in the treatment of mastitis-metritis-agalactia (MMA) syndrome, also known as Postpartum Dysgalactia Syndrome (PPDS).[1] Previous studies have shown that meloxicam administered to sows at the end of farrowing at 0.4 mg/kg IM is an effective treatment for PPDS, and the greater availability of milk and colostrum during the critical starting process of lactation resulted in healthier piglets (less scouring), less preweaning mortality, and heavier weaning weights.[2,3] Further studies documented similar performance benefits in subclinical cases of PPDS and/or as a result of the mitigation of pain associated with udder engorgement or uterine involution.[4-7] Due to the repeatability of the performance responses, internationally the administration of meloxicam to all gilts/sows post-farrowing has become a routine practice in many herds. It has been theorized that the noted performance benefits could be due to the anti-endotoxin activity as well as the anti-inflammatory properties of NSAIDs making the sow more comfortable and willing to nurse, less apt to crush her piglets, and have improved feed/water intake contributing to greater milk production.[8] The objective of this study was to see if similar performance benefits could be seen with a new compounded, drench suspension formulation of meloxicam.

Materials and methods

A commercial 10,000 head sow farm was the site of this study. 851 mixed parity females were randomly assigned to one of two treatment groups: Control = no treatment post-farrowing or Meloxicam = 90 mg (6 mL of a 1.5% Suspension) dosed via an oral drench. Sows were enrolled in the study after the farm crew had determined farrowing was complete and had set the litter. Each morning, sows that had farrowed overnight were treated with Meloxicam. Each afternoon, any sow that had completed farrowing that day was treated with Meloxicam before the crew left. Any sow with a rectal temperature > 103.5°F at 24 hours post-farrowing received a three-day therapy of Flunixin® and Polyflex® per the farm's standard operating procedure (SOP).

Data collected included:

1. Genetic line, parity, gestation length, total born, total nursed, litter weights at enrollment and just prior to weaning
2. Sow rectal temperatures at time of enrollment, and at 12 and 24-hours post-farrowing
3. Health events – any sow or litter treatments
4. Age, weight, and reason for any piglet deaths

The study design was a 2 × 2 factorial evaluating the effects of Meloxicam post-farrowing and additional medications at 24 hours using PROC Mixed in SAS (v. 9.0). The main effects of Meloxicam treatment, additional medication, parity class (P1 vs P2&3 vs $P > 3$) were used as main effects with litter birthweight quartile and number piglets nursed as covariates where appropriate. Significance was determined at $P < 0.05$, with $0.05 < P < 0.10$ considered a tendency. When significant, means separation was performed using a Tukey adjustment. Data are reported as means and standard error.

Results and discussion

The litter birth weight was significantly higher in Control litters compared to Meloxicam litters ($P = 0.02$) and number born alive tended to be higher ($P = 0.077$). However, these were artifacts as the treatments were not initiated until after farrowing was complete. It is important to note that despite the Meloxicam litters starting off lighter than the Control litters, they ended up with heavier weaning weights (Table 1).

Meloxicam treatment significantly reduced rectal temperatures at 12 hours post-farrowing compared to Controls ($P = 0.001$). The Meloxicam treated sows continued to have numerically lower temperatures at 24 hours, but the differences were no longer statistically significant at that time point.

Overall, scour rates were ~10% on this farm, and litters from sows treated with Meloxicam tended to have less scours ($P = 0.091$).

Although the magnitude of response was different from earlier findings, the trends followed the same patterns with Meloxicam treatment tending to reduce scours, numerically lower preweaning mortality, and have heavier litter weaning weights. Some of the benefits of the Meloxicam treatment may have been masked by the widespread administration of flunixin (another NSAID) and ampicillin the day following farrowing. 28% of the sows in this trial were treated with this combination of injectables per the farm's SOP. Sows receiving additional medications had significantly higher preweaning mortality, despite additional NSAID and antibiotic therapy. Because Meloxicam was so effective at reducing fever in the first 12 hours, further studies are warranted to investigate a longer duration of Meloxicam therapy without flunixin.

References

1. European Medicines Agency. Accessed October 20, 2022. **https://www.ema.europa.eu/en/medicines/veterinary/EPAR/metacam**

2. Hirsch AC, Philipp H, Kleemann R. Investigation on the efficacy of meloxicam in sows with mastitis-metritis-agalactia syndrome. J Vet Pharmacol Ther. 2003; 26:355–360.

3. Keller, F. Improved early piglet survival after Metacam® treatment of sows post farrowing. Proc 22nd IPVS Congress, Korea. 2012. 249.

4. Finestra A, Rubio S, Salleras JM. Post-farrowing treatment of sows with meloxicam or tolfenamic acid on the preweaning mortality rate of the low birth weight piglets in subclinical MMA. *Proc of 19th IPVS Congress*, Denmark. 2006. 477.

5. Andreoni, S, Vischi O, Persico F. Effects of routine administration of meloxicam (Metacam®) in sows after farrowing on piglets performance. *Proc 23rd IPVS Congress*, Mexico. 2014. 164.

6. Revilla E, Ubiergo A, Martinez P, et al. Post-farrowing treatment of sows with meloxicam on the preweaning weight gain and mortality rate of the low birth weight piglets in subclinical MMA. *Proc 19th IPVS Congress*, Denmark. 2006. Vol 2 475.

7. Guinez D, Grosse-Liesner V, Rouderque C, Ruiz A. Farrowing culling and therapeutic injections rates in litters from sows treated with Metacam®. *Proc 24th IPVS Congress*, Ireland. 2016. 630.

8. Tenbergen R, Friendship R, Cassar G. Investigation of the use of meloxicam post farrowing for improving sow performance and reducing pain. *J Swine Health Prod*. 2014;22(1):10-15.

Table 1: Fever and litter performance

Variable	Control		Meloxicam		*P*-Value
	Mean	STDERR	Mean	STDERR	
Temp at dosing, °F	102.38	0.0514	102.37	0.0485	0.2498
Temp at 12 hours, °F	102.17	0.0451	102.08	0.0418	0.0013
Temp at 24 hours, °F	102.01	0.0484	101.98	0.0443	0.1957
Number born alive	13.184	0.041	13.089	0.071	0.0769
Litter birth weight, kg	18.040	0.171	17.563	0.144	0.0203
Number piglets nursed	12.095	0.074	12.200	0.072	0.4828
Sow medicated (1/0)	0.289	0.022	0.284	0.022	0.5853
Scours (1/0)	0.116	0.015	0.091	0.014	0.0907
Sow removed (1/0)	0.039	0.009	0.027	0.008	0.2442
Litter weaning weight, kg	58.78	0.100	61.40	0.036	0.001
Adjusted weaning weight, kg	66.26	0.620	67.45	0.595	0.0401
Preweaning mortality, %	0.129	0.001	0.123	0.006	0.2685

https://doi.org/10.54846/am2023/78

Novel virulence-associated gene frequency across capsular serotypes and sequence types in *Streptococcus suis*

Joshua Elston, MB(ASCP); Jessica Peterson; Steve Tousignant, DVM, PhD
Newport Laboratories Inc., Worthington, Minnesota

Introduction

Streptococcus suis is a significant burden to the swine industry and an animal welfare concern.[1] Both commensal and pathogenic strains subsist in pigs, and identifying clinically relevant strains is important in autogenous vaccine development and disease control programs. Recently discovered Virulence-Associated Genes (VAGs), known as SSU_RS03100, SSU_RS09155, and SSU_RS09525, have been reported to be indicators of pathogenicity.[2] Our group aims to explore relationships between these VAGs among serotypes and sequence types (STs) that have previously been associated with pathogenic pathotypes,[1] with an end goal of further differentiating strains by Whole Genome Sequencing (WGS).

Methods

127 *Streptococcus suis* isolates were obtained during routine diagnostics in the Newport Laboratories bacteriology department from late 2021 to April 2022. Nucleic acids were isolated using the Kingfisher Core Kit (Applied Biosystems, Cat. # A32700) according to manufacturer's instructions. Extracts were subjected to library preparation using the Illumina XT Kit (Illumina, Cat. # FC-131-1096). Suggested library protocols were followed, except for a manual normalization post-amplification. Sequencing data which leveraged the Illumina MiSeq's paired-end chemistry was compiled over several sequencing operations. Indexed reads were assembled to VAG and capsular type references using SeqMan NGen(DNAStar). Sequence types were attained by submitting *de novo* assemblies to PubMLST.

Results

Frequencies of VAGs SSU_RS03100, SSU_RS09155, and SSU_RS09525 over the entire sample set were .89, .84, and .92, respectively. Previous research reported frequencies of .82, .85, and .91, respectively.[2] Frequencies of VAGs SSU_RS03100, SSU_RS09155, and SSU_RS09525 across STs previously reported to be associated with pathogenic pathotype[1] were observed to be ST1(1,.96,1), ST25(1,1,1), ST28(.93,1,1), ST29(1,.67,1), ST94(.67,.67,.67), ST373(1,1,1), and ST977(1,1,1), respectively. Frequencies of VAGs SSU_RS03100, SSU_RS09155, and SSU_RS09525 across serotypes previously reported to be associated with pathogenic pathotype[1] were observed to be 1(1,1,1), ½(.9,1,1), 2(1,1,1), 7(1,.86,1), 14(1,1,1), and 23(1,1,1).

Discussion

Our results suggest that serotypes and STs reported to be pathogenic align well with the presence of the novel VAGs. However, many serotypes and STs exhibited 100% prevalence of the novel VAGs though these were not associated with pathogenicity.[1] These isolates represented a small proportion of our sample set. Additional investigation of these less common serotypes and STs could further reveal trends with regards to the novel VAGs. We believe that each predictive approach has merit and their use in tandem can increase confidence in differentiating between commensal and pathogenic strains for autogenous vaccine development.

References

1. Estrada AA, Gottschalk M, Rossow S, Rendahl A, Gebhart C, Marthaler DG. Serotype and genotype (multilocus sequence type) of *Streptococcus suis* isolates from the United States serve as predictors of pathotype. J Clin Microbiol. 2019;57:e00377-e419. **https://doi.org/10.1128/JCM.00377-19**

2. Estrada AA, Gottschalk M, Gebhart CJ, Marthaler DG. Comparative analysis of *Streptococcus suis* genomes identifies novel candidate virulence-associated genes in North American isolates. Veterinary Research. 2022;53:23. **https://doi.org/10.1186/s13567-022-01039-8**

https://doi.org/10.54846/am2023/79

Developing a model protocol for tracking antibiotic use and AMR monitoring across a large-scale commercial swine production system: The IMAGINE Project

T. Spronk, DVM; R. Edler, MS; K. Havas, DVM, MS, PhD; S. Dee, DVM, MS, PhD
Pipestone Veterinary Services, Pipestone, Minneosta

Introduction

Antimicrobial use (AMU) in livestock is often implicated to be a source of antimicrobial resistance (AMR) in humans, despite a lack of data to support the hypothesis. This project is the first attempt to conduct AMR monitoring across a large-scale swine production system in the US and compare AMR in swine and public health pathogens with its respective on-farm antibiotic use.

Methods

To gain insight into this critical issue, Pipestone has partnered with USDA National Animal Health Monitoring System and South Dakota State University, with support from the National Pork Board and the Foundation for Food and Agricultural Research's International Consortium on Antimicrobial Stewardship in Agriculture. The project is in its third year and focusing on:

1. To conduct antimicrobial resistance (AMR) monitoring across pathogens of veterinary importance and food safety importance across a large-scale commercial swine production system.
2. To compare AMR data with antimicrobial use (AMU) data across study farms.
3. To develop and validate a model protocol of AMR monitoring for the US swine industry.

Results

The project has now been running for two years with several more years hopefully to come. Successful outcomes so far include public: private partnership team building, data collected from farm sites enrolled in the study and isolation of targeted pathogens, preliminary statistical analysis of AMU and AMR, use of both phenotypic and genotypic methods to evaluate AMR, and the continuous improvement of the protocol based on year 1 - 2 learnings. Preliminary observations (year 1 vs year 2) include completion of sampling across enrolled sites, similar recovery rates of targeted bacteria at the level of the pig and the environment, and no significant differences in AMR between sick and healthy pigs across all organisms tested. Regarding efforts in whole genome sequencing, the serovar prediction of *E coli* and Salmonella genomes revealed high diversity indicating that the no strains were endemic. Finally, we are beginning to identify outliers (problem farms), ie, farms with high AMU, or farms with a higher prevalence of monophasic *Salmonella typhimurium*. Clearly, we need to understand why these observations are occurring. Is it due to farm health status, animal flow, building design and environment, recent disease

Discussion

To better understand the relationship between AMU and AMR, the US swine industry must collect accurate data at the level of the farm. This project is the first attempt to accomplish this goal and the first set of observations are promising. As we develop our database, we will be able to tease apart these outliers, determine causation, apply interventions, and measure outcomes. Therefore, further sampling is critical. This is the innate value of this project, to identify problems and solve them for farmers and consumers. Due to its unique design and resource base, there is not another project that can do this. We look forward to further years of discovery and continuing to assist the industry by improving the understanding of AMU and AMR.

https://doi.org/10.54846/am2023/80

Virabloc impacts viability of an ASFV surrogate (*Emiliania huxleyi*) virus in feed: Implications for ASFV transmission in today's global supply chain

A. L. Wagner[1,2], PhD; C. Balestreri[3], DPhil; I. D. Girard[1], PhD; E. Groos[2]; K. Schuman[2]; D. C. Schroeder[3], PhD
[1] Probiotech International, Inc, Saint-Hyacinthe, Quebec Canada; [2] Essential Ag Solutions, Sioux Falls, South Dakota; [3]Department of Veterinary Population Medicine, University of Minnesota, St. Paul, Minnesota

Abstract

The threat of African swine fever virus (ASFV), a large double stranded DNA virus or megavirus, to the Americas is gaining concern among swine veterinarians, nutritionists and producers. With a high risk of transmission in today's global supply chain, a desired suitable megavirus ASFV-surrogate, Emiliania huxleyi virus (EhV), has been utilized for research purposes at the University of Minnesota to better understand the viability of ASFV under various conditions while reducing the threat of exposure to swine industry in North America. Virabloc is a commercially available feed additive for swine that is comprised of a proprietary blend of medium chain fatty acids and botanicals. Some of the compounds found in Virabloc have previously demonstrated efficacy for potential mitigation on positive single stranded RNA viruses, such as Seneca virus A, but had yet to be investigated on megaviruses. The research that will be presented will focus on the effectiveness of Virabloc to impair the viability of EhV in feed over a time course equivalent to that of minimum storage time from manufacturing to feeding on a farm.

Introduction

African swine fever (ASF) was initially identified more than a century ago in domestic pigs in Africa and is characterized as causing an acute hemorrhagic fever followed by death. Over the last century, transcontinental spread of ASF to Europe occurred in 1957 and 1960. From Europe, ASF was transmitted to the Caribbean and Brazil. However, in the mid-1990s eradication of the virus was achieved in all countries except Sardina and Africa. In 2007, another transcontinental spread occurred. The initial spread throughout Africa was determined to be due to the warthog (*Phacochoerus africanus*) population which often was infected with the soft tick (*Ornithodoros* spp) parasite; the warthog can carry the virus without showing any symptoms.[1] The wild boar (*Sus scrofa*) population throughout Europe and Asia are partially attributing to the spread and maintenance of ASF. Although ASFV, the virus responsible for African swine fever, has not yet been detected in regions such as the Americas, Australia and New Zealand, it has been detected in parts of Europe, Russia, Mongolia, Asia, and more recently in the Caribbean (**https://empres-i.apps.fao.org/**).

African swine fever virus is a large (200 nm in diameter) double stranded DNA megavirus that is structurally comprised of an outer envelope, capsid, inner capsule membrane, core shell, and inner core. Depending on the strain, the DNA is 170 - 190 kb which encodes 150 - 200 viral proteins[2-4] including over 100 non-essential proteins that reduce host defensive mechanisms to the virus.[2] These mechanisms include inhibition of type I interferon antiviral responses, impairing the main antiviral pathways that limit virus replication, programmed cell death.[5] The essential proteins are required for cytoplasmic replication and viral genome transcription.[2] During each of the aforementioned outbreaks over the last century, the strain of the virus has likely differed even though the method of infection of each strain remains the same.[2,6,7] The size of the virus enables it to enter the monocyte-macrophage via both clathrin-dependent endocytic pathway and macropinocytosis pathway.[7,8]

Despite discovery over a century ago, methods to eradicate the virus have been limited as there is not currently an effective vaccine option. The current options to mitigate the transmission of ASFV have been to quarantine and euthanize infected herds, increase biosecurity measurements, and restrict animal movement. All transmission mechanisms have not been elucidated; however, ASFV can be transmitted by both direct and indirect contact. Direct contact includes contact with infected animals or ingesting contaminated materials or meat/pork products from an infected source.[9,10] All of the potential mechanisms of indirect contact are still being investigated, and because the source of introduction during ASF outbreaks on commercial pig farms cannot be reliably identified,[5] it is possible that indirect contact may play role in ASFV entering isolated ASFV-free countries. Potential mechanisms of indirect contact for ASFV transmission include but are not limited to clothing, transport trucks, feed ingredients, and insects, such as soft tick vectors (*Ornithodoros* spp) and stable flies (*Stomoxys calcitrans*).[11-14]

Because feed is a potential source of indirect transmission for a number of viruses that infect domestic pigs,[11-17] the interest in ingredients that may mitigate viral particle viability in feed has been increasing in recent years. Medium chain fatty acids, monoglycerides, and botanicals have been previously investigated as potential ingredients for reducing Seneca virus A (SVA), Porcine reproductive and respiratory syndrome virus (PRRSV), and Porcine epidemic diarrhea virus (PEDV) in feed and ingredients.[18,19] Because these viruses are positive single stranded RNA viruses, the efficacy of these ingredients to reduce viral particles of a large megavirus like ASFV, requires investigation. With the high risk of ASFV transmission in today's global supply chain, a desired suitable double stranded DNA surrogate megavirus, Emiliania huxleyi virus (EhV), has been utilized for research purposes.[21] EhV is an ecologically important megavirus, which controls blooms of the marine unicellular phytoplankton *Emiliania huxleyi* [21,22] and shares many important features with ASFV.[23] Both ASFV and EhV are equally sensitive to time and temperature exposure. In fact, EhV has recently been shown to be one of the most thermally stable viruses known, with temperatures up to 100°C leaving a percentage of viable particle still intact.[23] It was also shown that EhV could be detected in all feed matrices tested and that no degradation in EhV viability was observed in after the 23-day transportation event across the US.[24] Given the similarities shared between ASFV and EhV, Balestreri et al proposed the use of EhV as a surrogate for ASFV.[23]

The number of feed ingredients commercially available to the swine industry with anti-viral properties is continually growing. However, the investigation of the components or a commercially available product on the megavirus EhV as surrogate for ASFV had not yet been demonstrated. Virabloc (Essential Ag Solutions, Sioux Falls, SD) is a commercially available feed additive that consists of a proprietary blend of medium chain fatty acids (MCFA), botanical extracts and monoglycerides. The individual ingredients in Virabloc have been previously tested for their anti-viral properties[18-20,25]; however, the commercially available product had not been examined on the megavirus EhV. The objective of this trial was to determine virus particle viability of EhV contaminated soybean meal following exposure to Virabloc for a period of time that would be relevant to feed distribution from manufacturing to animal consumption with hopes to have a better understanding of how a feed additive with anti-viral properties may impact a similar megavirus, ASFV, transmission in the global supply chain.

Materials and methods

Cell culture and EhV-86 stock

A culture of *Emiliania huxleyi* was grown in Alga-Gro® Seawater Medium (Carolina Biological Supply Company, Burlington, North Carolina) at 15°C with 18h/6h light/dark cycle (ca. 2400 lux) until the concentration of 2×10^5 cells mL-1 was reached. Isolate EhV-86 was added to *E huxleyi* at a multiplicity of infection (MOI) of 1 and grown in a 15°C incubator until lysis was observed, which was usually after 4 d.[21] The lysate was filtered through a 0.45 µm filter (Nalgene™ Rapid-Flow™ Bottle Top Filters, ThermoFisher Scientific, Waltham, Massachusetts) to remove cell debris. This filtration and infection procedure was repeated several times. The filtered lysate was divided into aliquots and kept in the dark at 4°C until use.

Feed matrix

Previous research results have shown that megaviruses, such as ASFV and EhV, can survive in various experimentally inoculated feed matrices such as conventional soybean meal (SBMC).[24] Therefore, the feed matrix used in this study was conventional solvent extracted, dehulled soybean meal (containing 1-2% oil and 46-47% crude protein).

Treatments

Aliquots of EhV filtrate (final concentration of 1×10^6 EhV per gram of SBMC or per 100 ul of treatment) was exposed to Virabloc (final concentration 0.05% per gram of SBMC or per 100ul) for hours or days at room temperature (22°C). After treatment, the virus was recovered from the SBMC filtrate as described in Palowski et al.[24] Briefly, elution took place using the Alga-Gro® Seawater Medium (Carolina Biological Supply Company, Burlington, North Carolina). The tube was vortexed repeatedly before incubating in a water bath set at 40°C for a total time of 30 min. The tube was centrifuged to collect the SBMC at the bottom of the tube. The supernatant (virus eluant) was removed and filtered through 0.22 µm syringe filter (MillexTM-GP Sterile Syringe Filters with PES Membrane, MilliporeSigmaTM, Waltham, Massachusetts) into an Amicon centrifugal tube (Millipore Amicon Ultra 15mL, MilliporeSigmaTM, Waltham, Massachusetts). The virus eluant was washed 3 times with 1X phosphate buffered saline (1 × PBS, ThermoFisher Scientific, Waltham, Massachusetts) via centrifugation as per manufacturer's instructions, and the final 200 µL volume obtained was split into two aliquots of 100 µL each. This process was repeated to create biologically independent replicate samples.

Viability qPCR

To the split virus aliquots, PMAxx dye (100 µM, final concentration, Biotium Inc, CA, US) was added, mixed and incubated in the dark at room temperature for 10 min. One of the aliquots was then exposed for 30 min to the light using a PMA-Lite device (Biotium Inc, CA, US) to cross-link PMAxx Dye to the DNA (V-qPCR). The second aliquot that was not light treated was used as a control for S-PCR. A QIAamp® MinElute® Virus Spin (Qiagen, CA, US) was used to extract DNA from all aliquots. Quantitative PCR was conducted using QuantiNova SYBR Green PCR kit (Qiagen, CA, US) using EhV MCP standards as described in Palowski et al.[24]

Results and discussion

Virus extraction efficiency of conventional soybean meal

Before performing multiple extractions from SBMC samples, we first evaluated the recovery rate of EhV using the SBMC elution protocol. Overall, viral load (total EhV genomic copies recovered) was successfully quantified using the V-qPCR assay (Figure 1). The virus measured in the eluant without SBMC (Figure 1, green) was similar ($\log_{10}$ 7.6 vs $\log_{10}$ 8) to the theoretical starting amount of virus added. The recovery rate of V-qPCR using the elution protocol on SBMC was similar (5% vs 20%) to the percentage reported by Palowski et al.[24] This data confirms that > 80% of the virus remains attached or associated to the feed matrix in some way.

Virus extraction after exposure to Virabloc

The final Virabloc formulation reduced EhV by more than 3 log (Figure 2, dark grey). More importantly, the 1 hr exposure time to Virabloc to EhV in SBMC gave a comparable log reduction (2.64 log reduction, Figure 2 light grey). Recent findings have also demonstrated that MCFA blends can reduce ASFV infectivity[26,27]; however, to the best of the author's knowledge a complex commercially available product has not been tested on ASFV particle viability. Additionally, a 10-day exposure resulted in a 3.6 log reduction in EhV numbers (preliminary data not shown). Further experiments are being carried out to extend the exposure times to 20 days.

Conclusion

Assuming the log reduction seen with the surrogate EhV will also be demonstrated with ASFV, then an ASFV infection would most likely not be observed based on current estimates of the minimum infective dose from feed.[28] Taken together, we show here the great potential of Virabloc to reduce the exposure risk of megaviruses (ASFV-like viruses) to swine. However, additional research is warranted with longer incubation times to further understand the impact on the global swine industry.

References

1. Plowright W, Parker J, Peirce MA. African swine fever virus in ticks (*Ornithodoros moubata*, Murray) collected from animal burrows in Tanzania. *Nature*. 1969; 221:1071-73.

2. Dixon LK, Chapman DA, Netherton CL, et al., African swine fever virus replication and genomics. *Virus Res.* 2013: 173:3-14.

Figure 1: *Emiliania huxleyi* virus (EhV) quantity as detected by V-qPCR. The EhV-86 eluant without feed (control, green) and from conventional soybean meal (SBMC). Bar = standard deviation from the mean.

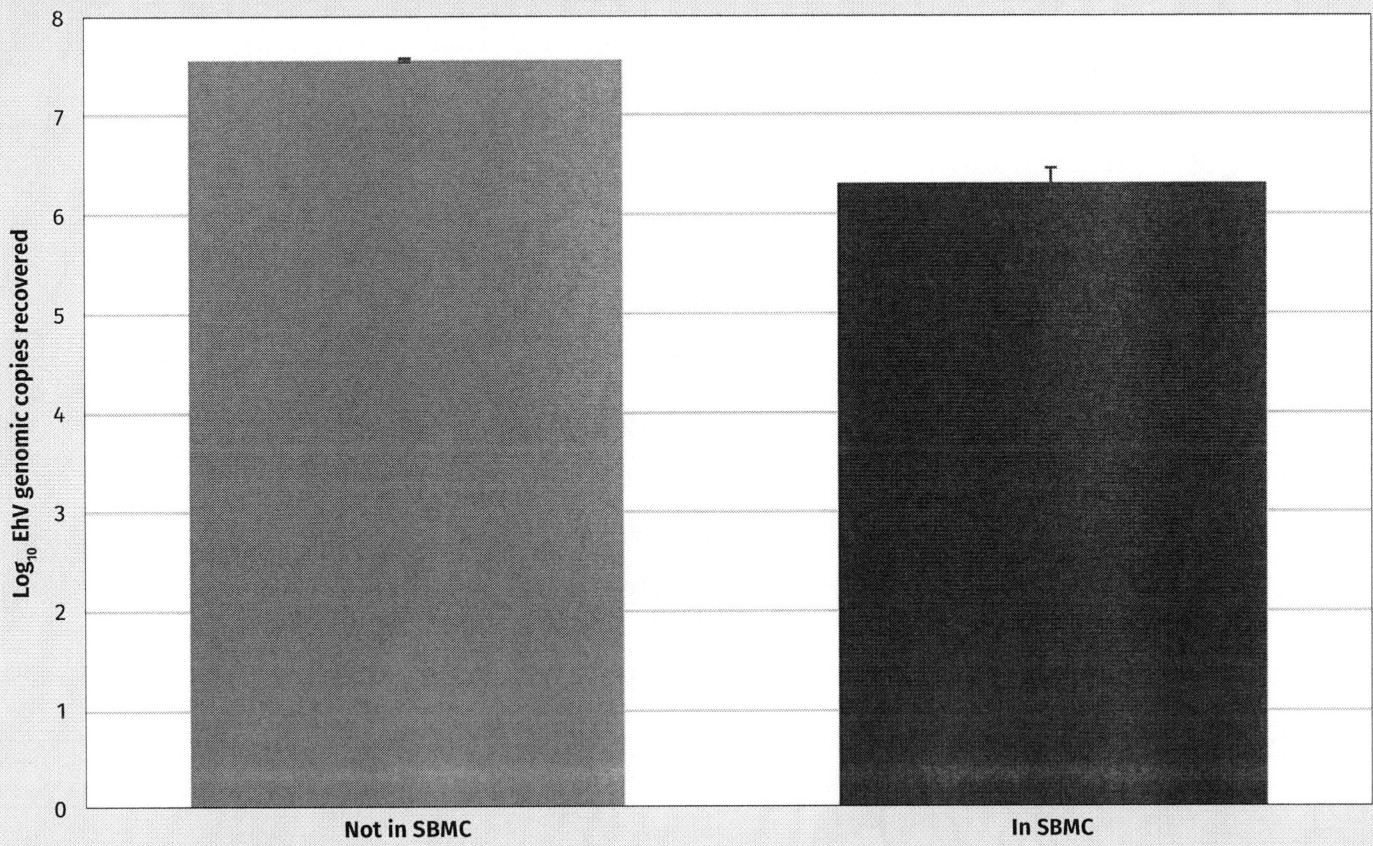

3. Alejo A, Matamoros T, Guerra M, et al., A proteomic atlas of the African swine fever virus particle. *J Virol.* 2018; 92:e01293-18.

4. Karger A, Perez-Nunez D, Urquiza J, et al., An update on African swine fever virology. *Viruses.* 2019; 11(9):864.

5. Dixon LK, Stahl K, Ferran J, et al. African swine fever epidemiology and control. *Annu. Rev. Anim. Biosci.* 2020; 8:221-46.

6. Quembo CJ, Jori F, Vosloo W, et al. Genetic characterization of African swine virus isolates from soft ticks at the wildlife/domestic interface in Mozambique and identification of a novel genotype. *Transbound. Emerg. Dis.* 2018; 65:420-31.

7. Gaudreault NN, Madden DW, Wilson WC, et al. African swine fever virus: an emerging DNA arbovirus. *Front. Vet. Sci.* 2020; 7:215.

8. Sanchez EG, Perez-Nunez D, Revilla Y. Mechanisms of entry and endosomal pathway of African swine fever virus. *Vaccines (Basel).* 2017;5(4):42.

9. Chenais E, Depner K, Guberti V, et al. Epidemiological considerations of African swine fever in Europe 2014-2018. *Porcine Health Manag. 2019; 5:6.*

10. Guinat C, Reis AL, Netherton CL, et al. Dynamics of African swine fever virus shedding and excretion in domestic pigs infected by intramuscular inoculation and contact transmission *Vet. Res.* 2014; 45:93.

11. Sánchez-Vizcaíno JM, Mur L, Gomez-Villamandos JC, et al. An update on the epidemiology and pathology of African swine fever. *J. Comp. Pathol.* 2015; 152:9-21.

12. De Carvalho Ferreira HC, Backer JA, Weesendorp E, et al. Transmission rate of African swine fever virus under experimental conditions. *Vet. Microbiol.* 2013. 165:296-304.

13. Stoian A, Petrovan V, Laconst A, et al. Stability of classical swine fever virus and pseudorabies virus in animal feed ingredients exposed to transpacific conditions. *Transboun. Emerg. Dis.* 2020; 67(4):1623-32.

14. Khanal P, Olcha M, Niederwerder M. Detection of African swine fever virus in feed dust collected from experimentally inoculated complete feed using quantitative PCR and virus titration assays. *Transbound, Emerg. Dis.* 2022; 69:97-102.

15. Dee S, Havas K, Spronk G. Detection of Senecavirus A in pigs from historically negative national swine herd and associated with feed imports from endemically infected countries. *Transbound. Emerg. Dis.* 2022; 1-3. **doi:10.111/tbed.14684**

16. Dee S, Clement T, Schelkopf, et al. An evaluation of contaminated complete feed as a vehicle for porcine epidemic diarrhea virus infection of naïve pigs following consumption via natural feeding behavior:proof of concept. *BMC Vet. Res.* 2014; 10:176.

17. Patterson G, Niederwerder M, Dee S. Risks to animal health associated with imported feed ingredients. *JAVMA.* 2019; 254(7):1-2.

18. Dee SA, Niederwerder MC, Edler R, et al. An evaluation of additives for mitigating the risk of virus contaminated feed using an ice-block challenge model. *Tranbound. Emerg. Dis.* 2020; 68:833-845.

19. Phillips FC, Rubach JK, Poss MJ, et al. Monoglyceride reduces viability of porcine epidemic diarrhoea virus in feed and prevents transmission to post-weaned piglets. *Transbound. Emerg. Dis.* 2022; 69:121-127.

20. Gebhardt JT, Woodworth JC, Jones CK, et al. Determining the impact of commercial feed additives as potential porcine epidemic diarrhea virus mitigation strategies as determined by polymerase chain reaction analysis and bioassay. *Transl. Anim. Sci.* 2019; 3:94-102.

21. Schroeder DC, Oke J, Malin G, Wilson WH. Coccolithovirus (Phycodnaviridae): Characterisation of a new large dsDNA algal virus that infects *Emiliania huxleyi*. *Arch. Virol.* 2002; 147, 1685–1698, **doi:10.1007/s00705-002-0841-3**

Figure 2: *Emiliania huxleyi* virus (EhV) quantity as detected by V-qPCR after exposure to Virabloc for 1 hr in the absence or presence SBMC (T). Non treated (NT) control is also shown. Bar = standard deviation from the mean.

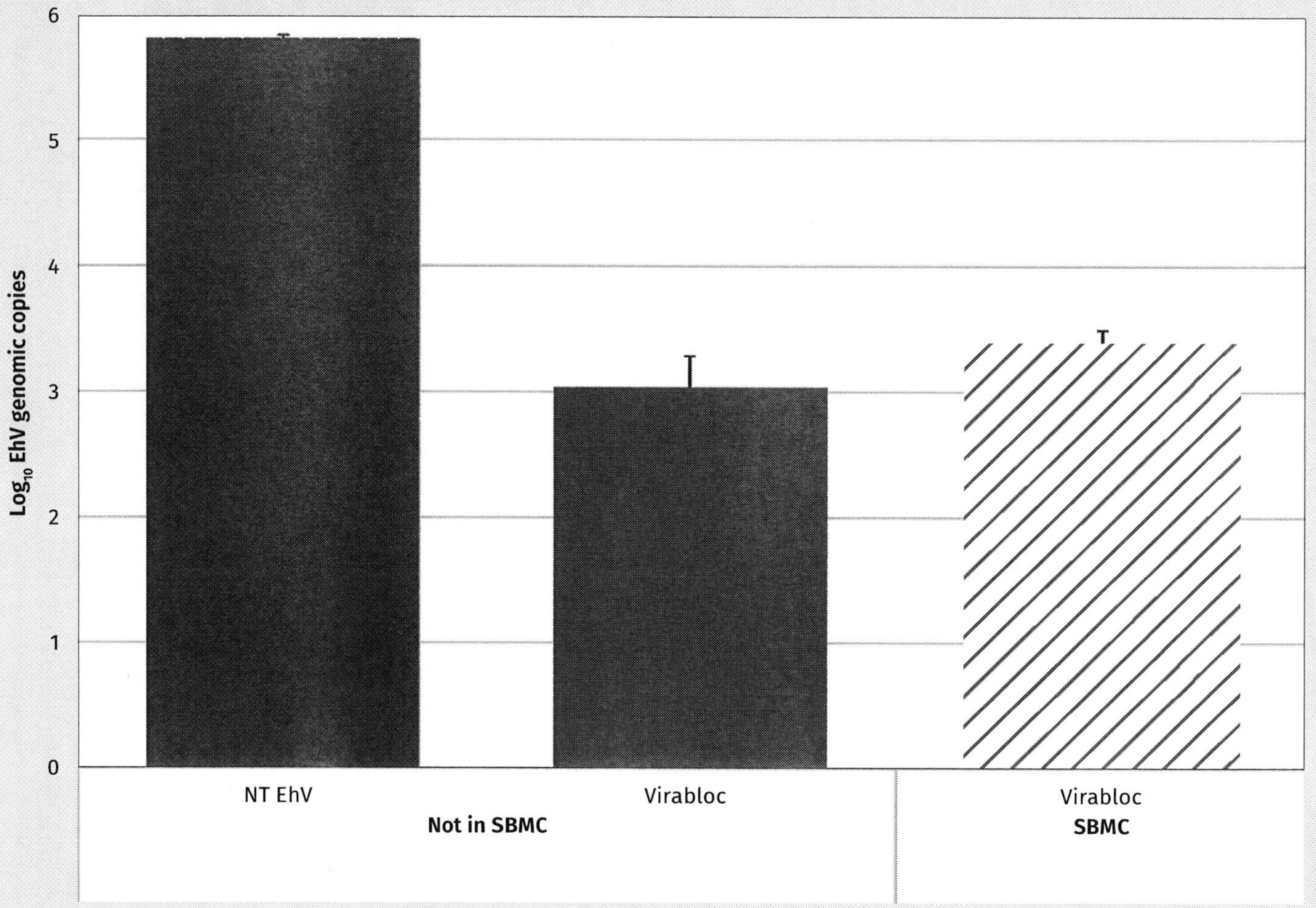

22. Schroeder DC, Oke J, Hall M, G. Malin et al. Virus succession observed during an *Emiliania huxleyi* bloom. *Appl. Environ. Microbiol*. 2003; 69, **doi:10.1128/AEM.69.5.2484-2490.2003**

23. Balesteri C, Schroeder DC, Sampedro F, et al. Unexpected thermal stability of two enveloped megaviruses, Emiliania huxleyi virus and African swine fever virus, as measured by viability PCR. *Frontiers in Microbiology*. 2022; in review.

24. Palowski A, Balestreri C, Urriola PE, van de Light JLG et al. Survival of a surrogate ASFV-like algal virus in feed matrices using a 23-day commercial United States truck transport model. *Frontiers in Microbiology*. 2022; in review.

25. Cochrane R, Dritz SS, Woodworth JC, et al. Assessing the effects of medium-chain fatty acids and fat sources on PEDV infectivity. *Transl Anim Sci*. 2019; 4(2):txz179. **doi: 10.1093/tas/txz179**

26. Jackman JA, Hakobyan A, Zakaryan H et al. Inhibition of African swine fever virus in liquid and feed by medium-chain fatty acids and glycerol monolaurate. *J. Anim Sci Biotech*. 2020; 11:114, **doi:10.1186/s40104-020-00517-3**

27. Niederwerder MC, Dee S, Diel DG, et al. Mitigating the risk of Africna swine fever virus in feed with anti-viral chemical additives. *Transbound Emerg Dis*. 2021; 68:477-486

28. Niederwerder MC, Stoian AMM, Rowland RRR, Dritz SS, et al. Infectious dose of African swine fever virus when consumed naturally in liquid or feed. *Emerg. Infect. Dis*. 2019; 25, 891–897, **doi:10.3201/eid2505.181495**

https://doi.org/10.54846/am2023/81

Sample quality management in swine surveillance

E. Tapia, DVM; S. Zimmerman, DVM
IDEXX Laboratories, Inc., Westbrook, Maine

Immunity, biosecurity, and surveillance are the basis of disease control and prevention. Among the three, surveillance may be the most important because its function is to verify that the other two (herd immunity and biosecurity) are functioning correctly.[1]

When designing a surveillance program, the most frequent question and point of discussion concerns the number of samples, frequency of sampling, and ages to be sampled. Would it be better to monitor with ELISA or PCR? What type of specimen - serum, oral fluids, or something else?

PCR is widely used in swine surveillance because it is considered to be diagnostically sensitivity, diagnostically specific, and provide for early detection. The objective herein is to touch on some concepts that are frequently missed when considering the use of a PCR in a swine surveillance program.

Internal sample control

The performance of qPCR can be affected by testing procedures, ie, extraction and/or amplification steps, primers and probes, technique, reagents, equipment, software, etc, but detection may also be affected by sample handling. Was the sample chilled properly after collection to protect the analyte? Did it undergo multiple freeze/thaw cycles? Thus, prior to testing, PCR targets in diagnostic samples may be subjected to a variety of adverse conditions in the course of collection, storage, transport, and processing. How do these conditions affect the PCR results?

IDEXX Laboratories developed a qPCR procedure that includes an indicator of sample quality. The internal control for RealPCR™ swine tests is based on the detection of endogenous porcine DNA or RNA sequences present in the host sample and is referred to as the "internal sample control" (ISC). Inherent to the sample, the ISC is subjected to the same conditions as the target. Thus, detection of endogenous DNA or RNA in swine samples is a quality control for sample addition, extraction, and amplification.

Primers and probe for detection of the ISC are included in the RealPCR™ target-specific mix and detection of the ISC indicates both that the test procedure was conducted correctly and that the sample was of good quality. Failure to detect the ISC indicates that something was not right either with the sample or the test.

Research findings

Several investigations have been done to characterize the effect of sample handling on the integrity of the sample.[2,3] Using a RT-qPCR kit (IDEXX Laboratories, Inc.) designed to detect both PRRSV and ISC RNAs simultaneously, it was shown that the temperature and time at which the specimen is held can impact results. For serum samples, both PRRSV and ISC results were unaffected if the samples were maintained at ≤ 20°C ... even for up to 168 hours (Table 1). Oral fluid specimens showed a decline in ISC and PRRSV even when maintained at 4°C for 24 hours. Therefore, the study recommended that oral fluid and fecal samples should be frozen in a non-self-defrosting freezer and submitted frozen to the laboratory for testing.

Table 1: Daily (24 hrs) ECq decline by temperature[1]

Target	Specimen	4°C	10°C	20°C	30°C
PRRSV	Serum	0.00[2]	0.00[2]	0.00	0.11
	Oral fluid	0.03	0.05	0.07	0.08
	Feces	0.05	0.06	0.05	0.05
ISC	Serum	0.00	0.00	0.07	0.10
	Oral fluid	0.03	0.05	0.07	0.08
	Feces	0.06	0.05	0.08	0.10

1 Data analyzed using a mixed-effects regression model

2 No decline in ECq over time ($P > 0.05$; mixed-effects regression model). Cq were re-expressed as "efficiency standardized Cqs" (ECqs)

Conclusions

The use of an ISC requires understanding the ISC response to interpret the results. Studies performed to date indicate that the impact of less-than-optimum environmental conditions on RT-qPCR PRRSV and ISC results are similar in serum, oral fluid, and fecal samples. Therefore, the use of an ISC is a useful and important addition to quality management in routine PRRSV RT-qPCR testing.

References

1. Munguía-Ramírez B, Armenta-Leyva B, Giménez-Lirola L, Wang C, Zimmerman J. 2022. Surveillance on swine farms using antemortem specimens. In: Maes D, Segalés J (editors). Optimising Pig Herd Health and Production. Burleigh Dodds Science Publishing, Ltd., Cambridge UK, (in press).

2. Munguía-Ramírez B, Armenta-Leyva B, Henao-Díaz A, Ye F, Doolittle K, Zimmerman S, Giménez-Lirola LG, Zimmerman J. November 2022. Effect of sample handling on PRRSV detection by RT-qPCR. Proc 2022 ISU James D McKean Swine Disease Conference, Iowa State University. Ames, Iowa, pp. 48-49.

3. Munguia-Ramirez B, Armenta-Leyva B, Henao-Diaz A, Ye F, Doolittle K, Zimmerman S, Gimenez-Lirola LG, Zimmerman J. October 2022. Use of a porcine endogenous reference gene (internal sample control) in a PRRSV RT-qPCR. Proc 65th Ann Meet American Association of Veterinary Laboratory Diagnosticians, Minneapolis, MN, p. 81.

https://doi.org/10.54846/am2023/83

Overview of the benefits of Furst Protect in sows and nursery pigs

F. Sandberg, PhD; K. Soltwedel, MS, PhD; R. Fischer, MS, PhD; W. Kwon, MS, PhD
Furst-McNess Company, Rockford, Illinois

Introduction

The correct nutritional supplementation during health, and times of disease challenge, can be hard, as the requirements of the animal changes between the health and disease state, as well as during the time course of a disease challenge. The animal either in utero, post-birth or post-weaning can be fed to support correct and adequate development of essential organs, including immune organs. If the animal has developed adequately prior to a challenge, the animal has greater resilience in surviving the challenge through a more efficient innate immune response, with the least amount of impact. Furst Protect is a combination of fatty acids in the form of highly available monoglycerides, lysophospholipids, natural antioxidants, and additional phytonutrients that all play an important role in immune system development and function.[1] In the work summarized by[2] various compounds including Furst Protect, and organic acids, to medium chain fatty acids, to monoglycerides and essential oils were found effective too varying degrees at preventing infection of nursery pigs from feed artificially contaminated with PEDV, PRRSV and SVA viruses. Other research has shown that Furst Protect™ supplementation for the first 3 weeks in the nursery, reduced mortality and improves growth performance in nursery pigs beyond the supplementation time frame, and even into the finishing phase.[3,4] Further research has demonstrated that supplementation of Furst Protect via drinking water, can generate the same benefits of improved immunity, post supplementation.[4] Presented here are two additional trials, firstly a meta-analysis of sow farm trials with Furst Protect. Secondly, a controlled study of Furst Protect supplementation via drinking water in a commercial research farm. The overall body of information is supportive that Furst Protect is a significant tool for reducing mortality in swine production systems through novel nutritional supplementation of the innate immune system.

Methods

Furst protect sow farm mortality

A large number of controlled on farm sow trials were summarized by accessing data through their record keeping systems (PigKnows, MetaFarms) and analyzed within the farm, and across all farms to create more adequate power of assessing sow mortality (defined as both dead sows, and sows destroyed) effects. The breakdown of mortality was either not available or deemed sufficiently accurate, and therefore was not analyzed. For the sow data, multiple sow farms were collected and analyzed as individual farms and collectively as multiple farms together to get further information on the effects of Furst Protect on sow survivability. Multiple trials were pooled together (sows deads+destroyed by week as the replicate) by taking equal control periods, and treatment periods. For example, if a farm implemented Furst Protect for an 8-week period we would contrast the performance of the 8 weeks prior to implementation. To do this a single time is meaningless, as sow farm mortality is variable and seasonally driven. However, when this is repeated multiple times, across various times of the year, the collective data allows for statistical evaluation to more adequately determine whether Furst Protect impacted sow mortality. To protect farm identities, the farms were given identification numbers, and 9 farms took part, representing 32,050 sows ranging in farm sizes from 1,200 to 10,000 sows. Data was analyzed using GLM in Minitab with week as the replicate.

Furst Protect via water, nursery performance and mortality

A controlled study was conducted at a commercial research facility, with 26 pigs (DNA female, DNA 600 sire) per pen for 12 pens assigned by weight and gender to control, and 12 pens assigned by weight and gender to Furst Protect supplementation via water for the first 7 days post-weaning. Pigs averaged 11.95 lbs at weaning. Data was analyzed first for outliers, using the formal Grubb's procedure in Minitab, and then using GLM in to determined whether a trend ($P < 0.10$), or statistically significant ($P < 0.05$).

Results

Furst Protect sow farm mortality

In Table 1 is the summary of the Meta-analysis of 9 farms, consisting of a total of 32,050 sows. Sows supplemented with Furst Protect™ had a reduction ($P = 0.001$) in sow mortality from 14.91% to 11.62% which equals 3.29% points, or a 22% reduction. In Figure 1 is a graphic display of the data, showing that Furst Protect reduced the weeks during which high sow mortality occurred (upper bar), with a slight reduction in the lowest mortality level (lower bar).

Furst Protect via water, nursery performance and mortality

Furst Protect supplementation via drinking water for the first 7 days only, tended to improve day 0-7 ADG (0.19 vs 0.15 lbs/d; $P = 0.07$) and FCR (1.33 vs 1.70; $P = 0.07$). Overall ADG from 0 to 42 days tended to be improved by Furst Protect (0.80 vs 0.78 lbs/d; $P = 0.10$). Overall FCR from 0-42 days was numerically improved by Furst Protect (1.52 vs 1.54; $P = 0.14$). In Table 2 is the effects on mortality by early supplementation with Furst Protect Direct. As shown in previous trials[3,4] early supplementation with Furst Protect, impacts mortality positively in later nursery phases. This effect appears independent whether via feed or water supplementation. In this trial, mortality tended towards a reduction ($P < 0.10$) from 1.85% to 0.31%.

Discussion

Furst Protect has shown across multiple trials, in both sows and nursery pigs to support improved livability. Further research is warranted to elude the mechanisms by which this occurs. In terms of sow mortality, unfortunately, we can report that the reduced mortality is not from a reduction of prolapses, but through

Table 1: The effect of Furst Protect™ supplementation on sow livability across multiple sow farms

Sow farm ID	Sows	Weeks compared	Control (% mortality)	Furst Protect™ (% mortality)	SEM	*P*-value
FMC001	6,000	8 vs 8	10.87	6.36	0.93	0.004
FMC002	10,000	12 vs 12	15.17	12.69	1.04	0.107
FMC003	3,200	8 vs 8	15.01	15.65	2.36	0.851
FMC004	1,200	10 vs 10*	14.05	11.84	2.38	0.578
FMC005	1,300	20 vs 20*	11.47	13.64	1.99	0.469
FMC006	1,200	10 vs 10*	18.80	13.89	0.41	0.014
FMC007	2,100	12 vs 12	20.31	14.36	2.21	0.069
FMC008	2,950	12 vs 12	21.30	10.70	2.04	0.001
FMC009	4,100	24 vs 24	10.51	9.89	1.09	0.680
	32,050	Overall	14.91[a]	11.62[b]	0.71	0.001

* Refers to batched 5 week groups ie, 10 weeks would be 2 × 5 week batches.

[a,b] Different superscripts within the same row are statistically different ($P < 0.05$).

Figure 1: The effect of Furst Protect supplementation on the 95% confidence interval for sow livability across 9 farms.

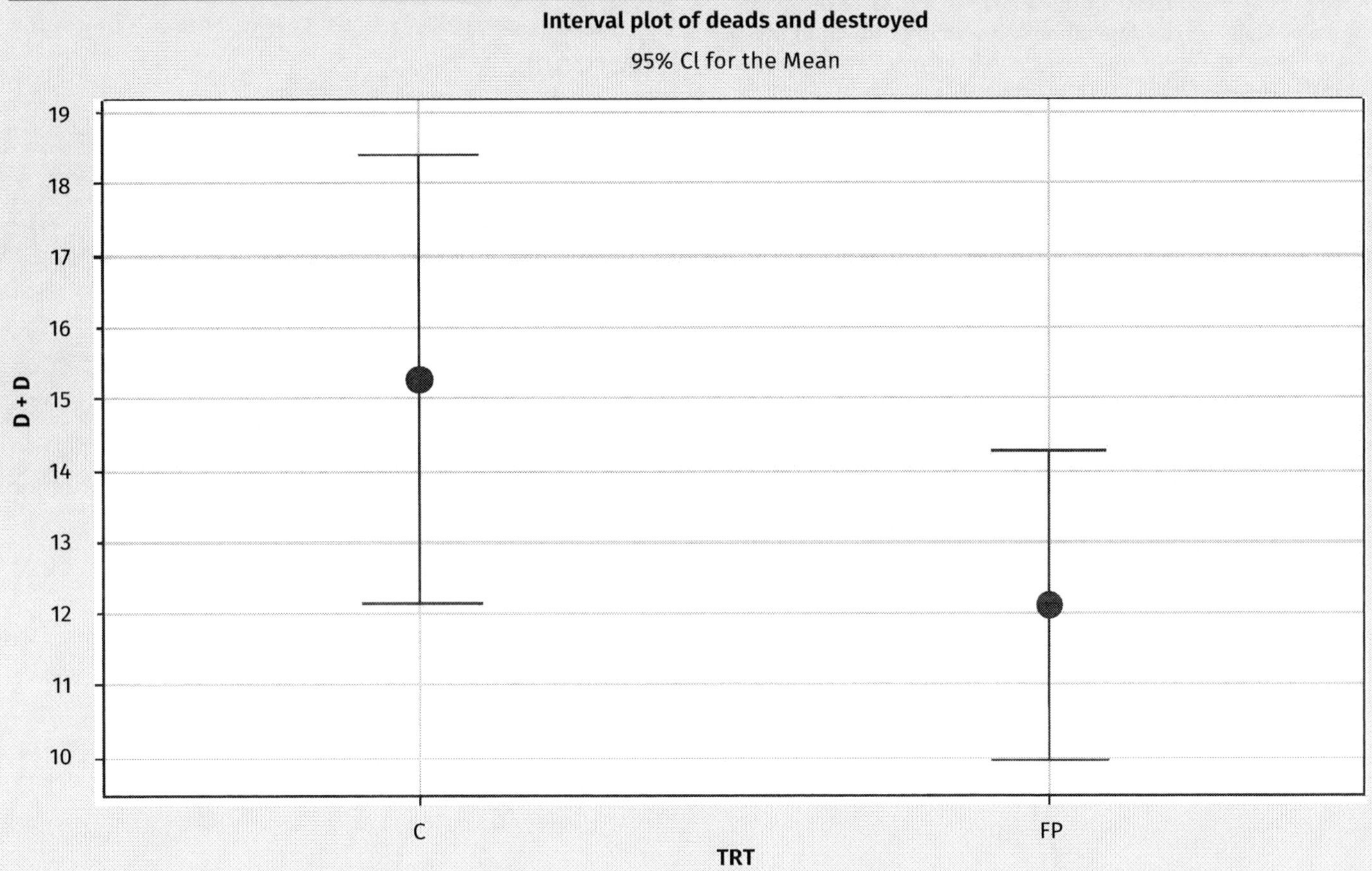

Table 2: The effect of Furst Protect supplementation via drinking water for first 7 days post weaning (FPD, n = 11) vs no supplementation (control, n = 12) on mortality throughout the nursery phase

Parameter	Control	FPD	SEM	*P*-value TRT
Mortality 0-7	0.617	0.000	0.470	0.353
Mortality 7-21	0.309	0.309	0.332	1.00
Mortality 21-42	0.926[c]	0.000[d]	0.379	0.093
Mortality 0-42	1.852[c]	0.309[d]	0.603	0.080

[c,d] Different superscripts within the samw row tended towards a reduction $P < 0.10$.

reductions of other causes. We believe that adequate supplementation of the innate immune system (in its full definition) can successfully reduce mortality in sows and piglets, thereby enhancing performance and animal welfare. Reducing mortality (through enhanced health) improves producer profitability and relieves pressure on the labor force as healthier animals require less time being given supplemental care and management.

From a financial perspective, cost of Furst Protect for a sow farm is approximately $0.30 to $0.60 per pig depending on feeding level, and according to recognized industry sources (Ron Ketchum, through SMS records; Dr Ken Stalder, ISU, through simulation model) 1% in sow mortality increases cost of production of weaned pigs by $0.30-$0.50, or that 1 dead sow has a cash cost to the farm of $1000 ± $300 depending on when in production life mortality occurs. To summarize in simple terms, a 1000 head sow farm produces 28,000 piglets, and for each 1% mortality 10 sows incur a cost of $10,000 to the farm, which across 28,000 piglets translate to $0.36/pig weaned. In the trial summarized here, Furst Protect reduced sow mortality by 3.29% generating a reduction in production cost of $1.18 per pig, at a cost of $0.40 leading to a 3 to 1 return; or stated differently a $0.78 per pig weaned net margin improvement.

From a financial perspective, cost of Furst Protect via drinking water is approximately $0.20 per pig. 1000 pigs cost $40,000 and with initial feed, treatment and medications carry a cash cost of $43 per hd by the end of the first week increasing towards $55 towards the end of the nursery as they consume more feed. 10 dead pigs therefore incur a cost of on average $500, which becomes additional carrying cost across the remaining 990 pigs to be sold of $0.51 per pig. In addition, the sales revenue of those 10 pigs are lost, which typically on average is around $50 for a feeder pig. Putting the overall cost of 1% nursery mortality at $1.02 per pig leaving that nursery. Trial reported here showed a 1.54% reduction in nursery mortality, with a return of 7.9 to 1; or stated differently a $1.37 per pig net margin improvement.

References

1. Radzikowska *et al.* (2019) Nutrients 11:2290.
2. Dee *et al.* (2020) Transboundary and Emerging Disease: 00: 1-13.
3. Sandberg *et al.* (2021). Leman Swine Conference.
4. Sandberg *et al.* (2021). Leman Swine Conference.

https://doi.org/10.54846/am2023/84

Evaluation of the vaccination during lactation with ERYSENG® PARVO/LEPTO in the follicle genesis, stress-related parameters, and analytes vs a non-vaccinated control group

J. Miguel[1], DVM; I. Ballarà[1], DVM; S. Ramírez[1], DVM; M. Baratelli[1]; J. Roca[2]; C. Gil[3]
[1]HIPRA HQ, Amer (Spain); [2]Department of Medicine and Animal Surgery, University of Murcia, Campus de Espinardo, Murcia (Spain); [3]Carracaballar SA, Segovia (Spain)

Introduction

It has been previously demonstrated that the administration of some vaccines including antigens against Erysipelas, Parvovirus, and *Leptospira* spp. during lactation can cause side effects.[1,2] This type of vaccine is common on most pig farms around the world to prevent reproductive problems.[3,4] The fever induced by vaccination derives in a reduced feed intake having an impact on milk production and consequently the piglet weight at weaning. Previous studies have already reported the negative impact of feeding restriction during lactation on ovarian development, as well as on ovulation rate and total number of piglets born.[5,6]

The objective consisted of evaluating if the vaccination at 10 days of lactation with ERYSENG® PARVO/LEPTO comparing vs a non-vaccinated control group, has an impact on the follicles genesis and consequently could impair the subsequent reproductive performance. Other parameters that can be affected by the vaccination were also evaluated, such us rectal temperature, piglet weight at weaning and analytes stress-related (cortisol and α-amylase).

Methods

The study was conducted on an experimental 300-sows size breeder swine farm (genetic line CG36 Choice Genetics) located in Spain during the period of April to June of 2022. Healthy gestating sows (n = 60) were recruited during 3 consecutive batches and divided into two groups: Group 1, vaccinated with ERYSENG® PARVO/LEPTO and group 2, non-vaccinated group and injected with a placebo (PBS) (Table 1). The animals were managed following the standard procedures of the farms and the different groups of sows were housed together in the same farrowing barn.

Factors such as the farrowing cycle, previous reproductive performance, and quality of the follicles before the vaccination were considered to establish the groups (Table 2).

The sows in Group 1 and Group 2 were injected at 10 days after farrowing. The rectal temperature and the clinical signs were measured just before vaccination, 6 hours post-vaccination (hpv) and 24hpv. Saliva samples were collected twice on the day of vaccination (before and after 30 minutes), and 3 days after vaccination to assess two analytes (cortisol and α-amylase) related to stress level. The ovarian follicles were studied at 10 days after farrowing (before vaccination) and at 21 days after vaccination, coinciding with weaning. The piglets' body weight was evaluated the day of vaccination and 11 days after.

The saliva was collected using Salivette® tubes (Sarstedt, Aktiengesellschaft & Co, Nümbrecht, Germany). For sample collection, a small sponge, held by a surgical clamp, was inserted into the sow's mouth, allowing the sow to chew on the sponge for 20-40 seconds. The saliva-impregnated sponges were then placed in Salivette® tubes which were centrifuged at 3000 × g for 10 min. The particle-free supernatant from the centrifugation was aspirated and deposited in Eppendorf tubes which were transported at 5°C to the Animal Andrology Laboratory of the University of Murcia where they were stored at -80°C until the cortisol and α-amylase analytes were analyzed. Cortisol concentration was quantified using a solid-phase competitive chemiluminescent enzyme immunoassay according to the method described by Escribano.[7] The level of α-amylase activity was quantified using the commercial α-Amylase test (OSR6182, Beckman Coulter, Inc., Kraemer, California, USA) that had been validated in porcine saliva by Fuentes et al.[8]

Table 1: N° of sows and piglets by group

Group	N° sows	N° piglets	Treatment
1	29	340	Vaccination with ERYSENG® PARVO/LEPTO
2	28	329	Vaccination with placebo

Table 2: Average liveborn piglets

	Average liveborn
ERYSENG® PARVO/LEPTO	15.21 ± 2.95
Group 2	15.63 ± 2.51

Results are represented as average and standard deviation. No statistically signifigant difference was observed.

The ovaries were visualized with a portable ultrasound scanner (LOGIQ Book XP; GE Healthcare) incorporating a 4-10 MHz transrectal linear probe (i739-RS, GE Healthcare). The protocol to be followed was that described by Lopes.[9] The probe was introduced via the rectum, between 35 and 45 cm beyond the anal sphincter, a position that allowed visualization of both ovaries. In each sow, both ovaries were scanned in different planes and all images were recorded. In the data collection, anechogenic spherical structures, with thin borders and posterior reinforcement that could show irregular contours at times, were counted as follicles. The largest diameter of the follicles, as a reference for size, was measured using the calibrated measurement functions of the ultrasound software. Anechogenic structures with diameters greater than 1.1cm were considered cysts (Figure 1). Circular, homogeneous, hypoechogenic structures were identified as corpora lutea. The diameters of four follicles were recorded for each sow, two for each ovary, and all follicles in both ovaries were counted. The number was only counted if the follicles were larger than 0.5 cm in diameter (Figure 2). At smaller sizes it is considered an inaccurate task.

The statistical analyses were performed with the Prism 9 program (GraphPad Software Inc., California, USA). The ultrasound data (number and largest diameter of the follicles) and cortisol and α-amylase levels were statistically analyzed comparing the data between experimental groups using the nonparametric Kruskal-Wallis test since the data did not meet the criteria for normal distribution so the results were expressed as median and percentiles. Significant differences were considered significant for a $P < 0.05$. Whereas, data collation, initial analysis and graphs generation were performed using Excel (Microsoft) and GraphPad Prism 9. As the data met the criteria for normal distribution, ANOVA was used, with Tukey's test along with Chi-square with Bonferroni correction.

Results

The average rectal temperature (RT) of the studied animals was ranging between 38.33°C and 38.59°C before vaccination. Fever was considered as a T greater than 39.3°C. A mild and transient increase of RT was observed in both groups after injection, at 6 hours and 1-day post vaccination which was resolved within 24 hours (Figure 3). No statistically significant differences were

Figure 1: Follicles of more than 1.1 cm at weaning considered as cysts.

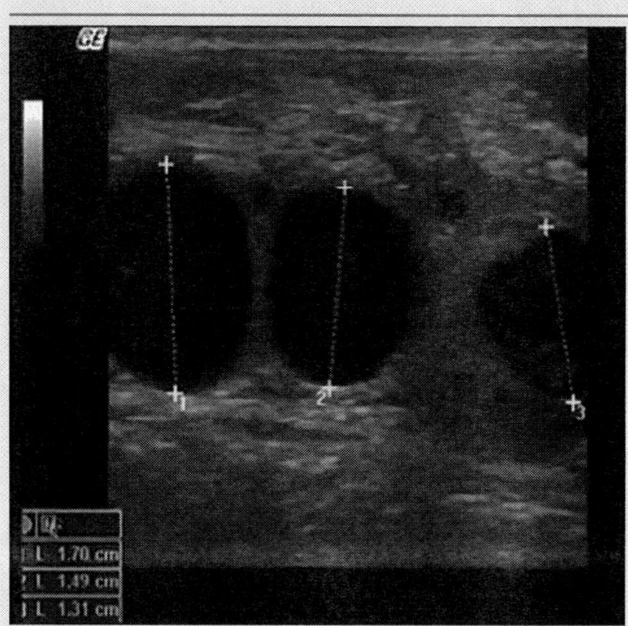

Figure 2: Image of mature follicles close to the ovulation.

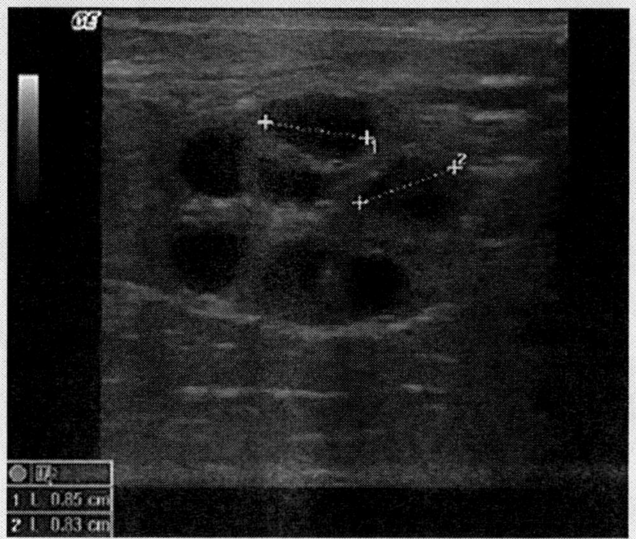

Figure 3: Rectal temperature of the sows at different timepoints.

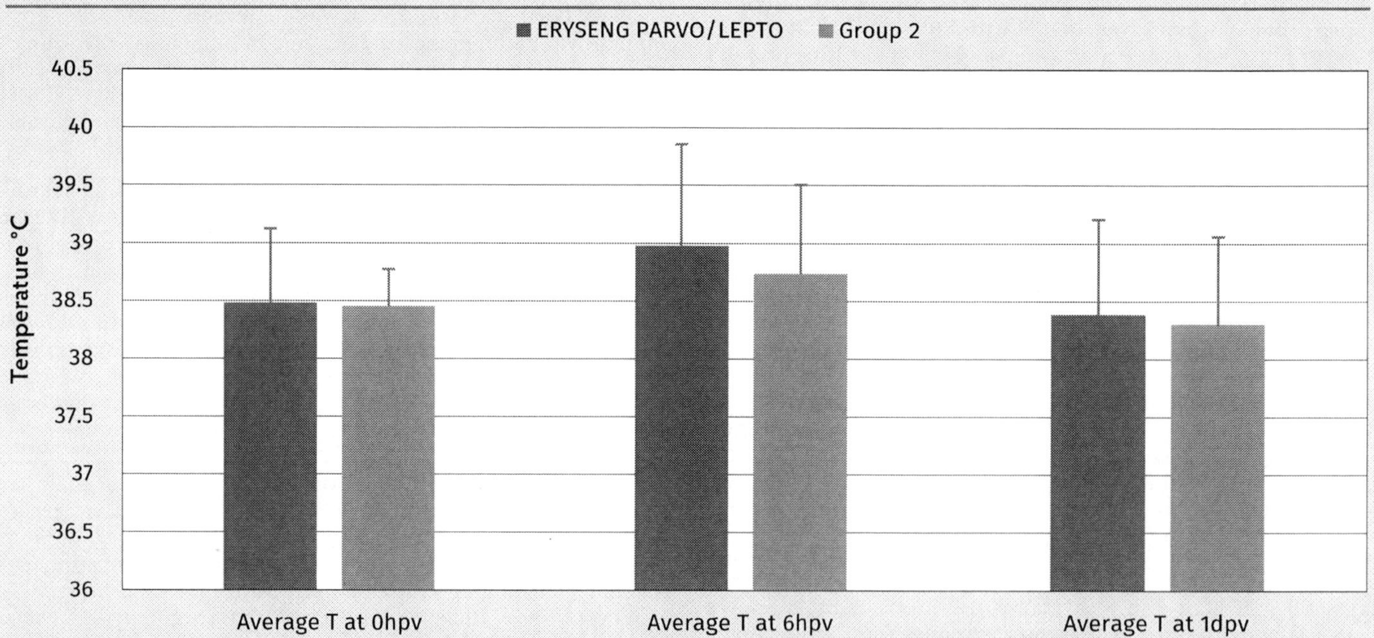

Results are represented as average and standard deviation. No statistically signifigant difference was observed.

observed between both groups at any of the evaluated time points. General and local reactions were monitored during the vaccine administration and none of these reactions were observed during the entire study.

The piglets recruited for the study were weighed on the day of sow vaccination (10 days of life) and 11 days after (21 days of life). A total of 340 and 329 piglets were recruited respectively in groups 1 and 2. The body weight at birth of the recruited pigs was similar between groups, no statistically significant differences were observed (Both the average piglet weight and the weight per litter). The body weight at day 21 of group 1 was 4 g higher than group 2; these differences between groups were not statistically significant (Table 3).

Regarding the evaluation of the follicular development, the average diameter of the follicles at the day 10 of the trial (Figure 4), showed no statistically significant differences between groups (Table 4). 21 days later, the number and the size of the follicles was the normal number for weaned sows (Table 5). There were no statistical differences between groups for the total number of follicles and the size.

The cortisol and α-amylase data at each evaluation moment are observed in the Table 6. There are not any statistical differences between groups at any of the sampling moments. But, for the cortisol results, the values are statistically higher in the moments 1 and 2, than in the 3 in both groups.

Discussion

The observed follicular diameter of lactating sows was as expected at 5 - 12 days after farrowing, but the number of follicles was not possible to be evaluated as the size of most of them was < 5 mm. The results of follicle size at weaning indicated that the vaccination with ERYSENG® PARVO/LEPTO did not influence the ovarian follicular development as it was similar for both groups and the results were the same as to be expected for reproductively active newly weaned sows.

With regard to the cortisol and α-amylase and the differences of the results found between the moments 1 - 2, and 3, the differences could be explained by assuming the stress caused by the transrectal ultrasound effect carried out three times in both groups. Despite the vaccination, the results were the same in both groups. The results of the other parameters evaluated such as the RT and the piglet weight at weaning did not show differences between groups either.

From the results obtained, it can be concluded that ERYSENG® PARVO/LEPTO has not influenced the sows' ovarian follicular development.

References

1. Rodriguez-Ballara, I.; Ibanez, G. Safety comparative study of two reproductive vaccines administered during lactation. 49th Annual Meeting of the American Association of Swine Veterinarians (March 3-6, 2018) – 223.

2. Balderrama, V., Tribó F., et al. 2019. 50th Annual Meeting of the American Association of Swine Veterinarians (Orlando; March 9-12, 2019) – 209.

3. Adler, B. and De LA Peña Mocrezuma, A. 2010. Vet Mcribiolo. 140-292-4.

4. Mengeling, WI., et al. 1979. Am J Vet Research. 40 (2):204-7.

5. Soede, N. M., et al. 2020. Consequences of negative energy balance on follicular development and oocyte quality in primiparous sows. Biology of Reproduction, 2020, 102(2), 388-399.

Figure 4: Follicles at day 10 of lactation.

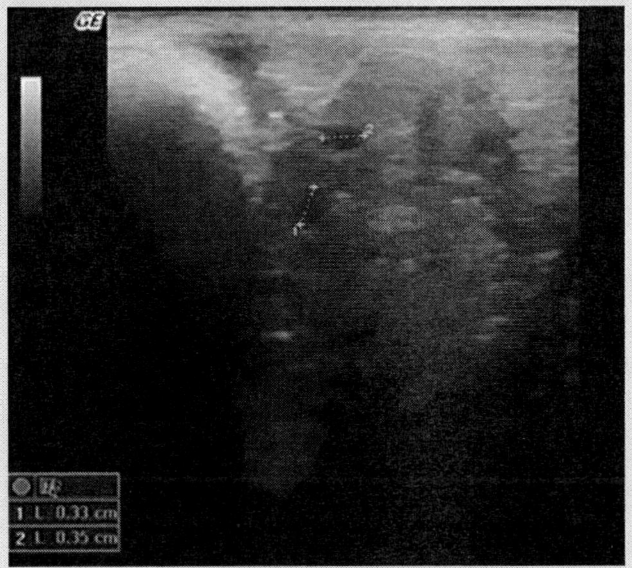

Table 3 : Average litter weight at 21 days of life and ADG

	Average piglet weight 10 dol	Average piglet weight 21 dol	ADG
ERYSENG® PARVO/ LEPTO	2.73 ± 0.71	6.14 ± 1.15	310.28 ± 80.62
Group 2	2.69 ± 0.74	5.77 ± 0.90	279.28 ± 59.19

Results are represented as average and standard deviation. No statistically signifigant difference was observed.

Table 4: Follicles size at day 10 of lactation

	Median (mm)	Percentile 25 and 75 (mm)
Group 1 (ERYSENG® PARVO/LEPTO)	2.8	0.22 - 0.33
Group 2 (PBS)	2.6	0.13 - 0.34

Results are represented as median and percentiles 25 and 75.

Table 5: N° of follicles and size at weaning

Parameter	Group	Median	Percentile 25 and 75
N° of follicles	Group 1 (ERYSENG® PARVO/LEPTO)	17.5	15.3 - 20.8
	Group 2 (PBS)	15	13.5 - 18.0
Size of follicles (mm)	Group 1 (ERYSENG® PARVO/LEPTO)	4.9	3.8 - 5.7
	Group 2 (PBS)	5.2	4.3 - 5.7

Results are represented as median and percentiles 25 and 75.

Table 6: Cortisol and α-amylase at different time moments

Analyte	Sample	Group 1	Group 2
Cortisol (ng/mL)	1	143.1[a] (39.8 - 288.8)	127.9[a] (64.5 - 195.6)
	2	185.3[a] (96.2 - 266.9)	132.2[a] (74.2 - 215.5)
	3	59.1[b] (32.4 - 149.4)	78.1[b] (36.9 - 99.6)
α-amylase (UI/mL)	1	366.3 (161.9 - 781.3)	323.0 (160.5 - 767.9)
	2	234.4 (100.6 - 397.1)	238.3 (119.0 - 484.7)
	3	177.7 (58.9 - 274.5)	219.4 (81.0 - 594.8)

Average values and percentile 25 and 75 (within parentheses) of the cortisol and α-amylase levels Results are represented as median and percentiles 25 and 75. [a,b] indicate significant statistical differences (Kruskal-Wallis test; $P \leq 0.05$).

6. Quesnel, H., Prunier, A. 2000. Influence of the nutritional status on ovarian development in female pigs. Animal Reproduction Science 60-61 (2000) 185-197.

7. Escribano, D., et al 2012. Validation of an automated chemiluminescent immunoassay for salivary cortisol measurements in pigs. Journal of Veterinary Diagnostic Investigation, volume 24, Issue 5, September 2012.

8. Fuentes, M., et al 2011. Validation of an automated method for salivary alpha-amylase measurements in pigs (Sus scrofa domesticus) and its application as a stress biomarker. Journal Veterinary Diagnostic Investigation, 2011 Mar;23(2):282-7.

9. Lopes T.P., Sánchez-Osorio J., Bolarín A., Martínez E.A., Roca J. Relevance of ovarian follicular development to the seasonal impairment of fertility in weaned sows. The Veterinary Journal. 2014 Mar;199(3):382-6. **doi: 10.1016/j.tvjl.2013.11.026. Epub 2013 Dec 2. PMID: 24461203**

https://doi.org/10.54846/am2023/85

Effects of an additional iron dextran injection administered to piglets on hemoglobin, performance, and carcass responses

T. B. Chevalier[1], BS, MS; J. W. Lyons[2], BS, DVM; D. B. Paczosa[1], BS;
G. K. Rentfrow[1], BS, MS, PhD; M. D. Lindemann[1], BS, PhD
[1]Department of Animal and Food Sciences, University of Kentucky, Lexington Kentucky;
[2]Pharmacosmos Inc., Watchung, New Jersey

Introduction

It is common practice to provide newborn piglets with supplemental iron to prevent iron deficiency, usually through an intramuscular (IM) injection of an iron complex, because piglets are born with very low iron stores and receive little through the sow milk.[1] Moreover, it has been suggested that the iron requirement of piglets during lactation is mainly dependent on the growth rate of the piglet during the lactation period.[1-3] Modern swine production has undergone improvements in genetics that have resulted in increased litter sizes and growth rates of nursing piglets. Given these advancements, recent work has shown that the industry standard iron injection (100-200 mg Fe) administered to piglets shortly after birth is not sufficient to maintain the iron status of the pig throughout the lactation period.[4,5] Unsurprisingly, faster-growing pigs in a litter are most susceptible to becoming iron deficient at weaning creating potential postweaning problems.[6]

Recently, Chevalier et al[7] reported that pigs administered 200 mg iron at birth had peak hemoglobin (Hb) concentration on d 17 that subsequently declined to weaning at d 22. Furthermore, pigs that only received 100 mg iron at birth had peak Hb at d 11 with a decline in Hb thereafter. These results agree with previous literature which suggests a possibility of pigs having sub-optimal hemoglobin concentrations (generally < 11 g/dL)[8] at weaning. Interestingly, it has been demonstrated that pigs with optimal hemoglobin concentrations at weaning have greater growth performance in the subsequent postweaning periods.[9,10] Furthermore, Olsen[11] demonstrated that an additional iron injection given to piglets 4 to 6 days after the initial iron injection resulted in a much larger percentage (89 % vs. 20 %) of pigs having optimal Hb concentration (≥ 11 g/dL) at weaning and a 454 g increase in body weight (BW) 42 d postweaning. Thus, there is a continued need to reevaluate iron supplementation practices as there may be a potential benefit to administration of an additional iron injection before weaning. The current experiment evaluated a second iron injection and also assessed growth performance and carcass responses past the nursery period through the finishing period.

Methods

Animals and experimental design

A total of 144 crossbred pigs (Yorkshire x Landrace x Large White; 72 barrows and 72 gilts; initial BW = 2.86 ± 0.01 kg) were assigned to either a control (CON) or added-injection treatment (+Fe) at d 6-8 of age. At enrollment (d 6-8), pairs were created by pairing two same-sex siblings with a BW difference < 0.23 kg. Within a pair, one pig was assigned to the CON and the other to +Fe group. All pigs received an initial 200 mg intramuscular (IM) iron injection (Uniferon, Pharmacosmos, Inc.) < 24 h following birth. Pigs assigned to the +Fe group received an additional 200 mg Fe IM injection at d 6-8. All pigs were weaned at d 22-25 to a nursery facility and housed 6 pigs/pen. Once weaned, both CON and +Fe pigs received a common corn-soybean meal diet formulated to meet or exceed the requirement estimates relative to body weight until slaughter. The common diet fed for the duration of the experiment was formulated to supply an added 100 mg/kg iron as ferrous sulfate. Pigs were moved from the nursery to a finisher facility about 5 wk after weaning. All pigs had ad libitum access to feed and water throughout the entire experiment.

Response measures

Body weight (BW) and feed disappearance were recorded every 2 wk to determine average daily gain (ADG), average daily feed intake (ADFI), and gain to feed ratio (G/F). Hemoglobin (Hb) was measured at birth, enrollment, weaning, and the end of the nursery and slaughter periods by using a HemoCue Hb 201+ analyzer (HemoCue America, Brea, California). Pigs (1 pig/pen) were slaughtered at ~ 18 weeks of age under USDA inspection. The hot (HCW) and cold carcass weight (CCW), carcass length, backfat depth at four locations
(1st rib, last rib, 10th rib, and last lumbar), longissimus dorsi muscle area (LMA) were measured. Primal cut weights were also determined.

Statistical analysis

Data analyses were performed in SAS 9.4 (SAS Inst. Inc., Cary, NC, USA) by least-squares analysis of variance using the generalized linear model as a randomized complete block design. The individual pig served as the experimental unit for BW, ADG, carcass measures, and Hb data whereas the pen served as the experimental unit for ADFI and gain/feed ratio data. The statistical model included sex, treatment, sex × treatment, and allotment pair for BW, ADG, Hb, and carcass characteristics. For ADFI and gain/feed ratio, the statistical model only included effects for sex, treatment, and sex × treatment. Statistically significant differences were established at $P \leq 0.05$; tendencies were established at $P \leq 0.10$.

Results

There were expected sex effects for ADG and some carcass measures but no sex × treatment interactions for any response measures. The +Fe pigs had a greater Hb at weaning (13.1 vs. 10.7 g/dL, $P < 0.0001$) and end of the nursery (12.1 vs. 11.7 g/dL, $P = 0.01$) compared to CON pigs (Table 1). During the finisher period, +Fe pigs had a greater ADG ($P = 0.05$) compared to CON pigs (Table 2).

Table 1: Effects of an additional iron injection administered at d 6-8 on hemoglobin (Hb, g/dL) concentration[1]

Time point	Treatment		SEM	*P*-value
	CON	+Fe		
Birth (d 0)	9.9	9.9	0.214	0.93
Enrollment (d 6-8)	9.1	8.9	0.122	0.13
Weaning (d 22-24)	10.7	13.1	0.152	< 0.0001
End of nursery[2]	11.7	12.1	0.129	0.01
End of study	11.9	12.0	0.136	0.66

1 Data represent 144 pigs or 72 pigs/treatment. All pigs received an initial iron injection (200 mg Fe) at birth; the +Fe group received an additional iron injection (200 mg Fe) at enrollment.

2 Nursery period was about 5 weeks postweaning.

Table 2: Effects of an additional iron injection administered at d 6-8 on pig bodyweight and growth performance[1,2]

Items	Treatment		SEM	*P*-value
	CON	+Fe		
Bodyweight, kg				
Birth (d 0)	1.53	1.55	0.01	0.53
Enrollment (d 6-8)	2.87	2.85	0.01	0.44
Weaning (d 22-25)	7.44	7.40	0.06	0.67
Wk 2	10.51	10.85	0.11	0.03
Wk 4	19.08	19.96	0.22	0.01
Wk 5	27.63	28.15	0.27	0.17
Wk 7	40.10	40.77	0.35	0.18
Wk 9	53.69	54.72	0.49	0.13
Wk 11	67.28	68.49	0.60	0.15
Wk 13	81.03	82.70	0.76	0.11
Wk 15	91.77	93.63	0.83	0.11
Wk 17	103.89	106.88	0.97	0.03
End of study	112.79	115.77	1.03	0.04
Average daily gain, kg				
Preweaning	0.29	0.29	0.004	0.73
Nursery	0.55	0.56	0.01	0.10
Finisher	0.91	0.94	0.01	0.05
Wean to end of study	0.81	0.83	0.01	0.04
Average daily feed intake, kg[3]	1.88	1.90	0.04	0.74
Gain/feed ratio[3]	0.43	0.44	0.004	0.33

1 Data represent 144 pigs or 72 pigs/treatment. All pigs received an initial iron injection (200 mg Fe) at birth; the +Fe group received an additional iron injection (200 mg Fe) at enrollment.

2 Preweaning period included enrollment to weaning; nursery period included weaning to about Wk 5; finisher period included Wk 5 to slaughter.

3 Average daily feed intake and gain/feed ratio are for the entire wean to end of study period.

Table 3: Effects of an additional iron injection administered at d 6-8 on carcass traits[1]

Items	Treatment CON	Treatment +Fe	SEM	*P*-value
Slaughter weight, kg	113.13	114.72	1.75	0.53
Hot carcass weight, kg	82.75	84.85	1.49	0.33
Cold carcass weight, kg	80.93	82.99	1.48	0.34
Carcass length, cm	81.73	82.37	0.66	0.50
Backfat depth, cm				
First rib	3.37	3.43	0.22	0.84
Last rib	2.16	2.60	0.23	0.19
10th rib	1.82	1.85	0.15	0.88
Last lumbar	1.57	1.50	0.14	0.76
Loin muscle area, cm^2	44.06	45.32	1.69	0.60
Absolute primal cut, kg				
Boston butt	3.72	3.71	0.08	0.96
Picnic shoulder	3.90	3.98	0.13	0.65
Loin	9.95	10.67	0.23	0.04
Belly	6.34	6.30	0.16	0.87
Spare rib	1.55	1.63	0.04	0.14
Ham	9.04	9.28	0.18	0.35

1 Treatment means are reported as least squares means from 12 pigs per treatment; data is based on the left side of the carcass. All pigs received an initial iron injection (200 mg Fe) at birth; the +Fe group received an additional iron injection (200 mg Fe) at enrollment.

Overall, pigs receiving the added iron injection had an ~ 4% increase in ADG ($P = 0.04$) from weaning to the end of study. The cumulative improvement in ADG from weaning to end of study resulted in +Fe pigs having a heavier BW at the end of the study (115.77 vs. 112.79 kg; $P = 0.04$). Carcass response measures (Table 3) of the selected slaughter pigs demonstrated that +Fe pigs had ~ 2.5% numerically heavier HCW and CCW ($P = 0.34$) and significantly heavier trimmed loin (10.67 vs. 9.95 kg; $P = 0.04$) compared to the CON pigs.

Discussion

The mean hemoglobin value for CON pigs at weaning was 10.7 g/dL that is below the oft-recognized target of 11 g/dL and demonstrated that greater than 50% of the pigs would be considered anemic. The added iron injection resulted in an increase in mean hemoglobin to 13.1 ($P < 0.0001$). The elevated hemoglobin was maintained through the end of the nursery period ($P < 0.01$). During this period of elevated hemoglobin status, the BW was improved at Wk 2 and Wk 4 in the nursery ($P < 0.03$) and, while not significant at the end of the nursery at Wk 5, was an advantage of about 500 grams. The BW difference between the two treatment groups continued to grow with each biweekly weighing, becoming significant ($P < 0.04$) again by the end of the study at which time the difference between treatments was about 3.0 kg. The current nursery growth performance response agrees with previous work[2] which reported that pigs administered a second iron injection one week prior to a weaning at 28 days had an increase in daily BW gain (380 g vs. 362 g) through the following three weeks in the nursery as well as a report[12] that observed an increase in nursery ADG of 28 grams (526 vs. 498 g/d; $P = 0.033$) when a second injection was given 4 days before weaning at d 18-24. The continuation of the pigs used herein through the finishing stage allowed a broader assessment of the impact of that improved nursery growth. The selection of one pig/pen for slaughter and carcass measures resulted in a smaller treatment difference (that would be an artifact of the selection process) being only about 50% of the BW difference of the total group of pigs for which growth performance was measured. In general, the +Fe pigs had numerically greater carcass weight and primal cuts that were reflective of the slaughter weight improvement. However, there was a significant increase in loin weight ($P < 0.04$) that was probably the result of the slight increase in BW coupled with the numerical increase in carcass length. In conclusion, administering an additional iron injection resulted in greater Hb at weaning and the end of the nursery as well as improved growth performance from weaning to slaughter weight and improved carcass measures commensurate with that increased body weight.

References

1. Venn, J., R. McCance, and E. Widdowson. 1947. Iron metabolism in piglet anaemia. *J. Comp. Pathol. Ther.* 57: 314-325. **doi :https://doi.org/10.1016/S0368-1742(47)80037-2**

2. Kamphues, J. K. Männer, and C. Netzer. 1992. Effects of a 2nd iron injection in suckling pigs on iron retention and performance before and after weaning. *Proc. 12th IPVS Congress.* 601.

3. Van Gorp, S., H. Segers, and C. Von der Recke. 2012. Preventing iron deficiency by avoiding an iron gap in modern pig production. *Proc. Am. Assoc. Swine.Vet.* 43: 407-408.

4. Bhattarai, S., J. P. Nielsen. 2015. Early indicators of iron deficiency in large piglets at weaning. *J. Swine Health Prod.* 23: 10–17.

5. Perri, A. M., R. M. Friendship, J. C. Harding, and T. L. O'Sullivan. 2016. An investigation of iron deficiency and anemia in piglets and the effect of iron status at weaning on post-weaning performance. *J. Swine Health Prod.* 24:10-20.

6. Jolliff, J., and D. Mahan. 2011. Effect of injected and dietary iron in young pigs on blood hematology and postnatal pig growth performance. *J. Anim. Sci.* 89:4068-4080.

7. Chevalier, T. B., H. J. Monegue, M. D. Lindemann. 2021. Effects of iron injection dosage to newborn piglets on hematological measures, growth performance pre and postweaning, and postweaning tissue mineral content. *J. Swine Health Prod.* 29: 189-199.

8. Thorn, C. E., 2010. Hematology of the pig, In: Weiss, D. J., and J. K. Wardrop, 6th ed. Shalm's Veterinary Hematology, Wiley-Blackwell., Ames, IA, p. 843-851.

9. Bhattarai, S., and J. P. Nielsen. 2015b. Association between hematological status at weaning and weight gain post-weaning in piglets. *Livestock Sci.* 182:64-68. **doi: https://doi.org/10.1016/j.livsci.2015.10.017**

10. Fredericks, L., C. Olsen, A. Maschhoff, and C. Shull. 2018. Evaluation of the impact of iron dosage on post-weaning weight gain, and mortality. *Proc. Am. Assoc. Swine. Vet.* 49: 315.

11. Olsen, C. 2019. The economics of iron deficiency anemia on US swine production: An annual impact of 46-335 million US dollars. *Proc. Am. Assoc. Swine. Vet.* 50: 351-352.

12. Chevalier, T. B., and M. D. Lindemann. 2019. Effects of an additional iron injection administered to piglets before weaning. *J. Anim. Sci.* 97 (Suppl. 3):227. **doi.org/10.1093/jas/skz258.462**

Poster 01

https://doi.org/10.54846/am2023/86

Comparison of high-performing and low-performing suckling pig fecal microbiome

E. Snezek[1], MPH; S. Krantz[2], DVM; M. Costa[3], DVM, PhD
[1]College of Veterinary Medicine, University of Georgia, Athens, Georgia; [2]Tosh Pork, Henry, Tennessee;
[3]Large Animal Clinical Sciences, University of Saskatchewan

Introduction

The gastrointestinal tract (GIT) microbiota has recently become of interest in animal health. Composition of the GIT microbiota has been linked to changes in digestibility, feed intake, growth performance, immune system development, and prevention of some diseases in swine.[1-3] Changes in the composition of the gut microbiome in piglets has been studied in relation to the sow gut microbiome and the environment while nursing.[4] Despite published correlations of specific microbes to performance, there is currently a knowledge gap comparing the microbiome of suckling piglets to individual pre-weaning performance. Though some studies show that sow microbiome modulation influences entire litter performance,[1] little has been described at the level of the individual suckling pig. Our hypothesis is that there are differences in the GIT microbiome composition between high-performing and low-performing suckling pigs. The primary objective of this cohort study was to compare the fecal microbiome between high-performing suckling piglets to low-performing suckling piglets.

Materials and methods

This study was conducted in a high-health, 6,000 head, commercial multiplication sow herd. A convenience cohort of nine individually ear-tagged litters were observed and sampled throughout the suckling period for retrospective enrollment and analysis. Litters were farrowed the same day and were distributed across first (n = 2), second (n = 3), and third or greater (n = 4) parity sows which were considered healthy and devoid of antibiotic therapy at farrowing. Sows remained untreated, and cross-fostering of piglets was not allowed for the duration of lactation. One-hundred and thirty-six piglets were initially enrolled, received identical processing procedures, and were weighed on days 1 and 21. Sterile, polyester tipped swabs were used to collect individual piglet fecal samples on days 1, 3, 7, and 21. All samples were transported in a cooler and promptly stored at -80°C until enrollment and testing. To reduce the influence of day 1 weight variation, individual suckling pig performance was analyzed by the percent average daily gain compared to day 1 weight (%ADG), using the calculation (weight gain/birth weight)/wean weight. Fifty-seven of the 136 piglets sampled were enrolled as follows: the highest and lowest 25% of all piglets based on %ADG, and the 2 highest and lowest %ADG pigs in each litter. Each litter was represented by at least four piglets, with 228 total piglet samples processed. Fecal microbiome was profiled using 16s rRNA sequencing. Samples were processed using QIIME2 and analyzed using R Bioconductor package.

Results and discussion

Two-hundred and twenty-eight piglet fecal samples were submitted as well as 27 total sow fecal, colostrum, and vaginal samples. Of the samples sent, 10 were from first parity gilts, 18 were from second parity sows, and 29 were from third parity or higher sows. Analysis of 16s rRNA sequencing revealed a significant difference in alpha diversity at the phylum level between high-performing and low-performing pigs at day 21 ($P = 0.02$), characterized by increased phyla diversity in low-preforming pigs (Figure 1A). There was no significant difference in alpha diversity between the performance groups for days 1, 3, or 7. The difference in alpha diversity seen at day 21 suggests that development of a more diverse microbiome alone through the nursing period is not necessarily beneficial to the pig. Beta diversity was not significantly affected by performance status but was impacted by age (Figure 1B). Beta diversity increased from day 1 to 21, indicating that microbiota diversity within each performance group increased with age. This is consistent with previous studies examining the changes in piglet microbiome from birth to weaning. Based on the results, increased alpha diversity is correlated with low-performance and increased beta diversity is correlated with age in suckling pigs.

References

1. Monteiro, M., Poor, A., Muro, B., et.al. The sow microbiome: Current and future perspectives to maximize the productivity in swine herds. *J Swine Health Prod*. 2022;*30*(4), 238–250.

2. Uryu, H., Tsukahara, T., Ishikawa, H., et. al. Comparison of productivity and fecal microbiotas of sows in commercial farms. *Microorganisms*. 2020;*8*(10), 1469.

3. Wang, X., Tsai, T., Deng, F., et. al. Longitudinal investigation of the swine gut microbiome from birth to market reveals stage and growth performance associated bacteria. *Microbiome*. 2019;*7*(1).

4. Liu, H., Zeng, X., Zhang, G., et. al. Maternal milk and fecal microbes guide the spatiotemporal development of mucosa-associated microbiota and barrier function in the porcine neonatal gut. *BMC Biology*. 2019;*17*(1).

Figure 1: (A) Comparison of day 21 phyla alpha diversity in low-performing (LP) and high-performing (HP) piglets. **(B)** Differences in beta diversity based on age of pig when sample was taken.

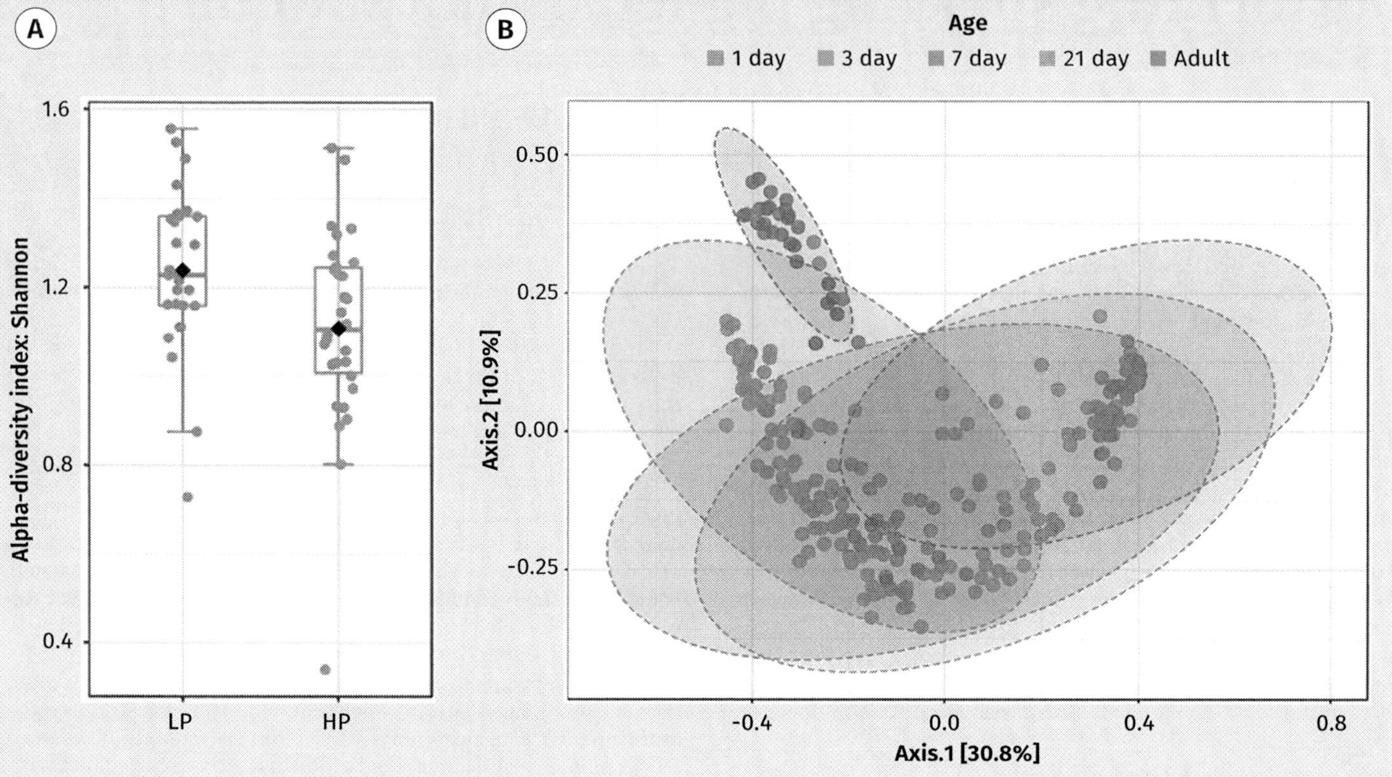

https://doi.org/10.54846/am2023/87

An investigation of the number of pigs in gestation pens that will chew on a rope during oral fluid collection

B. Werner[1], BS; M. Hood[2], DVM; D. Reicks[2], DVM; J. Morgan[3], DVM

[1]University of Minnesota, College of Veterinary Medicine, St. Paul, Minnesota; [2]Reicks Veterinary Research & Consulting, St. Peter, Minnesota; [3]Boehringer Ingelheim Animal Health USA Inc., Duluth, Georgia

Introduction and objective

As gestation pens become more common in the swine industry, oral fluids will remain a convenient sampling strategy. Understanding the factors affecting the percentage of sows that chew on the rope would increase confidence in low disease prevalence situations. A previous study found that 80% of growing pigs in a pen interact with a rope in 60 minutes.[1] A similar study was conducted for sows in pens with electronic sow feeders. The study found that with 4 ropes hung, approximately 20 sows in a group would contact each rope in a pen of 175 sows.[2] Similar studies have not been conducted in sow farms containing small group gestation pens, housing 12 sows per pen. The objective of this study was to determine how many sows or gilts in a gestation pen of up to 12 animals will chew on a rope within 45 minutes and to analyze various factors affecting these results.

Materials and methods

Three sow farms were enrolled in the study. Farms were selected based on the presence of small group gestation pens. One-hundred gestation pens (groups) totaling 1181 animals were observed in the study. Cull and sick pens were excluded. Each group had either a dried, previously used 1.59 cm cotton diameter rope or new 1.59 cm cotton diameter rope hung at eye level (approximately 15.24 cm off the floor) for 45 minutes. The time the rope was hung was recorded. During the 45 minutes, each time a new sow or gilt chewed on the rope, the animal's ID number and the time were recorded. If the rope was observed to be in danger of being eaten or falling, the time of observation was noted, and the rope was removed. The parity and ID number for each sow was recorded, as well as if they had been treated. Each group had the rope hung on two consecutive days, once in the morning (7 am-10 am) and once in the afternoon (12 pm - 3 pm).

Results

The total percentage of animals that chewed on the rope was 52.9%. 53.9% of the animals chewed on the rope in the morning, and 51.9% of the animals chewed in the afternoon. On day 1 of exposure, 51.1% of animals chewed versus 54.7% on day 2. New ropes had a higher percentage of animals that chewed at 57.2% than reused ropes at 49.2%. Farm 1 had an average parity of 4, and 47.1% of animals chewed. Farm 2 had an average parity of 1, and 65.3% of animals chewed. Farm 3 had an average parity of 0, and 50.9% of animals chewed. A difference in the percentage of animals that chewed was noted between parities with younger animals chewing more often compared to older animals. See figure 1 for parity data. 35.9% of animals chewed within 15 minutes versus 47.0% within 30 minutes and 52.9% within 45 minutes. Thirteen out of two hundred of the ropes had to be removed early due to being in danger of being eaten or falling in the pit.

Figure 1: The percentage of sows that chewed the ropes by parity.

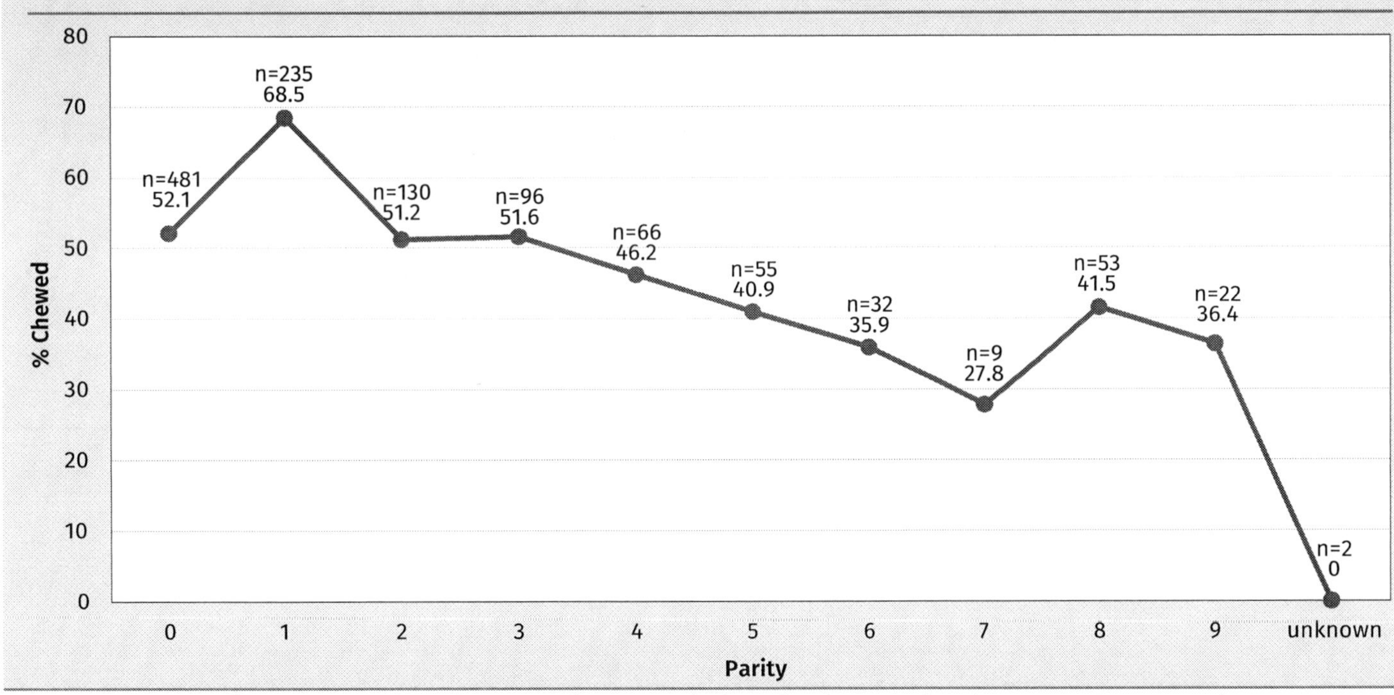

Discussion

In this study, approximately half of sows or gilts will chew on an oral fluid rope in a small gestation pen. Time of day and training the day prior resulted in no significant difference in the percent chewed. Hanging the rope for more than 30 minutes would not be recommended due to the increased risk of the rope being eaten or falling in the pit. In order to maximize this sample size, it would be recommended to use a new rope in pens with younger animals and hang the rope for approximately 30 minutes.

References

1. Seddon, Y. M., J. H. Guy, and S. A. Edwards. "Optimising oral fluid collection from groups of pigs: effect of housing system and provision of ropes." *The Veterinary Journal* 193.1 (2012): 180-184.

2. Pierdon, Meghann K., Amy L. Martell, and Thomas D. Parsons. "Use of ropes to collect oral fluids from gestating sows housed in dynamic groups and fed via electronic sow feeder." *Journal of Swine Health and Production* 24.5 (2016): 246-252.

https://doi.org/10.54846/am2023/88

Evaluation of ozone viricidal effects on porcine reproductive and respiratory syndrome virus within semi-truck cabs

A. Bishop[1], BS; D. Boykin[2], DVM; D. Holtkamp[1], DVM, MS; J. Zhang[1], MD, MS, PhD; C. Puls [3], PhD; R. Evelsizer [3], DVM; A. Cosgrove [3], PhD

[1]Iowa State University, Ames, Iowa; [2]Cacus Family Farms, Osceola, Iowa; [3]Elanco Animal Health, Glenwood, Indiana

Introduction

While there are many biosecurity protocols in place on hog sites and for trailer clean-out, semi-truck and tractor cab biosecurity protocols are limited. The objective of this research was to determine if ozone proved beneficial for decontaminating semi-truck cabs to reduce the spread of PRRSV.

Materials and methods

A negative control, positive control, and treatment group were evaluated for five materials. Materials include: metal (aluminum), cloth, leather, vinyl, and rubber. Five coupons (replicates) were used for each treatment and control groups. For the positive control and treatment groups, each coupon was contaminated with 4 mL of modified live PRRSV vaccine (Prevacent, Elanco Animal Health).[1] Each coupon in the negative control groups were contaminated with 4 mL of MEM. All groups were subjected to a 1-hour dry time, followed by varying contact times. Contact times were 15-minutes, 30-minutes, 1-hour, 1.5-hour, and 2-hours. A truck cab was simulated using plastic and a metal chair. Coupons were placed in the simulated cab based on their typical location in a semi-truck cab: metal under the seat, rubber on the floor, cloth and leather on the seat, and vinyl elevated in front of the seat. Both negative and positive controls were placed in a similar fashion. After the contact time, a 30-minute down-time was implemented to mimic the time allowed for ozone to leave the cab before entry then samples were collected.[3] The collection protocol included elution of the coupon with 2 mL of MEM ten times. This was completed over a petri-dish to collect media run-off from the coupon. The coupon surface was then swabbed with a sterile polyester tipped swabs, and the tip was placed in a 5 mL snap cap tube along with the elution media that remained in the petri-dish. Samples were immediately placed on dry-ice and submitted to a diagnostic lab for virus isolation and titration.[2]

Results

Due to inconsistent positive control results for the leather and cloth coupons, this data has been removed from the study. Table 1 compares the titration levels of the 2-hour positive controls to the ozone treated coupons. If a sample was positive for virus isolation, but negative for virus titration this is indicated by a red dash (-). Other results demonstrated that as time of exposure to ozone decreased, there was a decrease in effective virus elimination. Within the 15-minute exposure time it was noted that 5/5 metal coupons, 4/5 vinyl coupons, and 3/5 rubber coupons were positive on virus isolation.

Table 1: Comparison of the titration levels from the 2-hour contact time of both the positive controls to the ozone treated coupons

Coupon Type	Positive control Virus titration TCID50/mL	Ozone Virus titration TCID50/mL
Metal 1	3.16 × 10^3	5.62 × 10^3
Metal 2	3.16 × 10^3	Negative
Metal 3	5.62 × 10^3	3.16 × 10^2
Metal 4	-	Negative
Metal 5	3.16 × 10^2	Negative
Vinyl 1	3.16 × 10^5	-
Vinyl 2	5.62 × 10^3	Negative
Vinyl 3	3.16 × 10^5	3.16 × 10^3
Vinyl 4	1.78 × 10^5	Negative
Vinyl 5	3.16 × 10^3	-
Rubber 1	3.16 × 10^3	5.62 × 10^4
Rubber 2	Negative	3.16 × 10^3
Rubber 3	1.78 × 10^2	5.62 × 10^3
Rubber 4	5.62 × 10^3	Negative
Rubber 5	Negative	Negative

Discussion

While results from this study did not demonstrate complete viricidal effects against PRRSV using ozone, there was some reduction in the virus titers relative to the positive control on metal and vinyl at the 2-hour contact time. Regarding ozone efficacy against PRRSV studies may be required to better understand parameters of contact time and dose.

References

1. C. Grace Elijah, D., Cassandra K. Jones, P., et al., (2022, March 1). *Quantification of decontamination strategies for semi-truck cabs.* 53rd Annual Meeting of the American Association of Swine Veterinarians.

2. Scott Dee, J. D. (2004, July). *An assessment of sanitation protocols for commercial transport vehicles contaminated with porcine reproductive and respiratory syndrome virus.* Retrieved from National Library of Medicine: **https://www.ncbi.nlm.nih.gov/pmc/articles/PMC1142141/**

3. Understanding ozone disinfection technology with STERISAFE. (2020, August 28). Retrieved from Health Europa: **https://www.healtheuropa.com/understanding-ozone-disinfection-technology-with-sterisafe/102447/#:~:text=The%20STERISAFE%20PRO%20uses%20an,nitrogen%20from%20the%20ozone%20production.**

https://doi.org/10.54846/am2023/89

Serial dilution of serum samples to assess the relative antibody levels generated by ELISA serotests for common swine pathogens

E. Larsen[1], MS; E. Russell[1], MS; B. O'Brien[2], MS, VMD; B. Thacker[1], DVM, PhD, MBA
[1]College of Veterinary Medicine, Lincoln Memorial University, Harrogate, Tennessee;
[2]Merck Animal Health, Lenexa, Kansas

Introduction

Enzyme-linked immunosorbent assays (ELISAs) are commonly used to measure serum antibody levels in swine. A limitation of ELISAs is that the level of antibody in a sample is a relative value that does not yield a "titer." Traditional serological tests such as hemagglutination inhibition (HI) or indirect immunofluorescent antibody assay (IFA) yield a titer based on testing serial dilutions of the sample and the reported titer is the highest dilution that shows a positive result. With ELISAs, the sample is tested at a single dilution and the optical density value that is generated by the assay is compared against a standard control sample.

The advantage of tests that generate a titer is that the results can be used to determine the age when maternally derived antibodies (MDA) wane, which may impact duration of passive protection or potential interference with pig vaccination. The half-life of MDA in pigs is approximately 2 weeks so a two-fold decline in titer occurs every 2 weeks of age.

The objective of this study was to 1) determine the antibody "titers" to porcine circovirus type 2 (PCV2), swine influenza virus (IAV-S), *Mycoplasma hyopneumoniae* (Mhyo) and *Lawsonia intracellularis* (Li) by serial dilution of positive serum samples, testing each dilution by ELISA and assigning an endpoint titer based on the highest positive dilution and 2) estimate the weeks to seronegative based on the starting ELISA antibody level.

Methods

Archived serum from experimental vaccination/challenge studies were used. Samples had antibody levels that were considered to be moderate to high. For IAV-S, Mhyo and Li, 11 sera were used; 22 sera for PCV2. The test sera were diluted two-fold with pre-vaccination, seronegative serum with the resulting dilutions ranging from 1:2 to 1:128. All testing was performed at the Merck Animal Health Diagnostic Laboratory (Lawrence, Kansas) using commercial ELISA kits. The results shown include the ELISA endpoint dilution titer (ELISA titer) and the average ELISA antibody levels of the original sample and the subsequent dilutions. For PCV2 and Mhyo, both indirect ELISAs, results are shown as the S/P ratios. For IAV-S and Li, both competitive ELISAs, the results are shown as S/N and percent inhibition, respectively.

Results

The results are shown in Table 1. In addition to the ELISA antibody levels by ELISA titer, the estimated weeks until seronegative based on a 2-week half-life is presented. As expected, the antibody level measured by each ELISA declined with each serial dilution. We report only the average values; the range of values overlapped between adjacent dilutions but not 4-fold dilutions (data not shown).

Discussion

The main utility of this data is for assessing Mhyo MDA decay to avoid vaccination interference. Vaccination interference and/or duration of passive immunity are potentials concerns for PCV2 and IAV-S but IFA and HI, respectively, are typically used to assess MDA levels. For Li, less is known about the importance of MDA and the percent inhibition differences between dilution titers of 1:16 to 1:64 were small and overlapping, making it difficult to interpret the data and therefore estimate the weeks to seronegative.

Table 1: Average ELISA antibody levels by ELISA titer based on two-fold dilutions and estimated weeks to negative

		Average antibody levels by ELISA titer			
ELISA titer	**Weeks to negative**	**PCV2 S/P**	**IAV-S S/N**	**Mhyo S/P**	**Li % Inh.**
0	0	0.241	0.852	0.170	10.0
2	2	0.405	0.467	0.494	38.0
4	4	0.515	0.326	0.710	56.7
8	6	0.674	0.189	0.997	68.6
16	8	0.813	0.115	1.222	77.7
32	10	1.064	0.090	1.461	82.7
64	12	1.280	0.081	1.734	79.4
128	14	1.628	**No samples exceeded an ELISA titer of 1:64**		

Poster 05

https://doi.org/10.54846/am2023/90

Investigation of factors associated with the receipt of an etiologic diagnosis for suckling piglets at a veterinary diagnostic laboratory

Morgan Johnson[1]; Giovani Trevisan[1], DVM, PhD; Gustavo De-Sousa-E-Silva[1], DVM, PhD; Eric Burrough[1], DVM, PhD; Daniel Linhares[1], DVM, PhD; Rebecca Robbins[2], DVM, PhD; Edgar Diaz-Estrada[3], DVM, PhD
[1]Department of Production Animal Medicine, Iowa State University, Ames, Iowa;
[2]R. C. Robbins Swine Consulting Services PLLC, Amarillo, Texas; [3]Ceva Animal Health, Lenexa, Kansas

Introduction

Among 416 enteric tissue diagnostic submissions from suckling pigs received between January 2020 and June 2022 at Iowa State Veterinary Diagnostic Laboratory (ISU-VDL), 22.67% did not receive a confirmed etiological diagnosis (VDL data). At the ISU-VDL, diagnosticians assign diagnostic codes (dx code) to each tissue case to track the physiologic system, insult type, lesion, and etiology. With many cases not receiving an etiological diagnosis, it begs the question, how can submission quality be improved? The objective of this study was to assess factors in the diagnostic submission process associated with cases assigned an etiologic diagnosis.

Materials and methods

This was a prospective observational study conducted at ISU-VDL. A survey was used to collect daily information from direct observation of suckling piglet enteric tissue cases submitted with histology, requested from 06/07/22 to 07/29/22 (n = 100). The primary outcome variable was the etiologic diagnosis specified (yes or no). Exploration of physical characteristics included the presence of fluid leakage, visually apparent autolysis, putrid smell, and the inclusion of fresh or formalin-fixed intestinal sections. Intestinal tissue temperature was taken with a non-contact infrared thermometer (Equate™). Temperature values were categorized into quartiles. Discretionary submission information was collected, such as the presence of necessary samples, the number of animals sampled, request for diagnostician discretion, and the inclusion of a clinical description. Associations between each of the aforementioned variables and receiving etiology were assessed with Chi-square, and the risk ratio was reported as a measure of association with 95% confidence intervals (OpenEpi v.3.01). Statistical significance was set at P-value < 0.05 and statistical trend for significance 0.05 - 0.25. A risk ratio greater than one and a significant P-value indicate a high likelihood of a specified etiologic diagnosis.

Results

Of the 100 cases, 30.98% had a "not specified" etiology code. Intestinal quartile temperature results were 47.6°F, 52.0°F, 61.7°F, and 72.1°F, respectively. Table 1 describes the risk ratio of various submission factors. Including both fresh and fixed small intestine in the submission was a significant factor.

Discussion

This is the first report on the investigation of factors impacting the establishment of an etiologic diagnosis. Factors tending to be associated with a greater probability of obtaining an etiologic diagnosis were the presence of fixed and fresh tissue, specifically large and small intestine, receiving ≥ 2 animals, absence of leaking fluid (e.g., blood, formalin), and allowing discretion of diagnostician for testing. High temperatures at receiving were associated with the presence of visible autolysis; however, temperature and etiologic diagnosis were not related perhaps due to no extreme temperature cases observed in this study. These associations indicate that to an extent, pathologists can identify lesions despite some degree of autolysis. Understanding associations for receiving etiologic diagnoses is relevant for veterinarians and producers as it assists in time and resource allocation when submitting to a VDL.

Table 1: The risk ratio of various submission factors

Characteristic by dx code	Risk ratio	*P*-Value	% of Cases
Discretion of diagnostician	1.147	0.1529	48%
No leakage	1.435	0.1179	93%
Fixed and fresh small intestine	**1.465**	**0.049**	**86%**
Fixed and fresh colon	1.103	0.2406	62%
Animals submitted ≥ 2	1.166	0.1884	79%

Poster 06

https://doi.org/10.54846/am2023/91

Tissue submission quality of nursery age enteric cases and effect on receiving a diagnosis

L. Tidgren[1], BS; R. Robbins[2], DVM, PhD; E. Diaz[2], DVM, MS; G. Silva[1], DVM, MS, PhD;
D. Linhares[1], DVM, MS, PhD; G. Trevisan[1], DVM, MBA, PhD
[1]Iowa State University, College of Veterinary Medicine, Ames, Iowa; [2]Ceva Animal Health LLC, Lenexa, Kansas

Introduction

From an estimated 80,000 porcine cases submitted to the diagnostic laboratory (VDL) yearly, around 20% do not receive a final diagnostic (Dx) code. A portion of this percentage is attributed to either autolysis of tissues or the absence of key samples. Undiagnosed cases provide no answers or application points for swine producers and veterinarians but proceed to accrue hefty costs. It is hypothesized that proper sample collection and shipping techniques greatly diminish the number of nonsensical results for a higher percentage of case diagnoses. This would allow for faster access to data for an effective swine health intervention plan. The objective of this study was to pinpoint if various necropsy and shipment characteristics were associated with receiving a definitive Dx code for nursery aged swine enteric tissue cases.

Methods and materials

A convenience sample was taken from a VDL receiving laboratory. A recorder followed cases as they were unpackaged and all swine cases received between 6/6/22 to 7/14/22 containing intestinal tissue samples or submission forms noting enteric issues occurring in pigs between 35 and 70 days of age were enrolled (N = 104). Data was documented on intestinal cuts present and if they were fresh or fixed, age of the animal, if the sample was cool (≤ 15°C) or warm (> 15°C) on arrival in the laboratory, potential leaking, shipment type, time between collection and receiving, tissue odor, observable quality, and if the diagnostician was given discretion for testing. The tissue odor scale was ordinal ranging from 1 indicating absence of a putrid smell, 2 moderate putrid smell, and 3 for severe putrid smell. As cases were finalized by diagnosticians, additional information including presence of autolysis noted by the pathologist, tests performed, and if a Dx code was assigned. Factors potentially affecting autolysis were charted in contingency tables and analyzed using the chi-square test. Associations were described as significant ($P < 0.05$) or a trend (0.05 - 0.20). Effect was reported as a risk ratio and 95% confidence interval.

Results and discussion

The presence of autolysis significantly lowered the risk of a case receiving a Dx code by 62%; therefore, factors believed to influence autolysis became the focus. A smell of 1 and a cool temperature on receipt (≤ 15°C) were found to have statistically significant *P*-values. A smell of 1 and a cool temperature were found to be protective, associated with a lower risk of having autolysis noted on the case reports. A smell of 1 was associated with a 70% lower risk of finding autolysis on the case reports. There was a trend for association when looking for the temperature. Samples received with a cool temperature had a 53% lower risk of being received with autolysis. Thus, to increase chances of obtaining a Dx code when collecting samples on a farm, autolysis should be minimized. This can be done by observing odor quality at the time of collection, as well as keeping samples ≤ 15°C through the use of ice packs and insulated transport containers. Results are reported in Table 1.

Table 1: Tissue sampling factors associated with autolysis and the risk of autolysis receiving a diagnostic code

Variable by dx code rendered	Risk ratio	Confidence interval	*P*-value
Autolysis noted on case report	0.38	0.19, 0.75	0.0001
Variable by autolysis	**Risk ratio**	**Confidence interval**	***P*-value**
Smell of 1	0.29	0.11, 0.73	0.004
Cool temperature	0.46	0.21, 1.00	0.07

Poster 07

https://doi.org/10.54846/am2023/92

A field study examining the effects of a novel maternal pheromone on performance and livability of weaned pigs

B. Steidley[1]; R.C. Robbins[2]; K. Wilson[3]; J. McGlone[3]
[1]Oklahoma State University, Stillwater, Oklahoma; [2]R.C. Robbins Swine Consulting Services, Amarillo, Texas; [3]Texas Tech University, Lubbock, Texas

Introduction

Pigs experience unavoidable stress in commercial systems at weaning. A novel maternal pheromone (MP) was recently reported that may benefit weaned pigs.[1] The MP is the odor of the mother that is typically absent in the post-weaning environment. When the MP was applied in a controlled post-weaning environment, pigs grew faster and fought less than control pigs.[1] The objective of this study was to determine the effects of MP on performance and livability among weaned pigs in a commercial nursery with conventional production methods, wean ages and health challenges.

Materials and methods

Four sow sources filled 8 rooms, with a total of 9,556 pigs, over a 2-week period and they remained for about 47 d; rooms were blocked by sow source and randomly assigned to treatment (MP) or control treatments. Sow sources had similar yet variable health status (PED 2fvi, PRRS 2fvi, mycoplasma and influenza positive). Weaned pigs averaged 21.9 days of age (range 18 to 25 d). Fence-line feeders were stainless steel, 1.2 m in length, and a trough divided with 6 feeder holes per pen. Three evenly spaced plastic strips infused with MP (0.0005% myristic acid and skatole) were placed 14 to 16 inches above the feed trough for 48 hours. Control rooms were managed as usual for this farm (CON). An investigator was on site collecting gross pathology data for 19 days post-weaning. Descriptive statistics were used to compare key performance indicators including ADG, end weight or adjusted 70 day weight, ADFI, F:G and livability between treatments.

Results

Two nursery rooms were identified as health challenged (N = 2 rooms; 2,206 pigs) and summarized separately from the conventional groups (N = 6 rooms; 7,350 pigs). Key performance indicators were approximately equal between MP and CON treatments among relatively healthy pigs (Table 1). Within the health challenged (HC) population, numerical differences between treatment groups were observed. The MP-HC group had a 92.06% livability while CON-HC group had 87.86% (+ 4.2% difference). The MP-HC group had improved ADG (+ 10%), higher final weight adjusted to a common age (+ 4.9 lb./pig), lower feed intake (- 25%), and improved feed efficiency (+ 12%) compared with CON-HC pigs. Gross pathologic lesions were assigned to 3 categories: respiratory/systemic (68%), gastrointestinal (18%), and no gross lesions/unknown (14%). In addition, stomach fill amongst necropsied pigs was classified as full, partial, or empty. The MP group had 63% full stomachs, 18% partially filled, and 18% empty. The CON group had 54% full stomachs, 27% partial, and 19% empty.

Discussion

Due to barns being confounded by health status, our descriptive summary was stratified by health status. Among conventional health rooms, data were variable, and we observed no differences in key performance or health measures between treatments. The use of MP improved key performance indicators among health-challenged weaned pigs. Nursery pig performance and health may benefit from the use of MP, especially among pigs or barns with Livability below about 97%. Research is needed to better understand how to efficiently apply the MP, the economic value of use of a MP, and the mechanism of how MP may change pig behavior, performance, and health in different phases of production.

Reference

1. Archer C, Garcia A, Henderson M and McGlone JJ (2022) Olfactory enrichment using a maternal pheromone improved post-weaning pig performance and behavior. Front. Vet. Sci. 9:965370. **https://doi.org/10.3389/fvets.2022.965370**

Table 1: Descriptive statistics of key performance indicators

Room ID	A	B	C	D	E	F	G	H
Treatment	Control	MP	MP	Control	MP	Control	Control	MP
Health status	Challenged	Conventional	Conventional	Conventional	Challenged	Conventional	Conventional	Conventional
Number started	1211	1233	1260	1254	995	1293	1077	1233
Number ending	1064	1182	1212	1227	916	1277	1045	1224
All-cause mortality	12.14%	4.14%	3.81%	2.15%	7.94%	1.24%	2.97%	0.73%
Wein weight (lbs.)	11.21	14.85	13.79	12.27	12.22	13.14	14.97	13.16
End weight (lbs.)	38.52	50.345	52.25	49.56	39.31	47.68	51.78	41.49
Avg. days in phase	50	49	49	49	45	45	44	43
ADG (lbs.)	0.55	0.72	0.78	0.76	0.60	0.77	0.84	0.66
Adj. Wt. @70 days (lbs.)	49.44	65.56	68.73	65.53	54.36	66.87	73.52	59.29
Total feed intake (lbs.)	47,590	67,130	67,830	68,030	35,710	61,830	56,030	49,550
F:G	1.64	1.6	1.46	1.49	1.44	1.40	1.46	1.43

MP = maternal pheromone

Poster 08

https://doi.org/10.54846/am2023/93

Pharmacokinetics of extravascular maropitant citrate in swine

J. Gebert[3]; J. Smith[1]; L. Ebner[3]; K. Bennett[1]; R. Collins[1]; C. Hampton[1]; S. Kleine[4]; P-Y. Mulon[1]; C. Smith[4]; R. Seddighi[1]; G. Bussieres[4]; J. Mochel[2]; H. Knych[5]

[1]Large Animal Clinical Sciences, College of Veterinary Medicine, University of Tennessee, Knoxville, Tennessee; [2]Biomedical Sciences, College of Veterinary Medicine, Iowa State University, Ames, Iowa; [3]College of Veterinary Medicine, Lincoln Memorial University, Harrogate, Tennesse; [4]Small Animal Clinical Sciences, College of Veterinary Medicine, University of Tennessee, Knoxville, Tennessee; [5]K. L. Maddy Equine Analytical Chemistry Laboratory, School of Veterinary Medicine, University of California-Davis, Davis, California

Introduction

Swine are at risk of emesis from clinical conditions and in the peri-operative period. As such, it would be advantageous to have an anti-emetic for these scenarios. Maropitant citrate, a neurokinin-1 receptor antagonist, has been used in small animals to decrease emesis, motion sickness, and increase return of normal appetite post-operatively.[1] Currently no prospective studies describe the use of maropitant in pigs. The objective of this study was to estimate the pharmacokinetics of a single intramuscular (IM) injection of maropitant at a dose of 1.0 mg/kg. A secondary objective was to estimate the pilot pharmacokinetics of a single oral (PO) dose of 2.0 mg/kg of maropitant.

Methods

For the initial objective, maropitant was administered IM to six commercial pigs (Landrace/Yorkshire cross; 45.9 ± 2.5 kg, 98-99 days of age) at a dose of 1.0 mg/kg. Plasma samples were collected over a 72-hour period. For the secondary objective, two commercial pigs from the first study were selected to receive 2.0 mg/kg of maropitant PO after a washout period and plasma samples were also collected over a 72-hour period. Plasma concentrations were measured via liquid chromatography/mass spectrometry (LC-MS/MS) using a non-compartmental analysis to derive the pharmacologic parameters.

Results

After IM administration, the maximum concentration (C_{max}) observed was 412 ± 132.0 ng/mL. The time to maximum concentration (T_{max}) ranged from 0.083 to 1.0 hours. Elimination half-life was 6.6 ± 0.82 hours and mean residence time was 6.1 ± 1.7 hours. Area under the curve (AUC_{inf}) was 1336 ± 132.0 hr*ng/mL. The secondary objective observed the relative bioavailability after PO administration to be 15.5 and 27.2% in the two pigs compared to the IM route. Figure 1 depicts time vs concentration of maropitant of IM administration and the two PO administration pigs.

Discussion

The C_{max} observed after IM administration was higher than observed in studies evaluating subcutaneous (SC) administration in dogs and rabbits.[2,3] Further investigation is needed to evaluate if this maximum concentration is efficacious to provide anti-emetic effects to swine. The time to maximum concentration was also faster than observed in SC injections of dogs and rabbits.[1,2] The area under the curve after IM dosing in pigs was greater than observed in dogs and rabbits after a SC injection.[2,3] Volume of distribution was higher (15.9 L/kg) than rabbits and horses.[2,4] No adverse side effects were observed during IM or PO administration. Further research is needed to identify if these differences are due to route differences or species bioavailability. Additional research on withdrawal recommendations is necessary.

References

1. Hay Kraus BL. Spotlight on the perioperative use of maropitant citrate. Vet Med (Auckl). 2017;8:41-51 **https://doi.org/10,2147/VMRR.S126469**

2. Ozawa SM, Hawkings MG, Drazenovich TL, Kass PH, Knych HK. Pharmacokinetics of maropitant citrate in New Zealand White Rabbits (Oryctolagus cuniculus). American Journal of Veterinary Research. 2019;80(10):963-8.

3. Benchaoui HA, Cox SR, Schneider RP, Boucher JF, Clemence RG. The pharmacokinetics of maropitant, a novel neurokinin type-1 receptor antagonist, in dogs. Journal of Veterinary Pharmacology and Therapeutics. 2007;30(4):336-44.

4. Berryhill EH, Knych H, Edman JM, Magdesian KG. Pharmacokinetics of single doses of maropitant citrate in adult horses. Journal of Veterinary Pharmacology and Therapeutics. 2019;42(4):487-91.

Figure 1: Time versus concentration data for maropitant in pigs after intramuscular (IM) administration, n = 6; black circles) as well as 2 pilot pigs after oral administration (orange square and grey triangle).

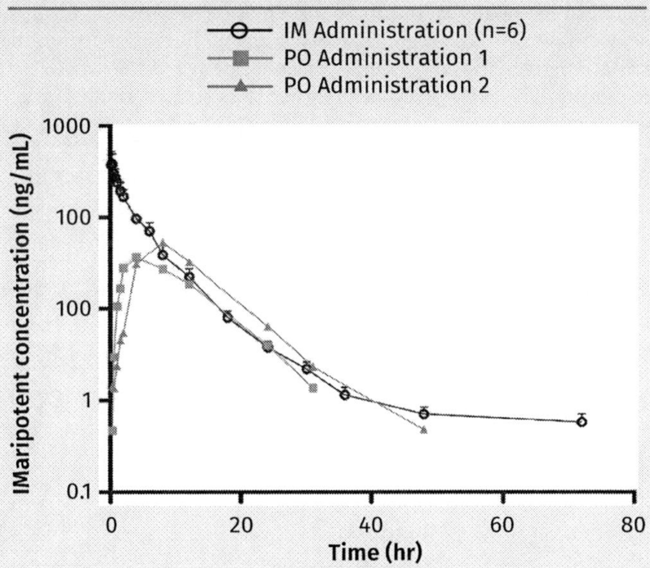

https://doi.org/10.54846/am2023/94

Evaluating the economic impact of an autogenous influenza vaccine in the finishing phase with a pre-post analysis utilizing closeout data

Kaci Way[1], BS; Jessica Higgins[2], DVM; Todd Wolff[3], DVM; Derald Holtkamp[4], DVM, MS; Lynette Holman[2], DVM
[1]The Ohio State University, Columbus, Ohio; [2]Kalmbach Swine Management, Upper Sandusky, Ohio;
[3]Phibro Animal Health, Teaneck, New Jersey; [4]Iowa State University, Ames, Iowa

Introduction

Influenza A virus (IAV) is a growing concern among swine practitioners. This is due to swine influenza's large economic impact on the industry, its zoonotic potential, and its ability to rapidly mutate.[1] Following the implementation of a vaccination program for IAV, it is important to assess whether the expected economic benefit is being achieved and whether vaccination should be continued based on the evaluation. The objective of this study was to evaluate the economic impact of an autogenous influenza vaccine used in finishing pigs with a pre-post analysis utilizing closeout data.

Materials and methods

The study took place in a medium sized system that is *Mycoplasma hyopneumoniae* negative with variable porcine reproductive and respiratory syndrome virus (PRRSV) status located in the Midwest United States. Closeout data utilized for this analysis included multiple lots of pigs started from October 1, 2020, until February 28, 2022. The pre-vaccination group included lots of finishing pigs started from October 1, 2020, until February 28, 2021. The post-vaccination group included lots of finishing pigs started from October 1, 2021, until February 28, 2022. To address the potential confounding effect of seasonality, only lots started between October through February were included in this study. There were 47 lots of finishing pigs included in the pre-vaccination data and 40 included in the post-vaccination data. The pre-vaccination group was vaccinated with a single dose of a commercial killed influenza vaccination given at weaning or 8 - 9 weeks of age at the labeled dose per pig. The post-vaccination group was given a single 1 mL dose of a five-strain autogenous influenza vaccination formulated specifically for finishing pigs at 8 - 9 weeks of age.

Results and discussion

The closeout data was extracted and analyzed using Excel. Potential confounders and outliers were explored using box and whisker plots of the average weight at placement, average finishing weight, and days to market. Frequency distributions and box and whisker plots were created for key performance indicators (KPIs), including average daily gain (ADG), feed conversion (FCR), mortality rate, and feed per head. The economic analysis was performed to estimate a relative benefit: cost ratio using a fixed time-to-market production and economic model. Average finishing mortality was 7.52% for the pre-vaccination group and 5.66% for the post-vaccination group. ADG was 1.96 for the pre-vaccination group and 2.02 for the post-vaccination group. The pre-vaccination FCR was 2.85, and the post-vaccination FCR was 2.70. Administering an autogenous influenza vaccine to pigs at 8 - 9 weeks of age that costs $0.52 per pig placed was beneficial to this swine production system with a $10.54 increase in profit per pig placed for the post-vaccination group and a benefit-to-cost ratio of 20.3:1 (Table 1). In conclusion, the use of an autogenous influenza vaccine in growing pigs demonstrated that it is not only beneficial to health and production metrics but also proved to be economically favorable.

Reference

1. Influenza viruses - diseases of swine - Wiley Online Library. (n.d.). Retrieved November 16, 2022, from **https://onlinelibrary.wiley.com/doi/10.1002/9781119350927.ch36**

Table 1: Summary of economic analysis by treatment group

	Pre-vaccination	Post-vaccination	Difference
Number of pigs placed (pigs/group)	1,000	1,000	0
Cost: Cost of intervention ($/pig)	$0	$0.52	**$0.52**
Feed price ($/ton)	$365	$365	$0.00
Total costs ($/pig placed)	$243.81	$243.37	$0.44
Number pigs marketed (pigs/group)	924	943	19
Average live weight at market (pounds/pig)	279.6	286.5	6.9
Market pig price ($/pound)	$1.15	$1.15	$0
Total annual revenue ($/pig placed)	$223.02	$233.12	-$10.10
Benefit: Net profit ($/pig placed)	- $20.79	- $10.25	**- $10.54**

Poster 10

https://doi.org/10.54846/am2023/95

Evaluation of market trailer hygiene and transmission of porcine enteric coronaviruses

Sarah Albers[1]; Kate Dion[2], DVM; Jana Morgan[3], DVM
[1]University of Wisconsin, Madison, Wisconsin; [2]The HANOR Company, Webster City, Iowa;
[3]Boehringer Ingelheim Animal Health USA Inc., Duluth, Georgia

Introduction and objective

Market trailer transportation serves as potential source of transmission for porcine enteric coronaviruses (PEC) from contamination at the harvesting facility or prior farms visited. One study found 5.2% of trailers tested positive for PEDV[1] and another found 3% positive for PEDV and 5% for PDCoV on environmental swabs.[2] Due to cost, time, and truck wash availability, proper trailer hygiene is not always maintained during marketing, causing the potential for compromised biosecurity and spread of PEC. In the production system this study was completed in, pigs are typically marketed in four cuts. For cuts one through three, all trailers are expected to have a wash prior to arriving at the farm. During the fourth cut, trailers may be unwashed only if they have transported pigs from this system on the prior load. The objective of this study was to determine if farms with introductions of PEC during marketing have more unwashed market trailers arrive at the farm compared to the farms that were not suspected of having PEC during marketing.

Materials and methods

A retrospective case control study was completed from February 2022 to May 2022 utilizing transportation schedules and invoices to evaluate wash status of trailers that arrived during marketing. Cases were defined as finishing farms that tested positive for PEC during marketing via RT-PCR. For each case, market trailer hygiene was evaluated from the first date of marketing to the day prior to the PEC diagnosis. Invoice records from transportation companies were utilized to verify wash status. A cut off of $85 was used to define an adequate wash and any invoices totaling less than $85 were considered unwashed trailers.

The cases were matched with a negative control farm, that was not suspected of having PEC, based on proximity (within 90 km), number of pig spaces, and fill time (within 3 weeks of the case farm). The same number of trailers were evaluated for wash status in the controls as defined by its matched case.

Results and discussion

19 cases were identified that tested positive for PEC during marketing and were matched with 19 controls. 33 unwashed trailers arrived at the case farms compared to 42 unwashed trailers at the control farms. 45% of the unwashed trailers arrived 7 days or less prior to the diagnosis of PEC for the cases compared to 38% of the controls. 70% of the unwashed trailers were expected to be washed for the case farms compared to 76% of the negative controls. Although the percentage of unwashed trailers was similar between the cases and controls, the fact that the trailers were expected to be washed highlights a concern for biosecurity and potential spread of PEC. This project showed that it was possible to assess trailer wash status utilizing records already available in the production system. To improve biosecurity and decrease the spread of pathogens, there should be further procedures in place to verify trailers are being washed when expected and provide feedback to communicate biosecurity expectations.

References

1. Lowe J, Gauger P, Harmon K, Zhang J, Connor J, Yeske P, Loula T, Levis I, Dufresne L, Main R. Role of transportation in spread of porcine epidemic diarrhea virus infection, United States. Emerg Infect Dis. 2014 May;20(5):872-4. **https://doi.org/10.3201/eid2005.131628**

2. Vazquez G, Dion K, Sponheim A. Assessment of market trailer contamination with porcine coronaviruses and Senecavirus A, In: Proceedings of the 53rd AASV Annual Meeting. American Association of Swine Veterinarians; 2022:71.

https://doi.org/10.54846/am2023/96

Evaluation of chemical treatments to deactivate PRRSV in water sources

Austin Janssen[1], BS; Brent Sexton[2], DVM; Kevin Eggers[3], DVM
[1]Iowa State University, Ames, Iowa; [2]The Maschhoffs, LLC, Carlisle, Illinois;
[3]Zoetis Animal Health, Parsippany, New Jersey

Introduction

In the US swine industry, the annual cost of PRRSV associated disease is estimated to be $664 million.[1] Due to these economics, intensive biosecurity measures are often enacted to mitigate PRRSV introduction via environmental sources. Notably, there are limited studies evaluating the viability of PRRSV in water and its ability to serve as a source of infection. The objective of this study was to evaluate PRRSV viability in water after applying a chemical treatment model. The results would be evaluated using two different diagnostic assay methods, specifically bioassay and virus isolation (VI) testing.

Materials and methods

Bioassay

Thirty-two wean pigs, sourced from a PRRSV-naïve sow herd were individually penned in a 2,400 head commercial barn without direct nose-to-nose contact. Pigs were allocated into one of eight treatment groups. Treatment groups 1-6 each contained 4 pigs and were assigned one of six possible chemical water treatments. Group 7 consisted of 4 pigs acting as PRRSV-positive controls. Group 8 consisted of 4 pigs acting as PRRSV-negative controls. Water treatments consisted of commercially available products with varying active ingredients: bleach, hydrogen peroxide, chlorine dioxide, organic acids, hypochlorous acid, and chlorinated water. Respective treatments were applied to PRRSV-contaminated water per label indication. After five minutes, pigs in treatment groups 1-6 were orally inoculated with a ten-milliliter dose of treated contaminated water at TCID(50) of $1 \times 10^{5.3}$.[2] Individual pig serum samples were collected at 4- and 7-days-post-inoculation (dpi) for subsequent PRRSV PCR testing.

Virus isolation

As in the bioassay procedure, water was contaminated with PRRSV-positive serum to achieve a PCR CT of 20-21 driven by VI testing capabilities. Chemical treatments were applied at the same rate and concentration as previous. After five minutes, a neutralizing agent was added to each treatment. A sample was taken from each replicate, frozen, and submitted for PRRSV testing via PCR and VI. VI cell cultures were subsequently screened for cytotoxicity.

Results

Reference Table 1 for complete results.

Discussion

Due to the unexpected and variable treatment outcomes, no definite determination of these chemical treatments can be made. Future testing should incorporate improved bioassay biosecurity measures and a better understanding of the minimum PRRSV oral infectious dose. The inability to achieve consistent results from either assay method highlights the challenges of experimental design in assessing the viability of PRRSV in water.

References

1. Holtkamp DJ, Kliebenstein JB, Neumann EJ, et al. Assessment of the economic impact of porcine reproductive and respiratory syndrome virus on United States pork producers. *J Swine Health Prod.* 2013;21(2):72-84.

2. Hermann JR, MuÑoz-Zanzi CA, Roof MB, et al. Probability of porcine reproductive and respiratory syndrome (PRRS) virus infection as a function of exposure route and dose. *Vet Microbiol.* 2005;110:7-16.

Table 1: Bioassay PCR results and VI results

Treatment	Bioassay 4-dpi (+/total)	Bioassay 7-dpi (+/total)	Virus isolation (+/total)
Bleach	(4/4)	(4/4)	(4/4)
H_2O_2	(4/4)	(4/4)	(2/4)
ClO_2	(4/4)	(4/4)	(2/4)
Organic acids	(3/4)	(4/4)	(1/4)
HOCl	(2/4)	(4/4)	(1/4)
Chlorinated water	(4/4)	(4/4)	NA
Positive control	(3/4)	(4/4)	(1/4)
Negative control	(1/4)	(4/4)	NA

https://doi.org/10.54846/am2023/97

Evaluation of the relative PCR cycle times (Ct) of rotavirus A and C in natural planned exposure (NPE) materials, colostrum deprived piglets, and environmental samples

Alexis Berte[1], BS; Kayla Blake[2], DVM, MAg.; Kimberlee Baker[2], DVM, MS; Jessica Davenport[2], DVM; Darin Madson[2], DVM, PhD, DACVP
[1]Iowa State University; [2]JBS Live Pork

Introduction

Rotavirus is a non-enveloped RNA virus that causes blunting of the intestinal villi, diarrhea, increased mortality, and poor weight gain in neonatal piglets. Rotavirus is resistant to many disinfectants, and environmental build-up is a significant source of infection. NPE is used in gestating females to ensure adequate lactogenic immunity. The purpose of this observational study was to compare the relative amount of rotavirus in different NPE materials and in various environmental samples on a sow farm with clinical rotavirus.

Materials and methods

A 2500-head multiplier sow farm with a two-year history of rotavirus diarrhea in suckling piglets was used for this study. NPE materials were created using an intestinal homogenate (IH) harvested from two to six-day old, clinically diarrheic piglets. The farm NPE protocol targets 7 mL IH per sow. The following NPE materials were prepared and spiked with IH at the following dilution rates: feed slurry (1:5), water from a PVC pipe (1:250), water from a medicator drawing up an IH stock solution (1:672), and a filtered IH solution for use in a Maximus sprayer system (1:2). IH served as a positive control, and farm water served as a negative control. Immediately after preparation (T0) and 12 hours later (T12), all solutions were tested in triplicate for rotavirus A, B, and C by PCR. Three snatch farrowed, colostrum deprived piglets were fed per NPE solution and necropsied 12 hours later to serve as bioassays for rotavirus viability in NPE materials. For the environmental samples, nine farrowing crates were sampled before and after washing via a sterile gauze soaked in PBS. They were swabbed in the same 91 square cm area on the back floor of the crate. Skin swabs were taken from the rump of the same nine sows before washing, after washing, and after a vigorous scrub. Vaginal vault swabs were collected with a double guarded culture swab after a sterile scrub, and rectal swabs were collected with a sterile swab placed in PBS. Pit samples were collected from four areas beneath the nine test farrowing crates.

Results and discussion

Rotavirus C was detected from diarrheic piglets and all NPE materials. All environmental samples were positive for rotavirus A and C, except sow rectal swabs and farm well water. All bioassay piglets including the negative controls were positive for rotavirus indicating environmental contamination, thereby invalidating all the other bioassays. Various steps including snatch farrowing, piglet specific feeding syringes, separate totes per test group, disinfecting the necropsy station between each pig, and changing gloves between every pig, were taken to try and prevent contamination of the bioassay piglets. The results of this observational study confirm that rotavirus is extremely easy to propagate in colostrum deprived piglets, and that the farm environment is a significant source of contamination, despite standard sanitation protocols. Most interestingly, the negative rectal swabs suggest sows were not actively shedding rotavirus which may be due to immunization through repeated NPE use. It was also discovered that the Maximus sprayer system was the least labor-intensive form of NPE tested. In future studies, it would be beneficial to use piglets that have only have one strain of rotavirus to determine if there is a compounding effect between rotavirus A and C. It would also be beneficial to push the testing times out to 24 hours for each of the NPE solutions.

Table 1: Average rotavirus Ct values from PCR

Samples	Rota A	Rota C	T0 samples*	Rota A	Rota C
Crate before wash	30	36	Spiked water pipes	31	23
Crate after wash	32	36	Medicator water	35	25
Pit beneath crates	24	33	Spiked feed slurry	31	20
Skin before wash	30	35	Sprayer solution	31	14
Skin after wash	28	34	IH (+ control)	25	15
Skin after scrub	32	36	Well water (- control)	> 36	> 36
Rectal swab	> 36	> 36	*Data not shown for T12 due to minimal change in Ct values versus T0		
Vaginal swab	30	35			

IH = intestinal homogenate

Poster 13

https://doi.org/10.54846/am2023/98

Antibiotic use for the treatment of lameness in sows

B. Cincotta; M. Pierdon, VMD
University of Pennsylvania School of Veterinary Medicine

Introduction

Documented to be a common medical issue afflicting sows and leading to their premature exit from the herd, lameness can have a severe negative impact on the welfare of animals and economic success of swine farms. To counteract such consequences of lameness, antibiotics, among other medications, are administered to lame sows. The extent to which antibiotics are used as a treatment for lameness and other sow medical issues is of great interest given growing concerns about antibiotic resistance amongst the medical community and beyond. Here, we look to examine the extent to which antibiotics are employed to treat lameness versus other medical issues afflicting sows and which categories of antibiotics are utilized for such treatments. It is hypothesized that lameness will be the main medical indication for antibiotic use.

Methods

Treatment records were collected from 3 sow farms using pen gestation, including a small farm with 258 sows and two larger farms with 5,232 and 5,273 sows respectively. From paper or electronic records kept by barn staff, information regarding medical condition, medication(s) type and dosage, and date of treatment, was input into Excel according to farm and sow ID. For each of the larger farms, records from a full year were provided, while records from 600 days were provided by the smaller farm. Descriptive statistics were performed using Excel (Microsoft Corporation, USA). Total antibiotic use was first calculated as the total milligrams of product used. Antibiotic use rate was then calculated as the total milligrams of antibiotics used per average head of inventory per day to standardize the measure across the farms with different herd sizes. The proportion of antibiotic use for each type of antibiotic and for each treatment reason were calculated as the total milligrams for each, out of the overall total milligrams of antibiotic used. No feed or water medication use was reported at these facilities.

Results

Over one-year period, 0.746 mg/sow/day of antibiotics were used across all farms for individual treatments. The smallest farm used the most antibiotics, using 0.992 mg/sow/day, while the two larger farms used 0.672 mg/sow/day and 0.855 mg/sow/day respectively. Lameness was responsible for more antibiotic use than any other treatment across all three farms, accounting for 72.7% of the total milligrams of antibiotics used (Figure 1). The second most common reason for antibiotic use was reproductive problems, representing 24.5% of milligrams used. Other issues accounted for 2.7% of antibiotic usage. Lincomycin was the most used antibiotic, making up for 69.7% of the milligrams of antibiotic used. Oxytetracyclines were the next most commonly used antibiotic, accounting for 23.8% of antibiotics used. Ceftiofurs (9.5%) and macrolides (0.1%) were the other classes used. With regards to lameness specifically, 93.7% of the milligrams of antibiotic used was lincomycin and 6.2% were ceftiofurs.

Figure 1: Proportions of antibiotic use according to reason for individual treatment on 3 sow farms over the course of a year.

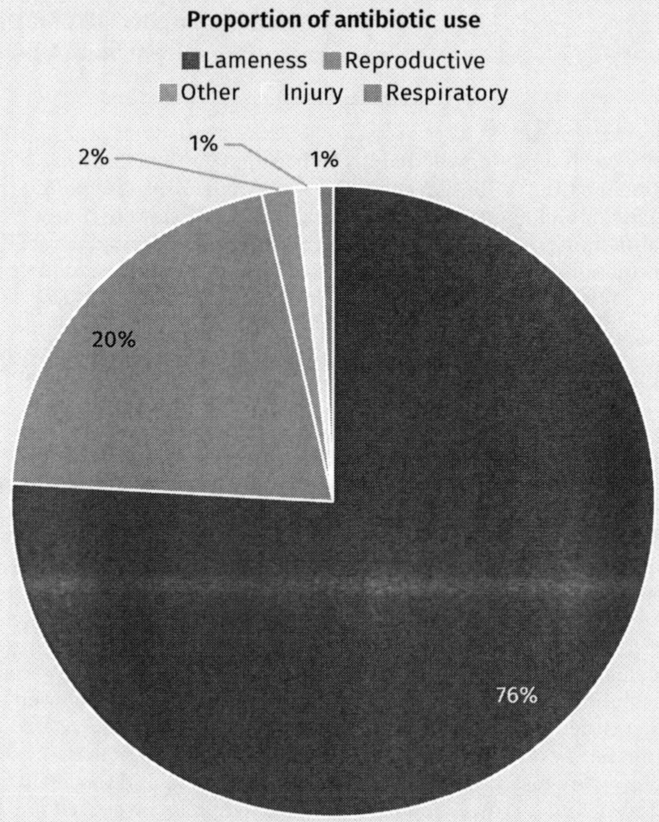

Discussion

As hypothesized, lameness was the top reason for antibiotic use. However, while antibiotics were frequently employed to treat lameness, the total amount used for the individual treatment of animals was relatively low compared to the amounts that would be used if all animals in a herd were treated. The two antibiotics that were most used to treat lameness, lincomycin and ceftiofurs, are both classified by OIE as second level (of three) antibiotics. This indicates they should be used with care due to the potential threat of antimicrobial resistance. Given this level of concern, it is worth finding ways to prevent lameness to preclude their use. However, since a minimal amount of these drugs are being used for individual treatments of lameness, which is a significant animal welfare and economic concern on swine farms, the use of such antibiotics is justified. Future research should evaluate the effectiveness and timing of such treatments.

Poster 14

https://doi.org/10.54846/am2023/99

Lessons from swine biosecurity outreach programs directed to small producers and show pig exhibitors

J. F. Hernández Cuevas; J. Nolting, PhD, MS; A.G. Arruda, DVM, PhD, MS
Department of Veterinary Preventive Medicine, College of Veterinary Medicine, The Ohio State University, Columbus, Ohio

Introduction

The emergence and rapid appearance of foreign animal diseases (FADs) is a constant threat to animal health. Specifically, the US swine industry is threatened by the introduction of African swine fever (ASF) from nearby areas with confirmed cases. Even though several national and local efforts occurred in the past years to support producers in preventing and detecting ASF and other FADs, efforts have been largely dedicated to large commercial swine producers and not small-scale or exhibition swine. The first objective of this study was to use data from small producer outreach sessions to identify FAD- and biosecurity-related knowledge gaps within this population and assess the session's effectiveness. The second objective was to describe FAD preparedness in the exhibition swine sector, including basic aspects of their site biosecurity.

Methods

For the first objective, participants were invited to attend outreach sessions organized by state industry partners. During the program, participants were asked to voluntarily and anonymously complete two paper-based surveys: one before and one after the session. Important aspects covered in the surveys included knowledge of FADs and biosecurity concepts. For the second objective, participants were selected on a voluntary basis using an outreach educational booth at two national exhibition fairs: The Exhibition (Des Moines, Iowa) and Summer Spectacular (Louisville, Kentucky). Participants were engaged to create biosecurity plans using the Secure Pork Supply (SPS) framework and take home a modified biosecurity SPS plan and map. A few selected aspects of their biosecurity plans were captured for analysis (e.g. number of animals, whether a biosecurity plan was previously completed, etc.). Data from both efforts was analyzed using descriptive analysis and non-parametric tests, as applicable, using STATA.17.

Results and discussion

For objective 1, 31 participants attended five outreach sessions, the average (± SD) age of participants was 43.36 (19.62), and the mean number of pigs owned by participants was 51.2 (126.61). Approximately 48% of participants reported having a veterinarian-client relationship, but only called the veterinarian if needed. A large part of the participants (97%) indicated they had heard of ASF, however, only 29% of participants felt that they understood important aspects of the disease before attending the session. This is a concern for the US swine industry because it indicates that there could be potential delays in early identification of the disease and timely implementation of control measures in an outbreak situation. After the seminar, 71% of participants reported understanding important aspects of ASF, which was confirmed by analyzing specific knowledge-based questions. These initial results indicate a positive impact of the outreach program in the small producer population. For the second objective, 49 participants from 14 US states created their modified SPS plans and map. The mean number of pigs owned was 60.69 (188.34), and the mean number of buildings housing pigs was 1.88 (1.30). Approximately 98% of the population reported having a veterinarian whom they could call if their animals were sick; however, only 16% of the population said they had a biosecurity plan on-site, and 10% of the population had biosecurity plans developed with the help of a veterinarian. Furthermore, 6% of the participants reported facing challenges if needing to implement a cleaning and disinfection station and 24% of participants reported using off-site carcass disposal. The lack of biosecurity plans and oversight by professionals in this population is a concern for the swine industry, highlighting that FAD and biosecurity education is needed to control disease spread and implement control measures in an outbreak situation.

https://doi.org/10.54846/am2023/100

Impact of foot trimming on gilt productivity and foot health

A. Patev; M. Pierdon, VMD, DACAW
University of Pennsylvania School of Veterinary Medicine

Introduction

Lameness is a welfare issue for sows and a common reason for premature removal from the herd. This presents a significant economic issue for the producer. There is little data on the impact of hoof trimming in sows. There are, however, studies that indicate that preventative hoof trimming can improve the gait of the animals, which could correlate to decreased lameness. The objective of this study is to randomly allocate gilts to either receive hoof trimming or no hoof trimming to determine the impact of preventative foot maintenance on productivity, longevity, and foot health.

Materials and methods

This study was conducted over the course of 6 weeks at a 5000-sow farm utilizing pen gestation with large pens and electronic sow feeding stations. Select gilts (n = 348) were allotted to either treatment (trimming n = 177) or control (no trimming n = 171). Enrolled gilts were brought into a single large pen in the gestation barn each week and treatment and controls allotted from that pen. Gilts were walked to a foot trimming chute (Zinpro Corp, MN, USA) and their feet scored prior to trimming using a modified version of the Zinpro® FeetFirst® scoring system receiving a 0 for no lesion, a 1 for mild to moderate lesion, and 2 for severe lesions. Lesions scored include heel overgrowth and erosion, heel-sole crack, white line cracks, and cracked walls. Toe length was also recorded. Lameness and any injuries or abnormalities to the hoof, dewclaws, or coronary band were also noted. Treatment animals were trimmed to correct heel overgrowth and any long toes. Control animals were scored and released without trimming. Sows are being scored again when they entered farrowing. Records for breeding, pregnancy, and herd removal were obtained for all enrolled gilts. Foot scores were added together to calculate a total score for each animal at entry and at farrowing. A t-test was used to look for differences between follow up scores in treated and control animals. Logistic regression was performed for each binary production outcome including treatment and lameness at allotment as fixed effects.

Results

Of the 177 gilts in the treatment group, 127 gilts had one or more hooves trimmed. The average total foot score at entry was 4.77 ± 2.08 and the average toe length was 37.4 mm ± 3.2. The average total foot score at farrowing was 6.3 ± 3.6 for the controls and 5.5 ± 2.8 for the trimmed sows ($P = 0.12$) There was no impact of treatment on the odds of being bred (0.83, 95% CI 0.39 - 1.8, $P = 0.63$) with 92% of control animals bred and 91% of treatment gilts bred. There was also no impact on the odds of a gilt becoming pregnant (0.85, 95% CI 0.49 - 1.5, $P = 0.57$) with 83.0% of control gilts confirmed or presumed pregnant and 81.0% of treated gilts. There was no difference in the odds of removal for treated sows compared to control sows (1.1 95% CI 0.58 – 1.9, $P = 0.86$). There were 3 control sows and 5 treated sows removed for lameness. All production parameters measured are shown in Table 1.

Discussion

Overall, foot scores were low and average toe length was short relative to the trimming length recommended by the Zinpro® FeetFirst® scoring system which advises to trim toes once they exceed 55-60 mm. Preventative hoof trimming could be beneficial. However, it is important to consider hoof trimming will not prevent injury associated with the environment, inter-animal aggression, or existing issues, but may prevent future hoof health issues such as wall or white line cracks. Regular hoof trimming could help compensate for problems caused by conformation but will not fix conformational deformities. There was some difference in the follow up foot scores where the trimmed sows were numerically lower. Future follow up should include monitoring their foot scores for differences over time, future productivity, and eventual time to removal. Since the economic value of the gilt is regained as she farrows more litters, the economic benefit of foot trimming can be best ascertained by whether it improves her longevity and her productivity.

Table 1: Foot health and production parameters in gilts that were either control gilts or allocated to foot trimming at entry into the herd

Treatment	Entry to first service	Percent bred	Percent pregnant	Percent farrowed	Percent removed	Follow up score
Control (n = 171)	14.3 ± 13.9	92%	83%	53%	13%	6.3 ± 2.6 (n = 97)
Trim (n = 177)	15 ± 13.8	91%	78%	53%	14%	5.5 ± 2.8 (n = 95)

Poster 16

https://doi.org/10.54846/am2023/101

The effect of dietary zinc on *Clostridioides difficile* colonization and pathogenicity in neonatal piglets

G. DeMers[1], BS; J. Zackular[2], BS, MS, PhD; M. Pierdon[1], VMD, BA; L. Redding[1], VMD, BA, PhD
[1]Department of Clinical Studies—New Bolton Center, School of Veterinary Medicine, University of Pennsylvania, Kennett Square, Pennslyvania; [2]Department of Pathology and Laboratory Medicine, Perelman School of Medicine, University of Pennsylvania, Philadelphia, Pennsylvania

Introduction

Clostridioides difficile is a spore-forming anaerobic, gram-positive bacillus that is a significant enteric pathogen capable of causing severe and sometimes fatal diarrhea in both humans and livestock, including neonatal swine. The source of infection is most likely the dam, as even low levels of *C difficile* in the sow appear sufficient to result in colonization of piglets.[1] Transition metals such as zinc play an essential role in gut microbiota ecology and alter susceptibility of the host to enteric infections.[2] Zinc is included in the diet of swine at all stages of life as an essential trace element; it has also been consistently found to increase pig performance during the post-weaning period and to reduce the incidence and severity of post-weaning diarrhea due to *Escherichia coli* infection.[3] In mice however, excess dietary zinc has been shown to increase susceptibility to and severity of *C difficile* infection.[4] It is unknown whether dietary zinc supplementation affects rates of *C difficile* in sows and piglets or enhances virulence.

Objectives

The objectives of this study were to determine the effect of feeding high levels of dietary zinc on the prevalence of *C difficile* in gestating sows and piglets, the genetic variation in *C difficile* in sows and piglets, the phenotypic characteristics of *C difficile* associated with these animals, and the impact on piglet health and sow productivity.

Materials and methods

At loading into farrowing, 73 sows were randomized to either high Zn (2,500 ppm) ($n = 35$) or low Zn (125 ppm) ($n = 38$) diets ≥ 5 days prior to parturition. Fecal samples were collected from the sow at allocation and after parturition. Fecal swabs were collected from neonates within 3 days of birth at which point they also received a scour score. To detect *C difficile* colonization, anaerobic culture was performed, and strain DNA was isolated for comparative genomics. Toxin-specific cytotoxicity assays for *C difficile* were performed as well as growth dynamics, motility, and biofilm formation capacity of isolates determined. Risk factors for carrying *C difficile* other than diet (eg, maternal age, parity, early farrowing) were also recorded. Production characteristics were compared between the high and low zinc sows using a t-test for normal data and a Mann Whitney U-test for non-normal data.

Results

There was no difference in piglet scour scores between the high zinc sows (0.50 ± 0.98) and the control sows (0.34 ± 0.70) ($P = 0.39$). There was also no difference in piglet livability between high Zn (0.77 ± 0.32) and low Zn diets (0.92 ± 0.59) ($P = 0.36$). Likewise, stillborns were not different between high Zn (0.81 ± 1.0) and low Zn (1.1 ± 1.8) ($P = 0.89$). Culture results are pending—final results will be shared at the meeting.

Discussion

While productivity and scour score did not vary between the two treatments, further work is needed to determine the impact of the diets on *C difficile* prevalence and characteristics. The findings of this study will help further elucidate factors that shape gut health of piglets and the characteristics of pathogens that affect them. This work may serve to guide nutritional recommendations given by practitioners for pre-parturient swine considering the resulting health affects it may have on neonatal piglets.

References

1. Grześkowiak Ł, Zentek J, Vahjen W. Determination of the extent of *Clostridium difficile* colonisation and toxin accumulation in sows and neonatal piglets. *Anaerobe*. 2016;40:5-9.

2. Palmer L, Skaar E. Transition metals and virulence in bacteria. *Annual Review of Genetics*. 2016;50(1):67-91.

3. Johanns V, Epping L, Semmler T, et al. High-zinc supplementation of weaned piglets affects frequencies of virulence and bacteriocin associated genes among intestinal *Escherichia coli* populations. *Front. Vet. Sci*. 2020;7.

4. Zackular J, Moore J, Jordan A, et al. Dietary zinc alters the microbiota and decreases resistance to *Clostridium difficile* infection. *Nature Medicine*. 2016;22(11):1330-1334.

https://doi.org/10.54846/am2023/102

A survey of rectal temperatures immediately after transport and body weight gain for the first 17 days post placement for weaned pigs transported an extended distance during the summer months

H. Lathom[2]; T. Bauman[2]; V. Utrera[1]; D. Shaffer[1]; D. McKilligan[1]; M. Almeida[3]; K. Coulson[1]
[1]TechMix Global Inc., Stewart, Minnesota;[2] The Maschhoffs; [3]Veterinary Diagnostic Laboratory, Iowa State University

Introduction

Previous studies have demonstrated that the exposure to chronic heat can cause decreased growth performance in the nursery and finishing phases of pig production. The stress experienced by piglets during the weaning process has also been well documented in literature. This survey was conducted to measure rectal temperatures immediately post weaning and body weight gain over 17 days after placement for pigs hauled 7 - 8 hours during the summer months.

Materials and methods

This investigation was a longitudinal survey using 16 trailer loads of pigs delivered to two, 7,200 head, wean to finish (WTF) sites with a common sow farm source. The morning of weaning, 25 pigs (n = 400) were systematically selected and tagged. All pigs were loaded in the same compartment located on the bottom deck of each trailer. Rectal temperatures and body weights were collected immediately after unloading at the WTF site and body weights were collected again at 17 days post placement. The trailers were equipped with internal sensors on each deck and external sensors to measure ambient temperature and humidity. The average length of travel was 7 - 8 hours. Regression analyses were done using the Proc Mixed procedure of SAS. All average daily gain (ADG) data were analyzed with the random effect of room and the fixed effects of gender and post-transport weaning weight as well as the linear and quadratic effects of the variable of interest. Post-transport rectal temperature data were analyzed with the fixed effects of gender and post-transport weight and the linear and quadratic effects of the variable of interest.

Results and discussion

Mean trailer temperatures ranged from a minimum of 80.1°F to a maximum of 90.9°F. The mean rectal temperature post-transport was 105.5°F with a range from 103.0°F to 107.7°F. There was a slightly positive linear relationship ($R^2 = 0.05$) between mean bottom deck trailer temperature and piglet rectal temperature. There was a quadratic relationship ($R^2 = 0.14$) between mean bottom deck trailer temperature and ADG during the first 17 days post placement. There was no significant ($P > 0.05$) relationship between weaned pig rectal temperature during unloading and ADG the first 17 days post- weaning. These results suggest relatively weak relationships between temperature inside the trailer and pig growth immediately post-weaning and no relationship between piglet rectal temperature at unloading and subsequent pig growth rate. Future work should be conducted across a wider range of ambient temperatures.

Poster 18

https://doi.org/10.54846/am2023/103

Evaluation of the administration cost of feedback

Paul McDonald[1], BS; Marjorie Schleper[2], DVM; Laura Dalquist[2], DVM; Amanda Sponheim[3], DVM
[1]Purdue University, West Lafayette, Indiana, [2]Christensen Farms, Sleepy Eye, Minnesota,
[2]Swine Vet Center, St. Peter, Minnesota, [3]Boehringer Ingelheim Animal Health USA Inc., Duluth, Georgia

Introduction

Feedback is a common on-farm practice with the goal of exposing gilts and sows to farrowing house pathogens like Rotavirus to boost the quality of colostrum and subsequently decrease piglet disease. Feedback is historically viewed as less expensive than vaccine alternatives because it is "produced and manufactured" on farm. Batch to batch and farm to farm preparation, administration, and pathogen inclusion variation is expected to occur. Therefore, this study was aimed at determining the presence of pathogens of interest and administration cost of feedback in four commercial sow herds within one system.

Materials and methods

Sow farms were identified based on feedback administration frequency, prefarrow Rotavirus C vaccination status, and negative PRRSV status. Farm managers were interviewed to determine farm specific feedback practices and the number of man hours and equipment used to make feedback. Feedback cost per dose was calculated by multiplying the number of man hours required to mix and deliver each batch by the system starting wage per man hour and divided by the number of sows that receive each batch of feedback. The life of a blender and freezer was then divided over the number of doses in the lifespan of the appliance (based off the system average) and added to the cost per dose. One farm administered feedback once per week, one administered feedback three times per week, and two farms vaccinate for Rotavirus C prefarrow (Farms 2 and 3) and administered feedback twice per week (Farms 2 and 3). Farm staff mixed feedback using a standard protocol and collected one 50mL tube of feedback material per batch of feedback for three weeks. Tubes were labeled and frozen until submitting to the diagnostic lab for PRRSV, PCV2/3, and Rotavirus A, B, and C PCR. Samples were also anaerobically cultured to identify *Clostridium* species.

Results

All samples were PRRS and PCV2 PCR negative. Results are summarized in Table 1.

Discussion

This study demonstrates the variation in pathogen detection in feedback between and within system sow farms. Feedback is administered in the study system primarily for Rotavirus C. If Rotavirus C is not detected by PCR, the value of feedback should be questioned, especially if unintended pathogens are identified. It is interesting to note that the frequency of administration did not necessarily increase detection of Rotavirus C by PCR in this study. The cost of feedback administration is estimated to be between \$0.10 and \$0.17 per head per dose in this study, which can be used to compare to alternative methods for preparing sows for lactation.

Table 1: Feedback administration program cost estimate and diagnostic data by sow farm

	Feedback administration data				Diagnostic data (# positive/# tested)		
Farm	**Frequency per week**	**Man hours per batch**	**Sows per batch**	**Cost per head**	**Rotavirus C PCR**	**PCV3 PCR**	***Clostridium perfringens* culture**
1	1	2.66	405	\$0.17	0/3	0/3	3/3
2	2	1.16	297	\$0.11	6/6	5/6	1/6
3*	2	1.5	354	\$0.10	0/10	10/10	3/10
4	3	2.25	400	\$0.13	0/9	3/9	7/9

* Two samples per batch of feedback were submitted for the first two weeks of the study.

https://doi.org/10.54846/am2023/104

The effect of nucleoprotein reassortment on influenza A virus transmission in swine

Megan N. Neveau[1]; Brianna Cowan[2]; C. Joaquin Caceres[2]; Daniela S. Rajao[2]; Daniel R. Perez[2]; Phillip C. Gauger[1]; Amy L. Vincent Baker[2]; Tavis K. Anderson[2]
[1]Department of Veterinary Microbiology and Preventive Medicine, College of Veterinary Medicine, Iowa State University, Ames, Iowa; [2]Department of Population Health, College of Veterinary Medicine, University of Georgia, Athens, Georgia; [3]Virus and Prion Research Unit, National Animal Disease Center, USDA-ARS, Ames, Iowa

Introduction

The influenza A virus (IAV) genome consists of eight single-stranded, noncontiguous RNA segments known to encode 10-17 proteins. The segmented structure allows for reassortment of genes when multiple IAV strains concurrently infect the same host. In 2019 there was an increase in the detection frequency of H3 C-IVA (3.1990.4.A) clade IAVs in US swine. These increasingly detected C-IVA IAV were not antigenically distinct from previous strains from the same hemagglutinin (HA) clade but instead were characterized by the acquisition of an H1N1pdm09 (pdm) lineage nucleoprotein (NP) gene instead of a TRIG lineage NP.[1] The NP binds to the viral genome for RNA transcription, replication, and packaging. We hypothesized that the acquisition of the pdm lineage NP conferred a selective advantage over prior circulating H3 viruses containing TRIG lineage NP.

Methods

To investigate the role of NP reassortment in virus replication and transmission, we used phylogenetic methods to identify representative strains from two contemporary C-IVA hemagglutinin (HA) clades that contained the different evolutionary origins of the NP gene. The major clade, NC/19, contained the pdm NP; and the minor clade, MN/18, contained a TRIG NP. Reverse genetics was used to swap the pdm and TRIG lineage NP genes between the two strains, resulting in the following four viruses: wildtype-NC/19-pdmNP, NC/19-trigNP, wildtype-MN/18-trigNP, MN/18-pdmNP. The recombinant virus genes were confirmed by sequencing, and an *in vivo* transmission study was conducted in 65 conventional nursery-aged pigs.

Results

All viruses successfully infected the 10 primary pigs and transmitted to 5 contact pigs. The wildtype-MN/18-trigNP virus had delayed transmission to the contacts, with nasal shedding not detected until 5 days post contact (dpc), compared to 3 dpc for the other three viruses. Additional assays with experimental samples (titer kinetics over time and microscopic evaluation of pathology) and data analysis are pending. We will utilize these assays to determine if the pdm NP affects the transmission efficiency or associated lung pathology of the C-IVA viruses.

Discussion

The initial results suggest that reassortment to acquire a pdm lineage NP gene may improve transmission efficiency compared to similar IAV with TRIG-NP, but this is likely a multigenic trait. This study supports the use of whole genome sequencing for IAV as these data can detect novel reassortments that may influence the transmissibility and pathology of IAV, thereby impacting influenza epidemiology across pig populations. Additionally, integrating whole genome information with dominant HA/NA may improve the prediction of emerging strains to include in vaccines since our data suggest that viruses undergoing evolutionary change by reassortment can affect the expansion and contraction of IAV HA clades.

Reference

1. Neveau M., Zeller M., Kaplan B., et al. Genetic and Antigenic Characterization of an Expanding H3 Influenza A Virus Clade in U.S. Swine Visualized by Nextstrain. *mSphere* 2022;7(3). **doi:10.1128/msphere.00994-21**

https://doi.org/10.54846/am2023/105

Serologic profiling and viral monitoring of influenza in a large production system

A. Prowant[1], BS; B. Mason[2], DVM; K. Castevens[2], DVM; J. Maurer[3], VMD; M. Inskeep[3], DVM; E. Byers Taylor[3], DVM
[1]Iowa State University College of Veterinary Medicine, Ames, Iowa;
[2]Country View Family Farms, Middletown, Pennsylvania; [3]Merck Animal Health, Lenexa, Kansas

Introduction

Influenza A virus in swine (IAV-S) is a respiratory disease affecting pigs with variable levels of morbidity and mortality, resulting in negative economic impacts. Due to variable maternal antibody levels, the presence of virus, and currently available vaccines, vaccination of growing pigs has been largely limited as an IAV-S control method in the industry.[1] The need for implementation of more suitable vaccination strategies is crucial for disease control. The primary objective of this study is to analyze and interpret the serologic profile, presence, and progression of virus in pigs across different sow farm flows in a large commercial production system. Findings from this study can be used to establish future IAV-S control programs in both dams and pigs.

Materials and methods

The project consisted of sampling pigs from 4 flows: 2 serial pig sampling (Flow A and Flow B) and 2 cross-sectional sampling (Flow C and Flow D). In all flows, pigs were weaned and moved to a nursery at 3 weeks of age (WOA). None of the pigs in the study were vaccinated for IAV-S. In both Flows A and B, 45 pigs were individually ear-tagged and enrolled at 3 days of age (DOA). In both Flows C & D, 180 different pigs were sampled – 45 each from a different age group: 3 DOA, and 3, 5, and 8 WOA. Pigs sampled and enrolled at the sow farms in the 4 Flows were all from different litters with the dam parity distribution by group (P1, P2/P3, > P3) reflective of their farm. The 5 and 8 WOA pigs sampled were selected at random.

Serum was collected from selected pigs at 3 DOA, and 3, 5, and 8 WOA. Individual nasal swabs were collected at 3 WOA and oral fluids were collected from pens of a sub-set of the selected pigs at 5 and 8 WOA.

All serum samples were evaluated using hemagglutination-inhibition (HI) assay, nucleoprotein (NP) ELISA, and neuraminidase-inhibition (NI) assay. Both nasal swabs (pooled in groups of 3) and oral fluids (individual) were submitted for IAV RT-PCR and subtyping; sequencing was conducted on select positive samples.

Results

The RT-PCR and NP ELISA results are shown in Table 1. RT-PCR results were negative at every sample age in Flow A and Flow C. Samples from Flow B were positive at every time point and all positive samples were subtyped as H3N2. NP ELISA results varied: the number of positive samples in Flow A decreased over time; Flow B had its largest number of positives at 8 WOA; Flow C had its largest number of positives at 3 DOA; Flow D had its largest number of positives at 3 DOA and 8 WOA. At the time of writing the proceedings, only the NP ELISA results were available with HI and NI assay results to be featured in the final poster.

Discussion

NP ELISA results can reveal previous pig and/or dam viral exposure and/or whole virus inactivated vaccine use. In this study, the growing pig results of the NP ELISA and RT-PCR tests when evaluated together most likely indicate: 1) normal decay of maternal antibody with no flu introduction(s) in growing pigs in in both Flow A and Flow C, 2) a prolonged exposure event in the nursery phase in Flow B, and 3) previous or resolving flu exposure in Flow D.[2] Both HI and NI titer results can be indicative of influenza protection, but high antibody levels can interfere with whole virus inactivated vaccination success in growing pigs. The final data generated from the project will serve as an important foundation for future work on growing pig vaccine options and implementation in a large commercial operation.

References

1. Detmer, S., Gramer, M., Goyal, S., Torremorell, M., Torrison, J. (2012). Diagnostics and Surveillance for Swine Influenza. In: Richt, J., Webby, R. (eds) Swine Influenza. Current Topics in Microbiology and Immunology, vol 370. Springer, Berlin, Heidelberg. **https://doi.org/10.1007/82_2012_220**

2. Tan, M.P.; Tan, W.S.; Mohamed Alitheen, N.B.; Yap, W.B. M2e-Based Influenza Vaccines with Nucleoprotein: A Review. *Vaccines* 2021, *9*, 739. **https://doi.org/10.3390/vaccines9070739**

Table 1: IAV-S RT-PCR and NP ELISA results

Influenza RT-PCR results			
Sow flow (sampling type)	**Rate positive (avg positive Ct)**		
	Age of pigs		
	3 weeks*	**5 weeks**	**8 weeks**
Flow A *(Serial)*	0/14	0/6	0/6
Flow B *(Serial)*	14/14 (24.1)	4/6 (29.2)	6/6 (31.5)
Flow C *(Cross -sectional)*	0/15	0/6	0/6
Flow D *(Cross-sectional)*	0/15	1/6 (28.8)	0/5

*samples are nasal swabs; all others are oral fluids

NP ELISA results				
Sow flow *(sampling type)*	**% positive (# positive/total samples)**			
	Age of pigs			
	3 days	**3 weeks**	**5 weeks**	**8 weeks**
Flow A *(Serial)*	93.3% (42/45)	52.4% (22/42)	26.2% (11/42)	2.4% (1/42)
Flow B *(Serial)*	54.3% (25/46)	7.1% (3/42)	36.6% (15/41)	87.5% (35/40)
Flow C *(Cross -sectional)*	65.2% (30/46)	37.8% (17/45)	0% (0/45)	6.7% (3/45)
Flow D *(Cross-sectional)*	77.8% (35/45)	44.4% (20/45)	48.9% (22/45)	86.7% (39/45)

https://doi.org/10.54846/am2023/106

Evaluation of sanger sequencing detection of PRRS virus in *in-vitro* co-infections

Joaquín Álvarez-Norambuena; Mariana Kikuti; Albert Rovira; Cesar A. Corzo
Veterinary Population Medicine, University of Minnesota, Saint Paul, Minnesota

Introduction

Porcine reproductive and respiratory syndrome (PRRS) virus is known to cause great economic losses in the swine industry.[1] Modified-live vaccines (MLV) have been used in the field to mitigate impact and reduce wild-type (WT) virus shedding upon exposure.[2] Vaccines have also been used during outbreaks as control strategy.[3,4] The main diagnostic test to confirm a new viral introduction is PCR together with ORF5 sequencing. However, interpretation of these results can be complex when two strains are present in the same sample since sequencing may better represent one strain instead of the other since the most frequent nucleotide in each position is reported.[5] Therefore, the aim of this study is to assess the sequencing of in-vitro samples spiked with two PRRS virus strains at different concentrations.

Materials and methods

Two strains were used for this study, a wild-type RFLP 1-7-4 lineage 1A virus and a vaccine strain RFLP 2-5-2 lineage 5 (Ingelvac PRRS MLV,Boehringer Ingelheim Animal Health, Duluth, Georgia, USA). Each strain was subjected to quantitative PCR (qPCR) at the University of Minnesota Veterinary Diagnostic Laboratory (UMNVDL), and subsequently diluted to obtain three different concentrations of RNA copies per mL (10^6, 10^5, 10^4). Samples comprised a mix of all possible concentration combinations of WT and MLV to create 9 different groups with 3 replicates each (Table 1, data not shown). Samples were then RT-PCR tested and ORF5 sequenced.

Results

An MLV-like sequence (RFLP 2-5-2) was most frequently obtained when the vaccine had a concentration equal or higher to the wild-type strain (in 5 out of 9 groups, 55.6%). A wild-type sequence (RFLP 1-7-4) was most frequently obtained in 1 out of 9 (11.1%) groups when concentration of wild-type was 2 logs higher than the concentration of vaccine. Interestingly in group 4 one sequence was MLV-like (RFLP 2-5-2) and 2 sequences were wild-type (RFLP 1-7-4 and 1-10-4). Only one sequence was generated in group 8 (vaccine at 10^4 and wild-type at 10^5), which was classified as wild-type (RFLP 1-7-4), and none of the replicates for group 9 yielded a sequence. One to three nucleotide differences from either the original MLV or wild-type sequences to which a higher concentration was present in the tested sample were observed in 8 samples, as shown in the percent nucleotide identity in Table 1 (data not shown). This resulted in changes in the RFLP pattern of the consensus sequence for two samples.

Conclusions

A vaccine-like sequence was more often retrieved when WT and MLV were present at the same or higher concentrations, pointing towards Sanger sequencing potentially favoring amplification of this strain in detriment of the WT strain assessed. Whether these results hold true with larger sampling sizes and different PRRS viral strains remains to be elucidated. Still, it is important to consider that the vaccine strain may mask wild-type infections particularly in herds undergoing WT elimination.

References

1. Holtkamp DJ, Kliebenstein JB, Neumann EJ, et al. Assessment of the economic impact of porcine reproductive and respiratory syndrome virus on United States pork producers. *J Swine Health Prod.* 2013:72-84.

2. Linhares, D. C., Cano, J. P., Wetzell, T., Nerem, J., Torremorell, M., & Dee, S. A. (2012). Effect of modified-live porcine reproductive and respiratory syndrome virus (PRRSv) vaccine on the shedding of wild-type virus from an infected population of growing pigs. *Vaccine, 30*(2), 407-413.

3. Dee, S. A., Joo, H. S., Henry, S., Tokach, L., Park, B. K., Molitor, T., & Pijoan, C. (1996). Detecting subpopulations after PRRS virus infection in large breeding herds using multiple serologic tests. *Swine Health and Production*, 4(4), 181.

4. Linhares, D. C. L., Cano, J. P., Torremorell, M., & Morrison, R. B. (2014). Comparison of time to PRRSv-stability and production losses between two exposure programs to control PRRSv in sow herds. *Preventive veterinary medicine*, 116(1-2), 111-119.

5. Kikuti, M., Sanhueza, J., Vilalta, C., Paploski, I. A. D., VanderWaal, K., & Corzo, C. A. (2021). Porcine reproductive and respiratory syndrome virus 2 (PRRSV-2) genetic diversity and occurrence of wild type and vaccine-like strains in the United States swine industry. *PloS one*, 16(11), e0259531.

Table 1: PRRS ORF5 sequencing results by group according to RFLP classification and percent nucleotide identity to the original vaccine and wild-type strain

Group	Vaccine concentration	Wild-type concentration	RFLP	MLV percent nucleotide identity	Wild-type percent nucleotide identity
1	10^6	10^6	2-5-2	100	87.6
			2-5-2	100	87.6
			2-5-2	100	87.6
2	10^6	10^5	2-5-2	100	87.6
			2-5-2	100	87.6
			2-5-2	100	87.6
3	10^6	10^4	2-5-2	100	87.6
			2-5-2	100	87.6
			2-5-2	100	87.6
4	10^5	10^6	2-5-2	100	87.6
			1-10-4	88	99.6
			1-7-4	87.9	99.7
5	10^5	10^5	2-5-2	100	87.6
			1-5-2	99.7	87.6
			2-5-2	99.8	87.4
6	10^5	10^4	2-5-2	100	87.6
			2-5-2	100	87.6
			2-5-2	99.8	87.4
7	10^4	10^6	1-7-4	87.8	99.8
			1-7-4	87.7	99.8
			1-7-4	87.7	99.8
8	10^4	10^5	NA	NA	NA
			1-7-4	87.7	99.8
			NA	NA	NA
9	10^4	10^4	NA	NA	NA
			NA	NA	NA
			NA	NA	NA

NA = not available

Poster 22

https://doi.org/10.54846/am2023/107

Fine-scale analysis within owners of influenza A virus in swine reveals persistent clade circulation and limited clade diversity

B. Arruda[1], DVM, PhD; M. Neveau[2]; A. L. Vincent Baker[1], DVM, PhD; T. K. Andersen[1], PhD; G. S. Silva[2], DVM, MS, PhD; P. C. Gauger, DVM, MS, PhD[2]

[1]Virus and Prion Research Unit, National Animal Disease Center, Agriculture Research Service, US Department of Agriculture, Ames, Iowa; [2]Veterinary Diagnostic and Production Animal Medicine, College of Veterinary Medicine, Iowa State University, Ames, Iowa

Introduction

Influenza A virus (IAV) remains a dominant cause of morbidity in swine with continuous expansion of genetic and antigenic diversity due to incursions of human IAV and evolution in swine. The aggregated national USDA IAV swine surveillance system data documents significant numbers of co-circulating genetic clades; yet, less is known about the diversity present at finer spatial scales or within epidemiologically linked populations of swine within an owner.

Methods

To identify the amount and duration of diversity shared among sites with the same owner, we conducted a fine-scale analysis of IAV at the owner level using large scale data mining of nearly 8,000 hemagglutinin (HA) sequences collected and submitted between January 2013 and June 2021 to a veterinary diagnostic laboratory.

Results

Twenty-five HA clades were detected in the data set, and fourteen of these clades were detected more than twenty times. Of the top twenty-three contributing owners, clinical clade count ranged from 2/25 (8%) to 9/25 (36%). Within the top 6 contributing owners, the average length of time a clade was detected was 3.6 years, most clades were detected for longer than 2 years, and some clades were detected for at least 8 years. A clade was commonly detected within an owner across multiple farm types, including growing, breeding, and isolation stock. Evaluation of H3 antigenic motifs identified four motifs that made up nearly 60% of the detections.

Discussion

Fine-scale analysis demonstrated that the number of detected HA genetic clades and H3 predicted antigenic diversity at the owner level is much less than the national level, supporting the use of autogenous and custom vaccines for control at the owner level. For the owners we evaluated, the duration of circulation of a specific clade detected indicated that a vaccine update every 3 to 4 years may be sufficient to facilitate control and preventative measures unless a lateral introduction of a new clade is detected. Continuous introduction of susceptible animals including gilts and newborn piglets may contribute to persistent clades that circulate for multiple years within an owner, suggesting vaccination of incoming gilts and sows may achieve a more homogenous and less susceptible population.

Poster 23

https://doi.org/10.54846/am2023/108

Detection and localization of atypical porcine pestivirus in the reproductive tract of a persistently infected boar

Gyula Balka[1], DVM, PhD; Lukas Schwarz[2], DVM; René Brunthaler[3], DVM; Sandra Högler[4], DVM; Lilla Dénes[1], MS
[1]Department of Pathology, University of Veterinary Medicine, Budapest, Hungary; [2]University Clinic for Swine, Department for Farm Animals and Veterinary Public Health, University of Veterinary Medicine Vienna; [3]Institute of Pathology, Department for Pathobiology, University of Veterinary Medicine Vienna; [4]Institute of Pathology, Unit of Laboratory Animal Pathology, Department for Pathobiology, University of Veterinary Medicine Vienna

Introduction

Atypical porcine pestivirus (APPV, *Pestivirus K*) belongs to the genus *Pestivirus* within the family of the *Flaviviridae*. Recently, APPV has been identified as the causative agent of congenital tremor (CT) type A-II. The disease is a neurological disorder that affects newborn piglets and is characterized by generalized trembling of the animals. However, little is known about the epidemiology of the infection, transmission and spread of the virus between herds. It has already been confirmed, that similarly to other Flaviviruses, APPV can be shed by semen. Previously, we have identified the interstitial Leydig cells, the peritubular myoid cells and the smooth muscle cells of medium-sized arteries as the target cells of APPV in the testicles of newborn, CT-affected, congenitally infected piglets.[1]

Methods

In our study, we examined FFPE tissue samples obtained from a persistently infected 6-months-old, sexually mature boar, that was born with CT, and was still shedding 2.1×10^9 GE/mL APPV in its semen at the time of the study. RNAscope *in situ* hybridization was used to detect APPV genome in the tissues.

Results

We detected viral genome in the T-cell zones of the reproductive lymph nodes and the MALT of the ileum, periarteriolar lymphoid sheaths of the spleen, and in undifferentiated cells of the crypts in the ileal glands. We also identified positive cells among the neurons of the cerebral cortex, in the molecular and granular cell layer of the cerebellum, including the Purkinje-cells and less intense signal in the neurons of the spinal cord. We found weak signal in the colon, in the cells of the zona arcuata of the adrenal cortex and none in the lungs and the liver. In the reproductive organs, APPV was detected in the Leydig cells and peritubular myoid cells of the testicles, as well as in the cells beyond the Sertoli cell barrier (Sertoli cells and germ cells) and in the epididymis. We found positive cells in the prostate and the bulbourethral gland as well (Figure 1).

Discussion

Our results show that APPV – similarly to bovine viral diarrhoea virus – can cause persistent, congenital infection in pigs and such boars can actively shed the virus in their semen as a result of the infection of cells beyond the Sertoli cell barrier of the testicle and the accessory sex glands. Further research is needed to highlight the underlying mechanisms of the persistent infection. Our results indicate the boar studs should monitor their animals for APPV.

Reference

1. Dénes L, Ruedas-Torres I, Szilasi A, Balka G. Detection and localization of atypical porcine pestivirus in the testicles of naturally infected, congenital tremor affected piglets. Transbound Emerg Dis. 2022 Jul;69(4):e621-e629. **doi: 10.1111/tbed.14355**

Figure 1: Representative microscopic images of the reproductive organs obtained from a persistently infected APPV-positive boar after RNAscope ISH labelling of the virus. The arrows show the positive cells.

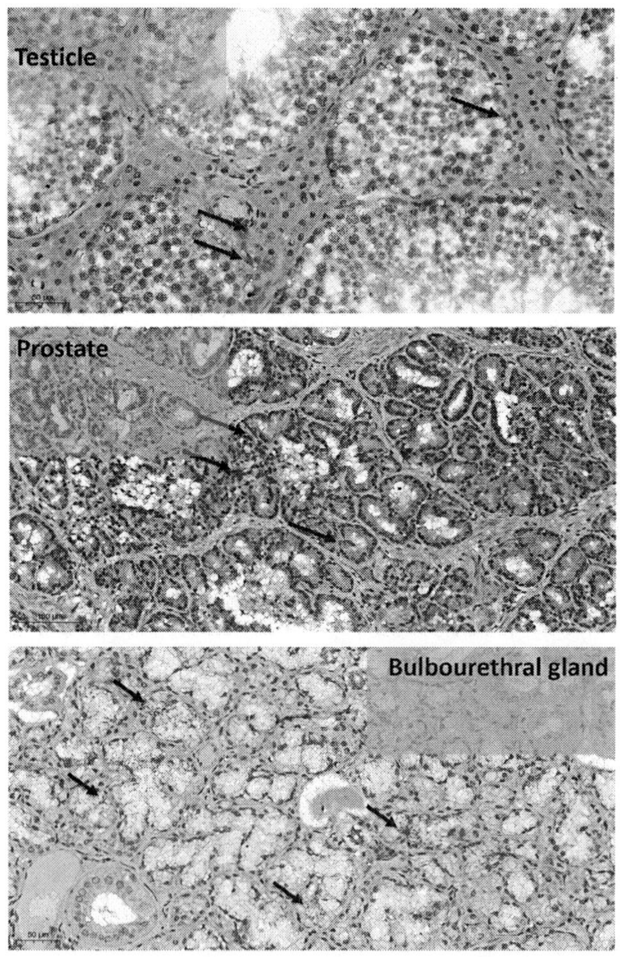

https://doi.org/10.54846/am2023/109

Impact of wean age on *Streptococcus suis* and *Mycoplasma hyorhinis* colonization dynamics during nursery period

Alyssa Betlach[1], DVM, PhD; Mark Schwartz[2]; Marcelo Gottschalk[3], DVM, PhD;
Maria Pieters[4], DVM, PhD; Paul Yeske[1], DVM, MS
[1]Swine Vet Center, St. Peter, Minnesota; [2]Schwartz Farms Inc., Sleepy Eye, Minnesota; [3]University of Montreal, St-Hyacinthe, Québec; [4]University of Minnesota, College of Veterinary Medicine, St. Paul, Minnesota

Introduction

Streptococcus suis (*S suis*) and *Mycoplasma hyorhinis* (*M hyorhinis*) are considered two prevalent bacterial causative agents for nursery pig disease.[1,2] Despite their long-standing presence in the industry, minimal information is known on the epidemiology of these pathogens, thus hindering the optimization of control methods. Antimicrobials are frequently utilized for the treatment and control of *S suis* and *M hyorhinis* infections. However, development of alternative methods for disease control is needed, considering antimicrobial resistance and societal pressures. Impact of production (eg, wean age) or pig management on epidemiology and disease control has been a long-standing question, especially when such modifications also influence other production parameters. Therefore, the objective of this study was to assess the effect of wean age on *S suis* and *M hyorhinis* colonization dynamics during the nursery period.

Materials and methods

An observational field study was performed in a porcine reproductive and respiratory syndrome virus and *Mycoplasma hyopneumoniae* negative nursery herd sourced from a sow from with a history of clinical disease associated to *S suis* serotypes 1 and 14 and *M hyorhinis* in downstream herds. Thirty, 25- and 18-day old pigs (n = 60 total) were randomly selected for longitudinal sampling throughout the nursery period. At weaning, each pig was ear-tagged with a unique identification number. Selected pigs were pen-housed with pigs of similar wean ages (ie, 25 or 18 days of age) within one room. Tonsil swabs were collected from each individual pig at three time points: 1) weaning, 2) three weeks-post weaning, and 3) nine weeks post-weaning. Sample size was calculated to estimate a 75% apparent prevalence based on a 95% confidence level and 1/5 precision level of desired prevalence, assuming perfect test accuracy. Swabs were individually tested using *S suis* quantitative real-time PCR (sub-species and serotypes 1, 14) and *M hyorhinis* real-time PCR. Differences in *S suis* and *M hyorhinis* prevalence, incidence, and bacterial load based on wean age were descriptively analyzed. Prevalence and incidence were calculated using 'epiR' package in R 3.6.1 (R Core Team, 2019).

Results and discussion

Prevalence of *S suis* serotypes 1, 14 among 18 day-old weaned pigs was fairly consistent throughout the nursery period, compared to 25 day-old weaned pigs, which a numerical reduction (24.6%) in prevalence was detected over time. Finding suggests that an older wean age may reduce colonization of *S suis* at end of nursery phase. Weaning age may be an effective means of control in problem herds. In comparison, *M hyorhinis* prevalence increased with time and was similar across wean age treatments. *Streptococcus suis* incidence rates (per week) were 0.052 and 0.043 for 18 day-old and 25 day-old weaned pigs, whereas *M hyorhinis* incidence rates were 0.284 and 0.305, respectively. Bacterial load for *S suis* and *M hyorhinis* were numerically similar across the wean age treatments. Such information provided insight towards potential risk factors associated with *S suis* and *M hyorhinis* colonization dynamics. Additional research related to *S suis* and *M hyorhinis* epidemiol is essential to advance disease control efforts.

References

1. Pieters, M.G., Maes, D., 2019. Mycoplasmosis. In: Zimmerman, J.J., Karriker, L.A., Ramirez, A., Schwartz, K.J., Stevenson, G.W., Zhang, J. (Eds), Diseases of Swine, 11th Ed., John Wiley & Sons Inc, New Jersey, USA, p. 863-883.
2. Gottschalk, M., Segura, M., 2019. Streptococcosis. In: Zimmerman, J.J., Karriker, L.A., Ramirez, A., Schwartz, K.J., Stevenson, G.W., Zhang, J. (Eds), Diseases of Swine, 11th Ed., John Wiley & Sons Inc, New Jersey, USA, p. 934-950.

https://doi.org/10.54846/am2023/110

Longitudinal study of atypical porcine pestivirus infection dynamics in pigs from birth to market reveals persistent infection in piglets with congenital tremors

A. Buckley[1]; J. Mora-Diaz[2]; A. Van Hulzen[2]; F. Matias Ferreyra[2,3]; S. Falkenberg[4,5]; L. Gimenez-Lirola[2]; B. Arruda[1]
[1]Virus and Prion Research Unit, NADC, ARS, USDA, Ames, Iowa; [2] Vet Diagnostic and Production Animal Medicine, CVM, ISU, Ames, Iowa; [3]Current address: Kansas State Veterinary Diagnostic Laboratory, KSU, Manhattan, Kansas;
[4]Ruminant Disease and Immunology Research Unit, NADC, ARS, USDA, Ames, Iowa;
[5]Current address: Department of Pathobiology, CVM, Auburn University, Auburn, Alabama

Introduction

Atypical porcine pestivirus (APPV) has been found to be associated with piglets demonstrating congenital tremors (CT) and clinical presentation of CT in piglets was reproduced after experimental challenge. The virus has been identified in swine of all ages globally; although, little is known about the pathogenesis and epidemiology of APPV. The objective of this research was to perform a longitudinal study following two cohorts of pigs, those born in litters with piglets exhibiting CT and those born in litters without CT, to analyze the dynamics of APPV infection from birth to market.

Materials and methods

A commercial sow farm experiencing an outbreak of CT in piglets was identified. Five litters with at least one piglet exhibiting CT were selected for the study in addition to three cohort litters not demonstrating CT based on parity of sow and date of farrowing. At weaning previously enrolled piglets were co-mingled upon placement at a wean-to-finish barn. Once a month, animals were bled and observed for clinical signs of CT. When pigs were market ready, four pigs from CT-positive litters and four pigs from CT-negative litters were necropsied for tissue collection to evaluate viral distribution. Monthly serum samples were tested for APPV RNA by PCR and the antibody response determined by ELISA. Tissue samples were tested for APPV RNA by PCR and localization was determined by *in situ* hybridization.

Results

There was a wide range in the percentage of affected piglets (8 - 75%) within CT-positive litters. APPV nucleic acid was detected in serum of both affected (100%) and unaffected (31%) piglets from CT-positive litters, while only a single pig was PCR positive from a negative CT litter at the first sampling timepoint. After comingling at weaning, most piglets from CT-negative litters tested PCR positive for APPV in the serum. Overall, a greater percentage of pigs exhibiting CT remained PCR positive throughout the growing phase compared to unaffected piglets from CT litters, while piglets from CT-negative litters had the shortest duration of viremia. CT-negative litters had evidence of maternal immunity; however, some CT-positive litters did as well. The majority of unaffected piglets from CT-positive litters and piglets from CT-negative litters developed an antibody response during sampling, while very few CT-positive piglets had a detectable antibody response after maternal immunity waned. APPV nucleic acid was present in multiple tissues from pigs born in both CT-positive and CT-negative litters at marketing especially the cerebellum, tonsil, and mandibular salivary gland.

Conclusion

This study has shed light on infection dynamics of APPV in swine and the impact immune status and timing of infection have on the persistence of APPV in serum and tissues, which can aide in the development of control measures for the industry.

Poster 26
https://doi.org/10.54846/am2023/111

Use of a viability PCR assay for the detection of *Mycoplasma hyopneumoniae*

Albert Canturri[1]; Lucina Galina-Pantoja[2]; Kimberly Vonnahme[2]; Maria Pieters[1,3,4]
[1]Department of Veterinary Population Medicine, College of Veterinary Medicine, University of Minnesota, St. Paul, Minnesota; [2]Zoetis Inc., Parsippany, New Jersey; [3]Veterinary Diagnostic Laboratory, University of Minnesota, St. Paul, Minnesota; [4]Swine Disease Eradication Center, University of Minnesota, St. Paul, Minnesota

Introduction

Due to its intrinsic difficulty for isolation, microbiologic culture is not routinely performed to identify *M hyopneumoniae* in clinical specimens.[1] Thus, the detection of *M hyopneumoniae* in clinical samples is usually determined by PCR targeting bacterial DNA. However, DNA molecules are known to be very stable in the environment and resistant to degradation, including resistance to pathogen inactivation. Hence, there is no direct relationship between viability of bacteria and their detection by PCR, which can lead to diagnostic uncertainty and interpretation issues in certain epidemiologic scenarios. Therefore, the objective of this study was to develop a PCR assay for the detection of viable *M hyopneumoniae*.

Materials and methods

Development and validation of the new assay followed the guidelines suggested by the Laboratory Technology Committee of the American Association of Veterinary Laboratory Diagnosticians.[2] To determine the analytical sensitivity of the assay, a standard curve spanning several orders of magnitude was prepared from serial 10-fold dilutions of a quantified targeted synthetic oligonucleotide. The limit of detection (LOD) was calculated based on the mean of the Ct value at the lowest copy number wherein 100% of the replicates were positive. The LOD was established based on the results of different replicates in independent runs. To define the analytical specificity of the assay, a panel comprised of several *M hyopneumoniae* reference and field strains, other Mycoplasma species, and bacterial and viral species commonly present in the respiratory tract of pigs was assembled. Repeatability was measured testing replicates of various concentrations of the target in separate runs. In order to determine the assay's selectivity and diagnostic characteristics, different clinical specimen matrices, including secretions from the upper and lower swine respiratory tract were collected and tested. The above-mentioned clinical specimens were obtained both from *M hyopneumoniae* experimentally infected and naive pigs. Side-by-side comparisons were performed to evaluate the detection of viable *M hyopneumoniae* and genetic material after bacterial inactivation using various methods, including chemical and thermal inactivation. Genetic material was extracted prior to and at 1, 2, 6 and 12 hours, and at 1, 2, and 3-days post-inactivation. All samples were analyzed by means of a commercially available real-time PCR and the viability PCR assay. The decay dynamics of viable *M hyopneumoniae* was compared to the standard detection and was evaluated by culture inactivation immediately prior to and 1, 2, 3, 4, 5, 10, 15, 20, 25 and 30-minutes post-inactivation.

Results and discussion

The analytical LOD was 1 genome copy number/μL (8 genome copy numbers/reaction). The assay met all the analytical specificity requirements, as all tested *M hyopneumoniae* strains were detected and the genetic material of other pathogens was not amplified. PCR amplification was obtained in clinical specimens from all tested swine matrices. As for repeatability, there was very low intra and inter-assay variation, as all SDs were < 1.0.

Mycoplasma hyopneumoniae DNA detection by the real-time PCR (current PCR assay) was consistent over time, irrespective of the viability status up until the last time it was measured. In contrast, viable *M hyopneumoniae* was no longer detected either at 25 minutes or immediately after, depending on the method used for inactivation. This study describes, for the first time, the development of a quantitative and rapid method for *M hyopneumoniae* viability assessment *in vitro*. This assay can aid to circumvent the current inability to differentiate viable from non-viable bacterial cells. Lack of viability differentiation has been a key limiting factor for the diagnosis of *M hyopneumoniae*, especially when assessing the efficacy of control interventions in the field. Important scenarios in which this assay can be applied may include the evaluation of the effects that antibiotic treatments have on the clearance of *M hyopneumoniae* from the respiratory tract of pigs, and the detection or pathogen genetic material in the tail end of disease eradication.

References

1. Kobisch M, Friis NF. Swine mycoplasmoses. *Rev Sci Tech.* 1996;15(4):1569-1605.

2. Toohey-Kurth K, Reising MM, Tallmadge RL, et al. Suggested guidelines for validation of real-time PCR assays in veterinary diagnostic laboratories. *J Vet Diagnostic Investig.* 2020;32(6):802-814.

https://doi.org/10.54846/am2023/112

Prevalence and risk factors for *Lawsonia intracellularis* detection in vaccinated and non-vaccinated swine sites

T-Y. Cheng[1], DVM, PhD; M. Campler[1], PhD; J. Angulo[2], DVM, PhD; L. Van De Weyer[2], DVM, MSc; A. Arruda[1], DVM, PhD
[1]The Ohio State University, Columbus, Ohio; [2]Zoetis, Parsippany, New Jersey

Introduction

Proliferative enteropathy (PE) associated with the Gram-negative obligate intracellular bacterium *Lawsonia intracellularis* remains a challenge for swine producers globally. Oral fluids (OF) have recently been used for *L intracellularis* antibody detection with a reported 100% specificity and between 84.6-88.5% sensitivity for IgA and IgG, respectively. However, although the initial reports are promising, additional work is needed to ensure that OF can be used as a reliable detection tool for *L intracellularis*. Thus, our objectives were to 1) compare the detection rate of *L intracellularis* between fecal and oral fluid samples and 2) investigate risk factors for subclinical ileitis in vaccinated and non-vaccinated farms.

Materials and methods

Forty Canadian swine sites located in Manitoba (n = 20), Ontario (n = 18), and Quebec (n = 2) were recruited for the study. Of those, 20 sites were actively vaccinated with a *L intracellularis* vaccine during the study period. From June to October 2021, Swine fecal and oral fluid samples were collected from 3 randomly selected pens at their mid- and late-finishing phases, yielding 6 fecal and 6 oral fluid samples from each farm. An in-house PCR (Biovet Inc.) was used to detect *L intracellularis* DNA. A questionnaire was used to collect farm demographics and herd health information. A logistic mixed model was used to investigate the association between *L intracellularis* detection and specimen type controlling for farm demographics. In particular, the model fit *L intracellularis* detection as a binary outcome under binomial distribution with the following independent variables: specimen type, the use of *L intracellularis* vaccine and other vaccines and production phase at sample collection, with pen and farm identifications, and the province where the farm located as random effects. In addition, to investigate potential farm-level risk factors of *L intracellularis*, two Poisson regression models were constructed separately for farms with and without active vaccination for *L intracellularis*. These models contained the number of samples positive to *L intracellularis* (outcome), the number of collected samples (offset), the detection of common swine pathogens (eg, *Mycoplasma hyopneumoniae*, porcine reproductive and respiratory syndrome virus, porcine circoviruses) and the history of ileitis and vaccination.

Results and discussion

Overall, 235 fecal and 205 oral fluid samples were tested via PCR for *L intracellularis*. Results show 14.5% of fecal samples (34/235) and 22.9% of oral fluid samples (47/205) were positive. Oral fluids were 2.5 times more likely to detect *L intracellularis* than fecal samples. Among non-vaccinated farms, those positive for porcine circoviruses (odds ratio = 4.99) and showing gastrointestinal signs (OR = 10.00) were more likely to be positive for *L intracellularis*. In contrast, for vaccinated farms, those positive for *Mycoplasma hyopneumoniae* (OR = 4.41), ileitis positive in the past two years (OR = 3.08), vaccinating once with 1/2 dosage (OR = 4.55), and vaccinating for ileitis before 2021 (OR = 9.85) were more likely to be positive to *L intracellularis*. The results showed the presence of *L intracellularis* in swine oral fluids and its potential as a diagnostic specimen. In addition, the history of ileitis and the vaccination strategy regarding dosage and boosting was associated with subclinical ileitis in *L intracellularis* vaccinated herds.

https://doi.org/10.54846/am2023/113

Study of *Mycoplasma hyopneumoniae* gene mutations associated with antimicrobial resistance in field samples

Cipriano De Abreu[1]; Mark Schwartz[1,2]; Paul Yeske[3]; Maria Pieters[1,4,5]

[1]Veterinary Population Medicine Department, College of Veterinary Medicine, University of Minnesota, St. Paul, Minnesota; [2]Schwartz Farms, Inc., Sleepy Eye, Minnesota; [3]Swine Vet Center, St. Peter, Minnesota; [4]Veterinary Diagnostic Laboratory, College of Veterinary Medicine, University of Minnesota, St. Paul, Minneosta; [5]Swine Disease Eradication Center, College of Veterinary Medicine, University of Minnesota, St. Paul, Minnesota

Introduction

Mycoplasma hyopneumoniae (*M hyopneumoniae*) is a slow growing and hard to isolate bacterium,[1] that causes respiratory disease in swine operations worldwide.[2] Antimicrobial therapy is one of the main strategies used to control infections in the field.[3] Different antimicrobial classes, including macrolides, tetracyclines, fluoroquinolones, aminoglycosides, lincosamides and pleuromutilins are used to control *M hyopneumoniae* infections.[4] Various studies have reported the detection of antimicrobial resistance (AMR) mutations in *M hyopneumoniae* isolates.[6-9] Mutations in five different genes have been assessed and correlated with increased AMR in *Mycoplasmas*.[6] In this study, the detection of mutations, in the form of single nucleotide polymorphisms (SNPs), previously correlated with *M hyopneumoniae* AMR was evaluated in field specimens.

Materials and methods

An *M hyopneumoniae* positive 5,000-sow farm in which elimination by herd closure and medication had been attempted, was selected for the study. Deep tracheal secretions were collected from gilts and sows with clinical signs suggestive of *M hyopneumoniae* infection and from a cluster of neighboring sows (n = 251). Specimens were refrigerated and transported to the laboratory where they were processed for DNA extraction and *M hyopneumoniae* species-specific PCR. Specimens with Ct values ≤ 30 were selected for molecular characterization of the previously reported AMR genes

(n = 51). Gene segments coding for DNA gyrases (gyrA, gyrB), DNA topoisomerases (parE, parC) and 23S rRNA were amplified.[7-9] Resulting PCR products were purified and submitted for Sanger sequencing. Forward and reverse sequences for each gene were assembled to obtain a consensus sequence using in BioEdit 7.0.5. Consensus sequences were aligned to a reference sequence of *M hyopneumoniae* J strain (GenBank accession number AE017243.1) using ClustalW function in MEGA 11.0 software. The assembled sequences were evaluated for the presence of SNPs by manual examination.

Results and discussion

Sequencing success rate was greater in samples with Ct values ≤ 20 than in samples with greater Cts. A total of 46 gyrA, 28 gyrB, 47 parC, 44 parE and 17 23S RNA gene sequences were obtained out of the 51 samples tested. Sequences shared a homology ranging from 98.88% to 99.83% with those of the *M hyopneumoniae* reference strain. Various mutations, both silent and missense, were detected in the evaluated genes. A single SNP previously correlated to increased AMR was identified in parC and 23S RNA genes. In the parC gene, the SNP was detected in 45/47 of the analyzed sequences and represented a base substitution in position 248 (C248A), previously correlated to increased resistance *in vitro* to fluoroquinolones in *M hyopneumoniae*.[8] The SNP detected in the 23S RNA gene, was present in all analyzed sequences (17) and represented a base substitution in position 2071 (A2071G), previously associated with increased *in vitro* resistance to macrolides and lincosamides in *M hyopneumoniae*.[9] Both mutations have been reported in other *Mycoplasma* species.[5] Correlation between previous antimicrobial treatments applied at the farm and the detected mutations was not evaluated in this study. Findings of this research reinforce the need to apply culture-independent assays to detect AMR in *M hyopneumoniae*, which could aid improving disease control strategies.

References

1. Marois C et al., 2007. *Veterinary Microbiology*. 120. 1–2.
2. Pieters M & Maes D. 2019. In: Diseases of Swine. 863-871.
3. Maes D et al., 2020. *Veterinary Journal (England)*, 259-260.
4. Hannan P et al., 1997. *Antimic. Agent Chemot*. 41:2037–2040.
5. Gautier-Bouchardon A. V, 2018. *Microbiology spectrum*, 6(4).
6. Vicca J et al., 2007. *Microb Drug Resist*. 13:166–170.
7. Stakenborg T et al., 2005. *Microb Drug Resist*.11:290–294.
8. Le Carrou J et al., 2006. *Antimic. Agent Chemot*. 50:1959–1966.
9. Felde O et al., 2018. *PLoS ONE* 13(12): e0209030

Poster 29

https://doi.org/10.54846/am2023/114

Rethinking PCV2 surveillance: Revealing the value of dynamics over time

Gillian Greaves, MSc; Fabian O. Chamba Pardo, PhD, DVM; Courtney Werth, MSc; Clint Lichty, DVM
South West Ontario Veterinary Services, Stratford, Ontario, Canada

Introduction

Recently, cases of clinical porcine circovirus associated diseases (PCVAD) have been reported throughout all stages of production, despite successful porcine circovirus type 2 (PCV2) vaccination programs existing in North America for many years. To date, the longitudinal relationship between viral level, production parameters, and health status (eg, PRRSV) has not been well understood, specifically how this information may guide PCV2 control measures.

Methods

A longitudinal study was initiated in November 2021 to surveil 25 commercial flows across Canada with varying clinical PCV2 historical diagnostics. A flow-based sampling protocol was developed that monitors 13 cohorts of pigs per flow. A cohort is defined as 4-weeks of pig production, where the processing fluid marks the middle of the cohort. Samples are collected in each cohort at 4 strategic time points during the production cycle (processing fluids and oral fluids in downstream nurseries and finishing sites). Processing fluid samples contain 1-week worth of pooled processing tissues. Oral fluids are collected at 3 time points; at the start of nursery (~2 weeks post-weaning), end of nursery (~6 weeks post-weaning), start of finishing (~2 weeks post-placement) and consist of 3 pooled rope samples. Existing literature on sample type and quantity, pooling of samples, collection time, and environmental contamination was considered to develop this sampling strategy, while also balancing practicality and costs associated. PCV2 levels in enrolled flows are assessed using PCV2 PCR Ct values. In addition to diagnostics, survey information containing PCV2 vaccination protocols, herd and flow demographics, other disease statuses, and genetics are collected for each flow. Production parameters are gathered from all stages and will be associated with PCV2 PCR Ct values.

Results

To date, 15 flows in Ontario, five in Western Canada, and five in Quebec have been enrolled, representing > 30,000 sows with varying PCV2 historical diagnostics. Enrolled flows range in size from 185 to 3000 sows, from three different genetics suppliers. Approximately 75% of flows are multi-site production, the remainder are farrow-to-finish sites, and 2 herds are genetic multipliers. The majority of flows vaccinate gilts for PCV2, few herds vaccinate sows for PCV2, and all flows vaccinate nursery-aged pigs with 5 commercial PCV2 vaccines. PRRSV positive, stable, and negative health statuses are represented.

PCV2 PCR Ct values are variable between different flows enrolled, but also within-herd variability exists amongst cohorts (Figure 1). In some flows, the dynamics remain consistent. While in other flows, PCV2 levels change over time. As an example, a multiplier is enrolled with some cohorts that have low circulation of virus and others that have high levels of PCV2 (low Ct values). Viral levels of PCV2 fluctuate over time, with some cohorts showing increased circulation in later stages as compared to their initial processing fluid sample. Sixty one percent of PF PCV2 PCR Ct values from 92 cohorts, were between 37 and 40 (negative) and 33% of Ct values were below 35. Approximately 27% (19/71) of PCV2 PCR Ct values for early nursery oral fluids and 29% (17/58) and 34% (13/38) of late nursery and early finishing oral fluids were below 35, respectively. Production parameters including mortalities, mummies, and farrowing rate will be associated with PCV2 PCR Ct values.

Conclusion

Our data suggests that strategically timed longitudinal sampling of PCV2 within an entire production flow may be necessary to understand the changing dynamics of the virus over time. Future analyses will examine how PCV2 PCR Ct values are correlated with production parameters, to advance PCV2 control and management.

Figure 1: PCV2 PCR Ct value dynamics, by cohort, in 20 flows including processing fluids and oral fluids at the start and end of nursery (~2 and 6-weeks post-weaning), and start of finishing (~2 weeks post-placement).

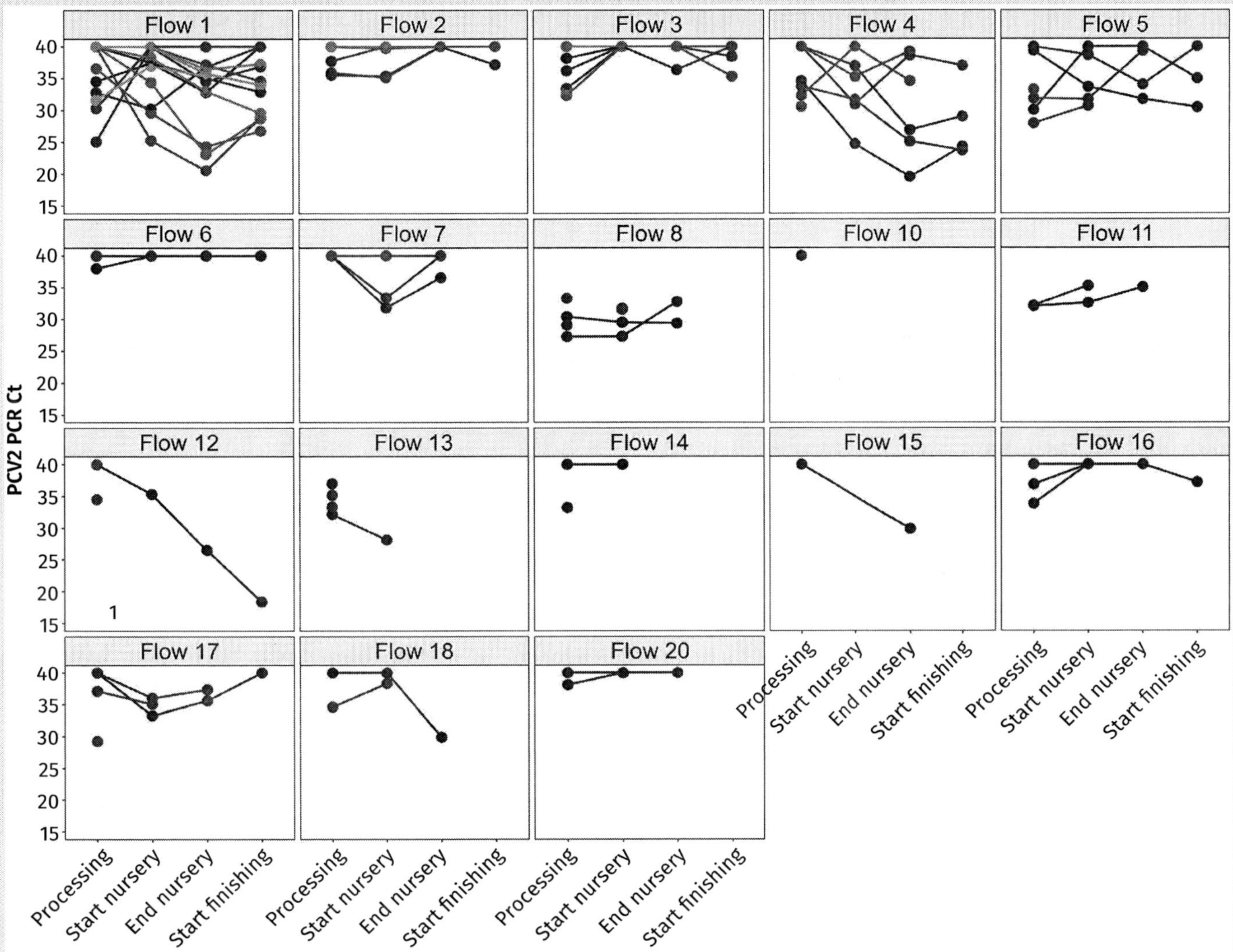

https://doi.org/10.54846/am2023/115

Evaluation of the virucidal activity of Intervention® against PRRSV 1-4-4 L1C and PRRSV 1-7-4

Erin Kettelkamp[1], DVM; Alyssa Betlach[1], DVM, PhD; Angie Quinonez-Munoz[2], BS; Sagar M. Goyal[2], PhD
[1]Swine Vet Center P.A., St. Peter, Minnesota;
[2]University of Minnesota, College of Veterinary Medicine, St. Paul, Minnesota

Introduction

Porcine reproductive and respiratory syndrome virus (PRRSV) continues to cause significant economic and production losses in the swine industry. Emergence of virulent PRRSV strains shown to be resistant to high temperature environments[1] raises concerns for current time and temperature supply quarantine protocols used on swine farms. Reliable methods for decontamination of fomites are necessary to mitigate the introduction and spread of pathogens. Previous research has elucidated disinfectant efficacy against modified live PRRS vaccine.[2] However, validation of current disinfectants is warranted in the face of more environmentally stable PRRSV field strains. Due to the risk of spreading novel pathogens, an in-vitro study was performed to measure disinfectant effectiveness. The aim of this study is to compare the effectiveness of accelerated hydrogen peroxide (Intervention®, Virox Animal Health; IVN) for inactivation of PRRSV 1-4-4 1C and PRRSV 1-7-4 on non-porous surfaces and to assess the time necessary for virus inactivation on simple surfaces following treatment with Intervention®.

Materials and methods

Two viruses (ie, PRRSV 14-4 lineage 1C and PRRSV 1-7-4) were propagated and titrated in MARC 145 cells. Two 24 well, plastic plates (Costar® REF 3524) were labeled per strain and 100 µL of virus was applied to seven wells of each plate respectively. Plates were dried in a biosafety cabinet at room temperature (20°C) for one hour. Then, 200 µL of IVN at 1:256 dilution was added to six wells of each plate followed by 800 µL of sterile beef extract eluent after five and ten minutes. Three replicates were conducted per time point. For each plate, one well was only eluted with BE elution buffer to serve as a positive control. Serial 10-fold dilutions were prepared of the eluates in MEM with 4% fetal bovine serum and antibiotics in 96-well plates using three wells/sample. Dilutions were inoculated in monolayers of MARC 145 cells prepared in 96-well microtiter plates. Plates were incubated at 37°C under 5% CO_2 and examined daily for the appearance of cytopathic effect (CPE). The 50% end points were calculated after seven days of incubation. Virus titers were calculated using the Karber method (1931) and expressed as $\log_{10}$ $TCID_{50}$ per 100µl. Moreover, indirect immunofluorescence was performed for four random samples to confirm the absence of cytopathic effects. Differences in viral detection between internal positive control and each treatment were determined by conducting a Fisher's exact test. A *P*-value of < 0.05 was used for significance.

Results and discussion

Cytotoxicity was present at 10^{-1} dilution for all treated samples. Additionally, indirect immunofluorescence results were negative for all samples tested. Compared to the positive control titers, absence of viral growth was observed for each treatment (*P*-value = 0.035; Table 1), resulting in complete viral inactivation. (No differences were observed between 5- and 10-minute treatments, or between PRRSV strains (Table 1). In conclusion, IVN at 1:256 dilution was able to inactivate > 99% of PRRSV 1-7-4 and 1-4-4 L1C. Using IVN, at this rate, may be an effective way to decontaminate non-porous surfaces or fomites entering swine farms when allowing for adequate time and temperature quarantine is not achievable to inactivate highly virulent strains of PRRSV.

References

1. Quinonez-Munz, A., & Goyal, S. (2022). *Comparative survival of different strains of porcine reproductive and respiratory syndrome virus*. SDEC Partners Research Update. Retrieved June 2022, from Swine Disease Eradication Center.

2. Dee, S. (2006). *An assessment of 3 sanitation protocols for PRRSV-positive transport vehicles*. Swine Health Research Report. Retrieved February 8, 2022, from **https://www.porkcheckoff.org/wp-content/uploads/2021/02/04-182-DEE-UofMN.pdf**

Table 1: Inactivation of PRRSV by IVN at 1:256 dilution

	PRRSV1-7-4		PRRSV 1-4-4 1C	
REPLICATES	5 min	10 min	5 min	10 min
1	0.00	0.00	0.00	0.00
2	0.00	0.00	0.00	0.00
3	0.00	0.00	0.00	0.00
Only BE buffer	2.83 Log_{10} $TCID_{50}$/100µl		2.50 Log_{10} $TCID_{50}$/100µl	
Positive control	4.50 Log_{10} $TCID_{50}$/100µl		2.83 Log_{10} $TCID_{50}$/100µl	

https://doi.org/10.54846/am2023/116

Progression of the newly emerged PRRSV L1C 144 variant in breeding herds

M. Kikuti, DVM, MPH, PhD; C. M. Mellini, DVM, MS; S. Rossow, DVM, PhD;
A. Rovira, DVM, PhD; C. A. Corzo, DVM, MS, PhD
Department of Veterinary Population Medicine, University of Minnesota, Saint Paul, Minnesota

Introduction

The Morrison Swine Health Monitoring Project (MSHMP) is a national voluntary initiative currently monitoring PRRSV in ~50% of the US sow population.[1] MSHMP monitors weekly sow farm PRRSV status and also ORF5 sequences, when available. Here we describe how the newly emerged lineage 1C RFLP 144 (L1C144) variant[2] has affected the monitored breeding herds.

Methods

The PRRSV L1C144 variant is being monitored amongst MSHMP participating systems through their sequence submissions to tree main veterinary diagnostic laboratories. Submissions were linked to a participant site using premises ID, farm name, production system, and/or location. Date of the first detected L1C144 variant sequence was linked to the site's reported PRRS status if the reported change to a new break occurred within ± 30 days of the sequence detection. Amongst sites affected, changes in status as well as time to stability and percentage of sites that reached stability within one year were described. Production data of sites that reported a rebreak associated with a L1C144 variant were also described

Results

As of July 2022, 1,030 sequences characterized as the L1C144 variant were identified, of which one-fourth (269/1030) originated from breeding herds. We were able to link sequences to 64 different sites from 13 production systems. However, sequence detection was only linked to a change in the reported PRRSV status for 49 sites. Of those, up to August 2022, 20 (40.8%) remained as status 1 since the first detection of this variant and 8 (16.3%) reported a new viral introduction (i.e. 1.2 status). The remaining 10, 3, 1, and 7 herds changed from the intial status 1 due to the detection of this variant to status 2, 2fvi, 2vx and depopulation, respectivelly. Amongst the sites that remained as status 1 throughout, they have remained as such for a median of 55 weeks (min. 4, max. 86, IQR: 32-63). For the sites that achieved stability, the median time to stability was 43 weeks (min. 18, max. 61, IQR: 29-58). Sites that depopulated did so in a median of 4 weeks from the outbreak detection (min. 2, max.37, IQR: 2-21). Sites that reported another viral introduction before reaching stability (1.2 status) did so in a median of 31 weeks (min. 6, max. 47, IQR: 23-43). Amongst those, other L1C 144 variant virus was identified within ± 30 days of status change in two sites. The minimum nucleotide identity between all within-farm L1C144 variant sequences of the first site was > 98%, indicating that in fact the same the same strain was circulating. A second increase in total sows removed (death and euthanized), percent stillborn, and total abortions was observed around the 1.2 change in status for this site.The within-farm nucleotide identity of L1C144 variant sequences for the second site was as low as 97.5%. The 1.2 change in status accompanied an increase in percent mummies and percent stillborn.

Discussion

Sites that were affected by this variant and were able to achieve stability did so within the previsouly described average of PRRS stability (~41 weeks).[3] In 2018 and 2019, before the emergence of this variant, 31% and 28% MSHMP sites that did not reach stability within one year, respectivelly. Still, excluding the depopulations and sites that detected this variant for the first time less than one year ago, 19 of the L1C 144 variant affected sites (58%) did not reach stability within one year. Here, rebreaks were considered according to change in status to 1.2, which usually represents a different PRRS sequence was identified. However, because many sites are detecting similar L1C144 variant sequences to the initial outbreak, some might opt to report a new outbreak due to the clinical manifestations whiles others may not. Many rebreaks might have gone undetected and need to be further investigated.

References

1. Perez AM, Linhares DCL, Arruda AG, et al. Individual or Common Good? Voluntary Data Sharing to Inform Disease Surveillance Systems in Food Animals. *Front Vet Sci*. 2019;6. **doi:10.3389/fvets.2019.00194**

2. Kikuti M, Paploski IAD, Pamornchainavakul N, et al. Emergence of a New Lineage 1C Variant of Porcine Reproductive and Respiratory Syndrome Virus 2 in the United States. *Front Vet Sci*. 2021;8. **https://doi.org/10.3389/fvets.2021.752938**

3. Sanhueza J, Vilalta C, Corzo C, Arruda A. Factors affecting Porcine Reproductive and Respiratory Syndrome virus time-to-stability in breeding herds in the Midwestern United States. *Transbound Emerg Dis*. 2018;Dec 6. **doi:10.1111/tbed.13091**

Poster 32

https://doi.org/10.54846/am2023/117

Tongue tips fluids: A new specimen for PRRSV RNA detection from dead pigs

Isadora F. Machado[1], DVM; Edison S. Magalhaes[1], DVM, MS; Ana Paula P. Silva[1], DMV, MS, PhD; Daniel C. A. Moraes[1], DVM, MS; Guilherme A. Cezar[1], DVM, MS; Mafalda P. Mil-Homens[1], DVM, MS; Onyekachukwu H. Osemeke[1], DVM, MS; Rodrigo Paiva[1], DVM, MBA, MS; Cesar A. A. Moura[2], DVM, PhD; Luis G. Gimenez-Lirola[1], DVM, MS, PhD; Phillip C. Gauger[1], DVM, MS, PhD; Giovani Trevisan[1], DVM, MBA, PhD; Gustavo S. Silva[1], DVM, MS, PhD; Daniel C. L. Linhares[1], DVM, MBA, MS, PhD

[1]Department of Veterinary Diagnostic and Production Animal Medicine, Iowa State University, Ames, Iowa;
[2]Iowa Select Farms, Iowa Falls, Iowa

Introduction

Porcine reproductive and respiratory syndrome virus (PRRSV) remains one of the most challenging pathogens for swine production.[1] Therefore, characterizing the activity of PRRSV in herds over time is critical to guiding biosecurity and disease management practices. Even though serum and processing fluids sampling are the most common method for PRRSV surveillance in breeding herds, other methods, such as family oral fluids, are also used depending on the desired sampling objective.[2,3] Recently, a novel alternative method using tongue tips from dead piglets was described to effectively detect PRRSV RNA via RT-qPCR.[4] The objective of this study was to investigate the use of tongue tip fluids for PRRSV monitoring in terms of PRRSV RNA and antibody detection and open reading frame five (ORF5) sequencing, as well as compare tongue tip sampling to other conventional methods.[5]

Methods

Three PRRSV RNA-positive unstable breeding herds (Farms A, B, and C) were selected for sampling. On each farm, tongue tips (= 2 cm) from dead piglets were collected in disposable bags (n = 16) from three age groups: newborn (< 24h of age), processing (2 - 7 days of age), and weaning (18 - 21 days of age). Blood samples (n = 45) were randomly collected from one piglet per litter from the same room and ages from which tongue tips were collected. Processing fluids (n = 3) and family oral fluids (n = 25) were collected from processing and weaning age pigs, respectively. Tongue tips, processing fluids, and family oral fluids were tested individually, serum samples were tested in pools of five for PRRSV RNA, and the positivity rate of RNA detection in tongue tips was compared to serum samples, processing fluids, and family oral fluids for each age group. Moreover, three tongue tip exudate samples from farm B were individually tested for ORF5 sequencing and ELISA. The ELISA test was based on the IDEXX PRRSV X3 ELISA previously adapted for isotype-specific (IgG, IgA, IgM) antibody detection in processing fluids.

Results

In farms A and B, PRRSV RNA was detected in 100% and 95.2% of tongue-tip samples, respectively, and pooled serum samples were positive in all age groups in different percentages for these farms (Table 1). In particular, for these two farms, the cycle threshold was 4.7 and 0.4 points, respectively, lower in tongue tips (29.3 and 26.4) compared to serum samples for the newborn age category (34 and 26.8). Additionally, PRRSV RNA was detected in processing fluids at farm A (100% positive samples) and B (50% positive samples), while family oral fluids were positive in all three farms, in different percentages (farm A 11.7%, farm B 22.8%, and farm C 43.4%). At farm C, no PRRSV RNA was detected in tongue tips, and pooled serum samples were PRRSV RNA-positive only at the weaning age. Moreover, three PRRSV RNA-positive tongue tip samples from farm B were also positive for PRRSV antibodies with S/P ratios of 1.25, 1.70, and 1.54, respectively. The ORF5 analysis detected a wild-type PRRSV of RFLP 1-1-2 of Lineage 1A for the three samples.

Discussion

Risk-based sampling using tongue tips from dead animals provided a good estimate of the population's health status. Further, no PRRSV RNA was detected in tongue tips when serum samples from the same group tested PRRSV RNA-negative. Collecting tongue tips from dead piglets was practical, suitable, and time-efficient for farm personnel under field conditions. Together, these results demonstrated the potential value of tongue tips for PRRSV diagnosis, monitoring, and surveillance.

References

1. Holtkamp DJ, Kliebenstein JB, Neumann EJ, et al. Assessment of the economic impact of porcine reproductive and respiratory syndrome virus on United States pork producers. *J Swine Health Prod.* 2013;21(2):72-84.

2. Lopez W, Angulo J, Linhares D, and Zimmerman J. 2018. Porcine reproductive and respiratory syndrome monitoring in breeding herds using processing fluids. *Journal of Swine Health and Production,* 26(3):146-150.

3. Almeida, M. N., Rotto, H., Schneider, P., Robb, C., Zimmerman, J. J., Holtkamp, D. J., ... & Linhares, D. C. L. (2020). Collecting oral fluid samples from due-to-wean litters. *Preventive veterinary medicine,* 174, 104810.

4. Baliellas, J., Novell, E., Enric-Tarancón, V., Vilalta, C., & Fraile, L. (2021). Porcine Reproductive and Respiratory Syndrome Surveillance in breeding Herds and Nurseries Using Tongue Tips from Dead Animals. *Veterinary Sciences,* 8(11), 259.

5. Machado, I. F., Magalhães, E. S., Poeta Silva, A. P. S., Moraes, D. C., Cezar, G., Mil-Homens, M. P., ... & Linhares, D. C. (2022). Porcine reproductive and respiratory syndrome virus RNA detection in tongue tips from dead animals. *Frontiers in Veterinary Science,* 1356.

Table 1: PRRSV RNA detection by sample type and age group in three commercial breeding herds

Farm	Age group	Serum		Tongue tips		Processing fluids		Family oral fluids	
		Percentage positive	Ct average	Percentage positive	Ct average	Percentage positive	Ct average	Percentage positive	Ct average
A	Newborn	33.3%	34.0	100%	29.3	NA	NA	NA	NA
	Processing	53.0%	24.4	100%	27.8	100%	32.2	NA	NA
	Weaning	33.3%	33.5	100%	36.3	NA	NA	11.7%	34.8
B	Newborn	23.0%	26.8	90.9%	26.4	NA	NA	NA	NA
	Processing	25.0%	19.3	100%	27.8	50%	21.9	NA	NA
	Weaning	37.5%	21.6	100%	29.3	NA	NA	22.8%	32.4
C	Newborn	0%	-	0%	-	NA	NA	NA	NA
	Processing	0%	-	0%	-	0%	-	NA	NA
	Weaning	60.0%	34.8	0%	-	NA	NA	43.4%	33.1

NA = not applicable

Poster 33

https://doi.org/10.54846/am2023/118

Which factors are relevant to identifying high nursery mortality groups under field conditions?

E. S. Magalhães[1]; C. A. A. Moura[2]; P. Thomas[2]; G. Trevisan[1]; J. J. Zimmerman[1]; D. J. Holtkamp[1]; D. C. L. Linhares[1];G. S. Silva[1]

[1] Veterinary Diagnostic and Production Animal Medicine Department, Iowa State University, Ames, Iowa;
[2] Iowa Select Farms, Iowa Falls, Iowa

Introduction

Swine nursery mortality is an important metric in the swine industry and it is caused by a complex interaction of factors involving the animals, environment, and pathogens. The impact of these factors is different between and within farms over time, creating a complex scenario when the objective is to analyze retrospective data and identify the major factors increasing swine nursery mortality. Despite knowledge of the various possible causes of nursery mortality (eg, birth weight, weaning age, PRRS, number of sources, ventilation, and stocking density), determining the most important factors for a determined population is difficult. Therefore, the objective of this study was to evaluate the use of a machine-learning model to identify high-nursery mortality groups and rank the major factors associated with groups of pigs within one US swine production system.

Methods

This study utilized retrospective data from 1,825 groups of pigs marketed between July 2021 to July 2022 in one US swine production system. Nursery mortality of each nursery group was the outcome of the analysis, which was converted into a dichotomous variable (High vs Low mortality) by categorizing the top 25% (n = 457) of groups with the highest nursery mortality averages to be categorized as "High" nursery mortality, while the remaining groups (75%; n = 1,368) were classified as "Low" nursery mortality. Along with the outcome, a total of 19 factors from the pre-weaning phase and the nursery phase for each group of pigs (breeding-to-nursery) were available in the dataset, and the objective was to rank these factors by importance for classifying groups as High nursery mortality. For this purpose, the total data was initially divided into a training set (80% of the data = 1,460) and a test set (20% = 365 groups) where a random forest (RF) machine learning model was applied to explore the association of the 19 factors and the groups with high nursery mortality using the "caret" package[1] on R software. The RF model was initially trained on the training set of data using a 10-fold cross-validation process and later tested on the test set of data by classifying the outcome of the groups in that dataset. The model performance was then assessed by calculating the accuracy (proportion of groups correctly classified), specificity, and sensitivity. Furthermore, the variables were ranked by importance using the Gini index values, where larger values represent factors with higher relevance for nursery mortality.

Results

The mean nursery mortality of the top 25% quartile among the study population was 10.2%, while the remaining 75% had 2.55% average mortality. Across the 19 factors ranked using the RF model (Figure 1), the most important factors were pre-weaning mortality, farrowing rate, sow farm size, weaning age, stocking density in the nursery, average parity at farrow, PRRS status at weaning, and antibiotic application at weaning. Partial dependence plots highlighted the importance of the following characteristics for higher nursery mortality: high pre-weaning mortality in weaned groups, young weaning age, large farms, high stocking density, young average parity at farrow, and epidemic PRRS. Furthermore, the RF model developed for this study classified high nursery mortality groups on the test training set with 92,3% accuracy, 92,9% sensitivity, and 89,8% specificity.

Discussion and conclusions

The approach utilized in this study provides an automated and structured method to identify and rank the importance of multiple factors impacting nursery mortality. Also, it demonstrated the importance of factors in the breed-to-wean phase of production on the downstream survivability of nursery groups. Notably, this analysis helps producers prioritize the most important factors contributing to nursery mortality which may be used to guide the allocation of resources to reduce nursery mortality.

Reference

1. Kuhn, M. (2008). Building predictive models in R using the caret package.Journal of Statistical Software,28,1–26.

Figure 1: Variable importance rank performed by random forest machine-learning model.

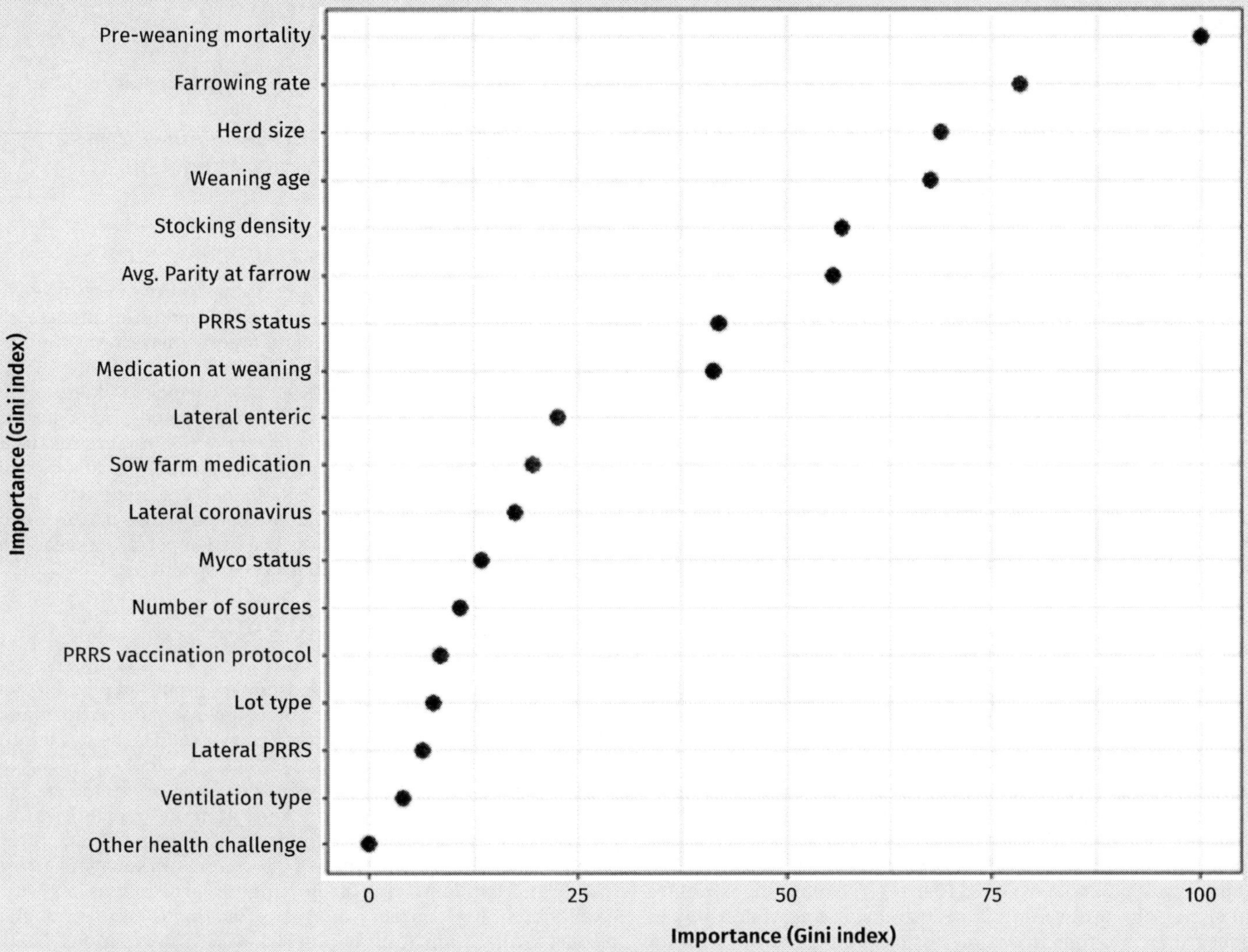

https://doi.org/10.54846/am2023/119

Forecasting swine nursery mortality under field conditions

E. S. Magalhães[1]; D. Zhang[2]; C. Wang[1, 2]; C. A. A. Moura[3]; P. Thomas[3]; G. Trevisan[1]; J. J. Zimmerman[1]; C. J. Rademacher[1]; D. J. Holtkamp[1]; G. S. Silva[1]; D. C. L. Linhares[1]
[1] Veterinary Diagnostic and Production Animal Medicine Department, Iowa State University, Ames, Iowa;
[2] Department of Statistics, College of Liberal Arts and Sciences, Iowa State University, Ames, Iowa;
[3] Iowa Select Farms, Iowa Falls, Iowa

Introduction

Swine nursery mortality is a key performance indicator (KPI) in the swine industry and refers to the aggregated mortality of a group of pigs throughout the nursery phase. Notably, this outcome accounts for almost 50% of all growing phase mortality in a short period (~ 49 days of 170 days). Even though swine nursery mortality is a metric concerning the post-weaning phase of production, previous authors described that the retrospective performance of the piglets during the pre-weaning phase and nursery stocking conditions are highly correlated with post-weaning survivability.[1,3] Although the importance of the pre-weaning phase on the downstream nursery performance is known, this information is not yet being utilized in the swine industry to predict the nursery performance. Thus, the objective of this study was to build a forecasting model to predict the nursery mortality for individual nursery groups of pigs based on their retrospective performance in the pre-weaning phase and stocking conditions.

Methods

Algorithms were developed using SAS software to automatically import and integrate data from multiple sources within one large swine production system in the Midwestern USA, creating a system-specific master table. Retrospective information for 1831 nursery groups stocked between January 2021 – July 2022 in this company was available in the master table, containing variables concerning the pre-weaning phase and stocking conditions in nursery sites. Multiple linear regression (MLR), Neural Network (NN) models, and Random Forest (RF) models were investigated for predicting nursery mortality using the explanatory variables collected in the master table, which consisted of 31 variables from the pre-weaning phase and 12 related to stocking conditions. The outcome was nursery mortality, defined as the number of piglets that died in the nursery phase as a percentage of the pigs placed. The forecasting methods were implemented using the "caret" package.[2] The parameters were tuned with 3 repetitions of the 5-fold cross-validation. In order to evaluate the prediction performance on the dataset, for each record, the training set was the dataset excluding that record. Therefater, the trained model was then used to predict the nursery mortality of for the record excluded previously. The predictive performance of the models (MLR and RF) was compared with the actual values for nursery mortality, being evaluated in terms of root mean square error (RMSE), and coefficient of determination (R2), and mean absolute error (MAE).

Results

This study population's mean nursery mortality percentage was 4.48%. After the cross-validation and parameter tuning process, the model with the best overall prediction performance was the Random Forest in terms of Root Mean Squared Error (RMSE = 0.3973), Mean Absolute Error (MAE = 0.2995), and coefficient of determination (R^2 = 0.7390). Hence, the random forest model could explain 73% of the variability in the groups where the nursery mortality values were predicted. Figure 1 shows the average predicted vs. actual mortality over time using the RF model. The prediction performance for MLR in terms of RMSE, MAE, and R^2 was 0.5569, 0.4316, and 0.4873, respectively. Lastly, the prediction performance for the NN model was 0.4571, 0.3291, and 0.6545, respectively.

Discussion and conclusions

This study demonstrated the importance of analyzing all retrospective information available for nursery groups of pigs at placement in nursery sites to predict their downstream performance through this phase. Among the nursery groups, the actual and predicted mortality were higher for cohorts stocked during the winter period, while for the entire data, the predicted average monthly mortality followed the trend of actual mortality of the groups in the same period. These results can help to guide future decisions within swine production systems in terms of future trends and by identifying/projecting the groups with a greater likelihood of high nursery mortality, leading then to strategical resources allocation in support of risk-based interventions. Lastly, it is important to note that no causal claim can be made from these results, and the external validity for other systems under different conditions is limited.

References

1. Agostini et al. 2015. Managing variability in decision making in swine growing-finishing units. Ir. Vet. J. 68.

2. Kuhn, M. (2008). Building predictive models in R using the caret package.Journal of Statistical Software,28,1–26.

3. Main et al 2004. Increasing weaning age improves pig performance in a multisite production system. J. Anim. Sci. 82, 1499–1507.

Figure 1: Actual vs predicted (RF) nursery mortality using a random forest ML model.

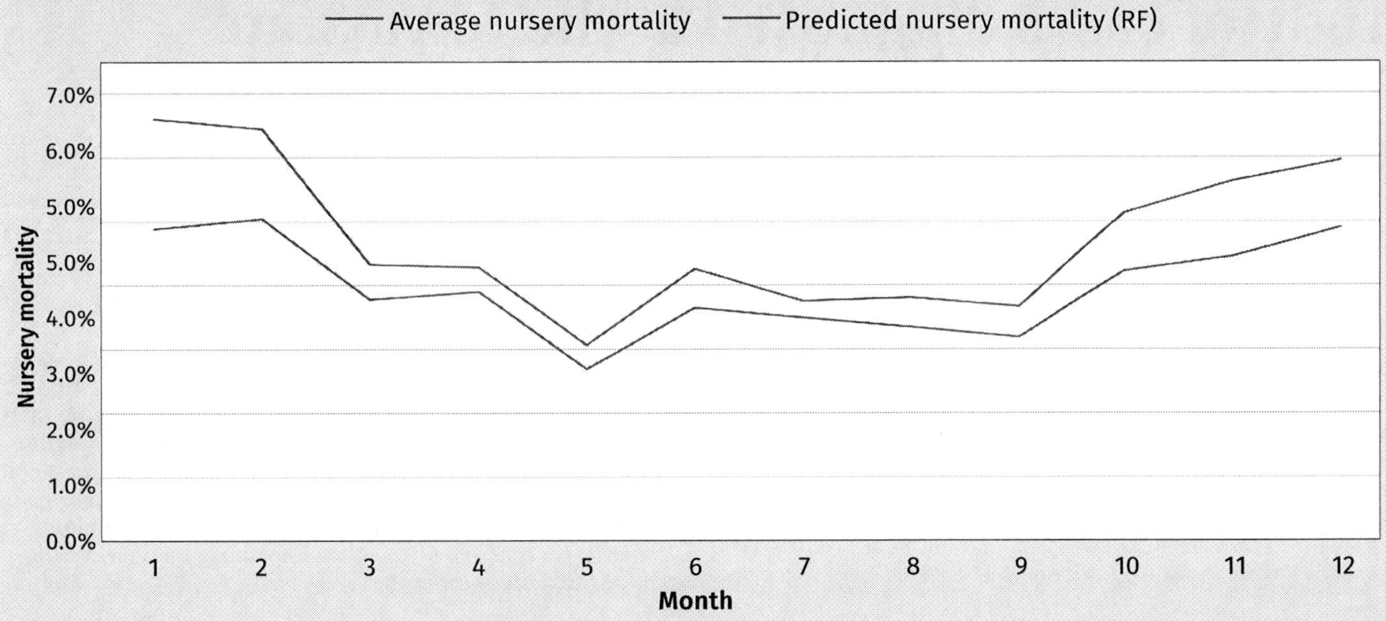

Poster 35

https://doi.org/10.54846/am2023/120

PRRSV encephalitis and myelitis: A retrospective study of cases diagnosed at the Iowa State University Veterinary Diagnostic Laboratory from 2016 to 2022

M. Mainenti, DVM, DACVP; P. C. Gauger, DVM, MS, PhD; M. C. Rahe, DVM, MPH, PhD, DACVP; C. Siepker, DVM, DACVP; A. Michael, DVM, PhD, DACVP; P. Pineyro, DVM, MVSc, DVSc, PhD; M. Almeida, DVM, MS, PhD; D. R. Magstadt, DVM, MS
Department of Veterinary Diagnostic and Production Animal Medicine, Iowa State University, Ames, Iowa

Introduction

Non-suppurative encephalitis is a sporadic presentation of porcine reproductive and respiratory syndrome virus (PRRSV) reported in both natural and experimental infections.[1,2] Recently, the Iowa State University Veterinary Diagnostic Laboratory (ISU VDL) has observed an increased number of diagnostic cases with neurologic disease due to PRRSV encephalitis and/or myelitis, rising from an average of 1.8 per year in 2007-2020 to 17 cases in 2021.

Methods

A total of 40 cases comprising 82 pigs diagnosed with PRRSV encephalitis/myelitis at the ISU VDL from January 2016 to July 2022 were retrospectively evaluated to report common clinical and pathologic features as well as identified viral strains. The diagnosis of PRRSV encephalitis and/or myelitis was based on a combination of histologic evaluation and viral detection by rRT-PCR and immunohistochemistry in lung and central nervous system (CNS). Sequencing of the ORF5 region was performed in 35 cases.

Results

Neurologic signs were reported in 73% of the cases; respiratory signs in 63%. Paddling was the most frequent neurologic sign (41%), followed by ataxia (28%) and lateral recumbency (17%). Other signs include head tilt, trembling, rear paralysis, nystagmus, stargazing, high fever, lethargy, purple ears, and increased mortality (up to 12% rate when reported). On histology, cerebrum was most consistently affected (100%), followed by cerebellum (97%) and brainstem (88%). Spinal cord was submitted from 30 animals and affected in 90%. Interstitial pneumonia consistent with PRRSV infection was present in 54/57 animals (95%) where lung was available.

The average cycle threshold (Ct) value of PRRSV detection by rRT-PCR in lung and CNS was 16.4 (median 16). Only lineage 1 PRRSV was detected, consisting of 5 sublineages (L1A, L1C, L1D, L1E, L1H) and 8 RFLP patterns. Almost 60% of the sequences were 1-4-4 L1C variant (29%) or 1-7-4 L1A strain (29%). Coinfections were present in 60% of cases and mainly consisted of bacterial meningoencephalitis (23%; mainly *Streptococcus suis*), polyserositis (20%; mainly *Glaesserella parasuis* and *Mycoplasma hyorhinis*), and synovitis (18%; mainly *Mycoplasma hyorhinis*).

Discussion

PRRSV involvement of the CNS should be considered as a differential in neurologic cases, especially when occurring concomitantly with respiratory signs or in farms with a history of ongoing PRRSV infection. Sequencing data suggest that PRRSV neurologic disease is not strain-dependent. While exclusive PRRSV detection by PCR in the CNS does not fully support causation, histologic and immunohistochemical evaluation of the brain, spinal cord, and other tissues can aid in differentiating this infection from primary bacterial meningoencephalitis and other causes of neurologic disease.

References

1. Cao J, Li B, Fang L, et al. Pathogenesis of nonsuppurative encephalitis caused by highly pathogenic Porcine reproductive and respiratory syndrome virus. *J Vet Diagn Invest.* 2012; 24(4):767-71.

2. Thanawongnuwech R, Halbur PG, Andrews JJ. Immunohistochemical detection of porcine reproductive and respiratory syndrome virus antigen in neurovascular lesions. *J Vet Diagn Invest.* 1997; 9:334-37.

Poster 36

https://doi.org/10.54846/am2023/121

Addressing biocontainment through environmental contamination assessment in farms housing PRRS Lineage 1C 1-4-4 positive pigs

Claudio Marcello Melini[1]; Mariana Kikuti[1]; Montse Torremorell[1]; Laura Bruner[2]; Chase Stahl[3]; Brian Roggow[3]; Paul Yeske[2]; Brad Leuwerke[2]; Matt Allerson[4]; Katie O'Brien[4]; Mark Schwartz[5]; Cesar A. Corzo[1]

[1]Department of Veterinary Population Medicine, University of Minnesota; [2]Swine Vet Center; [3]Fairmont Vet Clinic; [4]Holden Farms; [5]Schwartz Farms

Introduction

The newly emerged porcine reproductive and respiratory syndrome virus (PRRSV) L1C 1-4-4 strain disseminated throughout southern Minnesota during the fall of 2020, spring and summer of 2021. Unfortunately, the strain continues to reach susceptible farms in the Midwest. Given the speed at which this strain was disseminating, questions regarding transmission routes became evident as farms with robust biosecurity measures and some that were strategically located were testing positive for this strain. Therefore, an exploratory study was designed to assess whether PRRS virus RNA could be detected in surfaces outside of barns housing positive pigs as a proxy of dissemination and to understand whether biocontainment interventions needed to be implemented.

Methods

A total of eight (seven PRRS positive and one PRRS free?) conveniently selected farms (3 breeding and 5 growing pig farms) were included in this study. A sampling protocol that included wiping of specific surfaces was developed in conjunction with swine practitioners. A surface sampling protocol previously validated for PRRSV and influenza[1,2] was used. Briefly, sterile cloths moistened with 20 ml of PBS were used to sample a 1ft. × 1ft. area. Once the sample was collected, the cloth was deposited in a sterile plastic bag and fluid was extracted and poured into Corning Falcon™ tubes (Fisher scientific), labeled according to sampled surface and farm. Samples were then refrigerated and submitted for PRRSV RT-PCR testing at the University of Minnesota Veterinary Diagnostic Laboratory.

Results

Out of 169 environmental samples collected at seven farms housing PRRSV L1C 1-4-4 positive pigs, 19 (11.24%) samples from six farms yielded a positive RT-PCR result. Eight (42%) of the positive samples originated from exhaust fan housings in four farms with a Ct value ranging from 30.29 to 35.98. Main door entry doorknobs in four farms tested positive in four occasions with Ct values ranging between 31.68 and 36.35. The remaining positive samples originated from surfaces such as ante-room floors and mortality carts with Ct values ranging from 25.41 to 35.38.

Conclusions

Overall, the detection of viral genetic material on surfaces on the outer areas of a farm housing PRRSV-positive pigs is possible using a targeted surface sampling. This exploratory study provides evidence that biocontainment efforts to prevent the spread of PRRSV from infected farms need to be reviewed.

References

1. Garrido-Mantilla, J., Alvarez, J., Culhane, M., Nirmala, J., Cano, J.C., Torremorell, M. (2019). Comparison of individual, group and environmental sampling strategies to conduct influenza surveillance in pigs. BMC Veterinary Research. 15(1), 61.

2. Vilalta, C., Sanhueza, J., Garrido, J., Murray, D., Morrison, R., Corzo, C., Torremorell., M. (2019). Indirect assessment of porcine and respiratory syndrome virus status in pigs prior to weaning by sampling sows and the environment. Veterinary Microbiology, 237, 108406.

https://doi.org/10.54846/am2023/122

Evaluation of infectiousness of the newly emergent and virulent Minnesota PRRS virus variant

Claudio Marcello Melini[1]; Mariana Kikuti[1]; Montserrat Torremorell[1]; Kimberly VanderWaal[1]; Stephanie Rossow[2]; Jerry Torrison[2]; Cesar A. Corzo[1]
[1]Department of Veterinary Population Medicine, University of Minnesota;
[2]Veterinary Diagnostic Laboratory, University of Minnesota

Introduction

In the swine industry, porcine reproductive and respiratory syndrome virus (PRRSV) is a costly disease due to losses from reproductive failure (eg, abortions, stillbirths, premature farrowing), increased mortality and poor growth performance in growing pigs.[1] The estimated yearly loss for US producers can reach up to $664 million.[2] During the fall of 2020 there was an increase in wean-to-market mortality that was associated to a newly emerged variant of PRRSV classified as L1C 1-4-4 in over 91% of the cases.[3] The epidemic subsided in early 2021; however, during late spring and summer another wave of transmission occurred infecting multiple sow and growing pig farms leading the industry to question why this virus was spreading at such a fast pace and breaking through current biosecurity practices. A study was designed to compare this newly emergent strain with two other strains, a L1A 1-7-4 and L9 1-4-2 to assess the minimum infectious dose (ID_{50}), viremia magnitude, shedding pattern, and histopathological lesions in experimental infected pigs.

Methods

For each strain, thirty-six PRRSV-negative four-week-old barrows were randomly allocated to five different challenge groups and one control group. The PRRSV inoculum was titrated, and through 10-fold dilutions were five different infectious concentrations were obtained starting from a 10^4 $TCID_{50}$/mL. Five days after arrival at the University of Minnesota Veterinary Isolation Facility and once both PRRS ELISA and PCR confirmed that pigs were PRRS negative, pigs were intranasally (ie, 1mL per nostril) challenged according to the groups selected dose (ie, 10^4 to 10^0 $TCID_{50}$). The control group was sham-inoculated with minimum-essential media. Blood and nasal-oral-rectal swabs were collected at day post-inoculation (DPI) 0, 1, 2, 4, 7, 11, 16, 21, 26 and 30 to monitor for the presence of the virus using RT-PCR for PRRSV. On DPC 11 two pigs from each group were euthanized and tissues were collected for histopathological assessment.

Results

Preliminary results show that at day 4 post inoculation PRRSV L1C 1-4-4 had 26/30 positive pigs, meanwhile L1A 1-7-4 had 19/30 positive pigs, and L9 1-4-2 had 15/30 positive pigs. Histopathological results and RT-PCR for the swaps to assess shedding are still pending.

References

1. Fraile, Lorenzo. "Control or Eradication? Costs and Benefits in the Case of PRRSV." *Veterinary Record* 170, no. 9 (March 2012): 223–24. **DOI: 10.1136/vr.e1386**

2. Holtkamp, D., Liebenstein, J., Neumann, E., Zimmerman, J., Rotto, H., Yoder, T., Wang, C., Yeske, P., Mowrer, C., Haley., C. (2013) Assessment of the economic impact of porcine reproductive and respiratory syndrome virus on the United States pork producers. Journal of Swine Health and Production, 21(2), 72-84.

3. Kikuti, M., Paploski, I., Pamornchainavakul, N., Picasso-Risso, C., Schwartz, M., Yeske, P., Leuwerke, B., et al. Emergence of a New Lineage 1C Variant of Porcine Reproductive and Respiratory Syndrome Virus 2 in the United States. *Frontiers in Veterinary Science* 8 (October 18, 2021): 752938. **DOI: 10.3389/fvets.2021.752938**

Poster 38
https://doi.org/10.54846/am2023/123

Ten-year retrospective analysis of bacterial pathogens associated with influenza infections in pigs (2012-2021)

M. Meneguzzi, DVM, MS; C.M. Melini, DVM; N. Noyes, DVM, PhD; M. Torremorell, DVM, PhD; F. Vannucci, DVM, MS, PhD
Department of Veterinary Population Medicine, University of Minnesota, Saint Paul, Minnesota

Introduction

Influenza A virus (IAV) is a common and important primary pathogen of pigs worldwide. Influenza infections are often accompanied by opportunistic bacteria that take advantage of the mechanisms of virulence of IAV to establish secondary co-infections, contributing to a significant enhancement of clinical signs, mortality, and greater economic losses.[1] For instance, the cost of IAV co-infections has been estimated to range between $3 and $10 per pig with higher costs attributed to co-infections with multiple pathogens.[2] Although IAV and bacterial co-infections cause great economic impact and impairment of animal health, comprehensive analyses of trends are still scarce, especially for bacterial pathogens prone to changes in antimicrobial resistance. Diagnostic laboratory cases can provide relevant information on the occurrence of viral-bacterial co-infections. Therefore, the purpose of this study was to perform a ten-year (2012-2021) retrospective analysis of diagnostic cases to identify bacterial agents associated with IAV co-infections in pigs.

Methods

Diagnostic laboratory data from 19,455 porcine lung samples from the United States with clinical history of respiratory illness were received at the Minnesota Veterinary Diagnostic Laboratory (MVDL) between January 2012 and December 2021. Lung samples were tested for IAV by PCR. Subsequently, the majority of the IAV PCR-positive cases were confirmed by virus isolation and subtypes of the IAV isolates were determined by PCR. Most of the lung samples were also tested for the presence of bacteria by aerobic culture. In addition, for some cases, infection with *Mycoplasma hyorhinis* and/or *Glaesserella parasuis* was detected by PCR test. Since both pathogens can cause systemic infections, specimens from pericardium, peritoneum, pleura, and bronchial swabs were considered.

Results

Of the 19,455 total porcine lung samples received, 16.45% (n = 3,202) were PCR positive for IAV with the most common IAV subtypes being H1N1 (34.07%), H1N2 (22.02%), and H3N2 (19.43%). Other subtype profiles were observed but in lower quantities. Out of the 3,202 IAV PCR-positive cases, IAV infections alone were observed in 824 (25.73%) cases, while for most of the cases (n = 2,378; 74.27%), IAV infection was observed together with one or two other bacterial co-infections. The most common bacteria in co-infection with IAV were *Pasteurella multocida, M hyorhinis, G parasuis,* and *Streptococcus suis*. We did not observe an association between the bacterial co-infection profile and a specific IAV subtype.

Discussion

Influenza causes damage to the respiratory epithelium facilitating interaction with secondary bacterial infections. Our study indicates that bacterial coinfections are prevalent in diagnostic cases of IAV infected pigs. The common bacteria associated with IAV infection is in accordance with what has been previously described.[3] Our study also emphasizes the need for research on microbial coinfections to enhance our understanding of IAV ecology and disease dynamics with other agents. The knowledge obtained by continued research efforts on IAV coinfections will provide useful information to further devise novel, effective approaches for swine disease prevention and control programs.

Acknowledgments

We would like to thank you University of Minnesta Veterinary Diagnostic Laboratory for kindly sharing the data for the analysis.

References

1. Morens DM, Taubenberger JK, Fauci AS. Predominant role of bacterial pneumonia as a cause of death in pandemic influenza: Implications for pandemic influenza preparedness. J Infect Dis. 2008;198(7):962-970. **doi:10.1086/591708**

2. Dykhuis Haden C, Painter T, Fangman T, Holtkamp D. Assessing production parameters and economic impact of swine influenza, PRRS and *Mycoplasma hyopneumoniae* on finishing pigs in a large production system. In: Proceedings of the AASV Annual Meeting. 2012:75-76.

3. Choi YK, Goyal SM, Joo HS. Retrospective analysis of etiologic agents associated with respiratory diseases in pigs. Can Vet J. 2003 Sep 1;44(9):735–737.

Poster 39

https://doi.org/10.54846/am2023/124

Neurosensory lesions in field cases of porcine reproductive and respiratory syndrome virus (PRRSV) infection

Alyona Michael, DVM, PhD, DACVP; Marcelo Almeida, DVM, MS, PhD; Chris Siepker, DVM, DACVP; Marta Mainenti, DVM, DACVP
Department of Veterinary Diagnostic and Production Animal Medicine, Iowa State University, Ames, Iowa

Introduction

PRRSV is a primary global driver of swine morbidity across all stages of production. Although neural lesions are reported with various PRRSV type 2 strains,[1] CNS disease was not a common feature in field outbreaks until recently. In 2021, there was an abrupt spike in cases diagnosed with PRRSV encephalitis at ISU-VDL, coinciding temporally with the emergence of the virulent PRRSV 1-4-4 L1C variant.[2] The SARS-CoV-2 pandemic has drawn scientific and laypublic attention to the myriad manifestations of viral neurotropism, including adverse impacts on quality of life associated with sensory neuropathies, such as anosmia and ageusia. Recently, recognition of unexpectedly widespread neurotropism in field cases of PRRSV received ISUVDL (internal data) raises concerns regarding impacts of viral encephalitis on pig health due to damage of sensory pathways unamenable to routine clinical evaluation.

Methods

Our objective was to evaluate neurosensory pathology in pigs with PRRSV encephalitis. Clinical signs of fulminant PRRSV encephalitis include shivering, opisthotonos, ataxia, nystagmus, circling and paddling, indicative of variable but potentially widespread involvement of the brain and spinal cord. Cribriform plate, eyes, and tongue from pigs with confirmed PRRS encephalitis (n = 10) were retrospectively trimmed and stained with H&E.

Results

Viral lesions, including mononuclear perivasculitis, malacia, and glial nodules were present in olfactory and optic tracts of pigs with PRRSV encephalitis (9/10). Olfactory nerves and rhinencephalon were most frequently affected (9/10), followed by iridoretinitis and optic neuritis (3/10). Tongues featured perivascular lymphocytic cuffing (5/10), however, neuritis was not observed in available sections. In several cases (2/10), there was additional presence of suppurative meningoencephalitis indicative of bacterial coinfection.

Discussion

In addition to overt ambulatory deficits and loss of consciousness, olfactory and audiovisual damage observed in PRRS-infected animals is strongly supportive of potential sensory deficits that may have lasting impacts on performance and welfare in convalescent animals. These results suggest a need for further investigation of PRRS neurotropism, as it relates to clinical phenotype, strain-dependent risk for sensory impairment, husbandry and management considerations of potentially sensorially impaired survivors in the face of a PRRS outbreak.

References

1. Halbur PG, Paul PS, Meng XJ, Lum MA, Andrews JJ, Rathje JA. Comparative pathogenicity of nine US porcine reproductive and respiratory syndrome virus (PRRSV) isolates in a five-week-old cesarean-derived, colostrum-deprived pig model. J Vet Diagn Invest. 1996 Jan;8(1):11-20.

2. Kikuti M, Paploski IAD, Pamornchainavakul N, Picasso-Risso C, Schwartz M, Yeske P, Leuwerke B, Bruner L, Murray D, Roggow BD, Thomas P, Feldmann L, Allerson M, Hensch M, Bauman T, Sexton B, Rovira A, VanderWaal K, Corzo CA. Emergence of a New Lineage 1C Variant of Porcine Reproductive and Respiratory Syndrome Virus 2 in the United States. Front Vet Sci. 2021 Oct 18;8:752938.

Poster 40

https://doi.org/10.54846/am2023/125

Use of family oral fluid to detect influenza A virus in the breeding herd

D. C. A. Moraes[1]; P C. Gauger[1]; I. F. Machado[1]; G. Cezar[1]; O. H. Osemeke[1]; R. Paiva[1]; C. A. Moura[2]; M. Putnam[1]; J. Smallfield[3]; A. Ramirez[4]; G. S. Silva[1]; D. C. L. Linhares[1]

[1]Department of Veterinary Diagnostic and Production Animal Medicine, Iowa State University, Ames, Iowa; [2]Iowa Select Farms, Iowa Falls, Iowa; [3]Kansas State University, Manhattan, Kansa; [4]University of Arizona, Oro Valley, Arizona

Introduction

Sampling options used to detect influenza A virus (IAV) may vary based on the sensitivity of the sample type and the convenience to perform sample collection. Family oral fluids (FOF) specimens have demonstrated the ability to be a practical and effective population-based sample type for porcine reproductive and respiratory syndrome virus (PRRSV) RNA detection at the weaning age.[1] Recently, udder wipes have been described for IAV detection in swine as a population sample with promising diagnostic sensitivity.[2] Nasal wipe is an individual sampling method to minimize stress on sows and pigs, and is easier sample type compared to nasal swabs.[3] However, studies have not evaluated the detection of IAV in breeding herds under low prevalence, lower than 15%, using FOF, and information is lacking regarding the successive use of FOF over time as a method to monitor the presence and epidemiology of IAV in commercial herds. Therefore, this study aims to compare different sample types (FOF, udder wipes, sow nasal wipes, and drinker wipes) to detect IAV RNA in breeding herds.

Methods

Three breeding herds were screened in the farrowing room at weaning age (17 - 21 days) using udder wipes to ensure evidence of IAV circulation prior to sample collection of FOF. Eligible herds had at least 15% (6/35) of samples testing positive, and sample collection took place within 48 hours (about 2 days) of screening. Samples were collected at the breeding herd (n = 3,500 sows) using matched sets (n = 57) of FOF, udder wipes, and nasal wipes from sows, drinker wipes (from sows and respective litters), and all 2 - 3-week-old pigs (close to weaning) within the respective litter for testing RT-rtPCR for IAV RNA. FOF sampling followed the methodology described by Almeida et al. (2021). All samples were tested at the Iowa State University Veterinary Diagnostic Laboratory for IAV RNA by RT-rtPCR following standard and previously validated protocols with the VetMAX- Gold SIV Detection Kit. A sample was considered positive with a Ct value < 38.

Results

One of the three breeding herds tested IAV RNA positive (6/35, 17.1%) at the screening with udder wipes. This farm was selected for the sample collection 48 hours later. A total of 57.9% (33/57) FOF samples tested positive, 49.1% (28/57) tested positive on udder wipes, 28.1% (16/57) tested positive on nasal wipes collected from sows, and 15.8% (9/57) tested positive in drinker wipes (Table 1).

Discussion and conclusion

This study assessed the use of different sample types to detect IAV in one commercial breeding herd. Detection of IAV can be challenging with prevalence ranging from 7 to 57% in breeding herds.[4] Based on these results, FOF may be a resourceful alternative population-based sample type for IAV in the breeding herd. Population-based sampling (ie, udder wipes, FOF) allows cost reduction and potential gains on the probability of detection by increasing the coverage of pigs, pens, rooms, and/or sites sampled. The number of positive udder wipes (77.2%) in room A agrees with the previous reports[2]. Providing different sampling approaches with alternate sample types for IAV detection allows veterinarians to make informed decisions on IAV monitoring in swine populations.

Table 1: IAV detection by FOF, udder wipe, nasal wipes in sows, and drinker wipes at weaning age

		FOF		Udder wipes		Sow nasal wipes		Drinker wipes	
Room	**Age**	**Samples positive % (n)**	**Ct value average (min-max)**	**Samples positive % (n)**	**Ct value average (min-max)**	**Samples positive % (n)**	**Ct value average (min-max)**	**Samples positive % (n)**	**Ct value average (min-max)**
A	Weaning	86.3% (19/22)	29.0 (24.4-33.9)	77.2% (17/22)	32.5 (27.5-37.4)	40.9% (9/22)	36.4 (33.1-37.9)	27.2% (6/22)	35.8 (33.7-36.9)
B	Weaning	54.1% (13/24)	32.9 (25.0-37.9)	45.8% (11/24)	33.4 (27.5-36.1)	29.1% (7/24)	37.4 (37.3-37.8)	12.5% (3/24)	36.5 (35.1-37.5)
C	Weaning	9.0% (1/11)	34.5	0 (0/11)	-	0 (0/11)	-	0 (0/11)	-

FOF = family oral fluids

Acknowledgments

American Association of Swine Veterinarians Foundation for funding this study, and the production system for allowing to perform the field study.

References

1. Almeida, M.N. et al. Longitudinal piglet sampling in commercial sow farms highlights the challenge of PRRSV detection. Porc Health Manag 7, 31 (2021).

2. Garrido-Mantilla et al. Comparison of individual, group and environmental sampling strategies to conduct influenza surveillance in pigs. BMC Vet Res 15, 61 (2019).

3. Nolting JM et al. Nasal Wipes for Influenza A Virus Detection and Isolation from Swine. J Vis Exp. 2015 Dec 4;(106):e53313.

4. Chamba Pardo FO et al. Influenza Herd-Level Prevalence and Seasonality in Breed-to-Wean Pig Farms in the Midwestern United States. Front Vet Sci. 2017 Oct 11;4:167.

https://doi.org/10.54846/am2023/126

Effect of storage temperature and time on PRRSV RNA detection by RT-qPCR

B. Munguía-Ramírez[1]; B. Armenta-Leyva[1]; A. Henao-Díaz[3]; F. Ye[2]; K. Doolittle[4]; S. Zimmerman[4]; L. Giménez-Lirola[1]; J. Zimmerman[1]
[1]Veterinary Diagnostic and Production Animal Medicine; [2]Department of Statistics, Iowa State University, Ames, Iowa; [3]PIC Latinoamérica, México; [4]IDEXX Laboratories, Inc.

Introduction

Prior to testing, PCR targets in diagnostic samples are subjected to a variety of adverse conditions in the course of collection, storage, transport, and processing. However, the effect of these conditions on PCR results is largely unexplored. The objective of this study was to evaluate the effect of storage temperature x time on the detection of PRRSV RNA and a pig-specific "internal sample control" (ISC) in serum, oral fluids, and fecal samples. Note: the ISC used in this study is a pig-specific RNA target present in all pig-derived specimens and, therefore, is a measure of sample quality. The RT-qPCR kit used in the study (IDEXX Laboratories, Inc.) detected both PRRSV and ISC RNAs simultaneously.

Methods

Serum specimens (n = 5) used in the study were from pigs experimentally inoculated with contemporary wild-type PRRS viruses. Oral fluids (n = 5), and feces (n = 5) were from individually housed pigs vaccinated with a PRRSV MLV (Boehringer Ingelheim Vetmedica, Inc.). Each sample was divided into 28 aliquots (500 uL) and each aliquot was subjected to one combination of temperature (4, 10, 20, or 30°C) by time (24, 48, 72, 96, 120, 144, or 168 hours). After completing all treatments, samples were tested by RT-qPCR and results reported for PRRSV and ISC as quantification cycles (Cq). Cqs were the re-expressed as "efficiency standardized Cqs (ECqs)" and the effect of storage temperature × time on ECqs was analyzed using a mixed-effects regression model (MRM) (R 4.2.1, R core team, 2022).

Results

In serum, PRRSV RNA was stable 4, 10, and 20°C (MRM: $P > 0.05$), but a negative effect on detection was discernible at 30°C (MRM: $P < 0.05$), with an estimated ΔECq of -0.11 every 24 hrs. For the ISC, a decrease in detectable RNA was observed only at 20 and 30°C (MRM: $P < 0.05$). In oral fluids, both PRRSV and the ISC were affected at all temperature × time treatments (MRM: $P < 0.05$), with an estimated ΔECq -0.03, -0.05, -0.07, and -0.08 for PRRSV RNA, and -0.02, -0.04, -0.05, and -0.05 for ISC RNA at 4, 10, 20, and 30°C, respectively. In feces, PRRSV and ISC RNAs were also affected by storage temperature and time (MRM: $P < 0.05$). The ΔECq for PRRSV RNA was -0.05 and -0.06 at 4°C and 10°C respectively, and -0.09 at 20 and 30°C. For ISC RNA, the ΔECq at 4°C and 10°C was -0.06; at 20 and 30°C the ΔECq was -0.08 and -0.10, respectively (Table 1).

Conclusions

This study showed that the effect of storage temperature × time on PRRSV and ISC RNAs was specimen dependent. Storing serum at 4, 10, or 20°C had no detectable impact on PRRSV ECqs for up to 168 hours, but storage at 30°C resulted in loss of detectable PRRSV RNA. The ISC RNA was stable in serum at 4 and 10°C, but storage at 20 and 30°C was detrimental to detection. In contrast, oral fluid and feces showed a constant decrease in PRRSV and ISC RNAs at all temperature × time treatments. Based on these data, serum samples should be stored at ≤ 20°C to optimize PRRSV RNA detection; oral fluid and fecal samples should be frozen in a non-self-defrosting freezer and submitted frozen for testing.

Table 1: Daily (24 hrs) ECq decline by temperature[1]

Target	Specimen	4°C	10°C	20°C	30°C
PRRSV	Serum	0.00[2]	0.00[2]	0.00[2]	0.11
	Oral fluid	0.03	0.05	0.07	0.08
	Feces	0.05	0.06	0.05	0.05
ISC	Serum	0.002	0.00[2]	0.07	0.10
	Oral fluid	0.03	0.05	0.07	0.08
	Feces	0.06	0.05	0.08	0.10

[1] Data analyzed using a mixed-effects regression model
[2] No decline in ECq over time ($P > 0.05$; mixed-effects regression model).
ISC = internal sample control

Poster 42

https://doi.org/10.54846/am2023/127

The effect of freeze-thaw on PRRSV RT-qPCR results

B. Munguía-Ramírez[1]; B. Armenta-Leyva[1]; A. Henao-Díaz[3]; F. Ye[2]; K. Doolittle[4]; S. Zimmerman[4]; L. Giménez-Lirola[1]; J. Zimmerman[1]
[1]Veterinary Diagnostic and Production Animal Medicine; [2]Department of Statistics, Iowa State University, Ames, Iowa; [3]PIC Latinoamérica, México; [4]IDEXX Laboratories, Inc., USA

Introduction

Re-testing samples to confirm qPCR results or to test for additional pathogens is a common practice in both diagnostic and research settings. This requires exposing samples to repeated freeze-thaw cycles. However, the effect of freeze-thaw cycles on nucleic acid detection by qPCR is largely unexplored. Therefore, the aim of this study was to evaluate the effect of freeze-thaw cycles on the detection of PRRSV RNA and a porcine internal sample control (ISC) in serum, oral fluids, and fecal samples using a commercial RT-qPCR (RealPCR*NA PRRS Types1-2 RNA Mix, IDEXX Laboratories, Inc.).

Methods

Serum samples (n = 10) used in the study were from pigs experimentally inoculated with wild-type PRRSV (n = 5) and age-matched negative control pigs (n = 5). Oral fluids (n = 10; 6 PRRSV pos; 4 PRRSV neg), and fecal samples (n = 10; 5 PRRSV pos; 5 PRRSV neg) were from individually housed 14-week-old pigs vaccinated with a PRRSV modified-live vaccine (MLV) (Ingelvac® PRRS MLV, Boehringer Ingelheim Vetmedica, Inc.) and sampled from -7 to 42 days post vaccination (DPV).

To test the freeze-thaw effect, oral fluid, serum, and fecal samples were subjected to 2, 5, 10, or 15 complete freeze-thaw cycles and then tested for PRRSV and ISC RNA. Thereafter, the freeze-thaw effect was analyzed in a mixed-effects regression model (MRM) using R 4.2.1 (R core team, 2022).

To conduct the experiment, samples were aliquoted (1 mL) into 2 mL tubes to create 4 complete sample sets (one per freeze-thaw "treatment"). Samples were submitted to freeze-thaw cycles by freezing at -80°C and then thawing at 4°C overnight. After completion of all treatments, samples (n = 40 per specimen) were stored at -80°C before completing the designated number of cycles. Samples were then tested using a commercial PRRSV RT-qPCR that included an ISC. RT-qPCR results reported as quantification cycles (Cq) were re-expressed in function of the PCR efficiency as "Efficiency standardized Cqs (ECqs)"

Results

PRRSV was detected in all known PRRSV-positive samples at all freeze-thaw treatments (n = 64). Similarly, the ISC was detected in all samples subjected to freeze-thaw treatments (n = 360). In serum, no freeze-thaw effect was detected in either PRRSV or ISC ECqs ($P > 0.05$; MRM). In oral fluids, the slope of the regression line showed a decrease of 0.027 (95%CI: 0.020, 0.033), and 0.052 (95%CI: 0.018, 0.085) ECqs per freeze-thaw cycle for PRRSV and ISC, respectively ($P < 0.05$; MRM). In feces, freeze-thaw cycles did not affect either PRRSV or ISC ECqs ($P > 0.05$) (Figure 1).

Conclusions

This study showed that freeze-thaw cycles had no detectable impact in serum and fecal samples on PRRSV or ISC RNAs, ie, the expected ECq change per freeze-thaw cycle was negligible. In oral fluids, a freeze-thaw effect was discernable, ie, a decrease in target concentration was observed as a function of the number of freeze-thaw cycle. It should be noted that these results apply to PRRSV and that further studies are needed to address the effect of freeze-thaw on other pathogens and other matrices.

Figure 1: Effect of freeze-thaw cycles on PRRSV and ISC RNAs indicated by the slopes of the mixed-effects regression model.

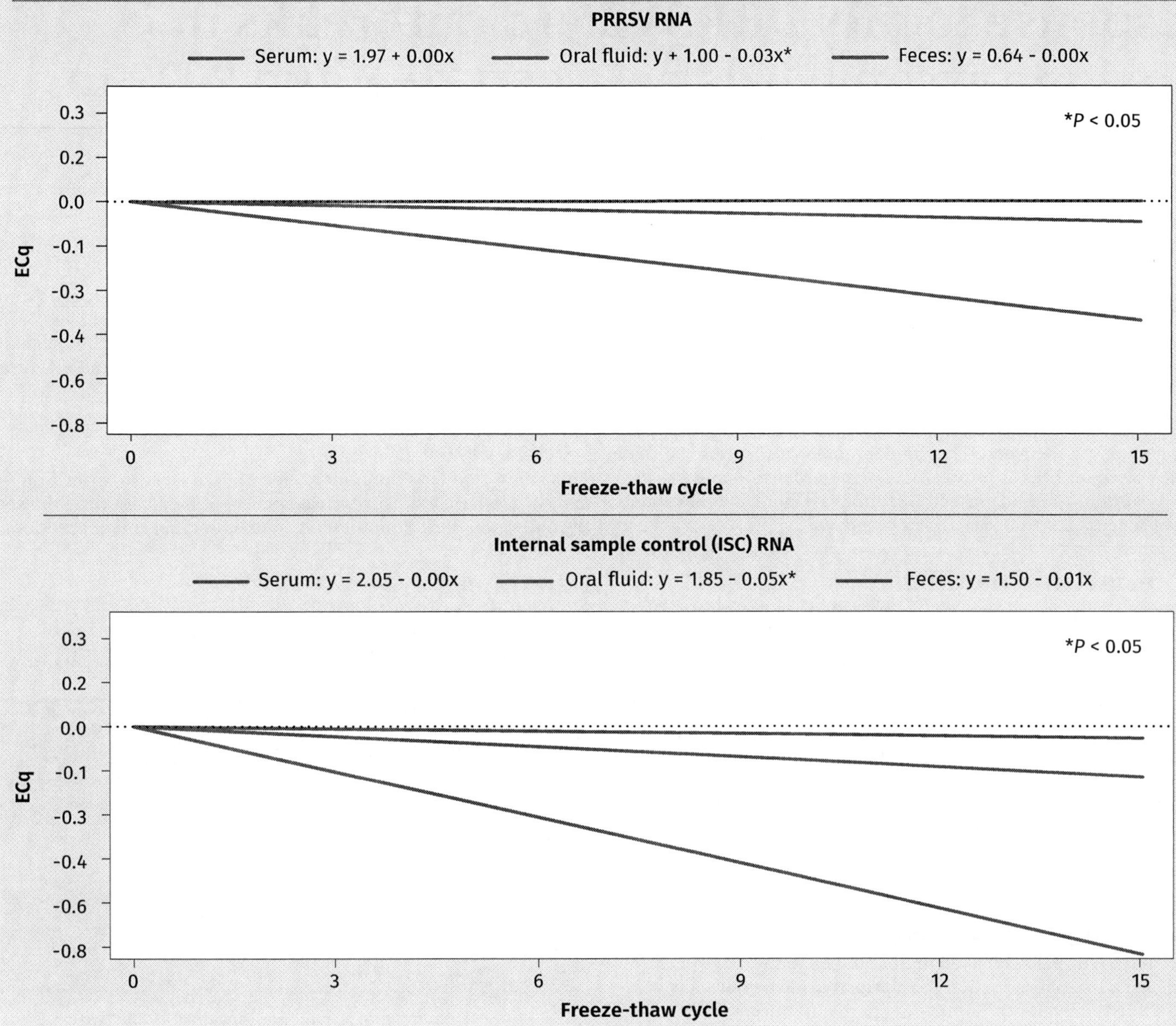

https://doi.org/10.54846/am2023/128

The impact of management practices in response to fifty PRRSV herd outbreaks in breed-to-wean herds between 2018 and 2022 in the midwestern US

Rodrigo Paiva[1]; Cesar A. Moura[2]; Pete Thomas[2]; Gustavo Silva[1]; Giovani Trevisan[1]; Daniel C. L. Linhares[1]
[1]Department of Veterinary Diagnostic and Production Animal Medicine,Iowa State University, Ames, Iowa;
[2]Iowa Select Farm Inc, Iowa Falls, Iowa

Introduction

Only 40% of the US breeding herd is negative for porcine reproductive and respiratory syndrome virus (PRRSV). Veterinarians report a longer time to produce PRRSV-negative pigs and a higher impact on productivity following outbreaks compared to 10-20 years ago.[1] The objective of this study was to assess the impact of key management practices implemented in response to fifty PRRSV herd outbreaks in breed-to-wean herds from one production system in the Midwestern US and describe whole-herd metrics of success associated with key practices.

Materials and methods

This was a retrospective cohort study following fifty PRRSV herd outbreaks during 2018-2022. Herd demographic characteristics and practices implemented as part of disease management were captured. Time-to-stability (TTS) was calculated as the number of weeks (wks) the farm took to achieve 8 consecutive weeks of PCR-negative results on processing fluids (n = 1 sample/week) or piglet serum testing (n = 30 piglets/week). Time-to-baseline-productivity (TTBP) was calculated based on an exponentially weighted moving average and 26 weeks of baseline prior to the outbreak. Total loss (TL) per 1,000 sows was calculated as the sum of piglets weaned below expected from the outbreak until reaching TTBP. Descriptive statistics compared TTS, TTBP, and TL.

Results and discussion

Most herd outbreaks followed in this study had the target to bring herds back to full productivity as a priority, and achieve low wild-type PRRSV prevalence in breeding herds, not necessarily bringing herds to the AASV status 4 (negative). Therefore, the relatively longer TTS in this study (median 77 wks) compared to the previous cohorts reported by Linhares et al,[2] and Trevisan et al,[3] 27, and 36 weeks respectively, was not surprising. Herd outbreaks where herd closure was implemented had a 40% percentage points higher success rate of achieving stability compared to those not implementing closure. In agreement with previous reports,[2,3] shorter TTS was achieved in herd outbreaks with exposure to the MLV vaccine before the outbreak and reporting whole-herd exposure sooner after the outbreak. Herd outbreaks (63%) with no reported live virus exposure (MLV or LVI) for ≥ 3 months and with no detection of PRRSV wild type before the outbreak had a higher total loss after the outbreak, which may reflect a lack of sufficient prior immunity, also reported in the previous cohort studies. *Mycoplasma hyopneumoniae* (Mhyo)-negative herds (32%) achieved TTBP sooner compared with Mhyo-positive herds. MLV vaccination after the outbreak, and sow feed medication (antibiotic week 0 - 4 after outbreak), were associated with less severe productivity impact (lower TL and shorter TTBP). The most frequent PRRSV strain detection changed from L1A RFLP 1 - 7 - 4 in 2018, 2019, and 2020 to L1C variant 1 - 4 - 4 in 2021 and 2022 respectively. The TTS, TTBP, and TL for L1C variant 1 - 4 - 4 were 59, 28, and 5,471 compared with 45, 21, and 4,814 for L1A RFLP 1 - 7 - 4 respectively.

Conclusions

This study confirms the importance of herd closure, prior immunity, whole-herd exposure, strategic use of medication right after the outbreak, PRRSV monitoring, and bio-management practices.

References

1. Yeske et al., AASV, 2022. p.350-351.
2. Linhares et al. *Prev Vet Med*, 2014. *116* (1-2), 111-119.
3. Trevisan et al., 20-109 IPPA, 2022. **https://www.fieldepi.org/POMP**

Table 1: Variability of TTS, TTBP, and TL (piglets not weaned per 1,000 sows) for the 50 herd outbreaks during 2018-2022

Percentiles	TTS	TTBP	TL
10th percentile	36	7	945
25th percentile	51	12	2,167
Median	77	22	3,524
75th percentile	113	29	6,642
90th percentile	141	45	10,475

TTS = time-to-stability; TTBP = time-to-baseline-productivity; TL = total loss

https://doi.org/10.54846/am2023/129

Assessment of the risk factors associated with weekly sow mortality rate under the condition of a production system in the midwestern United States during 2019-2021

Rodrigo C. Paiva[1]; Cesar A. Moura[2]; Pete Thomas[2]; Ben Haberl[2]; Laura Greiner[1]; Christopher J. Rademacher[1]; Daniel C. L. Linhares[1]; Gustavo S. Silva[1]
[1]Department of Veterinary Diagnostic and Production Animal Medicine, Iowa State University, Ames, Iowa; [2]Iowa Select Farm Inc, Iowa Falls, Iowa

Introduction

Sow mortality has significantly increased in the last three years in the US swine industry. The execution of hypothesis-driven research related to this topic is difficult, in part due to the lack of fundamental understanding of the root of the causes contributing to this problem. Thus, the aim of this observational study was to identify and measure the major risk factors associated with sow mortality in a large swine production system located in the Midwestern United States, a high swine-dense region.

Materials and methods

An observational study was performed using production and health data from one production system with an inventory of 240,000 sows in 48 sow farms. Data between July 2019 and December 2021 were used with a total of 6,279 herd-weeks records (total deaths sows/week). The eligibility criteria to enroll in this study consisted of all breed-to-wean herds from this production system that had all variables to be included in the analysis. Herd demographics, weekly productivity data, selected practices implemented as part of the daily management, and disease status were gathered. A multivariable Poisson mixed regression model was built and the outcome was the mortality rate/1,000 sows per week, and the sow farm was included as a random intercept. The multivariable model was obtained using a forward stepwise approach. First, univariate analyses were performed and variables with a *P*-value < 0.20 were retained and tested in the second step (multivariable). Differences in means were compared using a t-test with Tukey-Kramer adjustment and *P*-values < 0.05 were considered significant. All analyses were performed using the R program version 4.2.0.

Results

Table 1 describes the risk factors associated with sow mortality (multivariable model). The infectious causes such as breeding herds facing porcine reproductive and respiratory syndrome virus (PRRSV) outbreaks (epidemic), herds PRRSV epidemic and positive for *Mycoplasma hyopneumoniae* (Mhyo), PRRSV epidemic herds with no feed meds added, Seneca Valley Virus (SVA) positive herds were associated with higher sow mortality rate. The interaction between the PRRSV epidemic and Mhyo-positive herds was associated with a higher mortality rate. Non-infectious causes such as non-medicated feed and open pen gestation farms were associated with higher weekly sow mortality rate/1,000 sows.

Discussion and conclusions

The sow mortality in this observational study has significantly increased in the past three years as has been reported in the past years by the US swine industry. The disease status of the breed-to-wean herds, barn characteristics, and whole-herd feed medication were associated with the sow mortality rate. Herds epidemic for PRRSV, positive for Seneca virus, and the interaction of two diseases (*Mycoplasma hyopneumoniae*-positive herds and epidemic for PRRSV) showed a higher mortality rate compared to farms with single disease status. This study also documented the negative impact of the gestation type on sow mortality and showed that whole-herd medication can reduce sow mortality. The identification of risk factors is key to guiding producers and veterinarians in their decision-making process to establish proper interventions to reduce sow mortality.

References

1. Ross, Jason W. 2019. II Industry Summary 17-224.
2. Stock, Joseph D et al 2017. *Open Journal of Veterinary Medicine* 7, 85-97.
3. Deen & Xue. 1999. "Sow mortality in the US: An industry-wide perspective." **https://hdl.handle.net/11299/148140**

Table 1: Final multivariable model describing the risk factors associated with weekly sow mortality rate/1,000 sows

Variables	Categories	Mortality rate per 1,000 sows	95% CI	*P*-value
Year	2019	3.00[a]	2.84 - 3.16	< .0001
	2020	3.58[b]	3.40 - 3.76	
	2021	3.93[c]	3.73 - 4.14	
Season	Spring	3.55[a]	3.36 - 3.76	< .0001
	Summer	3.33[b]	3.15 - 3.53	
	Fall	3.57[a]	3.38 - 3.77	
	Winter	3.47[a]	3.29 - 3.66	
PRRSV status	Negative	3.16[a]	3.00- 3.33	< .0001
	Endemic	3.31[b]	3.15 - 3.48	
	Epidemic	4.03[c]	3.81 - 4.26	
Feed medication	Medication	3.27[a]	3.12 - 3.43	< .0001
	Non-medication	3.70[b]	3.53 – 3.88	
Mycoplasma status	Negative	3.45[a]	3.24 - 3.67	0.4780
	Positive	3.51[a]	3.35 - 3.69	
Gestation type	Stalls	3.35[a]	3.13 - 3.59	0.0290
	Open pen gestation	3.61[b]	3.43 - 3.79	
SVA status	Negative	3.36[a]	3.23 - 3.50	0.0040
	Positive	3.60[b]	3.37 - 3.84	
Interaction between PRRSV status and Myco status	Negative*Negative	3.21[ab]	2.98 - 3.46	< .0001
	Negative*Positive	3.12[a]	2.92 - 3.33	
	Endemic*Negative	3.32[ab]	3.09 - 3.57	
	Endemic*Positive	3.30[b]	3.11 - 3.50	
	Epidemic*Negative	3.84[c]	3.52 - 4.19	
	Epidemic*Positive	4.22[d]	3.96 - 4.50	
Interaction between PRRSV status and season	Negative*Spring	3.12[ab]	2.91 - 3.35	< .0001
	Negative*Summer	3.17[abc]	2.96 - 3.39	
	Negative*Fall	3.11[a]	2.91 - 3.33	
	Negative*Winter	3.25[abcd]	3.03 - 3.48	
	Endemic*Spring	3.30[bcd]	3.09 - 3.52	
	Endemic*Summer	3.33[cd]	3.12 - 3.54	
	Endemic*Fall	3.20[abc]	3.01 - 3.41	
	Endemic*Winter	3.42[d]	3.21 - 3.65	
	Epidemic*Spring	4.35[f]	3.97 - 4.77	
	Epidemic*Summer	3.52[de]	3.20 - 3.87	
	Epidemic*Fall	4.56[f]	4.20 - 4.95	
	Epidemic*Winter	3.76[e]	3.50 - 4.05	
Interaction between PRRSV status feed medication	Negative*Medication	3.02[a]	2.84 - 3.21	0.0030
	Negative*Non-medication	3.31[b]	3.12 - 3.52	
	Endemic*Medication	3.14[c]	2.97 - 3.33	
	Endemic*Non-medication	3.48[d]	3.29 - 3.68	
	Epidemic*Medication	3.69[e]	3.45 – 3.95	
	Epidemic*Non-medication	4.39[f]	4.11 - 4.69	

Poster 45

https://doi.org/10.54846/am2023/130

Virulence comparison of PRRSV 1-4-4 L1C variant strain and other PRRSV strains in experimentally inoculated pigs

Gaurav Rawal[1], DVM, MS; Marcelo Almeida[1], DVM, PhD; Phillip Gauger[1], DVM, PhD; Jeffrey Zimmerman[1], DVM, PhD; Fangshu Ye[2], BS; Chris Rademacher[1], DVM; Jianqiang Zhang[1], MD, PhD
[1]Department of Veterinary Diagnostic and Production Animal Medicine, Iowa State University;
[2]Department of Statistics, Iowa State University

Introduction

A PRRSV 1-4-4 L1C variant strain, which emerged in Iowa and Minnesota swine farms in October 2020, has spread to many other states and caused high production losses based on field observations.[1,2] However, no unequivocal experimental data was available to determine the virulence phenotype of the 1-4-4 L1C variant. In this experimental pig study, the virulence and transmissibility of the 1-4-4 L1C variant strain was characterized in comparison with three other circulating 1-4-4 strains (L1C non-variant, L1A, and L1H) and one highly virulent 1-7-4 L1A strain.

Methods

Seventy-two 3-week-old PRRSV-naive pigs were divided into 6 groups with 12 pigs per group. Forty-eight pigs (8/group) were for inoculation and 24 pigs (4/group) served as contact pigs. After one-week acclimation, pigs were inoculated with the corresponding virus or negative media intramuscularly and intranasally. At 2 days post inoculation (DPI), contact pigs were added to the pen adjacent to the inoculated pigs in each room. Daily temperature and clinical signs were recorded. Serum and oral fluid samples were collected at 0, 2, 4, 7, 10, 14, 21 and 28 DPI. Pigs were necropsied at 10 & 28 DPI. Fresh and formalin-fixed tissues were collected at necropsy for examinations. For all analyses, SAS was used and a *P*-value ≤ 0.05 was considered significant.

Results

The 1-4-4 L1C variant-inoculated pigs became more anorectic and lethargic, had higher mortality, had higher percentage of pigs with fever (> 40°C) during 0 - 10 DPI, and had significantly higher average body temperature than other virus-inoculated groups at several DPIs. The 1-4-4 L1C variant virus-inoculated group had significantly higher viremia levels compared to all other virus-inoculated groups at 2 DPI. There were more severe gross lung lesions in the 1-4-4 L1C variant-inoculated group compared to other virus-inoculated groups except the 1-7-4 L1A group at 10 DPI. The differences of ADG, microscopic lung lesion score, IHC score, and RNA level in different tissues at 10 DPI were not statistically significant between virus-inoculated groups. The serum IFN-α level in 1-4-4 L1C variant-inoculated group was significantly higher compared to all the virus-inoculated groups at 2 DPI (except 1-7-4 L1A group) and 7 DPI measured using ProcartaPlex™ Porcine Cytokine & Chemokine Panel 9-Plex (ThermoFisher). Serum antibodies in all virus-inoculated groups were readily detected by the commercial PRRS X3 ELISA and the ISUP virus strain-based IFA antibody assays during 7-28 DPI. At 2 days post contact, 4/4, 2/4, 2/4, 0/4, and 2/4 Contact pigs in the 1-4-4 L1C variant, 1-4-4 L1C non-variant, 1-4-4 L1A, 1-4-4 L1H, and 1-7-4 L1A groups became viremic, respectively.

Discussion

The collective data from this study confirmed that PRRSV 1-4-4 L1C variant is highly virulent in weaned pigs and the findings align with what field veterinarians observed for the L1C variant outbreak. The higher number of contact pigs becoming viremic at 2 days post contact implies the L1C variant strain may have higher transmissibility than other PRRSV strains although it needs to be confirmed with a study involving more pigs.

References

1. Kikuti M et al., Emergence of new lineage 1C variant of PRRSV2 in the United States. Fronties in veterinary science 2021, 8, 752938-752938, **doi:10.3389/fvets.2021.752938**

2. Trevisan G et al., Complete coding genome sequence of a novel PRRSV2 restriction fragment length polymorphism 1-4-4 L1C variant identified in Iowa, USA. Microbiology Resource Announcements. 2021;10(21):e00448-21.

https://doi.org/10.54846/am2023/131

Bayesian latent class analysis of the sensitivity and specificity of a novel PCR and indirect ELISA for African swine fever virus antibody

Rachel Schambow[1,2]; Luis G. Giménez-Lirola[3]; Vu Duc Hanh[4]; Lai Thi Lan Huong[4]; Nguyen Thi Lan[4]; Pham Hong Trang[4]; Do Duc Luc[5]; Ha Xuan Bo[5]; Rolf Rauh[6]; William Nelson[6]; Juan Carlos Mora-Díaz[7]; Albert Rovira[2,8]; Marie Culhane[2,9]; Andres Perez[1,2]
[1]Center for Animal Health and Food Safety, University of Minnesota, St. Paul, Minnesota; [2]Department of Population Medicine, College of Veterinary Medicine, University of Minnesota, St. Paul, Minnesota; [3]Innoceleris LLC, Ames, Iowa; [4]Faculty of Veterinary Medicine, Vietnam National University of Agriculture, Trau Quy, Gia Lam, Ha Noi, Vietnam; [5]Faculty of Animal Science, Vietnam National University of Agriculture, Trau Quy, Gia Lam, Ha Noi, Vietnam; [6]Tetracore, Inc., Rockville, Maryland; [7]Department of Veterinary Diagnostic and Production Animal Medicine, College of Veterinary Medicine, Iowa State University, Ames, Iowa; [8]Veterinary Diagnostic Laboratory, College of Veterinary Medicine, University of Minnesota, St. Paul, Minnesota; [9]Secure Food Systems Team, University of Minnesota, St. Paul, Minnesota

Introduction

Traditionally, the accuracy of new or alterative diagnostic tests is determined by comparison to a gold standard reference test. For African swine fever virus (ASFV), no suitable gold standard reference exists, creating difficulty in evaluating new diagnostic assays especially in field locations where access to advanced diagnostic techniques is limited. Bayesian latent class analysis (BLCA) uses the results of multiple imperfect tests applied to an individual to estimate the diagnostic sensitivity and specificity (DSe and DSp) of each test, forgoing the need for a gold standard reference test.[1] The objective of this work was to assess the DSe and DSp of a novel indirect ELISA for ASFV p30 antibody (iELISA, Innoceleris LLC. and Tetracore Inc) and the VetAlertTM ASFV DNA qPCR assay (qPCR, Tetracore Inc.) on paired serum and oral fluid samples.

Methods

Two modeling approaches were used and compared: 1) assumption of disease status based on population and 2) BLCA in which disease status was not assumed. Using a cross-sectional design, 398 paired serum/oral fluid samples were collected from individual pigs in Vietnam, of which 100 were from 30 acutely ASF-affected farms, 98 were from 37 chronically ASF-affected farms and 200 from 20 ASF-unaffected farms. Samples were tested using both the iELISA and PCR assays. For the assumption of disease status, acutely or chronically affected farms were assumed to be positive, and ASF-unaffected farms were assumed to be negative. DSe and 95% confidence intervals were then estimated for acute, chronic, and both populations combined, and likewise DSp was estimated using the ASF-unaffected population data. For the BLCA model, DSe and DSp were estimated by fitting a four-test, three-population model assuming dependence between serum/oral fluids tested using the same assay. Priors were estimated from previous reports of the tests' performance. A sensitivity analysis was conducted to examine the effects of vague priors on the posterior distributions.

Results

When assuming disease status, qPCR DSe was considerably higher than iELISA in the acutely affected population only (Table 1), while iELISA DSe was higher than qPCR in the chronically affected population. DSp was nearly perfect for all test and sample types. In the BLCA model, median posterior estimates represented averages across the three populations and were nearly identical to those when assuming disease status in the combined population. The DSe of qPCR was higher for both serum and oral fluids than of the iELISA. DSe using oral fluids was slightly worse for each test than serum. DSp was high for each test and sample type. DSe of iELISA using serum and oral fluids were highly correlated. Posterior estimates were robust against changes in prior distributions.

Discussion

Both tests and sample types performed with high DSp, though DSe estimates indicate that these tests may perform better using herd-level detection strategies than when applied to individual animals. DSe here was likely influenced by timing of disease in individual animals and other factors, and further work is warranted to elucidate this dynamic trend.

Reference

1. Cheung A, Dufour S, Jones G, Kostoulas P, Stevenson MA, Singanallur NB, Firestone SM. Bayesian latent class analysis when the reference test is imperfect. Rev Sci Tech OIE (2021) 40. **doi: 10.20506/rst.40.1.3224**

Table 1: Estimates and accompanying 95% confidence intervals of diagnostic sensitivity and specificity (DSe and DSp) of iELISA for African swine fever virus (ASFV) p30 antibodies (Ab) and VetAlert™ ASFV DNA qPCR on paired serum and oral fluid samples when assuming disease status in ASF acutely affected pigs only, chronically affected only, and combined populations, and Bayesian posterior median estimates and 95% credibility intervals from Bayesian latent class model when not assuming disease status.

Test	Sample	Acute DSe (95% CI)	Chronic DSe (95% CI)	Combined acute & chronic DSe (95% CI)	DSp (95% CI)	Bayesian posterior DSe (95% CI)	Bayesian posterior DSp (95% CI)
iELISA for ASF Ab	Serum	16 (9.43-24.7)	70.4 (60.3-79.2)	42.9 (35.9-50.1)	100 (98.2-100)	46.2 (39.4-52.9)	99.6 (98.1-99.97)
	Oral fluids	11 (5.62-18.8)	56.1 (45.7-66.1)	33.3 (26.8-40.4)	99 (96.4-99.9)	36.0 (29.7-42.9)	98.5 (96.5-99.6)
PCR for ASFV DNA	Serum	74 (64.3-82.3)	56.1 (45.7-66.1)	65.2 (58.1-71.8)	100 (98.2-100)	70.0 (63.6-76.0)	99.6 (98.0-99.99)
	Oral fluids	69 (58.97-77.9)	34.7 (25.4-44.98)	52.0 (44.8-59.2)	100 (98.2-100)	53.9 (46.7-61.0)	99.6 (98.1-99.99)

https://doi.org/10.54846/am2023/132

Influence of barn fill rate on nursery pig exit weight

Haley K. Schwecke[1,4]; Mike D. Tokach[1]; Alyssa M. Betlach[3]; Julia P. Holen[1,3]; Jordan T. Gebhardt[2]
[1]Department of Animal Sciences & Industry, Kansas State University, Manhattan, Kansas;
[2]Department of Diagnostic Medicine/Pathobiology, Kansas State University, Manhattan, Kansas;
[3]Swine Vet Center, St. Peter, Minnesota; [4] College of Veterinary Medicine, University of Minnesota, St. Paul, Minnesota

Introduction

Swine barns are often filled with multiple groups of pigs over a period of numerous days. However, there is little research available to know if the second or subsequent group entering the facility have different performance than the first fill group.[1,2] Therefore, a binomial regression analysis was conducted to determine the influence of barn fill rate on nursery pig's exit weights.

Methods

The datasets used in this project were an accumulation of 20 trials conducted from 2017 to 2019 with a total of 37,539 nursery pigs. An excel-based spreadsheet was used to compile, visualize, and descriptively compare growth performance data across trials. Fill time ranged from one to eight days with barns filled up to four groups of pigs across four commercial research facilities located in central Minnesota. The following data were collected per trial at the pen-level: Fill rate including fill group (range of one to four groups and days between entry), starting body weight (BW), final BW, average daily gain (ADG), average daily feed intake (ADFI), mortality rate, feed to gain ratio (F/G), starting head count, and final head count. An equation, "AdjFinalWt," was generated to adjust final BW of pens to a common number of days on feed using growth rate during the last period of each experiment. Two linear mixed models were fit using a forward selection process to determine significant factors influencing adjusted final weight (AdjFinalWt; model one) and actual end BW (exit weight; model two). Statistical models included experiment as a random intercept. Data were analyzed with the GLIMMIX procedure of SAS (v. 9.4), and results considered significant at $P < 0.05$.

Results and conclusions

In model one, the following factors were significant ($P < 0.001$): starting BW, source, fill group, and days on feed. In model two, the following factors were significant ($P < 0.001$): days on feed, starting BW, source, days after barn began filling, and fill group. Both models revealed a difference in pig exit BW due to their fill group (Figure 1). Adjusted final BW of pigs in the first two fill groups were 3.5 and 3.3 lb heavier adjusted final BW, respectively, than pigs in the fourth fill group, with pigs in the third fill group being intermediate (2.0 lb heavier than fourth fill group) in model one. Similarly, when days on feed was included in model 2, fill group had a similar magnitude of impact with pigs in fill group 4 being 3.8, 3.5 and 1.7 lb lighter final BW than pigs in the first, second, and third fill groups, respectively. Both models support heavier exit weights of nursery pigs when two or fewer groups enter a facility and lower final BW for the third and fourth groups placed in the facility, regardless of days between entry. In conclusion, these results can help producers determine the economic impact of filling nursery barns with fewer weaning groups.

References

1. Faccin, J. E. G., M. D. Tokach, M. W. Allerson, J. C. Woodworth, J. M. DeRouchey, S. S. Dritz, F. P. Bortolozzo, and R. D. Goodband. 2020. Relationship between weaning age and antibiotic usage on pig growth performance and mortality. J. Anim. Sci. 98:1-10. **doi:10.1093/jas/skaa363**

2. Wu, F., J. Liao, M. D. Tokach, S. S. Dritz, J. C. Woodworth, R. D. Goodband, J. M. DeRouchey, C. I. Vahl, H. I. Calderon-Cartagena, and D. L. Van De Stroet. 2018. A retrospective analysis of seasonal growth patterns of nursery and finishing pigs in commercial production. J. Swine Health Prod. 27:19-33.

Figure 1: Weight differences across fill groups by models.

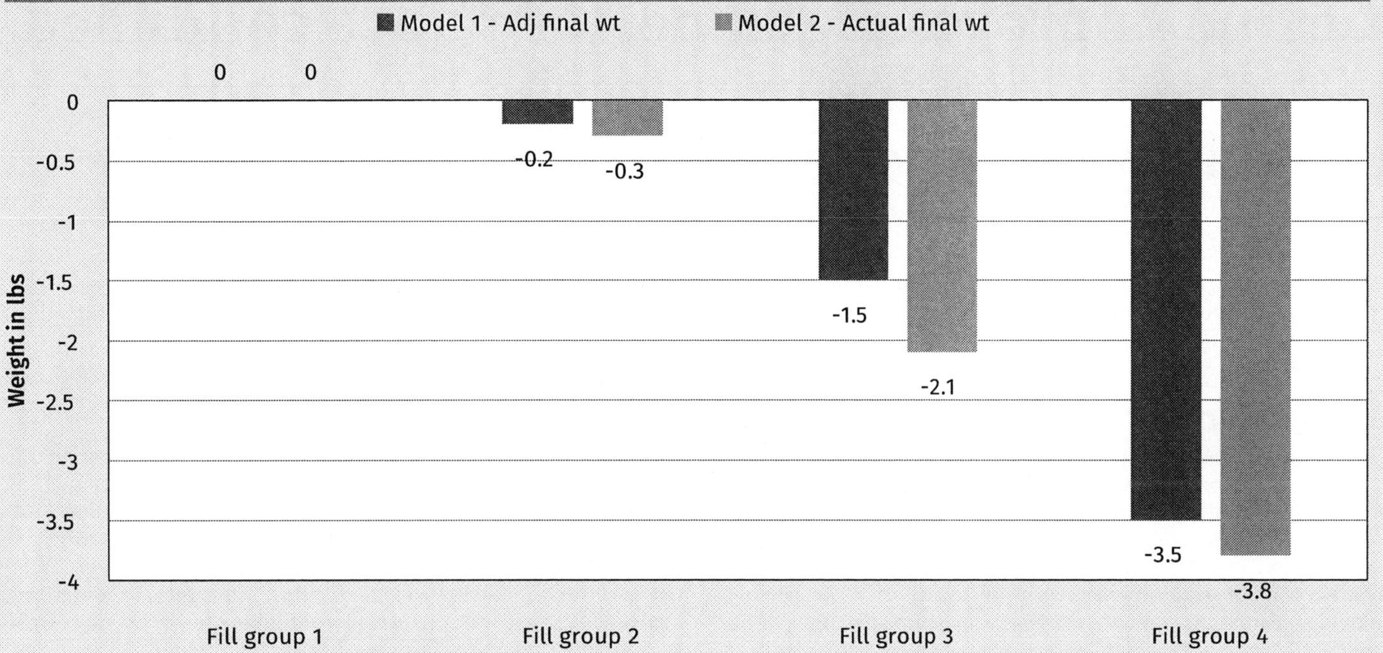

https://doi.org/10.54846/am2023/133

Vaccination with an *Escherichia coli* F4/F18 vaccine improves piglet performance combined with a reduction in antimicrobial use and secondary infections due to *Streptococcus suis*

Frédéric Vangroenweghe[1,2]; Mieke Boone[3]
[1] Elanco Animal Health Benelux, BU Swine & Ruminants, Antwerpen, Belgium;
[2]Ghent University, Faculty of Veterinary Medicine, Department of Internal Medicine – Reproduction – Population Medicine, Unit Porcine Health Management, Merelbeke, Belgium; [3]Medivet DAP, Waardamme, Belgium

Introduction

Post-weaning *Escherichia coli* (*E coli*) diarrhea (PWD) remains a major cause of economic losses for the pig industry.[1] Therapy to combat PWD typically consists of antibiotic treatment or supplementation of zinc oxide to the feed. Emergence of antimicrobial resistance to *E coli* and new EU regulations urge the need for alternative control strategies such as immunization.[2-5] The aim of the field study was to evaluate the effect of an oral live non-pathogenic *E coli* vaccine on piglet performance, health, and antimicrobial use in a farm with clinical PWD problems due to F18-ETEC.

Materials and methods

We evaluated vaccination with an oral live non-pathogenic *E coli* F4/F18 vaccine under field conditions in 10 consecutive batches against standard antimicrobial treatment in 10 historical control batches. Besides production data (weight, feed intake, mortality), health parameters were collected (TI_{100} – treatment incidence for a 100-day period during post-weaning phase, reason for death, weight of dead animals). Data were analyzed using JMP 15.0.

Results

Vaccine-treated groups demonstrated a significant improvement in feed conversion rate (1.67 vs. 1.61 in control and vaccine-treated group, respectively), mortality weight (7.80 vs. 5.46 kg in control and vaccine-treated group, respectively) and antimicrobial use, expressed as TI_{100} (69.43 vs. 0.13 days in control and vaccine-treated group, respectively). Other parameters (total weight gain and mortality rate) remained similar in both treatment groups. From a general health perspective, secondary infections due to *Streptococcus suis* (*S suis*) in the second part of nursery markedly reduced as indicated by the major reduction (- 91.5%) in amoxicillin use (Table 1).

Conclusions

In conclusion, the present study demonstrated efficacy of an oral live non-pathogenic *E coli* vaccine for active immunization of piglets against PWD under field conditions. Vaccine-treated groups showed an improvement in several economically important performance parameters while reducing overall antimicrobial use and infection pressure due to *S suis*. Therefore, vaccination against PWD may be considered a valuable alternative to consolidate piglet performance while meeting the new EU requirements concerning prudent use of antimicrobials in intensive pig production.[6]

References

1. Fairbrother JM, Nadeau É, Gyles CL. *Escherichia coli* in postweaning diarrhoea in pigs: an update on bacterial types, pathogenesis, and prevention strategies. Anim Heal Res Rev. 2005; 6: 17-39.

2. Boyen F, Vangroenweghe F, Butaye P, De Graef E, Castryck F, Heylen P, et al. Disk prediffusion is a reliable method for testing colistin susceptibility in porcine *E. coli* strains. Vet Microbiol. 2010; 144: 359-362.

3. Luppi A, Bonilauri P, Dottori M, Gherpelli Y, Biasi G, Merialdi G, et al. Antimicrobial resistance of F4+ *Escherichia coli* isolated from swine in Italy. Transbound Emerg Dis. 2013; 62: 67-71.

4. Nadeau E, Fairbrother JM, Zentek J, Bélanger L, Tremblay D, Tremblay C-L, et al. Efficacy of a single oral dose of a live bivalent *E. coli* vaccine against post-weaning diarrhoea due to F4 and F18-positive enterotoxigenic *E. coli*. Vet J. 2017; 226: 32-39.

5. Vangroenweghe F. Improved Piglet Performance and Reduced Mortality and Antimicrobial use Following Oral Vaccination with a Live Non-Pathogenic *Escherichia coli* F4/F18 Vaccine Against Post-Weaning Diarrhoea. Austin J Infect Dis. 2021; 8(2): 1048.

6. European Medicinal Agency. Questions and answers on veterinary medicinal products containing zinc oxide to be administered orally to food producing species. Outcome of a referral procedure under Article 35 of Directive 2001/82/EC (EMEA/V/A/118). 2017; EMA/394961/2017.

Table 1: Performance data of a comparative field trial with a historical control group with standard antimicrobial treatment and a vaccine-treated group using Coliprotec F4F18 (Elanco) on a farm with clinical problems of post-weaning diarrhea due to F18-ETEC and subsequent meningitis due to *S suis*. Significant differences are indicated by different superscript letters and their *P*-value.

Performance parameter	Control	Vaccine	*P*-value
Number of groups	10	10	-
Number of weaned piglets (± SEM)	2632 ± 63	2720 ± 98	0.181
Total weight of weaned piglets (kg ± SEM)	17475 ± 921	17895 ± 1088	0.386
Average weaning weight (kg ± SEM)	6.61 ± 0.22	6.53 ± 0.23	0.406
Number of sold piglets (± SEM)	2541 ± 68	2641 ± 69	0.157
Percentage sold piglets (± SEM)	96.48 ± 0.4	97.10 ± 0.2	0.094
Total weight sold piglets (kg)	60732 ± 1325	61183 ± 1567	0.414
Average sold weight (kg ± SEM)	23.97 ± 0.41	23.18 ± 0.26	0.063
Total daily gain (kg ± SEM)	17.36 ± 0.50	16.65 ± 0.34	0.129
# days in nursery (d ± SEM)	46.7 ± 1.2^{a}	49.6 ± 0.6^{b}	0.027
Mortality (# ± SEM)	91 ± 9	79 ± 6	0.133
(% ± SEM)	3.52 ± 0.38	2.90 ± 0.25	0.094
Weight of dead piglets (kg ± SEM)	7.80 ± 0.26^{a}	5.46 ± 0.42^{b}	0.00013
Feed intake per sold piglet (kg ± SEM)	28.98 ± 1.00	26.84 ± 0.56	0.052
Average daily feed intake (g ± SEM)	625 ± 27^{a}	540 ± 9^{b}	0.008
Average daily weight gain (g ± SEM)	375 ± 15^{a}	336 ± 6^{b}	0.018
Feed conversion rate (kg feed/kg gain ± SEM)	1.67 ± 0.02^{a}	1.61 ± 0.02^{b}	0.041
Antimicrobial treatment cost per piglet (€ ± SEM)	0.36 ± 0.11^{a}	0.05 ± 0.02^{b}	0.0086
Reduction in cost of antimicrobial treatment		86.1%	–
TI_{100} (d ± SEM)	69.43 ± 9.44^{a}	0.13 ± 0.13^{b}	0.000022
Reduction in antimicrobial use (%)		99.8%	–
Specific reduction in amoxicillin use (%)		91.5%	–
Month with TI_{100} = 0	0	9	–

Poster 49

https://doi.org/10.54846/am2023/134

Improving composting knowledge and competencies to create an impactful mortality management educational method

S. Probst Miller[1,2], DVM; M. Hutchinson[3]; A. Ramirez[4], DVM, MPH, PhD, DACVPM;
R. Robbins, DVM, PhD; S. Hornback[1,2], MBA
[1]AgCreate Solutions, Inc.; [2]Pork Avenue Training Portal; [3]University of Maine;
[4]University of Arizona College of Veterinary Medicine

Introduction

Using composting for effective livestock mortality management reduces spread of disease, provides a natural source of nutrient inputs, and contributes to environmental stewardship. Ensuring a competent labor force prepared to manage daily mortalities and potential depopulation catastrophes improves risk management. Management of mortality is a physical competency. In past studies, competency-based educational strategies improved physical competencies from a 25 - 60% level to an 85 - 95%+ competency level post-training.[1-3]

Methods

An advisory panel of national experts identified key knowledge points and physical competencies for both effective carcass decomposition and pathogen elimination via composting. These keys guided creation of two learning experiences: "Preparation for Composting" and "Building the Composting Pile." Pre- and post-evaluations were completed onsite by a third-party verifier asking staff in charge of composting to train them on how to prepare and build a compost pile. Knowledge points and competencies were recorded as present or not. After pre-assessment, participants completed two learning experiences administered via Pork Avenue Training Portal that followed the "See it. Do it. Teach it." educational methodology. These learning experiences consisted of online learning ("See it.") followed by on-farm mentoring administered by pre-trained farm staff ("Do it."), and lastly an opportunity to teach a third-party how to do the task ("Teach it."), which was also the post-training evaluation.

Results

Initially, 29 sites were evaluated with five sow farms and 24 wean-to-market barns. Post-training evaluations were unable to be completed at the five sow farms due to unexpected disease and visitor biosecurity requirements. Four wean-to-market sites were also unavailable due to illness and staffing issues. Data from these nine sites were excluded from analysis. Analysis was done using a paired t-test with alpha 0.05 in GraphPad Prism v.9.4.1. The average pre-training knowledge and competency score was 41% and average post-training knowledge and competency score was 74%. Knowledge and competency score across twenty farm staff participants increased by an average of 33 percentage points (Figure 1A). Knowledge scores increased from 40% to 82% post-training. Competency scores increased from 41% to 71% post-training. This educational style improved knowledge by 42 percentage points and improved physical competency by 30 percentage points (Figure 1A).

Conclusions

This study shows that competency-based educational strategies did have positive impact on both knowledge and physical competency but did not elevate competencies as high as previous studies (Figure 1B). Previous studies[1-3] looked at sow farm workers; this study evaluated contract growers. Assessing why these groups are different could aid in better understanding how to motivate new competencies in contract grower employees. Although both knowledge and competency were positively impacted, in this study there was a 12-percentage point gap between knowledge and competency (Figure 1A).

References

1. Probst Miller, S. (2020). Ensuring that a company's vision and mission and core values are living on farm via pre-training, targeted training, and post-training animal care competency measurements. **https://porkavenuetraining.com/site/wp-content/uploads/2020/12/Ensuring-that-a-companysvision-and-mission-are-living-on-farm-via-pre-training-and-post-training-animal-care-competency-measurements-Sarah-Probst-Miller-AgCreate-Solutions.pdf**

2. Probst Miller, S. and Krantz, S. (2019). Inspiring a verified culture of care post-identification of abuse on an undercover video using a baseline assessment of expressed competencies compared to post training measurement of expressed competencies. **https://porkavenuetraining.com/site/wp-content/uploads/2020/12/Inspiring-a-verified-culture.pdf**

3. Probst Miller, S. (2016). Impact of two training implementation plans on post-training competency expression of personnel on farm. **https://porkavenuetraining.com/site/wp-content/uploads/2021/06/Impact-of-Two-Training-Implementation-Plans-on-Post-Training-Competency-Expression-of-Personnel-on-Farm.pdf**

Figure 1: Pre- and post-training knowledge and competency scores. **A)** Average overall, knowledge, and competency scores pre- and post-training; N = 20. **B)** Overall knowledge and competency scores by participant compared to the average overall competencies from previous studies.

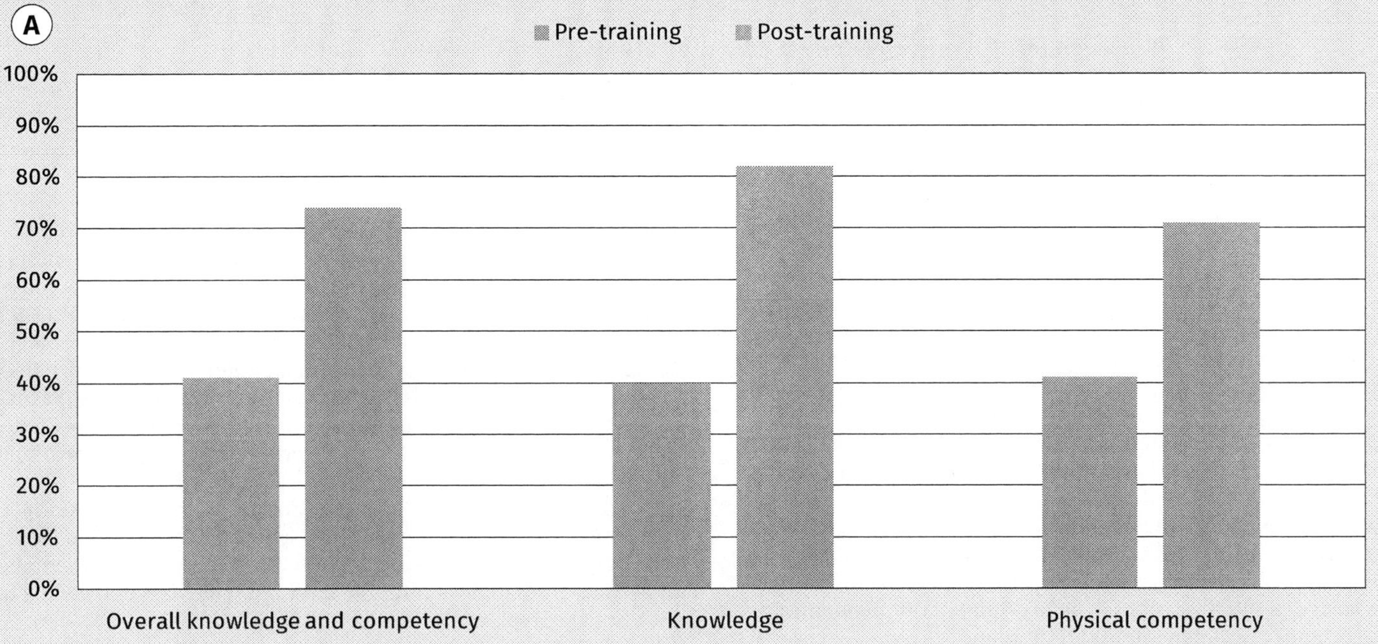

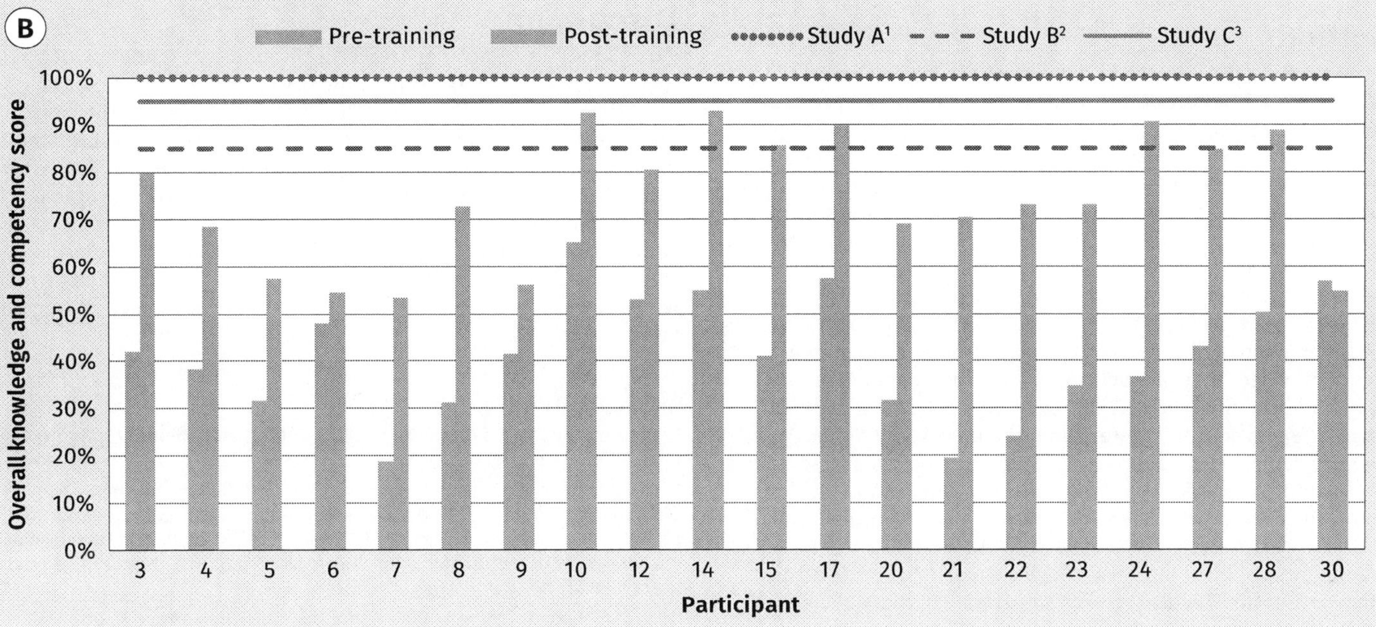

Poster 50

https://doi.org/10.54846/am2023/135

A survey of *Streptococcus suis* across sow production and neonatal piglets

S. Hernandez[1]; S. Burmeister[1]; D. Murray[2], DVM; A. Smith[1], PhD; E. Davis[1], PhD
[1]Arm and Hammer Animal and Food Production, Waukesha, Wisconsin;
[2]New Fashion Pork, Jackson, Minnesota

Introduction

Streptococcus suis is a major cause of mortality and morbidity in the swine industry causing systemic infection in young pigs and causing significant economic losses worldwide.[2] *S suis* resides asymptomatically in the gastrointestinal tract and genitals of pigs but the microbe is able to cross the dermal and mucosal layers and reach the bloodstream. Once the microbe is systemic, *S suis* causes infections in other areas of the lymphatic system causing diseases like arthritis and sepsis as well as crossing the blood-brain barrier causing meningitis and other pathologies.[3] *S suis* can also spread from swine to human and is the major cause of meningitis in other pork consuming countries. There are vaccinations that are being used worldwide, however these vaccine formulations target specific serotypes and are not effective against all strains equally.[2] The purpose of this survey was to determine the level of *S suis* in sow herds before and after implementation of CERTILLUS™ targeted microbial solution.

Methods

Three farms were selected to determine the levels and prevalence of *S suis*. Two timepoints were selected for this survey; 1) a sampling was completed for baseline levels at each farm (December 2021) and 2) sows from two farms were placed on a basal diet and sows from one farm were placed on a CERTILLUS targeted microbial solution (June 2022). Each sampling resulted in 45 samples per farm for a total of 135 samples per timepoint. Rectal and vaginal swabs were collected from the same lactating sow source and a rectal swab was obtained from a neonatal piglet in her litter. Swab samples were processed to determine colony forming units (CFU) per swab on Columbia CNA agar. A maximum of five isolates were selected from each sample and grown for 48 hours at 37° in Todd Hewitt Broth (THB), selected isolates were subjected to a multispecies PCR to determine the presence of *S suis*, *S parasuis*[1] and *S equi* subsp. *zooepidemicus*. Isolates were subjected to random amplified polymorphic DNA (RAPD) PCR to determine genetic diversity. Seven different Arm & Hammer™ *Bacillus* strains were utilized to screen 99 *S suis* isolates and a CERTILLUS targeted microbial solution was designed using the two strains (Arm & Hammer *Bacillus* strains 839 and 1999) that showed the best efficacy against the *S suis* populations.

Results

Across all timepoints and sample types, there were detectable levels of *S suis* found during this survey. *Streptococcus suis* counts at Farm 3 were decreased from the baseline to the post-CERTILLUS sampling, such that levels were 3.6 log CFU at baseline and 1.7 log CFU after the implementation of CERTILLUS ($P < 0.01$).

Streptococcus suis prevalence was similar across the farms fed a basal diet; on average 21% of samples had detectable *S suis* isolates. At Farm 3, 34% of the samples had detectable levels of *S suis* and this was decreased to 5% after the implementation of CERTILLUS ($P < 0.05$, Figure 1). Also observed at farm 3, *S suis* levels decreased numerically across the different sample types after CERTILLUS was implemented in sow diets, with the greatest decreases observed amongst the vaginal samples. *Streptococcus suis* was determined to be 4.0 log CFU in vaginal samples collected from Farm 3 at the baseline sampling and 1.8 log CFU after the implementation of CERTILLUS. A similar decrease was observed in the prevalence of vaginal samples that were positive for *S suis* at Farm 3 (45% vs. 5%). Furthermore, 67% of isolates obtained from vaginal samples had detectable *S suis* at the baseline sampling compared to 20% after the implementation of CERTILLUS, whereas the percentage of isolates from piglet samples decreased from 53% at baseline to 6% after the implementation of CERTILLUS.

Discussion

Farm 3 that was placed on CERTILLUS showed a reduction of *S suis* levels across the sample timepoint as well as the prevalence of *S suis* detected in the samples. The Farms that were on the basal diet showed no change in the levels of *S suis* detected in the samples or the prevalence of *S suis* that were detected in the samples. After CERTILLUS was implemented, the level of *S suis* within vaginal swabs revealed the largest decrease in prevalence of *S suis*. Our results indicate that sows placed on CERTILLUS targeted microbial solution had lower levels of *S suis* both vaginally and rectally which would lower the chances of her transferring potentially pathogenic *S suis* to her piglets.

References

1. Okwumabua O, O'Connor M, Shull E. 2003. A polymerase chain reaction (PCR) assay specific for *Streptococcus suis* based on the gene encoding the glutamate dehydrogenase. FEMS Microbiology Letters 218:79-84.

2. Arenas J., Zomer A., Harders-Westerveen J., Bootsma J., De Jonge M., Stockhofe-Zurwieden N., Smith H.E., De Greeff A. 2020. Identification of conditionally essential genes for *Streptococcus suis* infection in pigs. Virulence. Vol 11, No. 1: 446-464.

3. Rui L., Weiyi L., Yu M., Hong Z., Jiao Y., Zhe M., Hongjie F. 2018. The serine/threonine protein kinase of *Streptococcus suis* serotype 2 affects the ability of the pathogen to penetrate the blood-brain barrier. Cellular Microbiology. DOI:10.1111/cmi.12862

Figure 1: *S suis* (%) prevalence before and after the implementation of CERTILLUS; (a,b,c $P < 0.001$).

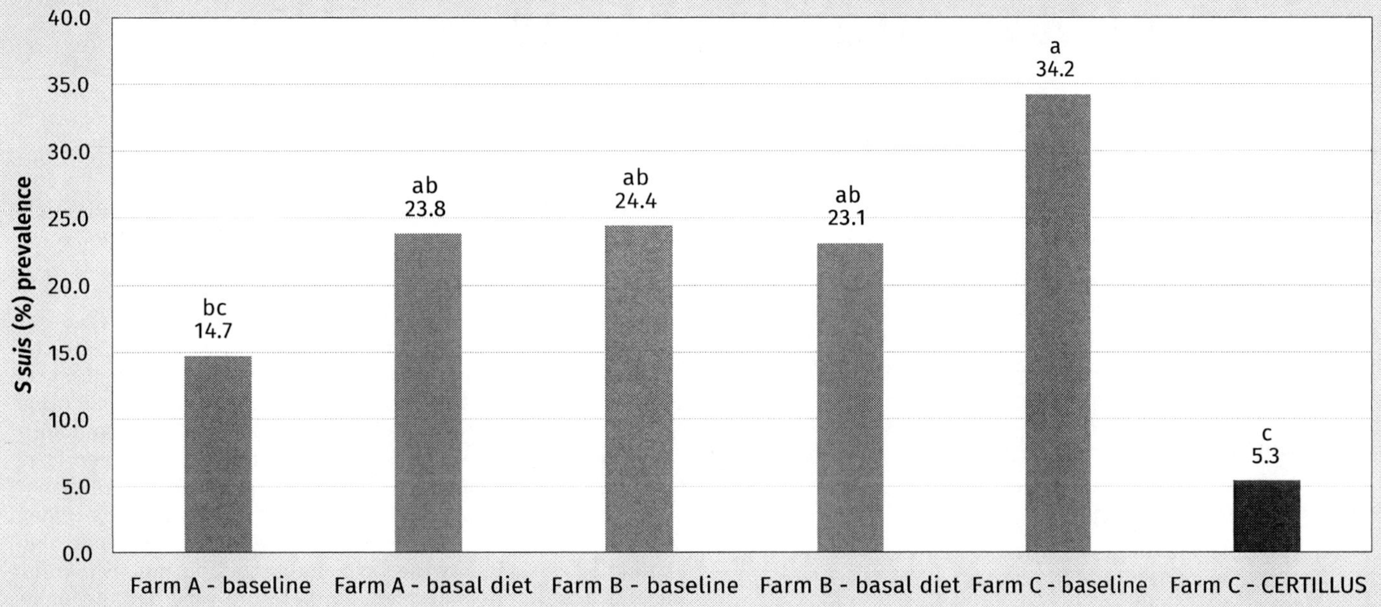

https://doi.org/10.54846/am2023/136

Assessment of the hyper-local ambient environment in large pig barns

D. Polson, DVM, MS, PhD
Boehringer Ingelheim Animal Health USA Inc., Duluth Georgia

Introduction

A potential influence on post-weaning live pig performance is ambient temperature. Temperatures levels and fluctuations within the direct animal environment can impact pig performance.[1] To describe and characterize ambient temperature at more granular (pen and zone) levels, the occasion of two consecutive disease challenge studies was used to incorporate a layer of pen-level ambient temperature monitoring in addition to the primary study protocol.

Materials and methods

As part of two different disease challenge studies spanning from September 2020 through July 2021, 72 pens across two rooms were equipped with individual in-line water meters to measure pen-level daily water usage, water temperature and ambient temperature (FlowIQ 2100 RF, Kamstrup Water Metering, Atlanta GA). Continuous external environmental data (eg, temperature, relative humidity, wind speed, wind direction) were also obtained from the nearest US national network weather station, located approximately 6 km from the research barn site.

Results and discussion

The external weather environment for the entire period of both studies was typical for the Upper Midwest United States and generally followed seasonal patterns. The average daily wind speed was 16.34 km/h during Study 1 and 15.47 km/h during Study 2. The daily average external temperature generally trended downward throughout the period of Study 1 with several cycles of falling and rising temperatures as weather patterns moved through the area. The daily average external temperature generally trended upward throughout the period of Study 2, also with several cycles of falling and rising temperatures as weather patterns moved through the area. The inside temperature also trended down over the Study 1 period as the study pigs aged and gained weight, showing a much less variable inside temperature average, maximum and minimum throughout the study period – an indication that the barn ventilation management was relatively consistent. During the periods of Study 1 (September 22 through December 22, 2020) and Study 2 (January 25 through July 13, 2021), deviations of the average temperature from the overall barn average were calculated for each of the 72 study pens (Figure 1). Considerable variation was recorded among individual pens throughout both rooms of the combined periods of both studies, with a pen-level temperature range of 6.84°C (-4.00 to +2.84) and 6.47°C (-3.75 to +2.72) within the West and East rooms, respectively. It is evident that use of an array of ambient temperature sensors assigned to all defined zones in a swine barn enable the continuous monitoring and assessment of ambient temperature variation throughout the rooms and barn.

Conclusions

Further research must be conducted to better assess the relationship of zone-level ambient temperature variation with measures of pig health, welfare and performance. As these relationships are further evaluated and better documented, the cost-effectiveness of operating pen-level ambient temperature sensors can be defined. Obviously important to the benefit:cost (specifically as the denominator) is the overall cost of purchasing, installing, connecting (to the cloud) and maintaining these meters/sensors. To be scalable and sustainable as a precision livestock monitoring method, it will be important that the per unit cost of cloud-connected combination water meters/temperature sensors decrease substantially.

Reference

1. Rauw W, Pena E, Gomez-Raya L, et al. (2020) Impact of environmental temperature on production traits in pigs. *Nature Scientific Reports* 10(2106).

Figure 1: Average pen-level temperature deviations from the overall barn mean ambient temperature (Celsius) for the entire evaluation period of Study 1 and Study 2, from September 22, 2020 through July 13, 2021.

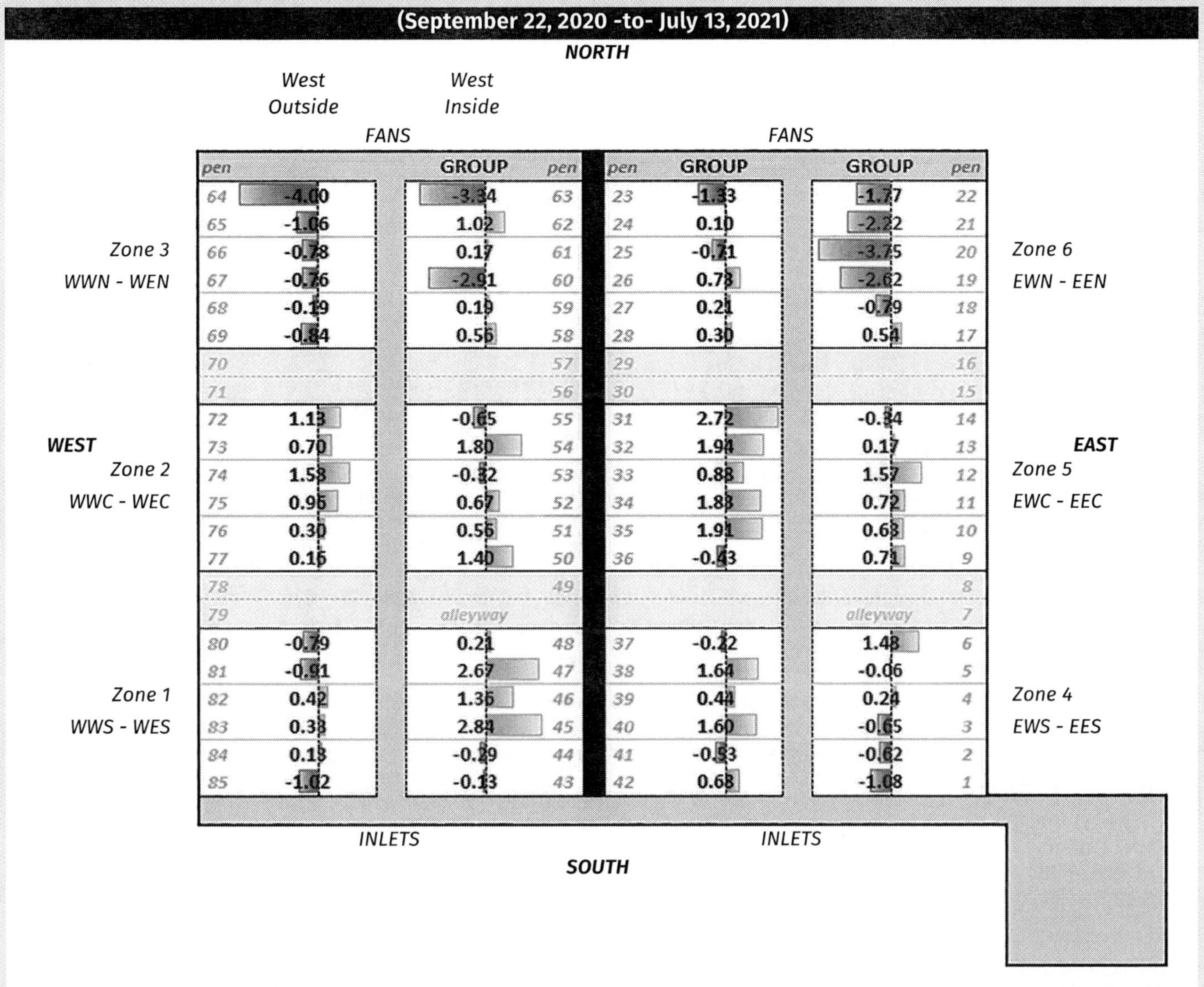

Poster 52

https://doi.org/10.54846/am2023/137

Control of cockroaches in swine facilities with new insect growth regulator Exhalt (Novaluron)

Gene Spellman, BS Central Life Sciences

Introduction

Large swine operations need good management programs in place to prevent insect pests from becoming a biosecurity and healthcare problem. Many different fly species can be found in swine barns, including but not limited to house flies, dark-eyed fruit flies, and drain flies. These pests are easily seen. Populations can be observed during daytime hours when workers are more often present. However, one group of insects may be present in large populations and not as noticeable - cockroaches. German cockroaches *Blatella germanica* are increasingly infecting swine facilities – barns and offices. Cockroaches require moisture, warmth and food. Today's swine facilities offer suitable environments.

Cockroaches prefer to move around and feed in the dark. After dark is the best time to sample for infestations. Inspecting areas where they may hide, including wall voids, cracks, crevices, and inside hollow crate panels and pipes often reveals hundreds or thousands of adult cockroaches and immature nymphs. Overnight placement of sticky cards can be an effective tool for monitoring cockroach activity. If cockroaches are seen in a facility during the daytime, it indicates a significant population present. Control steps should be taken to address the health threat adult and nymph populations pose to pigs and people for several following reasons.

Threat to pigs' health by breaking biosecurity plans

Swine producers spend tremendous resources on implementing biosecurity plans. Research proves cockroaches act as both a reservoir and vector of disease, such as *E coli*,[1] salmonella,[2] and *B hyodysenteriae*.[3] Unpublished work by Central shows German cockroaches tested PCR-positive for PED virus. Gregarious by nature, cockroaches feed on a variety of foods, including feces, animal feed and other animal sources. Inter-group clean up, sanitation and biosecurity programs do not eliminate cockroaches between groups of pigs. Cockroaches survive these biosecurity measures and can have contact with the next group of pigs. Combined with their habits and close proximity to swine, cockroaches pose a threat to pigs' health, undermining comprehensive biosecurity plans.

Possible health threat to people/ employees: A management headache

Research indicates cockroaches and their habits may create an increased environment for allergens and result in increased allergic reactions and/or allergies and asthma in people.[4] In heavily infested barns, cockroaches can spread into offices, break rooms and showering areas, and across secure and non-secure areas. This breaks biosecurity and causes concern for employees/ visitors/consultants coming into the facility. People fear "taking roaches home with them" in their clothes or footwear. This reduces people's interest in working in or visiting these facilities, a challenge for managers. Boots, coveralls, etc, must be shaken out thoroughly to avoid putting on infested clothing. "Bagging" clothes in sealed plastic bags while working/visiting severely infested facilities is necessary to prevent leaving with cockroaches. Controlling/preventing cockroach infestations will improve swine working environments for employees and visitors/ consultants.

Results

Control and prevent - the goal of Central's cockroach integrated pest management program (IPM).

Effectiveness of Exhalt IGR

Implementing an effective IPM (integrated pest management) program designed specifically for cockroaches should be a key component for all swine units as part of their biosecurity program. This should include inspecting for infestations, planning for the treatment program, implementation of the program, and evaluating the effectiveness of each treatment.

Central Life Sciences offers the following program which makes use of many different products, product forms and application methods to effectively combat cockroach infestations. This program uses adulticides (Lambda 9.7 CS Liquid Concentrate) and insect growth regulators Exhalt WDG (water disbursable granules) in a combination strategy to help achieve the goal of controlling the adults and helping prevent the immature nymph off-spring from becoming reproductive adults. Properly implemented, with good maintenance follow-up, Central's program helps prevent the immatures from becoming reproducing adults, with an eventual significant draw-down of the facility's cockroach infestation (Figure 1).

References

1. Ludek Zurek, Coby Schal. Evaluation of the German cockroach (Blattella germanica) as a vector for verotoxigenic Escherichia coli F18 in confined swine production. *Veterinary Microbiology Volume 101, Issue 4, 6 August 2004, Pages 263-267.*

2. Matthew Turner, Vincent Peta, Jose E.Pietri. New insight into the relationship between Salmonella Typhimurium and the German cockroach suggests active mechanisms of vector-borne transmission. *Research in Microbiology Volume 173, Issue 3, March–April 2022, 103920.*

3. R Blunt, K Mellits, E Corona-Barrera, P Pradal-Roa, S McOrist. Carriage of Brachyspira hyodysenteriae on common insect vectors. *Vet Microbiol. 2022 Jun;269:109417.*

4. Danh C. Do, Yilin Zhao, and Peisong Gao. Cockroach Allergen Exposure and Risk of Asthma. *Allergy. 2016 Apr; 71(4): 463–474.*

Figure 1: Exhalt insect growth regulator (IGR) greater than 95% effective kill against cockroach nymphs as they try to molt into adults. Exhalt sprayed onto various surfaces.

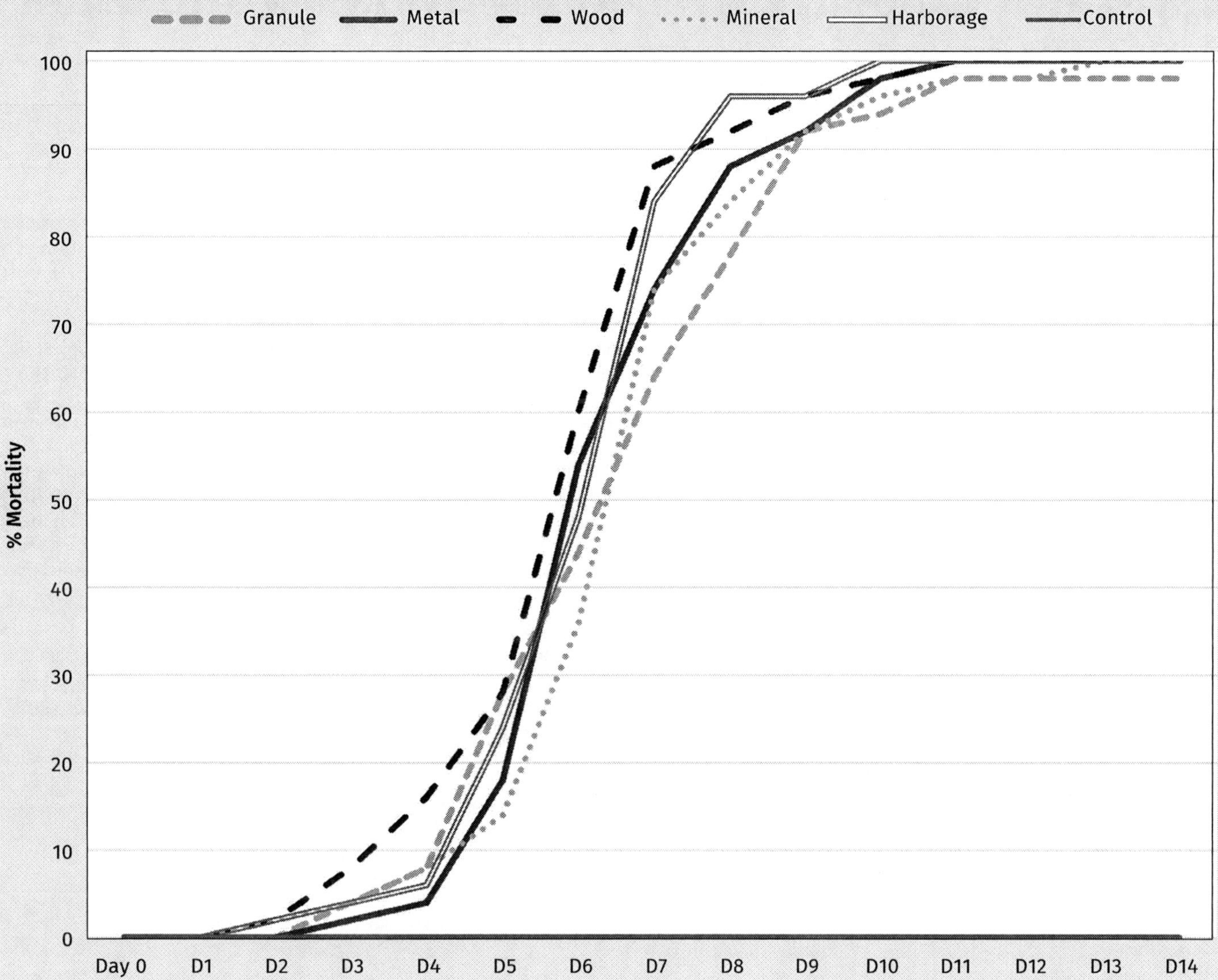

Poster 53

https://doi.org/10.54846/am2023/138

Utilizing lung lesion scoring to monitor swine respiratory disease in a commercial swine farm in the Philippines

Mary Grace Alba[1], DVM; Roselle Falconite-Cudal[1], DVM, DipPCSP; Karl Alexander Fabros[2], DVM
[1]Elanco Philippines Inc., Pasig City; [2]Elanco Malaysia Sdn. Bhd., Selangor

Introduction

Swine respiratory diseases in commercial swine herds remains challenging.[1] Chronic lung lesions and the lung lesion scoring (LLS) performed at the time of slaughter provide valuable information on the incidence and severity of swine respiratory pathogens. Chronic lung lesions like bronchopneumonia and pleurisy that correlate to farm data can complete the animal health program assessment. LLS provides clear information on changes in lung lesions pre and post medication. Simulation of disease cost further guided the veterinarian's decision towards treatment.

Methods

A commercial farm in the Philippines considered in-feed Pulmotil® medication to address respiratory disease in the herd. LLS was done across several months for a period of three years. Scoring systems used were Straw method and Slaughterhouse Pleurisy Evaluation System (SPES) for bronchopneumonia and pleurisy respectively.[2,3]

Results and discussion

Twelve LLS activities were completed. Baseline LLS1 was done before adding Pulmotil® (tilmicosin) to the grower feeds at 2 kg/ton for 21 days. Fewer lungs were classified as moderate and severe pneumonia from LLS2-4. LLS5 showed a resurgence of lesions. Pulmotil® in grower was used again with LLS6-7 which showed less lesions. It was then decided to transition the medication program and include Pulmotil® in the lactation diet one week before and two weeks after farrowing and monitoring. LLS10 showed bronchopneumonia lesions which were majorly of normal classification however with high pleurisy; hence, the farm opted to use a generic tilmicosin in the grower feed during the respiratory disease break. LLS11 showed an increase in severe pneumonia lesions. The use of Pulmotil® was returned and the use of grower feeds was eventually discontinued but remained in the lactating feed; LLS12 showed improved lung scores. The cost of bronchopneumonia was due to increased feed cost and reduced growth rates (Figure 1). Overall, LLS is an effective tool for monitoring swine respiratory diseases. It alerted the farm veterinarians on the efficacy of their medication and the cost of disease incursions.

References

1. Full Value Pig Global Survey. 2013. Elanco.

2. Straw, B. E., L. Backstrom, and A. D. Leman, 1986b: Examination of swine slaughter. Part II. Findings at slaughter and their significance. Compend. Contin. Educ. Vet. Pract. 8, 106–110.

3. Dottori, M., Evaluation of pig pleuritis at the slaughterhouse available on: Evaluation of pig pleuritis at the slaughterhouse - Articles - pig333, pig to pork community, 2013.

Figure 1: Cost of Bronchopneumonia

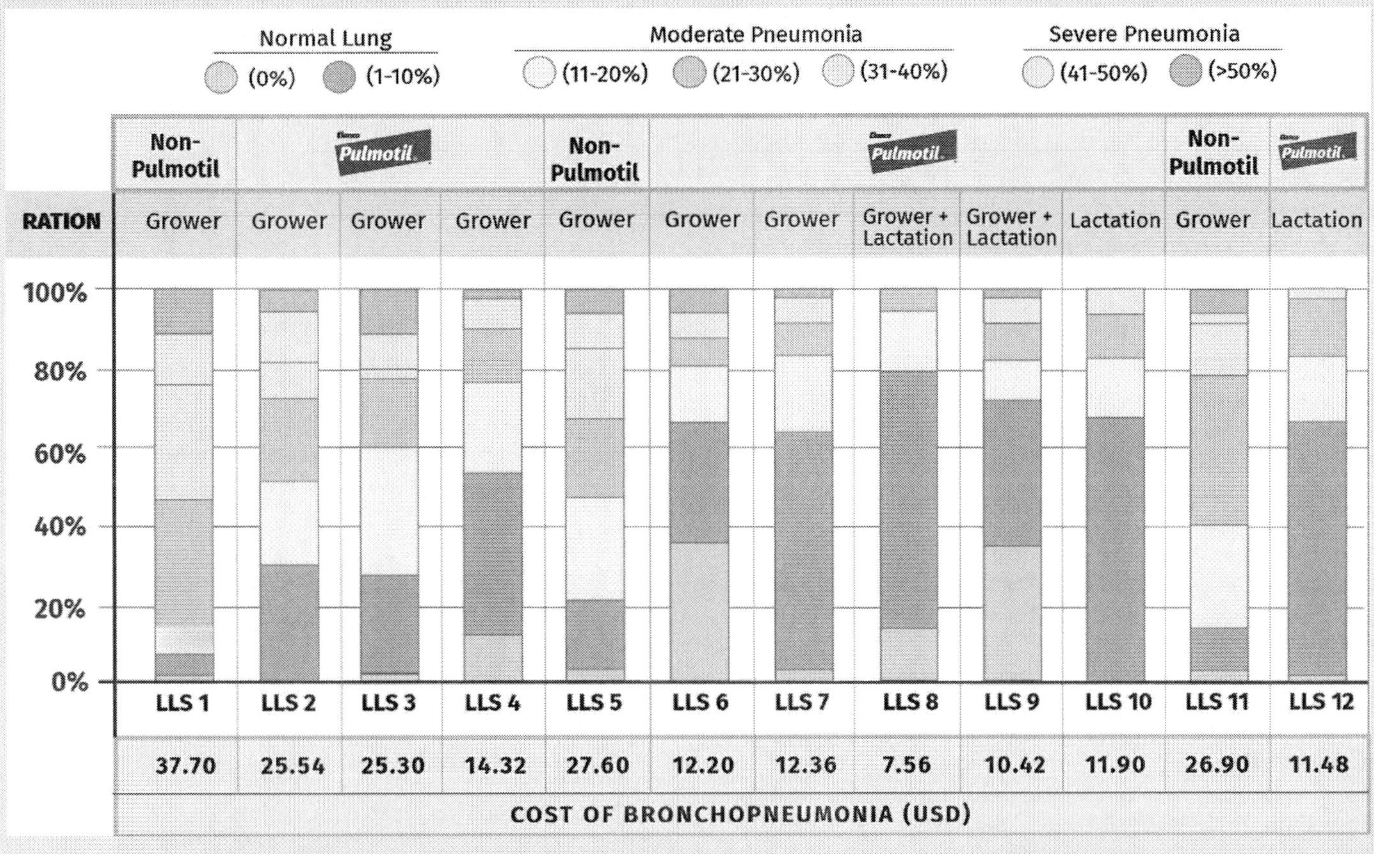

Poster 54

https://doi.org/10.54846/am2023/139

The first swine vaccine to show protection against ileitis, salmonellosis[1] and colibacillosis[2]

C. Cull[1], DVM, PhD, MBA; C. Stutheit[2]
[1]Midwest Veterinary Services, Inc., Oakland, Nebraska; [2]ENDOVAC® Animal Health, Columbia, Missouri

Introduction

Gram-negative bacteria have thousands of serotypes making it difficult to make effective commercial or autogenous vaccines that protect against a broad spectrum of gram-negative diseases. Thus, there is a paramount need for a single unique bacterin providing protection against virtually all gram-negative pathogens in swine.

Past research demonstrated ENDOVAC-Porci®, a core antigen and immune stimulant vaccine, provides piglets protection against the enteric and respiratory effects of gram-negative pathogens.[2]

Study Objective

This study expands the list of gram-negatives by showing protection against a *Lawsonia intracellularis* challenge and demonstrates ENDOVAC-Porci vaccine's performance as compared to a commercial ileitis vaccine.

Materials and methods

Study location

Midwest Veterinary Services, Inc. in Oakland, NE. Investigator; Charley Cull, DVM, PhD, MBA.

Study design

This study was a single-site, randomized, prospective, blinded, comparative placebo-controlled design. Individual "Animal" was the experimental unit. Number of Replicates was 30 per treatment group.

Statistical analysis performed by the "Center of Outcomes Research and Epidemiology" at Kansas State University.

Animals

90 healthy piglets weighing > 2.8 pounds were selected for inclusion in the study. Animals were commercially sourced from a PRRS-stable herd.

Vaccination

Group 1 and 2 piglets were vaccinated IM with 1 mL of either a saline placebo (group 1) or ENDOVAC-Porci from ENDOVAC Animal Health (group 2) at day 1 of age (study day 0) and boostered at weaning at 21 days of age (study day 21). The group 3 piglets were vaccinated at weaning (study day 21): IM with a 2mL dose of Porcilis® Ileitis from Merck Animal Health. The following vaccines were also given at study day 21: MLV PRRS, PCV2 and *Mycoplasma hyopneumoniae*.

Challenge

All piglets were challenged on study day 42, with gram-negative organism, *Lawsonia intracellularis*.[3]

Results

Body weight

Pig weight distribution at study day 0, at challenge study day 42 and at conclusion of study on day 70. See Table 1A.

Weight gain

At day 70 pigs vaccinated with ENDOVAC-Porci were 29% heavier than the control group and no statistical difference in weight gain compared to the Porcilis Ileitis vaccinated pigs. See Table 1B.

Morbidity

Daily observations for clinical and fecal scores began on study day 58 through the end of the study (study day 70). Clinical scores: % pigs with a score greater than 0. Based on body condition, respiration and behavior; 0 normal, 1, mild, 2 moderate and 3 severe. Fecal scores: % pigs with a score greater than 0; 0 normal, 1 soft stool, loose stool and 3 watery stools. See Table 1C.

Summary

The ENDOVAC-Porci vaccinated pigs were 29% (7.7 lb.) heavier than the control group and no statistical difference in weight gain compared to the Porcilis Ileitis vaccinated pigs at the end of the study (day 70). The ENDOVAC-Porci vaccinated piglets were healthier than the control pigs with 40.9% better clinical scores and 37.6% better fecal scores. The ENDOVAC-Porci vaccinated pigs also had better morbidity scores than the Porcilis Ileitis vaccinated pigs: 8.2% better clinical scores and 18.2% better fecal scores than Porcilis Ileitis vaccinated pigs.

Conclusion

ENDOVAC-Porci vaccinated pigs outperformed the control pigs and went head-to-head with Porcilis Ileitis. This study combined with past studies[1,2] provides compelling data to consider vaccination with ENDOVAC-Porci to control ileitis as well as other gram-negative diseases common to pig production systems.

References

1. Sprouse RF, Garner HE: Data on file at the University of Missouri-Columbia and USDA, Hyattsville, Maryland

2. Cull C. The effects of a core antigen bacterin with an immunostimulant on piglet health and performance outcomes when challenged with enteric and respiratory pathogens. In: Proceedings of the 53rd AASV Annual Meeting: 2022:167-169

3. Challenge isolate sourced from: Swine Services Unlimited, Inc., Rice, MN

Table 1: Results

A: Body weights

Day 0 by treatment group		
Treatment	Mean kg	Mean lb
Saline	1.67	3.68
ENDOVAC-Porci	1.60	3.53
Porcilis Ileitis	1.65	3.64
Effect of treatment (P = 0.48)		

Day 42 by treatment group		
Treatment	Mean kg	Mean lb
Saline	16.76	36.95
ENDOVAC-Porci	16.73	36.88
Porcilis Ileitis	17.05	37.59
Effect of treatment (P = 0.91)		

Day 70 by treatment group		
Treatment	Mean kg	Mean lb
Saline	28.93	63.78
ENDOVAC-Porci	32.43	71.50
Porcilis Ileitis	33.85	74.63
Effect of treatment (Saline vs ENDOVAC-Porci: P = 0.059; Saline vs Porcilis Ileitis: P = 0.046; ENDOVAC-Porci vs Porcilis Ileitis: P > 0.05)		

B: Weight gain

Day 42-70 by treatment group		
Treatment	Mean kg	Mean lb
Saline	12.17	26.83
ENDOVAC-Porci	15.70	34.60
Porcilis Ileitis	16.80	37.04
Effect of treatment (Saline vs ENDOVAC-Porci: P = 0.056; Saline vs Porcilis Ileitis: P = 0.047; ENDOVAC-Porci vs Porcilis Ileitis: P > 0.05)		

C: Clinical & fecal scores

Day 58-70			
Scoring	Saline	ENDOVAC-Porci	Porcilis Ileitis
Clinical	24.7[a]	14.6[b]	15.9[ab]
Fecal	27.4a	17.1[b]	20.9[ab]

Treatment means with different superscripts differ from each other (P < 0.05)

Poster 55

https://doi.org/10.54846/am2023/140

The effect of feeding Salmate DHA+® on sperm production and quality in boars

Darwin L. Reicks, BS, DVM; Tom Gall, MS

Introduction

Fish oil supplementation has been associated with increased sperm production in humans.[1] A complicating factor in implementation in boars has been freshness of the product under conditions of current manufacturing, storage, delivery, and usage conditions. Often, ingredients are stored for extended periods to reduce the risk of disease introduction through feed. A previous study had shown a significant increase in sperm production and reduction in rejected ejaculates when a commercial product high in DHA was fed to boars.[2] The purpose of this study was to evaluate a coated product under real world manufacturing and storage conditions.

Materials and methods

58 boars 6- 6 ½ months of age on arrival were randomly separated into treatment and control groups and breed lines split evenly. A total of 29 boars were allocated to the fish oil group and 29 to the control group. All boars were housed in the same room. Boars in the fish oil group received a supplement of 70 grams of a fish oil supplement (Salmate DHA+®) product top dressed daily at the time of feeding and data tracked for six months. Training commenced one week post arrival. The initial 3-4 ejaculates for each boar occurred in isolation and were discarded. Subsequent ejaculates were obtained in the main boar stud and data included for the study for six months post entry. Fish oil (Salmate DHA+) was also integrated into rations at three other studs in the spring of 2022. Data were collected and compared for 16 weeks prior to the inclusion of Salmate DHA+ in the ration, during a 6-week (one sperm production cycle) acclimation period, and for 16 weeks after the acclimation.

Results and conclusions

668 ejaculates were included in the data analysis. There was a significant difference in total sperm (Control = 90.4×10^9; Fish oil = 97.9×10^9; $P < 0.01$) and normal sperm (Control = 81.5×10^9; Fish oil = 88.9×10^9; $P < 0.01$). There was no difference in the % normal sperm (% of sperm with normal morphology) (Control = 91.5%; Fish oil = 92.2%) or Trash Rate (Control = 4.9%; Fish oil = 2.6%). For one of the studs implementing Salmate DHA+, 9200 ejaculates were included in the analysis. There was a significant difference in total sperm (Before fish oil = 84.2×10^9; After fish oil = 87.1×10^9; $P < 0.01$) and normal sperm (Before fish oil = 75.9×10^9; After fish oil = 79.1×10^9; $P < 0.01$). There was also a difference in the trash rate (Before fish oil = $5.8\% \times 10^9$; After fish oil = $3.1\% \times 10^9$; $P < 0.01$) and the average % of sperm with normal morphology (Before fish oil = 90.0×10^9; After fish oil = 90.7×10^9; $P < 0.01$). In the controlled head-to-head study, there was a benefit to using the Salmate DHA+ fish oil product in the ration for total sperm and normal sperm. When this was applied in the field to three other studs, the stud with data at the time of this paper submission also showed this same benefit in a before and after study of 9200 total ejaculates.

Results of two other boar studs conducting a before and after field study will be available for the AASV annual meeting.

References

1. Jensen TK et al. Associations of fish oil supplement use with testicular function in young men. JAMA Netw Open. 2020;3(1): e1919462. **doi:10.1001/jamanetworkopen.2019.19462**

2. Parsley, M.A., M.E. Wilson, T.J. Gall, and M.R.M. Ballard. 2021. Effect of stabilized fish oil on sperm quality and production of boars. Open J. of An. Sci., 11, 197-207.

Poster 56

https://doi.org/10.54846/am2023/141

Effects of Furst Strike Direct™ on mortality and injectable treatments in nursery pigs

F. Sandberg, PhD; K. Soltwedel, MS, PhD; R. Fischer, MS, PhD; W. Kwon, MS, PhD
Furst-McNess Company, Rockford, Illinois

Introduction

The first 7 days post-weaning are critical for establishment of water and feed intake, and to prevent damage to intestinal structure, and support innate immunity as piglets transfer from sows' milk to dry feed. Previous research has demonstrated that supplementation with monoglyceride forms of fat immediately post-weaning, can improve overall nursery performance.[1] Furst Strike Direct™ is a new nutritional supplement that is also provided via drinking water. Supplying high levels of monoglycerides, natural extracts and antioxidants immediately post-weaning, serve to support the pig during this transition period. Furst Strike Direct also provides novel natural flavorings that have been demonstrated to improve water consumption by 15-20%.[2] All prior research was conducted in controlled research facilities, and thus, it was decided to evaluate Furst Strike Direct™ in a larger scale 'live production' environment with a commercial wean to finish flow.

Methods

Ten wean to finish sites, ranging of 1500 to 3000 pigs, were used in the evaluation. There was a total of 21,885 pigs on the control and 21,966 pigs on Furst Strike Direct™ treatments. One room in the barn was randomly assigned as the control and the second adjacent room was assigned to receive Furst Strike Direct™ at a rate of a 16 oz bottle per 5 gallons of stock solution (metered at 1:128) continuously without break. Furst Strike Direct™ was used for the first approx 7 days starting from placement until the site was filled. Medical injectable treatments and mortality were measured for the first 7 weeks on feed. Barns were also designated into two separate groups – low health and high health. Low health were sites with a control mortality rate of greater than 3%. The low health sites had diagnosed presence of PEDV and streptococcus. Data were analyzed as a general linear model in Minitab. A statistical difference was observed where $P < 0.05$ and a statistical trend was $P < 0.10$.

Results

Furst Strike Direct tended to reduce mortality overall (2.43 vs 1.59%, $P < 0.10$), and reduced mortality (5.70 vs 2.35, $P < 0.01$) in groups with more significant health challenge. Figure 1 shows a boxplot summary of the mortality data. There was no statistical difference for number of treatments, however, numerically they changed by -21% overall, and -53% in health challenged groups when Furst Strike Direct was supplemented.

Discussion

Furst Strike Direct™ based on the results of this trial appears to support improved health, leading to reduced mortality, especially for health challenged flows, and potentially could be a tool to reduce antibiotic usage.

References

1. Sandberg F, Hartsook G, Soltwedel K and Bible M. Furst Protect Direct nourishes the immune system before a natural porcine epidemic diarrhea virus (PEDv) infection in nursery pigs. Leman Veterinary Conference 2021.

2. Sawyer AM, Carter SD, Aparachita P, Harshman JA, Bible MR and Sandberg FB. Evaluation of varying levels of a nutritional water supplement for first 7 days on growth performance of nursery pigs. National ASAS. 2019.

Figure 1: The effect of Furst Strike Direct supplementation on group mortality for 10 side-by-side comparisons in a commercial wean to finish flow (control = 2.43%; Furst Strike Direct = 1.59%, SEM 0.276, $P < 0.10$).

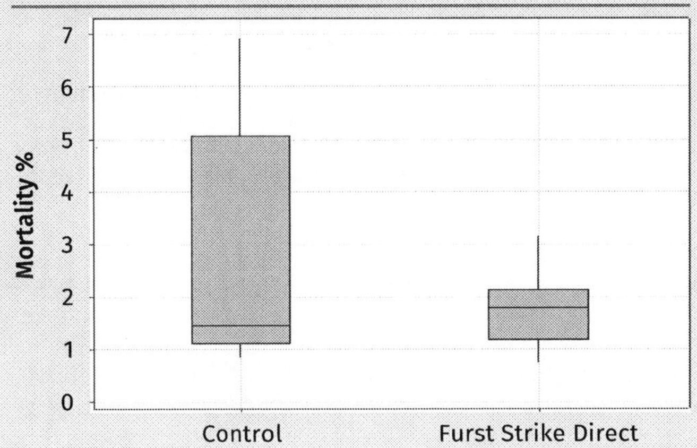

Poster 57

https://doi.org/10.54846/am2023/142

eCVI integration with AgView from National Pork Board streamlines data sharing

J. Tyler Holck, DVM, MS, MBA
Feed His People, LLC

Introduction

GlobalVetLink (GVL) is an industry leading software provider, with multiple solutions that streamline workflows and regulatory processes for swine veterinarians. These solutions include:

- Certificates of Veterinary Inspection (CVIs) - GlobalVetLink provides veterinarians with an easy-to-use, cloud-based software solution for digital CVIs for all species of animals. GVL CVIs are approved for all US states and territories and include automatic submission to state animal health officials.
- Veterinary Feed Directives (VFDs) - GlobalVetLink provides a trusted and proven electronic VFD platform. GVL VFDs are simple to create, comply with two-year record retention requirements, and streamline communication and treatment for veterinarians, feed distributors and producers.
- Veterinary Prescriptions - GlobalVetLink provides end-to-end management of digital veterinary prescriptions, helping ensure accuracy and compliance. GVL prescriptions provide clear instructions to clients, support VCPR, and streamline fulfillment and tracking of prescriptions.
- LabLink HIMS® - LabLink HIMS is a real-time diagnostic solution that enhances herd health management for disease monitoring, analysis, and pathogen response.

The GlobalVetLink platform supports the industry through providing easy to use solutions that streamline complex regulatory and compliance processes. GlobalVetLink also works to connect with the services that swine veterinarians use to further streamline their day-to-day processes.

Purpose of AgView

AgView, a free, opt-in technology solution from the National Pork Board, helps producers of all sizes and types provide disease status updates and pig movement data to state animal health officials. This solution promotes business continuity for America's pig farmers by uniquely making disease traceback and pig movement data available to the USDA and state animal health officials on day one of a foreign animal disease (FAD) outbreak.

Integration with AgView

GlobalVetLink has integrated with the AgView platform (Figure 1) to streamline the data sharing process. Producers who are registered with AgView can Opt In to automatically share their GlobalVetLink swine CVI data with AgView. Once a producer has created their free AgView account, they can follow these steps to share their GVL data with AgView:

- Go to **user.globalvetlink.com/agviewsignup**
- Enter your Email Address and the API Key and API Secret provided by AgView
  - Login to your AgView account to obtain this information
- Click Submit

This streamlined access to swine movement data supports efforts in swine disease traceability and prevention.

Streamlining data sharing

GlobalVetLink is proud to create connections across the industry to streamline data sharing, ensuring that veterinarians, animal health officials, laboratory managers and other key individuals have access to the data they need, when they need it. Some of these connections include:

- AgView - integration sharing Swine CVI Data
- Trace First and USAHERDS - integrations sharing CVI Data with State Animal Health Officials
- Prairie Systems and Seaboard Foods - integrations sharing VFD Data
- Wise Option - Practice Management integration for Equine Veterinarians

GlobalVetLink

GlobalVetLink is the leader in disease tracking and compliance documentation for animal health. The GVL Compliance Assistant SaaS platform collects, communicates, and stores animal health record information and complete compliance certificate records. Veterinarians for all species have access to automated creation and electronic submission of critical compliance documents, as well as having a real-time connection to an extended network of testing labs for submission and tracking of results. The GVL Compliance Assistant is used by over 10,000 production, equine, and companion veterinary clinics to accurately create, manage, and submit digital health certificates required by the USDA, state animal health offices, and international animal health agencies. GlobalVetLink uses artificial intelligence and blockchain technology to authenticate and securely store and communicate animal health records. For more information on GlobalVetLink, visit **globalvetlink.com**.

Reference

1. AgView from the National Pork Board. 2021. Accessed October 20, 2022. **www.agview.com**.

Figure 1: GlobalVetLink's integration with AgView streamlines the Swine CVI data sharing process for veterinarians and producers.

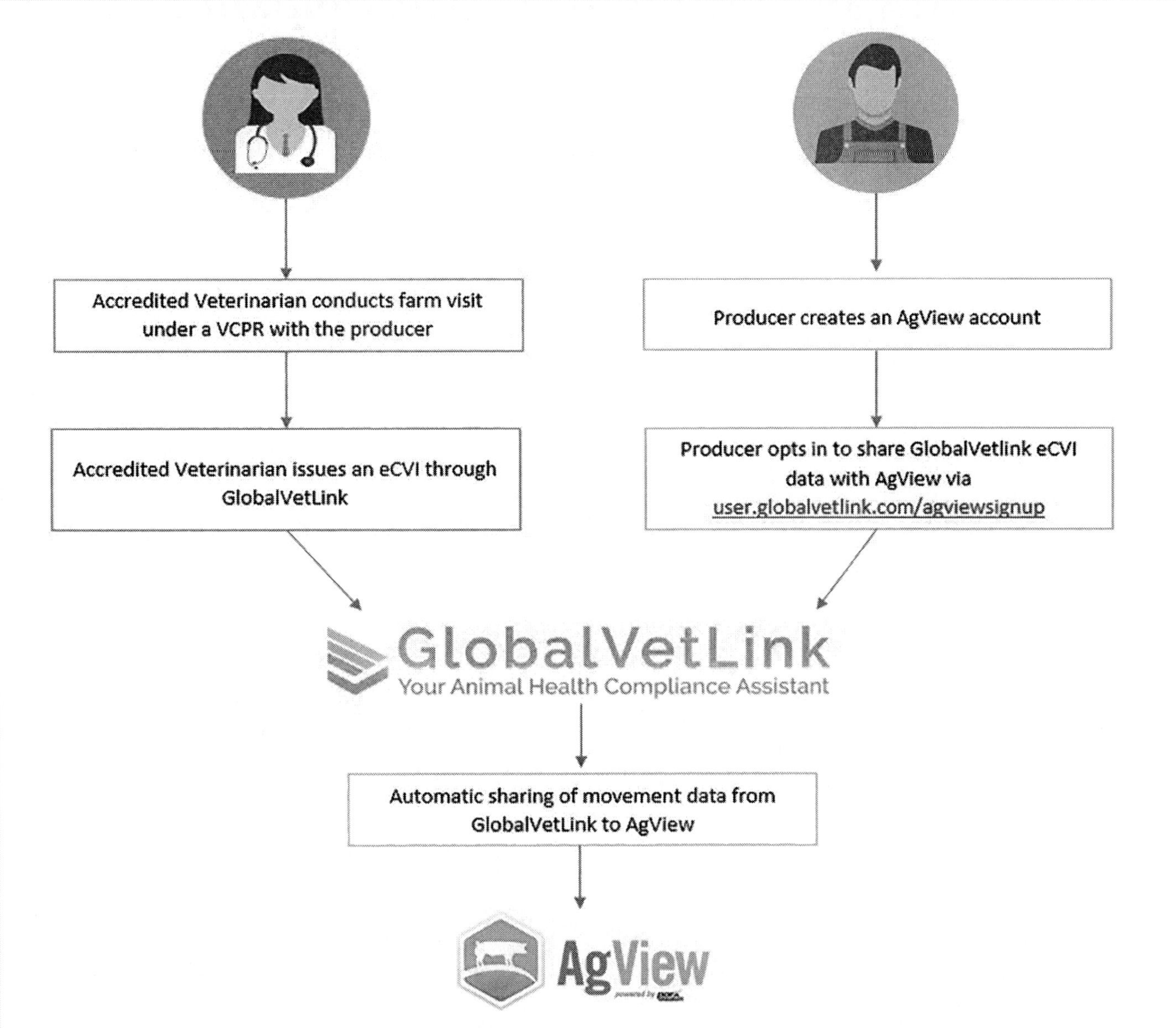

https://doi.org/10.54846/am2023/143

Comparative performances study of three commercial vaccines against *Mycoplasma hyopneumoniae* and porcine circovirus type 2

M. Bonneu[1]; S. Messier[1]; D. Bussieres[1]; M. De Groot[1]; R. Jordà[3]; L. Nodar[3]; N. Bernal[3]; I. Ballarà[3]; Y. Panaccio[2]; J. McKnight[2]
[1]Demeter Veterinary Service; [2]Hipra Canada Animal Health; [3]Laboratorios HIPRA S.A., Amer, Spain

Introduction

The objective of this study was to compare the production parameters and lung lesions between two intramuscular commercial vaccines and an intradermal needle-free commercial vaccine against *Mycoplasma hyopneumoniae* (M hyo) and PCV2, available in North America, in a Canadian farm.

Materials and methods

This study was conducted in a farrow-to-finish herd of 325 sows. The farm was Mhyo, PCV2 and PRRS positive and stable. The animals were monitored during the nursery period, from day 21 of age until day 45 of age, and on the fattening period, from day 45 until slaughter (approximately 6 months). The experimental design was: 418 piglets were weighed at weaning (day 0), tagged, and divided in 4 treatments, 104 piglets per treatment: Vaccine A (Fostera® PCV MH) by the intramuscular (IM) route, Vaccine B IM (Circumvent® PCV-M) by IM, Vaccine C (Mhyosphere® PCV ID) by intradermal (ID) route and control group (untreated). Blood samples were collected from 10 piglets in each group on day 0 for PCV2/PCV3 PCR, M hyo ELISA and PCV2 (IgG&IgM) and on week 26 for M hyo and PCV2 serology. Animals were weighted at day 0, 9 and 24 (end of nursery period), and then on day 50, 78, 107, 140 and 166. Feed consumption was monitored during the same days to assess the feed conversion rate (FCR) and average daily weight gain (ADWG). Lung lesions, mortality and antibiotic treatments used were also monitored.

Results

All animals were PRRS and PCV2/PCV3 PCR negative at day 0. Regarding serology at 10 weeks of age, an early circulation of M hyo was reported with more than 60% of positive animals on the control group. PCV2 circulation was confirmed at 26 weeks of age. Regarding the nursery phase, no bias was observed between groups on the weaning weight. However, statistically significant differences (P-value = 0.033) were observed on the ADWG between day 0 to day 9 (ADWG 0 - 9) comparing Mhyosphere® PCV ID vs Vaccine A and B (Figure 1). No differences were observed on body weights at the end of nursery period among different vaccine treatments neither with mortalities. Regarding the lung lesions, the highest % of lung surface with lesion was the control group (6,5%), and Vaccine A and Mhyosphere® PCV ID with the lowest (5,7% both treatments). However, no statistically significant differences between groups were found. Mhyosphere® PCV ID's group was the one with lowest incidence of pneumonia (55%) and with lowest adhesions (10%), however with no significant differences. In relation to global ADWG, Vaccine B (871 ± 40.2 g/day) and Mhyosphere® PCV ID (860 ± 27 g/day) had the highest ADGW, and control group (827 ± 50.7 g/day) the lowest. Regarding the global FCR, Mhyosphere® PCV ID had the lowest FCR (2.55 ± 0.16) compared to Vaccine A (2.69 ± 0.23) and Control group (2.68 ± 0.22) (P = 0.19). No significant differences were observed regarding mortality nor antibiotic treatments.

Conclusion

In nursery period, Mhyosphere® PCV ID was the only group without significant differences on ADWG 0-9 compared to control group, therefore, this vaccine demonstrated to be safer than the other ones. It is known that the intradermal administration has already demonstrated an advantage in terms of animal welfare,[1] moreover the formulation of the vaccine could influence the safety and so getting a better performance of piglets just after vaccination. In the fattening period, Mhyosphere® PCV ID presented the lowest incidence of M hyo-like lesions and the lowest % of lung surface, this was then correlated with the lowest FCR. This study confirms Mhyosphere® PCV ID as an efficacious vaccine in reducing M hyo lesions and controlling PCV2 infections.

Reference

1. A. Sanchez-Matamoros *et al.* 2021. *Welfare effect of intradermal needle-free vaccination on piglets.* ESPHM 2021. AWN-PP-08.

Figure 1: Average daily weight gain (g/d) per group between days 0 to 9 after vaccination (statistical test: Anova).

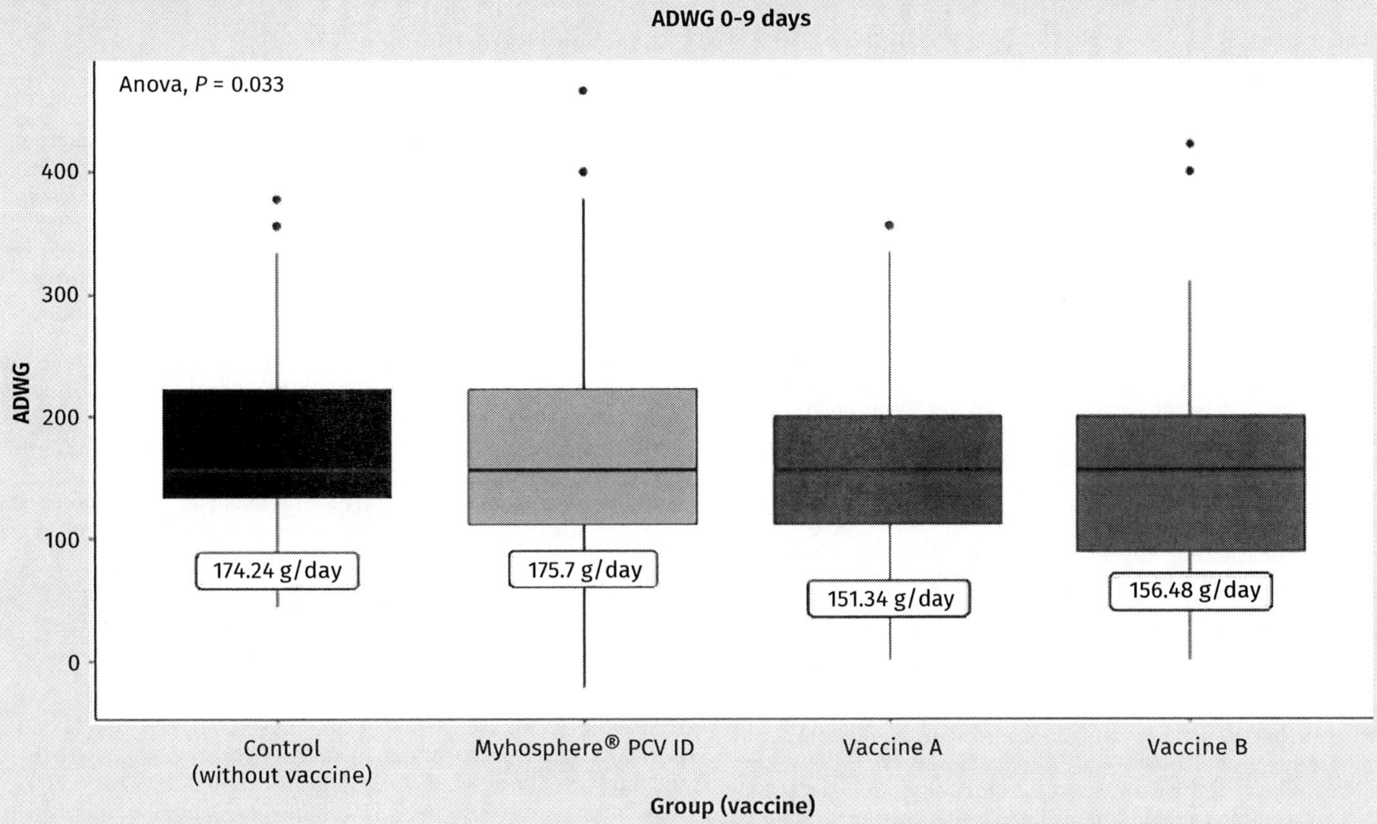

Poster 59

https://doi.org/10.54846/am2023/144

Monitoring PCV2 and PCV3 infection

A. Hidalgo, DVM, PhD; Zhaochun Li, DVM
IDEXX Laboratories, Inc., Westbrook, Maine

Introduction

Porcine circovirus type 2 (PCV2) is an economically important pathogen that has been associated with a broad range of clinical diseases including PCV2 systemic disease (PCV2-SD), respiratory and enteric disease, reproductive failure, and porcine dermatitis and nephropathy syndrome. PCV2-SD and subclinical infections are highly prevalent worldwide. Infection of pigs with porcine circovirus type 3 (PCV3) is thought to be linked to porcine circovirus disease-like scenarios.[1] PCV3 has been identified in most pig-producing countries and retrospective studies have shown that it has been circulating in swine populations for decades. While evidence of PCV3 involvement in some pathologies is expanding, the most frequent presentation of this viral infection is likely subclinical. A case definition for PCV3 systemic disease has been recently proposed, with main clinical signs similar to PCV2-SD.[2] Differentiation of PCV2 and PCV3 is needed in suspect cases of porcine circovirus systemic disease. This study reports on the monitoring of PCV2 and PCV3 infection in a commercial swine farm.

Materials and methods

A swine farm that had recently experienced problems compatible with porcine circovirus systemic disease in nursery pigs (wasting, weight loss, ill thrift and increase mortality) was selected for the study. Routine vaccination included immunization of piglets against PCV2 at 3 weeks of age. A total of 135 blood samples were obtained from sows, suckling piglets, and nursery pigs at 5 different sampling events from October 2021 to May 2022 (Table 1). Blood samples were individually tested with a commercial RT-PCR diagnostic kit for detection and differentiation of PCV2 and PCV3 (RealPCR PCV2/PCV3 Multiplex DNA Test, IDEXX). Quantification standards were used to produce standard curves and measure the PCV2 and PCV3 load present in the samples.

Results and discussion

Four different groups of sows, suckling piglets and nursery pigs were sampled during the study. PCV2 and PCV3 were both detected in 92% of the sampled groups (Table 1).

For PCV2, the percentage of positive animals in any of the groups of sows or suckling piglets was ≤ 55%. The average viral load in these animal groups was < 10^4 PCV2 DNA copies/mL. In nursery pigs, PCV2 was detected in 100% of the animals sampled from December 2021 to April 2022. The average PCV2 viral load in this group remained high from December 2021 to April 2022 ($10^{6.30}$ - $10^{7.23}$ PCV2 DNA copies/mL) and correlated with clinical signs of PCV2-SD and increased mortality of up to 10% in the nursery. A reduction in the percentage of positive nursery pigs to 50% and a marked decreased in the average viral load ($10^{4.57}$ PCV2 DNA copies/mL) was observed in May 2022, together with a reduction of clinical signs and mortality. This clinical improvement followed a review of vaccination practices and a change of PCV2 vaccine manufacturer in early 2022. PCV2 viral load has been correlated with porcine circovirus associated diseases previously and its use proposed for monitoring PCV2 vaccination efficacy.[3,4]

PCV3 was found in most of the groups sampled and its prevalence was higher among breeding animals (range 40 - 73.33%) than in suckling or nursery pigs. However, its average viral load remained < $10^{3.0}$ PCV3 DNA copies/mL in this animal group.

In summary, this study showed effective monitoring of relevant porcine circovirus infections in a farm with clinical disease in the nursery. The detection and quantification of PCV2 and PCV3 in different age groups over time helped to identify an increase in average PCV2 load in nursery pigs and improve vaccination practices.

References

1. Segalés et al. 2019. In: Zimmerman J. J. et al (Eds). Hoboken, NJ, USA: 473.

2. Saporiti et al. 2021. Transbound Emerg Dis.1–13.

3. Segalés et al. 2022. Vet. Sci. 9, 110.

4. Wozniak et al. 2019. Viruses 11:1135.

Table 1: Number of samples (n), percentage of PCV2 and PCV3 positive animals, average Ct values and average viral load (PCV2 or PCV3 DNA copies/mL) by sampling event and animal group.

	Sampling				
	Oct-21	Dec-21	Mar-22	Apr-22	May-22
Sows					
n	10	20	11		15
PCV2 % positive	50	55	27.27		40
PCV2 average Ct value	37.52	37.98	37.96		38.41
PCV2 average viral load	$10^{2.78}$	$10^{2.64}$	$10^{2.68}$	NT	$10^{2.88}$
PCV3 % positive	40	65	45.45		73.33
PCV3 average Ct value	39.70	34.11	35.67		34.49
PCV3 average viral load	$10^{1.70}$	$10^{3.35}$	$10^{2.93}$		$10^{3.26}$
Suckling piglets					
n	10	10	9	NT	10
PCV2 % positive	-	40	55		40
PCV2 average Ct value	-	35.48	37.75		38.71
PCV2 average viral load	-	$10^{3.39}$	$10^{3.07}$		$10^{2.80}$
PCV3 % positive	20	-	11		30
PCV3 average Ct value	26.59	-	44.84		35.21
PCV3 average viral load	$10^{5.58}$	-	$10^{0.36}$		$10^{3.06}$
Nursery pigs					
n		10	8	12	10
PCV2 % positive		100	100	100	50
PCV2 average Ct value		23.30	22.73	26.10	32.31
PCV2 average viral load	NT	$10^{7.0}$	$10^{7.23}$	$10^{6.30}$	$10^{4.57}$
PCV3 % positive		30	12.5	25	60
PCV3 average Ct value		28.44	33.10	30.41	35.42
PCV3 average viral load		$10^{5.04}$	$10^{3.65}$	$10^{4.41}$	$10^{3.0}$

NT = not tested

https://doi.org/10.54846/am2023/145

Long-term usage of a formaldehyde-based feed mitigant does not have a negative effect on sow performance metrics

D. Mellick[1]; D. Humphrey[2]; L. Greiner[2]; M. Bienhoff[1]; V. Mani[1]; V. Iseri[1]
[1]Kemin Industries, Inc., Des Moines, Iowa; [2]Department of Animal Science, Iowa State University, Ames, Iowa

Introduction

Feed has been proven to be a risk factor for viral infection in pigs.[1] Ingredients, such as formaldehyde, have been shown to mitigate this risk effectively.[2] Formaldehyde is an approved feed additive by the United States Food and Drug Administration, and multiple studies have previously been conducted confirming the safe inclusion in livestock and poultry diets.[3] Producers are continually looking to confirm the safe inclusion of long-term formaldehyde use in sow diets. The objective of this study was to compare sow performance prior to and after the use of Sal CURB® in three large US swine production systems over multiple years.

Methods

Historical sow farm data (number born alive per farrow, piglet and sow mortality) were obtained from three production systems representing approximately 550,000 sows per year. The data set represented 68 farms and was aggregated by month within each farm. Data from each system was standardized to start Sal CURB on the same month and truncated to be equal length, resulting in 51 months of data for number born alive and sow mortality, and 90 months of data for piglet mortality. A segmented regression of interrupted time-series analysis was used to evaluate the effect of Sal CURB-treated feed on sow performance using R v4.2.0. Outliers were removed on farm level data using statistical software (anomalize package in R v4.2.0). For number born alive, piglet mortality and sow mortality. 350 (5.66%), 375 (8.15%), and 320 (5.00%) data points were removed, respectively.

Results

For number born alive, there was not a significant change over time (slope of the line) prior to the usage of Sal CURB (Table 1). Immediately following Sal CURB inclusion, there was not a significant change in the number born alive compared to the period prior to Sal CURB usage. However, there was a significant positive increase in the slope for number born alive post-Sal CURB usage, compared to prior to Sal CURB usage. This increase was also observed in non-Sal CURB farms and likely the result of other factors. For piglet mortality, there was not a significant change over time prior to the usage of Sal CURB and following Sal CURB inclusion. Additionally, the slope of the regression line for piglet mortality post-Sal CURB usage was not different compared to prior to Sal CURB usage. For sow mortality, there was a significant increasing trend over time prior to the usage of Sal CURB, indicating a rising overall industry trend. There was not a significant change in sow mortality compared to the period pre-Sal CURB inclusion. Additionally, the increasing trend of sow mortality over time was not different post-Sal CURB usage compared to prior to Sal CURB usage.

Discussion

The interrupted time-series analysis proved to be an effective statistical model to analyze the effects of Sal CURB feed treatment on sow performance metrics. No negative effects on any of the performance metrics, number born alive per farrow, piglet mortality and sow mortality were observed in this long-term Sal CURB usage study.

References

1. Dee S, et al. An evaluation of contaminated complete feed as a vehicle for porcine epidemic diarrhea virus infection of naïve pigs following consumption via natural feeding behavior: Proof of concept. *BMC Vet Res.* 2014; 10:176.

2. Dee S., et al. An evaluation of a liquid antimicrobial (Sal CURB) for reducing the risk of porcine epidemic diarrhea virus infection of naïve pigs during consumption of contaminated feed. *BMC Vet Res.* 2014; 10:220.

3. EFSA Panel on Additives and Products or Substances used in Animal Feed (FEEDAP). Scientific Opinion on the safety and efficacy of formaldehyde for all animal species based on a dossier submitted by Regal BV. *EFSA Journal.* 2014; 12(2):3561.

Table 1: Parameter estimates, standard errors, and *P*-values for the segmented regression models of number born alive, piglet mortality, and sow mortality pre and post Sal CURB inclusion (n = 68 farms)

Parameter	Coefficient	Standard error[1]	*P*-value[2]
Number born alive			
Pre-Sal CURB level	12.671	0.066	**
Post Sal CURB level	12.770	0.078	NS
Pre-Sal CURB slope	-0.001	0.004	NS
Post Sal CURB slope	0.018	0.007	*
Piglet mortality (%)			
Pre-Sal CURB level	13.243	0.464	**
Post Sal CURB level	12.606	0.549	NS
Pre-Sal CURB slope	0.032	0.023	NS
Post Sal CURB slope	0.005	0.026	NS
Sow mortality (%)			
Pre-Sal CURB level	8.068	0.192	**
Post Sal CURB level	7.773	0.273	NS
Pre-Sal CURB slope	0.082	0.012	**
Post Sal CURB slope	0.094	0.019	NS

[1] Standard error for post Sal CURB level and slope coefficients are represented as change from pre-Sal CURB coefficients

[2] NS = $P > 0.05$; * $P \leq 0.05$; ** $P < 0.001$

https://doi.org/10.54846/am2023/146

Productivity profile in a boar stud in southern Brazil: Estimated breeding value (EBV) and collection interval

J. Pradieé[1], DVM, Dr; K. Calderan[1], DMV; L. P. Bevilacqua, DMV; R. Ausejo[2], DVM; P. I. De Quadros[1], DVM; M. M. Oliveira[3], Animal Science, Dr
[1]Bretanha Suínos, Passo Fundo, Rio Grande do Sul, Brazil; [2]Magapor, Ejea de los Caballeros, Zaragoza, Spain; [3]Morgado Experimental Statistic, Passo Fundo, Rio Grande do Sul, Brazil

Introduction

There are several factors that influence the semen production of swine breeding males. Among them are genetics, and within it the genetic index, estimated breeding value (EBV), beyond the age of animals, collection intervals (CI), nutrition, and health. It is understood that EBV provides an estimate of the genetic potential of the animal, which is expressed relative to the population average,[1] what is related to data of animal production as litter size or carcass, for example. However, regard to boar management, starting with the selection of young boars and early identification of sub-fertile boars, it is increasing importance for achieving the goal of reducing the of sperm per semen dose and/or the use of a single (instead of two or three) inseminations per estrus.[2] And thinking about the greater production of semen of the best males, the aim of the present analysis was to evaluate the productivity of the animals at a boar stud and correlating the groups of EBV 80 - 100; 120 - 140 and 140 - 160 with the collection intervals (CI3; CI5 and CI7).

Methods

The data variables were from Magavision® and Gesipor® software, from a boar stud of Southern Brazil from August 2nd, 2021, to July 21st, 2022. The data were all from animals it the doses were approved in the boar stud production. Variables examined from this database included: motility (CASA), ejaculate volume (mL), interval between collections (days), estimated breeding value (EBV), total sperm concentration per collection ($\times 10^9$) and viable sperm concentration ($\times 10^9$). EBV (at collection) was split into three arbitrary classes or groups: Group 1: 80 - 100 (n = 816 collections); Group 2: 100 - 120 (n = 1148 collections); Group 3: 120 - 140 (n = 2912 collections); Group 4: 140 - 160 (n = 2279). Ejaculates within each group were then filtered by collection intervals of 3, 5 and 7 days. Statistical analyses were performed using the R Core Team (2022).3 Variables were tested for ANOVA and media were compared by Tukey test ($P < 0.001$).

Results and discussion

We observed that most of the collections performed (72.55%), and thus, most of the doses produced come from animals with a high EBV (120 - 160), which corresponds to animals aged 1.15 years on average. Animals with a higher EBV (younger) even with a higher % of sperm pathology, do not exceed 30% of the target that discards the ejaculate from production, taking the desirable characteristics of the best EBV to the field. There was statistical difference between all EBV groups, within the CI3 (days) for the motility variable (Table 1). Animals with EBV between 100 and 140 presented higher means of sperm motility, remaining above 4.5. For sperm pathology variable, males with EBV between 100 and 120, in CI3, had a lower % of pathology (16.7%), and very young males, with EBV between 140 and 160, had a higher % of pathology (21.1 %), remaining below 30%, which determines failure. There was a lower volume in the CI3 interval than in the other intervals, which is explained by physiology, the less volume high concentration. And for the variables total sperm collection and viable concentration, they followed the pattern of the smaller the volume, the higher the concentration. We did not find references that corroborated or contradicted our results with EBV and interval collection. Within each interval collections and EBV there were statistical differences, for almost all variables. The objective was not to find a superior EBV to another but studied and measure productivity of boars observing variables: motility, pathology, ejaculate volume, total sperm collection and viable sperm concentration in different collection intervals and EBVs groups to optimize the semen production of boars. The intention was not to seek a "better" EBV, but rather to understand the productivity parameters within each EBV group and within each collection interval, to observe a profile.

References

1. Kodak O & Nagy I. Historical overview of the selection indices applied in pig breeding. Acta Agraria Kaposvariensis. 2019;23(1):22–31. **DOI: 10.31914/aak.2294**

2. Waberski D, Riesenbeck A, Schulze M., et al. Application of preserved boar semen for artificial insemination: Past, present and future challenges. Theriogenology. 2019;37: 2-7.

3. R: A language and environment for statistical computing. R Foundation for Statistical Computing, Vienna, Austria. URL **https://www.R-project.org/** Spencer Graves, Hans-Peter Piepho and Luciano Selzer with help from Sundar Dorai-Raj (2019). multcompView: Visualizations of Paired Comparisons. R package version 0.1-8. **https://CRAN.R-project.org/package=multcompView**. Pohlert T (2014). _The Pairwise Multiple Comparison of Mean Ranks Package (PMCMR)_. R package, <URL: **https://CRAN.R-project.org/package=PMCMR**>. Thorsten Pohlert (2022). PMCMRplus: Calculate Pairwise Multiple Comparisons of Mean Rank Sums Extended. R package version 1.9.4. **https://CRAN.R-project.org/package=PMCMRplus**

Table 1: Mean ± standard deviation for motility, pathology (%), ejaculate volume (mL), total sperm per collection (× 10^9) and viable sperm concentration (× 10^9) by EBV groups 80 - 100; 100 - 120; 120 - 140 and 140 - 160 between collection interval (3, 5 and 7) at Southern Brazil boar stud

	Motility	Pathology	Ejaculate volume	Total sperm por collection	Viable sperm concentration
EBV			**IC3**		
80-100 (n = 584)	4.59 ± 0.961^{c}	18.1 ± 6.58^{b}	258 ± 127^{a}	191 ± 78.8^{c}	35.7 ± 13.5^{a}
100-120 (n = 741)	4.74 ± 0.937^{a}	16.7 ± 6.45^{c}	229 ± 93.9^{b}	218 ± 97^{b}	37.4 ± 13.8^{a}
120-140 (n = 1726)	4.70 ± 0.947^{abcd}	17.4 ± 6.27^{b}	195 ± 86.5^{c}	222 ± 107^{b}	32.0 ± 13.4^{b}
140-160 (n = 1630)	4.66 ± 0.995^{b}	21.1 ± 5.85a	174 ± 61.2^{d}	256 ± 102^{a}	33.1 ± 13.8^{b}
EBV			**IC5**		
80-100 (n = 153)	4.73 ± 0.954^{a}	18.3 ± 5.99^{b}	253 ± 127^{a}	225 ± 92.5^{b}	41.8 ± 17.2^{ab}
100-120 (n = 338)	4.69 ± 0.956^{a}	17.4 ± 6.27^{b}	247 ± 102^{a}	240 ± 106^{b}	44.5 ± 16.8^{a}
120-140 (n = 605)	4.76 ± 0.929^{a}	17.8 ± 6.27^{b}	188 ± 89.9^{b}	261 ± 118^{a}	35.6 ± 14.7^{c}
140-160 (n = 506)	4.69 ± 0.971^{a}	21.6 ± 5.82^{a}	183 ± 64^{b}	301 ± 131^{ab}	39.7 ± 15.6^{b}
EBV			**IC7**		
80-100 (n = 79)	4.80 ± 0.955^{a}	20.4 ± 7.09^{a}	259 ± 118^{a}	273 ± 121^{b}	51.1 ± 22.2^{a}
100-120 (n = 69)	4.79 ± 0.896^{a}	18.4 ± 6.06^{ab}	255 ± 111^{b}	266 ± 124^{b}	48.6 ± 17.2^{a}
120-140 (n = 581)	4.80 ± 0.932^{a}	18.1 ± 6.61^{b}	160 ± 86.9^{b}	305 ± 164^{b}	34.7 ± 17.7^{b}
140-160 (n = 143)	4.8 ± 0.876^{a}	20.7 ± 5.67^{a}	183 ± 79.6^{b}	349 ± 170^{a}	46.3 ± 24.3^{a}

Different letters on the line differ statistically $P < 0.001$

https://doi.org/10.54846/am2023/147

Understanding IAV-S diversity over time using the SEQUIVITY® Dashboard

Channing Sebo, DVM; Jeremy Maurer, VMD; Ross Free, DVM
Merck Animal Health, Lenexa, Kansas

Introduction

Merck Animal Health has used the SEQUIVITY® Dashboard to assist customers with sequence management and prescription vaccine design since 2017. Sequence analysis is key to vaccine design and epidemiologic investigations for influenza A virus in swine (IAV-S) as well as other pathogens. Managing the disease impact and economic losses associated with influenza continues to be a challenge for the global swine industry. Prevention of IAV-S can be difficult given the ease of transmission, risk of zoonotic transmission, and high genetic mutation rate of the virus. Vaccination against IAV-S is particularly challenging due to the extensive diversity of hemagglutinin (HA) and neuraminidase (NA) proteins present in the field due to antigenic shift and drift.[1] Additionally, preferential pairing among specific HA and NA genetic clades has shown to be associated with genetic reassortment of co-circulating viruses.[2] Through the SEQUIVITY® Dashboard, Merck Animal Health can evaluate HA and NA clade pairing data offering insights to better understand strain dominance, virus recombination, geospatial spread, and ultimately success of vaccination.

Methods and results

An analysis using the SEQUIVITY® Dashboard was performed to evaluate HA-NA pairing data for the period January 2021 - November 2022 (Table 1). The data was comprised of both customer and publicly available data. Results are depicted as a percentage of the total number of sequences for the given HA clade in the database over the time period. Approximately 3,500 HA sequences and 1,400 NA sequences were analyzed. HA sequencing is more commonly performed alone. No H1 beta or N2 1998A sequences were detected in our database over this time period.

It is worth noting that some HA clade types displayed more variation in their NA clade pairing, while others almost exclusively paired with a specific NA clade.

One example of a regional HA-NA pairing difference occurs with the H1 alphas. The dominant H1 alpha in the eastern portion of the United States commonly paired with a N1 classic (11.7%) while H1 alphas in the Midwest commonly paired with a N2 2002B (12.4%). H3 human-like also displayed a more diverse pairing behavior. The majority paired with N2 2002B (26.6%), but also paired with N1 classic, N1 pandemic, N2 1998B, N2 2002A and N2 human-like NA clades.

HA clades found to more commonly paired with specific NA clades include the H1 gammas (pair with N1 classic, 52.3%; other NAs - 0.4%), H1 delta 2s (pair with N2 1998B -44.9%; other NAs -0.2%). The most dominant HA-NA pairs present in the United States over this time period was a H1 gamma paired with a N1 classic followed by a H3 cluster 1-4 paired with a N2 2002B.

Conclusion

As veterinarians and producers investigate IAV-S and hope to understand the complex pathogen dynamics, tools like the SEQUIVITY® Dashboard are critical to design successful vaccine intervention strategies. HA and NA pairings are complex and ever changing, but critical in our understanding of this pathogen. The SEQUIVITY® Dashboard is currently used extensively by US and Canadian swine veterinarians to monitor IAV-S and other pathogens to make proactive prevention strategies and mitigate financial and disease impacts.

Table 1: HA-NA pairings are depicted as percentages of the total number of sequences for that given HA type in the database within this time period. Over 3500 HA sequences and over 1400 NA sequences were analyzed and summarize within this table.

		H1							H3	
		Alpha	Gamma	Pandemic	Beta	Delta 1a	Delta 1b	Delta 2	Cluster 1-4	Human-like
N1	Classic	11.7%	52.3%	1.9%	–	–	1.4%	–	–	0.2%
	Pandemic	2.1%	0.1%	24.5%	–	–	–	0.2%	–	0.6%
N2	1998A	–	–	–	–	–	–	–	–	–
	1998B	2.1%	0.1%	0.3%	–	0.8%	–	44.9%	0.3%	0.8%
	2002A	2.8%	0.1%	0.5%	–	0.8%	29.4%	–	2.6%	6.2%
	2002B	12.4%	0.1%	0.8%	–	22.1%	10.5%	–	29.9%	26.6%
	Human-like	–	–	–	–	–	–	–	–	2.6%
NA pairing unavailable		68.9%	47.2%	72.0%	–	76.3%	58.7%	54.9%	67.2%	63.0%

HA = hemagglutinin; NA = neuraminidase

References

1. Vincent AL et al. Swine Influenza Viruses: A North American Perspective. *Adv V Res.* 2008: 127-154.

2. Zeller MA et al. Coordinated evolution between N2 neuraminidase and H1 and H3 hemagglutinin genes increased influenza A virus genetic diversity in swine. *Virus Evolution.* 2021.

Poster 63

https://doi.org/10.54846/am2023/148

Isolate, antigen concentration, and adjuvants are important when developing autogenous vaccines

Steve Tousignant, DVM, PhD; Andrea Pitkin, DVM, MS; Keith Wilson, DVM
Newport Laboratories Inc., Worthington, Minnesota

Introduction

Autogenous vaccines are good options for controlling endemic bacterial and viral pathogens of swine when no, or limited, licensed commercial vaccines are available. Similar to licensed commercial vaccines, autogenous vaccines are required to undergo product safety and sterility testing as outlined by the Code of Federal Regulation Title 9 Part 113.113 under the United States Department of Agriculture.[1] A series of studies were conducted to help understand the importance of isolate selection, antigen concentration and adjuvant selection in an autogenous vaccine program.

Methods

Study 1: Isolate selection – A study was designed to investigate the role of serotype on mortality consisting of heterologous and homologous vaccinated as well as non-vaccinated groups of pigs challenged with *Streptococcus suis* serotype 1/2 and 2.

Study 2: Antigen concentration – A study was designed to investigate the role three levels of antigen concentration on mortality in pigs vaccinated and subsequently challenged with a homologous *Strep suis* serotype 2.

Study 3: Adjuvant selection – A study was designed to investigate the role of two different adjuvants on mortality in pigs vaccinated and subsequently challenged with a homologous *Strep suis* serotype 2.

In each study, pigs were vaccinated at day D0, and D21, challenged at D28, and followed for 5 days post challenge.

Results

Study 1: Isolate selection – Mortality among the non-vaccinated controls was 57% and 100% in the serotype 1/2 and 2 challenges, respectively. In the both the serotype 1/2 and 2 vaccinated pigs, mortality was 0% in the homologous, and 13% in the heterologous challenge.

Study 2: Antigen concentration – Mortality among pigs vaccinated with the highest antigen concentration was 13%. Mortality was 63%, and 100% among pigs vaccinated with the mid, and low-levels of antigen, respectively.

Study 3: Adjuvant selection – Mortality among pigs vaccinated with adjuvant A was 70%, and adjuvant B was 0%.

Discussion

Autogenous vaccines are a good option for controlling various bacterial and viral pathogens of swine. Using *Streptococcus suis* as the pathogen of interest, a series of vaccine challenge studies confirmed three important factors when developing an autogenous vaccine. 1) Ensuring correct isolate selection for vaccine development is critical. Cross protection between isolates is limited at best, and proper diagnostic investigations can help refine selection. 2) Utilize the correct amount of antigen to elicit the desired response, without added financial costs. 3) Adjuvant selection is important to ensure the success of an autogenous vaccine program.

These three factors should be considered when developing an autogenous vaccine. They can also serve as starting points to investigate when less than favorable outcomes are documented with autogenous vaccine evaluations in the field.

Reference

1. Code of Federal Regulations, Title 9, Part 113.113. **https://www.ecfr.gov/current/title-9/chapter-I/subchapter-E/part-113/subject-group-ECFR275e24cb93d1fab/section-113.113**

Poster 64

https://doi.org/10.54846/am2023/149

Meta-analysis of field studies when using NutriQuest Swine Awake®

Jake Erceg[1], MS; Rob Musser[1], PhD; J. Tyler Holck[2], DVM, MS, MBA
[1]NutriQuest, Mason City, Iowa; [2]Feed His People, LLC, Gilbert, Iowa

Introduction

NutriQuest Swine Awake® is a novel, bio-active, non-antimicrobial protein derived from direct feed microbial technology. In pigs, NutriQuest Swine Awake has been associated with the induction of immunomodulatory cytokines thus providing improved immunity and performance in the health challenges.[1-3] Response to NutriQuest Swine Awake was initially observed to mitigate health challenges in commercial sow herds experiencing PEDV outbreaks.[4] NutriQuest Swine Awake has also been shown to improve ADG, reduce lung lesions and mediate the immune response in a PRRSV challenge model[5] and is being evaluated in an SIV challenge model as well.

Meta-analysis of 12 Swine Awake field studies in PRRSV-positive pigs

Twelve studies completed in commercial production facilities across the Midwest (IA, NE, MN, IL) and Southeast (NC), were combined to evaluate the effect of NutriQuest Swine Awake administered via water in PRRS positive pigs (121,463 pigs). Each farm, barn or room were treated as individual observations, and the difference between these two treatments was analyzed using a chi-square test. The data in Figure 1 show absolute mortality differences between treated (60,684 pigs) and untreated pigs (60,780 pigs).

Conclusions

Under the conditions of these field trials Swine Awake provided in the water for 21 consecutive days post-placement significantly reduced total cull and mortality by 2.3 percent compared to non-treated controls (25% reduction). In addition, Swine Awake treated pigs required less individual animal treatments compared to non-treated controls. The improvements in total cull/mortality and individual treatments were more dramatic at the sites which experienced more severe health challenges, both enteric and respiratory.

References

1. Brennan K et al. Interleukin 8. *xPharm: The Comprehensive Pharmacology Reference*, 1-4, 2007.

2. Lopez-Castejon G et al. Understanding the mechanism of IL-1β secretion. *Cytokine & Growth Factor Reviews*, 2011;22(4), 189–195.

3. Kajdaniuk D et al. Transforming growth factor β1 (TGFβ1) in physiology and pathology. *Endokrynol Pol*, 2013;64(5), 384–396.

4. Data upon request, NutriQuest.

5. Song, R et al. A novel, non-antimicrobial protein improves growth performance and immune responses of pigs experimentally infected with porcine reproductive and respiratory syndrome virus. J. Anim. Sci., paper in press.

Figure 1: Mortality differences between pigs treated with Swine Awake (60,684) and untreated pigs (60,780).

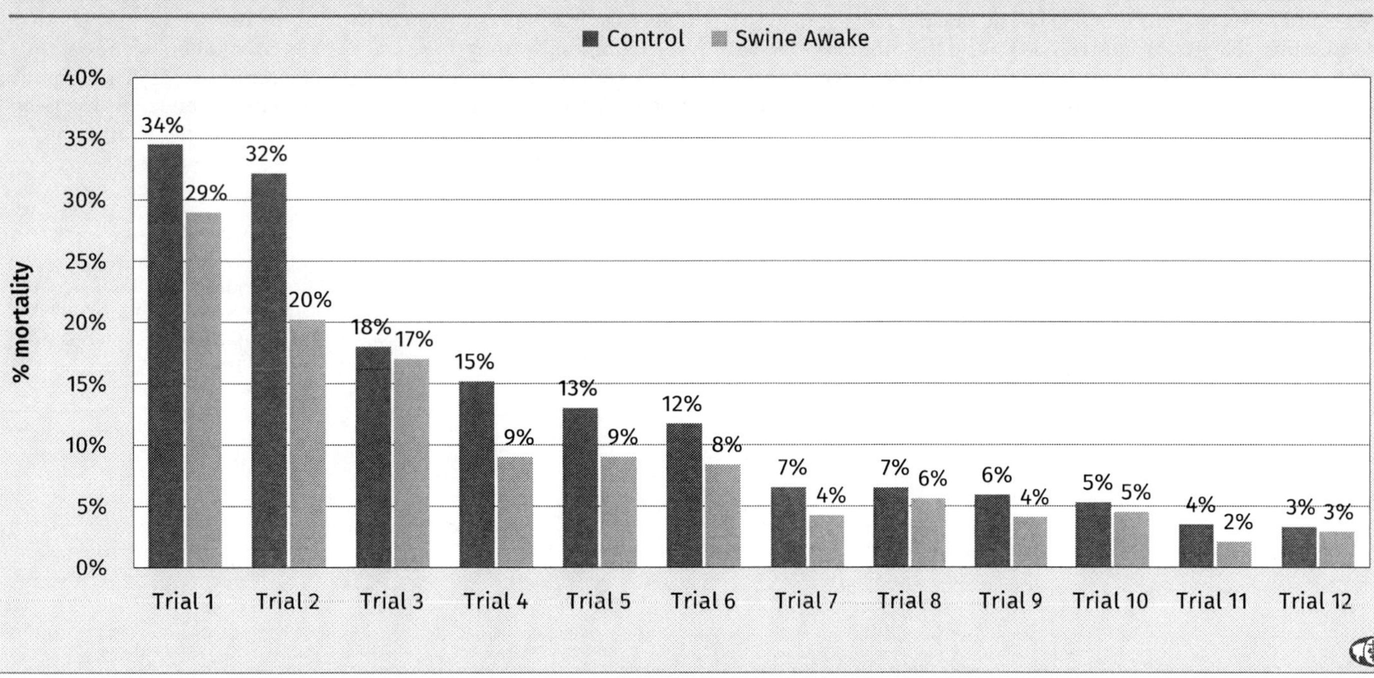

https://doi.org/10.54846/am2023/150

Effects of an additional iron dextran injection administered to piglets on differential gene expression at weaning

James L. Pierce[1,3], MS, PhD; J. Wesley Lyons[2], BS, DVM; Tyler B. Chevalier[3], MS; Duncan B. Paczosa[3], BS; Merlin D. Lindemann[3] , BS, PhD
[1]James L Pierce Consulting, Nicholasville, Kentucky; [2]Pharmacosmos Inc., Watchung, New Jersey; [3]Department of Animal and Food Sciences, University of Kentucky, Lexington, Kentucky

Introduction

Increased litter size combined with faster growing piglets consuming an all-milk diet increase the risk of developing sub-clinical or full-scale iron deficiency anemia.[1] An administration of 200 mg iron at 1-3 d of age has been demonstrated to meet the iron need of pigs until only approximately 4 kg BW due to the depletion of iron stores from birth and low levels of iron in sow milk (approximately 1 mg daily intake). The combination of larger litters of faster growing pigs can contribute to a potential "iron gap" before weaning.[2] Iron deficient anemia has been studied extensively since farrowing moved indoors and piglets no longer had access to the Fe in soil. There is, however, a dearth of information regarding molecular effects responsible for many effects of anemia beyond those which are clinically observable. Therefore, the objective of this experiment was to evaluate the differential gene expression at weaning between pigs receiving one or two 200 mg Fe dextran injections by sequencing the mRNA of liver and duodenal samples.

Methods

Six female littermate pairs were used in an experiment to evaluate the mRNA expression in tissues from piglets given one or two 1 mL injections of iron dextran (200 mg Fe/mL, Uniferon, Pharmacosmos Inc. Watchung, NJ). All pigs in the litter were administered the first 1 mL injection < 24 hours after birth. On d 7, pigs were paired by weight (1.72 ± 0.13 kg) and one pig from each pair was randomly selected as control (CON) and the other received a second injection (+Fe). At weaning on d 22 each piglet was anesthetized, and samples of liver and duodenum were taken from the anesthetized piglets and placed in DNA/RNA Shield (Zymo, Irvine, CA) for preservation until mRNA extraction. RNA-Seq libraries were sequenced to a depth of at least 30 million read pairs (150 bp paired-end sequencing) per sample. Differential Gene Expression data were analyzed with GeneSpring (Agilent, Santa Clara, CA) with a fold-change (FC) of |1.2| $P < 0.05$. Pathway analysis was conducted with Ingenuity Pathway Analysis (IPA QIAGEN, Redwood City, CA) software with Z-score cutoff of $P < 0.05$.

Results

In the duodenum 435 genes were significantly changed with a FC ≥ |1.2| $P < 0.05$. Most notably, Claudin 1 and Claudin 2 were inversely affected by the second dose of iron. Claudin 1 (CLDN1, FC = 4.48, $P = 0.0423$) was up regulated and is responsible for cell-to-cell adhesion in the epithelial cell sheets and serves as a barrier to prevent water and solutes from passing through the paracellular space. Claudin 2 (CLDN2, FC = -1.41, $P = 0.0097$) was down regulated because it is expressed in cation leaky epithelia. In the liver 362 genes were expressed with a FC ≥ |1.2| $P < 0.05$. The gene most affected by a second dose of 200 mg Fe was HAMP (Hepcidin Antimicrobial Peptide) with a FC of 40.8.

Discussion

HAMP is a liver-produced hormone that is the main circulating regulator of Fe absorption and distribution across tissues. It also controls the major flows of Fe into plasma by promoting endocytosis and degradation of ferroportin (SLC4A1). This leads to the retention of Fe in Fe-exporting cells and decreased flow of Fe into plasma. Metabolic pathway changes in the duodenum and liver not only supported the improved hematological measures observed with a second iron injection but also provide evidence for the improved feed conversion and growth rates in pigs given two iron injections pre-weaning. In the duodenum, there is a down regulation of gluconeogenesis ($P < 0.05$). Concurrently, there is a decrease in the production of urea in the liver ($P < 0.05$). These observations show that there is less need for gluconeogenesis, thus less urea production from deaminated amino acids. The genomic and pathway analyses provided empirical evidence linking gene expression with phenotypic observations of piglet health and growth improvements.

References

1. Gaddy, H.J. et al., 2012: "A review of recent supplemental iron industry practices and current usage of Uniferon® (iron dextran complex injection, 200 mg/mL) in baby pigs". In: Proc. 43rd AASV Annual Meeting. March 2012.

2. Van Gorp, S. et al., 2012: "Preventing iron deficiency by avoiding an iron gap in modern pig production". In: Proc. 43rd AASV Annual Meeting. March 2012.

Poster 66

https://doi.org/10.54846/am2023/151

The use of Ambitine® in late finishing diets and the effect on growth performance and carcass measurements

Stacie Crowder, PhD
PMI, Arden Hills, Minnesota

Introduction

Feed costs represent 70% of the input costs associated with pork production.[1] The late finishing phase also sees an increase in stressors such as increased intake, feeder space competition, environmental stress, and limited pen space. These stressors make feed intake erratic and limit performance potential. As pigs increase in weight and size, they become less efficient. This has led to the industry needing technologies to support sustainable growth and carcass efficiency. Under these stressful conditions the inclusion of functional components, such as plant extracts and organic acids, in the diets appear to be effective in improving the productive performance of pigs.[1,2] Two experiments were conducted to evaluate the use of Ambitine® in late finishing diets to improve growth performance and carcass measurements.

Materials and methods

Trial 1 was conducted using 90 pigs with an initial start weight of 226.5 lbs. Ambitine® was included at 0.10% of the diet. Growth performance measurements were collected at d18 and d33 with ADG, ADFI, and F/G calculated for each weigh period. Body composition using real time ultrasound technique was used to determine the carcass measurements.

Trial 2 was conducted using 594 pigs with an initial start weight of 231.4 lbs. Ambitine® was included at 0.10% of the diet. Growth performance measurements were collected at d29 and d43 with ADG, ADFI, and F/G calculated for each weigh period. Carcass measurements were collected at the packing plant on all pigs. Data was analyzed using the PROC Mixed procedure of SAS (SAS Institute, Inc., Cary, NC). Comparisons between treatments was considered significantly different when the probability is ≤ 0.05.

Results

Ambitine® fed to late finishing pigs in trial 1 resulted in 3.3 lb heavier pigs at d18 and 3.97 lb heavier pigs after 33 days. Pigs fed Ambitine® had improved ($P < 0.05$) ADG and F:G during the trial. Carcass weight was increased, and back fat thickness was decreased when Ambitine® was fed to late finishing pigs for 33 days. Ambitine® fed to late finish pigs in trail 2 resulted in 2.5 lb heavier pigs at marketing. Overall feeding Ambitine® to late finishing pigs resulted in heavier pigs at marketing vs the control fed pigs in both trials.

References

1. J. Anim. Sci. 2007. 85:1530–1539.

2. Tung, C. M. and J. E. Pettigrew, NA. Pork Checkoff. Critical review of acidifiers. Accessed: 09 September 2015.

Poster 67

https://doi.org/10.54846/am2023/152

Utilizing Feed Allocation System (FAS)™ data to populate AgView™ for foreign animal disease preparedness

J. Tyler Holck[1], DVM, MS, MBA; Heidi Vittetoe[2], BS; Brandee Graham[2], BS; Joel Stave[3], BA
[1]Feed His People, LLC, Gilbert, Iowa; [2]JWV Pork, Washington, Iowa; [3]Prairie Systems, Spencer, Iowa

Introduction

Feed Allocation System (FAS) is a web-based productivity tool designed to complement the day-to-day process of managing feed ordering and delivering feed. In addition to feed information, daily mortality and animal treatments can be easily entered using FAS Mobile™ in addition to herd visit reports and other on-farm information. FAS data is organized at the group/lot level with inventories and dates of pig placements and exits providing important inter-premises movement data. AgView is a technology solution supported by producer checkoff dollars that is available at no additional cost from the National Pork Board that helps participating producers of all sizes and types provide disease status updates and pig movement data to state animal health officials. This paper describes how an independent swine producer utilizes FAS to populate their AgView account with inter-premises movement data using a data-sharing integration between FAS and AgView.

JWV Pork

JWV Pork is a 4th generation family owned, farrow-to-finish farm operation in SE Iowa which raises 400,000 pigs annually. JWV Pork implemented FAS in Nov 2021 to automate their feed ordering process. They then expanded their use of FAS to include daily treatment and mortality reporting for all sites and organization of service reports and field documentation by staff and contract growers. JWV Pork implemented AgView in Nov 2021 and began utilizing the automated data share of inter-premise movement from FAS into AgView beginning in April 2021.

FAS group set-up and subsequent data collection in FAS Mobile

FAS subscribers organize their data by production sites during initial set-up, including location and premise identification (PIN) information and site capacity. Groups of pigs identified by a unique lot number are established based upon a placement event, either weaning or purchase. All pig movements are recorded in FAS and inventories are maintained real-time based upon daily mobile entries of mortalities. In addition, all feed deliveries are recorded with the potential to identify specific vehicle information via MyFeedWatch™. All FAS data is then available for download into AgView.

AgView data requirements for inter-premises movement data

AgView data includes 1) site information (premises identification numbers (PINs), type of farm, number of pigs); 2) inter-premise animal movement data (both in-state and out-of-state); and 3) diagnostic lab results from participating labs. This information can either be manually entered into AgView or downloaded automatically from FAS.

FAS – AgView data share

The AgView integration feature in FAS provides a portal to export animal movements related to inventory records to AgView electronically for inter-premise animal movement data reporting (Figure 1).

Conclusion

The AgView integration feature within FAS™ allows producers to seamlessly share their animal movement data real-time without

Figure 1: FAS integration with AgView streamlines the swine movement data sharing process.

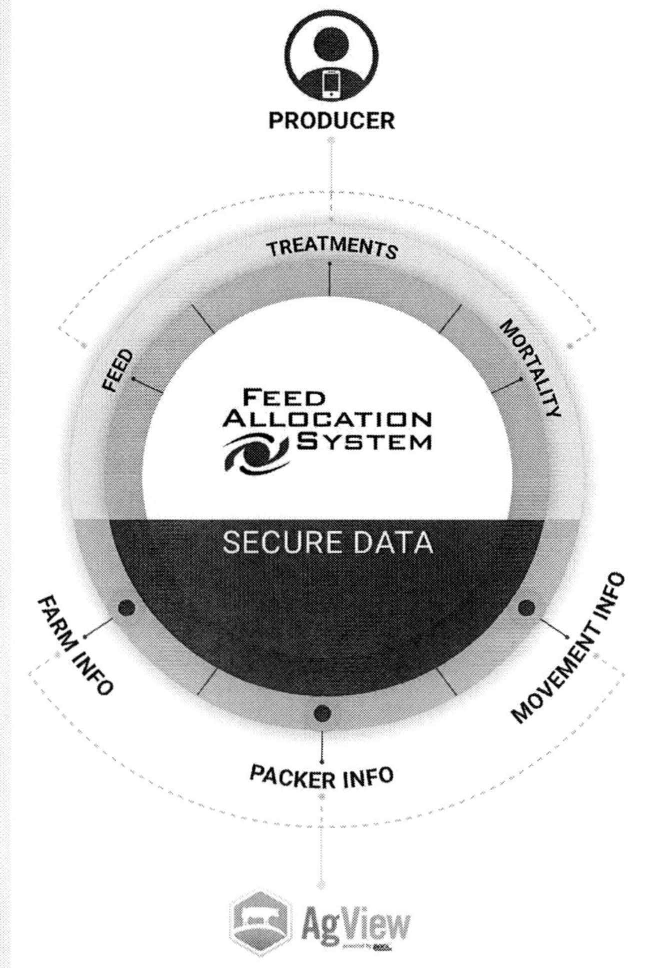

duplication of entry and risk of transcription error. This feature saves producers time and utilizes data collected daily in FAS to support foreign-animal disease preparedness and response.

https://doi.org/10.54846/am2023/153

Maternal genetics contribute to enhanced, natural robustness to PRRS in finishing pigs

C. A. Sevillano[1], PhD; J. R. Dunkelberger[1,2], PhD; E. Little[3], MSc; S. Dee[3], DVM, PhD; E. F. Knol[1], PhD
[1]Topigs Norsvin Research Center B.V., Beuningen, The Netherlands; [2]Topigs Norsvin USA, Burnsville, Minnesota;
[3]Pipestone Applied Research, Pipestone, Minnesota

Introduction

Results of previous PRRS challenge trials conducted by our group provide evidence of substantial phenotypic variation in robustness to experimental PRRS challenge in finishing pigs. The degree of phenotypic variation observed was due, in part, to the genetics of the host. In general, the maternal genetic contribution to finishing performance is overlooked, since in most designs, dam is nested within sire. Therefore, the objective of this study was to quantify the maternal genetic contribution to robustness to PRRS for crossbred finishing pigs using data collected from two experimental PRRS challenge trials.

Materials and methods

Data collected on n = 1,329 and n = 1,144 three-way commercial crossbred finishing pigs from Trials A and B, respectively, were used for this study. Pigs originated from 135 sows and 22 boars for Trial A and 94 sows and 47 boars for Trial B. Boars used for Trial B were selected based on extreme breeding values for mortality under challenge, using Trial A as the training population. To avoid confounding of dam by sire (as with Trial A), pooled semen doses were used for Trial B, consisting of two boars per pool.

Pigs were farrowed at a commercial sow farm and transported to a commercial research facility at weaning. Ear tissue was collected from each pig and submitted for genotyping using a custom single nucleotide polymorphism chip from Illumina. Pigs were vaccinated using a PRRS modified live virus vaccine within one week of placement and subsequently inoculated with PRRSV 1-7-4 via intramuscular injection. Pigs in Trial A were inoculated at four weeks post-vaccination and pigs in Trial B at six weeks post-vaccination. All pigs were followed from placement through slaughter and mortality was recorded throughout.

Mortality was analyzed as a binary trait (0 alive, 1 dead) using records collected between 0 and 28-days post-infection (dpi). Analyses were conducted using the logit function with either an Animal or Sire-Dam model. For the Animal model, sex, barn, and date on test were fitted as fixed effects and litter and pen were fitted as random effects. The Sire-Dam model was identical to the Animal model, except that litter was not included, due to confounding with the maternal effect. A genomic relationship matrix (GRM) was used to account for relatedness among individuals using all available genotypes for trial animals and their parents (n = 2,473). Analyses were conducted using ASReml-R 4.1.0.160 [1] and the GRM was constructed using ASRgenomics 1.0.0 package,[2] both implemented in RStudio.[3]

Results and discussion

Overall mortality rate from 0 to 28 dpi was 3.1%, 8,7%, and 6.1% for Trial A, B, and A/B combined, respectively. Results from this study validate previous findings of substantial, naturally occurring genetic variation in robustness to PRRS.[4] Whether using only data collected from Trial A,[4] Trial B, or both trials together (Table 1), mortality under PRRS challenge was a heritable trait. Results of the combined analysis demonstrate that both sire and dam contribute to the overall genetic variance in mortality, as expected (Table 1). Results also indicate that the use of pooled semen in Trial B helped to separate the overall genetic variation in mortality into the maternal vs. paternal components.

References

1. Butler DG, Cullis B, Gilmour A, et al. *ASReml-R reference manual*. 2017;4.

2. Gezan S, de Oliveira A, and Murray D. An R package with Complementary Genomic Functions. ASRgenomics: VSN International, Hemel Hempstead. 2021.

3. RStudio Team. *RStudio: Integrated development environment for R*. RStudio, PBC, Boston, MA. 2021.

4. Dunkelberger J R, Mathur PK, Little E, et al. Evidence of naturally occurring genetic variation in host response to challenge with a highly pathogenic PRRSV strain. In: *Proceedings, 12th World Congress of Genetics Applied to Livestock Production*. 2022.

Table 1: Variance components (SE) for analysis of mortality from 0 to 28 days following inoculation with PRRS virus

Estimate (SE)	Trial B[a]	Trials A/B combined[b]	
Variance component	Animal model	Animal model	Sire-dam model
Animal genetic	1.28 (0.71)	0.49 (0.39)	-
Sire genetic	-	-	0.42 (0.26)
Dam genetic	-	-	0.26 (0.22)
Maternal environmental	0.01 (0.22)	0.24 (0.24)	-
Pen	0.48 (0.30)	0.22 (0.15)	0.23 (0.15)
Heritability	0.26 (0.11)	0.12 (0.09)	0.32 (0.12)

[a] Trial B: One of two PRRS challenge trials conducted, including data collected from n = 1,144 commercial crossbred finishing pigs.
[b] Trials A/B Combined: Joint analysis of data collected from Trials A and B (n = 2,473).

https://doi.org/10.54846/am2023/154

Evaluation of NSAIDs on weaning/vaccination stress and a subsequent PRRSV challenge

B. Payne[1], DVM; D. Rosener[1], DVM; R. Edler[2], MS; T. Williams[2], DVM; E. Little[2], MS
[1]Veterinary Pharmaceutical Solutions, St. Peter, Minnesota; [2]Pipestone Research, Pipestone, Minnesota

Introduction

Non-steroidal anti-inflammatory drugs (NSAIDs) are widely used in swine medicine for their antipyretic, anti-inflammatory, and analgesic properties. However, there is limited information on their pharmacokinetics and efficacy, and even less information on comparative effectiveness. Acetylsalicylic acid (aspirin) continues to be widely used even though under both nursery and laboratory conditions the highest achievable dose is 3.07 mg/kg, which is just a fraction of the targeted dose of 48 mg/kg.[1,2]

It is common practice to administer vaccines at weaning, and while they are necessary for long-term health benefits, they frequently trigger a febrile response and a depression in feed intake, growth, and feed conversion. Earlier work has demonstrated that ibuprofen may help mitigate these negative side effects.[3]

The objective of this study was to evaluate a variety of NSAIDs on their antipyretic abilities and anti-inflammatory benefits in nursery pigs as measured by feed intake, growth, feed conversion, and mortality following weaning/vaccination and a subsequent 1-7-4 PRRSV challenge.

Methods

404 pigs were split into six treatment groups: negative control (CON), injectable flunixin (FLU), oral aspirin (ASP), oral sodium salicylate (SS), oral compounded ibuprofen/meloxicam (IBU/MEL), and oral compounded meloxicam (MEL). The pigs were placed in 8 pens within four rooms of a BSL2 facility where each room had two water lines – one for the medicated water and one for the CON and FLU pens. All pigs were vaccinated at weaning with a commercial PRRS vaccine as they were enrolled in the study on Day -14. Pig and feed weights were obtained on Days -14, 0, 21, and 42.

On Day 0, the groups were challenged with a highly virulent PRRSV lineage 1, 1-7-4 RFLP, via a seeder pig model. The seeder pigs were given a 2 ml injection of PRRSV with a titer of $10^{3.5}$ $TCID_{50}$ per mL.

Oral NSAIDs were administered for seven days starting on Day -14 (immediately following weaning/vaccination) and again starting on Day 3 (when clinical signs of PRRS were observed). FLU was administered once on Day -14 and again on Day 3. The oral NSAIDs were administered via water medicators calibrated at a 1:128 ratio. The stock solutions for the four oral NSAID groups were prepared as follows: ASP 4 oz., SS 16 oz., IBU 16 oz., and MEL 6.5 oz. per gallon of stock solution as a loading dose the first day, followed by six days with ASP 1oz., SS 10 oz., MEL 3.25 oz., and MEL 3.25 oz. per gallon of stock solution, respectively.

Due to the robust nature of the PRRSV challenge and continuing death loss, for animal welfare reasons all pigs were given an injection of enrofloxacin on Day 17, and a third pulse of the NSAIDs was initiated on Day 18 following the same dosage/duration as the two earlier pulses.

Results and discussion

This pilot study contained some limitations, primarily the lack of sample size which was exacerbated by the excessive death loss across treatment groups. Another likely confounder was the compensatory gains in the pens with higher mortality as they had more pen and feeder space. While there were few statistically significant differences between the treatment groups in this study, the lower mortality rates in the MEL and IBU/MEL treatment groups are notable (Table 1). Additional studies are needed to further elucidate the differences between NSAIDs.

References

1. Patterson A, Karriker L, Apley M. Determination of acetylsalicylic acid and sodium salicylate product stability and achievable serum concentrations under typical nursery conditions. *Proc 36th AASV.* 39-42.

2. Langston C, Apley MD, Boothe DM, Clark TP, Davidson GF, Dowling P, Kemp DT, Papich MG, Riddell MG, Riviere JE, Tubbs RC, Wilcke JR, USP Veterinary Pharmaceutical Information Monographs – Anti-inflammatories. *J Vet Pharmacol Ther.* 2004;27 (S1):4-14.

3. Fry, RS PhD; Williams S, Paton N, Cook D. Effect of administering non-steroidal anti-inflammatory drugs (aspirin or ibuprofen) through the water as possible interventions to mitigate the adverse effects of vaccination on nursery pig performance. *Proc 43rd AASV.* 135-137.

Table 1: Growth performance and mortality per treatment. Values in columns with different letters are significantly different at $P < 0.05$.

	Parameter	Control	Flunixin	Meloxicam	Ibuprofen/ Meloxicam	Aspirin	Sodium Salicylate
Day -14 to 0	ADG, lbs.	0.58^{abc}	0.57^{bc}	0.61^{ab}	0.68^{a}	0.61^{ab}	0.50^{c}
	Avg. daily feed intake, lbs.	0.79	0.77	0.78	0.69	0.78	0.85
	Feed/Gain	1.38^{ab}	1.37^{ab}	1.28^{a}	1.02^{a}	1.28^{a}	1.72^{b}
Day 0 to 42	ADG, lbs.	0.57	0.63	0.83	0.86	0.69	0.81
	Avg. daily feed intake, lbs.	1.79^{b}	1.83^{b}	1.63^{b}	2.01^{b}	2.48^{a}	1.91^{b}
	Feed/Gain*	3.17	4.56	2.13	2.57	4.35	2.78
Day -14 to 42	ADG, lbs.	0.58	0.63	0.76	0.80	0.68	0.71
	Avg. daily feed intake, lbs.	1.41^{b}	1.45^{b}	1.3^{4b}	1.56^{b}	1.85^{a}	1.53^{b}
	Feed/Gain**	2.94	2.38	1.82	2.02	3.47	2.31
	Mortality	35.8%	31.9%	25.4%	22.4%	33.8%	30.3%

* Pen 3 FLU, Pen 9 ASP, and Pen 11 CON omitted

** Pen 3 FLU omitted

Poster 70

https://doi.org/10.54846/am2023/155

Towards the identification of a marker of claw quality in sows using partial least squares analysis and classification and regression trees

T. Kramer[1], DVM, MSc; A. Klein[1], DVM; A. Cornelison[1], MS; W. Schweer[1], PhD; Z. Rambo[1], PhD; M. Socha[1], PhD; G. C. Alberton[2], DVM, PhD; L. Rodrigues[1], DVM, PhD
[1]Zinpro Corporation, Eden Prairie, Minnesota; [2]Federal University of Parana, Palotina Sector, Palotina, PR, Brazil

Introduction

Sow welfare and economic return are directly impaired by claw lesions and lameness. We developed a reliable and standardized method for the evaluation of claw lesions in sows. Towards identifying a marker of claw quality, Partial Least Squares Analysis (PLS) and Classification and Regression Trees (CRT) were used to reduce dataset dimension and solve classification problems, respectively.

Methods

A total of 18,458 sows (parities 1 - 7+; 10% of inventory size within each herd, up to a maximum of 100 sows per herd) were randomly selected and scored from 0 [(no lesions) to 3 (severe lesions)] for claw lesions in the hind legs after parturition.[1] Lesions included heel overgrowth and erosion (HOE), heel-sole crack (HSC), separation along the white line (WL), horizontal (CHW) and vertical (CVW) wall cracks, and overgrown toes (T), or dewclaws (DC). A lesion index (L-Index) was calculated as the sum of the higher scores from each lesion, per animal. Further, the L-Index was classified as Ideal (0 - 2), Acceptable (3 - 6), Worrisome (7 - 9), or Critical (10+). A PLS was used to predict L-Index using HOE, HSC, WL, CHW, CVW, T, and DC as independent variables (IV). Subsequently, the IV with highest Variable Importance in the Projection (VIP; above 1) were analyzed using CRT to identify the best predictors of claw quality.

Results

The average scores were 0.98 ± 0.89 for HOE, 0.17 ± 0.57 for HSC, 0.84 ± 1.16 for WL, 0.69 ± 0.81 for CHW, 0.30 ± 0.71 for CVW, 0.38 ± 0.64 for T, 0.66 ± 0.87 for DC, and 4.00 ± 3.14 for L-Index. The number of sows classified as either Ideal, Acceptable, Worrisome, or Critical were 6,862, 7,845, 2,680, and 1,071, respectively. The high VIPs of HOE (1.157 - 1.197), CHW (1.093 - 1.132), and WL (1.074 and 1.112) established these lesions as likely being dominant (associatively) in affecting the L-Index. The CRT analysis showed two main variables (CHW and WL) as critical predictors of the L-Index. Specifically, 50.8% of Ideal sows had CHW = 0 and 37.0% had CHW between 0 -1 and WL = 0 ($P < 0.01$). Moreover, 33.2% of Acceptable sows had CHW = 0, 16.1% had CHW > 1, and 21.2% had CHW between 0 - 1 and WL < 1 ($P < 0.01$).

Conclusions

Claw quality appears to be mostly impacted by CHW and WL. The model may have reliable and practical applications for the support and implementation of claw quality assessment in sows, particularly in large herds where assessment may be challenging.

Reference

1. Deen, J., Schuttert, M., Van Amstel, S., Ossent, P., & Van Barneveld, R. (2009). Feet first from Zinpro: lesion scoring guide. *Zinpro Corporation, Eden Prairie, MN, USA.*

Poster 71

https://doi.org/10.54846/am2023/156

Characterization of the metabolic conditions contributing to agalactia in sows

Jamie M. Studer[1], MS; Zoe E. Kiefer[1], MS; Lucas Koester[1], PhD; Erika Johnson[1], BS; Lance H. Baumgard[1], PhD; Stephan Schmitz-Esser[1], PhD; Laura L. Greiner[1], PhD; Aileen F. Keating[1], PhD; Attila Farkas[2], DVM; Lucina Galina Pantoja[3], DVM, PhD; Kimberly A. Vonnahme[3], PhD; Jason W. Ross[1], PhD

[1]Department of Animal Science, Iowa State University, Ames, Iowa; [2]Carthage Veterinary Services, Carthage, Illinois; [3]Zoetis, Parsippany, New Jersey

Introduction

Modern hyperprolific sows are susceptible to metabolic disorders such as post-partum dysgalactia syndrome (PDS) or mastitis-metritis-agalactia (MMA) during the periparturient period. These conditions affect the mammary glands and/or reproductive tract in sows after farrowing, and the predominant outcome is a reduction (dysgalactia) or cessation (agalactia) in milk production. Insufficient milk production by the sow results in substantial economic losses to the swine industry due to increased piglet mortality and premature sow removal. To characterize the metabolic conditions contributing to agalactia in sows, a case definition was established, and a research study was conducted on a commercial sow farm experiencing poor lactation performance. We hypothesized differences in immune cells, circulating analytes, and metabolic pathways would exist at farrowing in sows that subsequently developed clinical agalactia compared to healthy herd mates.

Methods

Thirty-six sows with clinical symptoms of agalactia (AGAL) were matched by parity and day of lactation with thirty-six healthy control (CON) sows that remained unaffected. Blood samples were collected at farrowing (timepoint 1) and at onset of clinical agalactia (timepoint 2). Blood was utilized to evaluate differences in circulating immune cell populations, metabolites, and inflammatory markers between agalactic and control sows. Clinical outcomes including sow rectal temperatures, piglet mortality, and piglet body weights were also collected.

Results and discussion

As expected, piglets from agalactic sows had lower ($P < 0.01$) average daily gain and higher mortality than piglets from control sows. At timepoint 1, an 8.9% increase in alanine aminotransferase (ALT; $P = 0.06$) and 26.6% increase in aspartate aminotransferase (AST; $P < 0.01$) was observed in AGAL sows compared to CON. Increased AST was also observed in AGAL sows at timepoint 2 ($P = 0.06$). Elevated levels of the enzymes ALT and AST may indicate degradation of muscle tissue for mobilization of energy reserves to support the metabolic demands of lactation.[1] Differences in concentrations of serum proteins were identified at timepoint 2, with total protein being 4% higher ($P = 0.01$) in AGAL sows compared to CON. Additionally, serum proteins globulin and albumin were 28.6% higher and 5.7% lower, respectfully, in AGAL sows compared to CON at timepoint 2 ($P \leq 0.01$). Circulating cholesterol was 19.5% higher ($P < 0.01$) in AGAL sows compared to CON at timepoint 2, which may indicate increased breakdown of fat to offset the nutrient losses induced by lactation.[2,3] Evaluation of the serum metabolome via liquid chromatography-mass spectrometry revealed differences in 253 and 322 molecular features between CON and AGAL sows at timepoint 1 and timepoint 2, respectively ($P < 0.10$). Further investigation is needed to determine which of the 253 molecular features identified at timepoint 1 may be biomarkers of agalactia to aid in early detection before clinical signs are apparent. In summary, sows experiencing agalactia exhibited altered metabolism compared to healthy herd mates as demonstrated by differences in circulating metabolites.

References

1. Verheyen AJM, Maes DGD, Mateusen B, et al. Serum biochemical reference values for gestating and lactating sows. *Vet J.* 2007;174(1):92-98. **doi:10.1016/j.tvjl.2006.04.001**

2. Ma C, Gao Q, Zhang W, Azad MAK, Kong X. Alterations in the Blood Parameters and Fecal Microbiota and Metabolites during Pregnant and Lactating Stages in Bama Mini Pigs as a Model. *Mediators Inflamm.* 2020;2020:1-13. **doi:10.1155/2020/8829072**

3. Patra MK, De UK, Kent Y, Rungsung S, Krishnaswamy N, Deka BC. Influence of seasonal variation on post-farrowing dysgalactia syndrome (PFDS) and serum biochemistry profiles in the periparturient sow. *Trop Anim Health Prod.* 2021;53(3):346. **doi:10.1007/s11250-021-02793-1**

https://doi.org/10.54846/am2023/157

Will swine veterinarians answer the call to meet the next generation challenges of our industry?

Gordon Spronk, DVM; Karyn Havas, DVM, PhD, MS, DACVPM; Scott Dee, DVM, MS, PhD, DACVM
Pipestone Veterinary Services, Pipestone, Minnesota

Adapted from:

Spronk G, Havas K, Patterson G, and Dee S. *Will swine veterinarians lead by meeting the Next Generation needs of our industry?* JAVMA (published). **https://doi.org/10.2460/javma.22.10.0480**

Executive summary

The US swine industry faces many challenges including labor supply, market access, threats of transboundary animal disease entry to the national herd, ie, ASF, and the relentless pressures of domestic diseases, ie, PRRS. The task of everyone in the industry including swine veterinarians is to meet industry needs in response to these challenges. Specifically, swine veterinarians can meet swine *disease* challenges by providing solutions including education, disease interventions and implementation of biosecurity on their customers local farms to mitigate disease risk. This paper reviews current industry challenges including specific needs that swine veterinarians: field practitioners, researchers, and government veterinarians may consider delivering to meet the needs of USA swine farms to achieve the goal of the swine industry to provide safe, affordable, and nutritious protein to a growing global population. Previous generations of swine veterinarians have led the US swine industry in response to the historical disease challenges resulting in the elimination of foot and mouth disease virus, classical swine fever virus, and pseudorabies virus from the national herd. Will the next generation of swine veterinarians answer the call these new challenges?

The current top ten challenges of the US swine industry

African swine fever virus, challenge; keep this virus out of our national herd.

As of date of this writing, the pathogen known as African swine fever virus (ASFV) has been on the island of Hispaniola in the Caribbean,[1] 721 miles from Miami, Florida since announcement of its presence on July 29, 2021, by USDA/APHIS. This reintroduction of ASF into North America raises the risk of introduction of ASFV to the US swine national herd. An ASF announcement on US soil would immediately stop the exportation of pork, with long term projections of $50 B in losses.[2] A recent paper by Carriquiry et al, summarized that should ASFV incursion to the US take place, the exportation of pork would immediately stop, resulting in a surplus of domestic supply and a series of negative impacts throughout the food supply system, specifically long term projections of $50 billion in losses, nationwide employment losses of 140,000 jobs, and downsizing of the swine industry.[2] The magnitude of this financial impact has already occurred in Belgium (2018) and Germany (2020) and we urge the reader to investigate further the financial impact in those countries as case studies of impact of ASFV on our swine herds and industry.

In summary, as of the time of this writing, Belgium has eliminated ASF due to very quick reaction and preplanning but is struggling to regain its export markets (Spronk G. personal communication). Germany is facing the decline of its industry due to many factors with ASFV introduction into both domestic pig and wild boar populations the most recent and most severe. Therefore, it is imperative that swine veterinarians at all levels do everything possible to keep ASFV out of the US. While great effort is being applied by the USDA Animal and Plant Health Inspection Service and US Customs and Border Protection to detect the importation of ASFV entry via smuggled meat and contaminated food products,[15] a recently identified risk factor that that has a low level of awareness in our industry is the importation of soy-based products from ASFV-positive countries.[16-18] The concern is that ingredients, such as soybean meal and soy oilcake, have been demonstrated to be highly protective for many viruses of veterinary significance, such as ASFV, foot-and-mouth disease virus (FMDV), classical swine fever virus (CSFV), Senecavirus A (SVA), PRRS virus (PRRSV) and pseudorabies virus (PRV) and are potentially contaminated with virus as entry to the US.[19-22] Over the period of January 2019 to July 2022, 6.4 M metric tons of soy products entered the US, with approximately 99.3% originating from four of 23 ASFV-positive countries: Russia, China, Ukraine, and India (Table 1). This is important because transmission of ASFV has been demonstrated following the feeding of diets formulated with soybean meal and inoculated with ASFV, and ASFV DNA has been detected in the commercial feed systems of endemically infected countries that import feed and feed ingredients to the US.[23,24]

To further describe and document this risk, a case study describing the initial introduction of Senecavirus A into a historically naïve national swine herd following importation of soybean meal from a known Senecavirus A positive country was recently published, the first case that links the entry of a new virus into a naïve national swine herd following feed importation.[25] This case study validates published data on feed risk generated at the laboratory level and raises awareness that this event could happen anywhere at any time. It also supports the practice of Responsible Imports protocols to minimize foreign animal disease entry into the US via feed imports.[26] Responsible Imports is a science-based, industry-driven approach to safely import feed ingredients that are not manufactured in the US, involving the auditing of manufacturing plants, shipment of product in one-time use tote bags transported to the US in locked containers, to designated facilities in the US. These products are then stored in quarantine rooms for 30 days at 25°C to enhance viral degradation prior to moving to feed milling facilities in rural areas.[27] This approach has been adapted throughout North America and since 2019, a national program of managing high risk feed imports such as grains and oil seeds from ASFV-positive countries has been in place in Canada.[28] Australia and Denmark have also developed national plans to manage the risk of feed imports from countries of

Table 1: Total soy imports in million metric tons into the US from 23 ASFV-positive countries over the period of January 2019 to July 2022. The top four countries, India, Russia, Ukraine, and China shipped 6,344,607 metric tons of soy-based products to the US over this time period, 99.3% of total imports.

Country	Year				Total	% of Total
	2022 (Jan-Jul)	2021	2020	2019	2019-2022	
India	229,703	1,009,127	1,706,487	1,584,008	4,529,324	70.87%
Russia	271,902	306,511	257,910	82,643	918,966	14.38%
Ukraine	191,162	200,505	224,389	160,581	776,637	12.15%
China	46,745	40,618	7,582	24,734	119,679	1.87%
All others*	2,693	5,822	10,091	27,860	46,466	0.73%
Grand Total	**742,205**	**1,562,583**	**2,206,459**	**1,879,826**	**6,391,073**	**100.00%**

* Austria, Belgium, Benin, Bulgaria, Burma, Cen African Rep, Cote d`Ivoire, Dominican Rep, Estonia, Germany, Hong Kong, Indonesia, Iran, Italy, Laos, Moldova, Nepal, Nigeria, Poland, Romania, Rwanda, Singapore, Slovakia, South Africa, South Korea, Taiwan, Thailand, Vietnam.

lower health status (Dahl J, personal communication). While Responsible Imports efforts are voluntary and industry-driven, initiatives such as the US Swine Health Improvement Plan (US SHIP) incorporate multiple standards, including traceability, surveillance, premise biosecurity, as well as a feed biosecurity standard, with the goal being to validate freedom from ASF and CSF across participating farms.[29] US SHIP is modeled after the National Poultry Improvement Plan, a well-established means of protecting the poultry industry's ability to trade in the face of disease challenges, such as highly pathogenic avian influenza. Currently there is strong industry support, as the National Pork Board, the National Pork Producers Council and the National ASF Task Force are in favor of the fast-tracking of US SHIP.[30]

In conjunction with the Responsible Imports initiative is the opportunity to use feed additives to reduce the risk of pathogen (including ASFV) transmission via feed. Preliminary data document the ability of multiple products to reduce the risk of SVA, PEDV, and PRRSV in feed. In addition, the ability of select additives to reduce the risk of ASFV transmission through feed has been published.[10,31] However, this is "off-label", as none of these products have been approved by the FDA for use against viruses.

In conclusion, fast-tracking the implementation of US SHIP, application of Responsible Import procedures in swine production companies, and the strategic use of validated feed additives at the mill provide a science-based, comprehensive means to reduce, if not eliminate, the risk of viral transport in feed and feed ingredients in our national herd. Clearly, this is excellent progress with a great deal of collaboration across all aspects of the industry with government; therefore, it is our opinion that keeping ASFV out of the US is not only technically possible but creates the most value for the entire industry if we choose to work together. However, it will depend on clear collaboration and communication of industry needs with the appropriate government authorities including USDA/APHIS to accomplish this goal.

Porcine reproductive and respiratory syndrome (PRRS), challenge; elimination of this virus from our national herd.

The endemic viral disease known as porcine reproductive and respiratory syndrome (PRRS) is annually costing the domestic swine industry approximately $663 M in production losses and increased costs,[3] based on introduction of the virus from the sow farm to WTM sites. Recently a new PRRS behavior of lateral proximity spread in WTM groups has been observed in the field with costs up to $6 per pig following lateral introduction of PRRSV into local sites (Spronk G., 2022, unpublished data) in groups from negative sow farms. Therefore, swine veterinarians, including field practitioners, researchers and government officials, have the opportunity to work together to biosecure their local farmers from the effects of PRRSV through virus elimination and prevention of re-infection. As the long-term goals for managing these viruses are different (prevent entry vs. elimination + prevent re-infection), different approaches to biosecurity (transboundary vs. domestic) are required to achieve success. Other national industries are already down this path; Hungary has completed their PRRS elimination program[4] and Denmark is restarting their efforts (Spronk, G. and Dee, S., personal communication). Starting with elimination of the virus and then educating the industry on science-based biosecurity principles to properly implement efficacious re-infection prevention strategies are an opportunity for swine veterinarians at the national, regional and local herd level.

Sow farm PRRS status, challenge; implement known sow farm biosecurity to keep this and other pathogens out of sow farms.

In the late 1980s, biosecurity in the swine industry consisted primarily of washing your hands, disinfection of footwear (rubber boots) and the changing of clothing (coveralls) between farms, with the occasional producer having a shower-in entry protocol and a quarantine and testing program for incoming animals (Spronk G., Dee S., personal experience). With the emergence of porcine reproductive and respiratory syndrome virus, many new routes of transmission were identified based on field observation and confirmed by research trials, including direct routes, ie, infected genetic material (infected pigs and contaminated semen) and indirect routes, ie, contaminated fomites, ie, incoming farm supplies, boots, coveralls, and transport, virus-positive aerosols, and contaminated feed.[5-9] The education and implementation of on farm biosecurity by the swine veterinarian took on greater importance, particularly at the level of the boar stud and the breed to wean sow herd, due to their position in the pork production pyramid. As each route of transmission of PRRSV was identified, interventions were investigated, leading to the development of science-based biosecurity protocols, including the purchase of breeding stock and semen from a PRRSV-naïve source in combination with quarantine and testing, personnel shower-in procedures, decontamination of transport and incoming farm supplies, air filtration, and feed mitigation.[10-14]

As it pertains to elimination of PRRSV from breed to wean farms, this is a well-documented topic that requires little discussion. Successful elimination of PRRSV from breeding populations was first described in 1997 using a technique known as Test and Removal.[32] Shortly thereafter, a more efficient, less invasive approach known as Herd Closure was developed and validated.[33] The current industry gold standard, Herd Closure is widely applied and highly efficacious at the sow farm level, with successful viral elimination documented at > 95 % (Pipestone Veterinary Services internal data).

Next Generation Biosecurity (NexGenBio) is defined as a comprehensive, science-based means to biosecure the breed to wean population, combining protocols designed to mitigate the direct and indirect routes of pathogen transmission.[24] Table 2 provides a summary of the four levels of NexGenBio organized by route of transmission, associated risk factors, and examples of interventions. Level 1 targets the direct routes (quarantine and testing of genetic material purchased from negative sources), followed by application of protocols targeting the indirect routes, first through the management of mechanical risk (level 2), followed by management of aerosol risk (level 3), and finally management of the feed risk (level 4). The cumulative effect of all four levels has been validated through published studies across both experimental facilities and field-based breed to wean populations.[9,34] The challenge facing swine veterinarians is that in the industry today is the wide degree of variation regarding the level of application across farms. For example, while the majority of commercial breed to wean farms in the US have successfully completed level 1 and 2, not all have advanced to levels 3 and 4. Obviously, air filtration is difficult to apply in areas of warm climates and to naturally ventilated facilities. In addition, the risk of viral transmission via feed is a new area of science, first described in 2014,[35] not all farmers and swine veterinarians are aware of the information and/or are prepared to deal with this risk. Fortunately, due to numerous peer-reviewed publications, as well as a recently published, open access Special Issue on the topic of feed risk,[36] information is readily available to veterinarians. Clearly, the application of NexGenBio protocols to breed to wean farms is technically possible and has brought about a positive effect in controlling PRRS. Using the combined approach of Herd Closure (to eliminate) and implementation of NexGenBio protocols (to prevent reinfection) are effective in keeping PRRS and other pathogens out of breed to wean farms.

PRRS in wean to market farms. Challenge; apply First Generation Biosecurity protocols to wean to market farms to prevent PRRS infection/reinfection.

An industry challenge that is rising in importance is transmission and circulation of PRRS in wean to market farms. This production area historically has limited biosecurity effort applied over time, therefore, the "First Generation" nomenclature. Typically, these populations have been determined to be of low risk or value as they are depopulated regularly following the sale of market weight animals and if groups were delivered to the site from a PRRS negative sow farm they generally maintained that status. For that reason, many wean to market farms have yet to apply known interventions to prevent pathogen transmission. However, since the ability of wean to market populations to become infected and transmit pathogens such as PRRSV, PEDV, and influenza A virus of swine has been documented,[37-39] it makes sense to determine what can be done to reduce area spread and protect surrounding breed to wean and wean to market farms. At first glance, it seems that the primary risk factors from pathogen spread associated with wean to market population are contaminated transport, fomites (such as equipment used for power washing), personnel (boots and coveralls),[11,40] feed, and air, suggesting that we could begin applying what we already know. If this is the case, a "First Generation" program could include the sanitation of transport vehicles following delivery to processing facilities (level 1), fomite-based interventions, such as stopping the movement of equipment between sites and the changing of boots and coveralls between sites (level 2), and the use of feed mitigants (level 3). Since we already understand how to mitigate these risks, it may be more productive to first educate, instill discipline, ie, "do what we know we need to do", provide evidence of cost: benefit, and then seek new information, as needed. The ability to apply First Generation biosecurity protocols to wean to market farms is technically possible and swine veterinarians have a key role to play in implementation to prevent PRRS transmission.

How we feed our pigs is changing. Challenge; keep feed costs the most competitive in the world.

As animal health improves, the nutritional needs of the pigs will change to meet the needs of a faster growing, more efficient pig. Additionally, the industry is challenged by changing demand for current energy and protein sources created by the 2005 renewable fuel standard (RFS) as soybean production will increasingly be used for biodiesel production. This challenge is best met with close collaboration of corn/soy producers, researchers, nutritionists, and herd veterinarians to solve nutritional challenges using different energy, protein, and amino acid sources all while meeting the growing global demand of sustainability interventions and measurement.[41-43] The USA

Table 2: A summary of the four levels of Next Generation Biosecurity organized by route of transmission, associated risk factors, and examples of interventions

Level	Route	Risk factors	Examples of interventions
4	Indirect	Feed	Feed additives, Responsible Imports
3	Indirect	Aerosols	Air filtration
2	Indirect	Mechanical (fomite-based)	Transport sanitation, personnel shower in with clothing change, controlled supply entry and equipment use
1	Direct	Genetics (pigs and semen)	Purchase from naïve source, quarantine and test

enjoys a competitive advantage in locally grown and processed feed ingredients, the challenge is to keep that competitive advantage.

Five more industry challenges

On-farm innovation to meet current and future needs. It is obvious that continuous improvement using technologies both new and existing to meet current and future production challenges to keep improving our industry. Field veterinarians and staff are in a unique position to observe on-farm production needs and connect innovators to the farm to meet that need. The skills of field observation, connecting and communicating these observations and potential solutions to innovators is a clear opportunity for all veterinarians.[47]

Continue and increase value creation with international trade. The value created by maintaining and increasing export volume is estimated to be over $8.1B per year or $60 per head.[44] Export value assumes that our products can cross international borders based on negotiated agreements. These agreements include how product will move under current WHOA regulations. Swine veterinarians may wish to be aware of the value created along with the needs of the industry to continue to move products across international markets. The value created will be shared by all by the successful continuation of export growth of our industry into more countries and meeting the demands of current international customers. In summary, maintaining and increasing export volume creates value and is good for the entire industry.

Application of technology via data collection and analysis, fourth-revolution agriculture. While the focus on 'smart farming' will continue to be on field/crop production, there are opportunities to apply the same technological process to manage a swine population better. The industry is challenged to continuous improvement in how we collect data (sow farm data, WTM production data, financial data, carcass data and over-all enterprise data) to improve how we manage our assets, make decisions and determine best daily operational direction.[50] The US has a unique advantage in the level of data sharing that already occurs across the entire production chain. Swine veterinarians have a unique 'place at the table' to contribute to the overall success of the farm by encouraging data collection, analysis and change in operational direction to improve all aspects of the production chain especially at the farm level and implement this fourth-revolution in the barn using analytics and organizational learning to better care for our livestock to produce safe, affordable nutrition food.

Attract, maintain, and train the next generation labor force. We all know the current challenges of a highly competitive environment and demand for staffing at all levels; owners, operators, managers, technicians, farm staff and veterinarians. We can all play a role in attracting the best and the brightest to our industry. Staff retention is key to long term success and swine veterinarians have a role to play in staff education, retention, and interactions on the farm.[48]

The challenge of value creation sharing. The percentage of pigs used for open market pig price discovery has been declining for the past decades (see daily USDA reporting here as example; **https://www.ams.usda.gov/mnreports/ams_2511.pdf**). All future industry owners are challenged with value sharing to maintain industry growth and competitiveness. The challenge for the next generation will be to maintain appropriate value sharing for all segments of the industry.[45,46]

The challenge of sustainability. The next generation swine industry is challenged to meet society demands of sustainable protein production.[49] How we care for our pigs, handle our manure, meter our water to minimize waste, use or generate renewable fuel are all part of the effort to be a good steward of the assets under our responsibility. The entire US industry is uniquely positioned to meet this global challenge as the most sustainable production system in the world. We can be the best farmers on the planet, we just need to set our minds to it.

Summary and conclusions

The goal of this session is to review current challenges and discuss ways for swine veterinarians to respond with solutions to these challenges. What can swine veterinarians apply at the farm immediately?

1. Keep ASFV out of the US

Plan: Encourage your customers to enroll their farms in the US SHIP program.[29] US SHIP is the best strategy of mounting a coordinated plan to prevent the introduction of trade limiting diseases, such as ASF. The initiative to fast-track implementation of US SHIP is gaining momentum with strong buy-in from producers around the US and has the support of state and federal veterinary agencies. Success or failure of the program is based on producer engagement. Ask your congressman to support funding of SHIP to keep our industry globally competitive.

2. Eliminate PRRSV from breed to wean farms

Plan: Conduct a survey of your current customer base and organize sow farms according to PRRS status. Prioritize elimination efforts based on likelihood of success, evaluating farm location, current level of biosecurity, facility design, and economic benefit. Once organized, start the process of elimination using published, validated methods of herd closure and call on experienced colleagues for advice as needed. Encourage your colleagues to set a goal of national elimination of PRRS and lead the industry to achievement of that goal. There is a long history of AASV veterinarians willing to collaborate and help each other out for the entire industry.

3. Apply Next Generation Biosecurity to breed to wean farms

Plan: In conjunction with your PRRS-based survey, evaluate the current biosecurity practices of your breed to wean customers. At what level are they operating at? Do they have the direct routes managed properly with interventions in place? What about the mechanical risks? Can an aerosol and/or feed biosecurity plan be implemented? Are audits results reviewed and protocols improved? Again, seek advice from experienced colleagues as needed. We all need to work together and help each other!

4. Apply First Generation Biosecurity to wean to market farms

Plan: As this is an emerging area, educational efforts will be important as new information comes forward. In the meantime, use the knowledge we already have and start discussions on implementing what we already know, such as washing market trucks and changing clothing and footwear between sites, stopping movement of equipment and begin to use composting for carcass disposal rather than using rendering trucks. This is very "common sense-based" decision making and the sooner we begin the educational process, the faster the adaptation across the industry.

5. Changing nutritional needs of high health herds

Plan: Work closely with your nutritional provider to collect correct data, analyze it and make appropriate changes as ingredient sources change. Additionally, cooperate with field research to confirm and challenge current thinking to make better decisions at all production levels.

6. Innovate

Plan: Many possible options here, find a field production need and work with your choice of innovators to find a solution, there are many. A great example, SoundTalks. This tool gives producers the opportunity to detect and respond to respiratory disease. It uses audio-based technology to detect and differentiate coughing pigs from other barn noises. It then turns that information into a respiratory distress index. The field technician can then use this information for earlier disease detection and intervention. This is just one example of good innovation, there are many others you can apply to your farms.

7. International trade

Plan: Educate yourself and your clients on the value of international trade. Increase your knowledge and stay involved to remove international trade barriers and offer to assist in any way possible. One option; increase the number of APHIS/FAD field report investigations to gain confidence of our trading partners. Go visit another country to learn more about the barriers to international trade.

8. Application of technology, the fourth-revolution at the farm level using data collection, analysis, and benchmarking

Plan: Encourage your clients to collect data from their farm. Any kind of data; water use, antibiotic use, electrical use as examples. All will be valuable in future decision making to continue to keep pork affordable, safe and nutritious.

9. Attracting new talent to our industry

Plan: Our industry faces a labor shortage at nearly if not all levels. Each of us can contribute to attract, train and keep the next generation of staff and leaders in our industry. We all can value people at every level of our industry from next generation owners to barn staff. One way to do this? Learn how to listen.

10. Sustainability

Plan: Participate in the National Pork Board Sustainability initiative (**www.porkcheckoff.org/pork-production-management/sustainability/**) or initiate your own. Educate yourself and your client by recording accurate data on all sites including electricity, water usage, manure usage and all other inputs into a site. This will contribute to an industry goal of 'meeting the needs of the present without compromising the ability of future generations to meet their own needs.'

Summary

In closing, these types of discussions with customers and working together as swine veterinarians: field practitioners, researchers, and government veterinarians, to meet current and future needs are effective ways to build strong, lasting relationships between the swine veterinarian and the farmer. If there is one thing common to swine veterinarians, it is the ability to seek to understand a problem, to meet a need, and to work to develop a solution to benefit the farmer. Therefore, "**Can swine veterinarians lead by meeting the Next Generation needs of our industry?**" is an affirmative **YES.** Now it's time to make your decision to make the US swine industry the 'best in the world'.

Acknowledgments

The authors wish to graciously acknowledge the contributions of Steve Henry, Jurgen Kramer, Dermot Hayes, Aidan Connolly, Mark Greenwood, Jerry Torrison, Steve Pollmann, Bill Evan Cameron Schmitt and many industry leaders willing to share their input and comments.

References

1. Rademacher C, Brown J, Karriker L. Impact of human behavior on the spread of African swine fever virus: what every veterinarian should know. *JAVMA*. 2022; 260(12):1413-1417. **https://doi.org/10.2460/javma.22.06.0250**

2. Carriquiry M, Elobeid A, Swenson D, Hayes D. Impacts of African Swine Fever in Iowa and the United States. Iowa State University Center for Agricultural and Rural Development; 2020; Working Paper 20-WP 600.

3. Holtkamp DJ, Kliebenstein JB, Neumann EJ, Zimmerman JJ, Rotto HF, Yoder TK, Wang C, Yeske PE, Mowrer CL, Haley CA. Assessment of the economic impact of porcine reproductive and respiratory syndrome virus on United States pork producers. *J. Swine Health Prod*. 2013; 21:72–84.

4. Szabo, I. The experience of the National PRRS eradication program in Hungary. *Proceedings of the 13th Annual Meeting European Symposium of Porcine Health Management*. Budapest, Hungary. 2022.

5. Wills RW, Zimmerman JJ, Yoon KJ, Swenson SL, McGinley MJ, Hill HT, Platt KB, Christopher-Hennings J, Nelson EA. Porcine reproductive and respiratory syndrome virus: a persistent infection. *Vet Microbiol*. 1997; 55(1-4):231-40. **doi: 10.1016/s0378-1135(96)01337-5**

6. Christopher-Hennings J, Nelson EA, Hines RJ, Nelson JK, Swenson SL, Zimmerman JJ, Chase CL, Yaeger MJ, Benfield DA. Persistence of porcine reproductive and respiratory syndrome virus in serum and semen of adult boars. *J Vet Diagn Invest*. 1995; 7(4): 456-64. **doi: 10.1177/104063879500700406**

7. Dee SA, Deen J, Rossow KD, Mahlum C, Otake S, Joo HS, Pijoan C. Mechanical transmission of porcine reproductive and respiratory syndrome virus throughout a coordinated sequence of events during cold weather. *Can J Vet Res*. 2002; 66: 232-239.

8. Dee SA, Deen J, Otake S, Pijoan C. An assessment of transport vehicles as a source of porcine reproductive and respiratory syndrome virus transmission to susceptible pigs. *Can J Vet Res*. 2004; 68:124-133.

9. Dee SA, Otake S, Oliviera S, Deen J. Evidence of long-distance airborne spread of porcine reproductive and respiratory syndrome virus and *Mycoplasma hyopnuemoniae*. *Vet Res*. 2009; 40(4)39.

10. Dee S, Niederwerder MC, Edler Roy, Hanson D, Singrey A, Cochrane R, Spronk G, Nelson E. An evaluation of additives for mitigating the risk of virus-contaminated feed using an ice block challenge model. *Transbound Emerg Dis*. 2021; (2):833-845. **doi: 10.1111/tbed.13749**

11. Pitkin A, Otake S, Dee SA. A one-night downtime period prevents the spread of PRRSV and *Mycoplasma hyopneumoniae* by personnel and fomites (boots and coveralls). *J. Swine Health Prod*; 2011; 19:345-348.

12. Dee SA, Torremorell M, Deen J, Thompson B, Pijoan C. An evaluation of the Thermo-Assisted Drying and Decontamination (TADD) system for the elimination of porcine reproductive and respiratory syndrome virus from contaminated livestock transport vehicles. *Can J Vet Res*. 2005; 68:208-214.

13. Dee SA, Deen J, Pijoan C. An evaluation of four intervention strategies to prevent mechanical transmission of porcine reproductive and respiratory syndrome virus. *Can J Vet Res*. 2004; 68:19-26.

14. Dee SA, Otake S, Deen J. Use of a production region model to assess the efficacy of various air filtration systems for preventing the airborne transmission of porcine reproductive and respiratory syndrome virus and *Mycoplasma hyopneumoniae*. Results of a 2-year study. *Virus Res*. 2010; 154:177-184.

15. Dixon LK, Sun H, Roberts H. African swine fever. *Antiviral Res.* 2019;165:34-41. **doi:10.1016/j.antiviral.2019.02.018**

16. Patterson G, Niederwerder MC, Spronk G, Dee SA. Quantification of soy-based feed ingredient entry from ASFV-positive countries to the United States by ocean freight shipping and associated seaports. *Transbound Emerg Dis.* 2021;68(4):2603-2609. **doi: 10.1111/tbed.13881**

17. Patterson G. An analysis of select swine feed ingredients and pork products imported into the United States from African swine fever virus affected countries. Transbound Emerg Dis. 2022;69(1):128-136. **doi: 10.1111/tbed.14265**

18. Blomme AK, Jones CK, Gebhardt JT, Woodworth JC, Paulk CB. Assessment of soy-based imports into the United States and associated foreign animal disease status. *Transbound Emerg Dis.* 2022. 69(1):137-148. **doi: 10.1111/tbed.14284**

19. Dee SA, Bauermann F, Niederwerder MC, Singrey A, Clement T, de Lima M Long C, Patterson G, Sheahan M, Stoian AMM, Petrovan V, Jones C, De Jong J, Ji J, Spronk GD, Minion Christopher-Hennings J, Zimmerman JJ, Rowland RRR, Nelson E, Sundberg P, Diel DG. Survival of viral pathogens in animal feed ingredients under transboundary shipping models. *PLOS ONE.* 2018; 13(3): e0194509.

20. Stoian AMM, Petrovan V, Constance LA, Olcha M, Dee S, Diel DG, Sheehan MA, Rowland RRR, Patterson G, Niederwerder M. Stability of classical swine fever virus and pseudorabies virus in animal feed ingredients exposed to transpacific shipping conditions. *Transbound Emerg Dis.* 2020; 67(4):1623-1632. **doi: 10.1111/tbed.13498**

21. Stenfeldt C, Bertram M, Crockett H, Hartwig E, Smoglia G, Niederwerder M, Diel D, Dee S, Artz J. The risk and mitigation of foot and mouth disease virus infection of pigs through consumption of contaminated feed. *Transbound Emerg Dis.* 2022; 69(1):72-87. **doi: 10.1111/tbed.14230**

22. Caserta L, Gomes C, Noll J, Singrey A, Niederwerder M, Dee S, Nelson E, Diel D. Stability of Senecavirus A in animal feed ingredients and infection following consumption of contaminated feed. *Transbound Emerg Dis.* 2021; 69: 88-96. **doi:10.1111/tbed.14310**

23. Niederwerder MC, Stoian AM, Rowland RR, Dritz SS, Petrovan V, Constance LA, Gebhardt JT, Olcha M, Jones CK, Woodworth JC, Fang Y, Liang J, Hefley TJ. Infectious Dose of African Swine Fever Virus When Consumed Naturally in Liquid or Feed. *Emerg Infect Dis.* 2019; 25(5):891-897. **doi: 10.3201/eid2505.181495**

24. Dee S, Spronk G. Feed: A new pathway for the domestic and transboundary spread of viral pathogens of veterinary significance (Special Issue Editorial). *Transbound Emerg Dis.* 2022; 69(1):1-3. **doi: 10.1111/tbed.14349**

25. Dee SA, Havas K, Spronk G. Detection of Senecavirus A in pigs from a historically negative national swine herd and associated with feed imports from endemically infected countries. *Transbound Emerg Dis.* 2022; **doi: 10.1111/tbed.14684**

26. Patterson G, Niederwerder M, Dee SA. Risks to animal health associated with imported feed ingredients. *JAVMA.* 2019; 254(7):790-791. **doi: 10.2460/javma.254.7.790**

27. Dee N, Havas K, Shah A, Singrey A, Spronk G, Niederwerder M, Nelson E, Dee SA. Evaluating the effect of temperature on viral survival in plant-based feed during storage. Transbound Emerg Dis. 2022. **doi: 10.1111/tbed.14546**

28. Calvin S, Snow A, Brockhoff E. African swine fever risk and plant-based feed ingredients: Canada's approach to risk management of imported feed products. *Transbound Emerg Dis.* 2022; 69(1):176-181. **doi: 10.1111/tbed.14281**

29. Proceedings of the US Swine Health Improvement Plan House of Delegates Meeting (US SHIP HOD), Bloomington, MN Sept 6-8, 2022.

30. Proceedings of the National ASF Strategy Working Group, US Pork Forum, Louisville, KY, March, 2022.

31. Niederwerder MC, Dee S, Stoian AMM, Constance LA, Olcha M, Petrovan V, Patterson G, Cino G, Rowland RRR. Mitigating the risk of African swine fever virus in feed with antiviral chemical additives. *Transbound Emerg Dis.* 2021; 68(2):477-486. **doi: 10.1111/tbed.13699**

32. Dee SA, Molitor TW. Elimination of PRRS virus using a test and removal process. *Vet Rec.*1998; 143:474-476.

33. Corzo CA, Mondaca E, Wayne S, Torremorell M, Dee S, Davies P, Morrison RB. Control and elimination of porcine reproductive and respiratory syndrome virus. *Virus Res.* 2010;154(1-2):185-92. **doi: 10.1016/j.virusres.2010.08.016**

34. Havas K, Brands L, Cochrane R, Spronk G, Nerem J, Dee SA. An assessment of enhanced biosecurity interventions and their impact on porcine reproductive and respiratory syndrome virus outbreaks within a managed group of farrow-to-wean farms, 2020-2021. *Frontiers* (under review).

35. Dee S, Clement T, Schelkopf A, Nerem J, Knudsen D, Hennings J and Nelson E. An evaluation of contaminated complete feed as a vehicle for porcine epidemic diarrhea virus infection of naïve pigs following consumption via natural feeding behavior: Proof of concept. *BMC Vet Res.* 2014; 10:176.

36. Dee S and Spronk G (guest editors). Feed: A new pathway for the domestic and transboundary spread of viral pathogens of veterinary significance. *Transbound Emerg Dis,* Special Issue. Volume 69, Number 1, pp.1-188, Jan 2022.

37. Corzo C, Gramer M, Dee SA, Morrison RB and Torremorell M. Airborne detection of swine influenza A virus in air samples collected inside, outside and downwind from swine barns. *PLOS One.* Vol 8,8, e71444.

38. Linhares D, Torremorell M, Cano JP, and Dee SA. Effect of modified live porcine reproductive and respiratory syndrome (PRRS) vaccine on the shedding of wild-type virus from an endemically infected population of growing pigs. *Vaccine.* 2012; 30:407-413.

39. Kikuti M, Drebes D, Robbins R, Dufresne L, Sanhueza JM, Corzo C. Growing pig incidence rate, control and prevention of porcine epidemic diarrhea virus in a large pig production system in the United States. *Porc Health Mgmt.* 2022; 8:23, **doi.org/10.1186/s40813-022-002689**

40. Otake S, Dee SA, Rossow KD, Deen J, Joo HS, Molitor TW, Pijoan C. Transmission of porcine reproductive and respiratory syndrome virus by fomites (boots and coveralls). *Swine Health Prod* 2002. 10(2): 59-65.

41. Sullivan CR, Thaler RC, Berg KM, De Jong JA, Neill CA, Ward JP, Frahm PF, Hanson DB, Boerboom MJ. Effects of Increasing Levels of Hybrid Rye on Growth Performance of 45 to 125 kg Pigs, *Journal of Animal Science* 2022. 100: **https://doi.org/10.1093/jas/skac064.065**

42. Bromm, J, Tokach, M, Woodworth, J, Goodband, R, DeRouchey, J, Gebhardt, J. Effects of Increasing Fat Levels in Diets Containing 40% Distillers Dried Grains with Solubles on Growth Performance and Carcass Characteristics in Commercial Finishing Pigs. *Kansas Agricultural Experiment Station Research Reports.* 2022. 8(10): **https://doi.org/10.4148/2378-5977.8386**

43. Hutchens, W, Tokach, M, Dritz, S, Woodworth, J, DeRouchey, J, Goodband, R, Calderón, H. Evaluating the Effects of Pharmacological Levels of Zinc Oxide, Diet Acidification and Dietary Crude Protein on Growth Performance of Nursery Pigs. *Kansas Agricultural Experiment Station Research Reports.* 2022. 6(10): **https://doi.org/10.4148/2378-5977.7988**

44. United States Department of Agriculture/USMEF; **www.usmef.org/downloads/Pork-2012-to-2021.pdf**

45. Campbell, T, Kay, A. Solution aversion: On the relation between ideology and motivated disbelief. *Journal of Personality and Social Psychology, 2014. 107*(5), **https://doi.org/10.1037/a0037963**

46. Parcell, J, Schroeder T. Development of Methodology for Calculating Net Price Information for Negotiated Barrows and Gilts. *AMS/USDA Agreement* 12-25-A-5729, May 2014.

47. Padhi, A, Batra, G., Santhanam, N. *The titanium economy: how industrial technology can create a better, faster, stronger America.* New York. Hachette Book Group. 2022.

48. Lee, Kate. Simple Ways to Make Your Team Feel Valued. *Harvard Business Review* July 19, 2022.

49. Bell, D, McLoughlin D, Beyersdorfer D, Hjortshoej M. Danish Crown, Feeding the Future. *Harvard Business School* Case 522-057, December 2021.

50. **https://sloanreview.mit.edu/audio/big-data-in-agriculture-land-olakes-teddy-bekele/** accessed November, 2022.

https://doi.org/10.54846/am2023/158

Lessons learned from the poultry industry for housing and health

Craig Rowles, DVM
Versova Management Company, Clive, Iowa

Introduction

The egg industry and the swine industry share several characteristics and challenges. Each industry provides animal derived protein food products to the consumer. Each industry is faced with headwinds associated with perceptions of animal welfare considerations. Each industry faces risk with the potential introduction of foreign animal disease. This paper will focus on how these challenges have impacted the egg industry, how the industry has responded, and provide suggestions for the swine industry.

Egg production in the United States

Egg production in the United States has followed a similar path as the swine industry. Fifty-100 years ago, small farms raised backyard flocks that produced eggs for their own family consumption with any excess production being sold in small towns and cities. "Mama's eggs" eventually developed into small 10,000-20,000 bird flocks with eggs being handled by hand. Off-site service trucks picked up eggs for transport to offline packaging and distribution.

In the 1970-1980s, further advancements in mechanization and egg handling drove further consolidation in the industry. Large, inline egg grading plants required larger flock size and farm size to maximize efficiency. These farms were often located close to population centers. Just as the swine industry moved toward consolidation in North Carolina during the 1980s, so too did the egg industry as production moved away from the Midwest and closer to the population centers.

In the 1990s, midwestern egg farmers began to compete by combining their capital, adapting the same technologies, and took advantage of their natural corn and soymeal basis. Large, modern egg production facilities were constructed to service both the shell and processed egg markets. Today, Iowa and Ohio rank as the number one and two egg production states in the country. (Figure 1)

Animal welfare considerations

In the 1980s the egg industry began to receive scrutiny over welfare related concerns related to the housing systems of egg laying hens. Around 2006, specific groups like the Humane Society of the United States (HSUS) began campaigns that challenged animal welfare issues across all species. In particular, the egg industry was a focus point of this scrutiny. Their approach took two major prongs.

Ballot initiatives

The first approach was to pursue animal welfare related ballot initiatives. Initiatives were passed in Florida and Arizona, but the first important ballot initiative was in the bellwether state of California. Prop 2 was placed on the ballot in 2008 and was passed in November of that year. That initiative prohibited the confinement of pregnant pigs, calves raised for veal, and egg-laying hens in a manner that does not allow them to turn around freely, lie down, stand up, and fully extend their limbs.[1] This led to further ballot initiatives in other states as well. (Table 1)

In 2018, Prop 12 was passed in California. This second ballot initiative established minimum space requirements based on square feet for calves raised for veal, breeding pigs, and egg-laying hens. Furthermore, it banned the sale in the state of veal from calves, pork from breeding pigs, and eggs from hens when the animals are confined to areas below the minimum square-foot requirements.[2] This has had a more direct impact on the swine industry as the vast majority of pork is produced outside the state of California. This initiative has been challenged by the pork industry due to commerce clause issues. At the time of this writing the case is under consideration at the supreme court of the United States with a ruling expected in the spring of 2023.

The egg industry did respond to the direct challenges they faced. In 2011 they worked in cooperation with industry stakeholders to create the Coalition for Sustainable Eggs.[3] The members included producers, retailers, universities which funded research into all the different available housing systems and provide science-based rankings. That research supported the use of colony cage systems with additional enrichments, including perches, scratch areas, and dedicated nest areas as the most long-term sustainable system.

The egg industry took that information and began a process of negotiation with HSUS. They agreed to work together to fashion an amendment to the 2014 farm bill. The intention of that amendment was to provide a designated roadmap and timetable to give the industry time to adopt these practices in a predetermined manner. That amendment was defeated as it did not receive widespread support from other members of the agricultural community. That defeat led to the next stage of lobbying efforts by HSUS.

Direct corporate lobbying

The second pathway pursued by animal rights activist groups was directed at the boardroom of large corporations. In 2015 the bellwether food corporation, McDonald's, announced they would transition to cage-free hens over the next decade with a goal of 100% transition to cage free by 2025. This initial announcement was quickly followed by a tidal wave of competing food producing corporations which followed with similar announcements and similar timelines. (Table 1)

Where does the industry stand today?

The most recent USDA NASS data as of January 2022 indicates that 34.1% or 111.1 million laying hens are now raised in cage-free environments.[4] This would include hens raised as organic (5.6%) as well as just cage free (28.5%). According to USDA's agricultural Marketing Service, approximately 66% of U.S. hens must be in cage free production by 2026 to meet projected demand.[4] In short, the industry has converted 24% over the past 10 years. To meet the demand predictions the industry would need to convert the next 35% in 4 years. This is a daunting task for the industry.

These expectations are driven by the assumption that all corporations will meet the commitments that were made in 2015. There is evidence to suggest that not all retailers will meet those dates.

Figure 1: Top states for egg production.

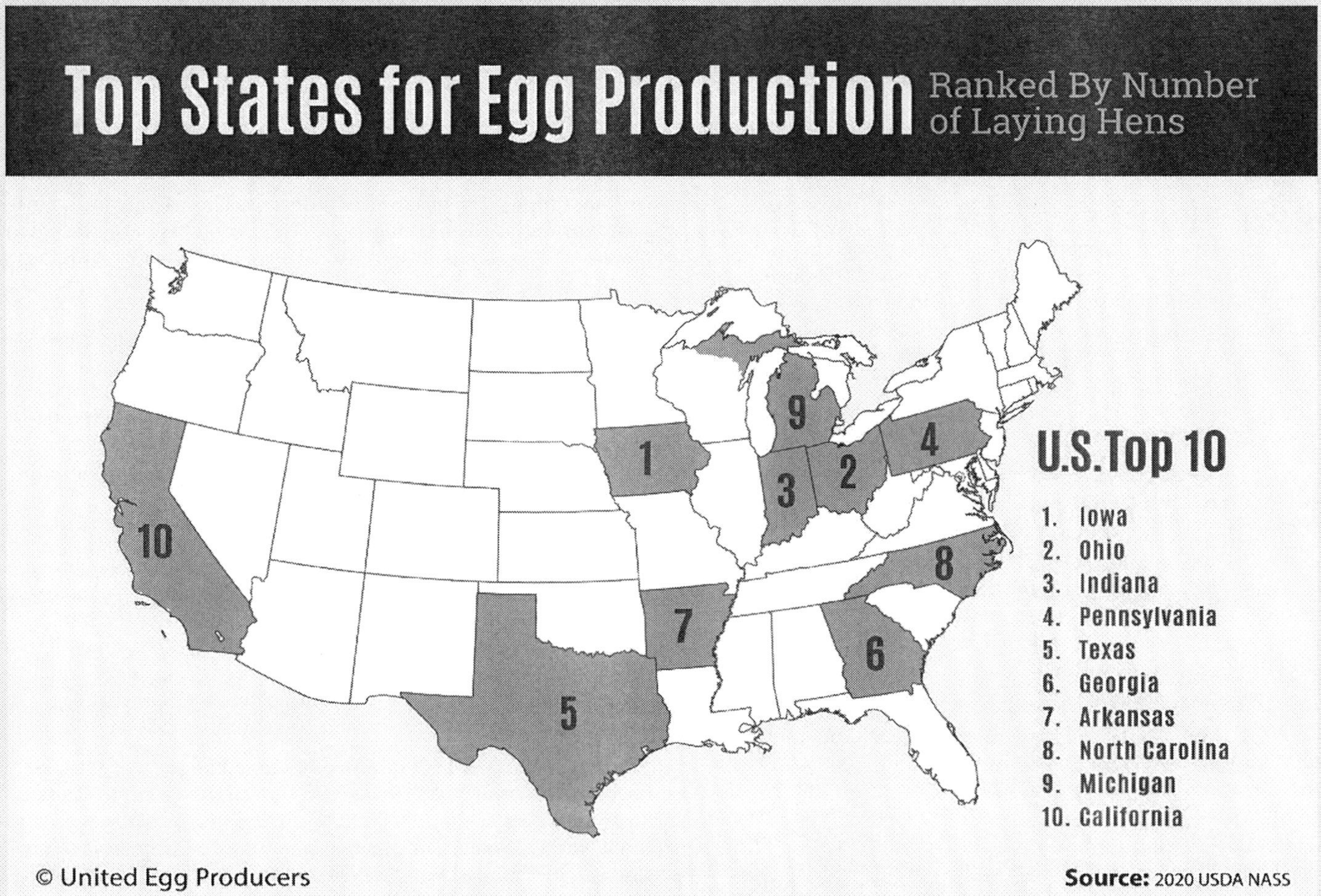

In the fall of 2022 both Kroger and Walmart announced they will not be able to supply 100% cage-free eggs by 2025. In their revised animal welfare statements, each grocer acknowledged the supply issues, the cost of production, and the strain that grocer cage-free pledges have put on the US egg industry. The grocers will continue to offer a variety of conventional and cage-free eggs.[5] It remains to be seen how this will be resolved.

To summarize this issue, the egg industry, and the swine industry both face challenges to adapt their methods of housing and production to meet the expectations of its customers.

Foreign animal disease

The egg industry and the swine industry also share a risk associated with foreign animal disease. In the case of the egg industry, that risk has been realized recently with the impact of highly pathogenic avian influenza (HPAI) in both 2015 and 2022. Historically, HPAI was a random event in the poultry industry. It usually started as a low pathogenic strain (LPAI) that would be transmitted from a wild bird to a domestic flock. It would then mutate or reassort into HPAI in that domestic flock and then spread laterally to other farms. Resources would be marshalled to depopulate the affected flocks and the outbreak would be brought under control. The source virus for 2015 and the 2022 outbreaks is unique in that it is an HPAI strain of H5N1 that is circulating directly in wild bird populations. As of October 25, 2022, 3,124 detections of the virus were found in samples from both live and wild bird mortality across the United States.[6] The impact on backyard and commercial flocks has been significant as well. As of October 26, 2022, there have been 569 confirmed cases with 248 commercial flocks and 321 backyard flocks with 47.73M birds affected.[7] These statistics include all poultry species, including turkeys, broilers, and ducks.

The take home message from these statistics is that the challenge for the industry has changed. We now have an HPAI virus circulating in multiple species of wild birds. This increases the risk of direct introduction into commercial flocks. It has focused the industry's attention on biosecurity and prevention of virus introduction. It has also highlighted the importance of a strong relationship between federal, state, and industry partners as they work to bring outbreaks under control.

An important asset in the relationship between poultry industry and regulatory bodies is the National Poultry Improvement Plan (NPIP). The NPIP is a voluntary State–Federal cooperative testing and certification program for poultry breeding flocks, baby chicks, poults, hatching eggs, hatcheries, and dealers. It became operative in 1935 with a three-pronged focus on certifying breeding stock, bird performance, and the elimination of bacillary white diarrhea (caused by *Salmonella pullorum*). The objective of the NPIP is to provide a cooperative State–Federal program

Table 1: State law summary

State	Requirements	Compliance date
California	All shell eggs and liquid egg products sold must be cage free and meet usable floor space defined by 2017 United Egg Producers (UEP) guidelines.	Current
Massachusetts	All shell eggs and egg products must be cage-free by UEP Standards.	Current
Nevada	All shell eggs sold must meet 144 square inches per hen. All shell eggs and egg products sold must be cage free by UEP Standards.	7/1/22 1/1/24
Oregon	All shell eggs and egg products from enclosures constructed or acquired before 1/1/12 must meet UEP Certified hen housing requirements. All shell eggs and egg products sold must be cage-free by UEP standards.	Current 1/1/24
Washington	All shell eggs and egg products sold from buildings built after 1/1/12 must meet American Humane enriched colony guidelines of 116.3 square inches per hen or greater. All shell eggs and egg products sold must be cage-free and meet UEP standards.	Current 1/1/24
Arizona	All shell eggs and egg products must meet one square foot of usable floor space per hen. All shell eggs and egg products sold must be cage-free by UEP standards.	1/1/23 1/1/25
Colorado	All shell eggs and egg products sold must meet one square foot of usable floor space per hen. All shell eggs and egg products sold must be cage-free by UEP standards.	1/1/23 1/1/25
Michigan	All shell eggs and egg products sold must be cage-free by UEP standards.	1/1/25
Utah	Hens housed in the state must meet cage free guidelines but does not address the sale of eggs within the state.	1/1/25
Rhode Island	Hens must meet the usable floor space of 2016 UEP guidelines	1/1/25

through which new technology can effectively be applied to the improvement of poultry and poultry products by establishing standards for the evaluation (testing) of poultry breeding stock, baby chicks, poults, and hatching eggs with respect to freedom from certain diseases.[8]

The National Poultry Improvement Plan was established in the early 1930's to provide a cooperative industry, state, and federal program through which new diagnostic technology can be effectively applied to the improvement of poultry and poultry products throughout the country.[8]

The NPIP was initiated to help diminish the spread of Pullorum Disease, caused by *Salmonella pullorum* which was rampant in the poultry industry and could cause upwards of 80% mortality in baby poultry. The program was later extended to include testing and monitoring for other poultry diseases. The program currently offers testing and monitoring for:[8]

- *Salmonella pullorum* (Causative agent of Pullorum disease)
- *Salmonella gallinarum* (Causative agent of Fowl Typhoid)
- *Mycoplasma gallisepticum*
- *Mycoplasma synoviae*
- *Mycoplasma meleagridis* (for turkeys)
- Avian influenza

NPIP played an integral role in the development of the poultry industry 14-point biosecurity standards. Those standards are a benchmark for producers to follow up protect their flocks from the introduction of avian influenza and maintain eligibility for indemnity in the face of an HPAI. More importantly, NPIP provides a formalized platform for industry, academia, federal and state regulators to determine technical standards, streamline flock and product transportation, and demonstrate freedom from disease to our national and international trading partners. We are aware of a move to provide a similar structure of the swine industry known as the Swine Health Improvement Plan (SWIP).[9] We would suggest that the swine industry and AASV consider supporting this effort. NPIP has provided strong benefits in building communication between all the stakeholders. We believe it could do the same for swine.

References

1. https://ballotpedia.org/California_Proposition_2,_Farm_Animal_Confinement_Initiative_(2008)
2. https://ballotpedia.org/California_Proposition_12,_Farm_Animal_Confinement_Initiative_(2018)
3. https://www2.sustainableeggcoalition.org/
4. https://unitedegg.com/facts-stats/
5. https://www.wattagnet.com/articles/45996-walmart-kroger-will-not-meet-2025-cage-free-commitments
6. https://www.aphis.usda.gov/aphis/ourfocus/animalhealth/animal-disease-information/avian/avian-influenza/hpai-2022/2022-hpai-wild-birds
7. https://www.aphis.usda.gov/aphis/ourfocus/animalhealth/animal-disease-information/avian/avian-influenza/hpai-2022/2022-hpai-commercial-backyard-flocks
8. https://www.aphis.usda.gov/aphis/ourfocus/animalhealth/nvap/nvap-reference-guide/poultry/national-poultry-improvement-plan
9. https://usswinehealthimprovementplan.com/

https://doi.org/10.54846/am2023/159

Stranger than fiction: PRRS in a remote farm

C. Fruge
The Maschhoffs LLC, Arnold, Nebraska

Introduction

Porcine reproductive and respiratory syndrome virus (PRRSV) is an RNA virus that causes reproductive failure in gestating sows and systemic disease in the growing herd. PRRSV is transmitted vertically, resulting in weaned piglets with respiratory disease, usually weak and underweight.[1] Lateral spread also occurs from area spread, transportation of infected stock, and biosecurity lapses.[2] With increasing PRRSV pressure in the state of Nebraska, The Maschhoffs LLC (TML) naïve sow herd was at high risk for exposure. During the late 2021 and early 2022 winter season, several sow breaks within the Nebraska system occurred with no obvious 'smoking guns' including viruses that did not percentage-wise match. This fueled a pursuit of increasing biosecurity execution and awareness within the region.

Investigation

Throughout the 2021 fall season, a 2500-head, PRRSV naïve sow farm had been experiencing an increase in abortions. All abort diagnostics submitted to Iowa State University Veterinary Diagnostic Laboratory (ISU VDL) in Ames, IA resulted in no major findings, confirming abortions at this time were induced by vaccination timing. PRRSV surveillance in the form of 100% weekly processing fluid submissions continued to test negative. On December 15, 2021, processing fluids tested PRRSV negative with no clinical signs reported. December 19, the farm staff reported widespread gilt litter scours and assumed to be due to gaps in acclimation of said animals. Processing fluids collected and submitted to ISU VDL on December 20 returned PRRSV positive on PCR (polymerase chain reaction) with a cycle time (Ct) of 22. December 21, abortions increased dramatically and sera from aborted sows returned PRRSV positive at a 23 Ct. Using ORF5 sequencing, the virus read as a 1-8-4 wild type. All recent breaks within the Nebraska pod were 1-8-4 viruses, however, were all greater or equal to 3% different on nucleotide and amino acid comparisons.

Area and weather risks

Health status of two finishing barns located seven miles north of the sow farm was unknown. The isolation (ISO) unit for the sow farm filled with gilts located a half mile to the west of the sow farm had not yet been tested. Regarding weather, a large statewide storm with reported wind gusts up to 64 miles per hour (mph), precipitation and a 44-degree temperature change occurred five days prior to reported clinical signs.

Animal movement risks

Gilts trucked from Illinois to Nebraska were moved into the isolation barn in late November. Several wean pig movements placed into western Iowa prior to the break. Recent transportation of cull sows out of the sow farm occurred on December 16.

Supply entry risks

Off-site delivery of monthly farm supplies on a clean truck from the distributor dropped at the farm manager's house. Manager moved supplies from house to farm via open truck bed 40 miles on high traffic route (multiple other swine producers observed on route).

Discussion

After multiple PRRSV breaks in the region with several percentage wise different 1-8-4 viruses (zero 100% matches), a multitude of biosecurity improvements were implemented stemming from found gaps. This includes, but is not limited to; perimeter fence required closed and locked during and after hours, visitor requests required 72 hours in advance with acceptance of vehicle cleaning and downtime, disposable booties put over shoes as all employees and visitors cross threshold into entryway, addition of extra disinfection step of supplies before entering into fumigation room, feed mill biosecurity and tier improvement, and live haul routes outlined to avoid high traffic areas. Regarding any future breaks, all confirmed positives to use whole genome sequencing (WGS) and expanding investigations out a month prior to break. Future biosecurity improvements involve mortality management practices, ventilation, and other facility upgrades. During all disease investigations, the main repeating message continues to ring true, never assume and opt to explore details and facts.

References

1. Karniychuk, U. U. and H. J. Nauwynck. 2013. Pathogenesis and prevention of placental and transplacental porcine reproductive and respiratory syndrome virus infection. Vet Res. 44(1):95. **doi: 10.1186/1297-9716-44-95**

2. Pileri, E. and E. Mateu. 2016. Review on the transmission porcine reproductive and respiratory syndrome virus between pigs and farms and impact on vaccination. Veterinary Research. 47(1):108.

https://doi.org/10.54846/am2023/160

An outbreak investigation of *Actinobacillus pleuropneumoniae* serotype 15 in Iowa in 2021-22

Isadora Machado[1]; Alyona Michael[1]; Mafalda Mil-Homens[1]; Ana Paula Silva[1]; Pete Thomas[2]; Levi Johnson[3]; Lori Feldmann[4]; Lauren Glowzenski[5]; Daniel Boykin[6]; Tyler Bauman[7]; Marcelo Almeida[1]; Daniel Linhares[1]; Gustavo Silva[1]; Derald Holtkamp[1]
[1]Department of Veterinary Diagnostic & Production Animal Medicine, Iowa State University, Ames, Iowa; [2]Iowa Select Farms, Iowa Falls, Iowa; [3]New Fashion Pork, Jackson, Minnesota; [4]Protein Sources, Mapleton, Minnesota; [5]TriOak Foods, Oakville, Iowa; [6]Cactus Family Farms, Osceola, Iowa; [7]The Maschhoffs, Carlyle, Illinois

Introduction

Porcine pleuropneumonia, caused by the gram-negative bacterium *Actinobacillus pleuropneumoniae* (App), economically impacts infected herds in pig production countries.[1,2] App is subclassified into 19 known serotypes, and the pathogenicity varies depending on the expression of virulence factors. Therefore, epidemiologic outcomes of field infections can vary widely, depending on the strain involved, degree of host immunity, and presence of coinfections.[3] Susceptible pigs affected by App suffer from a rapid onset of high fever (41°C), coughing, and respiratory distress. Gross pathologic findings in the lungs, such as consolidation and hemorrhagic areas, can be found up to 24 hours after App infection; resultant fatalities often present with characteristic bloody and foamy nasal discharge.[4] Transmission has traditionally been associated with direct oral or nasal contact between carrier animals and, to a lesser degree, aerosolized respiratory droplets over short distances, up to 2 meters.[5,6] Even though App is considered an obligate parasite with poor environmental persistence, its survival increases in organic material, wet environments, and cool temperatures.[7] Furthermore, the formation of biofilms and persistence in waterers has been documented.[8] Virulent strains of App associated with high mortality outbreaks are uncommon in the US. Also, vaccination is not routinely practiced in commercial operations in this country. Historically, App serotype 15 (App15) has been infrequently diagnosed at the Iowa State University Veterinary Diagnostic Laboratory (ISU-VDL) since 2010 and has not been associated with high mortality outbreaks in the US. The most common serotypes detected in the ISU-VDL are 7 and 8. However, during a 60-day-period, from November 25th, 2021, to January 24th, 2022, 20 grow-finish sites from nine unrelated companies in a small high pig-density geographic area in central Iowa reported acute mortality peaks. The cases were submitted to the ISU-VDL and diagnosed as porcine pleuropneumonia caused by App15. The cases during this period were interesting from an epidemiological perspective. Therefore, outbreak investigations were conducted to explore epidemiological aspects of the cases and potential routes of App15 transmission.

Methods

The case definition was applied to the site level. Thus, cases were eligible for this study if they met criteria based on clinical history, pathological assessment, and laboratory testing (App isolation and further characterization by gel-based serotyping PCR). Overall information was obtained through an epidemiological survey from the 20 cases conducted with herd veterinarians of each company. Seven companies agreed to share detailed information, in which one site per company was investigated through an outbreak investigation conducted by veterinarians from Iowa State University and members of the Swine Health Information Center's Rapid Response Team in collaboration with the herd veterinarians, site caretakers, and production management. The outbreak investigation was conducted retrospectively as an epidemiological investigation and biosecurity hazard analysis of the production processes to explore potential routes of App15 transmission and evaluate entry events related to swine movements, vehicles and deliveries, and people's movement over 14 days before the outbreak date.

Results and discussion

All cases occurred in commercial wean-to-market or finisher sites. The magnitude of death loss varied among the 20 cases, with an average of 25% total mortality. The most severe case reported 50% mortality in a single week. Among the clinical signs, sudden death was the most reported, followed by respiratory distress, dyspnea, cough, and cyanosis. The affected animals were generally medicated with antimicrobials through water and mass injections. Moreover, four of the seven intensively investigated sites reported stress events up to fourteen days before the outbreak date, such as energy and water outage.

Among the intensively investigated cases, gaps were detected in the biosecurity procedures, such as failure to follow a downtime period between farm visits, the inexistence of a bench entry system, and a lack of shower-in-shower-out procedure. Moreover, the knowledge of how third-party services providers operate was generally absent, with likely shared labor connections between production companies, such as carcass removal (rendering), load-out pig and vaccine crews, trailer washing, and feed delivery routes. Of the 20 growing pig cases over the 60 days, 18 used the same third-party services provider for rendering. The rendering company followed specific routes within a rendering collection point, with shared rendering trucks, routes, and drivers dedicated to the area, and the rendering route was unknown to the herd veterinarians. All 20 cases were located within the rendering collection point. The rendering company did not provide information on the routes and timing of pickups during the outbreak period. In most cases, the carcass disposal area was located at the site's entrance, and caretakers followed biosecurity procedures that may be insufficient to prevent the bacteria from being carried from the rendering area to the pigs in the barns.

This outbreak investigation demonstrated a heavy reliance on contracted services, such as rendering and load-out crew, that may have created a potential for operational connections between different production systems. Moreover, there is still a lack of knowledge and compliance on the basic biosecurity monitoring practices in the growing pig phase, which can expose the vulnerability of US swine production to emerging, re-emerging, and exotic diseases.

References

1. Hunneman, W. A. (1986). Incidence, economic effects, and control of *Haemophilus pleuropneumoniae* infections in pigs. *Veterinary Quarterly,* 8(1), 83-87. **doi.org/10.1080/01652176.1986.9694024**

2. Holmgren, N., Lundeheim, N., Wallgren, P. (1999). Infections with *Mycoplasma hyopneumoniae* and *Actinobacillus pleuropneumoniae* in fattening pigs. Influence of piglet production systems and influence on production parameters. *Journal of Veterinary Medicine,* Series B, 46(8), 535-544. **doi.org/10.1111/j.1439-0450.1999.tb01246.x**

3. Marois, C., Gottschalk, M., Morvan, H., Fablet, C., Madec, F., Kobisch, M. (2009). Experimental infection of SPF pigs with *Actinobacillus pleuropneumoniae* serotype 9 alone or in association with Mycoplasma hyopneumoniae. *Veterinary microbiology,* 135(3-4), 283-291. **doi.org/10.1016/j.vetmic.2008.09.061**

4. Baarsch, M. J., Foss, D. L., Murtaugh, M. P. (2000). Pathophysiologic correlates of acute porcine pleuropneumonia. *American journal of veterinary research,* 61(6), 684-690. **doi.org/10.2460/ajvr.2000.61.684**

5. Velthuis, A. G. J., De Jong, M. C. M., Kamp, E. M., Stockhofe, N., Verheijden, J. H. M. (2003). Design and analysis of an *Actinobacillus pleuropneumoniae* transmission experiment. *Preventive veterinary medicine,* 60(1), 53-68. **doi.org/10.1016/S0167-5877(03)00082-5**

6. Jobert, J. L., Savoye, C., Cariolet, R., Kobisch, M., Madec, F. (2000). Experimental aerosol transmission of *Actinobacillus pleuropneumoniae* to pigs. *Canadian Journal of Veterinary Research,* 64(1), 21. PMCID: PMC1189576

7. Assavacheep, P., Rycroft, A. N. (2013). Survival of *Actinobacillus pleuropneumoniae* outside the pig. *Research in veterinary science,* 94(1), 22-26. **doi.org/10.1016/j.rvsc.2012.07.024**

8. Loera-Muro, V. M., Jacques, M., Tremblay, Y. D., Avelar-Gonzalez, F. J., Muro, A. L., Ramírez-López, E. M., Guerrero-Barrera, A. L. (2013). Detection of *Actinobacillus pleuropneumoniae* in drinking water from pig farms. *Microbiology,* 159(Pt_3), 536-544. **doi.org/10.1099/mic.0.057992-0**

https://doi.org/10.54846/am2023/161

Biosecurity: Why do we do it and who do we do it for?

Brandi Burton, DVM
Suidae Health & Production, Algona, Iowa

Introduction

Biosecurity has been one of the most talked about topics of the swine industry since 2013 when PED wreaked havoc across the country. It became very clear at that time the attention towards biosecurity was not where it needed to be and great efforts have been made to increase awareness about its importance and its role to keeping the US swine herds healthy. These efforts aim to mitigate risk for US endemic diseases such as PED and PRRS as well as foreign animal diseases such as African swine fever (ASF) and foot-and-mouth disease (FMD). There has been added pressure to heighten biosecurity efforts the last few years with ASF cases on the rise and encroaching borders of US territories.

Although the industry agrees on the impact biosecurity can have on pig performance and profitability, there are no universal guidelines for which biosecurity measures should be taken and when. There is also a continual struggle getting caretakers and on farm-staff to completely buy into these efforts. Therefore, production systems, supervisors and veterinarians are always looking for protocols to cut out of the daily routine or to make these additional steps simpler while trying to be effective enough to get the job done. Or so that's the hope. It is also extremely difficult to audit biosecurity practices well. It's natural to do a job more thoroughly when that person knows he or she is being watched, but with biosecurity, it only takes one time of not following steps properly to infect a whole herd with a detrimental pathogen.

With all of that being said, it is important to revisit the basics of why biosecurity measures are taken, explore research targeting biosecurity processes and protocols and challenge the industry to re-evaluate who is it that we're doing all of these extra steps and measures for.

Biosecurity – why do we do it?

Definition

Googling the definition of biosecurity will produce several versions and definitions, but at one point, some of the major dictionary resources did not include this word in its vocabulary at all.[1] Saunders Veterinary Dictionary (1999) defines biosecurity as "the protection/security of susceptible animal herds from the introduction and transmission of infectious pathogens". Dr. Otake[2] uses this definition in his 2018 IPVS presentation titled "Prevention and control of swine diseases through biosecurity." He goes on to further define biosecurity with a set of requirements: science based, practically feasible (simple, organized), effective (cost vs benefit), committed to continue (execution), and measurable.

To summarize, biosecurity can be defined as a set of measures taken to reduce introduction of pathogens into a farm and to contain pathogens to a farm that should be backed by science, be easy, make economic sense, be continually executed, and be measured. But are these the things that dictate our biosecurity protocols and processes today?

Goals of biosecurity

In an ideal world, all the steps that are taken on farm results in no outside pathogen introductions, but realistically, these steps are taken to decrease the risk of these introductions. This is also why many times there are several layers of measures taken so there are built in safety measures for the nature of human error. However, there comes a time when too many steps become burdensome, and people are more likely to cut corners. It is even more important to very carefully select the number of steps that must be taken and focus on the ones that are most important. So which biosecurity steps are non-negotiable? A poll of veterinarians or industry experts would probably expose that there's no one answer to that question.

Backed by science

Biosecurity research has been and will continue to be very popular. The last several years, the most selected, attended, and successful student presentations at AASV and Leman Conference have been biosecurity based. The first ever Morrison Swine Innovator Prize was awarded to Megan Bloemer in 2018 with her project focusing on truck wash biosecurity. In 2019 and 2020, Zach Talbert was awarded that same prize for his presentation on creating a prototype device for fumigation rooms. Also in 2020, Erin Kettelkamp earned the title of top student presenter at AASV with her presentation titled "Effect of particle size and H_2O_2/PAA concentration on the efficacy of an aerosol decontamination system for items entering swine farms." Among the winners of these student presentations are even more student project focusing on various areas of biosecurity. Most recently, the Swine Health Information Center has put out a call for research proposals focusing on grow/finish biosecurity for sites and transport.

There is a lot to still learn on the biosecurity front, and this is even more evident the past two years with the emerging 1-4-4L1C variant PRRS strain that seemed to skirt around and through biosecurity measures that seemed to work extremely well in the past. However, with the past research that has been done and the future studies to still come, how much of that information are we using to make science-based decisions?

Biosecurity – who do we do it for?

When it comes to implementing or recommending biosecurity protocols across a system there are many things that are considered when making these decisions, and it all should go back to the basic requirements outlines by Dr. Otake: science based, practically feasible, effective, committed to continue, and measurable. It is very easy to make decisions that are practically feasible and effective, but it seems to be harder to make decisions that are science based and even harder to measure these processes objectively.

Biosecurity protocols – thoughts to ponder

There are many processes and protocols that can be debated but there are few that stick out as ones that are not agreed upon and do not have many, if any, published science backing.

Downtime

It is not unusual to have 3 nights or 72 hours down before going to certain farms; however, studies done at the University of Minnesota by Drs. Dee and Otake looking at transmission of PRRS and TGE observed the study protocols lacking a "downtime" element performed as well as the one study protocol requiring downtime. They conclude, "Findings of this study suggest that showering or mandatory animal avoidance periods are not always necessary to prevent mechanical transmission of even the most contagious swine pathogens[1]." With the increased labor shortage and increased need to use peoples' time most effectively and efficiently, is it time to re-evaluate excessive downtime requirements, specifically between high health herds? Who are these requirements for today?

Blue booties

It is common practice to expect anyone on farm, visitors or not, to wear blue booties or shoe coverings over their shoes while on site. Most of the time, this is even expected of those coming to the site but not going inside like feed mill truck drivers. In theory this seems like a well-accepted process to keep debris from the vehicle not in contact with the site itself and for the vehicle not to get contaminated with anything from the site. Is this process fool proof though? When putting on the bootie or taking off the bootie, is there risk of hands getting contaminated in that process? Are contaminated hands a higher risk of disease transmission or the bottom of a shoe? Is there science-based evidence to support the use and cost of this measure or is this one thing that can be visually appreciated and ensure people are following some sort of protocol?

Decontamination and disinfectant room protocols

There have been many studies completed on fogging D&D rooms – how many foggers, which products, how to best ensure all areas have been coated, etc. How often are farms being audited on compliance of these protocols or that these protocols are working the way they are intended? Most would agree cardboard boxes should not be brought into a farm, but how many times are there boxes in the farm even though the farm staff says they don't bring any in? How long are supplies supposed to sit in the D&D room after disinfection and where does that number come from?

Industry challenge

There are a lot of questions regarding what to do and what not to do when it comes to biosecurity and most of the time, there aren't clear answers. On the other hand, sometimes there are answers that people chose not to listen to or don't trust people enough to believe the results or that those results would be the same in their system. When that happens, decisions can be made from fear and skepticism, which ultimately does not address the root cause of the issue. At the end of the day, biosecurity decisions should be science based, practically feasible, effective, committed to continuing and measurable. Let's challenge one another and ensure this is the case when implementing biosecurity protocols and process and gear some energy towards research to answer some of these basic principles we so heavily rely on.

References

1. Pyburn, D. Biosecurity: How do you define it? AASV Library. November 11, 2009. Accessed November 10, 2022. **https://www.aasv.org/library/swineinfo/item.php?ISU/2001/034.PDF**

2. Otake, S. Prevention and control of swine diseases through biosecurity. IPVS 2018, Chongqing, China. Accessed November 13, 2022. **https://www.aasv.org/library/swineinfo/item.php?IPVS/2018/Keynotes%20(03-87)/59-Otake_S.pdf**

https://doi.org/10.54846/am2023/162

Farm Health Guardian: What we have learned so far

Brad Chappell, BSA; DVM; Mikala Sasley, BSc

Introduction

Farm Health Guardian (FHG) is a software platform that allows a visitor to a barn or premises to use an application on a hand-held device to sign in when arriving at the location, essentially replacing the paper log in at the entrance to the barn. The system utilizes a geofence around the premises, which is recognized by the app upon entry into the farmyard. The visitor is then asked a series of questions that pertain to the barn's biosecurity protocols, this information is reported to the barn manager, who can generate reports based on the information collected. Topigs Norsvin has been using the software since late 2020 and have learned many lessons as well as seen the value in the system.

The value of Farm Health Guardian

With biosecurity being an important priority, Topigs Norsvin supplies biosecurity SOPs to service providers and visitors when they are scheduled to enter a barn. The concern with relying on this approach is that there is no way to ensure that these protocols were communicated and understood by the person physically entering the barn. In addition to supplying biosecurity SOPs, using the FHG application ensures that the barn visitor is aware and in compliance with the SOPs by asking practical biosecurity questions. This can be done on the visitor's phone even before entering the controlled access zone. Some examples of the questions that are asked on the app: "Reason for visit", "Last Contact with Swine", "Are you co-habiting with any other person who has swine contact", "Are you entering with pork products", etc. If any of the answers fail to meet biosecurity standards then the visit will be "denied", and the person will not be allowed to enter the farm on this date. A notification will then be sent to those specified within the system. If the visitor has answered all the questions "correctly", they will receive approval for entry, the visitor can enter, and the barn manager is notified.

A downtime chart is programmed into the FHG app. Tracking all the user's barn visits allows the app to make the decision to "deny or approve" entry based on the amount of downtime that is required at each farm location. The FHG app will also ask the visitor if they are bringing any equipment or supplies into the barn. It is useful to record this information and use it to make decisions for the farm to potentially purchase the supplies, instead of bringing them in each visit. This can help to reduce potential "disease introductions" into the farm.

The information that is collected by the app allows administrators to generate reports based on the log in information. If there is ever a health status change at the farm, an outbreak report can be run, showing people that have visited the farm and how they are connected to the "industry".

Lessons learned

During the time that FHG has been put into operation at Topigs Norsvin farms, having one dedicated person to implement its use has been beneficial. This person is responsible for ensuring that the barns are equipped with the correct hardware such as having a tablet with the FHG app installed, as well as providing the training for the software.

Since the FHG application requires the use of location, some individuals may be reluctant to have the FHG app on their phone since they do not like the idea of being "tracked" by an app. For this reason and other potential concerns, it is beneficial to have a dedicated person to speak to service providers and explain the functionality of FHG and discuss any privacy concerns. One solution is that the FHG app can be set to only use the location when a geo-fence is nearby.

With the downtime chart programmed into the FHG app, any accidental forgetfulness of downtime is prevented.

Limitations

Currently the FHG software limits one premises per customer group, meaning Farm A can only be in customer A's database and not in customer B's database. This inability to communicate downtime between customer groups will cause unnecessary visit denials due to seemingly insufficient downtime requirements.

There is ongoing research into using technology to electronically register trucks that enter the controlled access zone, for example feed or livestock trucks. We are hoping to trial this technology over the next two or three months.

Conclusion

Farm Health Guardian has shown to be a useful tool in protecting the biosecurity of the farm. The app works as an electronic sign in for visitors and service providers, relaying the information provided to the barn manager. Tracking this type of information electronically makes the information more accessible in an outbreak scenario. FHG has shown its' value and shows lots of potential for future development.

https://doi.org/10.54846/am2023/163

Impactful investments: Measuring return on investment for biosecurity

Pete Thomas, DVM, MS; Katie Coleman, DVM
Iowa Select Farms, Iowa Falls, Iowa

Introduction

Disease prevention is a critical component of successful swine production operations. It has been well documented that porcine reproductive and respiratory syndrome (PRRS) costs the United States swine industry more than $600 million dollars annually.1 Two major pathogens that cause significant economic losses associated with disease include porcine reproductive and respiratory syndrome virus (PRRSV) and porcine epidemic diarrhea virus (PEDV). Estimated costs associated with PRRS in growing pigs range from $2.24 to $5.57.[1,2] High health breeding stock is typically negative for these two pathogens, but keeping pigs negative in commercial production settings can been difficult.

Iowa Select Farms (ISF) has most of its sow farms, gilt development units, and finishing sites located in swine dense regions of Iowa. In fact, many of the sow farms are in some of the most swine dense counties in the United States. Frequent introductions of novel PRRS strains and failure to maintain long-term PRRS stability has been a challenge due to these geographical concerns. Commercial wean-to-finish farms across the state see large seasonal increases in PEDV outbreaks during the colder months. Swine diseases such as PRRS and PED do not have vaccines that effectively prevent significant disease from occurring.

Disease prevention for these diseases, among many others relies on good biosecurity practices and investments. Between different production systems, diseases can have different economic impacts due to unique characteristics of that system and structure. Similarly, the cost of interventions can vary between production systems. I will give examples of measuring the cost of disease in the Iowa Select Farms production system, as well as some general examples of biosecurity investment costs. These same principles may be applied to different disease and investment strategies.

Wean-to-finish biosecurity investment

Different pathogens have different routes of infection. For the wean-to-finish example, I will use PEDV. PEDV is not thought to commonly move via aerosol transmission, but rather via fecal-oral exposure. Therefore, non-filtered facilities, should be able to invest in practices that are near 100% successful in preventing PED introduction.

Our system developed two different models to look at the impact of PEDV introduction into wean-to-finish populations. The first model looks at the cost of PED introduction during interim cuts due to dirty trailer or load-out crew contamination of a site. Prior to using this model for PED, it was used to evaluate the impact of ractopamine fed to pigs that were light on first cuts. During the winter of 2016/2017, the model found that pigs infected with PED during marketing weighed 1.98kg less at market and took an average of 4.6 days longer to reach market.[3] Additionally, 30% of groups marketed during October - April from 2018 - 2022 broke with PED during marketing.

Using the above information along with an average current hog price of $1.96 USD per kg and a typical wean-to-market barn contract cost of $0.115 per day, the total cost for a PED outbreak during marketing would be $4.41 per pig. With a 30% incidence rate, the cost per pig in the entire production system during the October - March time frame is $1.32, or $0.66 per pig annually.

We were able to calculate the re-break rate of wean-pig sites that placed pigs in a barn that had broken during marketing in 2018/2019 at 58%. A recent multi-variate model developed by Iowa State University includes 2 years of close-out data and demonstrates a 2.1% increase in nursery phase mortality for groups infected with PEDV in our system. Wean-pigs may have a different value to different producers whether they are buying pigs, producing their own, and contract agreements. For the sake of this analysis, we'll use a value of $40.00 per wean-pig. 17.4% of our wean-groups broke with PED during the October 2016 through November 2017 (0.58×0.30) timeframe due to the premise being infected at marketing and infectious virus remaining.

Using the above calculations and valuation, the mortality cost of a wean-pig break is $0.84 per pig. An additional conservative minimum impact of $4.41 on weight and days to market should be attributed per pig as well (may be quite a bit higher). A total cost of a wean-pig break with PEDV is at least $5.25 per pig. An impact of $0.91 per pig marketed as a system during the 6-month timeframe, and $0.46 for pigs marketed throughout the entire year.

Biosecurity investments considered, included: Washing all trailers prior to final cuts, hiring a crew supervisor, and white washing all sites. An estimate of the system per-pig cost of these interventions is:

- Increased trailer washing: $0.76 per pig
- Whitewash positive facilities: $0.03 per pig
- Hiring a crew supervisor + associated costs: $0.02 per pig

Based on the above economic analysis of the cost of disease outbreak we decided to focus on preventing PED re-breaks in wean pigs by white washing all positive facilities and hiring a crew supervisor to oversee load-crew biosecurity and wash inspections of positive sites.

In the first year, we saw a reduction in re-break rates from 58% to 19.6%. In the winter of 2021 - 2022 we saw a re-break rate of 11.5%. The current cost of PEDV re-breaks from positive finishing sites at marketing is $0.11 per pig marketed as a system. Overall, this resulted in a saving of $0.35 per pig on an investment of $0.05 per pig, a 7:1 return on investment.

Sow farm biosecurity investment

Iowa Select Farms has most of its sow farms and gilt development units located in swine dense regions of Iowa, which has contributed the challenges of keeping our sow herds healthy and minimizing disease introductions. Many of the sow farms within the ISF production system are 20 - 25 years old and re-investment in upgrading these facilities has been a priority. Production systems and large swine veterinary clinics have shared historical data of PRRS break incidence, demonstrating a significant reduction in frequency of breaks after air filtration.[4,5] It has also been demonstrated that there is an economic return on the investment in air filtration utilizing negative pressure designs when typical results can be achieved.[6,7]

Due to the age and design of the facilities and the develop of new technologies, it was determined that positive pressure filtration gave the best chance for success. Concerns associated with a negative pressure system were related to multiple avenues for unfiltered air to enter, and the immense amount of attention to continuously sealing the facility. Over the past 7 years, ISF has completed positive pressure filtration 32 sow farms and 10 GDU's. This represents about 185,000 sows.

Prior to filtration the annualized break-rate for PRRS in our swine dense sow regions was 86%, or an average of 1 PRRS break every 14 months. Since filtration we have reduced the break rate to 18%, or about a 4.8X reduction in new strain introductions on these farms. An average of 1 outbreak every 5.5 years per farm. Of course, some farms have done better, and some have done worse than the average.

The cost of a PRRS outbreak can be calculated by looking at the impact of wean-pig throughput and downstream performance after an outbreak. Fortunately, we have been able to partner with Iowa State University to develop models to look at both impacts in our production system. Analysis of > 2,000 close-outs in a multivariate model shows an increase of 4.6% nursery mortality for all pigs placed in the first 20 weeks after an outbreak. Additionally, an analysis of 50 PRRS outbreaks between 2018 - 2022 in our system demonstrates median losses of 3,524 pigs per 1,000 sows.[8] Total losses due to sow farm and nursery mortality total $155,000 per 1,000 sows inventoried using the $40.00 per wean-pig as before. This of course, does not consider any later mortality, reduced feed efficiency, or weight gain.

The cost to filter our facilities was not specifically captured due to extent of the remodels. Many of the components involved needed to be replaced (electrical, fans, inlets, ceilings, etc) due to the age of the facilities, and farms were being remodeled to pen gestation at the same time. For this example, I'll work backward on look at a very conservative return, and how it compares to the total remodel investment.

The current break rate of our filtered facilities is one break every 5.5 years. Prior to filtration, our break-rate was one break every 1.2 years. Our first farms have now been filtered for 6 years. On average over 6 years, we would've had 1.2 PRRS outbreaks per farm since filtering. Given the past break-rate, we would've expected 5 outbreaks without the investments made, or 3.8 more breaks on average. During this same timeframe sow farm throughput and nursery mortality due to a PRRS break cost us $155 per sow. By reducing the PRRS break rate we've saved (3.8 × $155) $589 per sow. The remodel cost to remodel a farm just for positive pressure filtration is unknown in our projects, and certainly overlaps with the other costs of the remodel; However, the health benefits achieved more than pays for the average total cost to remodel these facilities to pen gestation, positive pressure filtration, and update needed components and equipment of 20 - 25-year-old facilities.

This example is simplistic in the fact that it doesn't break out specifically the cost of the filtration in the remodel, calculate potential additional costs such as increased LP usage, and future filter replacements. Nor does it capture additional benefits of reducing PRRS outbreak incidence such as reduced medication, reduced grow-finish mortality, improved weight gain, and improved feed efficiency. But it does allow for a very conservative justification of a positive return on a capital investment 6 - 7 years after these projects were started.

Discussion

Biosecurity is an essential part of disease mitigation and production efficiency. It is critical to ensure that you know the costs and benefits associated with biosecurity decisions. Expensive investments can be made that have minimal impacts, and inexpensive investments/practices can be implemented that can have large benefits. In day-to-day production it can be difficult to know the true impact of an investment, since many factors can influence the outcome, and the impact can be difficult to quantify. Even in some of these instances you can get to some conservative estimates of benefits attributed or expected from the investment and ensure it is giving you a reasonable return on the investment you have made or are planning to make.

References

1. Holtkamp DJ, et al. Assessment of the economic impact of porcine reproductive and respiratory syndrome virus on United States pork producers. *J Swine Health and Prod.* 2013; 21(2):74-84.

2. Haden CD, et al. Assessing production parameters and economic impact of swine influenza, PRRS, and *Mycoplasma hyopneumoniae* on finishing pigs in a large production system.

3. Pavlovic, L. Experiences managing wean-to-finish PEDV in a production system. *Proc of the American Association of Swine Veterinarians annual meeting.* 2018:485-487.

4. Reicks D. Filtration: Ten years and 100 farms later...What have we learned? *Allen D. Leman Swine Conference.* 2014. St. Paul, MN.

5. Dee S, et al. A four-year summary of air filtration system efficacy for preventing airborne spread of porcine reproductive and respiratory syndrome virus and Mycoplasma hyopneumoniae. *J Swine Health and Prod.* 2011; 19(5):292-294.

6. Alonso C, et al. Economic review of air filtration systems in North American swine herds. *Proc of the American Association of Swine Veterinarians annual meeting.* 2013:65-66.

7. Reicks D. A financial evaluation of PRRSv introduction risk mitigation attributed to air filtration of pig production sites. *Proc Allen D. Leman Swine Conference.* 2011:37-42.

8. Rodrigo P, et al. The impact of management practices in response to fifty PRRSV herd outbreaks in breed-to-wean herds between 2018 and 2022 in the Midwestern US. *Proc of the American Association of Swine Veterinarians annual meeting.* 2023: Poster.

https://doi.org/10.54846/am2023/164

Water biology: The next frontier for biosecurity

G. Doughan[1], DVM; L. Karriker[1], DVM, MS, ACVPM; K. Mou[2], PhD
[1]Swine Medicine Education Center, Iowa State University College of Veterinary Medicine, Ames, Iowa;
[2]Iowa State University Veterinary Diagnostic Laboratory, Ames, Iowa

Introduction

It is well known that water can serve as a vector and reservoir for many bacterial, viral, and protozoal pathogens. In 2018, African swine fever (ASFV) was suspected to infect a breeding herd via river water transmission in Romania as similar outbreaks appeared in other counties near the Danube River.[1,2] Outside of this case, there is sparse information on the relative risk of water pathogens entering into swine barns through water lines or on the compositional makeup of biofilm inside of swine barn water lines, which may harbor such pathogens. Most swine water sources originate from privately owned wells with fewer incidences of surface water bodies or treated municipal water sources. Both privately owned wells and surface water bodies can be susceptible to contamination and they are not routinely monitored. The Iowa Department of Natural Resources (DNR) demonstrated that approximately 40% of wells sampled over a 15 year period in Iowa were contaminated with fecal coliform bacteria,[3] while other studies have demonstrated viral pathogen contamination (including norovirus, adenoviruses, and enteroviruses) in up to 30% of private wells and groundwater sources.[4-9] Within the swine industry, water treatment or water line disinfection is not traditionally considered in biosecurity mitigation planning. Often, it is only considered in cases where the water quality is known to be poor and disturbs water line flow or accessibility. Therefore, water is one of the largest inputs entering into swine farms that remains generally untreated and unmonitored for contaminants before it is consumed. Water is an underestimated and poorly understood biosecurity risk and should be evaluated to understand its potential for pathogen transmission and harborage within water line biofilms.

Groundwater contamination and transport mechanisms

In order to explain the scope and relative risk for pathogens entering into farms through groundwater sources, an introductory knowledge of regulatory and hydrogeologic factors is imperative. Public water sources are monitored by the United States Environmental Protection Agency (EPA) under the Safe Water Drinking Act (SWDA) and most of those sourcing from it can rely on a protected, regulated, and decontaminated water source. However, those who source water from private wells, an estimated 15% of the United States population, are not covered by these protections and may be more vulnerable to a contaminated water source than historically believed.[10] The 2013 Survey of Iowa Groundwater, which evaluated contaminants of emerging concern, estimated 300,000 Iowans are susceptible to unregulated and untested water sources for contaminants.[11] Groundwater contaminants can include antibiotics, pesticides, viral pathogens, coliform bacteria, antimicrobial resistance genes (ARGs), "forever chemicals" like polyfluoroalkyl substances (PFAS) that have gained recent public interest, and many others.[10,11] Today, water and groundwater contamination concerns have entered many environmental and socioeconomic agendas as a result of the recognition that these sources are more frequently contaminated with pathogens than previously understood.

Local and regional geologic factors influence the ability of contaminants to travel through watersheds and reach aquifers. Contaminants on the soil surface can be rapidly transported by precipitation events to deep depths through pore diameters that allow water passage, especially in the Fall and Spring during groundwater recharge.[12] Contaminant transmission can be influenced by drought, presence of tile drainage lines, bedrock fractures, soil porosity, and more. A figure described by Lawrence et al[14] and again by Olsen et al[13] summarizes the comparison of pathogen and soil matrix apertures, finding that apertures naturally occurring in the ground as well as sandstone pores can filter out generally most protozoa and some bacteria that are 1000 to nearly 1 micron in size. However, silt and limestone can filter out most protozoa, bacteria, and viruses, but leave a portion of viruses that are under one micron unfiltered.[13,14] For comparison, the average diameter of relevant swine pathogens such as porcine reproductive and respiratory syndrome virus (PRRSV), porcine epidemic diarrhea virus (PEDV) and ASFV are 0.06, 0.13, and 0.18 microns in size, respectively,[13] and in some cases an order of magnitude smaller than pore sizes. Risk assessments of the potential movement of PRRSV and PEDV through groundwater are needed. This is important to note as some wells may be protected and test negative for bacteria but could still contain smaller viruses.

Groundwater movement in geologic materials with low to high permeability can span from inches-per-year to feet-per-minute respectively,[12] creating opportunities for rapid spread of pathogenic viruses and bacteria, antimicrobials, and other contaminants.[10] For example, virus particles have been documented to penetrate through fractured bedrock in a matter of hours and contaminate municipal wells up to 300 meters deep.[15,16] Furthermore, mechanisms for underground storage of wastes such as manure pits, or sewage tanks that leak can also be sources for contamination and bypasses filtration from the above soil strata. Not only can contaminants penetrate to a protected aquifer, but they also can travel laterally throughout aquifers in which their borders can span thousands of square miles. Increased consumption of water from wells, such as power washing an empty barn or farrowing rooms, can aid in lateral transportation of contaminants by depleting local water sources and thus drawing water from other locations to the well.

When these hydrogeologic principles are extrapolated for biosecurity planning for swine farms, it's important to recognize the additional biosecurity risk of manure applications to surrounding fields. For example, fertilizing agricultural fields with manure containing PRRSV and or PEDV could have impacts on introducing the source farm or surrounding farms to those pathogens. If the farm of origin and surrounding farms are sourcing their water from an unregulated and untreated private well or water source, this potential contamination could not only affect subsequent turns of pigs, but also impact surrounding farms located within the shared aquifer. It is necessary to evaluate many of the pathogens that challenge the swine industry and understand if water transmission is a possible route of disease introduction.

Bacterial and viral contamination of wells and groundwater

There is a considerable amount of evidence documenting the presence of bacterial, viral and protozoal pathogens of human relevance present in privately owned well water and groundwater. For example, the Iowa Department of Natural Resources (Iowa DNR) completed a survey of Iowa's private wells and tested wells for coliform or fecal coliform bacteria and nitrates from 2002-2017.[3] The results for bacterial contamination found that an estimated 22,000 out of 55,000 wells tested positive for fecal coliform bacteria (roughly 40%), with the presence of at least one coliform bacterium present in a sample. In addition, the US EPA's water quality standard determined there is no safe level of coliform or fecal coliform bacteria in drinking water.[3] Viral pathogens have also been detected in these water sources, including adenoviruses (serotypes A2, A41, A31), rotavirus, norovirus[7] hepatitis A virus, and echovirus 3.[15,17] The prevalence of viral pathogens in wells and groundwater samples have varied. However, water samples from groundwater found that just over 30% of samples had the presence of at least one virus type, while the US EPA estimates that just under 30% of public wells will be contaminated with the presence of viral pathogens.[5, 15,18,19] Few viral pathogens of animals have been investigated in wells and groundwater. For example, during the outbreaks of highly pathogenic avian influenza virus in 2015, groundwater and well water and lagoons of affected barns in Iowa and Wisconsin were assessed for the presence avian influenza virus. The assessment found a few positive wells and evidence to support that avian influenza was able to penetrate into groundwater sources.[20]

Evaluating pathogen viability and transmission risk into swine barns

Pathogen presence and mechanisms of transport are meaningless without maintained infectious ability. However, there is evidence to support groundwater's ability to provide favorable conditions to facilitate and prolong pathogen viability. The rapid transport of pathogens from the surface to the aquifer allows for pathogens to maintain viability in groundwater. Active soil microbiomes just below the surface can inactivate pathogens if left too long in this layer. Once in the aquifer, groundwater is maintained at a lower temperature, provides a high organic and moisture content, and the absence of ultraviolet light increases pathogen survival.[13,21-23] Pathogen survivability is widely disputed as some reports estimate some pathogens to remain viable for days[22,24] and weeks,[25] while others report up to two to three years.[13]

Water can serve as a vector for many pathogens that can impact swine health. Waterborne pathogens that affect swine have been detected, including influenza A virus such as avian influenza,[26] African swine fever [2,27] rotavirus,[15] hepatitis E virus,[28] *Salmonella, Escherichia coli,*[29] *Actinobacillus pleuropneumoniae,*[30] *Leptospira,* and likely others. PRRSV nor PEDV have been detected as a naturally occurring waterborne pathogen. However, PRRSV has been successful in infecting swine from the positive control group where water lines were intentionally inoculated in a bioassay study.[31] In addition to known pathogens of concern, there may be unknown pathogens or organisms that predispose animals to other known pathogens. The swine industry has increasingly focused its attention on the persistent elevation in sow mortality due to pelvic organ prolapse (POP).[32] Although there are many contributing factors to the increase in sow mortality, the incidence of POPs was higher in farms with untreated well water than treated or municipal water.[32] Anecdotal evidence of suboptimal growth performance in the absence of infection with a recognized pathogen has also been reported to improve with disinfection of water lines. The impact of water transmission of microbes and biofilms in water lines warrants further investigation.

Current knowledge of water line biofilms in swine farms

Potential pathogen inoculation into swine water lines can be derived from two sources: through external water sources via contaminated water (as discussed previously), or through access points within the water system inside the barn. Access points can include the water dosing system (water medicator) and at each waterer, including cup, feeder, and nipple waterers. Bacteria from the environment, pig's mouth or snout, manure contamination, or poor sanitation of water medicator stock buckets can serve as sources of pathogen introduction to aid in the formation of biofilms.

Biofilms are developed by bacteria colonizing surfaces using adhesion factors. Once the bacteria have adhered to a surface, they utilize quorum sensing, which allows individual bacteria within the same and or different species to communicate using chemical signals, and execute colony-wide actions, such as the formation of biofilms.[33] Bacteria in biofilms form an extracellular polymeric substance (EPS) matrix that makes them difficult to remove even in the presence of chemical disinfectants and manual removal techniques.[34] Quorum sensing also allows for the transmission of antimicrobial resistance genes (ARGs) and other resistance mechanisms. Water quality challenges on farms such as water hardness, elevated total dissolved solids, and iron levels can provide a mineral framework for bacteria to quickly adhere, colonize, and begin biofilm formation. An established biofilm paired with commonly administered additives to the water lines, such as probiotics, live bacterial oral vaccinations, electrolytes, and antimicrobials, can create a favorable or selective environment for cellular growth. Mature biofilms have also been shown to incorporate viruses into their structures.[35] Once a biofilm has fully developed, biofilm shedding events can occur where fragments disperse into the main water circulation and are consumed by animals, utilized by farm personnel, or may inoculate other locations within the pipe.

In May of 2022, the authors submitted a three-inch section of water line pipe (termed "water line coupon") from a commercial, 2400 head wean to finish farm located in central Iowa for 16S rRNA sequencing and shotgun metagenomics sequencing to gain further knowledge on the composition of the biofilm in swine water lines. Next generation sequencing techniques were performed instead of traditional culturing and polymerase chain reaction (PCR) to identify a broader range of organisms cost-effectively.

The water line coupon was stored in 30 mL of sterile saline and frozen at -80° Celsius after collection until sample processing. The water line coupon was thawed, biofilm was removed with an inoculating loop, and a series of vortexes were applied to remove additional biofilm from the inside of the water line coupon. The supernatant was poured into a conical tube, vortexed twice for 30 seconds and allowed to settle for 30 minutes. The top layer of supernatant was collected and used for 16S rRNA and metagenomic shotgun sequencing. Following 16S sequencing, QIIME 2 pipeline was used for sequence data processing. Sequencing errors were removed using the DADA2 pipeline. Amplicon sequence variants (ASVs) with a frequency less than 0.01%, those that were classified as eukaryote, mitochondria, and chloroplast, and those that failed to be classified at least at the phylum-level were all removed from data analysis and reporting. Taxonomies with relative abundances > 1% were identified by 16S sequencing and confirmed with metagenomic shotgun sequencing (Figure 1). Although the *Enterobacteriaceae* family could not be resolved to the genus level, it was the most abundant taxon in the biofilm, making up 47% of the composition. The *Enterobacteriaceae* family contains important organisms such as *Escherichia coli, Salmonella, Enterobacter, Shigella, Proteus* and *Serratia* species. Other bacterial genera identified within the biofilm are commonly found in marine environments or wetlands,[36,37] wastewaters,[38-40] groundwater, soil environments,[41] and intestinal microflora[42] including intestinal microflora in sows.[43] Some of these genera are known to be iron oxidizers[37] and reducers,[36,41] nitrogen-fixing bacteria,[40] sulfur oxidizers and reducers,[36,38] and cellulose degraders.[39] It is uncertain if these organisms play a role in swine health and production, but their presence warrants further investigation. Taxonomies identified at the species level by metagenomic shotgun sequencing are being analyzed at this time. Antimicrobial resistance genes were also found within the biofilm using metagenomic shotgun sequencing (Table 1). The resistance genes included those that encode for multi-drug efflux pumps[44] and also resistances to antimicrobials used in the swine industry such as, tetracyclines,[45] aminoglycosides,[46] beta lactams,[47,48] quinolones,[49] and an unused, prohibited drug in swine, chloramphenicol.[50,51] Chloramphenicol has not been used in the swine industry in over 35 years. However, the selective pressure of using other antimicrobials may maintain genes that encode for multiple drug resistances over time, including the resistance genes for chloramphenicol. Eliminating biofilms from the inside of water lines is imperative not only from a disease standpoint, but also to preserve antimicrobial efficacy for the animals in the barn, and for future general use of antimicrobials.

Figure 1: Bacterial genera and families with relative abundances greater than 1% identified in water coupon biofilm using 16S rRNA sequencing.

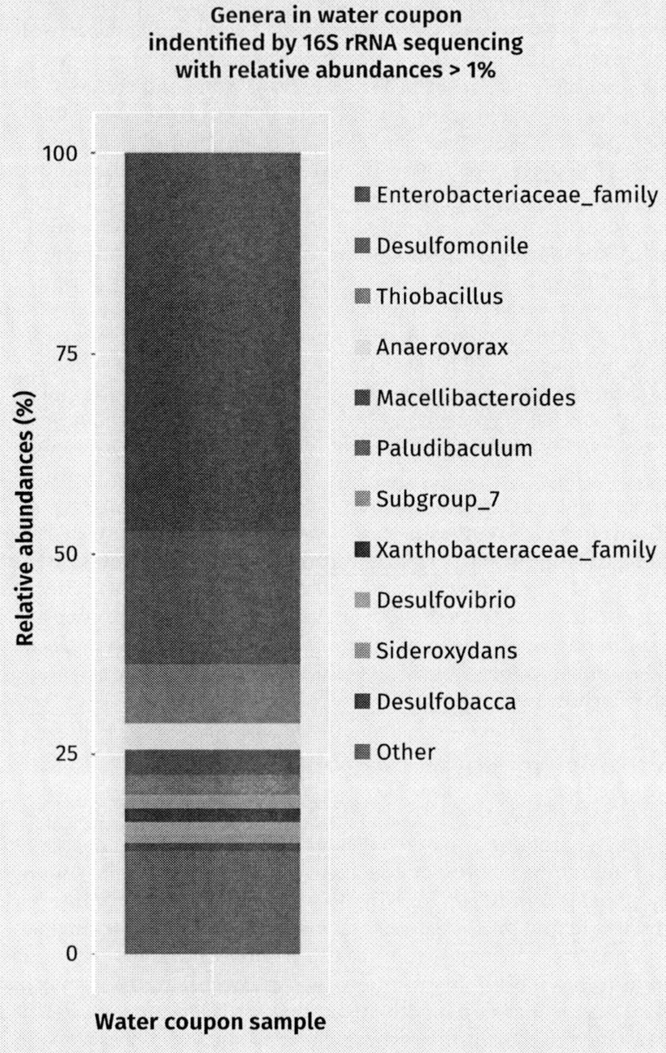

Table 1: Antimicrobial resistance genes found in water line coupon sample

AMR gene	Function
OqxB	Multi drug efflux pump
TetA	Tetracyclines
AacAad	Aminoglycosides
CatB2, cmlA	Chloramphenicol
qnrB	Quinolone
CARB, mrdA	Beta lactams

Conclusion

Groundwater is at risk for contamination and is no longer considered a protected source from bacterial, viral, and protozoal contaminants. The potential for pathogen transmission in swine farms has been missed and needs to be assessed not only as a mitigation step to secure swine farms from ASFV infection, but also may prove to be useful for reducing transmission of PRRSV and PEDV. Not only can pathogens enter into swine farms from the water source, but the design of current swine water distribution systems provide areas for potential contamination within the farm leading to the formation of biofilms. Without clear guidance on best practices on water line disinfection, biosecurity may be limited within and between groups of pigs. Understanding the bacterial, viral, and fungal composition of biofilms may allow industry insight into improving health, performance, and antimicrobial stewardship within swine farms.

References

1. Boklund A, Cay B, Depner K, Földi Z, Guberti V, Masiulis M, et al. Epidemiological analyses of African swine fever in the European Union (November 2017 until November 2018). EFSA Journal. 2018;16(11):e05494.

2. Niederwerder MC, Stoian AMM, Rowland RRR, Dritz SS, Petrovan V, Constance LA, et al. Infectious Dose of African Swine Fever Virus When Consumed Naturally in Liquid or Feed. Emerg Infect Dis. 2019;25(5):891-7.

3. Iowa Department of Natural Resources E. Iowa's Private Wells Contaminated by Nitrate and Bacteria 2019. Updated April 2019. Available from: **https://www.ewg.org/interactive-maps/2019_iowa_wells/#_edn12**

4. Borchardt MA, Bertz PD, Spencer SK, Battigelli DA. Incidence of enteric viruses in groundwater from household wells in Wisconsin. Appl Environ Microbiol. 2003;69(2):1172-80.

5. Borchardt MA, Bradbury KR, Gotkowitz MB, Cherry JA, Parker BL. Human enteric viruses in groundwater from a confined bedrock aquifer. Environ Sci Technol. 2007;41(18):6606-12.

6. Borchardt MA, Haas NL, Hunt RJ. Vulnerability of drinking-water wells in La Crosse, Wisconsin, to enteric-virus contamination from surface water contributions. Appl Environ Microbiol. 2004;70(10):5937-46.

7. Borchardt MA, Spencer SK, Kieke BA, Lambertini E, Loge FJ. Viruses in nondisinfected drinking water from municipal wells and community incidence of acute gastrointestinal illness. Environ Health Perspect. 2012;120(9):1272-9.

8. Borchardt MA, Stokdyk JP, Kieke BA, Muldoon MA, Spencer SK, Firnstahl AD, et al. Sources and Risk Factors for Nitrate and Microbial Contamination of Private Household Wells in the Fractured Dolomite Aquifer of Northeastern Wisconsin. Environ Health Perspect. 2021;129(6):067004.

9. Abbaszadegan M, Stewart P, Gerba C, LeChevallier M. A strategy for detection of viruses in groundwater by PCR (vol 65, pg 444, 1999). Appl Environ Microbiol. 1999;65(11):5192.

10. Centers for Disease Control and Prevention NCfEH. Healthy Housing Reference Manual2009. p. 1-12.

11. Libra RD, Gipp C, 2013 Survey of Iowa Groundwater and Evaluation of Public Well Vulnerability Classifications for Contaminants of Emerging Concern2013; (57). Available from: **https://s-iihr34.iihr.uiowa.edu/publications/uploads/2015-06-23_11-06-38_tis-57.pdf**

12. Prior JC, Boekhoff JL, Howes MR, Libra RD, VanDorpe PE, Vonk, JR. Iowa's groundwater basics a geological guide to the occurence, use, &vulnerability of Iowa's aquifers. Iowa Department of Natural Resources2003. Available from: **https://s-iihr34.iihr.uiowa.edu/publications/uploads/2014-08-24_08-08-21_es-06.pdf**

13. Olsen PC, Stone SJ, Groundwater & livestock production and husbandry, part 1, biosecurity. 51st Annual Meeting of the American Association of Swine Veterinarians; March 7-10th; Atlanta, GA: AASV; 2020. p. 384-98.

14. Lawrence AR, Macdonald D, Howard G, Barrett MH, Pedley S, Ahmed KM, et al. Guidelines for Assessing the Risk to Groundwater from On-Site Sanitation. 2001.

15. Bradbury KR, Borchardt MA, Gotkowitz M, Spencer SK, Zhu J, Hunt RJ. Source and Transport of Human Enteric Viruses in Deep Municipal Water Supply Wells. Environ Sci Technol. 2013;47(9):4096-103.

16. Levison JK, Novakowski KS. Rapid transport from the surface to wells in fractured rock: a unique infiltration tracer experiment. J Contam Hydrol. 2012;131(1-4):29-38.

17. Bradbury KR, Borchardt MA, Gotkowitz M, Spencer SK. Human viruses as tracers of wastewater pathways into deep municipal wells UW Extension Wisconsin Geological and Natural History Survey 2010. Available from: **https://wgnhs.wisc.edu/pubshare/WOFR2010-04a.pdf**

18. Abbaszadegan M, Lechevallier M, Gerba C. Occurrence of Viruses in US Groundwaters. Journal AWWA. 2003;95(9):107-20.

19. USEPA. Occurence and Monitoring Document for the Final Ground Water Rule. In: Agency USEP, editor. National Service Center for Environmental Publications (NSCEP)2006.

20. Borchardt MA, Spencer SK, Hubbard LE, Firnstahl AD, Stokdyk JP, Kolpin DW. Avian Influenza Virus RNA in Groundwater Wells Supplying Poultry Farms Affected by the 2015 Influenza Outbreak. Environmental Science & Technology Letters. 2017;4(7):268-72.

21. Pinon A, Vialette M. Survival of Viruses in Water. Intervirology. 2018;61(5):214-22.

22. Morris B. Pathogens and groundwater. Groundwateruk.org. Available from: **http://www.groundwateruk.org/downloads/Pathogens_and_groundwater_final%20version.pdf**

23. Gordon C, Toze S. Influence of groundwater characteristics on the survival of enteric viruses. J Appl Microbiol. 2003;95(3):536-44.

24. Givens CE, Kolpin DW, Borchardt MA, Duris JW, Moorman TB, Spencer SK. Detection of hepatitis E virus and other livestock-related pathogens in Iowa streams. Sci Total Environ. 2016;566-567:1042-51.

25. Sobsey M, Shields PA, Hauchman FH, Hazard RL, Iii LW. Survival and Transport of Hepatitis a Virus in Soils, Groundwater and Wastewater. Water Science and Technology. 1986;18:97-106.

26. Mateus-Anzola J, Martinez-Lopez B, Espinosa-Garcia AC, Ojeda-Flores R. Global subtype diversity, spatial distribution patterns, and phylogenetic analysis of avian influenza virus in water. Transbound Emerg Dis.12.

27. Dixon LK, Stahl K, Jori F, Vial L, Pfeiffer DU. African Swine Fever Epidemiology and Control. In: Lewin HA, Roberts RM, editors. Annual Review of Animal Biosciences, Vol 8, 2020. Annual Review of Animal Biosciences. 8. Palo Alto: Annual Reviews; 2020. p. 221-46.

28. WHO. Hepatitis E: World Health Organization; 2022. Updated June 24, 2022. Available from: **https://www.who.int/news-room/fact-sheets/detail/hepatitis-e**

29. Malgarin C, Takeuti K, de Lara A, Amaral A, Jühlich L, Barcellos D. Comparison of virulence factors of *Escherichia (E.) coli* isolated from nursery piglets with diarrhea and *E. coli* isolated from drinking water of the same herds. Allen D Leman Swine Conference 2015.

30. Loera-Muro VM, Jacques M, Tremblay YDN, Avelar-Gonzalez FJ, Muro AL, Ramirez-Lopez EM, et al. Detection of *Actinobacillus pleuropneumoniae* in drinking water from pig farms. Microbiology-(UK). 2013;159:536-44.

31. Garry M, Pitkin A, Stone S. Efficacy of commercial water treatment products to inactivate PRRS virus. AASV Annual Meeting: Our Oath in Practice: American Association of Swine Veterinarians; 2014. p. 333-4.

32. Chipman A, Rademacher C, Johnson C, Stalder K, Johnson A, Keating A, et al. Pelvic Organ Prolapse: An Industry-Wide Collaboration to Identify Putative Contributing Factors. ISU James D McKean Swine Disease Conference; November 1-22018. p. 57-60.

33. Windsor WJ. How Quorum Sensing Works American Society for Microbiology 2020. Updated June 12, 2020. Available from: **https://asm.org/Articles/2020/June/How-Quorum-Sensing-Works**

34. Fairchild B. Broiler Tip...Biofilms in poultry drinking systems. 2006 July:[1-3 pp.]. Available from: **https://poultry.caes.uga.edu/content/dam/caes-subsite/poultry/documents/archived-poultry-tips/biofilms-in-poultry-drinkers-jul-06.pdf**

35. Skraber S, Schijven J, Gantzer C, de Roda Husman A. Pathogenic viruses in drinking-water biofilms: a public health risk? Biofilms. 2005;2(2):105-17.

36. Ayala-Muñoz D, Macalady JL, Sánchez-España J, Falagán C, Courdeau E, Burgos WD. Microbial carbon, sulfur, iron, and nitrogen cycling linked to the potential remediation of a meromictic acidic pit lake. The ISME Journal. 2022.

37. Jain A, Coelho A, Madjarov J, Paquete CM, Gralnick JA. Evidence for Quinol Oxidation Activity of ImoA, a Novel NapC/NirT Family Protein from the Neutrophilic Fe(II)-Oxidizing Bacterium Sideroxydans lithotrophicus ES-1. mBio. 2022;13(5):e02150-22.

38. Zhang R-C, Chen C, Xu X-J, Lee D-J, Ren N-Q. The interaction between Pseudomonas C27 and Thiobacillus denitrificans in the integrated autotrophic and heterotrophic denitrification process. Science of The Total Environment. 2022;811:152360.

39. Liu J, Fu X, Yuan R, Chen Z, Zhou B, Chen H. Carbon sources derived from corncobs enhanced nitrogen removal in SBBR treating low C/N domestic sewage. Process Safety and Environmental Protection. 2022;166:628-37.

40. González-Cortés JJ, Valle A, Ramírez M, Cantero D. Characterization of Bacterial and Archaeal Communities by DGGE and Next Generation Sequencing (NGS) of Nitrification Bioreactors Using Two Different Intermediate Landfill Leachates as Ammonium Substrate. Waste and Biomass Valorization. 2022;13(9):3753-66.

41. Wu Q, Jiang X, Wu H, Zou L, Wang L, Shi J. Effects and Mechanisms of Copper Oxide Nanoparticles with Regard to Arsenic Availability in Soil–Rice Systems: Adsorption Behavior and Microbial Response. Environ Sci Technol. 2022;56(12):8142-54.

42. Kim M, Huda MN, Bennett BJ. Sequence meets function—microbiota and cardiovascular disease. Cardiovascular Research. 2021;118(2):399-412.

43. Gao L-m, Liu G-y, Wang H-l, Wassie T, Wu X, Yin Y-l. Impact of dietary supplementation with N-carbamoyl-aspartic acid on serum metabolites and intestinal microflora of sows. Journal of the Science of Food and Agriculture.n/a(n/a).

44. Liu B, Xie M, Gong Y, Dong Y, Zheng G, Wu H, et al. Prevalence, resistance phenotypes, and fluoroquinolone resistance genes of Salmonella isolates from raw milk of healthy dairy cows in Henan province, China. European Review for Medical and Pharmacological Sciences. 2022;26(18):6837-44.

45. Royer G, Ortiz de la Rosa J-M, Vuillemin X, Lacombe B, Chau F, Clermont O, et al. Reduced Chlorhexidine Susceptibility Is Associated with Tetracycline Resistance tet Genes in Clinical Isolates of Escherichia coli. Antimicrobial Agents and Chemotherapy. 2022;66(3):e01972-21.

46. Conco T, Kumari S, Awolusi OO, Allam M, Ismail A, Stenström TA, et al. Profiling of emerging pathogens, antibiotic resistance genes and mobile genetic elements in different biological wastewater treatment plants. Journal of Environmental Chemical Engineering. 2022;10(3):107596.

47. Jin J, Zhou Y, Zhang Z, Wang H, Hou W, Wang H, et al. Characteristics of Antimicrobial-Resistant Vibrio parahaemolyticus Strains and Identification of Related Antimicrobial Resistance Gene Mutations. Foodborne Pathogens and Disease. 2021;18(12):873-9.

48. Yang H, Wang F, Wang H, Lu X, Shen W, Chen X. Deleting mrdA and mrcB to significantly improve extracellular recombinant protein production in Escherichia coli. Biochemical Engineering Journal. 2019;143:185-95.

49. Taha SA, Omar HH, Hassan wH. Characterization of plasmid-mediated qnrA and qnrB genes among Enterobacteriaceae strains: quinolone resistance and ESBL production in Ismailia, Egypt. Egyptian Journal of Medical Human Genetics. 2019;20(1):26.

50. Zhang J, Yang C, Hu J, Zhang Y, Lai Y, Gong H, et al. Deciphering a novel chloramphenicols resistance mechanism: Oxidative inactivation of the propanediol pharmacophore. Water Res. 2022;225:119127.

51. Bischoff KM, White DG, Hume ME, Poole TL, Nisbet DJ. The chloramphenicol resistance gene cmlA is disseminated on transferable plasmids that confer multiple-drug resistance in swine Escherichia coli. FEMS Microbiology Letters. 2005;243(1):285-91.

https://doi.org/10.54846/am2023/165

Management and facility factors that impact biosecurity and biocontainment

Jordan Graham, DVM, MS
Swine Vet Center, Saint Peter, Minnesota

Introduction

Two challenges facing the achievement of effective biosecurity across the swine industry is the cost of biosecure facility design and the inconvenience of practicing proper biosecurity each day. Producers typically view these measures as a line item adding to their production costs per head. Successful producers understand that these costs are often necessary to maintain profitable production. However, the scale of biosecurity measures that could be implemented is essentially limitless. Finding balance between cost of interventions and adequate control over disease spread is critical, and often misunderstood.

Factors affecting area spread of disease

Biosecurity is broken down into two primary purposes: Bioexclusion and biocontainment. Bioexclusion is any practice or facility design implemented to prevent introduction of disease into a herd whereas biocontainment is the opposite, a design or practice put in place to prevent disease movement out of a positive or suspect herd. Within these two concepts, risks of disease introduction or dissemination revolve around 1.) Personnel/equipment 2.) Building design and location 3.) Live animal transport 4.) Mortality removal. This paper will focus on items 1 and 2.

Management decisions that affect biosecurity execution

Showering into and out of facilities, changing into site clothing, wearing shoe covers are examples of biosecurity practices we can easily implement and manage.

Real World Labor Example: A large multi-site production system has a team of grow/finish production managers that oversee contract labor crews handling loadout, vaccination, and power washing. The herd veterinarian implements a new vaccine requiring an additional vaccination at 3 weeks post-placement in the nursery. The same crew responsible for this vaccination, loaded market hogs at 5 AM and are scheduled to vaccinate this group at 11 AM. Key questions to ask in this scenario:

- Does the added value of this vaccine account for the added disease entry risk of the vaccination process?
  - Ways to mitigate these concerns:
    - Better working order realizing that pigs need to be loaded all times of the day.
      - May be possible to vaccinate before morning loads, realizing this depends on transportation and dock times.
    - Set up a method reduces the biosecurity risks of moving up stream to younger pigs.
      - Provide all equipment at each site ie, syringes, needles, etc.
        - Have a protocol for cleaning and be able to audit the protocol if this does occur.
      - Minimum change of clothes and wash hands if no shower available.[1]
        - In some cases, this would require 6-10 showers depending on size of crew and tasks to perform.
        - Understand that > than 4 showers a day is probably not sustainable.
- What expectations are given to contract labor crews?
  - Have the protocols been explained so that they can understand the process?
  - Have they been audited after training to ensure they truly understand the protocol?
- Who is accountable for their execution of biosecurity protocol?
  - Is this the production manager, site manager or where does the buck stop?
- Downtime – what does "1 night down from other pigs"[2] mean in a system that operates 24 hours a day?
  - A night of downtime needs to be defined by each system.
  - Does this refer to the procedures that have to be done before re-entering the system? If a shower, car wash/cab disinfection, and a clothes change is acceptable to restart the order if needed.

Facility factors that hinder biocontainment and bioexclusion

The 1990s saw the beginning of a dramatic change in the North American swine industry. The rapid shift from small farrow to finish operations up until that time to the 1,200 - 3,000-sow farrow-to-wean facilities moving on to combinations of nursery and finishers or wean to finish barns that are common in Northern Iowa and Southern Minnesota. This happened at same time veterinarians and producers were learning about the impact PRRSV and other highly contagious pathogens had on production, and how to control or prevent them. As a result, we have many facilities in our practice area that:

- Lack filtration
  - Was not available or considered at time of construction and difficult to remodel. Some of these curtain sites requiring significant remodeling to achieve filtration.
- Were built without isolation facilities and/or lack onsite gilt development
  - Isolation rooms that were connected to farrowing facilities (sharing manure removal, ventilation, etc) that would be difficult to truly maintain isolation.
- Closely located to other swine facilities
  - Cost of feed and access to crop acres for manure application are reasons the buildings were located as such, but also lead to the same concerns of cross trafficking in feed delivery and manure applications.

Filtration, onsite gilt development, and biosecure location are standard in new sow unit construction and have taken the farrow to wean facilities out of traditional production areas. Yet these older facilities will continue to operate as they can produce a far less costly pig than new construction, even with the health challenges. Veterinarians have found ways to manage these shortcomings, particularly using controlled exposure of replacement gilts to PRRSV and *Mycoplasma hyopneumoniae*. Yet the unintended consequences and cost of these programs manifest as areas becoming persistently PRRSV dense. This continued cycle will ensure future PRRSV spread within the area and likely viral mutation. The use of full genome sequencing suggests that this may result in more genetic sharing between virus and new strains emerging.[3] An improvement would be to add onsite gilt development to these facilities, yet lenders measure value of capital improvements through increased throughput. Gilt development does not increase throughput but does allow farms to complete herd closures[4] for disease elimination without disruption to parity structure or increased biosecurity risk through off-site gilt development. Veterinarians need to help lenders and producers assign value to management strategies like onsite gilt development.

Discussion

Biosecurity practices used within production systems are often specific actions that can be measured/managed. Yet we do not typically think of the ancillary impact that health or production practices will have on successful execution of biosecurity. Additionally, building design and capital improvements around disease control and replacement gilt flow need to be considered on a higher level, such as the area and viral recombination effect of continual pathogen exposure for disease control. Raising the awareness these practices have on biosecurity can help producers, lenders, and veterinarians make better informed decisions to reduce disease transmission.

References

1. Otake S, Dee SA, Rossow KD, et al. Transmission of porcine reproductive and respiratory syndrome virus by fomites (boots and coveralls). J Swine Health Prod. 2002;10(2):59-65.

2. Pitkin A, Otake S, Dee S. A one-night downtime period prevents the spread of porcine reproductive and respiratory syndrome virus and *Mycoplasma hyopneumoniae* by personnel and fomites (boots and coveralls). J Swine Health Prod 2011;19(6):345-348.

3. Trevisan G, Zeller M, Li G, Zhang J, Gauger P, Linhares D. PRRSV Whole-genome (WGS) Sequencing vs. Open Reading Frame-5 (ORF5). 2022 ISU James D. McKean Swine Disease Conference Proceedings.

4. Holtkamp DJ, Polson DD, Torremorell M, et al. Terminology for classifying swine herds by porcine reproductive and respiratory syndrome virus status. J Swine Health Prod. 2011;19(1):44–56.

https://doi.org/10.54846/am2023/166

Biosecurity from an international perspective

Eveline Willems, DVM
Regional veterinary manager Europe & NRAA
Topigs Norsvin

When thinking about biosecurity, most immediately think it is in reference to external biosecurity. Of course external biosecurity is a very important aspect of herd health management, as all the pathogens you are able to keep out of the farm, can surely not cause disease in the herd.

However, the focus of this presentation will not be external biosecurity as there have been so many topics on latest conferences on disease preparedness and other aspects of external biosecurity. Therefore only some specific topics will be highlighted where developments on this topic outside North America have been seen that perhaps could contribute to further improvements on external biosecurity in US.

An analysis by the Animal Health Service of the first 50 known outbreaks of PED in the Netherlands show us again the importance of hygiene of transport vehicles, as 65% of these outbreaks were related to transport. There are enormous differences in the easiness to clean different trailer types. When trailers are not easy to clean with holes, dipped floors, lots of corners and edges, the likelihood of not being cleaned properly increases. The trailer types that I have seen in North America would, in general, be considered to be 'difficult to clean'. From information of a third-party trailer inspection station we learned that, although it is known that the trailers will be inspected, the disapproval rate of supposedly clean trucks is still between 5 and 15%. In Europe it is much more common to use closed trucks with overpressure ventilation and virus filters for incoming air for transport of breeding stock as well as increasing amounts of high health animals from site I to site II. However much more important is that these trailers are developed to be easy to clean.

Another development is to design systems in a way to prevent people from being able to make mistakes on external biosecurity. Design of the showers for example should make it impossible to cheat the system and new farms are constructed in a way that it makes either the best way or the easiest way to do it, the preferably or only way to do it.

However when using the definition of Harvey for biosecurity "all the steps you could take to prevent your herd from disease", internal biosecurity, hygiene, vaccination programs and for example climate control would be as important as external biosecurity to protect your herd from disease. This part of biosecurity has become even more important with the pressure on lowering antibiotic use and for producing pigs with entire tails.

Legislative measures did trigger developments on such topics at a faster pace often supported by technology; precision farming. Knowing what is happening in the barn at all times. Making invisible aspects visible in a way turned out to be key for change in many farms as the issue becomes more tangible for the farm management. Following the poultry industry, new pig barns are being equipped with water meters per compartment. This makes it possible to measure the water consumption precisely and a sudden change in consumption is often an early indicator that something is wrong.

Furthermore, the advice on climate settings in the Netherlands was reviewed by specialists just two years ago; with our modern, fast-growing animals with a high feed efficiency, producing metabolic energy, the temperature settings were lowered by two degrees Celsius (3.6 degrees F) for nursery piglets and finishers. Since energy is expensive in Europe nowadays, it is in two ways beneficial to lower temperatures in the barns. Of course temperature is not the only aspect of a healthy climate for pigs. When a climate specialist visits the barn, he can only reflect on the situation on that very moment, therefore real-time monitoring is a real asset. This can be done with the healthy climate monitor. Such technology support farm-managers and specialists to define the actual situation of the climate during the whole day, week, year and early interference is made possible.

More focus on internal biosecurity aspects led to a significant decrease in the need of using antibiotics in the pork-industry in Europe and we experienced other unexpected benefits as well!

https://doi.org/10.54846/am2023/167

PRRS: The fight continues

Clayton Johnson, DVM
Carthage Veterinary Service, LTD, Carthage, Illinois

Is your farm and team prepared for PRRS season? PRRS is a frustrating and complicated disease. It demoralizes our teams at all levels, from the ownership to the newest technicians on the farm. We have numerous tools to help with PRRS prevention, control, stabilization and elimination but none of them individually represent a "silver bullet". Regardless of how challenging PRRS can be, we have to face the brutal reality that production systems that don't effectively manage PRRS cannot be competitive in today's marketplace. As such, we have to position our PRRS management resources strategically to optimally battle PRRS while being realistic that there will be times PRRS has the upper hand at our farm.

Where to start?

Biosecurity. Plain and simple, preventing wild type PRRS infection is the best possible outcome for a PRRS management program. Good biosecurity programs involve good biosecurity plans. Note the plural form of the term "plans" – PRRS can enter the farm through a number of different avenues and we must have specific biosecurity plans for each of these potential routes of introduction. While PRRS is infamous for its ability to "fly" into the farm via aerosols, the reality is very specific climatic conditions are required to facilitate PRRS transmission via aerosols. It's possible on specific days, but even in the winter most days in most pig producing regions don't have the appropriate climatic conditions to facilitate PRRS transmission via aerosols. Let's start with focusing on the areas PRRS can enter the farm every day.

People, supplies, equipment and animal introductions all represent huge areas of opportunity for PRRS introduction. Everything that comes into our farm can serve as a fomite and when contaminated, bring PRRS into our farm. Fomite management starts with defining "clean and dirty lines" (CDL). As my good friend Dr. Mark Thompson of The Maschhoffs taught me – Define and Defend these areas. The first CDL most of us encounter when coming to a farm is the bench entry. This area represents a CDL meant to remove our personal items most likely to be contaminated, specifically our personal shoes we wore to the farm. It is critical that we remove our shoes and swing our feet over the bench to the "clean" side without contaminating our socks. While this fact seems obvious, it's very common to find evidence at the bench that indicates our farm teams don't understand this concept. If you see shoes that are located farm away from the bench, you know those people had to contaminate their socks on the dirty side of the bench prior to crossing it. Shoe direction at the bench can also be a good tip of execution, if you see shoes facing the bench (toe side pointed toward the bench) that can also be a good indicator that there's a coaching opportunity for that farm. Benches represent a wonderful opportunity to teach people about the concepts of the CDL including:

1. There is a clean side and a dirty side
2. We don't want items that contact the dirty side to cross onto the clean side
3. For items that must come across the CDL onto the clean side, some sort of sanitation step should always occur prior immediately prior to crossing.

Supply and equipment entry represent our next key area of biosecurity focus. As a general rule of thumb, supplies and equipment coming into the farm should always be new and have never been on another pig farm. Historically, the pig industry has used "disinfection rooms" with "foggers" to aerosolize chemical disinfectant as our primary means of supply and equipment entry. Through a series of research projects Dr. Jim Lowe and his team at the University of Illinois have taught us that the aerosolization of chemical disinfectant represents an incomplete way to apply disinfectant to all surfaces of our supplies and equipment. Even when we work to space out supplies and place them on open bottom metal shelving, we still observe very poor disinfectant coverage of the sides of our supplies and equipment that aren't immediately facing the fogging machine. Simply put, replying solely on a fogging machine to disinfect your supplies and equipment is not going to result in complete sanitation of these incoming items. Fortunately, there are sanitation options outside of only chemical disinfection and even better, they're easy for our farm teams to execute! Many farms have transitioned to heated downtime as a means of supply introduction. We often use the same disinfection rooms that historically fogged our incoming items, except we had heat (if needed) to warm this room consistently to room temperature and place our incoming supplies in these rooms for 7 days before entering them into the farm. Compliance is easier as we don't have to space out all incoming supplies, stacking them in the room is acceptable. We obviously still need to use chemical disinfection on perishable items as well as emergency supplies needed in the farm right away, but I've found that compliance with manual spraying of chemical disinfectant on these "one off" introductions is much better than asking someone to manually disinfect every item that comes into the farm.

While our animal transport vehicles and mortality handling equipment don't come into the farm, I would be remiss to not include them in a discussion about PRRS prevention. Let's start with animal transport vehicles. These vehicles and their drivers represent an extremely unique and dangerous risk to our pig farms – they have to come to the farm and we know they go to other farms and animal concentration points which potentially have PRRS positive pigs. At the farm, we always consider the vehicles and drivers "dirty" and flawless execution of the CDL that separates the clean side of the farm from the dirty side of the vehicle is absolutely critical. It's easy for me to write an animal loading SOP that directs farm workers to load the truck without crossing the CDL, but I've loaded a lot of trucks in my life and I have the practical experience to understand that while the protocol is easy to write it's extremely difficult to execute. Asking our farm teams to load 1600 weaned pigs onto a trailer in 10°F temperatures with a 30-mph wind blowing directly into the chute without having some compromise of the CDL is an unrealistic expectation. As such, we lean on the principles of stages loading to create 2 CDLs. The first CDL is always at the back of the livestock vehicle and the end of your chute. The second CDL is the next or gate you encounter as you move into the clean side of your farm. We station one farm employee between these two CDLs and this area is considered dirty during the loading process. This employee has dedicated equipment (sort boards and other movement tools) as well as segregated coveralls and boots to wear in this area. Immediately after we load pigs, the employee working in between the CDLs immediately washes and disinfects this area, again using equipment dedicated for only this purpose. When items from between these CDLs need to be removed from the farm we remove these outside, directly to the dirty area outside the farm. Mortality handling equipment should follow the principles, with flawless execution of CDL at the mortality removal area being the focus. Ideally, we'd like to have an elevated platform to remove mortality across the CDL with the benefit of gravity helping move larger animals out of the farm. Where this option doesn't exist, consider using a rope system where you place the mortality immediately next to the CDL, tie a rope around the leg of the mortality and throw the other end of the rope outside. The employees working on the dirty side of the CDL will access this other end of the rope and use it to pull the mortality across the CDL.

If you're reading this article and satisfied that your farm and your team have all the tools and processes in place above, but you're still frustrated with your PRRS introduction rate, let's talk about air filtration. One of the more revolutionary tools the industry has developed, we all have to thank Dr. Satoshi Otake for his excellent work characterizing the risk of aerosol transmission as well as Dr. Scott Dee for his willingness to research and champion the early adoption of air filtration systems to mitigate this threat. As with all new technologies, the cost of air filtration has come down and a tool once generally reserved for high value assets such as boar studs is now often cost effective to adopt for commercial sow farms in pig dense locations. When coupled with positive pressure ventilation systems, air filtration systems give the farm a reasonably user-friendly tool to mitigate the threat of PRRS virus entering via contaminated aerosols. I think the most important consideration when considering air filtration is a thorough evaluation of your other potential PRRS introduction risks. If you have gaps in people entry, supply and equipment entry, transportation biosecurity or mortality handling we need to shore up those areas before consider investing in an air filtration system.

PRRS is complicated and frustrating, but if we optimally our biosecurity resources we can expect good success in preventing PRRS introduction into our farm. Excellent benchmarking tools exist through the Morrison Swine Health and Management Program (MSHMP) to help compare your PRRS outbreak rate to your peers in the industry. If you're looking for information on how to improve your biosecurity going into this winter, hopefully this article gives you somewhere to start. Remember, biosecurity plans need to be specific to your farm – work to define the optimal PRRS prevention plans for your team and operation.

Managing a PRRS outbreak on a breed to wean (BTW) farm

The immediate aftermath of a PRRS outbreak on a BTW farm is a chaotic situation. Emotions and ambiguity are both raging at high levels. Urgency to fix the problem and simultaneous frustration at the lack of control over the outcome will combine to cloud resource allocation. Providing a prognosis is always an important part of the diagnosis and veterinarians should share with their clients that key metrics such as born alive, pre-weaning mortality and pigs weaned will be dramatically impacted for 10 weeks and moderately impacted for another 10 weeks after that. In my experience it is best to set expectations fairly low, it is better to have better performance than your prognosis than to set an unrealistically positive future picture of what performance may do at the farm.

One of your primary areas of focus during the acute outbreak should be to contain the PRRS infection to just this specific farm. Ideally, biocontainment plans have been defined well before the PRRS outbreak so focus can be placed on executing the plan instead of defining it on the fly. Have a checklist of risk factors to review with each BTW PRRS outbreak. These risk factors should largely mimic the risk factors evaluated in a biosecurity audit with focus on risk factors that may connect the outbreak farm to other farms under your control. Focus efforts on dedicated previously shared assets to this farm with specific focus on livestock transportation and people movements. Even if you know the farm well, double check all potential risk factors with key stakeholders to ensure they are aware of the new PRRS status at the BTW farm and that your assumptions about people and supply movements are correct. Weaned pigs recently shipped from the farm should be tested for PRRS, even in the absence of clinical signs. It is very common to find infected batches of weaned pigs shipped over the last 2 - 4 weeks, even if there were no clinical signs at the BTW farm during that period. Additionally, neighboring farms should be contacted if possible to share your new PRRS status and inquire about potential cross traffic. Vehicle traffic patterns and areas where employees may overlap should be discussed and reviewed. Regardless of your relationship (or lack thereof) with the neighbor, you do not want their farm to become infected and serve as a risk for another PRRS outbreak after you have recovered from this outbreak.

PRRS monitoring

In breeding herds as well as growing pig farms, early detection of PRRS virus is critical to both best manage the impending disease consequences as well as to contain the PRRS outbreak to just this specific farm. While clinical signs of PRRS are often readily available soon after the outbreak, it's critical to continuously monitor our farms for clinical signs of the subacute phase of PRRS infection. In growing pigs, water consumption has shown to be a very effective objective measurement to detect leading indicators of the clinical signs of PRRS. More recently, audio surveillance via tools such as Sounds Talks have provided sensor-based monitoring to observe cough levels in pigs and identify farms in which additional investigation is warranted. In breeding pigs, abortions have been the primary objective measure to monitor, with abortions at a level greater than 1 abort per 1000 sows mated inventory signaling a need for further attention. Additionally, increases in stillborn rates and sows off feed should also provide a sense of urgency for veterinary examination.

Even with outstanding biosecurity plans, the reality is very few farms have a 0% PRRS introduction rate. Effective control of PRRS at the commercial breed to wean (BTW) level requires elimination of WT PRRS introductions into that farm. Initial diagnostics to diagnose the outbreak should include sequencing and if at all possible a whole genome sequence (WGS). After that initial set of diagnostics, future sampling focus should be on monitoring the PRRS stabilization and elimination program. How do we know if we're effectively progressing in our PRRS elimination program? We need to routinely monitor for the presence of PRRS in our piglet populations. We specifically monitor piglet populations as they represent the status of the entire herd and generally represent our most sensitive population for PRRS infection. In addition to understanding where we're at in our WT PRRS elimination programs, monitoring also provides confidence for routine management decisions such as gilt movements and introductions as well as comingling decisions for post-weaning pig placements.

Processing fluids (PF) represent an outstanding sample type to asses large number of piglets for the presence of PRRSV at the time of processing. We generally use these sample results to assess the incidence of in utero transmission from sow to piglet prior to birth. Once PF PRRS PCR results are consistently negative, then we progress to assessing older piglets in the farrowing house. Tongue tips represent a newer sample type to assess older piglets. Similar to processing fluids, we collect these tissues from all available piglets (in this case, piglets that become pre-weaning mortalities) and submit those for PCR. Family oral fluids (FOF) represent a sample that can be collected from piglets immediately prior to weaning. Ropes are hung in the farrowing crate in a location accessible to both the sow and her piglets, the sow teaches her piglets to chew on the rope, the piglets follow suite and in one hour after hanging the rope, we have a good sample to assess for PRRS via PCR. Serum may be our oldest diagnostic sample type in the farrowing house, but its still a very effective sample type. While routine collection from a large number of piglets may be too laborious in some situations, where it is possible to collect serum it is still valuable to do so as it gives you the ability to bias your samples toward the piglets you are most interested in testing.

Our general goal with PRRS monitoring focuses on assessing piglet populations over space and time. Piglet PRRS prevalence can be highly variable from room to room in the farrowing house at the end of the PRRS elimination program when prevalence is low. Constant monitoring with samples collected every day ideally and every week at minimum across all rooms in the farrowing house over time gives us the greatest confidence in our results. Regardless of sample type, we must provide direction to the diagnostic lab on how to pool our submitted samples. The cost of routine monitoring must be considered and pooling provides us the ability to control cost by minimizing the number of PCR assays requested. Pooling is a double-edged sword, while cost savings are obtained by intense pooling, we dilute our sample in this method and impact the likelihood of PRRS virus detection. The Field Epi team at Iowa State University has provided us with an online tool to calculate the impact of pooling on our chance of identifying PRRS virus – you can and should access this tool at **www.fieldepi.org/calc**. When we look at different pooling options across a different number of sample types, we consistently find that the following facts hold true:

1. More samples increase the probability of finding a positive
2. Higher prevalence increases the probability of finding a positive
3. If budget is a concern, collecting more samples and being more aggressive with pooling (same number of PCR tests but more samples pooled) increases the probability of finding a positive

When you have PF PCR positive results for too long or an unexpected drop in CTs on PF PCRs, it is recommended to collect tongue tips (TT) from stillbirths to verify if the virus is coming from in utero transmission via the sow herd. Additionally, requesting a WGS to compare to the baseline sequence can help to rule in or out the presence of a new PRRS introduction or recombination of multiple strains present on the farm. Should you have continued abortions beyond 4-6 weeks after the initial outbreak, it is recommended to submit serum from aborted sows as well as aborted piglets to evaluate the presence of other pathogens (Parvo, PCV) as well as PRRS.

https://doi.org/10.54846/am2023/168

Revenge of *E coli*, but wait, there's more…

D. Murray, DVM; L. Johnson DVM; A. Warner, MS
New Fashion Pork, Jackson Minnesota

Background

Many producers have seen a resurgence of clinically significant F18 *E coli* in the past 2 to 3 years. For our system these cases began in June of 2020, with several cases occurring that summer into late fall. These cases resulted in an average increase of 8% mortality. Some improvement was noted over winter but cases then resumed in late summer of 2021. These cases were unique due to their non responsiveness to F18 *E coli* vaccine (competitive inhibition), as historically vaccination had been very successful in preventing these outbreaks and even worked well in the face of an outbreak to slow down and stop mortality due to F18 *E coli*. Isolates from these cases also would typically show 4+ toxins associated with the F18 *E coli*. Although genetic resistance to F18 has been shown, genetic programs have not changed in the effected flows. What else had changed to contribute to these severe high morbidity/mortality cases? In June of 2021 similar clinical cases also began occurring in other non F18 positive flows that appeared grossly to be F18 but were ultimately diagnosed as *Streptococcus suis* (Figure 1). There was a suspected connection between the two clinical syndromes as grossly both cases would show a severe gastritis in addition to hemorrhagic small intestines. With similar gross presentations as well as disease onset timing, was something else underlying as a primary cause? If there was an underlying primary cause were *S suis* and F18 *E coli* were both secondary pathogens to this? In depth investigations occurred for both the F18 *E coli* as well as the *S suis* with whole genome sequencing performed on multiple isolates as well as isolates from cases which did not show gastritis lesions and would respond F18 *E coli* vaccination. These investigations did not reveal any new/unique isolates of either F18 or *S suis* to the system. Whole genome sequencing did not reveal unique differences in F18 *E coli* isolates on vaccine responsive cases vs non vaccine responsive cases. Vaccine compliance, titer, and purity was also verified in this investigation. With two apparent dead ends on "new" F18 and *S suis* introductions further investigation was pursued for an underlying primary cause and to determine when this clinical presentation was truly beginning.

Clinical presentation

Mortality timing in these cases can be variable however, sudden acute mortality in nursery pigs most frequently occurred at eight to ten days post wean. Necropsy findings (if no other underlying health issues are concurrent PRRSV, GPS, MHR etc..) revealed lungs which appeared grossly normal and would collapse after the chest cavity was opened but the color of the lungs was often unique. This unique color was a fluorescent pink and could be seen on found dead pigs. The lungs that appear grossly normal when the chest was opened also turned a fluorescent color after exposure to ambient air for several minutes (Figure 2). The heart may have mulberry heart lesions, but these lesions are not consistent on all necropsied pigs. The kidneys and spleen appeared grossly normal. The liver was typically normal but did show enlargement in some cases. The stomach would show a severe gastritis of the fundic region, and in rare cases also contain frank blood (Figure 3). The stomach wall was typically not thickened although in some cases could be. On euthanized pigs the gastritis lesion severity was more variable. Small intestines were grossly hemorrhagic with normal wall thickness. Colon was grossly normal. If the pig flow in question had underlying F18, this was ultimately diagnosed though routine tissue submissions on these cases. If the pig flow in question did not have underlying F18, *S suis* would typically be diagnosed though routine tissue submission in those cases. Occasionally other secondary pathogens were found as well such as salmonella in either the stomach or intestines. Although the onset timing of the first acute mortality was similar in both the F18 and *S suis* cases, on the F18 *E coli* cases it would rapidly spread throughout the site once the first barn became clinically ill. Acute mortality would then be found within 24 hours of nursery arrival with or without watery diarrhea and dehydration. Most cases involving F18 *E coli* loose stools were present prior to the first acute mortality but this was not always the case. In cases where F18 *E coli* was not involved there were very few clinical signs prior to the first acute mortality, and death could happen very rapidly. In all cases found dead pigs would have full stomachs of feed. In all cases mortality occurred for the remainder of the nursery turn and could also continue into the finishing turn as a combination of acute as well as chronic presentations. Average flow mortality attributed to this could range from two to ten plus% depending on secondary bacterial agents, stress, treatment timing, and other factors.

Diagnostics/investigation

Population dynamics

To determine if this issue was flow specific necropsies were performed in all flows throughout the system, in particular, on any acute mortality within two weeks of wean pig placement. Gastritis was found in all commercial flows as well as multiplication. Commercial flows include farms which wean directly into the nursery as well as farms that utilize a staging room to house wean pigs prior to nursery arrival. Transit times range from 30 minutes to 12 hours from sow farm to nursery arrival. To date one noncommercial flow is unaffected, and one commercial flow has a lower incidence with delayed timing. To determine if this condition occurred only after weaning, naturally occurring and quality control mortalities on suckling piglets from birth and to wean age were necropsied. Gastritis was found in piglets on the sow as young as 3 days of age with the majority of gastritis cases found day 15 of age to wean (Figure 4). Piglets had access to sow feed in farrowing but were not intentionally creep fed a specifically formulated ration. Sow farms with staging rooms utilize an early nursery diet in the staging rooms. Due to the widespread nature of the clinical issue it is unlikely that bulk feed ingredients such as corn or soybeans would be the root cause with multiple feed mills used in widespread geographic areas. However, VTM and other ingredients that have one source which are sent to each mill would have potential for contamination.

Diagnostics

Numerous full tissue work ups have been performed on affected pigs. Helicobacter species have not been found to date on these

Figure 1A: Good body condition pig euthanized day 8 post wean diagnosed as *Streptococcus suis*.

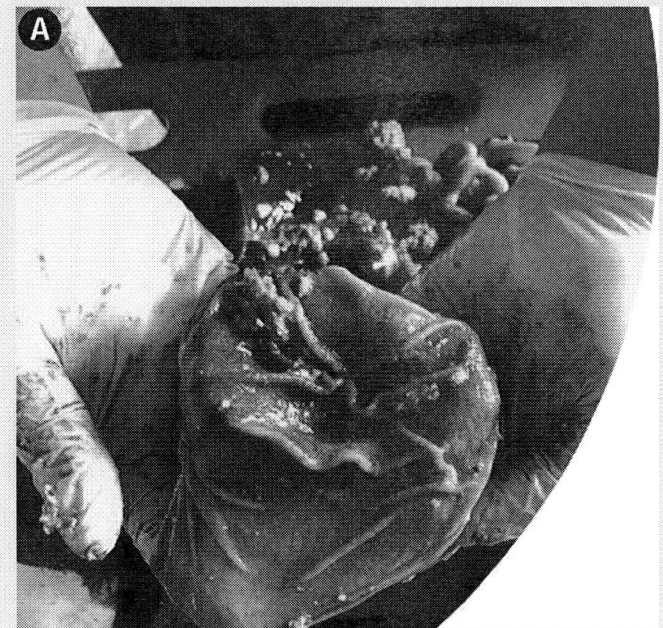

Figure 1B: Good body condition pig euthanized day 8 post wean diagnosed as *Streptococcus suis*.

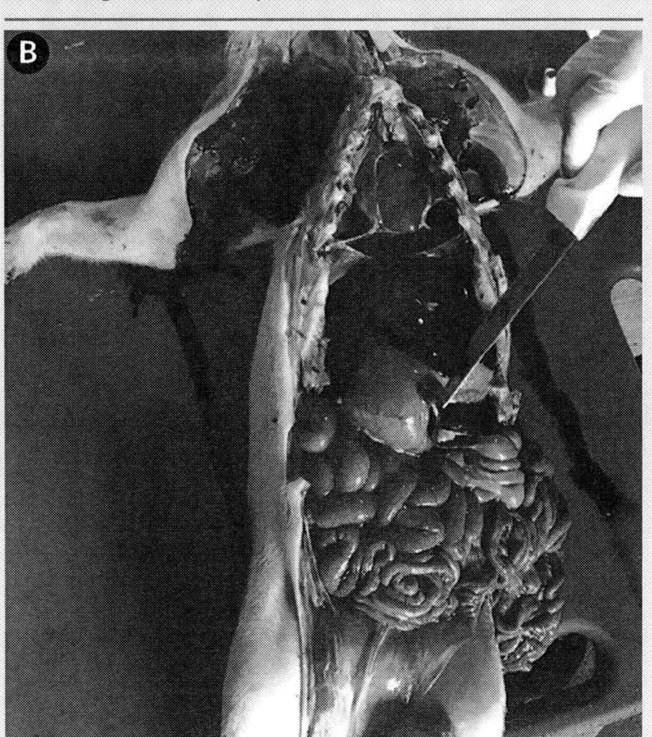

Figure 2: Two pictures of the same pig taken 30 minutes apart. **(A)** Normal lung. This picture shows the pig upon opening the chest.

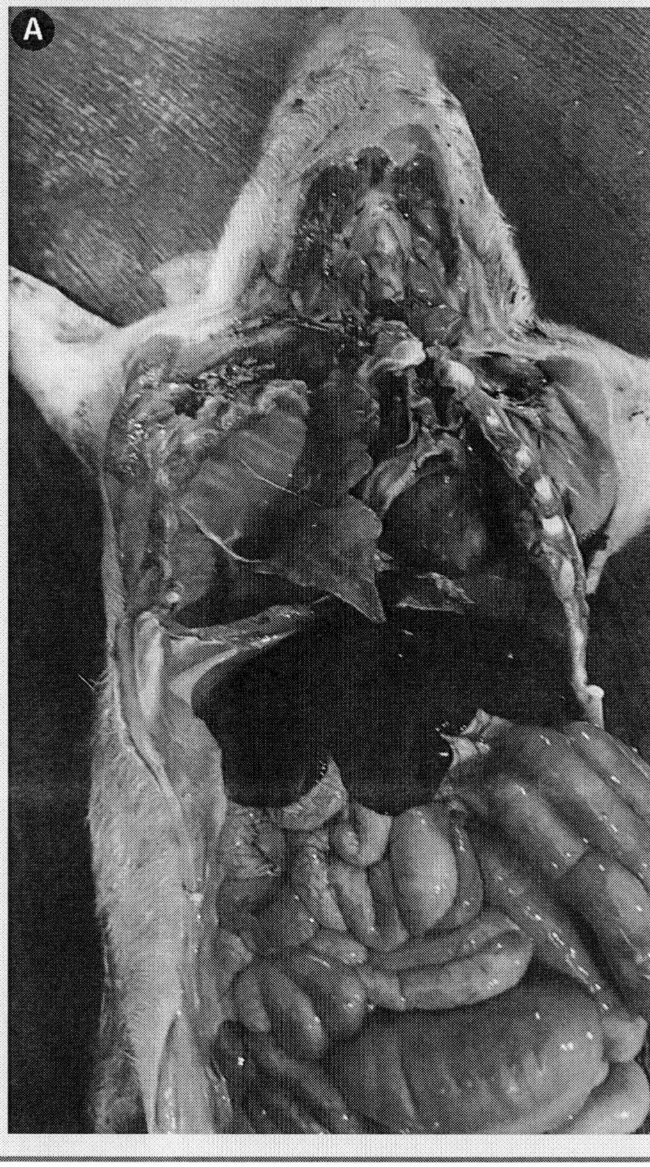

pigs and is not considered a differential at this time. Other pre-existing flow specific pathogens have been found but there have not been any new or unique pathogens detected to date. Histopathological classification of these lesions is typically described as mild to moderate gastritis with vascular congestion and hemorrhage of the superficial mucosa with little to no inflammation associated. These types of lesions may be better described as a gastropathy due to the lack of inflammation. Metagenomics have been performed on feces as well as stomach content in early nursery aged pigs as well as suckling piglets. No consistent or unique pathogens have been found to date on this testing. High reads of kobuvirus have been found on suckling piglets but the clinical significance of this remains unknown as this pathogen is considered ubiquitous. Ingredients which are common to all mills were investigated for the presence of *Bacillus cereus* or other unintentional bacillus species. Only commensal or probiotic bacillus

Figure 2: (B) Fluorescent lung. This picture shows the same pig as (2A) 30 minutes later. Pig location and lighting has not changed.

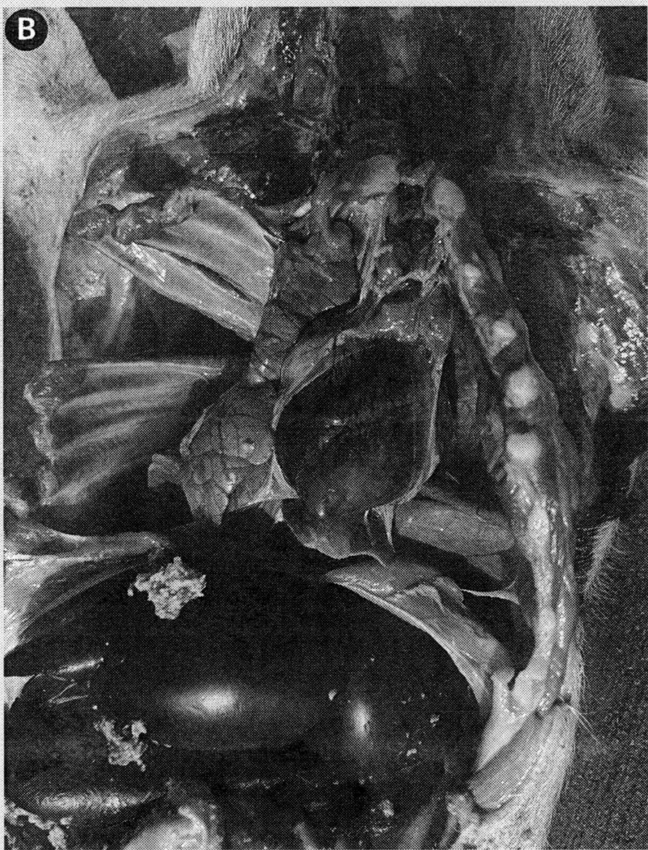

species were found including *Bacillus subtilis, Bacillus paralicheniformis, Bacillus velezensis,* and *Bacillus licheniformis* in above expected amounts. These Bacillus species would not be expected to cause gastritis lesions.

Investigations

To investigate potential sow farm causes field elimination trials and observations have occurred and are currently being pursued for known exposures such as drying agents used, vitamin supplementation, and vaccinations. Drying agent type has not made an impact on gastritis associated mortality. Pigs that did not receive any drying agent were still equally affected with gastritis associated mortality. Elimination of vitamin E and D_3 supplementation at birth has not decreased mortality associated with gastritis. Elimination of oral F5 (K88) *E coli* and oral salmonella vaccination to suckling piglets has also not decreased mortality associated with gastritis but does appear to increase nursery associated clinical cases of salmonella as well as *E coli*. Elimination of nursery vitamin D_3 or E and D_3 supplementation has also not decreased gastritis associated mortality, and anecdotally has potentially increased mortality slightly. Further evaluations are planned to investigate injectable vaccinations also.

Serology/toxicology

Several groups of nursery pigs as well as suckling piglets have been tested by serum chemistry, serum electrolyte, serum mineral, serum heavy metal, serum vitamin, liver mineral, liver vitamin. Due to supplementation serum vitamin D_3 and E levels are above age specific ranges, however toxic levels have not been found and no histologic lesions of toxicity have been found to date on any tissue submission. No overt mineral or heavy metal toxicities have been found as well. Further analysis of these results are currently underway.

Research

To better understand effect of vitamin E and D_3 nursery supplementation and diet a three x two factorial evaluation was performed. This evaluation consisted of three different diets with or without vitamin E and D_3 supplementation. To better understand incidence and correlation of gross lesions to culture/serum testing/histologic results, a subset (n = 48) of pigs were euthanized (evenly distributed between diets and supplemented groups) with full necropsies performed at day five and day ten post weaning. The results of the day five necropsied pigs showed 7/24 (29%) with a grossly normal stomach, 7/24 (29%) with a mild/moderate gastritis, and 10/24 (42%) with a severe gastritis (Table 1). The results of the day ten pigs showed 9/24 (38%) with a grossly normal stomach, 10/24 (42%) with a mild/moderate gastritis, and 5/25 (21%) with a severe gastritis (Table 2). In this evaluation pigs did develop a secondary F18 *E coli* with 4/24 culture positive by day five (3/4 positive culture pigs also had a severe gastritis), and 24/25 culture positive by day ten. Vitamin E and D_3 supplementation did not appear to have an impact on gastritis severity by day five as the distribution of lesions was nearly evenly spread between both groups. The day ten necropsied pigs which had vitamin E and D_3 supplementation had no severe gastritis, 6/12 with mild/moderate gastritis, and 6/12 grossly normal. Day 10 necropsied pigs which were not supplemented with vitamin E and D_3 had 5/12 with severe gastritis, 4/12 with mild/moderate gastritis, and 3/12 were grossly normal. Diet did appear to have an impact on evaluated pigs. Day 5 necropsy results by diet were as follows: Diet 1 had 2/8 grossly normal stomachs, 1/8 pigs with mild/moderate lesions, and 5/8 pigs with severe gastritis. Diet 2 had 2/8 pigs with grossly normal stomachs, 4/8 with mild/moderate gastritis, and 2/8 pigs with severe gastritis. Diet 3 had 3/8 pigs with grossly normal stomachs, 2/8 pigs with mild/moderate gastritis, and 3/8 pigs with severe gastritis. Day 10 necropsy results by diet were as follows: Diet 1 pigs had 4/8 grossly normal, 3/8 mild/moderate gastritis, and 1/8 severe gastritis. Diet 2 pigs had 2/8 grossly normal stomachs, 2/8 mild/moderate gastritis, and 4/8 severe gastritis. Diet 3 pigs had no severe gastritis with 3/8 stomachs grossly normal, and 5/8 mild/moderate gastritis. In other field evaluations of these diets, acute mortality due to gastritis appears to have been significantly reduced and nearly eliminated using similar diets to those evaluated in the study herein. The diets evaluated by this study consisted of a reduction in SID Lys, as well as an increase in total calcium and total dietary fiber with the use of sugar beet pulp and oat groats.

Success recommendations/things to avoid

As no known cause has been determined, controlling secondary infections, reducing stress, and a gut friendly diet have been the most successful in reducing morbidity/mortality in these cases. Pathogen reduction to help control secondary infections has been beneficial, which includes white washing barns as well as using a water line cleaner. Day to day management has been critical including but not limited to; early nursery pig care enabling pigs to get started correctly, environmental management, such as

Figure 3A: Early nursery-aged pig examples of severe gastritis.

Figure 3B: Early nursery-aged pig examples of severe gastritis.

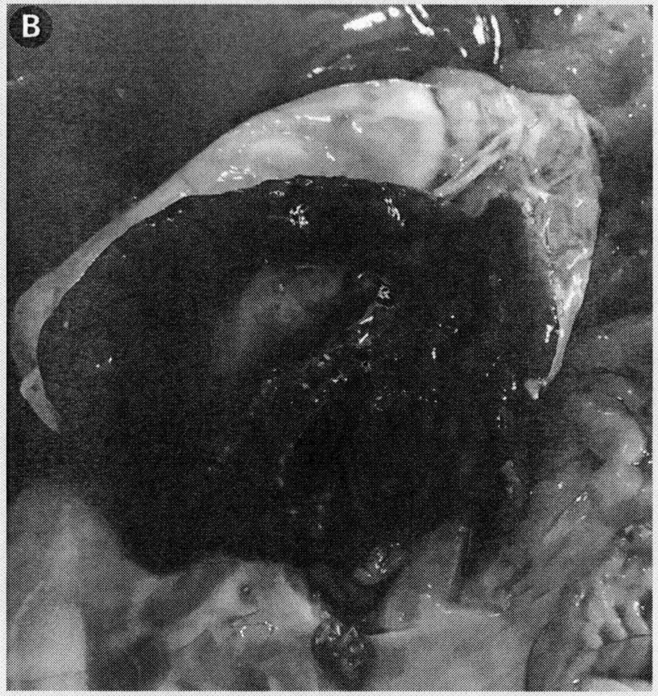

Figure 3C: Early nursery-aged pig examples of severe gastritis.

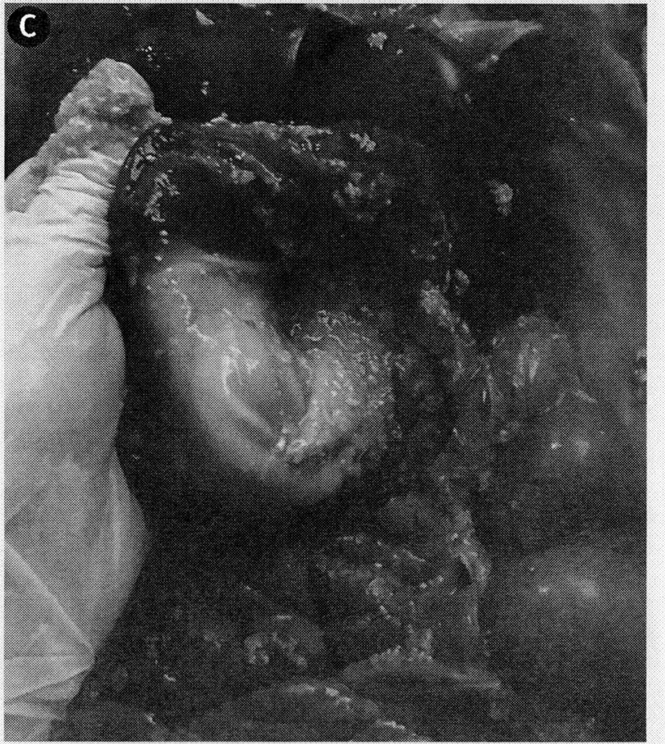

maintaining optimal barn temperature and humidity, and identifying at risk pigs to treat or pull early. Daily necropsies on all pigs, with very early intervention timing (such as with the first acute mortality with gastritis) has been essential. Nursery staff in most pig flows would currently necropsy all nursery mortalities, with pictures and information sent to production as well as veterinary service staff on a near daily basis. In conjunction with antibiotic therapy to control/treat secondary bacterial infections, rolled oat supplementation appears to be beneficial as well. Poor barn sanitation, poor environmental management, noncompliance on oral vaccines

(*E coli*/salmonella), late secondary infection control, and high protein/low fiber diets should be avoided. The magnitude of reduction and increase in these nutrients from a standard control should be evaluated. It appears that diet may have an influence on the prevalence of cases, which may be a reflection of the diet formulation.

Figure 4A: 20-day-old piglet with severe gastritis lesions.

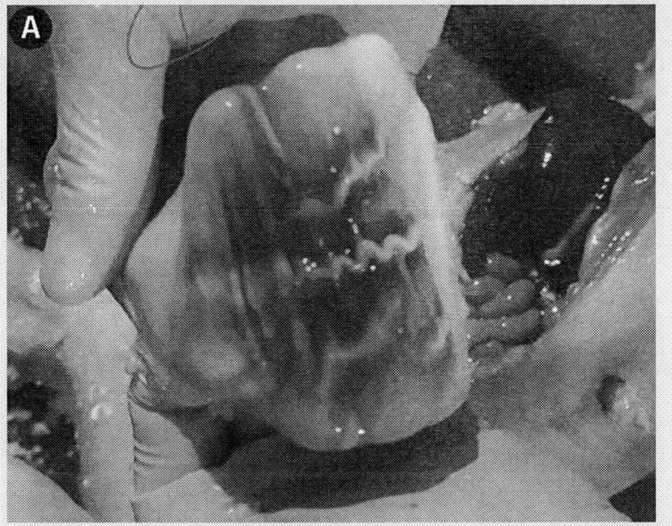

Figure 4B: 17-day-old piglet with mild/moderate gastritis lesions.

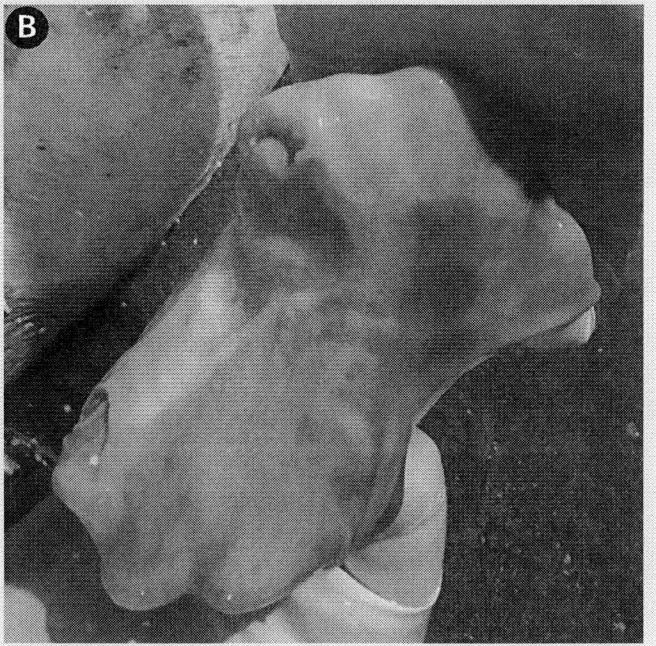

Table 1: Day 5 stomach lesions and aerobic culture results (*aerobic culture grew only F18 *E coli*)

Diet	Vitamin D/E supplementation	Stomach scores	Colon results *	Intestine results *	Stomach results *
1	no	0	No Sig. Growth	No Sig. Growth	No Sig. Growth
1	no	1	No Sig. Growth	No Sig. Growth	No Sig. Growth
1	no	2	No Sig. Growth	No Sig. Growth	No Sig. Growth
1	no	2	No Sig. Growth	No Sig. Growth	No Sig. Growth
1	**yes**	0	No Sig. Growth	No Sig. Growth	No Sig. Growth
1	**yes**	2	No Sig. Growth	No Sig. Growth	No Sig. Growth
1	**yes**	2	No Sig. Growth	No Sig. Growth	No Sig. Growth
1	**yes**	2	No Sig. Growth	No Sig. Growth	No Sig. Growth
2	no	0	No Sig. Growth	No Sig. Growth	No Sig. Growth
2	no	1	No Sig. Growth	No Sig. Growth	No Sig. Growth
2	no	1	No Sig. Growth	No Sig. Growth	No Sig. Growth
2	no	2	No Sig. Growth	No Sig. Growth	No Sig. Growth
2	**yes**	0	No Sig. Growth	No Sig. Growth	No Sig. Growth
2	**yes**	1	No Sig. Growth	No Sig. Growth	No Sig. Growth
2	**yes**	1	No Sig. Growth	No Sig. Growth	No Sig. Growth
2	**yes**	2	Moderate	No Sig. Growth	No Sig. Growth
3	no	0	No Sig. Growth	No Sig. Growth	No Sig. Growth
3	no	0	Moderate	Low	No Sig. Growth
3	no	2	No Sig. Growth	Low	No Sig. Growth
3	no	2	No Sig. Growth	No Sig. Growth	Few
3	**yes**	0	No Sig. Growth	No Sig. Growth	No Sig. Growth
3	**yes**	1	No Sig. Growth	No Sig. Growth	No Sig. Growth
3	**yes**	1	No Sig. Growth	No Sig. Growth	No Sig. Growth
3	**yes**	2	No Sig. Growth	No Sig. Growth	No Sig. Growth

Gross stomach lesion characterization 0 = normal 1 = mild/moderate 2 = severe

Table 2: Day 10 stomach lesions and aerobic culture results (*aerobic culture grew only F18 *E coli*)

Diet	Vitamin D/E supplementation	Stomach scores	Colon results *	Small Intestine results *	Stomach results *
1	no	0	High	Moderate	High
1	no	0	High	Low	No Sig. Growth
1	no	1	Low	Low	Couple
1	no	2	No Sig. Growth	Moderate	High
1	yes	0	High	Moderate	No Sig. Growth
1	yes	0	High	Moderate	No Sig. Growth
1	yes	1	Moderate	Moderate	No Sig. Growth
1	yes	1	Moderate	Moderate	Low
2	no	2	High	Moderate	Low
2	no	2	High / Couple	High	Low
2	no	2	High	High	Moderate
2	no	2	High	High	Low
2	yes	0	Moderate	High	Low
2	yes	0	Few	No Sig. Growth	Low
2	yes	1	No Sig. Growth	No Sig. Growth	Couple
2	yes	1	High	Moderate	No Sig. Growth
3	no	0	No Sig. Growth	No Sig. Growth	No Sig. Growth
3	no	1	Moderate	Moderate	No Sig. Growth
3	no	1	High	High	Couple
3	no	1	High	Moderate	Few
3	yes	0	High	High	No Sig. Growth
3	yes	0	No Sig. Growth	Moderate	No Sig. Growth
3	yes	1	Moderate	Moderate	No Sig. Growth
3	yes	1	High	Couple	Low

Gross stomach lesion characterization 0 = normal 1 = mild/moderate 2 = severe

https://doi.org/10.54846/am2023/169

Sapovirus: Diagnosis to solution ... a cooperative success

T. Petznick, DVM
ArkCare, Omaha, Nebraska

Introduction

From early 2017 to early 2019, a 4500-head farrow-to-wean operation experienced oscillating incidence of diarrhea, particularly in piglets that were 7 days of age and older. Typically, this diarrhea would self-limit and become unobservable by the end of the 20 – 24-day lactation period, however the expected performance of the pig was not being realized. While mortality due to diarrhea was minimal, the effect on morbidity was clear as it was rare to wean pigs over 12 pounds of bodyweight. The next two years would uncover an unexpected etiology of piglet diarrhea, establish that etiology as a sole and co-infective entity in several swine operations and result in a preventive solution due to the cooperative success of producer, veterinarian, diagnostic laboratory, and industry.

Pre-diagnosis

Over the two-year time period prior to the unique diagnosis, the farm experienced an ongoing presence of diarrhea in mid-to-late lactation piglets. The diarrhea was of a semi-liquid to pasty consistency with an off-color spectrum of yellow to grey. The diarrhea was non-responsive to antibiotic regimens. Since the diarrhea clinically resembled coccidiosis and fecal flotation had revealed positive samples, a regimen of administering an anti-protozoal product was established at day 3 of age. Early impressions were that there was some improvement, but it was not consistent nor complete. Further diagnostics revealed strains of rotavirus and lesions of atrophic enteritis to match it. A strain specific vaccination program for rotaviruses was implemented. Improvements observed were better than with coccidiosis regimen, but still incomplete and inconsistent. Due to the high level of frustration with resolving the diarrhea despite a preventative plan that had produced success in similar situations on other farms, a new set of diagnostics was ordered.

Sapovirus revealed

A diagnostic visit was arranged with the farrow-to-wean farm manager. Only new cases of diarrhea on the day of the visit were eligible for diagnostics. Typical presentation of the new cases was observed on the day of the visit and a full set of enteric tissues were submitted to the Iowa State University Veterinary Diagnostic Laboratory (ISU-VDL). The laboratory report confirmed atrophic enteritis lesions. Polymerase chain reaction (PcR) for the enteric coronaviruses and strains A, B, and C of rotavirus were negative. Fecal flotation for coccidiosis-related organisms were negative and no other histopathologic lesions were noted. Due to the observation of pathologic lesions, yet absence of etiology, next generation sequencing (NGS) was requested by the herd veterinarian. The herd veterinarian received a call from the diagnostic lab that NGS revealed porcine sapovirus (PSaV) on a high number of reads as the sole organism on feces that were submitted. PSaV was first identified in the United States in 1980, however was not differentiated as a definitive enteric pathogen in the field.[1]

Discovery

The results of the NGS initiated a cooperative effort as the evidence needed to be expounded upon. Researchers and diagnosticians at ISU-VDL were integral in helping to characterize the pathogen with development of diagnostic tools and a diagnostic survey of clinical vs. non-clinical litters. Diagnostic tools included development of a PSaV specific PcR and in situ hybridization (ISH) to establish presence of PSaV in enterocytes. Upon reviewing the significant results of the survey, Merck Animal Health (MAH) joined the collaboration and further invested expertise in developing and leading an additional clinical survey that indicated the negative effects of PSaV equated to an average of 1.47 less pounds of wean weight.[2]

In the meantime, other similar diarrhea cases within the practice's clients were reviewed with renewed scrutiny. This particular presentation of diarrhea was commonly noted within other herds, although it was on a more sporadic basis. Samples were obtained from several sites with the characteristic timing and character of diarrhea present. A variety of different outcomes were obtained, but essentially, the composite of those cases showed either coccidia organisms, rotavirus strains, or PSaV as the sole pathogen or co-pathogen involved. All, but one of those cases involved presence of PSaV. With the information gathered from the original operation, together with finding PSaV in several other similar diarrhea cases across multiple businesses and farms, it became clear that a solution was needed.

Search for a solution

Since addressing sanitation, environment, water, and co-infections as a means to abate the clinical signs was not effective, feedback was administered in an attempt to create some degree of maternal protection to the piglets. Unfortunately, this was unsuccessful at improving the clinical picture. Potential vaccine options were limited since the organism is very difficult to grow in cell culture. Therefore, a request was made to MAH to consider the possibility of developing an effective vaccine using their RNA-particle platform. The project went through several phases of trial and showed success with vaccine made available on a prescription basis to the industry in 2021.[2]

Cooperative success

The journey from problem identification to solution has truly been a cooperative success. In February of 2019, PSaV was far from a household name in the swine industry. Today it is part of the typical rule out list for mid to late lactation piglet diarrhea cases. PSaV is not a high mortality disease. In addition, pre-wean diarrhea interventions for the commonly understood diseases of coccidiosis and rotaviral enteritis can be frustrating to control. Considering those two facts, it would be very plausible that swine producers and veterinarians would still be naïve to the current understanding of PSaV, and its role were it not for relationship between producer, veterinarian, diagnostic laboratory, and industry.

Next steps

PSaV knowledge and understanding of its effects on piglets in lactation has disseminated over the last couple of years. More and more interest from fellow veterinarians and producers has been received with each passing day. PSaV vaccination throughout the industry is steadily increasing with similar outcomes to that experienced in the original herd of diagnosis. Considering the relatively rapid development of an effective vaccination solution, a lot has been left behind that needs to be understood with PSaV infections in the field, to include dynamics of infection in the nursery phase and a model of piglet exposure processes as well. There is no doubt that this can be accomplished with continued cooperation at all levels to ultimately do what is best for the pig itself.

References

1. Nagai, M., Wang, Q., Oka, T., Saif,L. 2020. Porcine sapovirus: Pathogenesis, epidemiology, genetic diversity, and diagnosis. Virus Res. **doi: 10.1016/j.virusres.2020.198025**

2. Thacker B., et al. 2022. Porcine sapovirus: Field experiences with a potentially emerging pathogen. 53rd Annual Meeting of the AASV, 2022. **doi:10.54846/am2022/72**

https://doi.org/10.54846/am2023/170

Remind me about *Actinobacillus pleuropneumoniae* (APP) again: APP experiences, control, and elimination

Jonathan Van Blarcom, VMD

Introduction

APP is a well-known swine pathogen that has historically been part of the swine respiratory disease complex. The pathogen has been largely eliminated in recent years from modern hog production in North America, especially in modern large scale production systems. While the bacteria is less prevalent than in the past, it persists within certain populations. APP can be transmitted from one herd to another without excellent biosecurity practices. With the aid of great team members, we have been able to eliminate the bacteria from over 100,000 sow spaces in different parts of the world. This presentation will be an attempt to review the pathogen, recent experiences treating within a herd and eliminating the bacteria from production systems.

The pathogen

APP is a gram negative, non-spore forming, non-motile, coccobacillary rod bacterium. There are two biotypes and at least 15 described serotypes worldwide. Serotypes 1,3,5,7 are historically the most prevalent in North America. The bacteria produces 4 exotoxins (ApxI, ApxII, ApxIII, ApxI) as well as endotoxin. The main route of transmission is vertical from sow to pig in the farrowing crate. Lateral transmission is also possible, especially via fomites. The majority of clinical signs are seen anywhere from late nursery to marketing age. Clinical signs include dyspnea, acute deaths, dog sitting, deep cough, blood-tinged snouts. Upon necropsy, findings include frothy blood tinged foam in the nose and respiratory tract of acutely dead pigs, trachea and bronchi filled with blood tinged mucus, adhesions between the lung and thoracic wall, fibrinohemorrhagic necrotizing pneumonia, pleuropneumonia, and severe pleuritis. More chronic pigs often have fibrous lung adhesions, pericarditis, and stunted growth. Mortalities can be dramatic in acute outbreaks in commercial finishing barns often reaching over 10-15% of the population in just a couple weeks. APP has a large economic impact through increased medication costs, elevated mortality, decreased meat quality, more condemns at market, reduced FCR and ADG, lower so productivity, and increased gilt cost.

Recent APP experiences

Several Midwest production systems have described lateral introductions into wean to market sites in recent years leading up to 2023. Reports from outbreak indicate similar disease pathogenesis to historic APP outbreaks. Investigations into the cause of the lateral transmission and herd introduction often lead to failures in routine biosecurity. The consensus from recent presentations and discussions is that fomites, people movement, equipment sharing, transportation, and rendering services are the main causes of the outbreaks.

I've had the opportunity early in my veterinary career to work in 2 different systems in 2 very different parts of the world that were APP positive. The first was in central Iowa with over 25,000 positive sows and the second system was in Russia with over 80,000 positive sows.

Acute APP outbreaks within known positive wean to market flows are often triggered by another pathogen, stress event, animal movement, marketing, or environmental changes. APP in combination with any of *Mycoplasma hyopneumoniae* (Mhp), PRRSV, or Influenza A often lead to dramatic acute outbreaks. The most elevated mortality in my experiences were in mid to late finishing in mycoplasma positive herds. APP breaks would be concurrent with Mhp circulation within an unstable finishing flow.

Control measures

Immediate action with antibiotics is critical to controlling the increasing mortality. Oral medications should be used in affected airspaces. Entire pens of pigs should receive an injectable antibiotic if a single pig within that pen is showing clinical signs for APP disease. Florfenicol has been a go to drug of choice in most outbreaks, but enrofloxacin, ceftiofur, amoxicillin, tiamulin have also shown to be effective at controlling disease. There are commercially available vaccines, but desired control was not accomplished in any of my experiences.

Elimination

Elimination cannot be accomplished with medication or vaccination. The positive source sow farm must undergo a depopulation (depop). Depopulating a sow farm and bringing in APP negative sows is a huge financial investment. Our team was pushed to find a way to control the costs of a depop by only having 2 weeks without weaning pigs from the sow farm. By using the methods described below, our teams have been able to accomplish a rolling depop of APP on 21 of the 21 attempted different sow farms.

Methods

- You must conduct an offsite breeding project that lines up to farrow immediately following the planned last wean from the APP positive sow farm.
  - The final wean of pigs from the APP positive sows should be done early (14-17 days of age) and with medication on board. You should refer to other medicated early wean strategies to ensure best possible nursery performance, but tulathromycin and ceftiofur were used in our eliminations.
  - Your nursery and nutrition must be set up and balanced to be ready for these small and young pigs.
- APP-positive wean sows should walk directly from farrowing rooms to a live haul trailer. They need to go through any withdrawal times at a different location.

- You will need to strategically set up 2 different areas in the farm for animals. You will want to have one area for the clean sows and one area for the APP positive sows that are to be depopulated. Theses areas must always be in different airspaces. They should not share the same building. You need two farm entry locations, two boot areas, two sets of equipment, two load out chutes, two shower/changing areas, etc.
- We chose to medicate APP positive sows to reduce bacterial shedding and reduce the risk of transmission. Tilmicosin in lactation feed and Tiamulin water soluble were both utilized.
- You will work your way through the buildings in a methodical way to walk the positive sows out of the barn.
- We used visual graphics as maps (Figure 1) to help the farm staff better understand where the different zones were. Theses are easy to create in excel using shaded colors and text labels. The production leader should work directly with the veterinarian to establish and move these zones as new areas of the farm are depopulated and cleaned. Your goal should be to make the APP-negative clean/APP positive-dirty demarcations as clear as possible for all levels of education.
- Thorough washing and disinfection needs to be performed once APP positive animals have been moved from a designated area in the farm. Wash and disinfect all areas immediately following sow removal and move your clean/dirty boundary accordingly.
- The clean/dirty areas should be separated with a physical barrier. Sows don't care about your planning or your graphics. If you have doors between buildings, you should close them, lock them, and bolt or screw them fast until you want them open. If you don't have doors between buildings, you should build a wall. A hammer drill, tap cons, some 2x4's and plywood work perfectly to build temporary walls in hallways separating

Figure 1: Simple graphic of sow farm with different airspaces outlined for separation. Graphics can be shared with staff and used as a guide to ensure teams can visualize the exact locations.

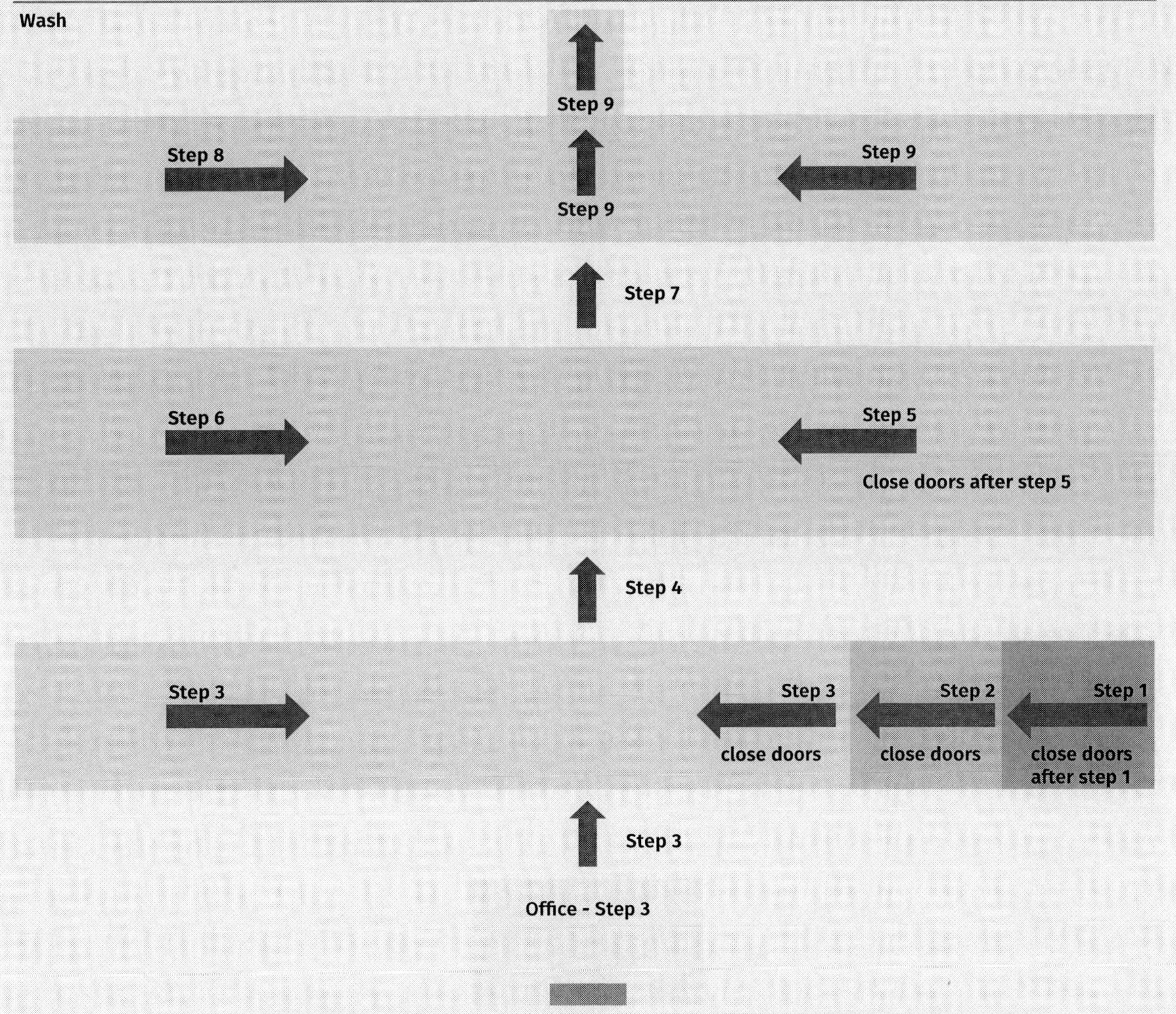

airspaces, and can be constructed in just a few minutes.

- Once entire airspaces have been completely washed and disinfected, they were given a week to dry and sit empty without pigs.
- We used a separate temporary/mobile loading chute to enter the APP negative sows into the clean building. The farm staff were split into two groups. They were fully separated with a completely different set of equipment, clothing, and footwear. As more airspaces on the farm become clean, you can bring in a larger number of your negative population. Keep in mind ideal pregnant sow movement times to avoid moving very heavy pregnant sows and sows early in gestation before embryo implantation occurs.

Conclusion

APP can be eliminated from a sow farm via a rolling depop. It requires careful planning, attention to detail, and a well-trained staff. Execution of the plan requires leadership and problem solving to ensure the plan can easily be followed. If you make it too difficult, the risk of failure is significant. However, if you make the "right way" also the easy way, your chance of success is quite high.

References

1. Swine Disease Manual 4th edition.
2. Diseases of Swine 10th edition.

https://doi.org/10.54846/am2023/171

A global update on African swine fever

K. A. Havas, DVM, PhD, MS, DACVPM (Epi)
Pipestone Research, Pipestone Holdings, Pipestone, Minnesota

Global disease distribution, epidemiology, and presentation

African swine fever was discovered in the early 1900s in Kenya after the introduction of European pigs to the region.[1] To date, 24 genotypes have been discovered based on sequencing of the p72 gene.[2,3] Since its discovery, it has been detected twice outside of Africa before starting regional and global panzootics. A genotype I strain found its way to Portugal in 1957 and was eradicated. It appeared again in 1960, most likely via plane and ship waste and then spread to Spain, Sardinia in Italy, where it is still endemic, and to other European countries.[4] It also reached Cuba in the 1970s and 1980s,[5] Brazil,[6] and Hispaniola.[4] It was eliminated from the Americas in the early 1980s and from Europe in the late 1995.[7]

Europe

A genotype II African swine fever virus has been circulating outside of Africa since 2007 when it was first recognized in the country of Georgia.[8] It quickly spread to neighboring countries, including into Russia[9] and then by 2014 it was found in numerous locations in Eastern Europe.[10] Five years later it was widespread across Eastern Europe and since has been diagnosed in Italy (outside of Sardinia), Germany, Belgium, and other countries in Europe. Figure 1 shows outbreaks from January 1st to November 11th, 2022.

Wild boars play a significant role in transmission dynamics in Europe.[11] Transmission occurred between domestic pigs and wild boar populations either through direct contact associated with poor biosecurity, or inappropriate disposal of dead carcasses. Wild boars maintained the infection in their population through direct contact but also via contact with carcasses[12,13] and contaminated environments.[13,14] The virus can likely persist in cold temperatures for many months in tissues and organs.[15] Wild boar migration and human activities spread the virus from Russia and across Eastern Europe. Some countries have only recorded cases in wild boar.[11]

Asia

In August 2018, ASF was detected in China for the first time.[16] Then it spread into Mongolia and took only an additional eight months to spread into eight countries (Vietnam, Cambodia, North Korea, Laos, Philippines, Myanmar, South Korea, and Timor-Leste.[17] Now it is found throughout Asia and Oceania (Figure 2). It is thought that a single introduction into the region occurred because of the high homology between viral isolates.[16,18] Further, the virus has high homology to isolates out of Europe, which confirms that emergence in this region is an extension of the European epizootic.[16] Today it is found widely across the region. The major source of transmission in China has been movement of infected pigs, feeding of swill, and contaminated fomites and vectors.[19] It is believed that low biosecurity on small farms, movement of infected pigs and pig products, and swill feeding contributed to spread in the region.[17]

Americas

ASFV was detected in the Dominican Republic in July 2021[20] with reports that disease was present as early as May 2021.[21] Soon thereafter, in September 2021, Haiti reported its first case. The USDA FADDL has published two research studies on the Hispaniola virus since this time. Whole genome sequencing of 73 viral isolates from 18 provinces collected between May and October 2021 found that the virus was most closely related to the Eastern European strains from between 2016 - 2018, but it seemed to be missing single nucleotide polymorphisms that have been reported in European strains since 2016. The DR isolates also having unique SNPs of their own.[21] Further, two genetic clusters seem to have diverged from a common ancestor since their introduction in the Dominican Republic. An unanswered question is whether the unique SNPs are related to the field observations related to reduced virulence. Yet, another study evaluated the low virulence field observations by assessing clinical presentation based on inoculation route when using an isolate from July 2021.[22] Three routes were evaluated: intra-muscular, intra-nasal, and via contact with infected pigs. They found that pigs infected via intranasal inoculation and contact with infected pigs had varying disease presentations with some animals experiencing acute disease and death, others prolonged disease that ended in death between 14 to 16 days post-infection and finally some pigs survived the 28-day observation period. All pigs infected by intramuscular inoculation experienced acute disease and died or were euthanized for animal welfare reasons within seven days.[22] The Dominican Republic now has a unique virus and that may be related to the change in disease severity, but that connection requires more investigation. Little additional information has arisen from Haiti, but at the US Animal Health Association conference during the Foreign and Emerging Disease Committee Meeting it was stated that Haiti has limited capacity to deal with the virus currently.

The emergence of ASFV in the Caribbean has led to collaboration in the region. A group called Peste Porcina Africana LaTam was formed to coordinate activities to prevent the spread of ASFV in the region. This group involves 22 organizations in 18 countries.[23] The International Services and Veterinary Services arms of the USDA Animal and Plant Health Inspection Services along with international organizations, such as Organismo Internacional Regional de Sanidad Agropecuaria (OIRSA), the Food and Agricultural Organization of the United Nations (FAO), and the World Organization for Animal Health (WOAH) continue to be involved in the ASF control effort.

Africa

While there have been 2 genotypes found outside of Africa, all 24 other genotypes have been found in Africa.[3] The first case was reported in Kenya in 1921.[1] Genotype I is found in the North and West of sub-Saharan Africa, IX and/or X in Central and East Africa (Democratic Republic of the Congo, Uganda Tanzania, and Kenya, and II on the East coast of sub-Saharan Africa and Madagascar. Southern Africa has a large variety of genotypes as well.[3] The most recent additions were genotype XXIII found in Ethiopia[24] and XXIV found in Mozambique.[2]

Figure 1: African swine fever outbreaks in domestic pigs (blue) and wild boar (red) between January 1, 2022 and November 9, 2022. Map was created using FAO EMPRES-i+, accessed on November 9, 2022. EMPRES-i+ reports data from national authorities, the World Organization for Animal Health and local FAO officers.

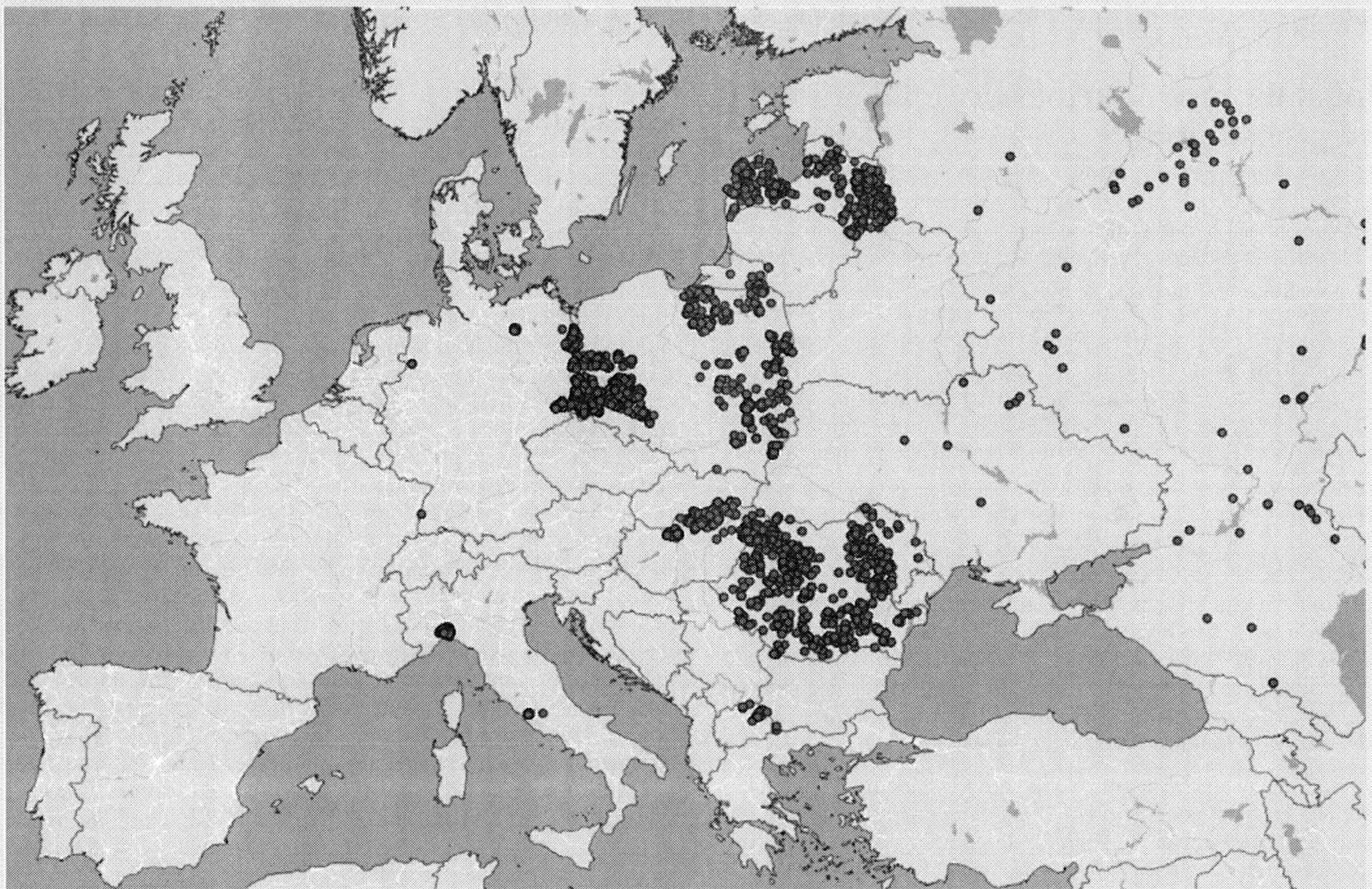

Clinical presentation of ASFV varies across the region with asymptomatic cases reported,[25-29] as well as acute disease.[30,31] Disease is transmitted through direct and indirect contact. Small scale farming with minimal biosecurity is common and contributes to disease spread, as is swill feeding and sale of animals in the face of an outbreak.[30,31] There is also a well-described sylvatic cycle involving wart hogs, their burrows, and *Ornithodoros* spp ticks. *Ornithodoros* ticks act as a competent biological vector for ASFV transmission. They are a threat worldwide but have documented transmission in sub-Saharan Africa.[32-34] Transmission of virus from Africa to other continents has been associated with airline and shipping waste that was improperly discarded.[4,6,8,35,36]

Scientific updates

Attenuated strains

African swine fever can present acutely as a hemorrhagic disease of pigs causing high morbidity and mortality. Infected pigs present as depressed, off feed, and febrile, with cyanotic ears, reddened skin and ecchymoses across the abdomen and back legs. It is also common to find conjunctivitis, diarrhea, and respiratory signs; neurologic signs have also been reported. Often the incubation period is 4 - 19 days and animals die within a week of starting clinical signs.[37,38] In recent years, attenuated ASFV genotype II strains have been reported in Europe[39] and the Dominican Republic.[22] While China reported attenuated genotype I[40] and II.[41] In China, the attenuated variants are thought to be both naturally occurring and from illegally produced and used vaccines.[42,43] From Pipestone's own experience in China, vaccines were originally widely used and were found to cause chronic reproductive issues in sows. After this was discovered, they were mostly used in finishing hogs, although it is believed that use has decreased although. Nonetheless, gene-deleted variants that resemble vaccine strains are still detected.[43] East Africa also reports attenuated strains of various genotypes.[25,26,29]

Attenuated ASFV strains are more difficult to detect. Reduced detection of viral nucleic acid occurs in blood,[44,45] oropharyngeal swabs,[44,45] fecal swabs,[44] tonsil scrapings,[45]and oral fluids.[45] Yet, they antibodies against attenuated strains are readily detectable.[44,45] This aligns with field observations that Pipestone has had in Chinese sites infected with attenuated strains.

Vaccine updates

In 2020, the Foreign Animal Disease Research Unit (FADRU) of the Agricultural Research Service, an arm of the United States Department of Agriculture reported on a gene-deleted ASFV vaccine candidate, ASFV-G-I177L. The deleted gene was previously undescribed, but its deletion resulted in a strain that did

Figure 2: African swine fever outbreaks in domestic pigs (blue), wild boar (red) and environmental samples (orange) between January 1, 2022 and November 9, 2022. Map was created using FAO EMPRES-i+ , accessed on November 9, 2022. EMPRES-i+ reports data from national authorities, the World Organization for Animal Health and local FAO officers.

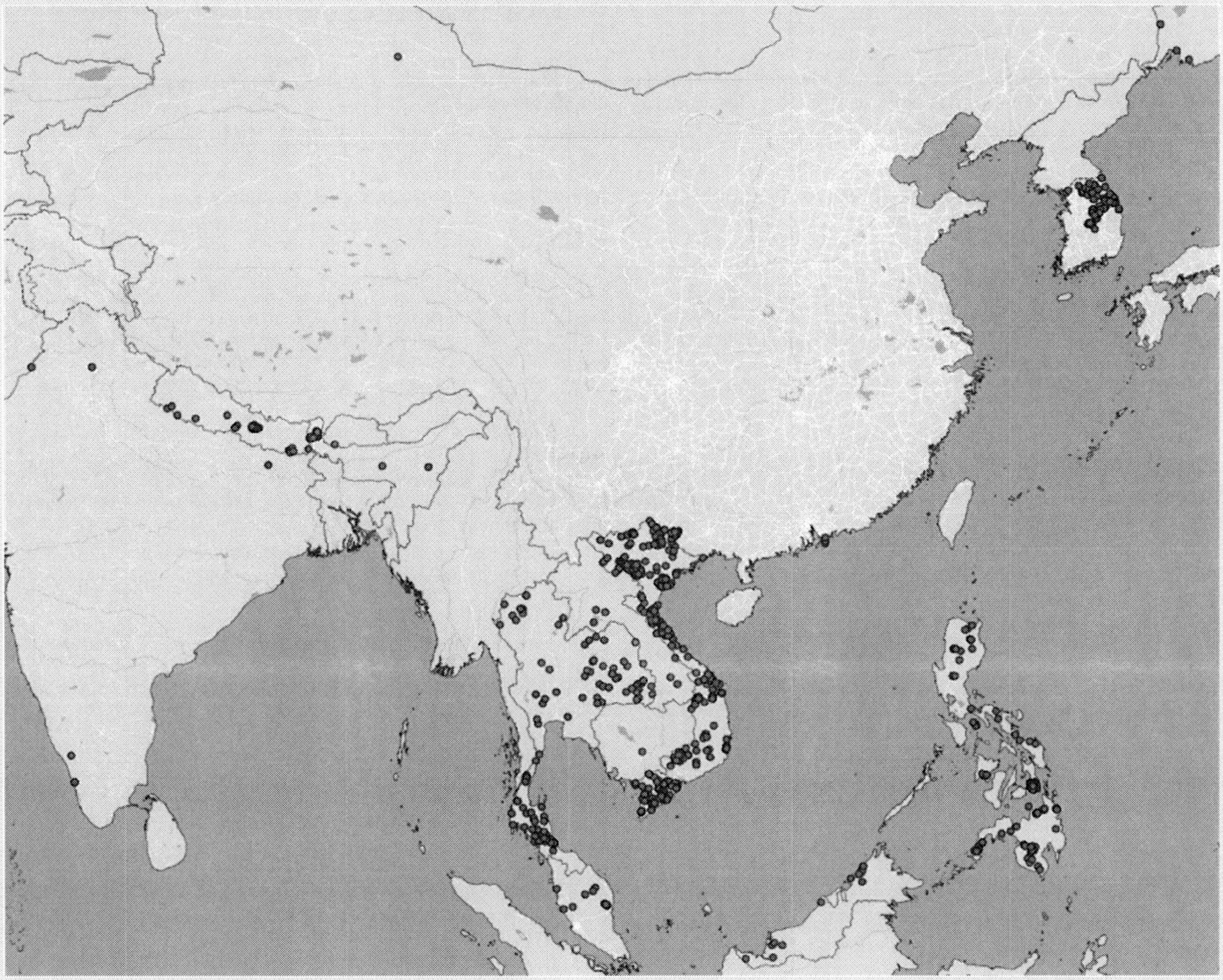

not cause disease or shedding. Inoculated pigs also showed a strong antibody response and sterile immunity at medium to high doses.[46] It was also reported that oronasal inoculation of this strain also provided immunity, suggesting it could be used as a bait vaccine.[47] ASFV is notoriously difficult to culture as it requires use of primary alveolar macrophages, a primary cell line. In 2021, FADRU reported the creation of the ASFV-G-I177L/LVR strain that could replicate in Plum Island porcine epithelial cells (PIPEC).[48] The next steps of evaluation occurred in collaboration with the National Veterinary Joint Stock Company (NAVETCO) in Vietnam. The studies found that when passaged the ASFV-G-I177L remained attenuated without any genetics changes, that the virus did not revert to virulence in serially infected pigs, it was safe to use at low and high doses without causing ASFV in European and Vietnamese pigs, and that in contact animals could be infected with the vaccine strain from infected animals.[49] A final study showed that European and Vietnamese pigs were protected from disease within four weeks post vaccination.[50]

The first commercial ASFV vaccine is now in use in Vietnam, but one cannot distinguish vaccinated from infected animals. Efforts have been made to identify a viral protein that elicits a strong humoral immune response while having no impact on virulence. The MGF110-5L-6L was such a candidate gene as it is highly conserved across variants, elicits a strong gene specific immune response along with the p30 gene that is often used in commercial indirect ELISA assays, and does not reduce virulence or replication when deleted from the virus The ASFV-G-I177L/MGF110-5L-6L gene deleted variant is a potential vaccine variant that could allow differentiation of vaccinated from infected animals.[51]

There were reports during the summer of 2022 that 730 vaccinated pigs of the more than 23,000 vaccinated had unexpectedly died. The vaccines used met testing and quality control, with quality defined as meeting sterility, safety, and potency. An investigation revealed that pigs that died were part of groups that were improperly vaccinated and could have been vaccinated after being exposed to African swine fever or other disease. The pigs that were properly vaccinated did not have excessive vaccine-associated deaths. Vietnam is moving forward with vaccination of 600,000 pigs.[52] There also reported the spread of ASFV due to sharing needles when pigs were improperly vaccinated.[53]

According to the website, the VACDIVA project (vacdiva.eu) is a European Union based consortium that is led by Dr. Jose Manuel Sanchez-Vizcaino, partnering with Ingenasa and Merck, and aims to develop three vaccines for wild boar and domestic pigs along with complementary DIVA test (accessed Nov 11, 2022). This program is developing vaccines using naturally attenuated strains: NH/P68, Lv17/WB/Riel, and ASF/ARRIAH/CV-1. The NH/P68 strain is an attenuated, non-hemagglutinating genotype I strain.[54,55] The NH/P68 has been evaluated with further gene deletions and was limited in its protectivity and retained some virulence.[56] The Lv17/WB/Reil strain has shown promise as an oral bait vaccine for wild boar[57] and safety studies showed promise but require more work.[58] The ARRIAH/CV-1 strain is a ASFV strain adapted to the CV-1 cell line,[59] a cell line from the kidney of African green monkeys (https://www.atcc.org/products/ccl-70; accessed Nov 11, 2022). It exhibited reduced virulence and protection against homologous viruses.[59] Other vaccines, such as subunit vaccines and inactivated vaccines are also being developed.[55]

References

1. Montgomery E.R. On a form of swine fever occurring in British East Africa (Kenya Colony) *J. Comp. Pathol. Therap.* 1921;34:159–191. **doi: 10.1016/S0368-1742(21)80031-4**

2. Quembo CJ, Jori F, Vosloo W, Heath L. Genetic characterization of African swine fever virus isolates from soft ticks at the wildlife/domestic interface in Mozambique and identification of a novel genotype. Transbound Emerg Dis. 2018 Apr;65(2):420-431. **doi: 10.1111/tbed.12700**

3. Njau, E.P.; Machuka, E.M.; Cleaveland, S.; Shirima, G.M.; Kusiluka, L.J.; Okoth, E.A.; Pelle, R. African Swine Fever Virus (ASFV): Biology, Genomics and Genotypes Circulating in Sub-Saharan Africa. *Viruses* 2021, 13, 2285.

4. Costard S., Wieland B., Glanville W., Jori F., Rowlands R., Vosloo W., Roger F., Pfeiffer D.U., Dixon L.K. African swine fever: how can global spread be prevented? *Philos. Trans. R. Soc. B Biol. Sci.* 2009;27:2683–2696. **doi: 10.1098/rstb.2009.0098**

5. Simeón-Negrín RE, Frías-Lepoureau MT. Eradication of African Swine Fever in Cuba (1971 and 1980. In: *Trends in Emerging Viral Infections of Swine*. Eds. Morilla A, Yoon KJ, Zimmerman JJ. 2002. 125-132.

6. Lyra TMP. The eradication of African swine fever in Brazil, 1978-1984. *Rev Sci Tech.* 2006; 25(1):93-103.

7. Costard S., Mur L., Lubroth J., Sanchez-Vizcaino J.M., Pfeiffer D.U. Epidemiology of African swine fever virus. *Virus Res.* 2013a;173:191–197. **doi: 10.1016/j.virusres.2012.10.030**

8. Rowlands, R.J.; Michaud, V.; Heath, L.; Hutchings, G.; Oura, C.; Vosloo, W.; Dwarka, R.; Onashvili, T.; Albina, E.; Dixon, L.K. African Swine Fever Virus Isolate, Georgia, 2007. *Emerg. Infect. Dis.* 2008, 14, 1870–1874.

9. Oganesyan AS, Petrova ON, Korennoy FI, Bardina NS, Gogin AE, Dudnikova SA. African swine fever in the Russian federation: spatio-temporal analysis and epidemiological overview. *Virus Res.* (2013) 173:204–11. **doi: 10.1016/j.virusres.2012.12.009**

10. Gallardo C, Fernández-Pinero J, Pelayo V, Gazaev I, Markowska-Daniel I, Pridotkas G, et al. Genetic Variation among African Swine Fever Genotype II Viruses, Eastern and Central Europe. *Emerg Infect Dis.* 2014;20(9):1544-1547. **https://doi.org/10.3201/eid2009.140554**

11. Sauter-Louis C, Conraths FJ, Probst C, Blohm U, Schulz, K, Sehl J, Fischer M, Forth JH, Zani L, Depner K, Mettenleiter TC, Beer M, Blome S. African swine fever in wild boar in Europe—A review. *Viruses.* 2021; 13(9):1717.

12. Probst C., Globig A., Knoll B., Conraths F.J., Depner K. Behaviour of free ranging wild boar towards their dead fellows: Potential implications for the transmission of African swine fever. *R. Soc. Open Sci.* 2017;4:170054. **doi: 10.1098/rsos.170054**

13. Pepin K.M., Golnar A.J., Abdo Z., Podgorski T. Ecological drivers of African swine fever virus persistence in wild boar populations: Insight for control. *Ecol. Evol.* 2020;10:2846–2859. **doi: 10.1002/ece3.6100**

14. Chenais E., Depner K., Guberti V., Dietze K., Viltrop A., Stahl K. Epidemiological considerations on African swine fever in Europe 2014–2018. *Porc. Health Manag.* 2019;5:6. **doi: 10.1186/s40813-018-0109-2**

15. Fischer M, Hühr J, Blome S, Conraths FJ, Probst C. Stability of African swine fever virus in carcasses of domestic pigs and wild boar experimentally infected with ASFV "Estonia 2014" isolate. *Viruses* 2020; 12(10):1118. **doi:10.3390/v12101118**

16. Ge S, Li J, Fan X, Liu F, Li L, Wang Q, Ren W, Bao J, Liu C, Wang H, Liu Y, Zhang Y, Xu T, Wu X, Wang Z. Molecular Characterization of African Swine Fever Virus, China, 2018. Emerg Infect Dis. 2018 Nov;24(11):2131-2133. **doi: 10.3201/eid2411.181274**

17. Lu G, Pan J, Zhang G. African swine fever virus in Asia: Its rapid spread and potential threat to unaffected countries. *J Infect.* 2020; 80(3): 350-371.

18. Le VP, Jeong DG, Yoon SW, Kwon HM, Trinh TBN, Nguyen TL, Bui TTN, Oh J, Kim JB, Cheong KM, Van Tuyen N, Bae E, Vu TTH, Yeom M, Na W, Song D. Outbreak of African Swine Fever, Vietnam, 2019. Emerg Infect Dis. 2019;25(7):1433-1435. **doi: 10.3201/eid2507.190303**

19. Wu K, Liu J, Wang L, Fan S, Li Z, Li Y, Yi L, Ding H, Zhao M, Chen J. Current state of global African swine fever vaccine development under the prevalence and transmission of ASF in China.*Vaccines (Basel).* 2020;8(3):531.

20. Gonzales W, Moreno C, Duran U, Henao N, Bencosme M, Lora P, Reyes R, núñez R, De Gracia A, Perez AM. African swine fever in the Dominican Republic. *Trans Emerg Dis.* 2021;68:3018-3019.

21. Lakin SM, O'Donnell VK, Xu L, Barrette RW, Barnabei J, et al. Whole genome sequencing and molecular epidemiology of the 2021 African swine fever virus outbreak in the Dominican Republic. *Trans Emerg Dis.* 2022. **doi:10.1111/tbed.14751**

22. Ramirez-Medina E, O'Donnell V, Silva E, Espinoza N, Velazquez-Salinas L, et al. Experimental Infection of Domestic Pigs with an African Swine Fever Virus Field Strain Isolated in 2021 from the Dominican Republic. *Viruses* 2022a, 14, 1090. **https://doi.org/10.3390/v14051090**

23. National Pork Producers Council (NPPC). National Pork Producers Council joins ASF Prevention America. *National Hog Farmer.* January 21, 2022. Accessed on November 9, 2022. Available at: **https://www.nationalhogfarmer.com/news/national-pork-producers-council-joins-asf-prevention-america**

24. Achenbach J.E., Gallardo C., Nieto-Pelegrín E., Rivera-Arroyo B., Degefa-Negi T., Arias M., Jenberie S., Mulisa D.D., Gizaw D., Gelaye E., et al. Identification of a New Genotype of African Swine Fever Virus in Domestic Pigs from Ethiopia. *Transbound. Emerg. Dis.* 2017;64:1393–1404. **doi: 10.1111/tbed.12511**

25. Atuhaire D.K., Afayoa M., Ochwo S., Mwesigwa S., Mwiine F.N., Okuni J.B., Olaho-Mukani W., Ojok L. Prevalence of African Swine Fever Virus in Apparently Healthy Domestic Pigs in Uganda. *BMC Vet. Res.* 2013;9:263. **doi: 10.1186/1746-6148-9-263**

26. Uttenthal Å, Braae UC, Ngowi HA, Rasmussen TB, Nielsen J, Johansen MV. ASFV in Tanzania: Asymptomatic Pigs Harbor Virus of Molecular Similarity to Georgia 2007. *Vet. Microbiol.* 2013;165:173–176. **doi: 10.1016/j.vetmic.2013.01.003**

27. Thomas LF, Bishop RP, Onzere C, Mcintosh MT, Lemire KA, de Glanville WA, Cook EAJ, Fèvre EM. Evidence for the Presence of African Swine Fever Virus in an Endemic Region of Western Kenya in the Absence of Any Reported Outbreak. *BMC Vet. Res.* 2016;12:192. **doi: 10.1186/s12917-016-0830-5**

28. Abworo EO, Onzere C, Oluoch Amimo J, Riitho V, Mwangi W, Davies J, Blome S, Bishop RP. Detection of African Swine Fever Virus in the Tissues of Asymptomatic Pigs in Smallholder Farming Systems along the Kenya–Uganda Border: Implications for Transmission in Endemic Areas and ASF Surveillance in East Africa. *J. Gen. Virol.* 2017;98:1806–1814. **doi: 10.1099/jgv.0.000848**

29. Chang'a J.S., Mayenga C., Settypalli T.B.K., Achenbach J.E., Mwanandota J.J., Magidanga B., Cattoli G., Jeremiah M., Kamigwe A., Guo S., et al. Symptomatic and Asymptomatic Cases of African Swine Fever in Tanzania. *Transbound. Emerg. Dis.* 2019;66:2402–2410. **doi: 10.1111/tbed.13298**

30. Mulumba-Mfumu LK, Saegerman C, Dixon LK, et al. African swine fever: Update on Eastern, Central and Southern Africa. *Transbound Emerg Dis.* 2019;00:1–19.

31. Penrith ML, Bastos A, Chenais E. With or without a Vaccine-A Review of Complementary and Alternative Approaches to Managing African Swine Fever in Resource-Constrained Smallholder Settings. Vaccines (Basel). 2021 Feb 2;9(2):116. **doi: 10.3390/vaccines9020116**

32. Plowright, W., Parker, J. A., & Peirce, M. A. (1969a). African swine fever virus in ticks (*Ornithodoros moubata* Murray) collected from animal burrows in Tanzania. *Nature. 221*:1071–1073. **https://doi.org/10.1038/2211071a0**

33. Plowright, W., Parker, J., & Peirce, M. A. (1969b). The epizootiology of African swine fever in Africa. *Veterinary Record. 85*:668–674.

34. Wilkinson P.J. The persistence of African swine fever in Africa and the Mediterranean. *Prev. Vet. Med.* 1984;2:71–82. **doi: 10.1016/0167-5877(84)90050-3**

35. Lubisi BA, Dwarka RM, Meenowa D, Jaumally R (2009) An investigation into the first outbreak of African swine fever in the Republic of Mauritius. Transbound Emerg Dis 56: 178–188.

36. Costard S, Jones BA, Martínex-López B, Mur L, de la Torre A, Martínez M, Sánchez-Vizcaíno F, Sánchez-Vizcaíno JM, Pfeiffer DU, Wieland B. Introduction of African swine fever into the European Union through illegal importation of pork and pork products. *PLoS ONE.* 2013b; **doi.org/10.1371/journal.pone.0061104**

37. Salguero FJ. Comparative Pathology and Pathogenesis of African Swine Fever Infection in Swine. Front Vet Sci. 2020;7:282. **doi: 10.3389/fvets.2020.00282**

38. World Organization for Animal Health. Chapter 3.9.1 African swine fever (infection with African swine fever virus). In: *Manual of Diagnostic Test and Vaccines for Terrestrial Animals.* 2022 Accessed November 14, 2022. Available at **https://www.woah.org/fileadmin/Home/eng/Health_standards/tahm/3.09.01_ASF.pdf**

39. Gallardo C, Soler A, Rodze I, Nieto R, Cano-Gómez C, Fernandez-Pinero J, Arias M. Attenuated and non-haemadsorbing (non-HAD) genotype II African swine fever virus (ASFV) isolated in Europe, Latvia 2017. *Tran Emerg Dis.* 2019; 66(3):1399-1404.

40. Sun E, Huang L, Zhang X, Zhang J, Shen D, Zhang Z, Wang Z, Huo H, Wang W, Huangfu H, Wang mW, Li F, Liu R, Sun J, Tian Z, Xia W, Guan Y, He X, Zhu Y, Zhao D, Bu Z. Genotype I African swine fever viruses emerged in domestic pigs in China and caused chronic infection. *Emerg Microbes Infect.* 2021a. 10:1, 2183-2193, **DOI: 10.1080/22221751.2021.1999779**

41. Sun E, Zhang Z, Wang Z, et al. Emergence and prevalence of naturally occurring lower virulent African swine fever viruses in domestic pigs in China in 2020. Sci. *China Life Sci.* 2021b; 64: 752–765. **doi.org/10.1007/s11427-021-1904-4**

42. Patton D. New China swine fever strains point to unlicensed vaccines. Reuters. January 21, 2021. Accessed on November 11, 2022. Available at **https://www.reuters.com/article/us-china-swinefever-vaccines-insight/new-china-swine-fever-strains-point-to-unlicensed-vaccines-idUSKBN29R00X**

43. Zhao K, Shi K, Zhou Q, Xiong C, Mo S, Zhou H, Long F, Wei H, Hu L, Mo M. The Development of a Multiplex Real-Time Quantitative PCR Assay for the Differential Detection of the Wild-Type Strain and the MGF505-2R, EP402R and I177L Gene-Deleted Strain of the African Swine Fever Virus. *Animals.* 2022; 12:1754. **doi.org/10.3390/ani12141754**

44. Gallardo C., Soler A., Nurmoja I., Cano-Gómez C., Cvetkova S., Frant M., Woźniakowski G., Simón A., Pérez C., Nieto R., et al. Dynamics of African Swine Fever Virus (ASFV) Infection in Domestic Pigs Infected with Virulent, Moderate Virulent and Attenuated Genotype II ASFV European Isolates. *Transbound. Emerg. Dis.* 2021;68:2826–2841. **doi: 10.1111/tbed.14222**

45. Havas KA, Gogin AE, Basalaeva JV, Sindryakova IP, Kolbasova OL, Titov IA, Lyska VM, Morgunov SY, Vlasov ME, Sevskikh TA, Pivova EY, Kudrjashov DA, Doolittle K, Zimmerman S, Witbeck W, Gimenez-Lirola LG, Nerem J, Spronk GD, Zimmerman JJ, Sereda AD. An Assessment of Diagnostic Assays and Sample Types in the Detection of an Attenuated Genotype 5 African Swine Fever Virus in European Pigs over a 3-Month Period. Pathogens. 2022 Mar 26;11(4):404. **doi: 10.3390/pathogens11040404**

46. Borca MV, Ramirez-Medina E, Silva E, Vuono E, Rai A, Pruitt S, Holinka LG, Velazquez-Salinas L, Zhu J, Gladue DP. Development of a highly effective African swine fever virus vaccine by deletion of the I177L gene results in sterile immunity against the current epidemic Eurasia strain. *J Virol* 2020 94:e02017-19. **https://doi.org/10.1128/JVI.02017-19**

47. Borca MV, Ramirez-Medina E, Silva E, Vuono E, Rai A, Pruitt S, Espinoza N, Velazquez-Salinas L, Gay CG, Gladue DP. ASFV-G-I177L as an Effective Oral Nasal Vaccine against the Eurasia Strain of Africa Swine Fever. *Viruses* 2021a, 13, 765. **https://doi.org/10.3390/v13050765**

48. Borca MV, Rai A, Ramirez-Medina E, Silva E, Velazquez-Salinas L, Vuono E, Pruitt S, Espinoza N, Gladue DP. A cell culture adapted vaccine virus against the current African swine fever virus pandemic strain. *J Virol* 2021b 95:e00123-21. **https://doi.org/10.1128/JVI.00123-21**

49. Tran XH, Phuong LTT, Huy NQ, Thuy DT, Nguyen VD, Quang PH, Ngôn QV, Rai A, Gay CG, Gladue DP, Borca MV. Evaluation of the Safety Profile of the ASFV Vaccine Candidate ASFV-G-ΔI177L. *Viruses.* 2022a;14(5):896. **doi: 10.3390/v14050896**

50. Tran XH, Le TTP, Nguyen QH, Do TT, Nguyen VD, Gay CG, Borca MV, Gladue DP. African swine fever virus vaccine candidate ASFV-G-ΔI177L efficiently protects European and native pig breeds against circulating Vietnamese field strain. *Transbound Emerg Dis.* 2022b;69(4):e497-e504. **doi: 10.1111/tbed.14329**

51. Ramirez-Medina E, Vuono E, Silva E, Rai A, Valladares A, Pruitt S, Espinoza N, Velazquez-Salinas L, Borca MV, Gladue DP. Evaluation of the deletion of MGF110-5L-6L on swine virulence from the pandemic strain of African swine fever virus and use as a DIVA marker in vaccine candidate ASFV-G--I177L. *J Virol.* 2022b; 96(14): **doi:10.1128/jvi.00597-22**

52. Thang B, Huan N. Special supervision of the African swine fever vaccine. Nong Nghiep Vietnam. Translated: Long NH. August 31, 2022. Accessed November 10, 2022. Available at **https://vietnamagriculture.nongnghiep.vn/special-supervision-of-the-african-swine-fever-vaccine-d331146.html**

53. Dinh T. Stricter control for the vaccination against African swine fever. Nong Nghiep Vietnam. Translated: Long NH. September 7, 2022. Accessed November 10, 2022. Available at **https://vietnamagriculture.nongnghiep.vn/stricter-control-for-the-vaccination-against-african-swine-fever-d331660.html**

54. Leitao A, Cartaxeiro C, Coelho R, Cruz B, Parkhouse RM. Portugal F, et al. The non-haemadsorbing African swine fever isolate ASFV/NH/P68 provides a model for defining the protective anti-virus immune response. *J Gen Virol.* 2001; 82: 513-523.

55. Urbano AC, Ferreira F. African swine fever control and prevention: an update on vaccine development. *Emerg Microbes Infect.* 2022; 11(1): 2021-2033.

56. Gallardo C, Sánchez EG, Pérez-Núñez D, Nogal M, de León P, Carrascosa AL, Nieto R, Soler A, Arias ML, Revilla Y. African swine fever virus (ASFV) protection mediated by NH/P68 and NH/P68 recombinant live-attenuated viruses. *Vaccine.* 2018; 36(19):2694-2704.

57. Barasona JA, Gallardo C, Cadenas-Fernández E, Jurado C, Rivera B, Rodríguez-Bertos A, Arias M, Sánchez-Vizcaíno JM. First oral vaccination of Eurasian wild boar against African swine fever virus genotype II. *Front Vet Sci.* 2019 6:137.

58. Barasona JA, Cadenas-Fernández E, Kosowska A, Barroso-Arévalo S, Rivera B, Sánchez R, Porras N, Gallardo C, Sánchez-Vizcaíno JM. Safety of African swine fever vaccine candidate Lv17/WB/Rie1 in wild boar: overdose and repeated doses. *Front Immunol.* 2021. 12:761753

59. Mazloum A, Igolkin AS, Zinyakov NG, Van Schalkwyk A, Vlasova NN. [Changes in the genome of African swine fever virus (Asfarviridae: Asfivirus: African swine fever virus) associated with adaptation to reproduction in continuous cell culture]. Vopr Virusol. 2021 Jul 9;66(3):211-216. **doi: 10.36233/0507-4088-50.** PMID: 34251158.

https://doi.org/10.54846/am2023/172

What's new? African swine fever vaccine update

Manuel V. Borca[1]; Elizabeth Vuono[1]; Elizabeth Ramirez-Medina[1]; Ayushi Rai[1,2]; Nallely Espinoza[1]; Edward Spinard[1]; Alyssa Valadares[1,2]; Ediane Silva[1]; Lauro Velazquez-Salinas[1]; Douglas P. Gladue[1]
[1]Plum Island Animal Disease Center, ARS, USDA, Greenport, New York;
[2]Oak Ridge Institute for Science and Education (ORISE), Oak Ridge, Tennessee

Introduction

African swine fever (ASF) is a contagious and often fatal viral disease of swine. The causative agent, ASF virus (ASFV), is a large, enveloped virus containing a double-stranded DNA (dsDNA) genome of approximately 190 kbp.[1] ASFV shares aspects of its genome structure and replication strategy with other large dsDNA viruses, including those of the Poxviridae, Iridoviridae, and Phycodnaviridae.[2] However, on a protein or amino acid level, there is little homology with the majority of the viral proteins, and very few ASFV proteins have been evaluated for their functionality or for their contribution to virus pathogenesis.

Currently, ASF is causing a pandemic that has recently spread to the western hemisphere on the island of Hispaniola. Historically ASFV was endemic in sub-Saharan Africa countries. This pandemic started with an outbreak in the Caucasus region in 2007, affecting Georgia, Armenia, Azerbaijan, and Russia. ASF has continued to spread uncontrollably across Eurasia, Sequencing of several contemporary epidemic ASFVs suggests high nucleotide similarity with only minor modifications compared to the initial 2007 outbreak strain, ASFV Georgia 2007/1, a highly virulent isolate that belongs to the ASFV genotype II group.[3]

Recently a live attenuated ASFV vaccine was approved for use in Vietnam, however there is no vaccine available for ASFV for worldwide use, and disease outbreaks are currently quelled by animal quarantine and slaughter. Attempts to vaccinate animals using other vaccine approaches including infected cell extracts, supernatants of infected pig peripheral blood leukocytes, purified and inactivated virions, infected glutaraldehyde-fixed macrophages, or detergent-treated infected alveolar macrophages failed to induce protective immunity.[4-7] Protective immunity develops in pigs surviving viral infection with moderately virulent or attenuated variants of ASFV, with long-term resistance to homologous, but rarely to heterologous, virus challenge.[8,9]

In this presentation, I will discuss the current efforts in commercialization of the ASFV-G-ΔI177L[10] vaccine and the possible introduction of a DIVA marker into current vaccine platforms.

Results

The ASFV-G-ΔI177L vaccine platform has been successfully reproduced in a commercial production facility in Vietnam. The commercially produced vaccine was tested and shown to be effective against a recent outbreak strain of ASFV that was isolated in Vietnam.[11] Importantly the vaccine was also shown to be effective in both local breeds and domestic breeds of swine.[11] The ASFV-G-ΔI177L vaccine has passed reversion to virulence tests,[12] and field trials that were conducted in both North and South Vietnam. Recently the department of Animal Health in Vietnam has approved the use of the vaccine under controlled settings.

The current ASFV-G-ΔI177L vaccine lacks a serological DIVA a component that could be helpful in deploying the vaccine to high risk and uninfected areas. To introduce a serological DIVA serum from ASFV vaccinated swine was used to determine novel ASFV proteins to make an additional deletion in ASFV vaccines. Two separate genetic targets were evaluated viral proteins E184L,[13] and MGF110-5L6L[14] in both cases the introduction of an additional deletion reduced vaccine efficacy.

Discussion

In the field of ASFV very little is known about individual viral protein function, due to this lack of information it is very difficult to predict how adding an additional deletion to an ASFV vaccine platform will affect vaccine production, or vaccine efficacy. An update to our on-going and future efforts towards DIVA markers and subunit vaccines as well as the current status of the deployment of the first generation ASFV vaccine in Vietnam will be further discussed.

References

1. Tulman ER, Delhon GA, Ku BK, Rock DL. 2009. African swine fever virus, p 43–87. In Van Etten JL (ed), Lesser known large dsDNA viruses, vol 328. Springer-Verlag, Berlin, Germany.

2. Costard S, Wieland B, de Glanville W, Jori F, Rowlands R, Vosloo W, Roger F, Pfeiffer DU, Dixon LK. 2009. African swine fever: how can global spread be prevented? Philos Trans R Soc Lond B Biol Sci 364:2683–2696.

3. Chapman DA, Darby AC, Da Silva M, Upton C, Radford AD, Dixon LK. 2011. Genomic analysis of highly virulent Georgia 2007/1 isolate of African swine fever virus. Emerg Infect Dis 17:599–605.

4. Coggins L. 1974. African swine fever virus. Pathogenesis. Prog Med Virol 18:48–63.

5. Forman AJ, Wardley RC, Wilkinson PJ. 1982. The immunological response of pigs and guinea pigs to antigens of African swine fever virus. Arch Virol 74:91–100.

6. Kihm UAM, Mueller H, Pool R. 1987. Approaches to vaccination.

7. Mebus CA. 1988. African swine fever. Adv Virus Res 35:251–269.

8. Hamdy FM, Dardiri AH. 1984. Clinical and immunologic responses of pigs to African swine fever virus isolated from the Western Hemisphere. Am J Vet Res 45:711–714.

9. Ruiz-Gonzalvo FC, Bruyel V. 1981. Immunological responses of pigs to partially attenuated ASF and their resistance to virulent homologous and heterologous viruses. Food and Agriculture Organization of the United Nations, Rome, Italy.

10. Borca, M. V., Ramirez-Medina, E., Silva, E., Vuono, E., Rai, A., Pruitt, S., Holinka, L. G., Velazquez-Salinas, L., Zhu, J., & Gladue, D. P. (2020). Development of a Highly Effective African Swine Fever Virus Vaccine by Deletion of the I177L Gene Results in Sterile Immunity against the Current Epidemic Eurasia Strain. *Journal of virology*, 94(7), e02017-19.

11. Tran, X. H., Le, T., Nguyen, Q. H., Do, T. T., Nguyen, V. D., Gay, C. G., Borca, M. V., & Gladue, D. P. (2021). African swine fever virus vaccine candidate ASFV-G-ΔI177L efficiently protects European and native pig breeds against circulating Vietnamese field strain. *Transboundary and emerging diseases.*

12. Tran, X. H., Phuong, L., Huy, N. Q., Thuy, D. T., Nguyen, V. D., Quang, P. H., Ngôn, Q. V., Rai, A., Gay, C. G., Gladue, D. P., & Borca, M. V. (2022). Evaluation of the Safety Profile of the ASFV Vaccine Candidate ASFV-G-ΔI177L. *Viruses*, 14(5), 896. **https://doi.org/10.3390/v14050896**

13. Ramirez-Medina, E., Vuono, E., Rai, A., Pruitt, S., Espinoza, N., Velazquez-Salinas, L., Pina-Pedrero, S., Zhu, J., Rodriguez, F., Borca, M. V., & Gladue, D. P. (2022). Deletion of E184L, a Putative DIVA Target from the Pandemic Strain of African Swine Fever Virus, Produces a Reduction in Virulence and Protection against Virulent Challenge. Journal of virology, 96(1), e0141921.

14. Ramirez-Medina, E., Vuono, E., Silva, E., Rai, A., Valladares, A., Pruitt, S., Espinoza, N., Velazquez-Salinas, L., Borca, M. V., & Gladue, D. P. (2022). Evaluation of the Deletion of MGF110-5L-6L on Swine Virulence from the Pandemic Strain of African Swine Fever Virus and Use as a DIVA Marker in Vaccine Candidate ASFV-G-ΔI177L. Journal of virology, 96(14), e0059722.

https://doi.org/10.54846/am2023/173

Who can test for foreign animal diseases? Certified Swine Sample Collectors

Pam Zaabel, DVM
National Pork Board

Introduction

Foreign animal disease (FAD) exercises conducted over the past several years have identified the need for additional human resources, specifically, the need for trained individuals on the farms to be able to collect diagnostic and surveillance samples during an FAD outbreak. It will be difficult for the few FAD diagnosticians (FADDs) and swine-focused veterinarians to adequately collect diagnostic and surveillance samples in a timely manner for the large number of swine farms involved. Currently in the United States, swine-focused veterinarians either collect or oversee the collection of samples from swine for diagnostic and surveillance testing for domestic diseases on production sites.

A group of collaborators including the National Pork Board, the American Association of Swine Veterinarians, the Multi-State Partnership for Security in Agriculture, and the Swine Medicine Education Center and the Center for Food Security and Public Health, both located at Iowa State University, received a grant through the USDA National Animal Disease Preparedness and Response Program to develop a training program. Objectives for the Certified Swine Sample Collector (CSSC) training program listed in the grant include developing the programs standards as well as the resources to be used in the training and conducting outreach to stakeholders. The goal of the CSSC program is to increase the number of producers, caretakers, and industry partners who are proficient in sample collection prior to an outbreak to ensure they are prepared to respond as an outbreak unfolds.

This national training program will assure state and federal animal health officials that individuals have been trained by category II accredited veterinarians working in the pork industry to correctly collect, handle, and submit samples, prior to an outbreak. During an FAD outbreak, state animal health officials (SAHOs) will determine when the trained individuals called Certified Swine Sample Collectors (CSSCs) will be authorized to collect samples in their state. Guidance on testing protocols will be provided by animal health officials.

Trainer qualifications and responsibilities

USDA category II accredited veterinarian with swine experience are the trainers, and they must either have a business relationship with the owner of the pigs on farms where individuals are trained or be requested to perform training by the site's category II accredited veterinarian under whom the collectors will be submitting samples or by the state animal health official. The trainers are responsible for conducting the training sessions, following the program standards, and recording and reporting the names and contact information of those who have completed training. Prior to training CSSCs, USDA category II accredited veterinarians are required to contact the animal health officials in the state(s) in which they plan to train or utilize CSSCs to confirm the state has rolled out the program and to discuss any additional requirements that might exist.

Trainee qualifications and responsibilities

Trainees must be approved by a category II accredited veterinarian, are encouraged to have a valid PQA Plus certification, and must successfully complete the training curriculum. These CSSCs are then responsible to properly collect, package, and submit samples when given permission to do so by the SAHO during an FAD outbreak. Samples can only be submitted under a category II accredited veterinarian in the state(s) in which the veterinarian is authorized or submitted under a state animal health official.

CSSCs can collect samples from any swine operation as requested by the accredited veterinarian or a state animal health official under whom the samples will be submitted. In addition, CSSCs can collect samples for multiple category II accredited veterinarians or as requested by a state animal health official. Certification stays with the CSSC if they change employment.

Curriculum

During the classroom instruction, trainers will review the videos and handouts covering the Tier 1 or Tier 2 sample types as well as the disease videos, and resources addressing additional topics such as biosecurity, animal restraint, and sample submission. Following the resource review, each trainee must pass a written examination. Once trainees pass the written examination, they will participate in hands-on training. For all sample types except processing fluid, vesicular fluid, and tracheal swabs, the trainer will demonstrate sample collection techniques. Trainees then will practice these techniques and demonstrate proficiency.

Recertification

Recertification is required annually with a 30-day grace period following the renewal date. Immediate recertification is required if a veterinary diagnostic lab informs the category II accredited veterinarian that submitted samples are deemed unacceptable. Any new curriculum needs to be taught with the same approach as the initial training. If the resources have not been updated since previous training, recertification consists of the CSSC demonstrating competency collecting samples as determined by their trainer.

Funding

Development of the CSSC training program was funded by USDA's Animal and Plant Health Inspection Service through the National Animal Disease Preparedness and Response Program for the two years of the grant period. Resources developed during the third year of the project and Spanish translation of program materials were funded by the Pork Checkoff.

Additional information

Program standards, resources, and a SAHO contact list for states which have rolled out the program are available at www.**securepork.org/cssc**

https://doi.org/10.54846/am2023/174

US Swine Health Improvement Plan, en route to USDA codification in 2024

R. Main[1], DVM, PhD; J. Roth[1], DVM, PhD; J. Christopher-Hennings[2], DVM, PhD; T. Holck[1], DVM, MS; J. Kraft[1], MS; L. Linhares[1], DVM, MS; J. Lowe[3], DVM, MS; C. Rademacher[1], DVM, MS; M. Torremorell[4], DVM, PhD; J. Torrison[5], DVM, PhD; G. Trevisan[1], DVM, PhD; J. Zimmerman[1], DVM, PhD
[1]Iowa State University, Ames, Iowa; [2]South Dakota State University, Brookings, South Dakota; [3]University of Illinois, Urbana-Champaign, Illinois; [4]University of Minnesota, Minneapolis, Minnesota; [5]Longhorn Vaccines and Diagnostics, LLC

Overview

The US Swine Health Improvement Plan (US SHIP) is being modeled after the National Poultry Improvement Plan (NPIP), a collaborative effort involving industry, state, and federal partners providing standards for certifying the health status of greater than 99% of commercial scale poultry and egg operations across the United States. US SHIP aims to establish a similar platform for safeguarding, improving, and representing the health status of swine across participating farm sites, supply chains, states, and regions (Figure 1).

The initial and principal objectives of US SHIP are to develop and implement an African swine fever (ASF)-classical swine fever (CSF) Monitored certification of US pork production operations (farm sites, live animal marketing operations, and slaughter facilities) modeled after the NPIP's H5/H7 Avian Influenza Monitored certification of US Commercial Poultry operations.

The US SHIP ASF-CSF Monitored certification aims to mitigate risks of disease introduction and provide a practical means for demonstrating evidence of freedom of disease (outside of foreign animal disease control areas) in support of ongoing interstate commerce. It also aims to provide a pathway towards the resumption of international trade over the course of a trade-impacting disease response and recovery period.

US SHIP is harnessing the energies and expertise of industry, state, and federal partners to establish an officially recognized system for safeguarding, certifying, and improving the health of

Figure 1: US Swine Health Improvement Plan. A proven platform for safegarding, certifying, and bettering animal health. An industry, state, and federal partnership.

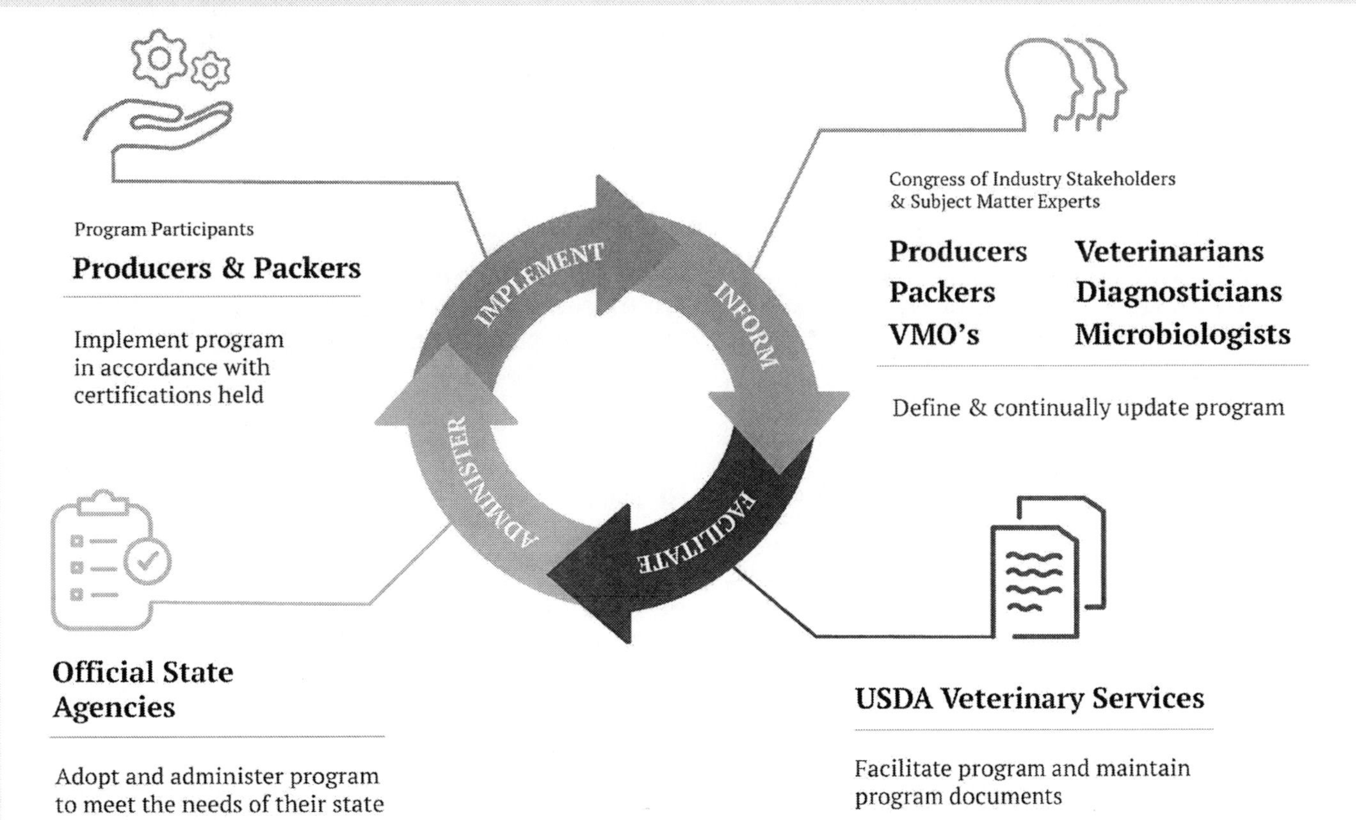

US swine and to improve the longer-term competitiveness of the US pork industry. US SHIP aims to incorporate a number of the foreign animal disease preparedness efforts into a singular USDA certification (ASF-CSF Monitored) of US swine production premises, live animal marketing operations, and slaughter facilities.

US SHIP aims to establish a national playbook of technical standards and associated certification that will be recognized across participating states—one that centers on disease prevention and demonstration of freedom of disease outside of control areas in support of animal health, commerce, and trade.

Transitioning from a pilot to a USDA Swine Health Program

There has been an ever-increasing level of understanding of, interest in, and support for the establishment of a US SHIP over the course of this program development period. During the past 12 months, the National Pork Producers Council (NPPC), National Pork Board (NPB), North American Meat Institute (NAMI), United States Animal Health Association (USAHA), American Association of Swine Veterinarians (AASV), and the American Association of Veterinary Laboratory Diagnosticians (AAVLD) each have come forward with motions and/or other words of support for expanding the resources to develop US SHIP. A joint industry "ASF Strategy Work Group" led by board members of NPB and NPPC in spring 2022 identified "expediting the development of US SHIP into a permanent USDA program" as one of the key industry priorities.

With such growing interest and support, the US SHIP Program Administration (pilot project investigators and staff) and USDA Veterinary Services Swine Health Staff have been working together to establish a four-year plan (2021–2024) for turning the US SHIP pilot project into an official USDA APHIS Swine Health Program (Figure 2). The US SHIP investigators and staff responsible for developing this program are working in partnership with USDA Veterinary Services Swine Health Staff and serving as the US SHIP Program Administration during the current start-up phase of this endeavor.

Developing US SHIP into a USDA APHIS Swine Health Program is essential for establishing credibility and recognition across states and ultimately US trading partners. NPIP's tripartite partnership (industry, state, and federal) and democratic approach to decision-making have a well-established record for delivering workable animal health assurance solutions for the US poultry and egg industries. Much of the responsibility for bringing forth, debating, and directly addressing species- and industry-specific animal health issues of industry-wide significance is deferred from the federal and state veterinary medical officials and associated agencies working in isolation to NPIP's formal congress of industry stakeholders and subject matter experts. This approach creates a sense of shared ownership of NPIP's officially recognized standards, definitions, policies, and health status classifications recognized across participants, participating states, USDA, and international trading partners.

Upon codification of US SHIP, similar to NPIP, the US SHIP Program Administration responsibilities are to be transitioned to a small team of swine interest USDA employees (ie, US SHIP Program Administrative Staff). It is envisioned that the US SHIP Program Administration office will be located in central Iowa. The Ames/Des Moines region is in close proximity to the center of US pork production and entities of national significance to US animal health and the US pork industry (eg, USDA's National Centers for Animal Health, NPB, NPPC, AASV, and Iowa State University).

The NPIP national headquarters and its five USDA APHIS employees are located in Conyers, Georgia. Locating and operating NPIP's small national administrative office in a poultry centric region of the country is an issue of importance to US poultry and egg industry stakeholders. Many international visitors and trading partners visit Georgia (eg, NPIP national headquarters, Georgia Poultry Lab Network, Poultry Diagnostic and Research Center at the University of Georgia, USDA Southeast Poultry Research Laboratory, and myriad commercial poultry operations) to learn and better appreciate the poultry health control programs and poultry and egg industries in the United States. The NPIP Senior Coordinator (the senior NPIP administrator) provides leadership for NPIP activities at the national level. The NPIP General Conference Committee consists of elected individuals (volunteers, one at-large member, and six regional representatives) who provide oversight and industry representation in the NPIP administration.

In the coming year, the US SHIP General Conference Committee (GCC) is being expanded to include a larger set of industry stakeholders appointed by NPB, NPPC, NAMI, AASV, and the State Pork Producer Associations. This group will be playing a critical role in developing and clarifying a number of program governance related items that are critical in guiding US SHIP towards codification. This expanded US SHIP GCC will serve in an interim capacity until the inaugural set of US SHIP GCC elections are held in 2023. Similar to the NPIP GCC, upon codification, the US SHIP GCC will play an active role in the US SHIP Program Administration and will serve in an officially recognized advisory capacity to the US Secretary of Agriculture on issues related to the health of US swine.

The initial slate of elected US SHIP GCC members is scheduled to be elected at the 2023 US SHIP House of Delegates (HOD). This timing will allow these individuals to be fully on-boarded and provide their input throughout the transition of US SHIP to an officially recognized USDA Swine Health Program in 2024.

US SHIP HOD meetings are anticipated to transition to an every-other-year rotation (Biennial Congress) upon transitioning to a USDA Swine Health Program in 2024.

House of Delegates meeting and initial progress across states

The second US SHIP HOD meeting was held in Bloomington, Minnesota, September 6–8th, 2022, with more than 250 people in attendance representing the interests of U.S. pork industry stakeholders across 30 states (Figure 3).

The US SHIP HOD is a democratic forum of US pork industry participants, representing each participating state, who determine the programs content, direction, and requirements for certification by majority vote. Each state expressing interest was allocated a specified number of voting delegates and the opportunity to invite up to two non-voting guests to attend the US SHIP HOD meeting. A formula was used to derive the number of voting delegates allocated to each state. The number of delegates includes a baseline allocation to each state and an allocation proportionate to the capacity (inventory) of the Breeding Herd and Growing Pig production sites enrolled in US SHIP in each state.

Delegates discussed and voted on a series of items to further inform the content and direction of this new USDA-sponsored program for swine health. Key topics included items related to

Figure 2: US SHIP development timeline to a USDA Swine Health Program.

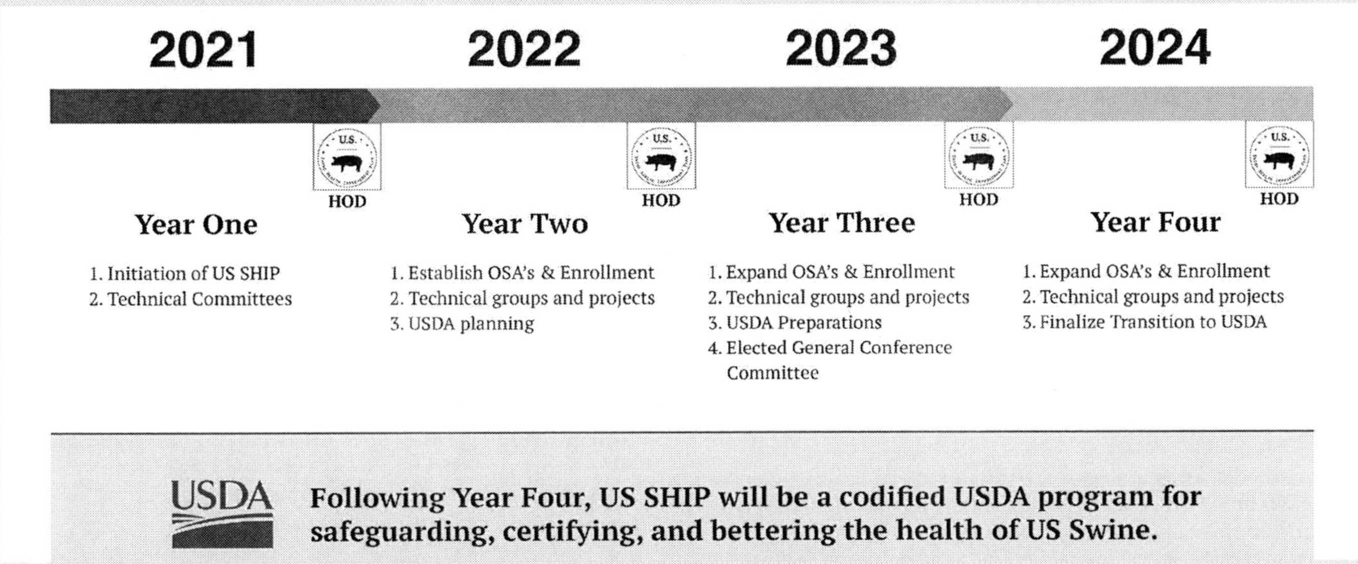

Figure 3: The second US SHIP House of Delegates was held in September 2022.

traceability, surveillance, feed biosafety, site biosecurity, live haul sanitation, and efforts to further define US SHIP's system of governance.

US SHIP Official State Agencies (OSAs) across the US began the process of enrolling sites in March and April 2022. In just the first few months, approximately 40 percent of the US breeding-herd and growing pigs across 31 states have enrolled in US SHIP. US SHIP aims to work with pork producers, live animal marketing operations, and packers across the country to greatly expand industry participation in this new USDA program for certifying the health of US swine and further develop the technical standards in the months and years ahead.

What is needed for US SHIP to be successful?

Answer = Participation

The US poultry and egg industries' NPIP has evolved over the past 85 years, and now 100% of primary breeders and greater than 99% of the commercial poultry (eg, meat-type chicken slaughter plants, meat-type turkey slaughter plants, and commercial table egg layers) in the United States participate in NPIP. This critical mass of participation across all 50-states is unquestionably a significant contributing factor toward NPIP's longstanding and proven track record of success. Such levels of participation have been critical towards providing US Commercial Poultry operations in states and regions not affected by an AIV event of significance (ie, HP-AIV or a lowly pathogenic AIV) an officially recognized mechanism for demonstrating freedom from disease. The H5/H7 Avian Influenza Monitored classification held by meat-type chicken and turkey slaughter plants, commercial table egg laying operations, and states has played a primary role in helping sustain export markets and interstate commerce from unaffected regions during times of an AIV outbreak of significance affecting commercial poultry in the US.

Obtaining a critical mass of participation in US SHIP is the foundation of being able to make progress towards protecting, improving, and representing the health status of all domestic pig production operations across supply chains, areas, states, and regions.

Unlike many endemic diseases, where the primary area of emphasis and economic return come from controlling or eliminating pathogens at the level of the breeding herds on a farm-by-farm basis, trade impacting diseases need to be kept out of and/or eliminated from the entire pork supply chain across areas, states, regions, and the country. Thus, a very different situation, and again, requiring a critical mass (if not universal) of participation and unified effort involving and relevant to all segments of the US pork industry.

How to participate and become US SHIP certified

Step 1. Enrollment

- Contact the US SHIP Official State Agency (OSA) in which the participating premises is located.
- Enroll live swine production and slaughter facility premises with the US SHIP OSA in the state in which the participating premises is located.
- Acknowledge understanding of and compliance with requirements for certification.
- Complete biosecurity survey.

Note: Participants can enroll sites within a state en-masse (ie, one Microsoft Excel or CSV file upload or verification of the premises-level demographic information, and one biosecurity survey for all the sites they are enrolling).

Step 2. Certification

- US SHIP OSA contacts US SHIP participant (swine owner) to initiate certification.
- US SHIP participants need to demonstrate competency in providing at least 30 days of movement information electronically in a common format (eg, a prescribed CSV file) to the US SHIP OSA in a timely manner.
- Boar stud, breeding herd, farrow-to-feeder, farrow-to-finish, and growing pig sites (US SHIP production site types) must be able to provide OSA access to a completed Secure Pork Supply Biosecurity Plan within 24 hours of the request.
- US SHIP OSA confers certification.

Step 3. Maintanence of certification

- Meet or exceed the requirements for the US SHIP certification(s) held.

Current program standards adequate starting point for USDA program

If broadly adopted and implemented across the US pork industry, the current program standards for US SHIP certification will be the basis for transformational change in the state of preparedness and in the ability to certify the health of the pigs across supply chains, states, and regions.

The current program standards position producers, live animal marketing operations, packers, and states to answer two fundamental foreign animal disease preparedness questions.

Question # 1: Do I know where the pigs are or where they came from?

Question # 2: Do I know something about their health status?

The traceability standards provide a functional and ongoing means of keeping the state's official database of swine premises up to date. The current traceability standards also require receivers and shippers of pigs to capture the premises indentification number (PIN) of origin and destination of all movements of swine into or out of their respective premises.

The sampling and testing requirements (which require a well-defined and recurring testing algorithm to be implemented in the event of an introduction into the United States) provide a tangible and objective means of knowing something about the pigs' health status, specifically concerning ASF/CSF status.

Producers, live animal marketing operations, packers, and states need to know the PIN of the premises from which they are receiving the pigs so that they know something about the pigs' health status (ie, certified as US SHIP ASF/CSF Monitored, or not).

This can be done in such a way that the details concerning the premises-level information records can remain locally with the US SHIP OSA of record. The US SHIP OSAs will be responsible for maintaining and depositing the PINs of the US SHIP-certified sites in their respective state to a centralized US SHIP Site Status Verification Database application.

This centralized US SHIP Site Status Verification Database can then be pinged with the PIN(s) interest by end users, and return the status of the US SHIP certification of the particular PIN(s) in question (ie, certified as US SHIP ASF-CSF Monitored, Yes or No).

Again, this process is straightforward and broadly scalable with the existing program standards and readily accessible technologies.

Closing

Now that the basic framework of this US SHIP endeavor is coming together, the principle need for making substantive progress is principally one of stakeholder participation, engagement, urgency, and resolve.

The swine veterinary profession and US pork industry have a long and well-recognized track record of continuous improvement and for adapting to meet new challenges. The ever-growing march of ASF across the globe presents a highly significant and long-term threat to export-centric pork producing countries throughout the world.

Establishing US SHIP aims to provide a sustainable and officially recognized platform for bringing industry, state, and federal partners together to address US swine health-related issues of high consequence over the course of time. Our partnering will safeguard, certify, and improve the health of US swine and increase the longer-term competitiveness of the US pork industry.

The time for such a national strategy is now!

US SHIP website = **https://usswinehealthimprovementplan.com**

https://doi.org/10.54846/am2023/175

What is targeted, science-based, African swine fever surveillance?

M. Culhane[1], DVM, PhD; P. Bonney[1], MS; S. Malladi[1], PhD; M. Remmenga[3], PhD; K.C. O'Hara[3], DVM, PhD; A. Ssematimba[1,2], PhD; B. Blair, DVM, PhD[1]; C. Corzo[1], DVM, PhD
[1]Secure Food Systems Team, College of Veterinary Medicine, University of Minnesota, St. Paul, Minnesota; [2]Department of Mathematics, Faculty of Science, Gulu University, Gulu, Uganda; [3]US Department of Agriculture, Animal and Plant Health Inspection Service, Veterinary Services, Center for Epidemiology and Animal Health, Surveillance Design and Analysis, Fort Collins, Colorado

Introduction

Outbreaks involving African swine fever virus (ASFV) are devastating to the swine industries of affected countries. Surveillance for ASFV is critically important because it facilitates timely detection and may limit the size and duration of the outbreak. Additionally, disease surveillance conducted pre-movement generates confidence that infected pigs are not being moved when permitted shipments are allowed to occur as part of business continuity efforts intended to mitigate outbreak-associated economic losses.

An ASFV surveillance system can be developed using robust heterogeneous disease transmission models that incorporate information not only about the disease triangle - virus (ASFV), host (pig), environment (barn layout) - but also include the best available information regarding diagnostic assay performance, pathogenesis of ASFV from laboratory studies, and disease outcomes from case studies of ASFV infected farms. With this information, a targeted surveillance system can be developed that is optimized for early detection with high probabilities of detection. Targeted surveillance increases the probability of detection (POD) by targeting sampling of those animals most likely to be infected at each time post-exposure and through using the appropriate sample type.

Methods

Using a stochastic within-herd ASFV transmission model[1] and an active surveillance simulation model, we predicted the probability of detection pre-movement for different targeted sampling schemes, simulating 10,000 iterations for each scenario. The model incorporated multiple transmission mechanisms such as direct contact between pigs in the same pen, nose-to-nose contact between pigs in adjacent pens, and people or fomite-mediated spread. The transmission model's predicted disease phase on various days post-exposure was used to simulate detection via active surveillance for a given diagnostic sensitivity of the PCR test on pools of five pig samples per test. Estimated rates of "normal" morbidity and mortality due to routine causes, ie, causes unrelated to ASF, were derived from North American swine industry data and informed predictions of the chances of sampling an ASFV-positive pig. The surveillance protocol options evaluated were all based on sample collection 24 hours pre-movement but with various prioritizations for targeted sampling of sick, dead, and/or apparently healthy (ie, randomly sampled live) pigs for testing.

For this work, we defined targeted individual pig samples as blood, spleen, or lymph nodes. We assumed that pig caretakers could easily sample dead pigs but only identify sick pigs with mild clinical signs 85% of the time. The clinical signs attributed to a moderately virulent ASFV strain included, but were not limited to, lethargy, anorexia, hyperemic skin, loss of appetite, red skin, and other mild symptoms.[2,3] Pigs were assumed to be exposed between one and 30 days prior to sample collection. Moderately virulent strains caused death in 40% of infected pigs. Highly virulent strains caused death in 100% of infected pigs.

Results

Within a 4,800-pig population housed in pens of 40 pigs per pen, active surveillance scenarios to detect ASFV via 20 individual pig samples collected 24 hours pre-movement had significantly different predicted probabilities of detection (PODs) based on the targeted sampling scheme. There was a 5% improvement in detection probability when the sampling was targeted to sick pigs, then dead pigs, then apparently healthy pigs compared to sampling dead pigs, then sick pigs, and then apparently healthy pigs when we assumed pigs were infected with a moderately virulent ASFV strain. However, if we assumed the pigs were infected with a highly virulent ASFV strain, there was a 7% improvement in detection probability when the sampling was targeted to dead pigs, then sick pigs, then apparently healthy pigs compared to prioritizing sick pigs, then dead pigs, then apparently healthy pigs. Interestingly, these significant improvements in POD were not seen for the different targeted sampling scenarios within a 1,200-pig population assumed to be infected with either a highly virulent or moderately virulent ASFV strain. Finally, regardless of barn size or ASFV strain virulence, targeting surveillance to include dead and sick pigs provided a much higher detection probability, usually an improvement of more than 15% compared to surveillance protocols that sampled only random live pigs. Interestingly, the improvement in POD was often more than 30% for surveillance in large, 4800-head barns, when targeted sampling of dead and sick pigs was conducted compared to simulations of surveillance that only sampled random live pigs. In conclusion, targeted sampling that includes the collection of specimens from sick and dead pigs is superior to random and/or convenience sampling of live (ie, apparently healthy) pigs (Table 1).

Discussion

The optimal, targeted priority surveillance scheme for ASFV will vary according to strain characteristics, barn/herd size, and the ability to identify sick pigs. Furthermore, in pig populations that routinely experience elevated levels of mortality and morbidity due to endemic diseases, it may be more difficult to find pigs that are sick or have died due to ASFV. Many of these factors can affect the ASFV infection prevalence in the population of dead pigs and pigs with clinical signs. If the goal of surveillance is to achieve a high probability of detection premovement, it is imperative to target sampling to include sick and dead pigs and also to incorporate targeted biosecurity practices to mitigate pathways of ASFV infection. Implementing both targeted biosecurity and targeted sampling decreases the likelihood of moving infected

Table 1: Overview of results from simulated surveillance scenarios for ASFV detection in growing pig populations using heterogeneous disease spread models within and between pens of 40 pigs

Total pig herd size	ASFV strain virulence	Which targeted sampling prioritization scheme had the highest probability of detection (POD) with 20 individual pig samples pooled by 5 for PCR?	Was there a 15% or greater improvement in POD with targeted sick and dead pig sampling compared to random live pig sampling?
4,800	Moderate	Sick then dead then apparently healthy (S > D > AH)	Yes
4,800	High	Dead then sick then apparently healthy (D > S > AH)	Yes
1,200	Moderate	Equivalent POD for both (D > S > AH and S > D > AH)	Yes
1,200	High	Equivalent POD for both (D > S > AH and S > D > AH)	Yes

S = sick; D = dead; AH = apparently healthy

but undetected pigs from a monitored premises in a control area when seeking to move pigs in an attempt to maintain a semblance of business continuity. Most importantly, producers, swine veterinarians, packers, academicians, state animal health officials, and federal regulatory agents can work together proactively as part of the risk assessment process. When all stakeholders work together, this helps ensure that both the qualitative estimates provided by workgroups and quantitative estimates from heterogeneous spread models are used. Together, these efforts help to more fully understand ASFV transmission risks, devise effective control strategies, and develop science-based, optimized surveillance protocols.

Acknowledgments and disclaimer

The authors would like to acknowledge the contributions and support of the entire Secure Food Systems team at the University of Minnesota, which also includes Catherine Alexander, Carol Cardona, David Halvorson, Myron Kebus, Michelle Leonard, Sylvia Martin, Rosemary Marusak, Miranda Medrano, and Margret Tuisaloo. Some of the project work presented here is part of a collaboration between USDA, APHIS, CEAH, and the UMN Secure Food Systems Team and conducted under a cooperative agreement (CON000000100634). Funding is also provided as part of research grants from USDA NIFA and with endowments. Production data was anonymously provided by swine veterinarians and pork producers. The findings and conclusions in this presentation have not been formally disseminated by the USDA and should not be construed to represent any agency determination or policy.

References

1. Ssematimba, A; Malladi, S., Bonney, P. J., Blair, B; St Charles, K. M., Boyer, T., Goldsmith, T., Corzo, C; Cardona, C. J., & Culhane, M. R. (2022) "African swine fever detection and transmission estimates using homogeneous versus heterogeneous model formulation in stochastic simulations within pig premises." *Open Veterinary Journal.* 12(6): 787-796 **DOI: 10.5455/OVJ.2022.v12.i6.2**

2. de Carvalho Ferreira, H. C., Weesendorp, E., Elbers, A. R., Bouma, A., Quak, S., Stegeman, J. A., & Loeffen, W. L. (2012). African swine fever virus excretion patterns in persistently infected animals: a quantitative approach. *Veterinary microbiology, 160*(3-4), 327–340. **DOI: 10.1016/j.vetmic.2012.06.025**

3. Malladi, S., Ssematimba, A., Bonney, P. J., St Charles, K. M., Boyer, T., Goldsmith, T., Walz, E., Cardona, C. J., & Culhane, M. R. (2022). Predicting the time to detect moderately virulent African swine fever virus in finisher swine herds using a stochastic disease transmission model. *BMC Veterinary Research, 18*(1), 84. **DOI: 10.1186/s12917-022-03188-6**

https://doi.org/10.54846/am2023/176

When does being prepared pay off? Japanese encephalitis - the Australian experience

C. Richards; J. van Dissel; R. Suter; M. Eastaugh; R. Dunlop; T. Harrison; J. Carr
Apiam Animal Health Ltd, East Bendigo, Victoria, Australia

Introduction

An outbreak of clinical disease caused by Japanese encephalitis virus (JEV) was observed concurrently across several piggeries in eastern Australia from February 2022. Clinical signs varied across farms and included prolonged gestation length, mummified and stillborn piglets, congenital abnormalities, and neurological abnormalities. Several veterinary groups across Australia identified these clinical signs over a 7-day period on piggeries that were located within the Murray Darling region, however in some cases geographically separated by thousands of kilometres. The Japanese Encephalitis virus was considered exotic to south-eastern Australia and as a result of the pathogen being confirmed, the Emergency Animal Disease Response Agreement (EADRA), a unique contractual arrangement between Australia's governments and industry groups to collectively reduce the risk of disease incursions and manage a response if an outbreak occurs, was enacted.

Japanese encephalitis virus

Japanese encephalitis virus (JEV) is a flavivirus, an enveloped positive strand RNA virus endemic in eastern and south-eastern Asia. This is a large group of viruses, often arthropod borne. JEV has been detected in sentinel pigs, mosquitos and a single human in the northern tropical tip of mainland Australia approximately 20 years ago, however, it had never previously been identified in south-eastern Australia, where the majority of the pig industry operates, or any other part of the country. JEV is a zoonotic mosquito-borne pathogen. Waterbirds are the reservoirs, and the virus can spill over into pigs which may act as amplification hosts. Humans and horses are also dead-end hosts. While vaccination of people is commonly practiced in Asia, the virus is notifiable in Australia and no vaccination policy in pigs or horses is practiced. Humans are usually vaccinated only as a precaution if travelling internationally to JEV endemic areas.

Clinical signs

Clinical signs in south-eastern Australia were first observed in February 2022 with over 80 piggeries reported over the following four months as having clinical disease confirmed by laboratory testing. Some sows failed to farrow by day 117 and in some cases extended to 126 days. Sows with extended gestation periods were induced and they passed large numbers of mummified and stillborn piglets (Figure 1). This clinical sign was observed in all parities of sows with variation from 10% to 100% of piglets affected in each litter. Mummified piglets appeared bloated with ascites (Figure 2) and there was a positive correlation between sows that tested serologically positive to JEV having entire litters of mummified piglets. Whether a sow within an affected herd showed clinical disease was dependent upon whether it has been infected with JEV, through mosquito bite and the stage of gestation when it was infected. Stillborn piglets presented with an array of congenital abnormalities including severe arthrogryposis and brachygnathia (Figure 3). Some piglets with these abnormalities were also born alive. Neurological abnormalities were a striking feature with cerebral and cerebella aplasia, hydranencephaly and meningocele (Figure 4). Some piglets were also born alive exhibiting congenital tremors.

Being prepared

With the spread of African swine fever (ASF) throughout Europe, south-east Asia and China, the peak industry pork producer body, Australian Pork Limited (APL) had formed an ASF Technical Panel consisting of industry veterinarians to monitor global disease spread, and consider the industries activities in relation to biosecurity, prevention and implications of the response detailed in AUSVETPLAN in the event of an ASF incursion. The panel also worked closely with both Federal and State veterinary officials to work through gaps identified from simulated outbreaks and challenges that occurred in ASF incursions in other countries that if addressed during 'peace-time' could result in an improved outcome for the industry in the event of an actual incursion. Activities included addressing gaps that had been identified as part of paper-based simulation exercises. These collaboration activities between the pork industry, veterinarians and government officials built a good relationship, also highlighting the main concerns of the pig industry; pig movements, continuity of semen supply, and destruction, disposal and decontamination limitations.

Lessons from the JEV outbreak

There were two key aspects that assisted industry in its response to the incursion of JEV into the Australian pork industry: the industry had an established taskforce that had undertaken simulated exercises in the event of a foreign animal disease (FAD) incursion, and that the industry taskforce and state government officials had an existing working relationship over several years discussing various aspects of a response to an FAD incursion. The collegiality of the Australian pig veterinarians also contributed to building an epidemiological picture and providing peer group support. There were significant resources provided by personnel (producers, abattoirs, laboratories) who collaborated with industry and government as part of working groups to address issues as they emerged. As the disease is zoonotic and vector borne, the involvement of the national vector management group, which included entomologists, agricultural chemical experts and industry was able to work together to establish mosquito management strategies. When the clinical signs become apparent in a piggery, the animals are likely to have been exposed to the pathogen over 60 days prior to the onset of clinical disease, therefore prevention of occurrence in subsequent periods relies on mosquito control programs through environmental control and application of chemicals to the pigs, and ongoing surveillance for the Japanese encephalitis virus through laboratory testing of mosquitoes.

Figure 1: Litter of mummified and stillborn piglets from an induced sow.

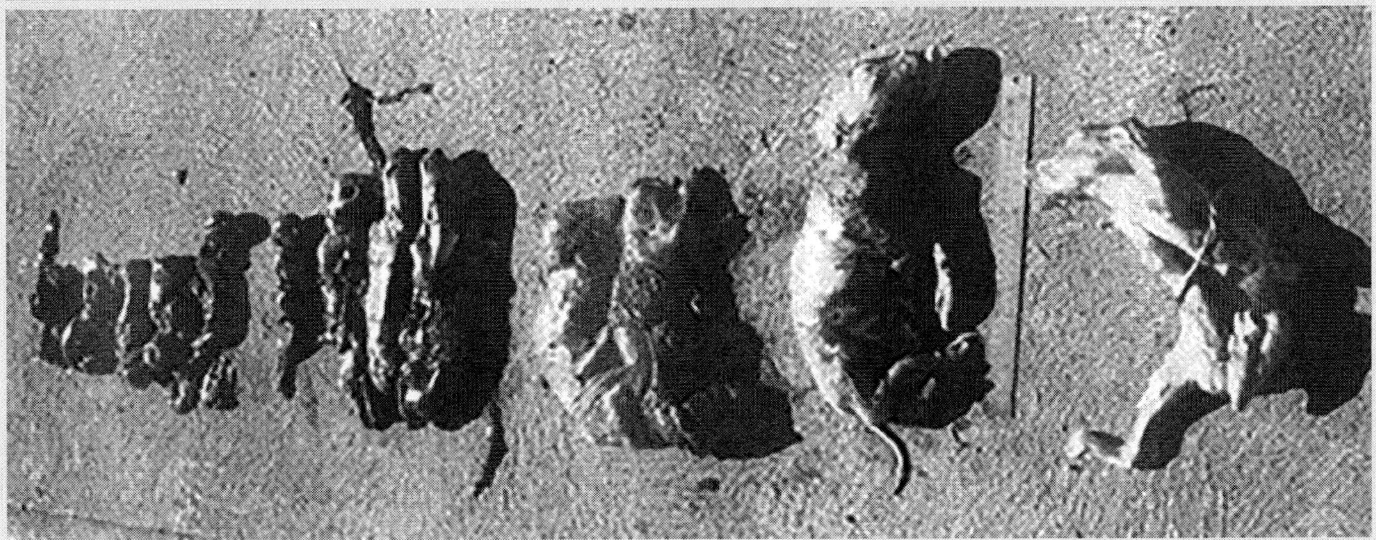

Figure 2: Mummified piglets appeared bloated with ascites.

Figure 3: Affected piglet with arthrogryposis.

Figure 4: Affected piglets had gross neurological abnormalities.

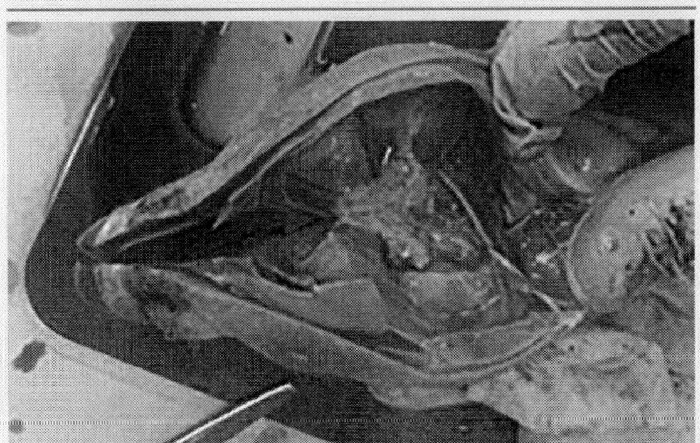

Conclusions

The ability of the Australian pig industry to respond to an incursion by a foreign animal disease such as Japanese encephalitis was enhanced by the previous planning that had been undertaken by industry for other exotic animal diseases, as well as the relationship that existed with state government officials. The event has also provided additional learnings that can be applied in the event of a future foreign animal disease incursion.

https://doi.org/10.54846/am2023/177

The time is now to eliminate porcine epidemic diarrhea virus (PEDV)

Paul Yeske, DVM, MS
Swine Vet Center, St. Peter, Minnesota

Introduction

Since the introduction of PEDV in 2013 to the US swine industry this disease has established itself and continues to remain in the swine population, as shown in both the MSHMP (Morrison Swine Health Monitoring Project) and SDRS (Swine Disease Reporting System).[1,2] Circulating in both the farrow to wean herds and wean to finish populations. There is a definite seasonal pattern to the disease with activity being the highest in winter months. This fall was no exception, with the break rate steadily increasing already.[1,2]

Success rate for elimination at the herd level is high, with many herds successfully eliminating PEDV. The time frame for elimination is relatively short compared to PRRSV and *Mycoplasma hyopneumoniae,* with most being done in 90-120 days. The principle of load close and homogenize followed by strict cleanup and good hygiene has worked well. We currently have the expertise to eliminate this disease from herds and we should have the same expertise to eliminate this virus from the North American industry.

An interesting observation is that the diagnostic labs have not shown a positive case for TGEV since March of 2021,[2] being effectively displaced from the industry by PEDV (Figure 1). Once this was a major concern for the industry and I think demonstrates that it can be eliminated from the industry.

Keeping herds positive is not easy as well. It requires a good area to do exposure and must have time and monitoring to establish exposure and cool down before going into herds. With some of these herds periodically circulating virus resulting in having to take measures to stabilize again. These also come with cost and risk like taking herds negative.

One of the biggest challenges is once PEDV starts circulating in the industry it does not take long for it to move. One factor is in transportation this was demonstrated early in the outbreak in 2013 the packing plants were contaminated quickly, and trucks had higher likelihood of become positive after being at the plants.[3] This results in market channels including cull sows become contaminated quickly and can potentially be sources of infection. Every year we see a limited number of sow farms becoming positive resulting in less than 9% of sow herd being positive with 2022 being one of the higher years.

Methods

The industry has the tools to successfully eliminate this virus. There are good diagnostic tools and the samples necessary to test are easily collected by veterinarians and producers. The incidence is relatively low for this agent compared to others that are in the industry see Figures 2 and 3. In looking at the recent history the number of positive samples has been above the 3-year predicted average see Figure 4. I believe we have the tools and the technical expertise that it can be eliminated for the entire industry with a concerted effort and cooperation from all producers. As an industry we need to establish the will to no longer deal with this disease by developing a national program to eliminate it. With the relatively low incidence rate, preparation, and timing could use the summer season to aid in the process. This would require herds that are doing ongoing virus exposure to stop and do eliminations in those herd. Planning timing and herd elimination to complete in the middle of summer. A program such as this would take cooperation, communication, and surveillance to be successful.

Discussion

My bias has always been to eliminate any disease that we can, so we have less to deal with as an industry. I believe that long term this only makes the industry more sustainable into the future. Although there are more severe economic diseases that impact herds every year. There is still a significant impact on the sow herds that are affected, and costs associated with cleaning up, stabilizing and going on to elimination. Herds that have chosen to live with the disease still have significant cost in keeping herds positive in on going monitoring, the effect on gilts that are exposed in new outbreaks that occur in the system because there is still active virus present. There is also risk to a larger segment of the industry as we see the virus being able to circulate especially in wean to finish. Wean to finish costs are not particularly high in market pigs and can sometimes go undetected, in groups that are starting however can be more difficult. These are all reasons that I believe the industry should take the initiative to eliminate this virus. I believe we have the knowledge, diagnostics, and abilities to accomplish this. It is a matter of resolve to do so. I think it is a good opportunity before we see another surge of the disease and more herds impacted to start the process. It will take a lot of preparation and organization and communication to get this done but I think it is the time and can be done in short period of time.

References

1. Corzo C. Morrison Swine Health Monitoring Project. 2022.

2. Linhares D. Swine Disease Monitoring System. 2022.

3. Lowe J. Role of Transportation in Spread of Porcine Epidemic Diarrhea Virus Infection, United States. Emerging Infectious Diseases 2014 (20) (5).

Figure 1: Demonstrates the number of TGEV positives from swine disease reporting system (SDRS) report.

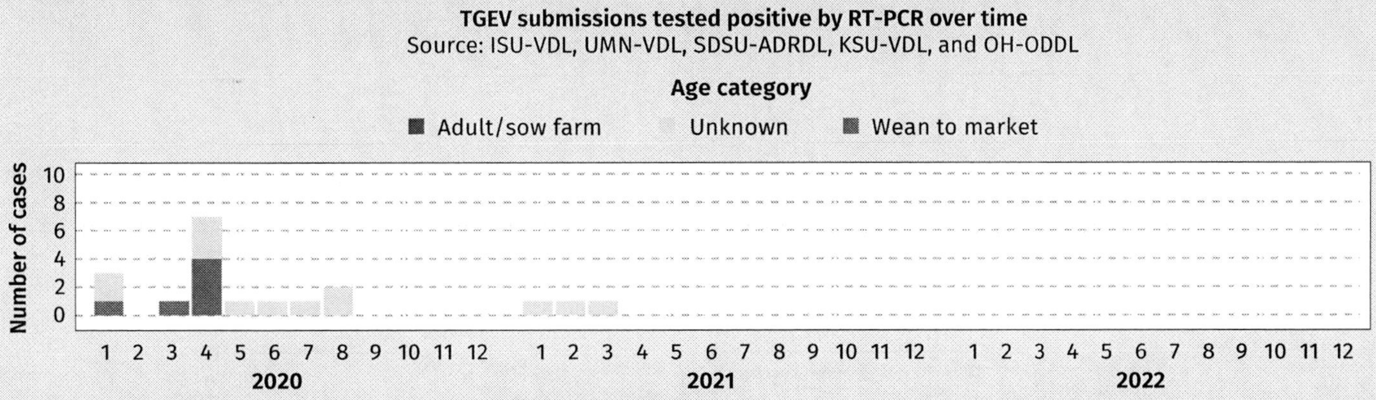

Figure 2: Cumulative incidence in sow herds.

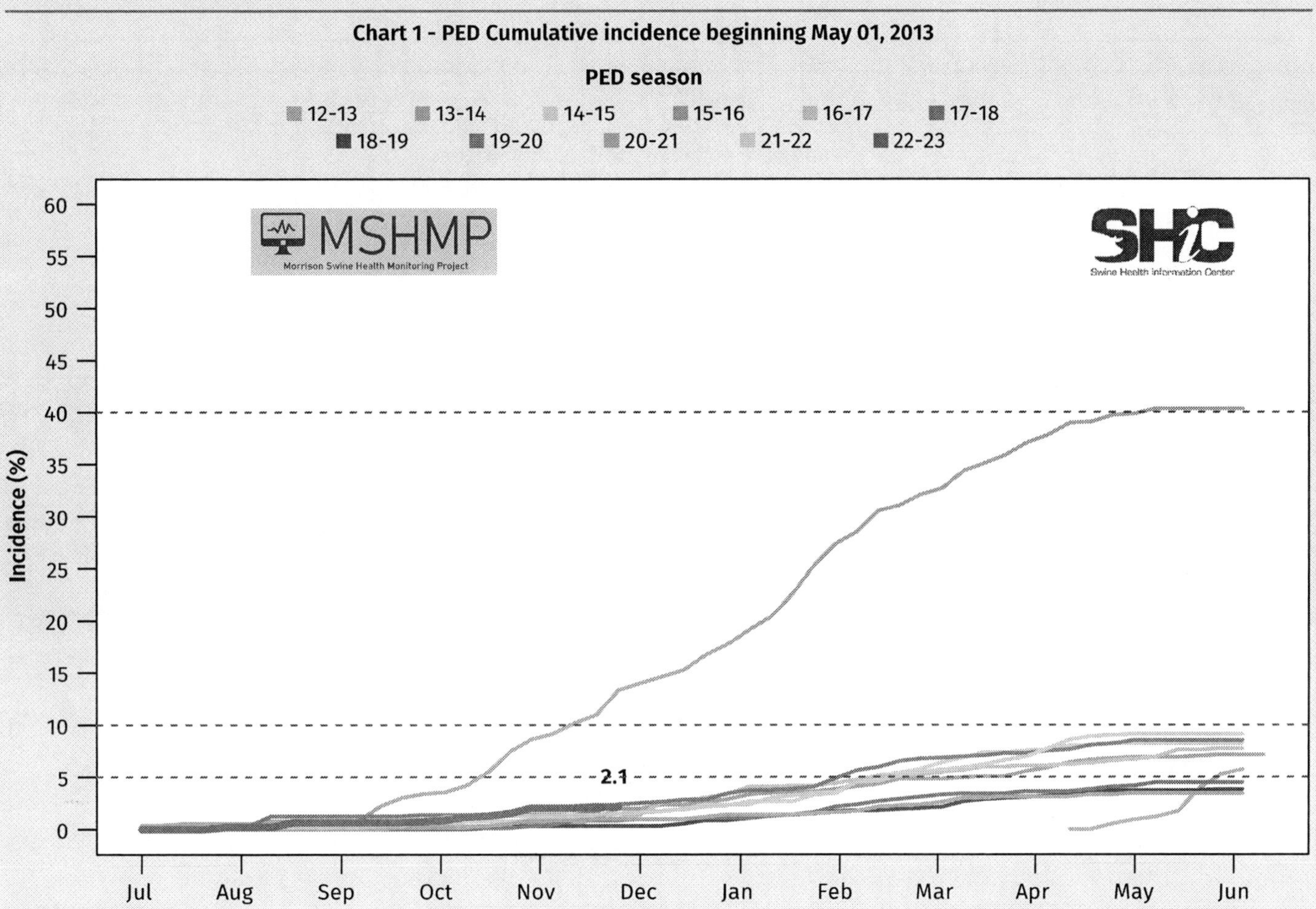

Figure 3: Percent positive samples by stage of production.

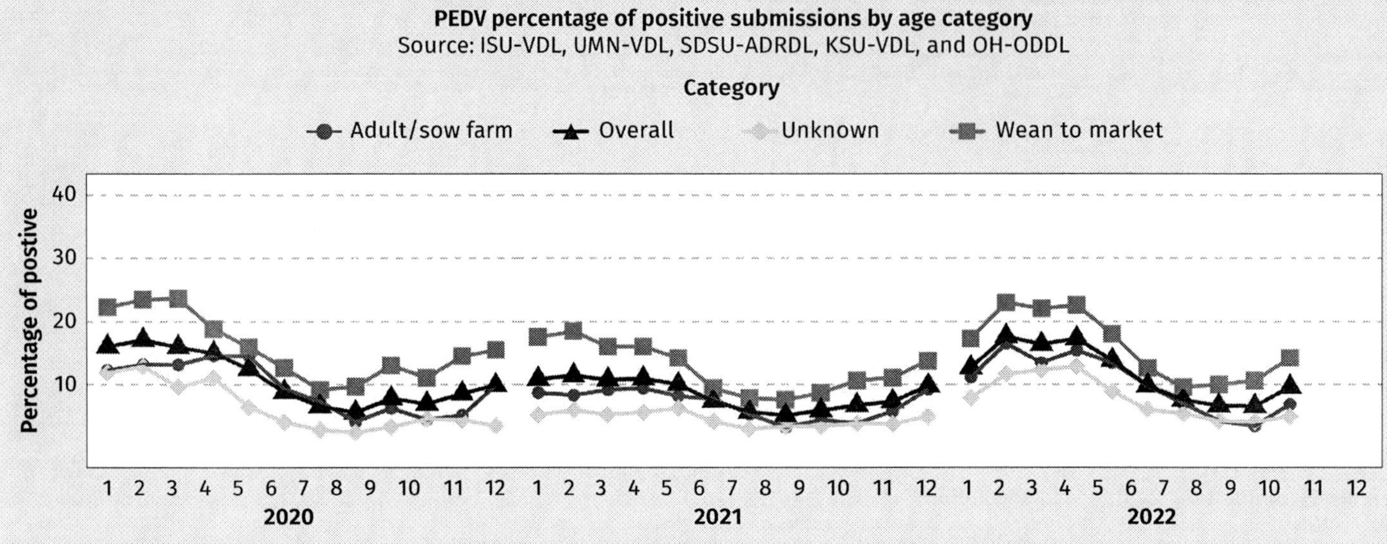

Figure 4: Predicted percentage of positives.

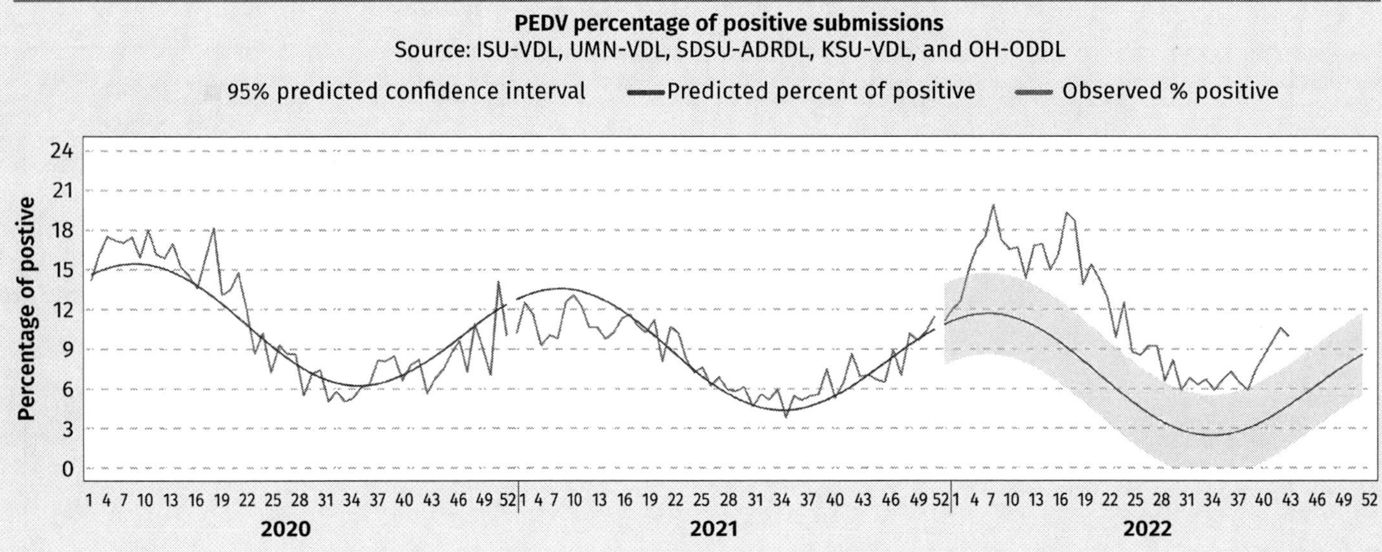

https://doi.org/10.54846/am2023/178

PED controlled exposure: A critical tool

Luc Dufresne, DVM
Seaboard Foods, Shawnee Mission, Kansas

To eradicate or not to eradicate, that is the question!

I vividly remember the summer of 2013. That is when PEDV went through a significant part of our system. Initially the virus went through our Colorado system (likely through infected feed ingredients as we were one of the index cases). Then through the Panhandle of Oklahoma and Kansas, where in a matter of 3 weeks it infected over 150,000 sows and an estimated 2,000,000 pigs from 3 different companies. The infection was first detected in a finisher site and then started moving quickly through multiple sites due to vector spreading and area spread. We were able to detect PED genetic material over a distance of 10 miles during the outbreak. We believe that the situation was worsened by several factors:

1. High pig density combined with sows and growing pigs in the same area
2. Very large finisher multi-sites with low biosecurity containment within said multi-sites
3. High wind speeds with no natural barriers
4. Wide use of natural ventilation

When we look back at the speed of transmission in the area and the inability we had of reducing the rate of infection during that initial outbreak, we decided that PED controlled exposure was the only way to minimize the risk of seeing that type of catastrophic loss repeat itself. There is a significant difference between a sow farm or system of production that is isolated from external growing pigs. In that situation PED elimination may be a viable option. Experience from other large systems that have experimented with elimination in swine dense areas shows that the approach can sometimes be ill advised due to the inability to eliminate the infection in growing pigs. Another area that potentially can be problematic is how long the virus can survive in slurry inside and outside of the barns. We have demonstrated that slurry from a lagoon remains infective 1 year after the infection was last diagnosed on a nursery site (internal trial done using bioassay).

Live inoculation

If the decision is to not eliminate PEDV from the sow herd, exposing gilts to PEDV is a key component to ensure the herd remains stable and continues to produce negative piglets.

- Live-virus exposure of gilts should be done in a manner to ensure all animals have time to not only develop immunity but also stop shedding the virus before they will be introduced into the sow herd.
- The exposure should be done in an off-site location to avoid reactivating the disease in the sow farm.
- The exposure should be done at least 10 weeks before entry to the sow farm.
- Exposure can be done effectively by administering inoculum through the waterline, gruel feeding, or oronasal spray.
- The process should be closely monitored to ensure clinical signs are present following exposure.
- Gilts should be tested by PCR before introduction in the sow farm to ensure they are not still shedding virus.
- A two-site system (one site for exposure and another site for cool-off) has been the best to prevent reactivation of PEDV in a stable sow farm. Biosecurity within site is important to minimize the risk of recirculation of the disease. We have seen previously exposed gilts shed again later in life when actively shedding gilts were placed in close proximity.
- Reducing direct and indirect contact between shedding and previously exposed animals remains very important, especially in groups that are ready to be shipped to the sow farms.

Pros of live virus inoculation

- Immunity matters! In a small internal trial, we showed that growing gilts exposed to PED live virus and vaccinated with PED killed vaccine 1 week pre-farrow were able to wean PED negative piglets when the animals were challenged at 5 days post-farrowing.
- While exposure to the indel strain reduced mortality and the severity of clinical signs, it did not prevent transmission from the sows to their piglets. In real life settings, we expect clinical recirculation in an exposed, immune herd to create 20% of the average loss when compared to a similar outbreak on a naïve farm.
- Most importantly the number of farms that break is greatly reduced. In at-risk areas where we detect high levels of circulation in growing pigs, we have found less than 20% of the sow farms in those regions would experience clinical rebreaks and those rebreaks in average experience 20% of the piglet losses that we experience in a naïve herd. Overall, the PED circulation in a specific area would create about 4% of the expected loss versus if the sow base in the region was naive.

Cons of live virus inoculation

- Problematic with onsite gilt development units.
- Problematic if done on sites where mature gilts are shipped to sow farms at the same time as exposure is being performed on the site.
- Does the inoculation maintain virus pressure in a region?

In the last 8 years, most of the PED rebreaks we've experienced were associated with high levels of regional PEDV circulation in growing pigs or associated with manure agitation or mishandling. If we don't find a way to effectively eliminate the risk of PED infection in growing pigs, it will be impossible to have farms remain naïve to the virus in pig dense areas. At this point, even with the knowledge we have acquired in a decade of PED circulation, I still don't see a way of effectively preventing what happened in the Oklahoma panhandle in the summer of 2013 with a naïve sow and growing pig population if the virus was to be reintroduced to the area. Based on these observations, PEDV live virus inoculation remains a critical tool to protect sow farm production in pig dense areas.

https://doi.org/10.54846/am2023/179

PEDV exposure, a tool for elimination?

Karine Talbot, DVM
HyLife Ltd

Introduction

Could controlled exposure of replacement gilts to porcine epidemic diarrhea virus (PEDV) be the solution to our recurrent and cyclic outbreaks in Manitoba? This question has been asked many times over the various PEDV outbreaks in Manitoba, but particularly during this last one...

Manitoba PEDV history

To understand our approach to PEDV, one needs to understand our history with this disease in Manitoba. The first case ever was in 2014, but it is in 2017 that things really got worst. Since then, it has been a 2-year cycle of PEDV outbreaks (Figure 1). Between each outbreak, we fully eradicate PEDV from all barns; until it comes back again...

PEDV is a provincially reportable disease in Manitoba (which means that all cases must be reported to our provincial government). Each case is counted as 1 (if a sow barn breaks and moves pigs to a nursery, it counts as 2). Our outbreaks are very localized to the most pig dense area of Manitoba, and although there have been a few cases out of the main high-risk area, most are located in the South-East (Figure 2).

We perform on-going surveillance for PEDV, even when all cases have been eradicated from barns; daily testing of each receiving dock at the two federally inspected slaughter plants, regular surveillance at assembly yards and provincial abattoirs, etc....

We are good at eradicating PEDV. Some barns have successfully done it 3 or 4 times now. But it keeps coming back.

Where are we at with PEDV now

In Manitoba, how PEDV is managed is not a single, individual producer's decision; it is a group decision. Whichever plan we chose has an impact on all: on producers dealing with PEDV year after year, on producers located in the high-risk area but that have been able to keep it out of their barn, but also on Manitoba producers that are outside the high-risk PEDV areas.

Future management of PED working group

In early 2022, a PEDV working group was created with the objective to figure out our approach to this disease in the future. The group consists of Pork Board representatives, veterinarians, integrators, independent producers, and the office of the Chief veterinarian (provincial government). What is the mandate of this group? Provide a PEDV management strategy for Manitoba. One of the first questions that had to be answered: Can we expose gilts to create immunity and prevent sow barns from re-breaking every 2 years?

Controlled exposure of gilts

Controlled exposure of gilts to acclimatize them before they are entered in sow barns located in the PEDV high-risk area sounds like a great plan at first. It could help stabilize the area, avoid multiple re-breaks, minimize severe financial loss, be more sustainable for a short to medium period of time until we can figure out how to eradicate PEDV in the long-term.

The group collaborated with researchers from the Western College of Veterinary Medicine in Saskatchewan who did a systematic review of studies assessing the efficacy of vaccines, biosecurity measures, feedback, PEDV shedding, financial impact of PEDV, and various PEDV control strategies (like controlled exposure of gilts).

Unfortunately, it was quickly realised that controlled exposure of gilts is not a solution for Manitoba producers dealing with PEDV. The logistics of doing it would impact all producers.

First, we would need to expose about fifty thousand gilts annually to PEDV to provide immunized replacement animals for the South-East sow barns (calculated using actual number of sows located in the South-East that were infected with PEDV during the 2021-22 outbreak). This number represents gilts from multiple production systems and independent producers, each with their own genetic and source of gilts. This would require multiple barns dedicated to exposing gilts to live PEDV, not just a few.

Second, acclimatized gilts would prevent major loss at the sow barns, but it does not guarantee the barns would not be infected again. If they do, we are then producing, once again, viremic piglets that must be placed in nurseries and thus exposing more barns to PEDV. It is a vicious cycle.

And third, but not least, constantly exposing animals to live virus would keep PEDV continually active in Manitoba, with the risk of contaminating our various contact points to the industry (rendering, transport, feed mills, service providers, manure pumping crews, etc). Where can you safely expose that many gilts, in that many different barns, without eventually impacting neighboring farms? Historically, we have not been particularly good at keeping PEDV contained to a few sites in Manitoba. How would we manage to expose thousands of gilts without leaking the virus to other farms?

Conclusion

After carefully reviewing studies and discussing with others that are exposing gilts as a mean to control PEDV, the Manitoba PEDV working group concluded that this strategy is not for us. Although it might be enticing at first look, the consequences of living with PEDV and purposely infecting naïve animals continuously is not the best strategy for the good of all. So, we are sticking together and keep working toward full eradication of PEDV!

Figure 1: Total cases of PEDV per year in Manitoba (as of November 15, 2022)

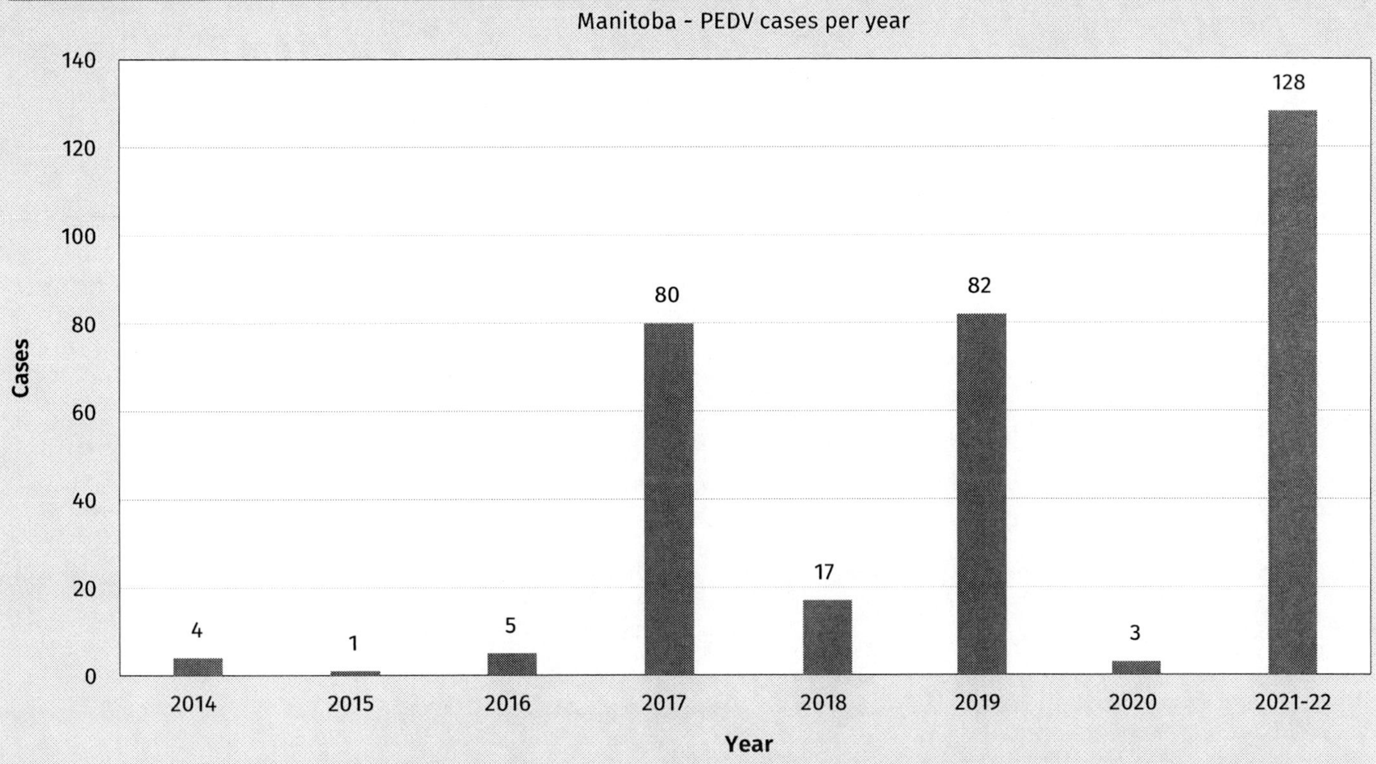

Figure 2: Location of PEDV cases in Manitoba and high-risk areas (from the Provincial Situation Report).

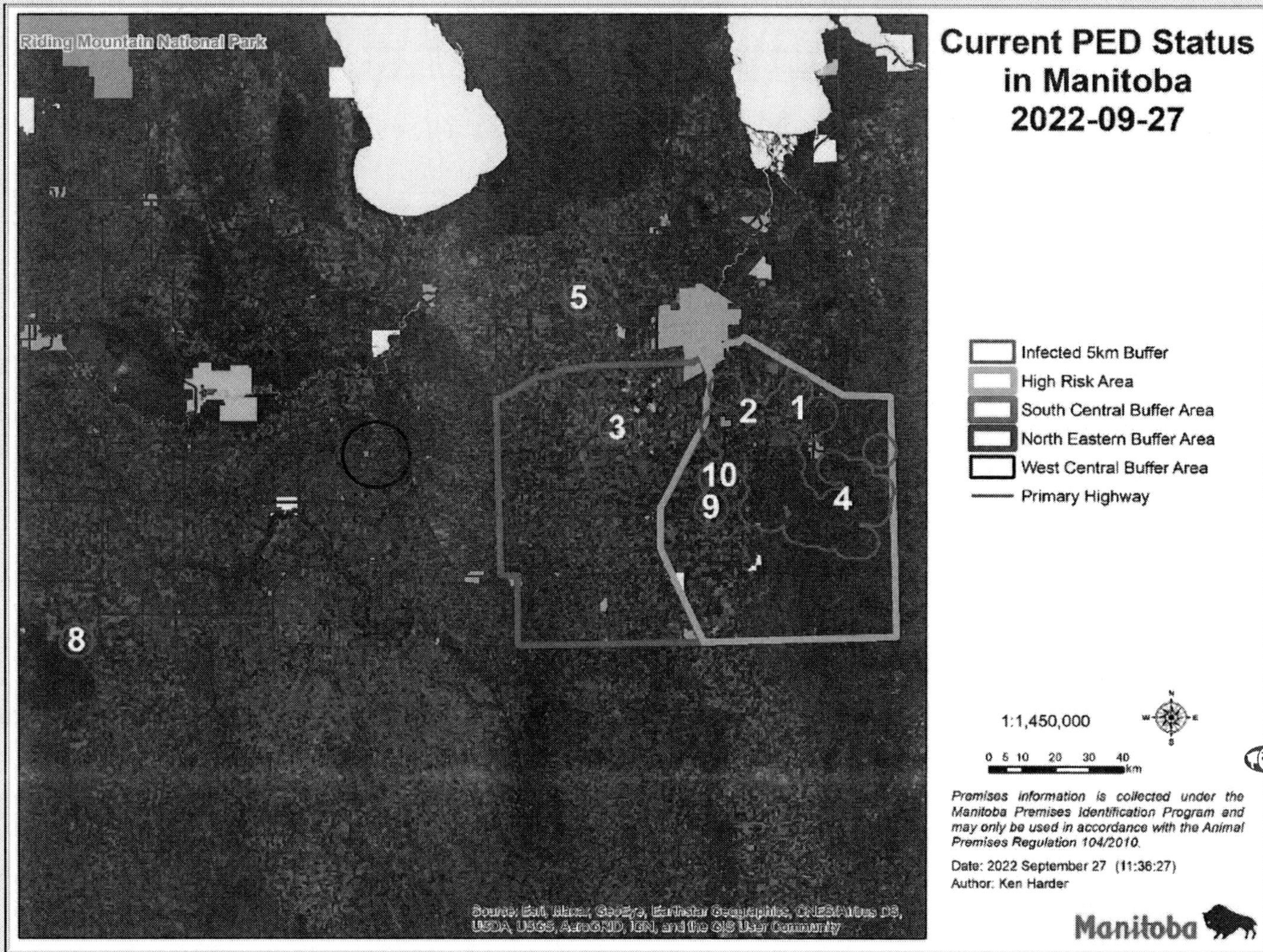

https://doi.org/10.54846/am2023/180

Experiences in removing planned PEDV exposure for acclimatization in a large commercial production system

L. Glowzenski, VMD
TriOak Foods, Oakville, Iowa

Introduction

Porcine epidemic diarrhea virus (PEDV) is an economically significant pathogen of concern to the United States' swine industry that became endemic following the initial outbreak in 2013. For nearly a decade, both growing pig sites and breeding herds have continued to circulate this enteric virus, posing an on-going risk of introduction to naïve sites.

Gilt acclimation is a tool utilized by swine practitioners to control and stabilize breeding herds for a variety of pathogens, including, but not limited to: Porcine respiratory and reproductive syndrome virus (PRRSV), *Mycoplasma hyopneumoniae* and porcine epidemic diarrhea virus (PEDV). The goal of planned exposure of PEDV to acclimate replacement gilts is to maintain a consistent supply of gilts, while at the same time reducing the significant risks associated with maintaining a PEDV naïve herd.[3] As long as the gilts are exposed at a young age, there have not been any recipient sow herds that have been infected through this process.[2]

PEDV acclimation has proven to be effective in reducing production and subsequently economic losses, by providing active immunity in previously naïve replacement animals. However, there are risks associated with live virus exposure, primarily, to the swine industry at-large by maintaining on-going, active viremia and circulation of a highly pathogenic organism within a population.

Removing planned exposure

Since the initial outbreak in 2013, our large, commercial, production system, comprised of over 70,000 sows across 22 locations, had routinely exposed all naïve replacement gilts with the US prototype strain of PEDV during off-site gilt development. This process proved to be easy to execute by administering oral feedback of PEDV inoculum on top of feed during the late nursery phase. The practice also proved to be effective, with none of the breeding herds suffering a PEDV outbreak during the period of planned exposure. At the time, all sow farms were receiving select gilts, over 26 weeks of age, that had ample time to 'cool down' and clear the virus following inoculation.

Our company's production model changed in recent years, with a transition from off-site gilt development and breeding herds receiving mature gilts to a model of on-site gilt development, with farms receiving 3-week-old, wean age gilts as breeding stock replacements. The on-site gilt development model posed a problem for our routine PEDV acclimation process. Since pigs need at least 6 - 10 weeks to clear PED virus after exposure and the fact there are naïve piglets constantly being born at the sow farm, our PEDV acclimation practice needed to end. There cannot be actively shedding pigs on the same site as naïve pigs – the risk of transmission is too great, being on the same site and location.

Removing planned PEDV exposure to acclimate gilts did not come without consequence. Since removing this practice from the system, we have endured several PEDV breaks in different geographical regions within the company, highlighting the benefit of this acclimatization practice in preventing outbreaks in breeding herds. Regardless of this consequence, our production model does not allow us to revert, and we instead must improve our biosecurity practices to prevent PEDV and other viral or bacterial diseases from entering our sites.

Discussion

Targeting specific pathogens for routine, planned exposure or acclimation, is a way to control, stabilize and prevent production losses on the farm. Although this is an effective practice, it is not always possible to implement and focus must be placed on prevention of disease, first and foremost.

Removing PEDV acclimation from a system is seamless – stop acclimating and begin entering naïve stock to the farm. The challenge with removing planned exposure is the biosecurity risk to the farm.

Regarding foreign animal diseases (FAD), like African swine fever (ASF) and foot-and-mouth disease (FMD), a focus on endemic swine disease eradication the United States needs to become a reality, now, to prepare us for the challenge of a FAD introduction and subsequent elimination efforts. If we cannot work together as a collective to eliminate a pathogen like PEDV, that is predominantly transmitted by fomites, we will not be able to eliminate a disease like ASF. Perhaps the time is now to challenge ourselves to eradicate an endemic virus, now, to prepare for our future challenges.

References

1. Ackerman M. Introducing high health animals into positive farms. AASV. 2011;449-450.

2. Connor J. Is it the time to eliminate PEDv, PDCoV, TGE & Mycoplasma from NA Swine - Opportunities and Challenges. Leman Swine Conference. 2019.

3. Thomas P. Experiences with porcine epidemic diarrhea virus (PEDV) acclimation of replacement gilts. AASV. 2016;260-362.

4. Turner MB. Review of Porcine Epidemic Diarrhea Virus Exposure Techniques. AASV. 2015;417-420.

5. Yeske P. School of hard knocks: Disease prevention, control and elimination. AASV. 2022;398-399.

https://doi.org/10.54846/am2023/181

Veterinary medicine and swine nutrition at AMVC

J. Hocker[1], DVM, MS; T. A. Kellner[2], PhD
[1]AMVC Management Services, Audubon, Iowa; [2]AMVC Nutritional Services, Audubon, Iowa

Introduction

The easy joke is that veterinarians are also nutritionists, so there are no challenges in the intersection of veterinary medicine and swine nutrition. The reality is that there are synergies and advantages of having intelligent individuals with different perspectives examining, evaluating, and questioning processes and procedures. Smart people rightly ask why things are done the way they are done. If these questions are embraced and used to make the organization better and more robust, there is no downside. That is our approach at AMVC, turn really smart solution-oriented people loose to look for areas of improvement and participate in that solution making process. The challenge in this thought process was eloquently described in Dr. Gregory Stevenson's Howard Dunne Memorial Lecture in 2015 where he pointed out, "Students today compared to those 25 years ago are much more likely to take offense by interpreting criticism of an idea as criticism or rejection of their person."[1] Substitute veterinarian or nutritionist for student in that sentence and the same aptly applies today. By understanding and rejecting the notion that questioning of ideas or processes is a criticism of an individual or profession, veterinary and nutrition professionals can have very rewarding interactions that yield great results.

What a veterinarian is responsible for?

Veterinarians are always immediately linked to the health and welfare of the pigs. This is the easiest point of reference and easiest to deal with from a practical standpoint. Create protocols, treatment regimens and recommendations and send them out into the swine world to support the health and welfare of pigs. But how does this actually happen? Without implementation, the best laid plans, protocols, recommendations, regimens, guidelines etc. are just words on paper. So, the veterinarian can often become the coach, the manager, the cheerleader, the confidant, the therapist, the human resources stand in, life coach, etc. Veterinarians are problem solvers by training. While the formal training is medicine in focus, the thought processes and mindset are ideal for complex problem solving where multiple systems are interacting ie, health, production, nutrition, management, employee interaction, logistics etc. Protocol and regimen development need to keep these aspects in mind to get adequate implementation. A one size fits all mindset will result in poor health program implementation and variables including farm layout, production staff understanding and tenure, current health challenges, etc need to be considered. Proper health management requires clear and concise protocols that can be executed by lay persons of any production experience level and the veterinarian needs to continuously evaluate the efficacy of the health management programs.

What a swine nutritionist is responsible for?

A swine nutritionist is responsible for oversight of the feeding program design, execution, and validation. Feed is greatest cost of raising a pig from farrow to finish. Thus, the nutritionist is responsible for selecting feed mill partners and locations, ingredient sourcing, procurement, and analysis, diet formulation, feed budgets and protocols, oversight of feed ordering and delivery, farm level feed protocol and usage execution, and economic and pig performance projections and validation. To be successful it is important that these responsibilities are not contained in a silo, but allow for input from veterinary, risk management, and production teams. It is important that the veterinary and production teams fully understand the feed program objectives and design.

How can a veterinarian support feed program design?

Take the time to understand the milling process and where the pinch points exist. If pigs do not eat feed, pigs do not grow, and welfare is negatively impacted. Evaluate all the steps necessary to get feed in front of the pig and work to minimize areas for failure. Are the health regimens being developed compatible with diet formulation in regard to days and pounds required for each ration? Can the treatment regimen exist in 12- or 24-ton increments? Do you have confidence in the feed delivery process? Do you know if there are tandem or single bins? Can the farmer or staff manage medication regimens that require split loads? Taking the time to analyze and understand some of these basic questions will result in better adoption and implementation of feed programs and feed medication regimens. Having the conversation about feed program goals and feed medication regimen goals with the nutritionist and formulation team will help both parties work better together in a cooperative fashion driving the production system to the goal of efficiently raising pigs and producing pork.

How can a swine nutritionist support medication and management protocols and practices?

It is important to have input and collaboration in protocols and practices from all perspectives involved in pork production. A swine nutritionist can help support in designing feed medication/additive programs, pig flow/space and marketing pig strategies, risk management strategies and sow farm strategies. The swine nutritionist and veterinarian should work together on the feed budget and how to integrate feed medication/vaccination programs into the feed budget. The key is to provide input, support, and collaborate and not dictate. Furthermore, a swine nutritionist can provide a unique perspective and support at the slat level. Nutritionists will look at feed usage, wastage, systems, and feeder adjustments, but can also provide another set of eyes on all management practices and production phases.

How can veterinary and nutrition teams work together with one unified purpose?

Communicate regularly with one another and embrace the differences in observational lenses. The different angle of perspective and observation will result in better solution development and help identify blind spots and biases that may not always be recognized. Often there are silos between nutrition, production and health. This is a mistake that costs the pig owner money and leads to counter productive challenges within organizations. When silos are created, monocular efforts to control individual spheres of influence lead to organizational dysfunction that ultimately negatively impacts pig health, welfare and overall performance.

In short, the veterinarians and nutritionist are an integral part of the system raising the pig. Not unlike the pig's internal organ system, where each component needs to be doing its job and working with the other components to support the pig's health, growth and welfare; the veterinarian and nutritionist need to work together synergistically toward the same goal – helping each pig grow as heathy and efficiently as possible.

Reference

1. Stevenson, Gregory. Howard Dunne Memorial Lecture. Because it's the right thing to do. Proceedings 2015 AASV Annual Meeting: Beyond Our Other: Integrity, Intensity, Professionalism. 2015. 11-26.

https://doi.org/10.54846/am2023/182

Veterinary medicine and swine nutrition at JBS

K. Coble, PhD; D. Madson, DVM
JBS Live Pork, LLC, Greeley, Colorado

Summary

When intra- and inter-personal surveys are completed, communication is most likely to be observed as a top opportunity in many organizations. Whether it is the type, form, or method of delivery, it is an aspect of teamwork that companies devote vast amounts of resources to, striving to create a better culture and stronger team. For some pork production systems, this process or interaction can be the most tenuous between veterinarians and nutritionists. To have a continued, harmonious, collection between those two technical teams is a rarity. This certainly is not all surprising when nutritionists are trained to think in terms of populations and requirements, whereas veterinarians are educated with individual minimum standards of care in mind.

Over the past years, *E coli* has been a challenge in many parts of the industry and our system was not immune. Both teams used their expertise, ultimately developing a world class model to understand this largely misunderstood health by nutrition interaction. This led to numerous research experiments being completed that went on to be successful field evaluations of which ultimately became new foundations of our nursery program. Another example would be when the severity of lameness spiked in one of our commercial sow flows. The veterinarian and nutrition teams were able to safely ask the other party to investigate causes, reaching across the aisle. A plan was put together for a large field examination where the two teams ultimately worked up the diagnostic causes of osteochondrosis in late maturing gilts. With a few nutritional changes, the two teams celebrated this as a win for both. This makes perfect since in our system to celebrate this as a win for our collective technical team rather than an instance where the feed was formulated incorrectly, or a disease misdiagnosed.

Communication can be the cause of diagnostic failures or production success when the entire team understands the process and that the communication is built on trust. Trusting that your teammate's work will be done timely and followed up on, or that your counter part is the expert in the room, serves as a basis for this successful atmosphere. While many examples have existed over the years within our system between our veterinary and nutrition groups, we continue to bring it back to the basics of trust and our JBS values.

Respecting your teammates knowledge and vantage point will lead to more successes for you than failures. No one accomplishes goals or pushes a production team forward when you stay on your own island, unwilling to learn and admit you don't know it all. Deferring to your teammate and asking them to teach you is not a weakness – it is a major ingredient in the recipe for success.

American Association of Swine Veterinarians
830 26th Street
Perry, IA 50220-2328
USA

ISBN 978-1-7138-7102-6

9 781713 871026

CURRAN ASSOCIATES INC.
proceedings.com